Communications in Computer and Information Science 361

Srija Unnikrishnan Sunil Surve
Deepak Bhoir (Eds.)

Advances in Computing, Communication, and Control

Third International Conference, ICAC3 2013
Mumbai, India, January 18-19, 2013
Proceedings

Volume Editors

Srija Unnikrishnan
Sunil Surve
Deepak Bhoir
Fr. Conceicao Rodrigues College of Engineering
Bandstand, Bandra (W), 400 050 Mumbai, Maharashtra, India
E-mail: {srija,surve,bhoir}@fragnel.edu.in

ISSN 1865-0929 e-ISSN 1865-0937
ISBN 978-3-642-36320-7 e-ISBN 978-3-642-36321-4
DOI 10.1007/978-3-642-36321-4
Springer Heidelberg Dordrecht London New York

Library of Congress Control Number: 2012956062

CR Subject Classification (1998): H.4.1-3, C.2.0-1, C.2.4-6, I.2.4, I.2.6-11, F.2.2, H.2.4, H.2.8, H.3.3-5, C.4, K.6.5, I.4.8-9

Typesetting: Camera-ready by author, data conversion by Scientific Publishing Services, Chennai, India

Printed on acid-free paper

Springer is part of Springer Science+Business Media (www.springer.com)

Preface

This conference proceedings volume contains the written versions of the contributions presented during the Third International Conference on Advances in Computing, Communication and Control (ICAC3), which is a biennial conference. As in the previous years, the conference was held at Fr. Conceicao Rodrigues College of Engineering, Mumbai, India, during January 18–19, 2013.

The conference provided a setting for discussing recent developments in a wide variety of topics including image processing, artificial intelligence, robotics, wireless communications, data warehousing and mining etc. The conference gave a good opportunity to participants coming from India, South Korea, Bangladesh, and Iran to present and discuss topics in their respective research areas.

Every submitted paper went through a rigorous review process; one non-blind review to check the suitability of the paper for the conference followed by three blind reviews out of which two were by experts and the third by peer authors. In case of contradictory reviews, additional reviews were commissioned. After this feedback was received, the three Program Committee members discussed any pending issues and provided a recommendation on acceptance and final decisions were made.

The 69 papers that were presented on the two days formed the heart of the conference and provided ample opportunity for discussion. The papers were split almost equally between the three main conference areas, i.e., computing, communication, and control. All presented papers are included in this proceedings volume published by Springer.

As organizers, we have many people to thank for their support in making the conference successful. The fruitful outcome of the conference was the result of a well-coordinated effort of the Program and Technical Committee members as well as the Advisory Committee. It is our pleasure to acknowledge the publication support from Springer. Special thanks to our Organizing Committee, who put immense effort into making the conference successful. We thank all the authors who submitted papers. It was a privilege as well as a great responsibility to oversee the reviewing of their papers. All in all, ICAC3-13 in Mumbai was very successful. The next ICAC3 will take place in Mumbai in 2015. We hope that it will be as interesting and enjoyable as its three predecessors.

January 2013

Srija Unnikrishnan
Sunil Surve
Deepak Bhoir

Preface

[illegible]

Organization

ICAC3 is a biennial conference organized by Fr. Conceicao Rodrigues College of Engineering, Mumbai, India

Executive Committee

Patrons	F. Diniz, Superior Victor, Director
Honorary Chairs	Munshi Srinivas
General Chair	Srija Unnikrishnan
Conference Chair	Sunil Surve
Conference Co-chair	Mahesh Sharma
Conference Steering Chair	Sapna Prabhu
Program Chair	Deepak Bhoir
Publication Chair	Shilpa Patil
Publicity Chair	Jagruti Save
Organizing Chairs	K. Narayanan Brijmohan S. Daga Merly Thomas Monica Khanore
Contact Chair	Prachi Patil
Finance Chairs	Vedvyas. S. Jorapur - Chair Roshni Padate - Co-chair Binsy Joseph - Co-chair

Advisory Committee

Vijay K. Kanabar	Metropolitan College Boston University, Boston, USA
Hemant Kekre	NMIMS, Mumbai, India
Sunil Survaiya	R&D Corning Cable Systems, Germany
Sameer Hemmady	SAIC, USA
Navdeep Singh	VJTI, Mumbai, India
Sudhir Sawarkar	Datta Meghe College of Engineering, Mumbai, India
Sanjay Gandhe	Sardar Patel Institute of Technology, Mumbai, India
Sriman Narayan Iyengar	School of Computing Science and Engg. V.I.T. University, Vellore, India
Mudasser F. Wyne	National University, USA

K.V.V. Murthy	Amrita Vishwa Vidyapeetham, Bengaluru, India
B.K. Lande	SAKEC, Mumbai, India
Deepak Garg	Thapar University, Punjab, India
Maulika Patel	G.H. Patel College of Engineering and Technology, Vallabh Vidya Nagar, India

Program/Technical Committee

Vijay K. Kanabar	Metropolitan College Boston University, Boston, USA
Mudasser F. Wyne	National University, USA
Sunil Survaiya	R & D Corning Cable Systems, Germany
Sameer Hemmady	SAIC, USA
Dinesh Sathyamoorthy	Science Technology Research, Institute of Defence, Ministry of Defence, Malaysia
Junwei Cao	Tsinghua University, China
Yudong Zhang	Research Scientist, Columbia University, USA
Ibrahim Kurcukkoc	University of Exeter, UK
Tole Sutikno	University of Ahmad Dahlan, Indonesia
K.V.V. Murthy	Indian Institute of Technology, Gandhinagar
N.M. Singh	VJTI, Mumbai, India
Sudhir Sawarkar	Datta Meghe College of Engineering, Navi Mumbai, India
S.T. Gandhe	Sandip Institute of Technology and Research Centre, Nashik, India
Chintan Modi	G.H. Patel College of Engineering and Technology, India
Deepak Garg	Thapar University, Punjab, India
Maulika Patel	G.H. Patel College of Engineering and Technology, Vallabh Vidya Nagar, India
Manik Lal Das	D.A.I.I.C.T, Gandhinagar, India
S.S. Sane	V.J.T.I., Mumbai, India
Srimannarayana Iyengar	V.I.T. University, Vellore, India
Tanuja K. Sarode	Thadomal Shahani Engineering College, Mumbai, India
Vivekanand Mishra	S.V. National Institute of Technology, Surat, India
Nikil Kothari	Dharmsinh Desai University, Nadiad, India
Vipul A. Shah	Dharmsinh Desai University, Nadiad, India
Rohin Daruwala	V.J.T.I., Mumbai, India
Subhasis Das	KIIT University, Bhubaneshwar, India
Chandan Chanda	Bengal Engineering and Science University, Shibpur, India
B.N. Chatterji	B.P. Poddar Institute of Management and Technology, WB, India

Debajyoti Mukhopadhyay	MIT Group of Institutions, Pune, Maharashtra, India
R. Bhavani	Department of CSE, Annamalai University, India
Susan Elias	Sri Venkateswara College of Engineering, Tamilnadu, India
Satya Sheel	Motilal Nehru National Institute of Technology, Allahabad, India
Sanjay Garg	Institute of Technology, Nirma University, Gujarat, India
J. Abdul Jaleel Jalaludeen	TKM College of Engineering, Kerala, India
Sudeep Thepade	Pimpri Chinchwad College of Engineering, Pune, India
Premila Manohar	MS Ramaiah Institute of Technology, Bangalore, India
Shivanand Handigund	Bangalore Institute of Technology, Banglore, India
S.S. Rathod	Sardar Patel Institute of Technology, Andheri, Mumbai, India
S.M. Khot	Fr. C. Rodrigues Institute of Technology, Mumbai, India
R.N. Awale	VeerMata Jijabai Technological Institute, Mumbai, India
Ramji M. Makwana	A.D. Patel Institute of Technology, Karamsad, Anand, Gujarat, India
Suprava Patnaik	XIE, Mahim, Mumbai, India
Shubhi Purwar	M.N.N.I.T., Allahabad, India
Shyamal Ghosh	Calcutta Institute of Engineering and Management, India
Vijaya K.	R.M.K. Engineering College, Tamilnadu, India
Archana Patankar	Thadomal Shahani Engineering College, Mumbai, India
Kishor Maradia	Government Engineering College Modasa, Gujarat, India
S. Nickolas	National Institute of Technology, Tiruchirappalli, India
Saurin R. Shah	Dharmsinh Desai University, Nadiad, India
Jayant Male	D.J. Sanghvi, Mumbai, India
Bhupendra Parekh	B.V.M. Engineering College, Vallabh Vidynagar, India
Surinder Singh	Sant Longowal Institute of Engineering and Technology, Longowal, Sangrur, Punjab, India
Rameshwar Kawitkar	Sinhgad College of Engineering, India

Local Organizing Committee

Swati Ringe
Hemant Khanolkar
Deepali Koshti
Supriya Kamoji
Tisha Jose
Ganesh Bhirud
Ajay Koli
Jagruti Nagaonkar
Heenakausar Pendhari
Sunil Chaudhari
Sarika Davare
Prajakta Dhamanskar
Girish Damle
Swapnali Makdey
Kalpana Deorukhkar
Sushma Nagdeote
Ashwini Pansare
Parshvi Shah
Vipin Palkar
Kranti Wagle
Sangeeta Parshionikar
Monali Shetty
Vaibhav Godbole
Dharmesh Rathod
Sonal Porwal
Anusha Jayasimhan

Table of Contents

Computing

Communication

Control

Others

k-QTPT: A Dynamic Query Optimization Approach for Autonomous Distributed Database Systems

Pankti Doshi[1] and Vijay Raisinghani[2]

[1] Department of Computer Science
Mukesh Patel School of Technology Management and Engineering
NMIMS Deemed-to-be University
Pankti.doshi@nmims.edu
[2] Department of Information Technology
Mukesh Patel School of Technology Management and Engineering
NMIMS Deemed-to-be University
rvijay@ieee.org

Abstract. Query processing in a distributed database system requires the transmission of data between sites using communication networks. Distributed query processing is an important factor in the overall performance of a distributed database system. In distributed query optimization, complexity and cost increases with increasing number of relations in the query. Cost is the sum of local cost (I/O cost and CPU cost at each site) and the cost of transferring data between sites. Extensive research has been done for query processing and optimization in distributed databases. Numerous search strategies like static, dynamic and randomized strategies are available for determining an optimal plan. However these search strategies are not suitable for the autonomous distributed database systems. These search strategies make certain assumptions (like all sites have same processing capability), which do not hold for autonomous systems. Mariposa, Query Trading (QT) and Query Trading with Processing Task Trading (QTPT) are the query processing algorithms developed for autonomous distributed database systems. However, they incur high optimization cost due to involvement of all nodes in generating optimal plan. We present our solution *k*-QTPT, to reduce the high optimization cost incurred by QTPT. In *k*-QTPT, only *k* nodes participate in generating optimal plans. We discuss implementation details of QT, QTPT algorithm and our solution *k*-QTPT. We evaluate *k*-QTPT through emulation. We show that the cost of optimization reduces substantially in *k*-QTPT as compared to QT and QTPT.

Keywords: Distributed databases, Query Processing, Query Optimization, Autonomous systems, Dynamic Query Optimization, Mariposa, Query Trading algorithm, Query Trading with Processing Task Trading, *k*-Query Trading with Processing Task Trading.

1 Introduction

In a today's business, data frequently resides on multiple sites inside an organization. This data might be managed by several database management systems for multiple reasons such as scalability, performance, access and management. Distributed

S. Unnikrishnan, S. Surve, and D. Bhoir (Eds.): ICAC3 2013, CCIS 361, pp. 1–13, 2013.

systems can be thought of as a collection of independent cooperating database systems. If database systems at different sites are of the same type, then such system is known as a homogenous DDBMS. In contrast to a homogenous system, there is a heterogeneous system, which has different types of databases at each site. The homogenous distributed database systems can be further divided into autonomous and non-autonomous systems. In non-autonomous DDBMS, there is a central optimizer, which decides at which node the query is to be executed. All nodes in non-autonomous system are aware about all other nodes. In autonomous DDBMS, there is no central optimizer, nodes itself decide whether to participate in query execution or not. All nodes in autonomous system are unaware about other nodes in the system. In a distributed database, the retrieval of data from different sites in a network is known as distributed query processing. The performance of a distributed query depends on how fast and efficiently it accesses data from multiple sites. In distributed query processing, a single query is divided into multiple sub queries, which are further allocated to the several sites depending upon the fragmentation schema and replication. There can be variety of methods for computing the result of a single query. For example, consider relations as shown in figure 1. The number of records and the size of each record are also shown in figure 1.

Site 1: Employee (Empno, Ename, Eaddress, Sex, Salary, Deptno) → 10,000 records, each record 100 bytes long

Site 2: Department (Dept_no, Dept_name, Dlocation) → 100 records, each record 35 bytes long

Fig. 1. Query execution example for DDBMS

It is assumed that relations employee and department are at site 1 and site 2 respectively. Assume that a query fired by user at site 3 is "For employee working in department HR, retrieve the employee and department details". Considering that there are 100 employees working in HR department the query execution can have one of the following approaches:

a. Transfer both the employee and department relations to the result site (site 3) and perform join at site 3. In this case total of (10,000*100) bytes from site 1 + (100*35) bytes from site 2 = 1003500 bytes must be transferred to site3.
b. Transfer the employee relation to site 2 (1,000,000 bytes need to be transferred), execute the join at site 2 and send the result ((100*100) + (100*35) = 13500) to site 3. Hence, total 1,000,000+13500 = 1,013,500 bytes need to be transferred.
c. Transfer one record of HR department (35 bytes) from department relation to site 1, execute the join at site 1, and send the result ((100*100) + (100*35) = 13500) to site 3. Hence, total 35 + 13500 = 13535 bytes need to be transferred.

Out of the three ways discussed above, (c) is the most optimal way in terms of data transfer to site 3. Finding an efficient way of processing a query is important. The objective of a distributed query optimizer is to find a strategy close to optimal and to avoid sub-optimal strategies. The selection of the optimal strategy requires the prediction of execution cost of the possible alternative plans. The execution cost is expressed as a sum of I/O, CPU and communication cost [1].

Extensive research has been done in the area of distributed query processing and optimization, several strategies like *static optimization strategies*, *dynamic optimization strategies* [2] and *randomized optimization strategies* [3] are available in literature. Query optimization is a difficult task in a distributed client/server environment as data location and size of data becomes a major factor, this is evident from the example discussed above. In a static optimization strategy, a given query is parsed, validated and optimized once. Plans for these queries are stored in the database, retrieved, and executed whenever the application program is executed. When initially determined parameters (like relation size) change, the plan stored in the database is invalidated and a new plan is generated. This approach cannot adapt to the changes, such as variations in the load of sites, thus the compiled plan show poor performance in many situations. To overcome this disadvantage of static approach, dynamic approaches like *dynamic programming* [4], *greedy algorithm* [5] and *iterative dynamic programming strategy* [6] were proposed. Dynamic optimization strategies work in bottom-up manner by building complex plans from simpler plans. Dynamic query optimization techniques have certain assumptions like 1. Static data allocation: objects like relations, records etc. cannot easily change sites to reflect changing access pattern and 2. Uniformity: all the sites have the same processing capabilities and memory. These assumptions are valid for non autonomous DDBMS.

In non-autonomous distributed database systems, there is a central query optimizer which decomposes the query and decides where to execute each of the sub queries [7]. This is possible because the central query optimizer is aware about all other nodes in a system and assumes that each node has same processing capabilities. In contrast to non autonomous DDBMS, in an autonomous distributed database system, nodes do not have complete knowledge of the system. The nodes are unaware of current load of other nodes. A node may accept or reject a query execution request based on its processing capabilities and current load. Before actual execution of the query, a node, where the query was initiated should ideally identify all other nodes which can participate. Run-time conditions, such as the existing load on the site and communication costs, would have a significant impact on the execution cost of the query. It is evident from the discussion above, that optimization strategies for non-autonomous distributed database systems are not suitable for autonomous distributed database systems. Hence, new techniques like *Mariposa* [8] and *Query Trading* [9] are proposed for autonomous distributed systems. In the next section, we discuss strategies for autonomous distributed database systems.

2 Optimization Strategies for Autonomus Distributed Database Systems

In this section we present, query optimization techniques for autonomous distributed database management systems. Here, we discuss Mariposa algorithm, Query Trading (QT) algorithm and variants of QT algorithm that use a new architecture for query optimization based on economic model [9]. We also discuss the shortcomings of each. A detailed review of Mariposa, QT and QTPT is given in [10].

A traditional distributed database system has a central optimizer which does not support total node autonomy. In order to support full node autonomy, in an autonomous distributed database system, each site has complete control over its resources. In an autonomous distributed database management system, all nodes are independent and unaware of physical schema, logical schema, data statistics etc. [10]. Here, nodes participating in query execution are not selected by the central optimizer. Nodes independently decide whether to participate or not depending upon the node's resource capacity and data availability. Hence, before actual query execution, all participating nodes are identified.

Fragkiskos Pentaris and Yannis Ioannidis [9] proposed an *economic model* for identifying all participating nodes and to support node autonomy in autonomous system. According to the economic model, as shown in figure 2, there are two types of nodes - buyer nodes and seller nodes. A node where a query is initiated (initiator node or buyer node) will request all other nodes (seller nodes) for query execution. It is assumed that the buyer node does not have sufficient resources to answer the query. The buyer node will send *Request for Bids* (RFB) to seller nodes, requesting cost of the sub-query that the seller node can execute. All seller nodes will send a reply to the buyer node, identifying the part of the query that they can execute, along with the cost of execution. Based on these replies, the buyer node will select the seller nodes with lowest possible cost and declare the selected nodes as *winners*. The query plan generator in the buyer will develop a query execution plan using these winning nodes.

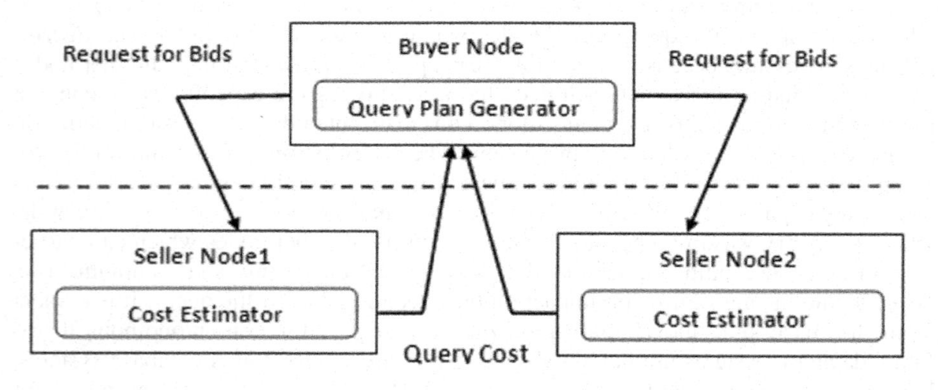

Fig. 2. Economic model for autonomous DDBMS

The Mariposa [8] and Query Trading algorithm [9] are based on this economic model. In general, both algorithms follow the process as stated below:

1. Buyer node prepares RFB for sub-queries that require cost estimation
2. Buyer node sends RFB to the seller nodes requesting cost for the sub-query.
3. Seller nodes calculate the costs for sub-queries and send replies back to buyer node.
4. Buyer node, based on replies, decides on an execution plan for the query; if required repeat steps 2 and 3.

A detailed review of Mariposa and QT is given in [10].

A query processing will involve data retrieval and left over processing on retrieved data (for example sorting, aggregation etc.). In distributed query optimization in QT, the processing can be performed either at seller nodes or at buyer nodes. One approach is that the seller nodes sends data available with them to buyer node and buyer node performs all the left over processing on the data; however this generates sub-optimal plans when the buyer node itself is resource constrained or when the buyer node is overloaded. To overcome this disadvantage, [9] proposed a new algorithm known as *Query Trading with Final Step of Processing Task Trading* known as *QTPT.*

QTPT algorithm is a variant of QT. It works in two phases. The first phase of QTPT works like QT in order to find an initial distributed query execution plan. In the second phase of QTPT, buyer again asks for bids from seller nodes for all processing tasks involved in plan and selects lowest bid nodes as winning nodes. Based on the bids received from seller nodes, initial query execution plan is revised at buyer node and processing tasks are distributed to the winning nodes. This generates more optimal plans as compared to QT; however the time required for optimization is higher as compared to QT, due to an additional phase. This disadvantage of QTPT is the motivation for our work. In next section, we analyze the performance of QT and QTPT.

3 Analysis of QT and QTPT

In autonomous systems, to increase local autonomy the optimizer consults the data sources involved in an operation to find the cost of that operation. Hence the dominant cost in optimization becomes the cost of contacting the underlying data sources. QT and QTPT algorithms construct optimal plans and support full node autonomy. However, QTPT has higher optimization time as it works in two phase. The key experimental results from [9] comparing QT and QTPT are as below.

a. **Time Delay:** As QTPT works in two phases, it spent lot of its time idle, waiting for replies to their bids in both phase. Hence, network and seller processing delay times are a large fraction of the total optimization time. Thus QTPT incurs high optimization time compare to QT.
b. **Startup Cost:** The QT algorithm has a start-up cost of a single round of message exchanges caused by the bidding procedure. The QTPT algorithm has a start up cost of two rounds of message exchange, one for finding initial query execution plan and second for requesting bids for processing tasks in query being optimized.
c. **Increasing Number of Partitions:** As the number of partitions is increased, the complexity requirements of all algorithms are raised. Experiments show that as number of partitions increases, all algorithms behave similarly and which heavily affects quality of plans produced.

As discussed above, QTPT generates optimal plans as compared to QT. However; QTPT has higher processing time compared to QT, as it works in two phases. To overcome the optimization cost in QTPT, due to two phases, we propose k-QTPT. In the next section, we discuss k-QTPT and its implementation.

4 *k*-QTPT

The proposed enhancement is to reduce amount of optimization time taken by QTPT.

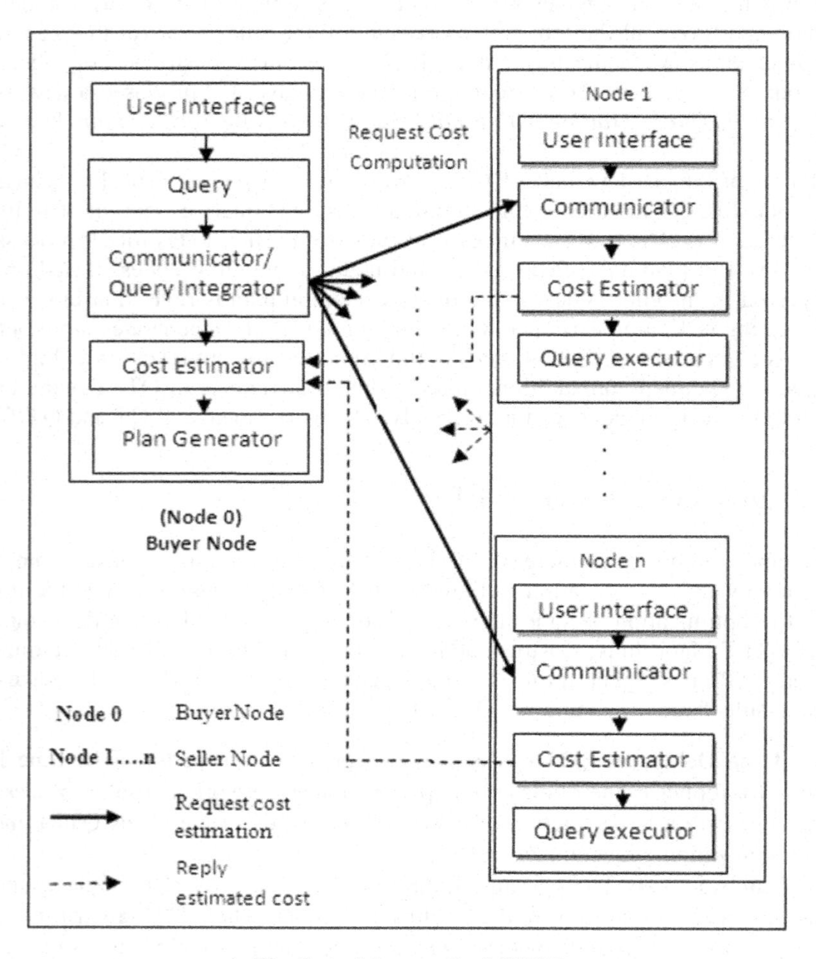

Fig. 3. Architecture for *k*-QTPT

Our proposed solution *k*-QTPT, also works in two phases like QTPT. The first phase in *k*-QTPT is same as that in QTPT, it determines initial query execution plan. In second phase, from the winners of first phase, buyer node buffers a list of *k* winning nodes. In second phase, RFBs for processing task will be requested from only these *k* nodes selected in first phase, instead of requesting from all nodes. This is expected to reduce optimization time substantially. The key assumption for selecting only *k* nodes is that *k* highest bidders of previous phase probably will be most suitable for next phase. Deciding appropriate value of *k* is one of the challenges for implementing *k*-QTPT. The proposed architecture as shown in figure 3, for the given solution is based on QT architecture. In next section, we present, experimental evaluation for QT, QTPT and *k*-QTPT. In case, the *n* nodes are grouped into R_i sets, where each set R_i has nodes which are replicas of each other, if we further assume *c* nodes per set, then *k* could be subset of these *c* nodes. We follow this approach in our experimental setup.

5 Experimental Setup

In order to do detailed performance study of QT, QTPT and *k*-QTPT we created a lab emulation of a DDBMS. In this section, we describe our emulation setup in terms of the system, workloads and algorithms emulated. Our system setup consists of 13 nodes, interconnected by a network. All nodes were connected using a LAN at a speed of 100 Mbps. Five tables were created in MySQL database. Each table had 6000 tuples and seven attributes. Horizontal fragmentation was done on each table. Each table had 5 fragments with 1000 to 2000 tuples each. Each fragment was replicated at 1 or 2 additional sites. The mirror sites of a fragment were grouped into a virtual set R_i. As shown in figure 4, there are five such virtual sets (R1, R2, R3, R4, and R5), of 2 to 3 nodes each.

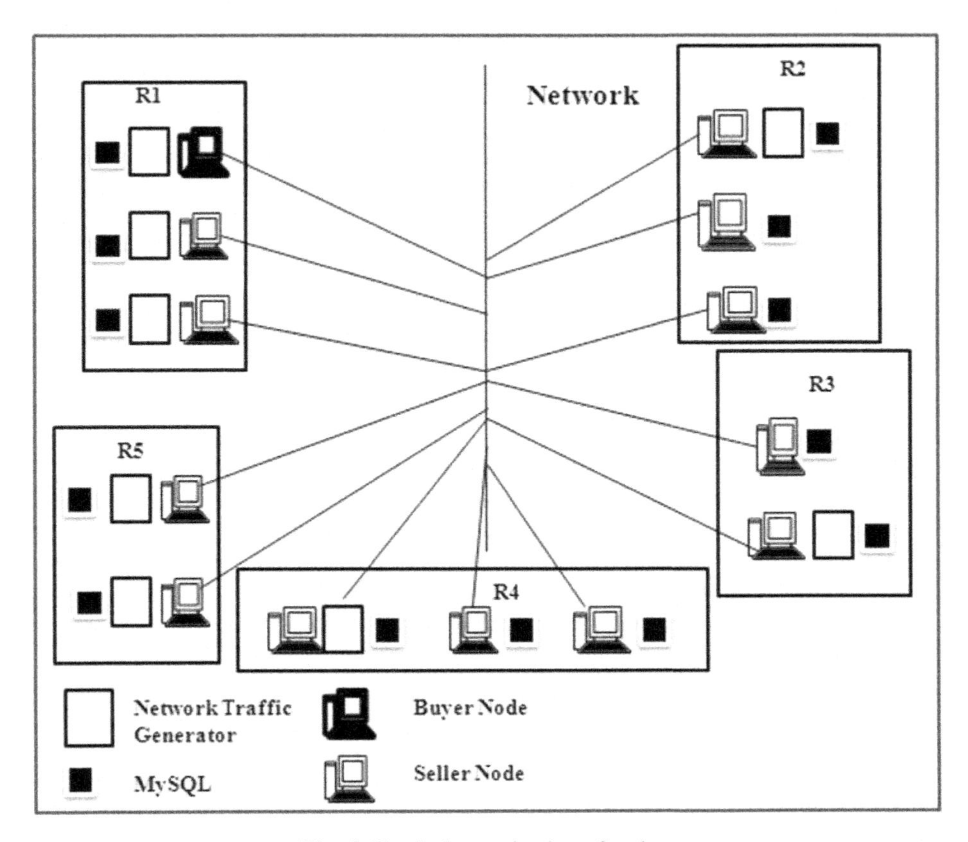

Fig. 4. Physical organization of nodes

We have used Java 1.6 and MySQL 5.1 to emulate a DDBMS. As shown in figure 4, each node is equipped with MySQL 5.1 The all 13 nodes in an emulation setup were having Microsoft Windows XP Professional version 2002, service pack 3 as operating system, 1.79GHz, 9 GB of RAM. Each node was equipped with NetBeans IDE 6.9.1, which includes support for JAVA FX SDK 1.3 and MySQL 5.1. In order to connect Java with MySQL, each node was installed with mysql-connector-java-5.1.14-bin.jar. In order to analyze results with varying load on the network, network cross traffic generator -- *LAN Traffic V2* is used to inject traffic to the network, from some of the nodes.

Challenges faced: We initially attempted to simulate the DDBMS. We studied tools such as Mckoiddb [11] simulator and CloudSim, but they were not suitable for our simulations. Later, we searched for open source database management system that supports distributed environment. We identified MySQL; however it was difficult to incorporate changes in optimization code due to limited documentation. Hence, we decided to emulate distributed environment over a LAN, by using Java and MySQL.

The second challenge, we faced was to estimate the cost for query execution at each node. In all variants of QT (including QTPT and *k*-QTPT), in each phase, it is necessary to estimate cost at each node, before actual query execution starts. We studied several database management tools, example Toad for MySQL but no tool supported this requirement. We identified database profiler in MySQL that provides cost estimate of a query before actual execution of the query. However, the profiler estimated the ideal cost, assuming that there was no load on the processor. To overcome the cost estimation problem, we used the heuristics based method proposed in the [12]. In this method, the cost estimation is done by abstracting certain parameters from the query and then running the query for multiple times, under different load conditions. We executed our sample query under different load conditions and measured the query execution time. The average costs were stored in tables within the MySQL database, at each node. These costs were then used by the seller nodes when sending replies to the buyer nodes.

We emulated the phases of QT, QTPT and *k*-QTPT by manually creating the sub-queries and programmatically invoking the required phase or multiple phases at the buyer node, with these sub-queries

The assumptions made for the emulation are as listed below:

- All nodes will run without any failure; system failure or hardware failure.
- Query cost and processing task cost are based on heuristics, due to unavailability of tool required to estimate cost of the query before actual execution of query. The load conditions of nodes are assumed and stored in a database. For simplicity, we have assumed that node will have either 80 percent load or 50 percent load. Experiments with current load conditions are planned for future work
- Ideal network conditions, example no packet losses, are assumed.

Based on the above experimental setup, we analyzed performance of *k*-QTPT and compared it with QT and QTPT. The evaluation approach taken to evaluate *k*-QTPT is stated as below:

- Implementation of QT
- Implementation of QTPT

- Implementation of *k*-QTPT, with different value of *k* and comparing the results with QT and QTPT.

In the next section, we present the results of our experiments, and our analysis based on the results.

6 Results and Analysis

To evaluate QT, QTPT and *k*-QTPT, we have measured optimization time, time taken to perform processing task at query initiating node (buyer node), time taken to perform processing task at query answering node (seller node) and total response time (total execution time). The main motive of our work, is to reduce optimization time of QTPT, hence our focus is on optimization time. We have taken results with fixed TCP data size and fixed inter packet delay. In our experiments, the number of rounds assumed to find best optimization plan is five and in each round we ran the experiment five times and took average of optimization time (in milliseconds). The results for optimization time comparison of QT, QTPT and *k*-QTPT is as shown in table1. Table includes, number of rounds (1 to 5), average optimization time and standard deviation for QT, QTPT, *k*-QTPT with *k*=1 and *k*=2. The graph plotted for optimization time versus number of rounds for results in table 1 is as shown in figure 5. The graph plotted is for 13 nodes (n). The average number of nodes (c) in each set (R_i) is 2.5. The number of winning nodes (w) buffered in first phase of *k*-QTPT is 5. According to *k*-QTPT, in second phase we request bids from *k* nodes of each set. Here, we have plotted graph for *k*=2 and *k*=1. The x-axis of graph represents number of rounds and y-axis of graph represents optimization time in milliseconds. As shown in figure 5, with decreasing value of *k*, optimization time of *k*-QTPT reduces compare to QTPT.

Table 1. Optimization time comparison for QT, QTPT and *k*-QTPT

	QT		QTPT		*k*-QTPT , *k*=2		*k*-QTPT , *k*=1	
Rounds	$\bar{X}$	∂	$\bar{X}$	∂	$\bar{X}$	∂	$\bar{X}$	∂
1	2659.4	102.4	3128	744.6	2862.4	180.2	2563.8	188.0
2	5290.4	1157.9	5524.8	797.9	5346.4	1088.9	4866	266.5
3	8259.6	1229.1	8474.8	1384.2	8084.4	1337.4	7690.6	1530.7
4	9881.2	1520.9	11987.8	1997.1	11287.4	3467.1	9565	1174.7
5	12503	2253.4	14518.4	3557.5	13986.8	1617.7	13496.8	1526.1
$\bar{X}$ - Average (Mean) optimization time ∂ - Standard Deviation								

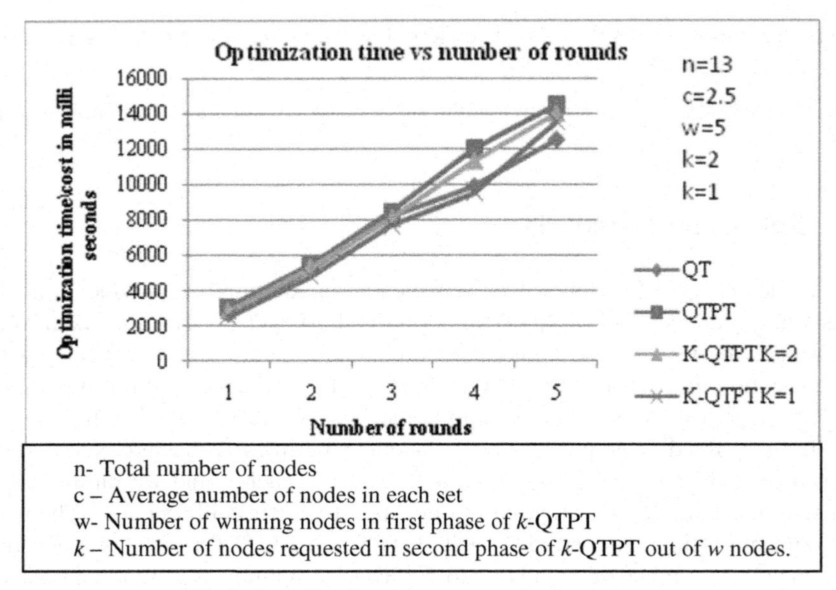

Fig. 5. Optimization time comparison for n=13

We also measured average time taken to query each node in each round. The readings for same are as shown in table 2. The graph for number of rounds vs. average time taken to query each node is as shown in figure 6.

Table 2. Average time taken to query each node for QT, QTPT and *k*-QTPT, *k*=2 and *k*=1

	QT	QTPT	*k*-QTPT, *k*=2	*k*-QTPT, *k*=1
Rounds	$\bar{X}$	$\bar{X}$	$\bar{X}$	$\bar{X}$
1	219.94	118.84	127.70	147.39
2	201.68	150.07	154.69	155.64
3	195.83	179.92	168.36	182.32
4	193.26	153.61	158.82	168.09
5	191.34	185.62	173.02	183.49
$\bar{X}$ - Average time taken to query each node				

As shown in figure 6, average time taken to query each node in QT, QTPT and *k*-QTPT is almost similar. Hence, we can conclude that optimization time depends on number of *k* nodes queried in second phase of *k*-QTPT. In figure 6, average time taken to query each node for QT is quiet higher compare to QTPT and *k*-QTPT. We believe that QT average time is higher compare to QT, QTPT and *k*-QTPT because of possibility of tables being queried first at a time of QT being cached. Due to caching, cache hit ratio is higher in QTPT and *k*-QTPT as being executed later. As QTPT in first phase is same as QT it is evident from QTPT, that if we run QT again that may reduce average time taken to query each node. However, further investigations are required for it.

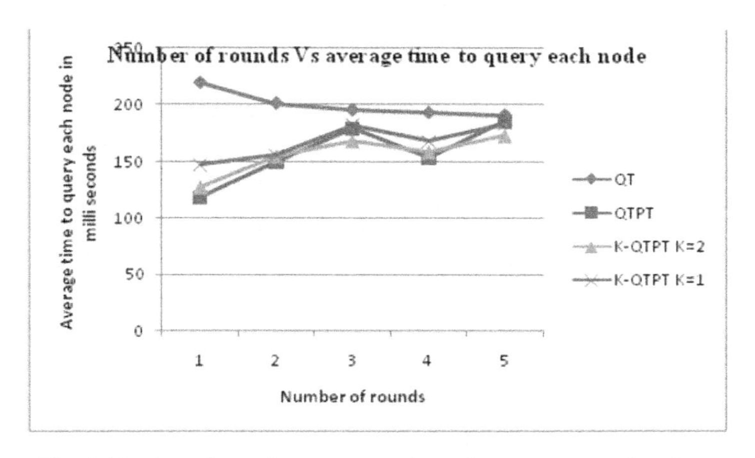

Fig. 6. Number of rounds vs. average time taken to query each node

In this section, we discussed experimental evaluation of QT, QTPT and *k*-QTPT. The results shows that optimization time of *k*-QTPT decreases substantially compare to QT and QTPT. The ideal value of k needs further investigation, which is our future work.

7 Conclusion and Future Work

The performance of a distributed database system depends on efficient query optimization. The dynamic strategies for query optimization produce optimal plans; however they are unsuitable for autonomous DDBMS. Mariposa, QT and QTPT are optimization strategies for autonomous systems. QTPT generates optimal plans compared to QT, but incurs high optimization cost. In our work, we have proposed *k*-QTPT, to reduce the high optimization cost of QTPT.

In QTPT, in both phases, costs are requested from all n nodes for all rounds (r).If we consider q as time taken to query each node then the optimization cost for QTPT is $q*r*n$ (for first phase) + $q*r*n$ (for second phase) i.e. $2(q*r*n)$.

From the explanation of *k*-QTPT in section 4, we can see that in the first phase the cost is $q*r*n$ (same as cost of first phase in QTPT) and in the second phase, it is $q*1*((n/c)*k))$, where n/c represents the average number of sets; and r=1.

From the above, we can easily see that *k*-QTPT will perform better when cost of *k*-QTPT is less than that of QTPT or $((q*r*n) + (q*1*((n/c*k)) < (2*q*r*)$ i.e. $k<r*c$.

Our experimental results show that with different values of k, optimization time of *k*-QTPT reduces as compared to QTPT.

We shall be conducting further investigations to study the impact of following parameters; CPU load at each node, effects of caching on different types of queries with local optimization and varying network conditions. Further, methods for accurately determining the query costs stored at the nodes needs to be investigated in detail.

References

1. Oszu, M.T., Valduriez, P.: Principles of Distributed database systems. Prentice Hall international, NJ (1999)
2. Aljanabv, A., Abuelrub, E., Odeh, M.: A survey of Distributed Query Optimization. The International Arab Journal of Information Technology 2(1) (2005)
3. Ioannidis, Y.E.: Query Optimization. In: Trucker, A. (ed.) The Computer Science and Engineering Handbook, pp. 1038–1054 (1997)
4. Selinger, P.G., Astrahan, M.M., Chamberlin, D.D., Lorie, R.A., Price, T.G.: Access Path selection in a relational database management system. In: ACM SIGMOD Conference on Management of Data, pp. 23–24. s.n., Boston (1979)
5. Palerno, F.P.: A database search problem. In: Tou, J.T. (ed.) Information Systems COINS, pp. 67–101. Plenum Press, New York (1974)
6. Kossmann, D., Stocker, K.: Iterative dynamic programming: A new class of query optimization algorithms. ACM Transactions on Database Systems 25(1) (2000)
7. Stonebraker, M., Aoki, P.M., Litwin, W., Pfeffer, A., Sah, A., Sidell, J., Stalien, C., Yu, A.: Mariposa: a wide area distributed database system. The VLDB Journal, 48–63 (March 1996)
8. Deshpande, A.V., Hellerstein, J.M.: Decoupled Query Optimization for Federated Database Systems. In: 18th International Conference of Data Engineering, pp. 716–792. IEEE Computer Society, Los Alamitos
9. Pentaris, F., Ioannidis, Y.: Query Optimization in Distributed Netwroks of Autonomous Database Systems. ACM Transactions on Database Systems 31(2), 537–583 (2006)
10. Doshi, P., Raisinghani, V.: Review of Dynamic Query Optimization Strategies in Distributed Database. In: International Conference on Network and Computer Science. IEEE Explorer, India (2011)
11. `http://www.mckoi.com/mckoiddb/index.html`
12. Zurek, T., DipperWaldrof, S., Na, K.: Gel. Data Query Cost Estimation. 7,668,803 US, Heidelberg (February 2010)
13. Ray, C.: Distributed Database Systems. Pearson Publication, s.l. (2009)
14. Jacobs, B.E., Walczak, C.A.: Optimization algorithms for distributed queries. IEEE Transactions on Software Engineering 9(1) (January 1983)
15. Ghaemi, R., Fard, A.M., Tabatabee, H., Sadeghizadeh, M.: Evolutionary query optimization for heterogenous database systems. World Academy of Science, Engineering and Technology 23 (2008)
16. Kossmann, D.: The state of art in distributed query processing. ACM Computing surveys 32(4), 422–469 (2000)
17. Bernstein, P., Goodman, N., Wong, E., Reeve, C., Rothine: Query Processing in a system for distributed databases (SDD-1). ACM Trasactions on Database Systems 6(4), 602–625 (1981)
18. Hass, L.M.: R*: A research project on distributed relational DBMS. Database Engineering 5 (1982)
19. Ono, K., Lohman, G.: Measuring complexity of join emumeration in query optimization. In: 16th International Conference on Very Large Databases (VLDB), pp. 314–325. s.n., Berkley (1990)

20. Ioannidis, Y.E., Kang, Y.C.: Randomized algorithms for optimizing large join queries. In: Proceedings of the ACM SIGMOD Conference on Management of Data, pp. 312–321. s.n., Atlantic city (1990)
21. Kirkpatrick, S., Gelatt, C.D., Vecchi, M.P.: Optimization by simulated Annealing. Science 220, 671–680 (1983)
22. Nahar, S., Sahni, S., Shragowitz, E.: Simulated Annealing and Combinatorial Optimization. In: Proceedings of the 23rd Design Automation Conference, pp. 293–299 (1986)

Market-Driven Continuous Double Auction Method for Service Allocation in Cloud Computing

Nima Farajian and Kamran Zamanifar

University of Isfahan, Department of Computer Engineering,
Faculty of Engineering
nimaff2000@yahoo.com, zamanifar@eng.ui.ac.ir

Abstract. In Cloud Computing, the computing resources, either software or hardware, are virtualized and allocated as services from providers to users. Each service requires various resources which all should be allocated for utilizing the service. Market-oriented approach is one of the most popular methods for managing Cloud services. In this paper a market-driven continuous double auction method (MCDA) for efficient cloud service allocation is presented which enables consumers to order various resources as a workflow for utilizing requested services. Also in the presented method consumers and providers make bids and requests based on market rivalry, time, opportunity, and eagerness factor. Ex-perimental results show that trading price depends on market supply and demand fairly and also proposed method is efficient in terms of successful allocation rate, resource utilization, and average connection time.

Keywords: cloud computing, continuous double auction, intelligent agent, market-driven agents, resource allocation.

1 Introduction

Cloud computing offers a new computing model in which resources such as computing power, storage and online applications are provided as 'services' over the internet [1] and is growing rapidly popular and appears suitable for resource sharing. Cloud computing is collection of several interconnected and virtualized computer which could be used as a one or more unified computing resource based on service-level agreements (SLAs) between providers and consumers [2].Cloud Resource management consists of finding and allocating required resources of each service through interaction among cloud participants. However this work couldn't be done easily because of: (i) the cloud environment is highly dynamic and resources could be added or removed frequently, (ii) consumers and providers make decision autonomously based on their status, (iii) consumers can acquire different multiple resources simultaneously and also compete with each other to access suitable resources [4]. It was noted [2] that a market-oriented approach for Cloud resource management is essential. It regulates supply and demand for resources; motivates the consumers to trade-off between deadline, budget, and the required level of quality of service; and provides an incentive for resource providers to participate in Cloud [3]. One of the

S. Unnikrishnan, S. Surve, and D. Bhoir (Eds.): ICAC3 2013, CCIS 361, pp. 14–24, 2013.

most challenging issues in market based models is pricing. Trading price should be fair, flexible, dynamic, and have high correlation to real market situation [2]. Two categories of market based models that are used for cloud resource management are commodities market models and auction models. In the commodity market model, resource prices are specified by resource providers, and consumers are charged according to the amount of the resource which they consume. In the auction model, each provider and consumer make decision autonomously and agree privately on the selling price.Auctions are used for products that have no standard values and supply and demand of the market affect the prices [3]. Generally there are 4 kinds of auction: English auction, Dutch auction, First and Second Price Auction and Double auction. The double auction model has a high potential for cloud computing. In a double auction model, consumers submit bids and providers submit requests at any time during the trading period. If at any time there are bids and requests that match or are compatible with a price, then a trade is executed immediately [12].

Three most popular double auctions are: Preston-McAfee Double Auction Protocol (PMDA) [5], Threshold Price Double Auction Protocol (TPDA) [6], and Continuous Double Auction Protocol (CDA) Kant and Grosu [7] showed that the CDA protocol provides high resource utilization in grid environments and is better from both the resource's and the user's perspective.

In this paper, a continuous double auction method for service allocation in cloud computing is presented in which consumers and providers can adjust the value of their bids and requests quickly to the acceptable prices of the market through time, competition, opportunity, and eagerness factors consideration. Because of close dependency between market conditions and the trading price, we call this method market-driven continuous double auction method (MCDA). The results illustrate that the trading price is fair, flexible, and adaptive to market variations and also the proposed method is efficient in resource utilization and successful allocation rate. The rest of the paper is organized as fallow. Section 2 describes some relevant works. The design details of this model and its mechanism are presented in Section 3. The evaluation and experimental results are shown in Section 4 and conclusion is provided in Section 5.

2 Related Works

Most of the studies and researches on management of resources and services in cloud computing is derived from methods and techniques which have been used in grid computing. The reason for this is high similarity between cloud and grid computing and also earlier emergence of grid computing.

Buyya et al. [3] used economic based concepts including commodity market, posted price modeling, contract net models, bargaining modeling for grid resource allocation. Schnizler et al. [8] introduced a double-sided combinational auction to allocate grid resources. Tan et al. [9] proposed a stable continues double auction, based on the more conventional CDA. It reduces the unnecessarily volatile behavior of the CDA, while maintaining other beneficial features. Amar et al. [10]

illustrated a comprehensive grid market model including a futures market and a centralized/decentralized spot market. Izakian et al. [11, 12] introduced a continuous double auction method in which resources are considered as provider agents and users as consumer agents. In each time unit, each provider agent determines its requested value based on its workload and each consumer agent determines its bid value based on the remaining time for bidding and the remaining resources for bidding. In [13] authors present a reverse VCG auction for cloud computing in which consumers hold auctions for required resources and providers submit their offers and winner determination is based on lowest offer. Fujiwara et al. [14] Presented combinational auction model for cloud computing and design spot and future auction market for provider to put their resource in it. Importantly, this method use combinational auction which auctioneer puts multiple different types of resources in auction and each consumer can bid for combination of them. Winners determination is based on maximum profit. Teymouri [15] proposed a new UCDA method based on CDA method in which bids are updated by auctioneer itself. Also providers determine the resource price based on their workload and users determine their bids based on jobs deadlines.

Sim [16] presents a market-driven negotiation mechanism for Grid resource management which includes (i) a market-driven strategy and (ii) a relaxed-criteria negotiation protocol. In market-driven strategy, agents consider opportunity and time factors to determine the value of prices. Sim [17] presented a resource management model for cloud computing based on market-driven agent which introduced in [16]. In this model there are provider, consumer and broker agents. Through a 4-stage resource discovery process (selection, evaluation, filtering, and recommendation), Broker agents match consumers' requests to suitable Resources and then consumer and provider agents negotiate for mutually acceptable resource time slots.

3 Market Mechanism and Problem Definition

3.1 Cloud Computing Market Model

We assume the cloud computing environment includes resource providers, consumers and an auction market. Consumers refer to cloud to use and consume its required services. Each service is composed of different resources which are provided by providers who sell them to make profit. The auction market consists of multiple auctioneers which each of them holds auction for one resource type. For each entity in cloud market there is one agent. Consumer agent works on behalf of the consumer and acquires required resources of the requested service by interaction with auctioneers, provider agent which works on behalf of the provider and sells its resource to consumers through related auctioneer and auctioneer agent which receives bids and requests from all consumers and providers and executes trade for all matching bids and requests. Fig 1 shows a cloud computing environment with the proposed mechanism.

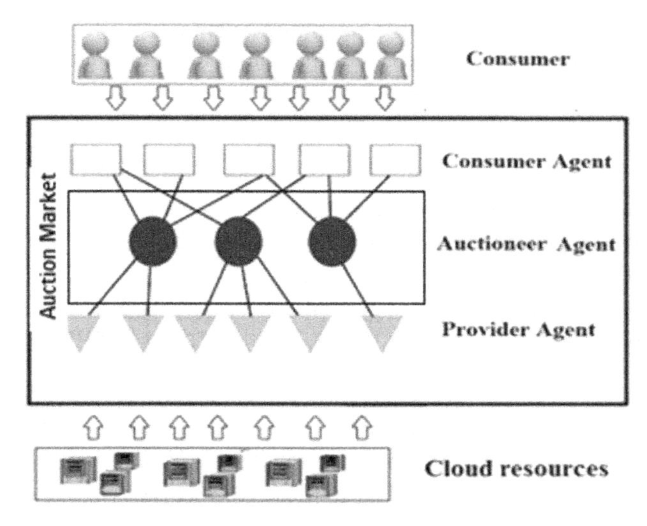

Fig. 1. Resource allocation schema in proposed method

3.2 Problem Formulation

Each resource R is characterized by five-tuple, Ri = (typ, spp, twp, rpp, mpp), in which typ is resource type(for example CPU, Hard Disk, Ram ...), spp is the resource specification (for CPU process speed , For RAM and Hard Disk capacity and ...), twp is current workload of resource and means that resource is busy until twp, rpp refer to lowest price for resource Ri(also called reserve price) and is expressed in form of Cloud units per second (C$ /S) . Also mpp is maximum price for resource Ri and is expressed in form of Cloud units per second (C$ /S). In this paper we assume that each resource is allocated by one consumer in each time unit.

Each service Ci is composed of some resource requests which each of them is characterized by five-tuple, Ci,j= (typ,spc,ts,td,mps), typ and spc determine type and specification of requested resource ,ts is amount of time that Ci need to utilize resource j and td is deadline and Ci must utilize resource j before td. mps represents the budget allocated to Ci,j.

3.3 Continuous Double Auction Method

Each consumer is looking for utilizing its requested service with minimum price before its deadline while each provider wants to sell its resource at maximum price and also reduces idle time of its resource. To utilize a service all required resources should be allocated and utilized before deadline otherwise service failed to utilize and consumer must pay penalty to providers for all other resources which is allocated. In the continuous double auction method at each time unit, consumers and providers submit bids and requests to the related auctioneer. An auctioneer maintains a list of the current bids and requests and matches the two offers when their prices are compatible. In the proposed method consumers and providers determine their bids and requests value autonomously based on market conditions and their interest. This method is inspired by the work presented in [12].

3.3.1 Determining Bid Values for Consumer Agents

Consumer agent in each time unit determines its bid price for each requested resource based on time, availability, competition and eagerness factors. By considering these factors consumer agent converges its bid value rapidly to the acceptable price of market.

Time factor: Since consumers are generally sensitive to deadline in acquiring re-quested resources, it is intuitive to consider time when formulating the bid price. Con-sumer agent time dependent bid price formula for resource j of service i at time t is determined in (1).

$$BidT(i,\mathrm{j},t) = \left[\frac{R_{min}}{mps} + \left(1 - \frac{R_{min}}{mps}\right)\left(1 - \frac{tre}{td}\right)^{\rho}\right] \times mps \quad (1)$$

Rmin is minimum reserve price between all providers and tre is time interval between current time (t) and td. $0 < \rho < \infty$ is the concession making strategy. Three classes of strategies are specified as follows: Conservative (ρ >1) which the consumer maintains a low bid value until current time gets close to td, Normal (ρ =1) which the consumer increases bid linear based on current time, and Conciliatory (0< ρ <1) which the consumer starts with a bid value close to *mps*. Fig. 2 describe the different convexity degrees of the curves with $\rho = 0.2,1,5$.

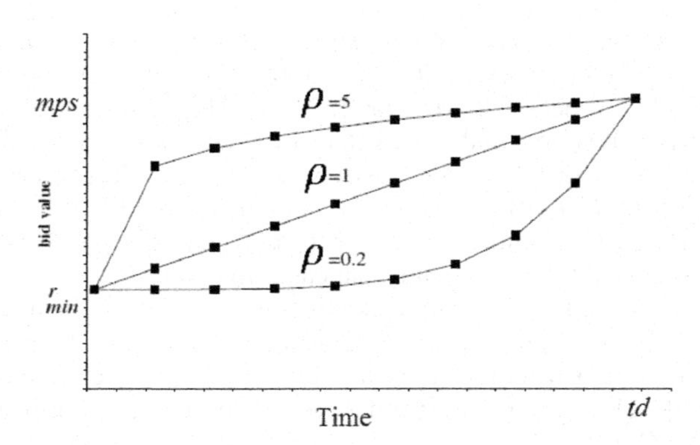

Fig. 2. Bid value based on time

Availability factor: Number of available resources is one of the important factors for consumers. Consumer agent bid price formula based on remaining resource is in (2).

$$BidR(i,\mathrm{j},t) = \left[\frac{R_{min}}{mps} + \left(1 - \frac{R_{min}}{mps}\right)\left(1 - \frac{N_i^t - 1}{N_{avg}^t}\right)^{\sigma}\right] \times mps \quad (2)$$

σ is the concession making strategy similar to ρ in (1). $\boldsymbol{N_{i,j}^t}$ is available resources for *Ci* in time t and $\boldsymbol{N_{avg}^t}$ is average number of available resource during the last **n** time units and is determined using (3). In cloud market resource number is constantly changing and resources are added and removed dynamically so we need to consider market conditions in terms of current number of resources compared to the market history.

$$N_{avg}^{t} = \frac{\sum_{T=t-n}^{t} N_{i,j}^{T}}{n} \tag{3}$$

It should be noted that for calculating $\boldsymbol{N_{i,j}^t}$ and $\boldsymbol{N_{avg}^t}$ only resources are considered which could be utilized for consumer request. The resource could be allocated to consumer request if

$$td - twp - ts \geq 0 \; and \;\; mps \geq rpp \;\; and \; spp \geq sps \tag{4}$$

Competition Factor: Since all consumers compete for acquiring resources and all providers compete to sell their resources, market rivalry and competition should also be considered. In a market with m consumers and n providers, a consumer agent $C1$ has m-1 competitors $\{C2... Cm\}$ and n trading partners $\{R1... Rn\}$. The probability that $C1$ is *not* the most preferred trading partner of *any* $Rj \in \{R1,..., Rn\}$ is $(m\text{-}1)/m$. So, the probability that $C1$ is *not* the most preferred trading partner of *all* $Rj \in \{R1,..., Rn\}$ is $\left[\frac{m-1}{m}\right]^n$. Consumer agent bid price formula based on market competition is in (5).

$$BidC(i,j,t) = \left[\frac{R_{min}}{mps} + \left(1 - \frac{R_{min}}{mps}\right)\left(\left[\frac{m-1}{m}\right]^n\right)\right] \times \;\; mps \tag{5}$$

Eagerness Factor: As mentioned before for utilizing a service, all required resources should be allocated before deadline, otherwise service is failed to utilize and in addition consumer must pay penalty. So consumer should make more effort and relaxes its bid price when some of its required resources are allocated. Eagerness factor represents how urgent it is for consumer to acquire necessary resources before deadline and is calculates as (6).

$$E(i,t) = \frac{\sum_{j=1}^{k} y_{i,j}}{k} \qquad where \; y_{i,j} = \begin{cases} 1 & if \; C_{i,j} \; allocated \\ 0 & if \; C_{i,j} \; unallocated \end{cases} \tag{6}$$

k is number of required resources for utilizing service Ci. Finally consumer agent make final bid price using (7), (8)

$$Bid_{tmp}(i,j,t) = \varphi \times BidT(i,j,t) + \mu \times \; BidR(i,j,t) \; + \; \gamma \; \times BidC(i,j,t) \quad \text{where} \;\; \varphi + \; \mu + \gamma = 1 \tag{7}$$

$$Bid\;(i,j,t) = Bid_{tmp}(i,j,t) + \left(mps - Bid_{tmp}(i,j,t)\right) \times E(i,t) \tag{8}$$

3.3.2 Determining the Request Value for Provider Agent

Each provider has two objectives: (i) make maximum profit for selling its resource, (ii) reduce idle time of its resource. Provider agent in each time unit determines its request value based on time and opportunity factors.

Time factor: Provider agent is sensitive to idle time of its resource and considers workload time for calculating price as in (9).

$$RequestT(i,t) = \left[\frac{\text{rpp}}{mpp} + \left(1 - \frac{\text{rpp}}{mpp}\right)\left(\frac{tst}{twp}\right)^{\tau}\right] \times mpp \tag{9}$$

tst is time interval between t and twp. Based on (8) each provider offers its reserve price when its resource is idle and after allocation makes price to mpp and with elapsing time decreases its price. Also τ is the concession making strategy like ρ *and* σ.

Opportunity factor: This factor expresses number of buyers for provider's resource. If the number of buyers decreases (increases), it means that market situation is unfavorable (favorable) for provider and it should decrease (increase) its price. According to number of buyers, provider calculates price formula in (10).

$$RequestC(i,t) = \left[\frac{\mathrm{rpp}}{mpp} + \left(1 - \frac{\mathrm{rpp}}{mpp}\right)\left(1 - \frac{C_i^t - 1}{C_{avg}^t}\right)^{\pi}\right] \times mpp \tag{10}$$

π is the concession making strategy similar to ρ, σ, τ . $\boldsymbol{C_i^t}$ is available buyer in time t and $\boldsymbol{C_{avg}^t}$ is average number of buyer during the last **m** time units and is determined using (11).

$$C_{avg} = \frac{\sum_{T=t-m}^{t} C_i^T}{m} \tag{11}$$

Provider agent finally makes its request price using (12).

$$Request\ (i,t) = \omega \times RequestT(i,t) + (1-\omega) \times RequestC(i,t) \quad \text{where } 0 \le \omega \le 1 \tag{12}$$

3.3.3 Auctioneer Agent

In each time unit, consumer and provider agents calculate their bid and request values and submit them to the auctioneer. The auctioneer sorts the bid values in increasing order and request values in decreasing order. Trade occurs when: (1) highest bid is more than or equal to the lowest request, (2) resource could be utilized by consumer as described in (3). When the above conditions are satisfied, the trade occurs at the average price of matching bid and request:

$$price = \frac{Highest\ matching\ Bid + Lowest\ matching\ Request}{2} \tag{13}$$

4 Experimental Results

In order to study the efficiency of the method presented in this paper, we developed a computational Cloud simulator using the java agent development framework (JADE) [18]. JADE is a middleware for developing multi agent systems and applications according to FIPA standards for intelligent agents. In our simulated system, consumers and providers are two kinds of agent and also for each resource type there is one auctioneer agent. Resource type and specification are shown in Table 1.

Table 1. Resource Specification

CPU Speed (GHz) : 1.5 ~ 3.5
Memory(MB):{512,1024,2048}
Disk Capacity(GB) : 50 ~ 1000

We use the system load concept i.e. the ratio of aggregated time and capacity of resource requests submitted to the cloud to the aggregated capacity that the cloud is capable of providing in the simulation period. The system load can be obtained using Eq. (14) which **m** is number of providers and **n** is total number of requests.

$$\rho_{Resource1} = \frac{\sum_{j=1}^{n}(spc_j \times ts_j)}{T_{total} \times \sum_{i=1}^{m} Spp_i} \quad (14)$$

In the simulated system there are 100 consumer agents and each service request of consumer agent requires 1 to 5 resources. Number of provider agent depends on system load and each provider agent provides one resource. For a fair comparison each provider set reserved price to 2$ and maximum price to 9$. The budget of each consumer is determined randomly between 3$ and 8$. Consumer agent utilization time of a resource (ts) is random integer number within range [2000, 5000] and deadline is from 2 to 5 times of ts and penalty fee for all resource is 5$. Also we set $\omega = 0.5$, φ =0.4 , $\mu = \gamma = 0.3$ and $\pi = \rho = \sigma = \tau = 1$.

To Evaluate efficiency of this method we consider success rate, average connection time, average resource utilization and average trading price (which is summarized in TABLE 2) and compared it with random continuous double auction (RCDA), continuous double auction method in [12] (CDA) and continuous double auction method in [15] (UCDA). In the RCDA consumers determine bids randomly between Rmin and mps and also providers set requests randomly between rpp and mpp.

Table 2.

Average connection time	$T_{avg} = \frac{1}{n}\sum_{j=1}^{n}(T_j^m - T_j^r)$
T_j^m	The time at which a request is matched to a resource
T_j^r	The time at which a consumer submits a request
n	The total no. of matches of requests to resources
Success rate	$Succ = \frac{R_{matched}}{R_{total}}$
$R_{matched}$	Number of requests that are matched to resource(s)
R_{total}	Total number of requests from consumers
Resource Utilization	$RU = \frac{T_{allocate}}{T_{total}}$
$T_{allocate}$	Total number of time slots allocated
T_{total}	Time window
Average Trading Price	$P_{avg} = \frac{1}{m}\sum_{j=1}^{m} P_j^{alloc}$
P_{trade}	Allocation price
m	The total no. of allocation

Fig 3 (a) illustrates comparison of successful allocation rate. In the proposed method, using the market-driven factors (especially eagerness factor) enables consumers to adjust their bid values rapidly to the acceptable prices of the market so consumers can acquire more resources before the deadline and as a result, the number of successful allocation is higher than other methods.

In fig 3(b) average connection time in different system loads is evaluated, in the presented method, the market-driven factors accelerate convergence speed of bid and request values, so average connection time is lower than other methods.

Fig 3 (c) shows resource utilization in different system loads and as shown in it, in MCDA resource utilization is more efficient than other methods especially in higher system load which is due to eagerness factor.

As shown in fig 3(d) when the system load increases (decreases), market demand increases (decreases) too and as a result, environment is more favorable (unfavorable) for providers and sell their resource with higher (lower) price. This figure illustrates that trading prices correlate to supply and demand of the market fairly and also is flexible to market variations so the proposed method motivates both consumers and providers to stay in the market and play their rules.

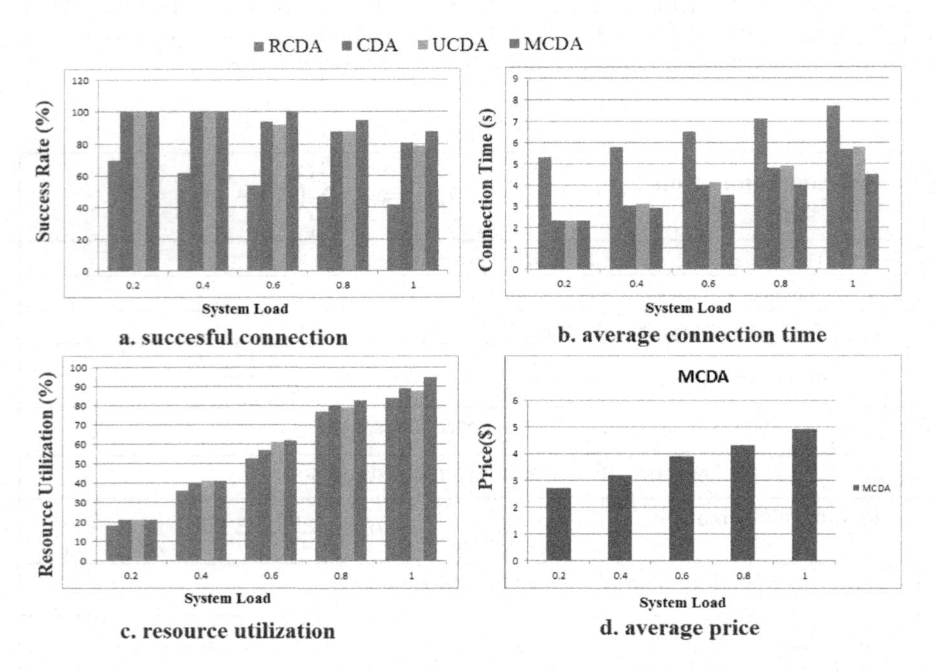

Fig. 3. Comparison on successful allocation, resource utilization, average connection time and average trading price

5 Conclusion

In Cloud Computing resource allocation is based on interaction and agreement between cloud consumers and providers. According to highly dynamic environment and also different policies of each provider and consumer, we need a method in which each participant can make decision autonomously based on market condition and its interest. In this paper a continuous double auction model is presented in which consumers and providers make bids and requests with time, opportunity, competition and eagerness consideration. Experimental results illustrate that proposed method is efficient and intensive for both consumers and providers and also trading price is fair, flexible and based on reality of the market.

References

1. Vaquero, L.M., Rodero-Merino, L., Caceres, J., Lindner, M.: A Break in the Clouds: Towards a Cloud Definition. ACM SIGCOMM Computer Communication Review 50, 39(1) (2009)
2. Buyya, R., et al.: Cloud computing and emerging IT platforms: Vision, hype, and reality for delivering computing as the 5th utility. Future Generation Computer Systems 25(6), 599–616 (2009)
3. Buyya, R., Abramson, D., Giddy, J., Stockinger, H.: Economic models for resource management and scheduling in Grid computing. The Journal of Concurrency and Computation 14(13-15), 1507–1542 (2002)
4. Sim, K.M.: A Survey of Bargaining Models for Grid Resource Allocation. ACM SIGECOM: E-commerce Exchanges 5(5), 22–32 (2006)
5. McAfee, R.P.: A dominant strategy double auction. Journal of Economic Theory 56, 434–450 (1992)
6. Yokoo, M., Sakurai, Y., Matsubara, S.: Robust double auction protocol against false-name bids. In: Proc. of the 21st IEEE International Conference on Distributed Computing Systems, pp. 137–145 (2001)
7. Kant, U., Grosu, D.: Auction-based resource allocation protocols in Grids. In: 16th International Conference on Parallel and Distributed Computing and Systems, pp. 20–27 (2004)
8. Schnizler, B., Neumann, D., Veit, D., Weinhardt, D.: Trading grid services - a multi-attribute combinatorial approach. European Journal of Operational Research 187(3), 943–961 (2008)
9. Tan, Z., Gurd, J.R.: Market-based grid resource allocation using astable continuous double auction. In: Proc. 8th IEEE/ACM Int. Conf. on Grid Computing (Grid 2007), pp. 283–290 (2007)
10. Amar, L., Stosser, J., Levy, E.: Harnessing migrations in a marketbased grid OS. In: Proc. 9th IEEE/ACM Int. Conf. on Grid Computing (Grid 2008), pp. 85–94 (2008)
11. Izakian, H., Ladani, B.T., Zamanifar, K., Abraham, A., Snasel, V.: A Continuous Double Auction Method for Resource Allocation in Computational Grids. In: IEEE Symposium on Computational Intelligence in Scheduling, pp. 29–35 (2009)
12. Izakian, H., Abraham, A., Tork Ladani, B.: An auction method for resource allocation in computational grids. Future Generation Computer Systems 26, 228–235 (2011)
13. Wu, X., et al.: Cloud computing resource allocation mechanism research based on reverse auction. Energy Procedia. 13, 736–741 (2011)

14. Fujiwara, I.: Applying Double-sided Combinational Auctions to Resource Allocation in Cloud Computing. In: IEEE 10th Annual International Symposium on Applications and the Internet (2010)
15. Teymouri, S., Rahmani, A.M.: A Continues Double Auction Method for Resource Allocation in Economic Grids. International Journal of Computer Application 43, 7–12 (2012)
16. Sim, K.M.: From Market-driven e-Negotiation Agents to Market-driven G-Negotiation Agents. In: Proc. of the IEEE Int. Conf. on e-Technology, e-Commerce and e-Services, Hong Kong, pp. 408–413 (2005)
17. Sim, K.M.: Agent-based Cloud Commerce. In: Proc. IEEE Int. Conf. on Industrial Eng. and Eng. Management, Hong Kong, pp. 717–721 (2009)
18. JADE, Java Agent Development Framework (2004), http://jade.cselt.it

Prediction Based Job Scheduling Strategy for a Volunteer Desktop Grid

Shaik Naseera[1] and K.V. Madhu Murthy[2]

[1] Department of Computer Science & Engineering,
Sreenivasa Institute of Technology and Management Studies, Chittoor, India
naseerakareem@gmail.com
[2] Department of Computer Science & Engineering
Sri Venkateswara University,
Tirupati, India

Abstract. Desktop grid is based on desktop computers owned and volunteered by individual users. Volunteer nodes donate their spare capacity like CPU cycles, memory and any other shared resources during their free time for public execution and withdraw from the public execution during their busy time. This is due to their high priority private jobs initiated by the node owner. Therefore, the volunteer nodes may join or leave the desktop grid at any instant of time because the volunteer nodes are not dedicated to public execution. Hence, the volunteer nodes availability time for public execution varies from one node to another node. In this paper, we consider the volunteer interferences as volunteer failures and duration of the interference (time to rejoin for public execution) as volunteer repair time. When a node fails during the execution of a job, the job might need to be resumed/restarted based on the fault tolerant capability of the node. This causes slowdown in the execution of the jobs at some nodes than the other nodes having the similar processing capability. This situation often leads to increased computational demands at some nodes. The desire to meet the increased computational demand at each node has influenced the interest on job scheduling policies that migrates local jobs for remote execution for the better turnaround time. In this paper we propose a prediction based job scheduling strategy (PJSS) that makes use of neural network load predictions combined with the node reliability parameters for making job scheduling decisions. The performance of the proposed method is compared against no-migration (NM) and resource exclusion (RE) algorithms. The simulation results show that the average turnaround time per job for the proposed method has got considerable improvement over no-migration and resource exclusion algorithms.

Keywords: Desktop Grid, job scheduling, migration, processing time, node failure, volatile.

1 Introduction

Advances in hardware and networking technologies made the availability of high speed computing devices manufactured and available at lower cost for desktop users.

S. Unnikrishnan, S. Surve, and D. Bhoir (Eds.): ICAC3 2013, CCIS 361, pp. 25–38, 2013.

Generally, the user desktops are not in use for majority of the time and hence plenty of idle computing cycles are available with the user desktops. Desktop grid is based on the desktop computers volunteered by the individuals. The objective of the desktop grid is to achieve high throughput computing by utilizing the idle computing cycles available with the user desktops.

Nodes volunteered for Desktop grid are located at different geographical locations and connected at the edge of internet. They possess different processing capability and storage. Therefore they are heterogeneous in natures. The nodes communicate with each other through internet service provider (ISP) server.

Desktop grid nodes are prone to two types of failures: volatile failure and interference failure [13]. Volatile failure is due to the network outages, link and crash failures that make system inaccessible to the desktop grid user. Interference failures are due to volunteer autonomic nature that the node owner can withdraw their participation from public execution due to high priority local jobs without giving prior notice. This causes nodes inaccessible to the desktop grid user as the Desktop grid should respect the volunteer autonomy. This leads to delay or entire loss of the job execution and this factor has influenced the interest on job scheduling policies that migrates local jobs for remote execution to get better turnaround time. It is found in [14] that the volunteer autonomic failures are much more frequent than link and crash failures. Therefore, the volunteer autonomic failures must be considered in job scheduling decisions in order to guarantee rapid turnaround time and reliable computation.

Job scheduling in a desktop Grid are mainly influenced by two factors: job transmission latency and node failure. Job migration to a remote node involves job transmission latency. By the time the job reaches to the remote node, the remote node might become busy and the selected target node may not be the best node for job scheduling. This is due to dynamic nature of the grid. This situation often demands the need for the prediction of remote node load statistics beforehand. This helps in making the job scheduling decisions by considering the change in load statistics of the remote node in the future interval of time. It is well known that artificial neural networks can be used to predict the future statistics based on the samples collected from the past history.

Node failure is another factor causes the slowdown in the execution of the scheduled jobs. This is due to node inferences caused by node owner's private jobs. Therefore, job transmission latency and node failures must be taken into account for job scheduling decision. In this paper we propose a prediction based job scheduling strategy (PJSS) that considers neural network predictions combined with node reliability parameters for selecting target node for scheduling the jobs generated in the Grid.

2 Related Work

Job scheduling algorithms are mainly classified into centralized, hierarchical and distributed algorithms [1]. In a centralized scheduling approach, there will be a central coordinator that takes scheduling decision for each job in the grid. This is a simple strategy that suffers with single point failure and scalability.

In a hierarchical scheduling algorithms [2], the nodes are divided into regions/sites and there will a local scheduler for each site and a global scheduler for the entire grid. The advantage of this approach is different scheduling policies can be adapted by the local scheduler and the global scheduler.

In a distributed scheduling [1], there is no central coordinator for making scheduling decisions. Each node possess its own local scheduler that takes scheduling decision for the locally generated jobs. The distributed scheduling algorithms are more complex compared to centralized and hierarchical job scheduling algorithms but they are highly scalable and fault tolerant.

A simple scheduling policy such as FIFO is applied in XtreamWeb[3]. X. He et al. have presented [6,7] Min-min algorithm and Max-min algorithms in which shortest job is scheduled on to a resource that offers minimum execution time and longest job is assigned to a node that offers minimum execution time respectively. Dinda P [8] has done significant work on the selection of resources based on the host load predictions.

Market driven, performance driven and trust driven scheduling strategies are proposed for workflow systems in the -grid context [10]. Dayi Zhou and Virginia lo [11] presented job scheduling strategies like migration immediate, wave migration, migration linger, wave linger and migration adaptive to achieve faster turnaround time for job scheduling algorithms in desktop grid computing environment. The nodes in [11] follow adaptive scheduling that identify night zone nodes and transfer the unfinished jobs that are delayed at any node.

Authors in [4,5] presented a scheduling algorithm that uses the percentage of CPU idle time as a metric for job scheduling. Algorithms like resource prioritization, resource exclusion, task replication, fixed threshold and host availability traces are proposed and shown that they have considerable performance for short lived applications. Derric Konda et. al have presented [12] a task scheduling algorithm that deals soft real-time applications that have deadlines. The approach proposed in [12] uses buffer space to improve task completion rate.

Most of the scheduling heuristics proposed in this section are based on the greedy choices that depend on the transitory completion times of the jobs. These methods do not consider the information about the changing environmental variables like node reliability and dynamically changing load conditions in the desktop grid when job transmission latencies are more. Some of the algorithms make scheduling decisions based on the periodically collected information available with the node. The collected information might become outdated by the time a new job is generated at the node. Considering this fact into account, we are proposing PJSS that takes future load predictions of the node into account for effective scheduling decision.

3 Modeling Issues

To make the model simple and realistic to the real desktop grid environment, the following modeling issues are considered. These modeling issues are chosen by taking into the account of field study conducted and found in the literature for the real grid environments.

- The jobs are generated at each node i on Poisson distribution with mean λ_i.
- The instruction and data size of the job at each node i follows exponential distribution with mean μ_i.
- The processing time of the job at each node i follows normal distribution with mean ψ_i and variance $\sigma^2_{\psi_i}$.
- The volunteer availability time between failures follows exponential distribution with mean $1/f_i$ and the duration of volunteer unavailability (time to rejoin or duration of the interference) follows exponential distribution with mean r_i.

4 List of Common Assumptions

To make the system simple and realistic to the real grid environment, the following assumptions are made. The assumptions are chosen from the field study conducted in real grid environments.

- There is a fixed bandwidth between two nodes (B is constant).
- The Congestion towards a node is chosen fixed (C is constant).
- The nodes communicate via ISP server located at equidistant to all the nodes, therefore the logical distance between nodes become constant (D is constant).
- The volunteer autonomic failures are more frequent compared to volatility failures. This assumption also confirms with the steady found in the literature [14].
- The size of the load packet that contains the remote node load information is assumed to be fixed for all the nodes. Therefore, *p is* constant.
- The jobs are independent. Each job has different size and processing time.
- For any job, the scheduling decision is taken only once. The decision policy of source node takes scheduling decision for each job.
- When a node decides to send a job for remote execution, it is dispatched completely to the remote node in its entirety.
- The local scheduler of each node schedules the jobs on FCFS basis.
- The size of the result for each job is same as the instruction/data size of the job.
- Each node after completing the execution of a job sends the result back to the source node where it is originally generated.
- The mean job arrival rate ($\bar{\lambda}$), mean node failure rate ($\bar{f}$), mean repair time between failures ($\bar{r}$), mean instruction/data size of the job ($\bar{\mu}$), mean job processing time ($\bar{\psi}$), mean clock speed ($\bar{c}$) for nodes in a desktop grid are chosen from sample space of normal distribution with corresponding mean and variance.

5 Problem Statement

"Let G be the desktop Grid consisting of N volunteer desktop nodes spread in a geographically different locations. The nodes are heterogeneous in nature and posses different clock speed c_i. The nodes are connected over internet and communicate with each other via Internet Service Provider (ISP) located at equidistance to all the nodes in the desktop Grid. The Desktop Grid user generates jobs at each node on Poisson distribution with mean λ_i. The mean instruction/data size of a job (bytes) for each node i follows exponential distribution with mean μ_i. The processing time of a job at node i follows normal distribution with mean ψ_i and variance $\sigma^2_{\varphi_i}$. The volunteer nodes in a desktop Grid encounter volunteer autonomic failures. Let f_i is the mean autonomic failure rate of the volunteer node, the volunteer availability time between failures for each node i follows exponential distribution with mean $1/f_i$ and the volunteer unavailability duration (mean time to rejoin/repair time) follows exponential distribution with mean r_i. The objective is to find a target node l for the scheduling of each job generated in the grid that minimizes the average turnaround time per job in the Grid ($\overline{G}^t$)."

5.1 Objective Function

The objective function is to Minimize $\overline{G}^t = \frac{1}{N}\sum_{i=1}^{N} \bar{T}_i^t$ where $\bar{T}_i^t = \frac{1}{m_i^t}\sum_{j=1}^{m_i^t} T_{ij}^l$ subject to

(i). $l: \hat{T}_{ij}^l < \hat{T}_{ij}^i$
(ii). $1 \leq l \leq N$ and $1 \leq i \leq N$

Where $\hat{T}_{ij}^l$ is the estimated turnaround time for job j of node i at node l. This is done using neural network predictions.

6 Proposed Technique: PJSS

The major overhead involved in the remote execution of the jobs is job transmission latency and remote node reliability. Job migration to remote node involves latency, by the time it reaches the remote node the node might become busy due to new local and remote job arrivals and the selected target node for job migration may not be the best node for job execution. This situation often demands the need for the prior estimation of change in load statistics of the remote node in the future interval of time. To address this issue, we employ artificial neural network (ANN) for each node. Each node train its ANN with the periodically collected load samples in the past using Back propagation neural network algorithm. The trained ANN is used for predicting the change in load statistics of the remote node for the future interval of time. In this way PJSS takes care of the transmission latency of the job into account for effective scheduling decision.

The design of artificial neural network used in this work is shown in Fig. 1. The input layer consists of 'M' input elements and output layer consists of 'O' output neurons. The input layer receives the external inputs and the output layer produces the predicted load value for the next interval. The input and output layer are separated by a hidden layer that captures the relationship between input and output values. The number of neurons in the hidden layer (H) is chosen using the following formula.

$$H = (M + O) \times \frac{2}{3} \quad (1)$$

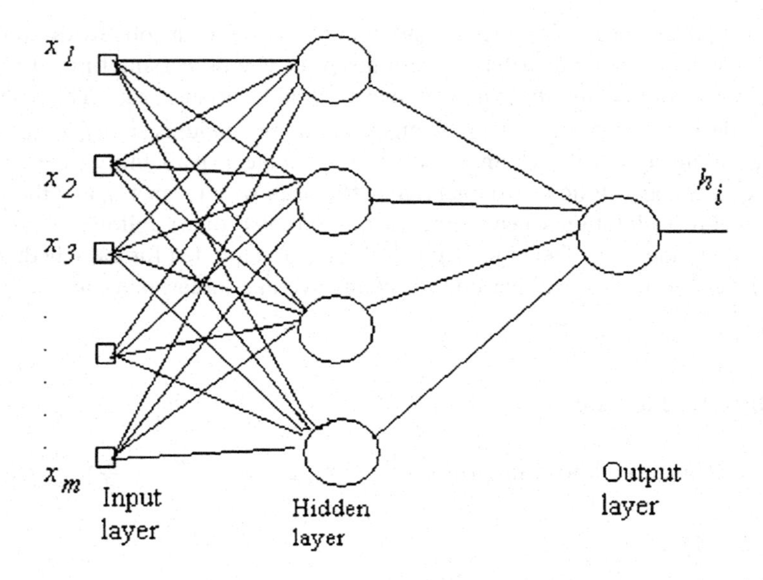

Fig. 1. A two layer feed forward Neural Network

The number of neurons in our design is chosen as 5 neurons in the input layer, one neuron in the output layer and 4 neurons in the hidden layer.

Each node train the neural network with periodically collected load patterns. The training pattern consists of set of input and output load samples generate in the past from the running node. Examples of training data sets are in the form of <input, output> vector pair. The input vector size and output vector size follows the design of neural network input and output elements respectively. The collected load patterns are normalized by using the following formula. LB represents the lower bound and UB represents the UB represents the upper bound of the normalization.

$$x_i = LB + \frac{x_i - x_{min}}{x_{max} - x_{min}} (UB - LB) \tag{2}$$

Where x_{min} and x_{max} are the minimum and maximum values of the load patterns available in the training patterns. The normalized load patterns are fed to the neural network and the neural network is trained with back propagation algorithm with a learning rate parameter η=0.7. During the training process the neural network adjust their interconnection weights and the summation done at each neuron i is as shown in Fig. 2.

When the network is run, each layer performs the calculation using the following formula and the result h_i is transferred to the next layer.

$$h_i = g\left(\sum_{j=1}^{m} h_j w_{ij} + b_i\right) \tag{3}$$

Where$g(y) = \begin{cases} \frac{1}{1+e^{-y}} & if\ hidden\ neuron \\ y & if\ output\ neuron \end{cases}$ and m is the number of inputs applied to neuron *i*.

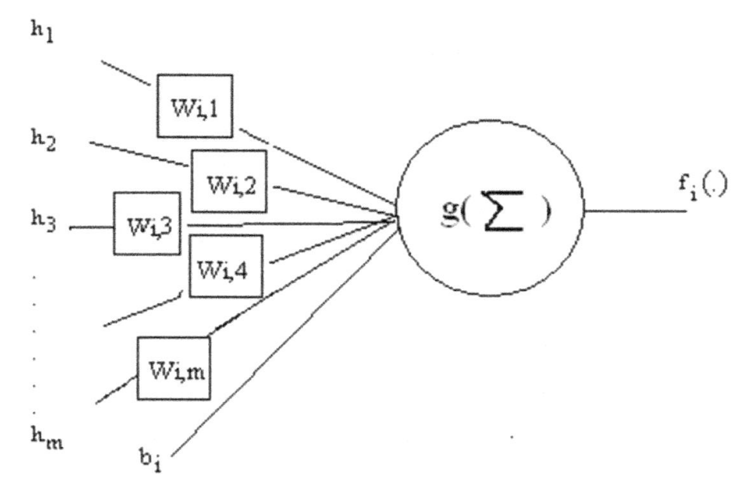

Fig. 2. The summation done at each neuron

The training process is stopped when the normalized mean squared error at the output layer is minimum i.e., 0.02. The error graph obtained for a sample node with learning rate parameter η=0.7 during training is shown in Fig. 3. The similar procedure is followed for all the remaining nodes in the Grid.

Each node predicts its own load statistics for the next interval from the trained ANN. These statistics are broadcasted to other nodes connected in the Grid. In this way every node not only maintains the local load statistics but also maintains the load statistics of the remote nodes in the Grid. When a local job is generated in a node, the node immediately takes scheduling decision based on the locally available remote node load statistics. Thus the job scheduling decision making delay is avoided and PJSS schedules the job to a remote node when the estimated turnaround time based on the predicted load at the remote node is less than the estimated turnaround time based on the current load at the local node.

Let $s_i(r), s_i(2r), s_i(3r), \dots, s_i(dr)$ be the average load traces of node *i* sampled at time $t = r, 2r, \dots, dr$ with periodic interval *r*. A sample $s_i(nr)$ is computed by the sum total of processing times of all the jobs (m_i^t) scheduled at node *i*.

$$s_i(nr) = \frac{1}{r}\sum_{t>(n-1)r}^{nr}\sum_{j=1}^{m_k^t} p_{kj}^i \text{ where } 1 \le k \le N \tag{4}$$

$h_i((d+1)r)$ is the average load predicted for node *i* for the next interval $t = (d+1)r$. It is obtained from *m* traces back from *t*. It is obtained by using the artificial neural network function$f_i(.)$ with *m* inputs and one output. It is given by,

$$h_i((d+1)r) = f_i(s_i(dr), s_i((d-1)r), \dots, s_i((d-m+1)r)) \tag{5}$$

The estimated waiting time $\widehat{w}_{ij}^{lt}$ for job *j* of node *i* at node *l* is given by,

$$\widehat{w_{ij}^{lt}} = h_l((d+1)r) \times (1 + \frac{1}{f_l} \times r_l) \quad where\ dr < t \le (d+1)r \tag{6}$$

The estimated turnaround time ($\widehat{T_{ij}^{l}}$) for job j of node i scheduled at node l is given by,

$$\widehat{T_{ij}^{l}} = L_{ij} + \widehat{w_{ij}^{lt}} + L_{ij} \tag{7}$$

A node i migrates job job j of node i to remote node l where

$$l = \{k | \widehat{T_{ij}^{k}} < \widehat{T_{ij}^{l}} \; where \; k = random(N)\} \tag{8}$$

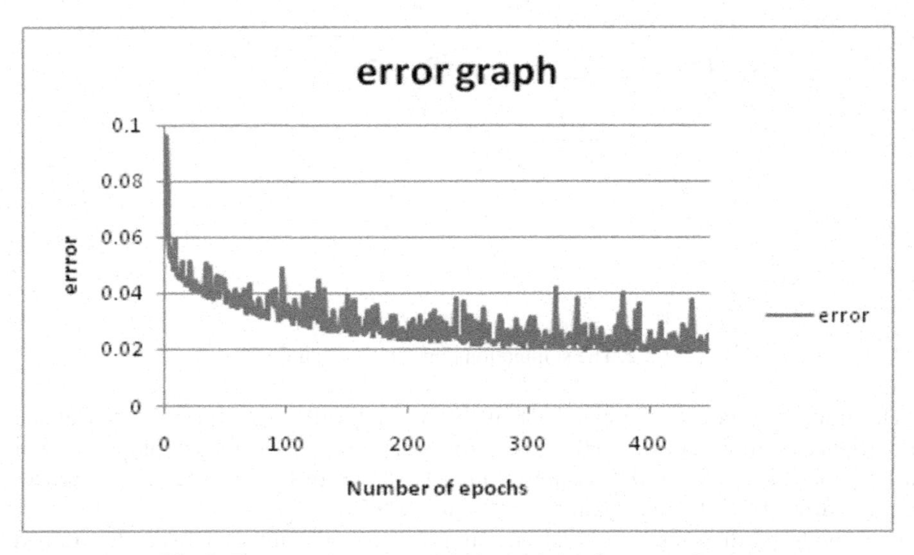

Fig. 3. Error graph generated during training for a sample node

7 Algorithm: PJSS

Algorithm PJSS
begin
Generate N nodes and perform the following steps concurrently;
Construct ANN for each node with I inputs elements, M hidden neurons and one output neuron;
Train the neural network using BP algorithm from the periodically collected average load samples;
Predict the local node load for the next interval of time from the trained ANN;
for each node *i* in the grid *G* **do**
Mark node up/down by following its failure f_i and repair time r_i distribu tion interval;
Perform the following steps when n node *i* is up
Generate job *j* for node *i* by following the job arrival time distribution parameter λ_i;
for each unscheduled job *j* **do**

```
            Randomly choose a node l from the quorum of 40% nodes in the
            Grid;
            Compute estimated turnaround time for job j at node l based on
            its predicted load;
            Compute estimated turnaround time for job j at node i based on
            its current load;
            if(estimated turnaround time at node l is less than estimated
               turnaround time at local node i)
            then
                 Schedule job j of node i on to node l;
                 Mark job j of node i as scheduled;
                 Schedule job j at local node i if j is unmarked and mark j as
                  scheduled;
            endif
        endfor
      for each periodic interval τ do
            Predict load for the next interval and broadcast to other nodes;
            Receive broadcasted predicted load from every other node k;
      endfor
  endfor
end;
```

8 Simulation Results

The simulation is done by varying the parameters like number of nodes (N), mean arrival rate of the job in the grid ($\bar{\lambda}$), mean size of the job in the grid ($\bar{\mu}$), mean processing time of the job in the grid ($\bar{\psi}$), mean clock speed of the node in the grid ($\bar{c}$), mean failure rate $\bar{f}$ and mean repair time between failures $\bar{r}$ of a node in the Grid. The effect of varying parameters on the performance of the algorithm PJSS is analyzed and is evaluated against Resource Exclusion (RE) and no-migration (NM) algorithm. The performance of PJSS is compared against RE with reference to average Turnaround Time ($\overline{G^t}$) with respect to number of jobs finished execution (u^t).

The graphs shown from Fig. 4 to Fig 8 shows that the performance of PJSS is found to be better than the RE and NM algorithms except in the case of varying mean clock speed ($\bar{c}$).

Fig .4 is drawn to study the effect of increasing arrival rate of the job on the performance of the PJSS with respect to NM and RE. The mean arrival rate ($\bar{\lambda}$) shown in braces in Fig. 4 is varied with fixed N=100, D=50Km, C=0.1Sec, $\bar{\mu} = 5KB$, $\bar{\psi}$=30Sec, $\bar{c} = 200MHz$, $\bar{f} = 65Sec$ and $\bar{r} = 10Sec$. It is observed that increase in the frequency of job arrivals impose more load on the nodes and makes them busy. Since PJSS works on the prediction based on the past load statistics, PJSS make effective scheduling decisions from the trained NN. Hence PJSS could achieve better performance than NM and RE for increasing arrival of the jobs in the Grid.

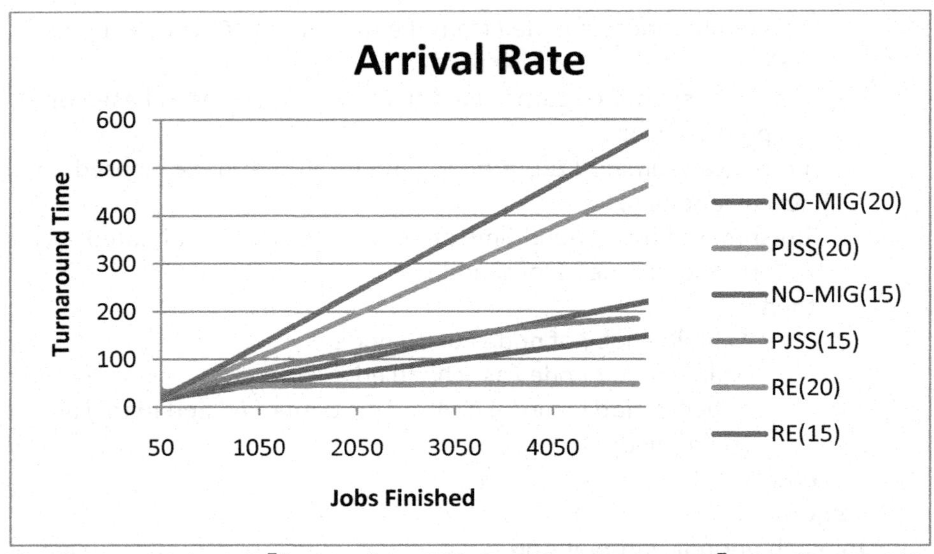

Fig. 4. The variation of $\bar{G}^t$ with u^t in PJSS for different values of $\bar{\lambda}$ with fixed N=100, D=50Km, C=0.1Sec, $\bar{\mu} = 5KB$, $\bar{\psi}$=20Sec, $\bar{c} = 200MHz$, $\bar{f} = 65Sec$ *and* $\bar{r} = 10Sec$..

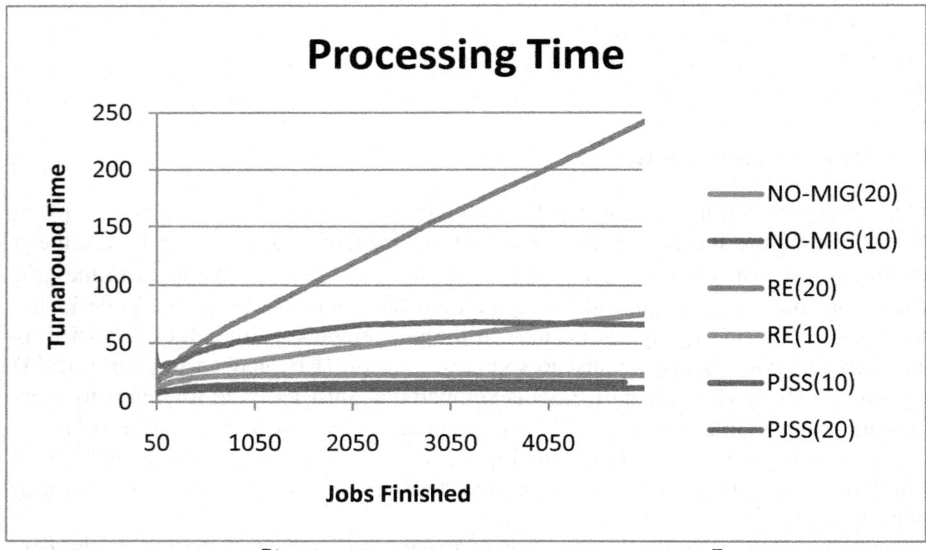

Fig. 5. The variation of $\bar{G}^t$ with u^t in PJSS for different values of $\bar{\psi}$ with fixed N=100, D=50Km, C=0.1Sec, $\bar{\lambda} = 30$, $\bar{\mu}$=5KB, $\bar{c} = 200MHz$, $\bar{f} = 65Sec$ *and* $\bar{r} = 10Sec$.

Fig. 5 is drawn to study the effect of increasing mean processing time of the job on the performance of the PJSS with respect to NM and RE. The mean processing time of the job ($\bar{\psi}$) shown in braces in Fig. 4 is varied with fixed N=100, D=50Km, C=0.1Sec, $\bar{\mu} = 5KB$, $\bar{\lambda}$=30Sec, $\bar{c} = 200MHz$, $\bar{f} = 65Sec$ and $\bar{r} = 10Sec$. It is observed that increase in the processing time of the job makes nodes longer time to complete the jobs and hence makes them much busier. Since PJSS works on the prediction based on the past load statistics, PJSS makes load estimations more accurately at the remote nodes from the trained NN and helps in making the effective scheduling decisions. Hence PJSS could achieve better performance than NM and RE for increasing mean processing time of the jobs in the Grid.

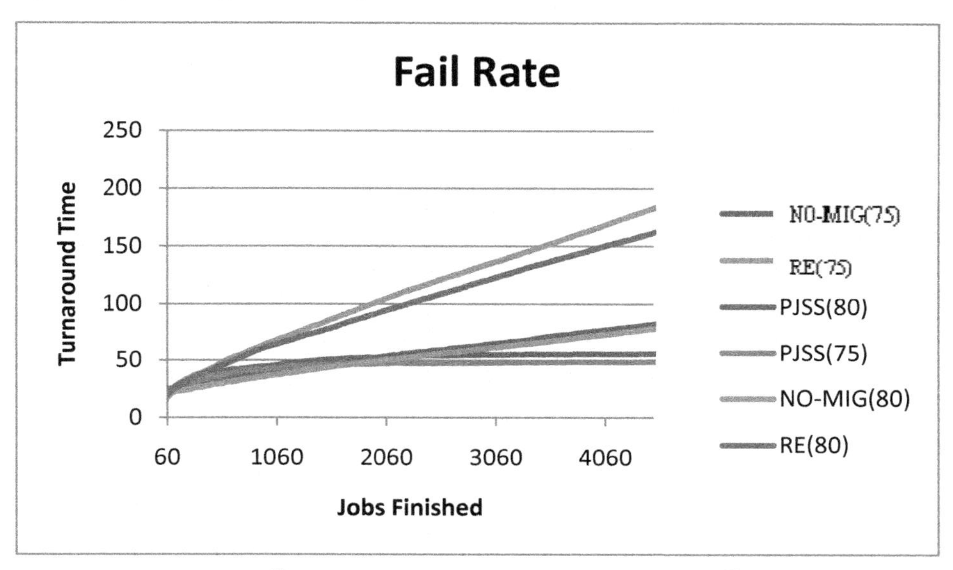

Fig. 6. The variation of $\bar{G}^t$ with u^t in PJSS for different values of $\bar{f}$ with fixed N=100, D=50Km, C=0.1Sec, $\bar{\lambda} = 30, \bar{\mu} = 5KB$, $\bar{\psi}$=20Sec, $\bar{c} = 200MHz\ and\ \bar{r} = 10Sec.$

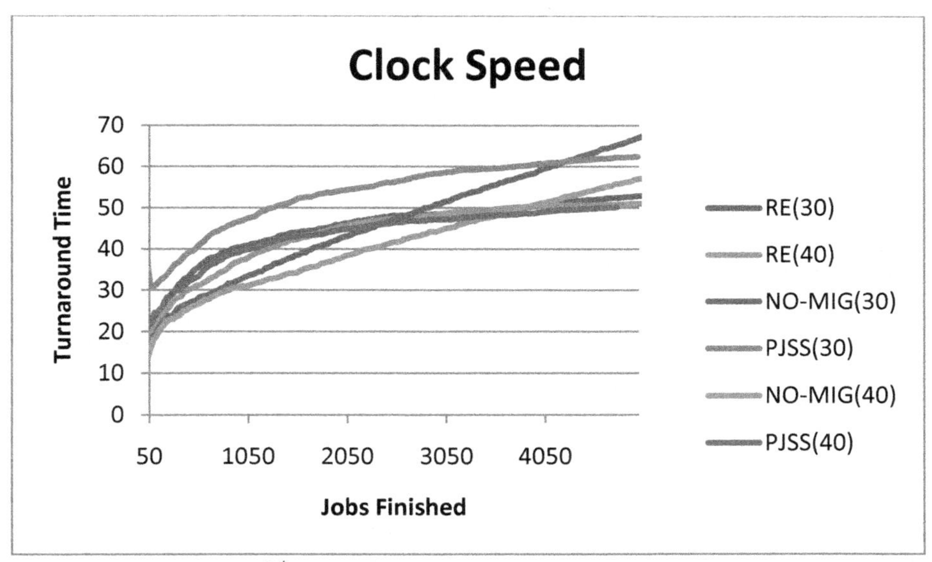

Fig. 7. The variation of $\bar{G}^t$ with u^t in PJSS for different values of $\bar{c}$ with fixed N=100, D=50Km, C=0.1Sec, $\bar{\lambda} = 30, \bar{\mu} = 5KB$, $\bar{\psi}$=20Sec, $\bar{f} = 65Sec\ and\ \bar{r} = 10Sec.$.

Fig. 6 is drawn to study the effect of mean failure rate of the nodes on the performance of the PJSS with respect to NM and RE. The mean failure rate of the nodes in the grid ($\bar{f}$) is shown in braces in Fig. 6 is varied with fixed N=100, D=50Km, C=0.1Sec, $\bar{\mu}$ = 5KB, $\bar{\lambda}$=30Sec, $\bar{c}$ = 200MHz, $\bar{\varphi}$=30Sec and $\bar{r}$ = 10Sec. As PJSS does not consider the failure rate of the nodes into account during the prediction, it could yield better performance on the lower node failure rates. It is observed from the simulation that PJSS is exhibiting better performance over RE and NM when average failure rate of the node in the grid is for every 65Sec.

Fig. 7 is drawn to study the effect of increasing clock speed of the nodes on the performance of the PJSS with respect to NM and RE. The mean clock speed of the job node ($\bar{c}$) scaled down to 1:10 is shown in braces in Fig. 7 is varied with fixed N=100, D=50Km, C=0.1Sec, $\bar{\lambda}$=30Sec, $\bar{\mu}$=5KB, $\bar{f} = 50Sec$ and $\bar{r} = 10Sec$. It is observed that increase in the clock speed of the nodes makes nodes more powerful and completed the instructions in the jobs at faster rate. This will become advantage for RE algorithm as it excludes the bottleneck in assigning slower nodes for processing longer jobs. Hence RE would exhibit similar performance in this case compared to PJSS and better performance than NM.

Fig. 8 is drawn to study the effect of increasing mean job size $\bar{\mu}$ in the grid. The effect of increase in job size $\bar{\mu}$ will be there on the job transmission latency in the Grid. Therefore PJSS will consider this into account while making the scheduling decisions along with remote node load predictions. Thus PJSS would exhibit better performance over RE and NM. The percentage of jobs initiated in this case will be comparatively less than the percentage of jobs migrated in the other cases.

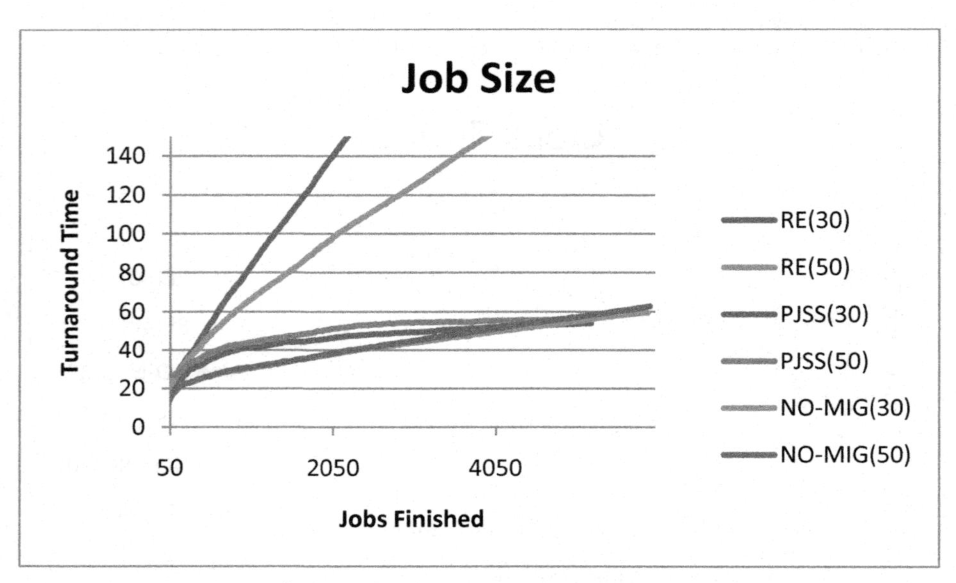

Fig. 8. The variation of $\bar{G}^t$ with u^t in PJSS for different values of $\bar{\mu}$ with fixed N=100, D=50Km, C=0.1Sec, $\bar{\lambda} = 30, \bar{c} = 200MHz, \bar{\psi}$=20Sec, $\bar{f} = 65Sec\ and\ \bar{r} = 10Sec$..

9 Conclusions and Future Scope

This paper investigate the feasibility and effectiveness of neural network predictions in the decision making process of job scheduling policy. The study is conducted to analyze the performance of PJSS for varying parameters. The simulation results by varying parameters are shown from Fig. 4 to Fig. 8. The results demonstrates that though PJSS makes decisions based on the load estimations from the periodically collected load statistics, It shows better performance than RE and NM in most of the cases.

It is found in the literature that trust is a firm belief about the competence of entity to behave as expected in the given environment. To address the dynamism in the network and improve the effectiveness in the scheduling decision, we would like incorporate trust metric into the scheduling decision as part of our future scope of research. In addition we would like to enhance the prediction performance by using the fast learning algorithms like Quick propagation and Resilient propagation algorithms.

References

1. Hamscher, V., Schwiegelshohn, U., Streit, A., Yahyapour, R.: Evaluation of Job-Scheduling Strategies for Grid Computing. In: Buyya, R., Baker, M. (eds.) GRID 2000. LNCS, vol. 1971, pp. 191–202. Springer, Heidelberg (2000)
2. Yagoubi, B., Slimani, Y.: Task Load Balancing Strategy for Grid Computing. Journal of Computer Science 3(3), 186–194 (2007)
3. Fedak, G., Germain, C., Neri, V., Cappello, F.: XtremWeb: A Generic Global Computing System. In: CCGRID 2001 (2001)
4. Casanova, H., Legrand, A., Zagorodnov, D., Berman, F.: Heuristics for Scheduling Prameter Sweep Applications in GRID Environments. In: HCW 2000 (2000)
5. Wolski, R., Spring, N., Hayes, J.: The Network Weather Service: A Distributed Resource Performance Forecasting Service for Metacomputing. In: Future Generation Computing Systems (1999)
6. He, X., Sun, X.-H., Laszewski, G.V.: QoS guided Min-min heuristic for grid task scheduling. Journal of Computer Science and Technology 18, 442–451 (2003)
7. Estiminani, K., Naghibzadeh, M.: A Min-min and Max-min selective algorithm for Grid task scheduling. In: The Third IEEE/IFIP International Conference on Internet, Uzbekistan (2007)
8. Dinda, P.: Online prediction of the running time of tasks. Cluster Computing 5(3), 225–236 (2002)
9. Entropia, Inc., http://www.entropia.com
10. Yu, J., Buyya, R.: A taxonomy of scientific workflow systems for grid computing. SIGMOD Record 34(3) (September 2005)
11. Zhou, D., Lo, V.: Wave Scheduler: Scheduling for Faster Turnaround Time in Peerbased Desktop Grid Systems. Presented at 11th Workshop on Job Scheduling Strategies for Parallel Processing In Conjunction with ICS 2005. The Cambridge Marriott-Kendall Square, Cambridge (2005)

12. Kondo, D., Kindarji, B., Fedak, G., Cappello, F.: Towards Soft Real-Time Applications on Enterprise Desktop Grids. In: Sixth IEEE International Symposium on Cluster Computing and the Grid (CCGRID 2006), pp. 65–72. IEEECS Press (May 2006)
13. Choi, S.J., Baik, M.S., Hwang, C.S., Gil, J.M., Yu, H.C.: Volunteer Availability based Fault Tolerant Scheduling Mechanism in Desktop Grid Computing Environment. In: The 3th IEEE International Symposium on Network Computing and Applications, Workshop Ouun Adaptive Grid Computing (NCA-AGC 2004), pp. 476–483 (August 2004)
14. Choi, S.J., Baik, M.S., Gil, J.M., Jung, S.Y., Hwang, C.S.: Adaptive Group Scheduling Mechanism using Mobile Agents in Peer-to-Peer Grid Computing Environment. Applied Intelligence, Special Issue on Agent-based Grid Computing 25(2), 199–221 (2006)

GRAF Password for More Secured User Authentication

Rohini Temkar, Geocey Shejy, and Dhanamma Jagli

Vivekananda Institute of Technology, Chembur, Mumbai
rohini_dighe@rediffmail.com, geocey@gmail.com,
dhana1210@yahoo.com

Abstract. There are many authentication schemes in the current state. Some of them are based on user's physical and behavioral properties, and some other authentication schemes are based on user's knowledge such as textual and graphical passwords. Moreover, there are some other important authentication schemes that are based on what you have, such as smart cards. GRAF password is an authentication scheme based on virtual environment by adding it as a response to actions performed on an object. Therefore, the resulted password space becomes very large compared to any existing authentication schemes.

Keywords: Authentication, entropy, virtual environment.

1 Introduction

Authentication is the process of validating who you are to whom you claimed to be. The password is a very common and widely authentication method still used up to now. In general, there are four human authentication techniques:

1. What you know (knowledge based)
2. What you have (token based)
3. What you are (biometrics)
4. What you recognize (recognition based)

1.1 Textual Authentication

Textual password is the most common authentication technique. Because of the huge advance in the uses of computer in many applications as data transfer, sharing data, login to emails or internet, some drawbacks of normal password appear like stolen the password, forgetting the password, weak password, etc.

1.2 Biometric Authentication

It consists of methods for uniquely recognizing humans based upon one or more intrinsic physical or behavioral traits. Each biometric recognition scheme is different considering consistency, uniqueness, and acceptability. Users tend to resist some biometrics recognition systems due to its intrusiveness to their privacy.

S. Unnikrishnan, S. Surve, and D. Bhoir (Eds.): ICAC3 2013, CCIS 361, pp. 39–44, 2013.

1.3 Graphical Authentication

Many graphical passwords schemes have been proposed. The strength of graphical passwords comes from the fact that users can recall and recognize pictures more than words.

2 Literature Survey

2.1 D´ej`a Vu: A User Study Using Images for Authentication [2]

Using D´ej`a Vu, the user creates an image portfolio, by selecting a subset of p images out of a set of sample images. To authenticate the user, the system presents a challenge set, consisting of n images. This challenge contains m images out of the portfolio, remaining n - m images are decoy images. To authenticate, the user must correctly identify the images which are part of her portfolio.

2.2 Graphical Password Authentication Using Cued Click Points [5]

It proposes and examines the usability and security of Cued Click Points (CCP), a cued-recall graphical password technique. Users click on one point per image for a sequence of images. The next image is based on the previous click-point. It presents the results of an initial user study which revealed positive results.

2.3 The Draw-A-Secret (DAS) Scheme [4]

A user is asked to draw a simple picture on a 2D grid. The coordinates of the grids occupied by the picture are stored in the order of the drawing. During authentication, the user is asked to re-draw the picture.

2.4 Pass Faces[6]

To log in, users select their Pass faces from a grid of faces displayed on the screen. This study uses the standard implementation of the Pass faces demonstration toolkit, requiring participants to memorize faces, and correctly selects all of them from grids of nine faces.

3 GRAF Password Scheme

The GRAF password presents a virtual environment containing various virtual objects. The user navigates through this environment and interacts with the objects. The GRAF password is simply the combination and the sequence of user interactions that occur in the virtual environment.

3.1 GRAF Password Workflow

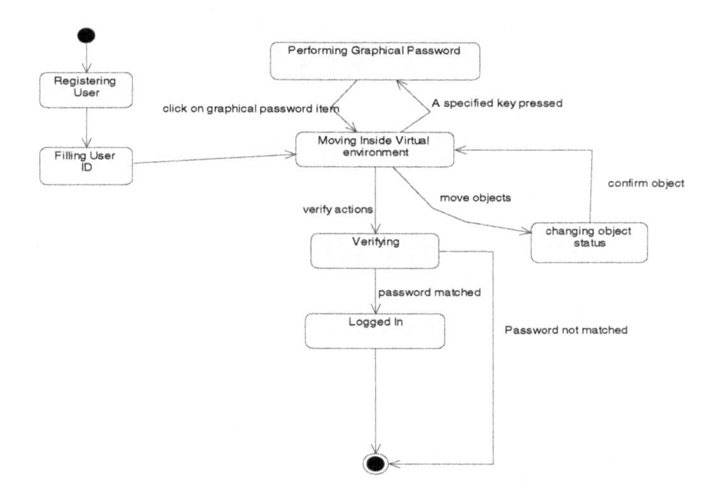

Fig. 1. State transition diagram for GRAF Password

3.2 GRAF Password Implementation Steps

Phase 1: 3-D Object Creation
Phase 2: Placing objects in 3-D environment
Phase 3: Registration of User

Tracking: At the time of registration, the actions and the interactions the user performs in the virtual environment have to be tracked using the locations and the properties of the involved objects.

Storing: The sequence of actions and interactions has to be maintained and stored in the database in the order of the input by the user. This sequence of routines becomes the user's GRAF password.

Phase 4: Authentication

3.3 GRAF Password Selection and Inputs

Any real life example of virtual environment is considered. Here the chessboard example is taken as an example.

The initial representation of user actions in the 3Dvirtual environment can be recorded as follows:

1. (15, 15) Action = Move Black Pawn to next square.
2. (15, 65) Action = Move White Pawn to next square.
3. (65, 15) Action = Move Black Knight to next square.
4. (15, 65) Action = Move White Knight to next square.

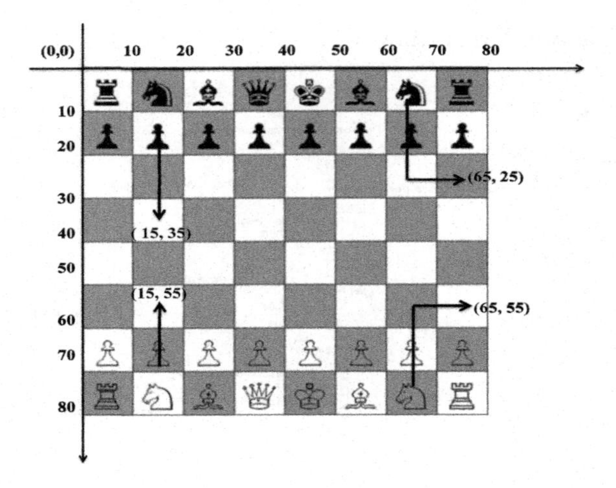

Fig. 2. Virtual environment of chessboard to generate GRAF Password

In order for a legitimate user to be authenticated, the user has to follow the same sequence and type of actions and interactions toward the objects for the user's original GRAF password.

4 Password Strength

4.1 Password Entropy

Entropy is the amount of randomness in a password or how difficult it is to guess.

$$\text{Entropy Per input} = \log_2(n) \tag{1}$$

$$\text{Password Entropy} = l * \text{Entropy per input} \tag{2}$$

Where $\boldsymbol{n}$ is the pool size of inputs and $\boldsymbol{l}$ is the length of the password.

4.2 GRAF Password Entropy

$$\text{The function m} = \sum f(O_i) \tag{3}$$

is the number of possible actions and interactions toward the object O_i and the function g(AC) counts the total number of actions and inputs toward the 3-D virtual environment.

$$\text{Entropy of GRAF Password} = \text{g(AC)} \times \log_2(\text{m}) \tag{4}$$

4.3 Security Analysis

Table 1. Entropy analysis of different graphical passwords

No of Inputs	Text Passwords	DAS (5X5 Grid)	DAS (10X10 Grid)	Pass Faces	GRAF Password
1	5.95	4.64	6.64	3.16	10.97
2	11.9	9.28	13.28	6.32	21.94
3	17.85	13.92	19.92	9.48	32.91
4	23.8	18.56	25.56	12.64	43.88
5	29.75	23.2	33.2	15.8	54.85

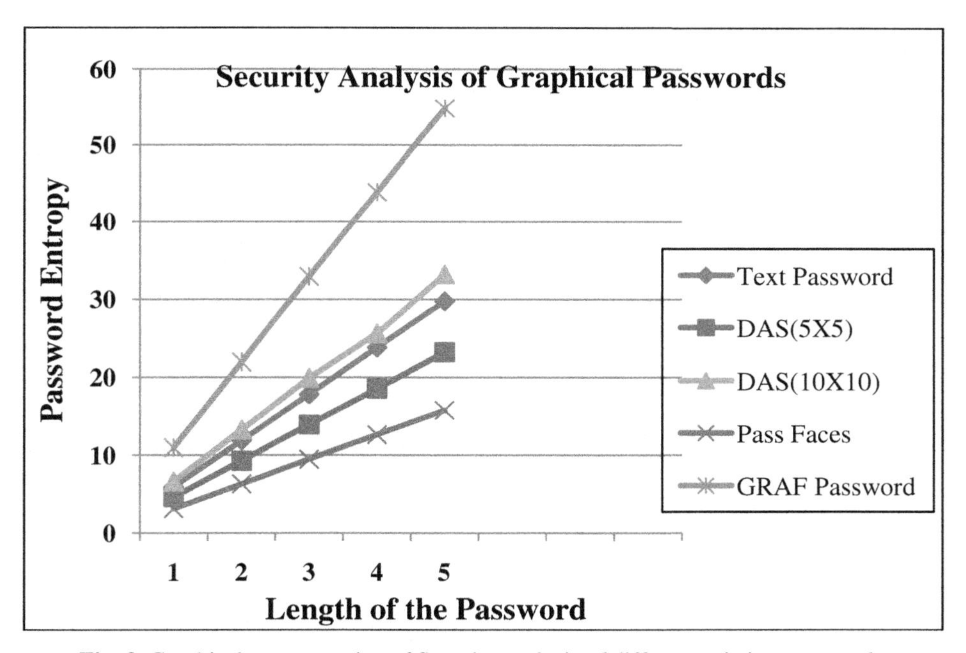

Fig. 3. Graphical representation of Security analysis of different existing passwords

5 Conclusion

GRAF password is an authentication scheme that includes various objects into a single virtual environment by adding it as a response to actions performed on an object. GRAF password reflects the user's preferences and requirements. Therefore, it is the user's choice and decision to construct the desired and preferred GRAF password. The resulted password space becomes very large compared to any existing authentication schemes. As compared to other graphical password methods and textual passwords the entropy of GRAF password is more which increases the security.

References

1. Klein, D.V.: Foiling the cracker: A survey of, and improvement to passwords security. In: Proc. USENIX Security Workshop, pp. 5–14 (1990)
2. Dhamija, R., Perrig, A.: Déjà Vu: A user study using images for authentication. In: Proc. 9th USINEX Security Symp., Denver, CO, pp. 45–58 (August 2000)
3. Wiedenbeck, S., Waters, J., Birget, J.-C., Brodskiy, A., Memon, N.: PassPoints: Design and longitudinal evaluation of a graphical password system. Int. J. Human-Comput. Stud. (Special Issue on HCI Research in Privacy and Security) 63(1/2), 102–127 (2005)
4. Jermyn, I., Mayer, A., Monrose, F., Reiter, M.K., Rubin, A.D.: The Design and Analysis of Graphical Passwords. In: Proceedings of the 8th USENIX Security Symposium (1999)
5. Chiasson, S., van Oorschot, P.C., Biddle, R.: Graphical Password Authentication Using Cued Click Points. In: Biskup, J., López, J. (eds.) ESORICS 2007. LNCS, vol. 4734, pp. 359–374. Springer, Heidelberg (2007)
6. Real User Corporation, The Science Behind Passfaces (October 2005), http://www.realusers.com
7. Suo, X., Zhu, Y., Owen, G.S.: Graphical passwords: A survey. In: Proc. 21st Annu. Comput. Security Appl. Conf., December 5-9, pp. 463–472 (2005)
8. Alsulaiman, F.A., El Saddik, A.: Three-Dimensional Password for More Secure Authentication. In: Proc. IEEE Transactions on Instrumentation and Measurement (September 2008)

Secured Mechanism for Dynamic Key Generation in ATM System

Praveen S. Challagidad[1,*] and Jayashree D. Mallapur[2]

[1] Dept. of CS & Engg. Basaveshwar Engineering College,
Bagalkot– 587102, Karnataka – India
praveesc07@gmail.com
[2] Dept. of E&C Engg. Basaveshwar Engineering College,
Bagalkot– 587102, Karnataka – India
bdmallapur@yahoo.co.in

Abstract. Automatic Teller Machines (ATMs) are used by millions of customers every day to make cash withdrawals from their accounts. However, the wide deployment and sometimes-secluded locations of ATMs make them ideal tools for criminals to turn traceable electronic money into clean cash. Existing ATMs do not have much security. There is only one security level provided in existing system that is password. Hackers can easily hack the password, and if intruder has the ATM card then he can easily access the money from ATM system. To prevent this we proposed secured mechanism for the ATM systems in this paper. First level security is the password. When the password is entered successfully next step is to generate four bit secured dynamic key (SDK) (i.e. a dynamic key generated using secured mechanism). The SDK is generated using efficient random selection algorithm and efficient encryption and decryption technique. This generated key is sent through GSM modem kit as an SMS to the customer who is presently using the ATM System to. This 4 bit SDK is authenticated as the access code to provide secured higher level security. The proposed work involves the implementation of the SDK generation mechanism. Further, using this implementation, we conducted experiments to study and compare the response times to send the dynamic key to the customer belonging to two different service providers.

Keywords: ATM, Security , Secured Dynamic key.

1 Introduction

A strong authentication mechanism is a fundamental requirement in managing fraud on the Internet [1]. Several authentication systems were proposed however; they are not effective enough to prevent attackers from performing illegal transactions over the Internet. The authentication system can be made more secured by using mobile technology.

S. Unnikrishnan, S. Surve, and D. Bhoir (Eds.): ICAC3 2013, CCIS 361, pp. 45–58, 2013.

1.1 Mobile Technology

Now-a-days, mobile technology has become very widespread in the world. Recent handsets have attractive features like internet access, video broadcasting like 3G and 4G mobiles. Since the mobile technology has penetrated into greater area across the world, we can use the technology for providing security in ATM services as well. Global System for Mobile Communication (GSM) modem kit can be used as a kind mobile technology.

1.2 GSM Modem Kit

GSM modem is a wireless modem that works with a GSM wireless network. A wireless modem behaves like a dial-up modem. The main difference between them is that a dial-up modem sends and receives data through a fixed telephone line, while a wireless modem sends and receives data through radio waves, like a GSM mobile phone [4], a GSM modem requires a SIM card from a wireless carrier in order to operate [13]. Figure 1 shows the application diagram of GSM Modem kit. This GSM modem kit can be connected to ATM system to provide more security as a is external device.

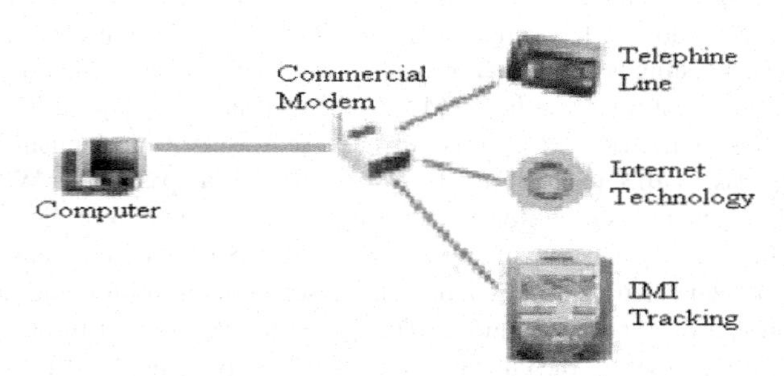

Fig. 1. Application diagram

1.3 ATM System

An ATM (Automatic or Automated Teller Machine) is a computerized machine designed to dispense cash to bank customers without need of human interaction. Many ATMs also allow customers to deposit cash or cheque, exchange currency and transfer money between their bank accounts [8]. ATM system is a safety critical system because its malfunctioning may lead someone to great economic loss [11], due less security.

1.4 Security in ATM

The security in ATM is done using the customer PIN (Personal Identification Number), which is the primary security level against fraud. Forgery of the magnetic stripe on cards is trivial in comparison to PIN acquisition. This proposed task intends to provide security for accessing ATM via GSM phone by sending SMS that consists of randomly generated 4 bit SDK to the user mobile that has to be entered to the ATM system to perform various banking transactions like withdrawal, transfer, balance check, etc[2],[8]. This SDK stream is used as a pin code for accessing ATM, at the same time a judgment must be made of the cost and effectiveness of the security measures.

1.5 Secured Dynamic Key

The dynamic key is the key which changes its values every time automatically. To provide security to any systems any time, key plays a very important role for authentication. If the key has some fixed values then it could be very easy for hackers to hack key values. To have more security, the key values can be changed automatically each and every time. By implementing some random selection algorithm the dynamic key can be generated. The generated dynamic key is more secured dynamic key (SDK). This secured dynamic key can be used to provide more security for the ATM system because the key is changing every time and is not possible hack this changing key easily by the hackers and fraudulent. Due to less security in present ATM system, the proposed system overcomes the drawback.

1.6 Need for Proposed System

In this proposed work, the mobile technology is used to send and receive SMS, which is done by GSM [5] modem kit. The proposed work is capable of preventing fraudulent ATM systems and Internet payments. Secured Dynamic Key Generation Scheme (SDKGS) that works together with advanced authentication technologies such as, smartcards and biometrics to prevent fraud payments over the ATM systems and Internet [9],[12].

This proposed work aims at providing more security than the existing system. In existing system, if card and password are known, then any one can carry out money transaction using the ATM card. But in our scheme it is not possible, where the key is generated dynamically and sent to the mobile number which is already registered with service providing bank. It uses a new SDKGS which is effective enough to prevent attackers, hackers from performing illegal transaction over the ATM system and Internet. The proper authentication by the user and correct user can be able to perform the transaction over the ATM system because of SDK/access code. Further system can be improved by using 3G and 4G technologies to make fast delivery of SMS to the ATM user. The ATM system should be kept where there is proper network coverage, and the user must have his/her mobile at the time of accessing the ATM. Following algorithm discuss SDKGS.

2 Algorithm to Generate Secured Dynamic Key

To generate secured dynamic key following steps are carried out:

1. Implement random selection algorithm which selects two characters randomly from A to Z and will have the corresponding values found in the following table.

Table 1. Alphabets from A to Z and their corresponding values

A	B	C	D	E	F	G	H	I	J	K	L	M
0	1	2	3	4	5	6	7	8	9	10	11	12
N	**O**	**P**	**Q**	**R**	**S**	**T**	**U**	**V**	**W**	**X**	**Y**	**Z**
13	14	15	16	17	18	19	20	21	22	23	24	25

2. Implement efficient encryption technique which generates two characters secured dynamic key from the randomly selected two characters (i.e. from first step).
3. Repeat second step to generate the two numerical secured dynamic key.
4. Combine two character dynamic key and two numerical dynamic key to generate final four alphanumeric dynamic key, which is SDK.
5. Implement efficient decryption technique for proper authentication of encrypted SDK.

The generated secured dynamic key is sent to the customer's mobile node, who is in the process of ATM access and authentication, which is the next level security check. Further, we could use our implementation to study the response time for the customer to receive the dynamic key on the mobile node with different service providers. For this purpose, we have used the four GSM service providers available in the location. Following topic discusses the proposed system operation.

3 System Operation

This works on authentication which uses dynamically generated access code (SDK) and retrieves the details pertaining to that specific person from the database. Here we use GSM modem to send SMS (Short Message Service). First level security is the password. Second level security is the 4bit SDK. ATM user should click on the GSM button which is provided in login form on ATM console. Once the GSM button is clicked, dynamic key is generated in ATM system using the steps discussed in the introduction part and is sent to GSM modem kit. This modem kit intern sends dynamic key to ATM user by communicating to the nearest BTS (Base Transceiver System). ATM user's mobile number is accessed by referring the database which is stored in ATM system. When ATM user receives the SMS, he/she has to enter SDK/access code in the login form of ATM system. After these two security levels are authenticated successfully then ATM user is allowed to make transactions over ATM. Architectural operation of system is shown below.

3.1 System Architecture

Figure 2 shows the architectural strategy of the proposed work.

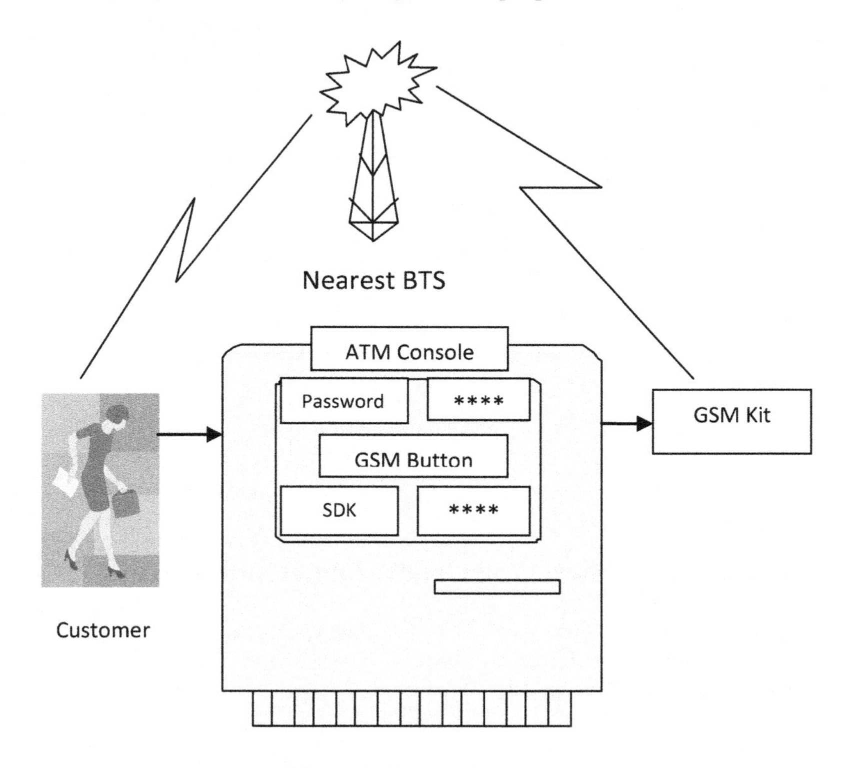

Fig. 2. Architectural strategy

Below figure 3 shows the block diagram of proposed work, which consists of login form which is the front end user interface and the database, random selection algorithm. SDK generation using encryption and decryption technique, SMS module, and authentication are the backend processes. These backend processes are activated once the GSM button is clicked. After completion of the transaction, ATM would be automatically directed to the login form.

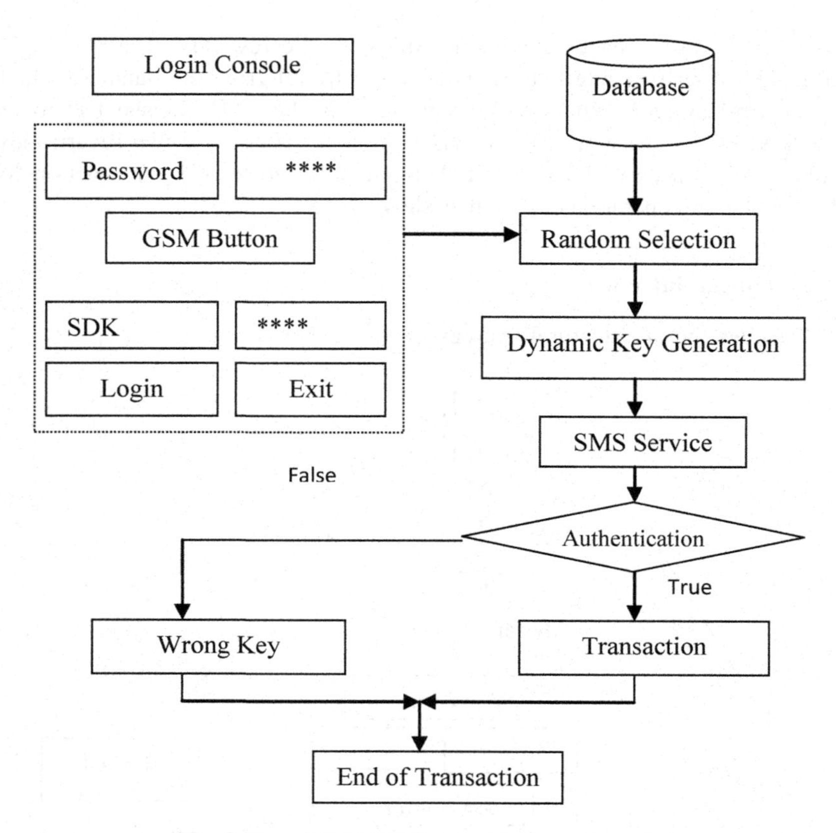

Fig. 3. Complete block diagram of proposed work

4 Mathematical Model for Generating Dynamic Keys

Here we have considered two examples, one encrypting and one decrypting, the alphabet is going to be the letters A through Z, and will have the corresponding values found in the table illustrated 1.

4.1 Encrypting

In this encrypting example, the plaintext to be encrypted is "AFFINE CIPHER" using the table mentioned above for the numeric values of each letter, taking to be 5, b to be 8, and m to be 26 since there are 26 characters in the alphabet being used. Only the value of a has a restriction since it has to be coprime with 26. The possible values that a could be are 1, 3, 5, 7, 9, 11, 15, 17, 19, 21, 23, and 25. The value for b can be arbitrary as long as a does not equal 1 since this is the shift of the cipher. General Encryption function is: $E(x) = (ax+b) \bmod m$.

Example: the encryption function for this example will be $y = E(x) = (5x+8) \pmod{26}$. The first step in encrypting the message is to write the numeric values of each letter.

Table 2. x values for AFFINCIPER

plaintext:	A	F	F	I	N	E	C	I	P	H	E	R
x:	0	5	5	8	13	4	2	8	15	7	4	17

Now, take each value of x, and solve the first part of the equation *(5x+8)*. After finding the value of *(5x+8)* for each character take the reminder when dividing t result of by the 26. The following table shows the first four steps of the encrypting process.

Table 3. Encrypted values for AFFINCYPER

plaintext:	A	F	F	I	N	E	C	I	P	H	E	R
x:	0	5	5	8	13	4	2	8	15	7	4	17
5x+8	8	33	33	48	73	28	18	48	83	43	28	93
(5x+8) mod	8	7	7	22	21	2	18	22	5	17	2	15

The final step in encrypting the message is to look up each numeric value in the table for the corresponding letters. In this example, the encrypted text would be IHHWVCSWFRCP. The table below shows the completed table for encrypting a message in the Affine cipher.

Table 4. Cipher text for AFFINCIPER

plaintext:	A	F	F	I	N	E	C	I	P	H	E	R
x:	0	5	5	8	13	4	2	8	15	7	4	17
5x+8	8	33	33	48	73	28	18	48	83	43	28	93
(5x+8) mod	8	7	7	22	21	2	18	22	5	17	2	15
ciphertext:	I	H	H	W	V	C	S	W	F	R	C	P

4.2 Decrypting

In this decryption example, the ciphertext that will be decrypted is the ciphertext from the encryption example. General decryption function is: $D(y) = 21(y\text{-}8) \bmod m$. The corresponding decryption function is, w $D(y) = 21(y\text{-}8) \bmod 26$ where a^{-1} is calculated to be 21, b is 8, and m is 26. To begin, write the numeric equivalents to each letter in the ciphertext, as shown in the table below.

Table 5. y values for cipher text

ciphertext:	I	H	H	W	V	C	S	W	F	R	C	P
y:	8	7	7	22	21	2	18	22	5	17	2	15

Now, the next step is to compute *21(y-8)*, and then take the remainder when that result is divided by 26. The following table shows the results of both computations.

Table 6. Decrypted values for cipher text

ciphertext:	I	H	H	W	V	C	S	W	F	R	C	P
y:	8	7	7	22	21	2	18	22	5	17	2	15
21(y-8):	0	-21	-21	294	273	-126	210	294	-63	189	-126	147
(21(y-8)) mod 26:	0	5	5	8	13	4	2	8	15	7	4	17

The final step in decrypting the ciphertext is to use the table to convert numeric values back into letters. The plaintext in this decryption is AFFINECIPHER. Below is the table with the final step completed.

Table 7. Plain text for cipher text

ciphertext:	I	H	H	W	V	C	S	W	F	R	C	P
y:	8	7	7	22	21	2	18	22	5	17	2	15
21(y-8):	0	-21	-21	294	273	-126	210	294	-63	189	-126	147
(21(y-8)) mod 26:	0	5	5	8	13	4	2	8	15	7	4	17
plaintext:	A	F	F	I	N	E	C	I	P	H	E	R

4.3 Entire Alphabets Encoded

To make encrypting and decrypting quicker, the entire alphabet can be encrypted to create a one to one map between the letters of the cleartext and the ciphertext. In this example, the one to one map would be the following:

Table 8. Entire alphabets encoded

Letter in the	**A**	**B**	**C**	**D**	**E**	**F**	**G**	**H**	**I**	**J**
number in the cleartext	0	1	2	3	4	5	6	7	8	9
(5x+8)mod(26)	8	13	18	23	2	7	12	17	22	1
Ciphertext letter	I	N	S	X	C	H	M	R	W	B

K	**L**	**M**	**N**	**O**	**P**	**Q**	**R**	**S**	**T**	**U**	**V**	**W**	**X**	**Y**	**Z**
10	11	12	13	14	15	16	17	18	19	20	21	22	23	24	25
6	11	16	21	0	5	10	15	20	25	4	9	14	19	24	3
G	L	Q	V	A	F	K	P	U	Z	E	J	O	T	Y	D

In below topic, results of proposed system is discussed.

5 Result and Discussions

The proposed experiment is conducted by using Nokia handset 5310 (music express) having Airtel SIM instead of GSM modem kit.

5.1 Testing Parameters

1. Time of test conducted: 8:05 pm (IST).
2. SIM card inserted in GSM modem kit: Airtel.
3. Location: Basaveshwar Engineering College, Bagalkot, Karnataka state, India.
4. Time taken: Time taken to deliver SMS in seconds from one subscriber to another subscriber. And average time taken in seconds is calculated.
5. Number of ATM users: Here we have considered10 ATM users.
6. Length of dynamic key: 4 (2 English capital alphabets and 2 decimal numbers)

5.2 Table of Test Results and Graphs

Table 9 show time taken to deliver SMS from Airtel to Airtel and their respective Dynamic Keys. Each time intervals are above 9.8 sec so we taken initial time as 9.8sec. The trial 8 (11:05 sec) indicates that network is busy at that time. *Average time taken to deliver the SMS = 10.51sec.*

Table 9. Time taken to deliver SMS from Airtel to Airtel

Airtel SIM/Trials	Subscriber	Time taken	Dynamic key
1	Airtel	10:59sec	PU63
2	Airtel	10:55sec	XF82
3	Airtel	10:25sec	AF958
4	Airtel	10:50sec	DG35
5	Airtel	10:02sec	GT11
6	Airtel	10:35sec	DB44
7	Airtel	10:45sec	TU33
8	Airtel	11:05sec	RS83
9	Airtel	10:03sec	UT84
10	Airtel	10:30sec	AB31

Figure 4 shows the number of user and time taken to deliver the SMS. Time is varying means that some time network is free and some time network is busy. The network traffic also involves other subscribers of the GSM service provider. In the worst case of traffic condition of the network during our experiments, the response time was 11.05sec.

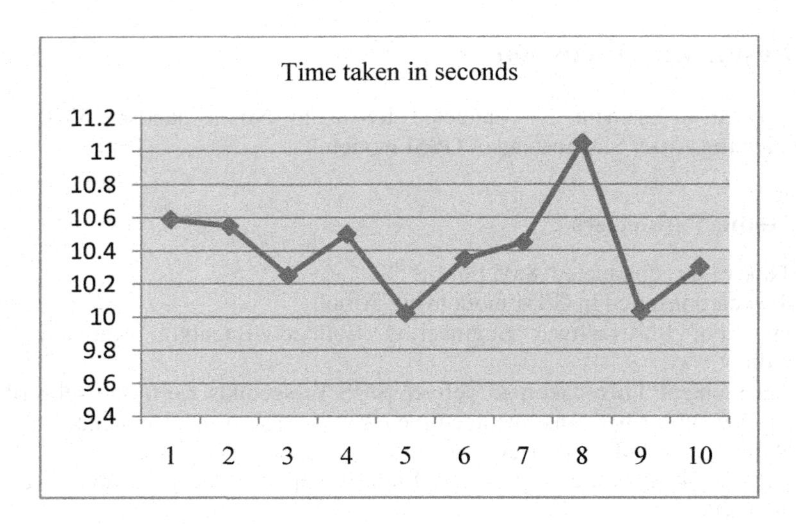

Fig. 4. SMS from Airtel to Airtel

Table 10. Time taken to deliver SMS form Airtel to BSNL

Airtel SIM/Trials	Subscriber	Time taken	Dynamic key
1	BSNL	11:27sec	PD21
2	BSNL	11:06sec	QG58
3	BSNL	10:47sec	GJ92
4	BSNL	11:20sec	AF58
5	BSNL	11:01sec	RE24
6	BSNL	10:20sec	NO95
7	BSNL	10:33sec	LO23
8	BSNL	11:32sec	SC91
9	BSNL	10:44sec	AP41
10	BSNL	11:15sec	CO51

Table 10 shows time taken to deliver SMS from Airtel to BSNL and their respective Dynamic Keys. As the response times are all above 10 sec so we taken initial time as 10 sec. The trials 1, 2, 4, 5, 8, and 10 have above 11 sec which indicates that network is busy at that time. *Average time taken to deliverSMS = 10.74sec.*

Figure 5 shows the number of user and time taken to deliver the SMS. By this we have shown that average time taken to deliver SMS from Airtel to BSNL is high compared to Airtel to Airtel.

Fig. 5. SMS from Airtel to BSNL

Table 11 show time taken to deliver SMS from Airtel to Vodafone and their respective Dynamic Key. Here we have taken initial time as 10sec. The trials 2, 3, and 6 have above 11 sec which indicate that network is busy at that time. *Average time taken to deliver SMS = 10.62sec.*

Table 11. Time taken to deliver SMS form Airtel to Vodafone

Airtel SIM/Trials	Subscriber	Time taken	Dynamic key
1	Vodafone	10:37sec	KM48
2	Vodafone	11.02sec	WM91
3	Vodafone	11:13sec	SD16
4	Vodafone	10:49sec	NF41
5	Vodafone	10:49sec	WP39
6	Vodafone	11:05sec	TA89
7	Vodafone	10:59sec	ST63
8	Vodafone	10:40sec	QT12
9	Vodafone	10:49sec	NT33
10	Vodafone	10:23sec	MA44

Figure 6 show the number of user and time taken to deliver the SMS. By this we have shown that average time taken to deliver SMS from Airtel to Vodafone is high compared to Airtel to Airtel, and less compared to Airtel to BSNL.

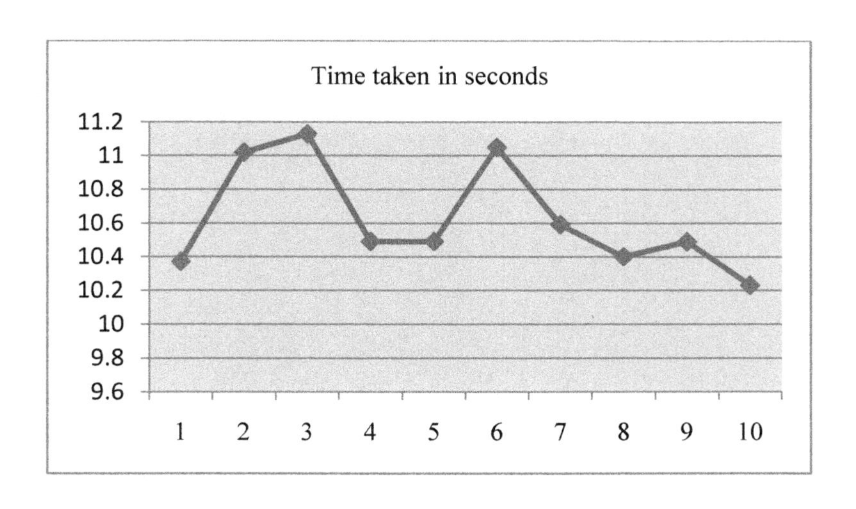

Fig. 6. SMS from Airtel to Vodafone

Table 12 shows time taken to deliver SMS from Airtel to Idea and their respective Dynamic Key. As the response times in the trials 2, 3, 4, 5, 6, and 7 are above 11 sec, the network is busy at that time. *Average time taken to deliver SMS = 10:98sec.*

Table 12. Time taken to deliver SMS form Airtel to Idea

Airtel SIM/Trials	Subscriber	Time taken	Dynamic key
1	Idea	10:41sec	UN82
2	Idea	11:52sec	SD75
3	Idea	11:22sec	CT36
4	Idea	11:31sec	PU93
5	Idea	11:03sec	XZ13
6	Idea	11:33sec	WP33
7	Idea	11:38sec	JP48
8	Idea	10:50sec	DP38
9	Idea	10:53sec	WD14
10	Idea	10:59sec	ND34

Figure 7 show the number of user and time taken to deliver the SMS. By this we have shown that average time taken to deliver SMS from Airtel to Idea is high compared to Airtel to Airtel, compared to Airtel to BSNL and Airtel to Vodafone.

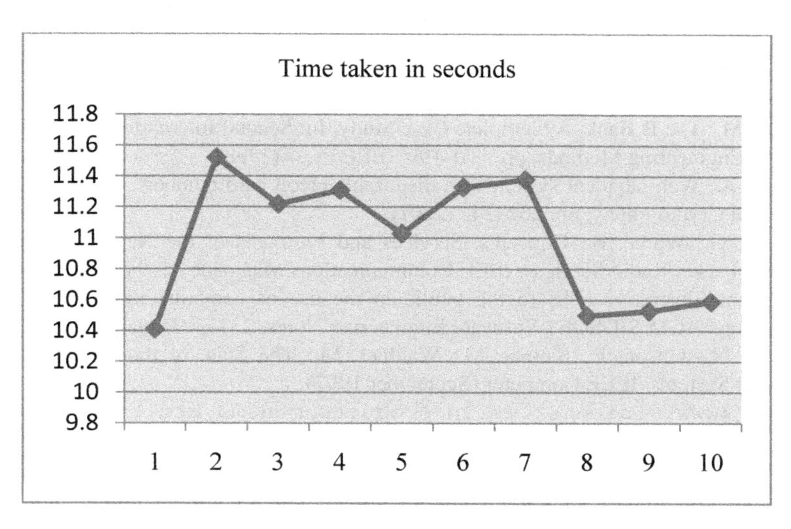

Fig. 7. SMS from Airtel to Idea

6 Conclusion and Future Enhancement

Finally our proposed system is able to provide efficient security using encryption and decryption techniques to generate secured dynamic key. It uses advanced authentication technologies and is well adapted with any possible future technology like 3G and 4G to make fast deliver of SMS. Moreover, it doesn't rely on fixed values where hacking one secret data will not destroy the whole system's security.

The various experiments prove the average time taken to deliver the SMS from one subscriber to another subscriber is below 11 sec, which is quite low and can be used for the security.

References

1. Dandash, O., Wang, Y., Le, P.D., Srinivasan, B.: A new Dynamic Key Generation Scheme for Fraudulent Internet Payment. In: International Conference on Information Technology, 0-7695-2776-0/07
2. Kanwal, S., Zafar, N.A.: Modeling of Automated Teller Machine System Using Discrete Structures. In: Proceedings of 8th IPMC (2007)
3. Li, Chen, Ma: Security in GSM (retrieved October 24th, 2004 from GSM-security)
4. Huynh, Nguyen: Overview of GSM and GSM security (retrieved October 25th, 2004 from Oregon State University, project)
5. ETS 300 608. Digital Cellular Telecommunication System (Phase 2), Specification of the Subscriber Identity Module-Mobile Equipment (SIM-ME) Interface. European Telecommunications Standards Institute (May 1998)

6. Stallings, W.: Practical Cryptography for Data Internetworks. IEEE Computer Society Press (1996)
7. Rubin, A.D., Wright, R.N.: Off-Line Generation of Limited-Use Credit Card Numbers. In: Syverson, P.F. (ed.) FC 2001. LNCS, vol. 2339, pp. 196–209. Springer, Heidelberg (2002)
8. Btuchi, M.: The B Bank: A Complete Case Study. In: Second International Conference on Formal Engineering Methods, pp. 190–199. IEEE Press (1998)
9. Shamir, A.: Web payment system with disposable credit card numbers. In: Proceedings of Financial Cryptography, pp. 232–242 (2001)
10. Brasche, G., Walke, B.: Concepts, Services and Protocols of the New GSM Phase 2+ General Packet Radio Services. IEEE Communications Magazine 35, 94–104 (1997)
11. Neumann: Illustrative risks to the public in the use of computer systems and related technology. ACM SIGSOFT Software Engineering Notes, 23–33 (January 1992)
12. Asokan, N., Janson, P., Steiner, M., Waidner, M.: The State of the Art in Electronic Payment Systems. IEEE Computer (September 1997)
13. http://www.visiontek.co.in/products/modems.htm

U-STRUCT: A Framework for Conversion of Unstructured Text Documents into Structured Form

Rajni Jindal[1] and Shweta Taneja[2]

[1] Dept. of Computer Engineering, Delhi Technological University
Formerly Delhi College of Engineering (DCE), Bawana Road, Delhi-42
rajnijindal@dce.ac.in
[2] Dept. of Computer Engineering, Delhi Technological University
Formerly Delhi College of Engineering (DCE), Bawana Road, Delhi-42
shweta_taneja08@yahoo.co.in

Abstract. The term Text Mining or Text Analytics refers to the process of extracting useful patterns or knowledge from text. The data in textual documents can be of two types, either it can be unstructured or semi-structured. Unstructured data is freely naturally occurring text, whereas web documents data (HTML or XML) is semi structured. Since the natural language text is not organized and does not represent context, it needs to be converted into structured form to perform data analysis and mine useful patterns from it. The field of text mining deals with mining useful patterns or knowledge from unstructured text.

In this paper, we propose a framework for the conversion of the unstructured text documents to a structured form. We present a generalized framework called U – STRUCT which translates unstructured text into structured form. This framework analyses the text documents from different views: lexically, syntactically and semantically and produces a generalized intermediate form of documents. Further, we also discuss the opportunities and challenges in the field of text mining.

Keywords: Text mining, data mining, Knowledge Discovery in Text (K.D.T) Process.

1 Introduction to Text Mining

The term Text Mining ([1], [10]) refers to the extraction of useful and non-trivial patterns or knowledge from different textual documents. Nowadays, the field of text mining has gained a lot of attention because of the huge amount of available text documents and the need to manage the information. Text mining [3] is an interdisciplinary field i.e.it is a confluence of many disciplines like Information Retrieval, Information Extraction, Statistics, Text Analysis, Natural Language Processing, Categorization, Topic Tracking, Visualization etc.

Text mining is a step of Knowledge Discovery in Text (K. D.T.) process [20]. As shown in figure 1, in this process firstly a Text Data warehouse is built. The data sources of text data warehouse are news stories, scientific research papers, emails, spreadsheets, contracts, warranties, medical and healthcare information, and so forth.

S. Unnikrishnan, S. Surve, and D. Bhoir (Eds.): ICAC3 2013, CCIS 361, pp. 59–69, 2013.

The concepts of data warehouse like Extraction, Transformation and Loading are used in designing the data warehouse. In the second step, Text Preprocessing is done to extract structured representations from raw unstructured data sources i.e. to prepare raw data for text mining. Text Preprocessing tasks include text clean up, stop words removal, stemming, tokenization, syntax check etc. The next step is Text Transformation. In this, Text document is represented by the words or features it contains and their occurrences. Two main approaches are used here: Bag of words [7] and Vector Space Model [8]. It also includes Feature Selection which deals with selecting a subset of features to represent the entire document. The next step is Attribute or Feature Selection which focuses on further Reduction of Dimensionality and removing irrelevant features. At this point, the text mining process merges with the traditional data mining process. Now we have a Structured Database on which the standard data mining techniques can be applied. The different data mining techniques are Classification, Prediction, Clustering Association Rules, and Visualization etc. and thus useful patterns can be discovered.

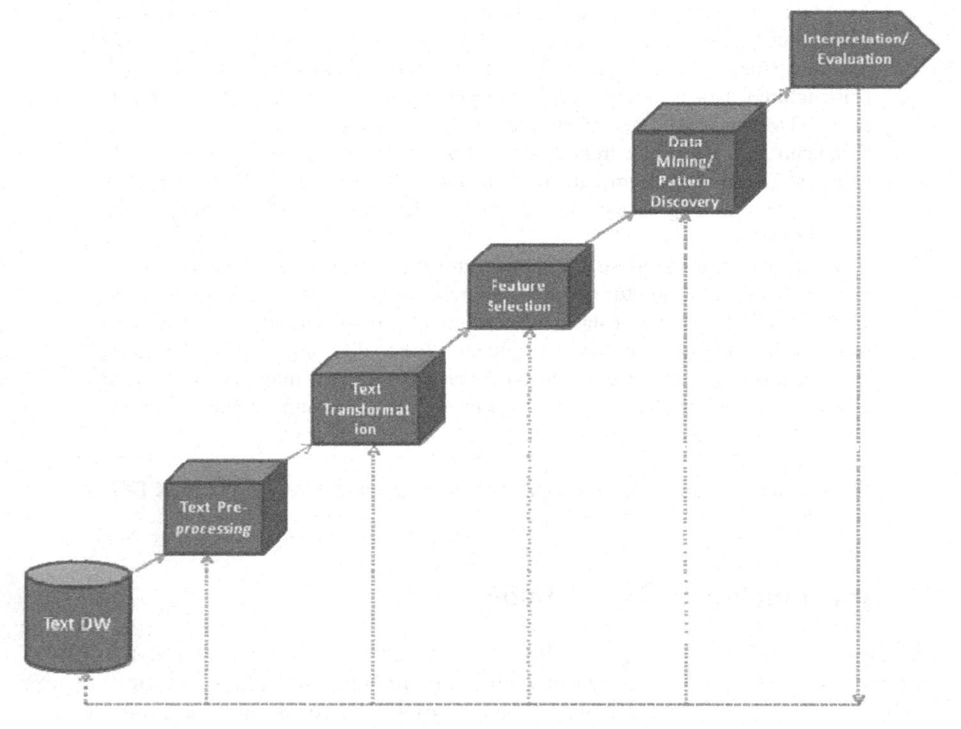

Fig. 1. Showing the process of KDT [12]

Text mining [6] differs from data mining in a way that data mining[9] concentrates on the discovery of knowledge from structured databases, whereas most of the available data present in text files is unstructured, for example journal articles which are available electronically. They comprise of some structured fields such as article title, author, publisher and date as well as some free text, like an abstract. Secondly, the preprocessing technique in data mining deals with cleaning and normalizing the data, as the data is already in a structured form, whereas in text mining, the preprocessing step converts the raw input unstructured data into structured form and prepares the data for text mining. Thirdly, due to unstructured nature of textual documents, text mining suffers with the problem of patterns overabundance which is more severe than in data mining field. So, some post refinement techniques are required to limit the number of patterns identified and select the relevant ones.

In this paper, we propose a generalized framework called U-STRUCT that converts unstructured data to a structured form in the text documents so that it can be used for making future predictions and analysis .The organization of the paper is as follows: Section 2 gives the need or motivation of the work. In Section 3, we give the related work done in this field. The next section 4 discusses the opportunities and challenges in preprocessing of text documents. Then in Section 5, we present the proposed framework U-STRUCT, describing its working or operation in the next section. Finally, we conclude and give the future scope.

2 Motivation of the Work

The field of text mining has become an emerging field nowadays. As most of the documents contain text, so text mining has gained a lot of importance. It has a lot of practical applications like evaluating public opinions (for example in review sites or blogs), spam filtering in emails, Automatic organizing of documents in digital libraries, Fraud detection by analyzing credit card data and so on.

In the field of text mining, there is a need for efficient preprocessing methods and techniques due to a variety of reasons like: Text documents contain free natural text which is unstructured and not organized, unstructured data does not follow any format, sequence or rules. It is totally unpredictable and is of less use. Secondly, due to a large number of text documents collection and its unstructured nature, different text mining methods generally produce a large number of patterns. This leads to pattern overabundance problem, which is more severe than in data mining applications. So, some efficient refinement techniques or preprocessing techniques are required so that useful knowledge can be obtained from free text. In this paper, we propose a generalized framework called U-STRUCT that converts unstructured data to a structured form which can be used for making future predictions and analysis. Till date, we could not find any standard method for text preprocessing which is more general and can be applied on all types of textual documents.

3 Related Work

In the field of preprocessing, few authors have given their contributions. In [15], authors have proposed methods for preprocessing of historical machine-printed documents. They have tried to improve the readability of old Sanskrit document by using efficient preprocessing methods like binarization operation. After scanning the paper document, noise is removed and then binarization is applied. The authors in [16] have analyzed the effect of preprocessing methods on Turkish texts. They combined two large datasets from Turkish newspapers using a crawler and performed methods like stemming, word weighing etc. In another work [17], authors have discussed the problem of dimensionality reduction in medical documents. They have made a comparison of dimensionality reduction techniques at the pre-processing stage and developed a new feature weighting scheme to improve the accuracy of classification. In the field of Arabic handwriting recognition, authors in [18] have proposed an orientation independent method for baseline detection of Arabic words. They also have given techniques for line and word separation, slant normalization and slope correction in handwritten Arabic documents. In [19], authors have discussed different methods used in the pre-processing of handwritten characters in a file. In the preprocessing stage, the input file is scanned, followed by the method of binarization, reference line estimation and thinning of the character image is done so that it can be used in the next stage of the feature extraction and recognition process.

Different researchers have contributed in the field of preprocessing of text documents, but either they have focused on Arabic Turkish or Sanskrit documents, Nobody has given attention to English documents from natural language point of view that is lexically, syntactically and semantically. In our paper, we have proposed a framework which converts an unstructured text document to a structured form, keeping in mind the constraints and challenges in natural text.

4 Opportunities and Challenges in Preprocessing of Unstructured Text Documents

Before discussing our framework, we present below the opportunities and challenges in pre-processing of text documents.

4.1 Data Is Not Well Organized and Labeled

The free natural language text is not well organized, either it is unstructured or semi structured .So, the first step in text mining is to organize or arrange the data in text documents.

4.2 Ambiguity in Data

The natural language text data contains the problem of ambiguities at many levels: at lexical level, parsing level or semantic level. Two kinds of ambiguities occur i.e. word ambiguity and sentence ambiguity. This problem should be solved so that useful information can be gathered and used in mining the textual data.

4.3 Lexical Analysis

The aim of lexical analyzer is to scan the documents and group the characters into meaningful tokens. In this case, problem arises if the documents are in different formats i.e. the number and time format of the documents should be consistent. Another problem is of acronyms and abbreviations which need to be converted into a standard form.

4.4 Semantic Analysis

To analyze the documents semantically is still an open challenge in the area of text mining. As the number of documents is huge, so there is a need to develop an efficient and cost effective tool for semantic analysis of the documents. The semantic analysis of the text documents gives us the important details like the relationships between objects or concepts in the documents.

4.5 Multilanguage Text Mining

Text Mining depends on the language used. Most of the Text Mining tools are based on Mining English language documents; it should consider the documents from other languages also and produce language independent intermediate form.

5 U- STRUCT: Proposed Framework

The figure 2 given below shows the proposed framework.

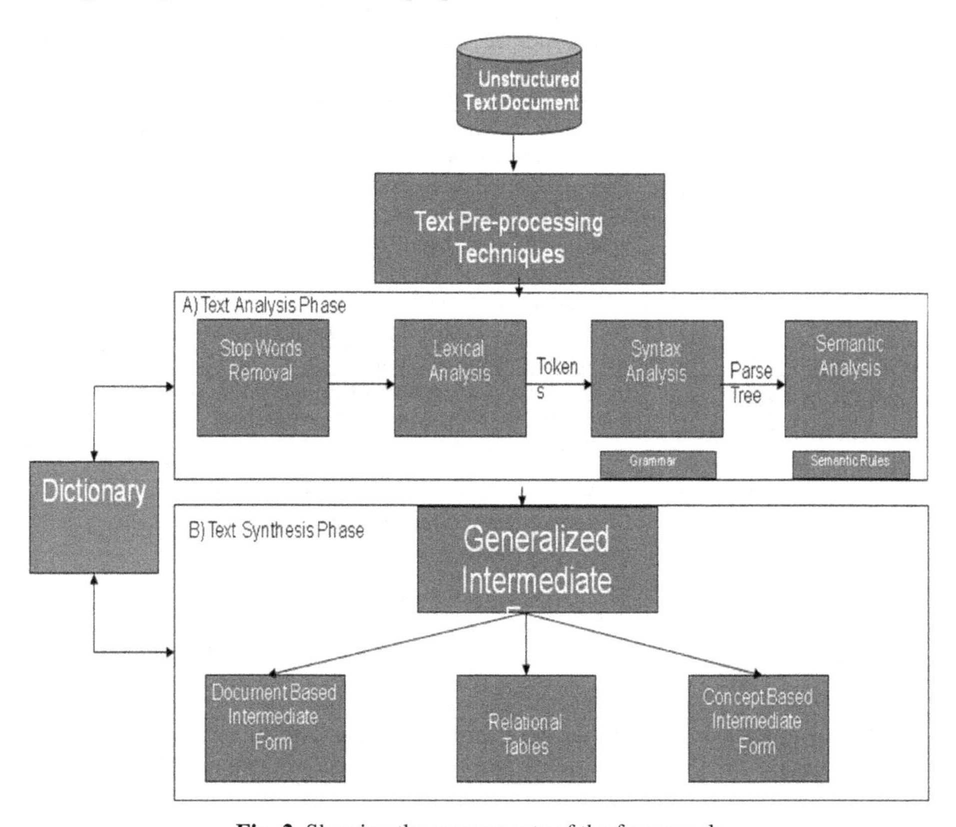

Fig. 2. Showing the components of the framework

5.1 Components of the Framework

The framework is divided into two phases: Text Analysis phase and Text Synthesis phase. In the Text Analysis phase, the raw unstructured text document undergoes text preprocessing or text interpreting. This phase consists of four steps: Stop words removal, Lexical Analysis or Scanner, Syntax Analysis or Parser and Semantic Analysis. The output of first phase is fed as input to the next phase that is Text Synthesis phase. In this phase, we generate a generalized Intermediate form which can be represented in three forms either it can be a document based intermediate form or in the form of Relational tables or it can be a concept based intermediate form. Along with these two phases, there is also a dictionary or a bookkeeping component which is used in both the phases.

Now, let us discuss the operation of the above mentioned phases and their respective components in more details:

Table 1. Components of the Framework

1. Text Analysis Phase	
Stop words removal	This is an optional step in the analysis of text document. Stop words are the commonly used words like 'a', 'the', 'of' etc. These are usually considered irrelevant and are ignored.
Lexical Analysis Step	In this step, the text document is scanned character by character, and grouped into tokens. Tokens can be Nouns, Verbs, Article, Adjective, Preposition, Number and special symbols.
Syntax Analysis Step	This step performs two functions, first it checks whether the incoming token is according to the specifications of the grammar or not. Then, if it is, it generates a syntax tree or a parse tree of a noun phrase, verb phrase, prepositional phrase, adjective phrase or a clause.
Semantic Analysis Step	In this step, a semantic action is executed which leads to intermediate representation of the document directly.
2. Text Synthesis Phase	
Document based Intermediate form	In Document based Intermediate form, each entity represents a document.
Relational tables	In this form, each entity represents a relational table.
Concept based intermediate form	In concept based Intermediate form, each entity represents an object or concept of interest in a specific domain.
3. Dictionary/Bookkeeping	
Dictionary	It is required mainly in Lexical Analysis step and Syntax Analysis step to store the tokens and check for spellings of the words etc.

5.2 Framework Operation

The input fed to the system is a raw unstructured text document. This document can be of any form like a news story, a business or a legal report, an email etc. First the document undergoes text preprocessing using the preprocessing techniques.

5.2.1 Text Analysis Phase

In the Text Analysis phase, the input text document is analyzed and few operations are applied on it.

5.2.1.1 Stop Words Removal. Stop words are the words like 'a', 'an', 'the' etc. These words are considered irrelevant and are therefore removed from the document.

5.2.1.2 Lexical Analysis / Scanner. Lexical Analyzer is also known as Scanner. In this Step, the text document is scanned character by character and grouped into meaningful tokens like chapters, sections, paragraphs, sentences, words. The most frequent method used in text mining systems involves breaking the text into sentences and words, which is called tokenization. Another function of Lexical Analyzer is to extract token features. These are simple functions of the tokens describing some specific property of the sequence of characters that make up the token. Among these features is inclusion of digits, punctuation special characters, types of capitalization etc.

Another function of Lexical Analysis is PoS Tagging. It stands for Part-of-Speech Tagging. PoS tagging is the tagging of tokens with the appropriate PoS tags based on the context in which they appear. PoS tags divide words into categories depending on the role they play in the sentence in which they appear. PoS tags provide information about the semantic content of a word. Nouns usually denote "tangible and intangible things," whereas prepositions express relationships between "things."Most PoS tag sets make use of the same basic categories. The commonly used set of tags [21] contains seven different tags (Proper Noun, Article, Noun, Adjective, Verb, Preposition and Number). Some systems contain a much more elaborate set of tags. For example, the complete Brown Corpus tag set [13] has more than 87 basic tags. These PoS tags capture syntactic category or variable like noun, verb, adjective etc .and they can be used for the identification of noun phrase, verb phrase or other parts of speech in the next Syntax Analysis phase. The output of Lexical Analysis is a sequence of tokens which is passed on to the next phase.

Challenging Issues in Lexical Analysis

- The main issue in Lexical Analysis is the ambiguity of words. Natural language words are ambiguous in nature. Ambiguity means the ability of a word to be understood in two or more possible senses or ways. Ambiguity gives a natural language its usability and flexibility and thus, it cannot be completely eliminated from the natural language. This ambiguity is termed as Lexical Ambiguity. For example, According to WordNet [14], the lexical pen, has five meanings which can be written as pen= {a writing implement with a point from which ink flows, an enclosure for confining livestock, a portable enclosure in which babies may be left to play, a correctional institution for those convicted of major crimes, female swan}.This issue needs to be resolved.

- Another challenge is in identifying limits for sentence boundaries in the English language .That is differentiating between a period that marks the end of a sentence and a period that is part of a previous token like Dr., Mr. or an email id - abc.def@yahoo.com and so on.

5.2.1.3 Syntax Analysis/ Parser. The input to Syntax Analysis or Parser step is the sequence of tokens .The Syntax Analysis performs two steps: firstly it checks whether the incoming tokens are according to the specified Grammar or not. So, the concept of Grammar comes into picture. And, if the tokens are in accordance with specifications of the grammar then a Parse tree is drawn else, there is an error. A Grammar or a Natural Grammar (G) is a formal specification of the language used, and is represented as

G = (VN T,VT, P, S) ,where VNT denotes a set of non terminal symbols, VT a set of terminal symbols, P is a set of production rules, and S represents a sentence. A non terminal symbol is a symbol that does not appear in an input string, but is a syntactic category or a variable in G. Examples of VN T are NP (Noun Phrase),VP (Verb Phrase), PP (Preposition Phrase), and so on. A terminal symbol is a symbol that represents a class of basic and indivisible symbols in input strings; it represents a part-of-speech symbol. Examples of terminal symbols are Noun, Verb, Adjective, Preposition etc. A production is a rule of the form $\alpha \rightarrow \beta$, where α is a non terminal symbol, and β represents a set of terminal symbol or non terminal symbols. When a sentence is successfully parsed, a structure so called a parse tree which represents an actual structure of a sentence is generated. For example, a parse tree of the sentence 'He holds bat with his hand 'is shown in the figure 3 given below:

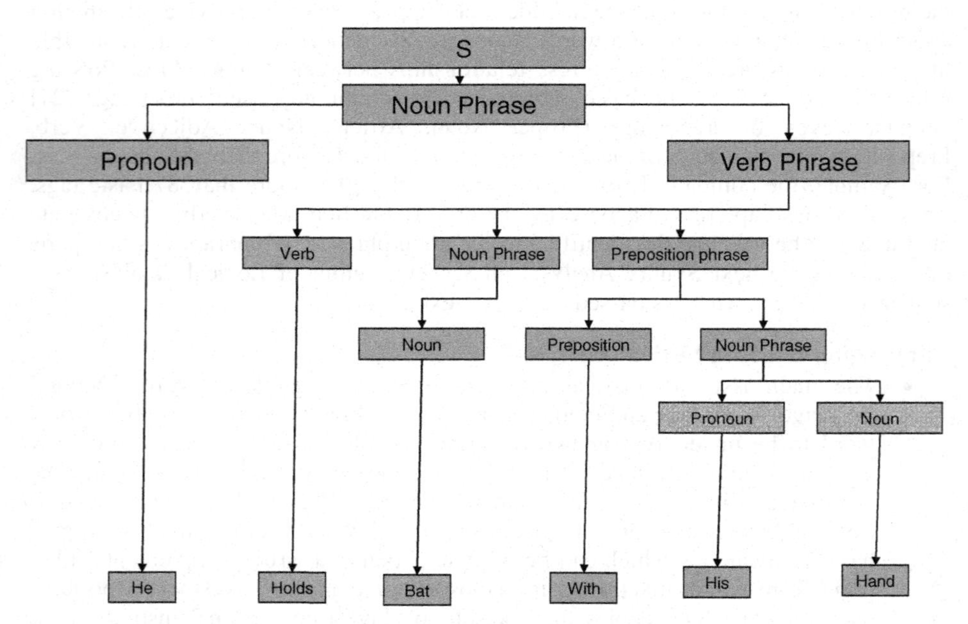

Fig. 3. Showing parse tree of the sentence 'He holds bat with his hand'

The major work in syntactic analysis is sentence parsing. During this, a sentence is broken down into phrases and sub phrases until an actual grammar structure of a sentence is recognized.

Challenging Issues in Syntax Analysis

- In Syntactic Analysis, the order in which words appear is an important issue which should be considered. The parsing of a sentence should start either from the beginning of a sentence or from the end or sometimes it can even start from the main verb.
- A syntactic analysis may lead to Collocations Discovery in the text. Collocations are phrases like "head ache", which make sense if they are taken as a whole. That is the meaning of the whole is different or greater than the meaning of its parts. So, such type of phrases should be detected in Syntax Analysis phase.
- To resolve Syntactic Ambiguity, a Knowledge base consisting of the meanings of all the ambiguous phrases that the author has used in the past has to be maintained.

5.2.1.4 Semantic Analysis. This is the next step after Syntax Analysis; it takes as input Parse tree generated in the previous phase. Semantic Analysis is a process of converting a parse tree into a semantic representation that is unambiguous and clear .It can be done by two methods: Context independent and Context dependent. Context independent deals with the meaning of words in a sentence without considering the context in which they are applied. Context dependent deals with the context of the sentence that is the situation in which the sentence is used, the immediately preceding sentences and so on. In this step, a Semantic Action is associated with each production rule of the grammar. For example, $\alpha \rightarrow \beta$ { α1.sem, α2.sem….. α n.sem},where the expression in { } is the semantic action. The semantic action is a piece of code that generates the intermediate form of text directly.

5.2.2 Text Synthesis Phase

The Text Synthesis Phase generates the intermediate form of text. This intermediate form can be document-based where each entity represents a document, or concept based where each entity represents an object or concept of interests in a specific domain or in the form of relational tables. A document-based Intermediate Form discovers patterns and relationships across documents. Document Visualization, Clustering and Categorization are techniques that can be applied on document-based Intermediate Form .Concept-based Form derives patterns and relationships across objects or concepts. Operations such as predictive modeling and association discovery come under this category. In Relational Tables, Categorization, Clustering techniques can be used to infer useful knowledge.

Challenging Issue in Text Synthesis Phase

- Complexity of the intermediate form is an important issue, which needs to be kept in mind.

5.2.3 Dictionary /Bookkeeping Component

The dictionary or Bookkeeping component performs two functions. Firstly it stores the tokens found in Lexical Analysis and Syntax Analysis step so that there is no redundancy of the tokens .Secondly, it helps to search any token, if required by other components. This will also check for any errors in the tokens like misspellings of words etc.

6 Conclusion and Future Work

In this paper, we have proposed a novel framework U - STRUCT that converts unstructured text document to a structured form. This is a generic framework which can be applied on all forms of text documents like a news story, an email, a business report or a scientific research paper. At present, we have presented the framework for a single document, in future it can be extended for multiple text documents as well. And also, till now we have worked on only English language, it can be applied on other languages also.

References

1. Kroeze, J.H., Matthee, M.C., Bothma, T.J.D.: Differentiating between data-mining and text-mining terminology. In: SAICSIT 2003, pp. 93–101. ACM Digital Library (2003)
2. Chen, H.: Knowledge management systems: a text mining perspective. Knowledge Computing Corporation (2001)
3. Stavrianou, A., Andritsos, P., Nicoloyannis, N.: Overview and Semantic Issues of Text Mining. SIGMOD Record 36, 3 (2007)
4. Han, J., Kamber, M.: Data Mining Concepts and Techniques, 2nd edn. Morgan Kaufmann. The University of Illinois at Urbana-Champaign (2006)
5. Pujari, A.K.: Data Mining Techniques. University Press (2002)
6. Gupta, V., Lehal, G.S.: A Survey of Text Mining Techniques and Applications. Journal of Emerging Technologies in Web Intelligence 1(1) (2009)
7. Berry, M.W., Browne, M.: Understanding Search Engines: Mathematical Modeling and Text Retrieval. SIAM, Philadelphia (1999)
8. Lee, S., Song, J., Kim, Y.: An Empirical Comparison of Four Text Mining Methods. In: 43rd Hawaii International Conference on HICSS, pp. 1–10 (2010)
9. Chen, M.S., Han, J., Yu, P.: Data Mining: A Overview from a Database Perspective. IEEE Transactions on Knowledge and Data Engineering 8(6) (1996)
10. Hearst, M.A.: Untangling text data mining. In: Proceedings of the 37th Annual Meeting of the Association for Computational Linguistics on Computational Linguistics, ACL 1999, pp. 3–10 (1999) ISBN:1-55860-609-3
11. Witten, I.H.: Text mining. In: Practical Handbook of Internet Computing, pp. 14-1–14-22. Chapman & Hall/CRC Press, Boca Raton, Florida (2005)
12. Fan, W., Wallace, L., Rich, S., Zhang: Tapping the power of text mining. Communications of the ACM - Privacy and Security in Highly Dynamic Systems 49(9), 76–82 (2006)
13. `http://en.wikipedia.org/wiki/Brown_Corpus`
14. `http://wordnet.princeton.edu/`

15. Balakrishnan, K., Sreedhanya, S., Soman, K.P.: Effect Of Pre-Processing On Historical Sanskrit Text Documents. International Journal of Engineering Research and Applications (IJERA) 2(4), 1529–1534 (2012) ISSN: 2248-9622
16. Torunoglu, D.: Analysis of preprocessing methods on classification of Turkish texts. In: International Symposium on Innovations in Intelligent Systems and Applications (INISTA), pp. 112–117 (2011)
17. Sagar, S., Imambi, S.T.: Pre processing of Medical Documents and Reducing Dimensionality. Advanced Computing: An International Journal (ACIJ) 2(5) (2011)
18. Farooq, F., Govindaraju, V., Perrone, M.: Pre-processing methods for handwritten Arabic documents. In: Proceedings of Eighth International Conference on Document Analysis and Recognition, vol. 1, pp. 267–271 (2005)
19. Suliman, A., Sulaiman, M.N., Othman, M.: Chain Coding and Pre Processing Stages of Handwritten Character Image File. Electronic Journal of Computer Science and Information Technology (eJCSIT) 2(1) (2010)
20. Feldman, R., Dagan, I.: Knowledge Discovery in Textual Databases (KDT). In: Proceedings of KDD 1995 (1995)
21. Shatkay, H., Feldman, R.: Mining the Biomedical Literature in the Genomic Era: An Overview. Journal of Computational Biology 10(6), 821–855 (2003)

Performing Web Log Analysis and Predicting Intelligent Navigation Behavior Based on Student Accessing Distance Education System

Shivkumar Khosla[1] and Varunakshi Bhojane[2]

[1] Department of Computer Engineering, Mumbai University, India
shivkumar_khosla@yahoo.co.in

[2] Department of Computer Engineering, Mumbai University, India
varunakshi_k@yahoo.com

Abstract. We have introduced a concept of capturing different web log file, while the user is accessing the Distance Education System website and provide the user with the intelligent navigation behavior on his browser. Web log file is saved in text (.txt) format with "comma" separated attributes. Since the Log files consist of irrelevant and inconsistent access information, therefore there was a need to perform Web log preprocessing which includes different techniques such as field extraction, data cleaning, and data summarization. Preprocessed information is been given to intelligent navigation module and it allows the student to have most frequently viewed subject at the top of the list, which allows them to have easy access to the tutorial or chapter within the subject. The analysis allows enhancing the personalization services in distance education system and making the system much effective.

Keywords: Intelligent Navigation, Personalization services, Web Usage Patterns, Web log information.

1 Introduction

Most of the student or users are moving towards the internet technology for learning or exchanging of the information. Distance Education System provides a medium through which student can learning the subject and exchanging the information with each other. In traditional distance education system, when the student enters the teaching websites he access the subject or the tutorial available of his interest by clicking on the navigation menu available and finally logout.

Traditional distance education system is unable to provide students with a variety of personalization services. In order to meet the requirements of different students, we should improve the topology & navigation structure in which the pages are being accessed. The paper presents an implementation procedure for the intelligent navigation. We have used web log capture techniques to acquire the student access information and improve the navigation procedure in the distance education system.

S. Unnikrishnan, S. Surve, and D. Bhoir (Eds.): ICAC3 2013, CCIS 361, pp. 70–81, 2013.

2 Log Information

2.1 Intelligence Navigation Log

Information Navigational log information is used to capture information such as SubjectID, SubjectName, SubjectCount, Position, Visibility, Chaptername etc. This information captured provides a provision for the navigation link while the user is accessing the distance education system site.

(Table 1) StudentID is of character data-type which is given to the users when the user (student) registers in the distance education system. StudentID is being referred from Student registration table. Subject-id is of integer data-type which is referred from subject table. Subject-count is of integer data-type which is used to keep the record count of each subject when the student accesses the respective subject. Subject-count is of integer data-type which is used to keep the record count of each subject when the student accesses the respective subject.

Table 1. Change Log Table

Attributes	Description
StudentID	Uniquely identify the user
SubjectID	Uniquely identify the subject
SubjectCount	Count of subject user access
Position	Position of the each subject in the navigation list

(Table 2) SubjectID is of integer data-type which is the unique identifier of the subject. SubjectName is of character data-type which consist of the subject name which will be displayed when the student access the sites. Position is of integer data-type which give an initial position of the subject to the user. Visible is of Boolean data-type which consist of 1 and 0 values where in 1 indicates subject is visible to the student while the student is accessing distance education sites and 0 indicates subject is not visible to the student.

Table 2. Subject Information Table

Attributes	Description
SubjectID	Uniquely identify the subject
SubjectName	Name of the Subject
Position	Position of the each subject in the navigation list
Visible	Describe whether the subject is available to the user

(Table 3) ChapterID is of character data-type which is the unique identifier of the chapter associated to the respective subject. SubjectID is referred from the subject table. ChapterName is of character data-type which assigns a name to the chapter which will be displayed on the student screen. Position is of integer data-type which gives a default position to each chapter. Visible is of Boolean data-type were in 1 represent the chapter is visible and 0 represent chapter is not visible to the student. Content is of Text data-type which consist of information that is to be displayed on the student screen whenever student click on the respective subject.

Table 3. Chapter Information Table

Attributes	Description
ChapterID	Uniquely identify the Chapter for the subject
SubjectID	It is being referenced from subject table
Chapter-Name	Name of the chapter within the subject
Position	Position of the each chapter in the navigation list
Visible	Describe whether the chapter is available to the user or not
Content	Consist of the information in each chapter

2.2 Web Log Information

In Distance Education system students may have different access modes, but for long-term statistical representation and also to make system as stable as possible we need to have only the useful information of the students. So, Web usage pattern establishes a set of intelligent way of extracting only the useful information. Access Information of the student is recorded as per the web log file information shown (Fig. 1.)

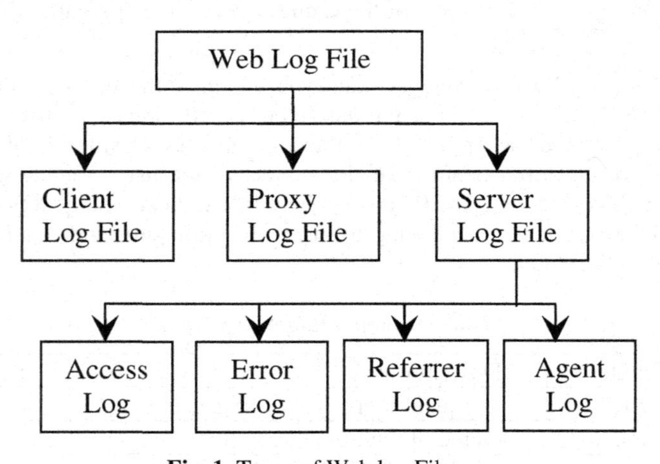

Fig. 1. Types of Web log Files

Client Log Files mostly consist of authentic i.e. username and password but it does not consist of any of the browser information.

Proxy Log File used to capture the user access data i.e. it capture the pages that are being accessed by the user. Proxy server is in many-many cardinality since there are many users accessing many pages.

Server Log Files are in relationship of many to one since there is only one web server response to many users. Server log file do not record the cashed pages requested.

Different types of Server Log File include:

1. Referrer Log
   Referrer Log file contains information about the pages that is being referrer.
2. Error Log
   Error Log File records the errors of web site especially page not found error (404 File not found).
3. Agent Log
   Agent Log File records the information about the website user's browser, browser version & Operating System.
4. Access Log
   Access Log file records all the click, hits and accesses made by the user to the website.

3 Data Preprocessing

Preprocessing converts the captured log data in to the valuable information which can be given for further pattern discovery. In this phase main steps includes are [1]:

1. Extraction the attributes from the web log which is located in web server
2. Cleaning the web logs and removing the redundant and irrelevant information.
3. Manage the data and put it in relational database or data warehouse.

3.1 Field Extraction

In Field Extraction we extract the attributes from the log file which is separated by character ','. The attributes that are extracted is stored in the log table which is relational database. Process flow is (shown in Fig.2).

3.2 Data Cleaning and Data Filtering

Data cleaning eliminates irrelevant or unnecessary records stored in log table. Since website will be accessed by millions of users. Data Cleaning is used to remove the records while analyzing the data. It removes all the jpg, gif and css files which the user has accessed & also the failed HTTP status code (404 Page Not Found). Data Filtering provides with the more better cleaning by removing the repeated pages that the user have accessed in the referrer string and also filtering help to reduce the path's accesses of the page by splitting the Referrer String and getting valuable information.

By performing Data Cleaning and Filtering errors files, inconsistent data, missing data, repeatable data will be detected and removed to improve the quality of data. Process flow is (shown in Fig.3).

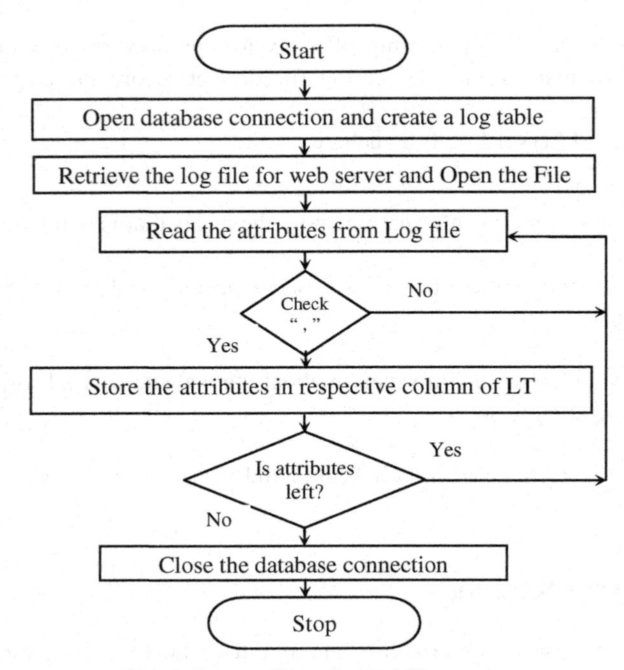

Fig. 2. Process Flow for Field Extraction

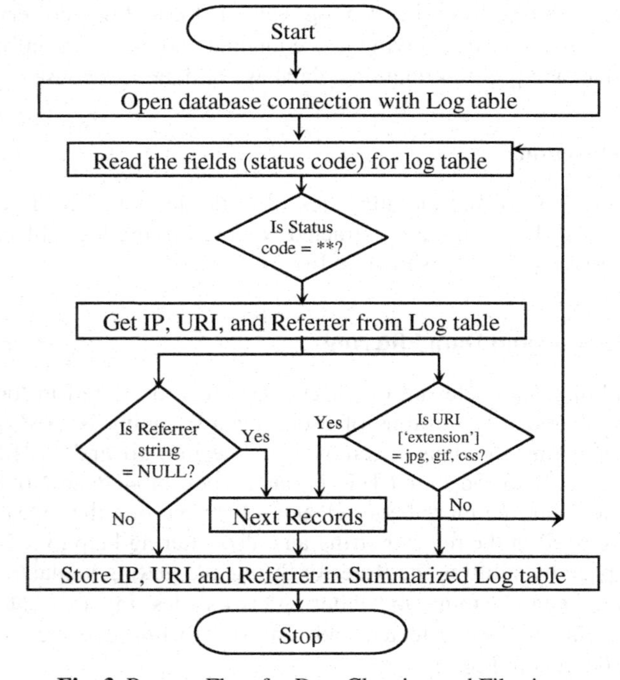

Fig. 3. Process Flow for Data Cleaning and Filtering

3.3 User Identification

User Identification is based identify the user with the User Id which help to predict the user behavior uniquely i.e. pages the user have accesses along with the IP address of the user through which user was accessing the sites.

User Id of the user also helps in Session Identification i.e. if the session is set then the user is registered user and can start recording the pages the user accessing and also provide the user with dynamism of the pages but if session is not set the user is allow to access the pages but do not record any of the log information of the user which in turn reduces the log information of the non-registered user.

3.4 Data Summarization

In this paper data summarization is describe by the graphical representation of the log that are being captured. Since, useful and consistent records remain in the Summarized log table which helps in further pattern discovery process.

Summarized log table when compared with the log table provides information such as Total no. of records, Total no. of registered User, Total no. of URL, Visited count of the registered user.

4 Intelligent Navigation Module Design

In traditional model of distance education system the student communicate with the web server through the internet and whenever student is accessing the subject there is no dynamic in the navigation of the pages. I.e. the link available to each user remains fixed on the left navigation menu. Intelligent navigation provides a procedure to have dynamism in the navigation menu while the user is accessing the distance education system (as shown in Fig. 4).

4.1 Intelligent Navigation Modules Included Are as Follows

Log Capture Module:

We have considered Client Log Files which mostly consist of Client IP address & authentication information. The Client Log files helps to uniquely identify the users and the subject which the user will be accessing. So after capturing information in the log file (.txt format), we then converted it into the table with the attributes such as ClientIP, UserID, Subjectid, SubjectCount, position, visibility.

Intelligent Navigation Module:

Step1: User (Student) is supposed to register first for Intelligence Navigation

Step2: After Registration create all subject entry in database for that user with subject-count =0

Step3: Allow the user to login to distance education system sites

Step4: Record for the click stream of the user on each subject

Step5: Update the subject-count on each access of the user to specific subject

Step6: Update the position with respect to the subject count available in the database.

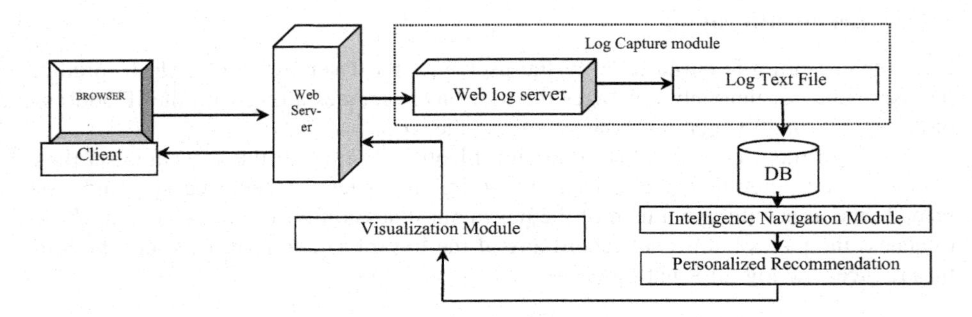

Fig. 4. Architecture of the Navigation system

Visualization Module
Visualization Module consists of a menu representation for the user. Menu consist list of SubjectName and the assigned chapter to each subject. We have used relational database technology (SQL) to carry out selection of the SubjectName which the user request and display it in the navigation section of the user browser.

5 Methodology

The Client request to the web server through a web browser for accessing distance education system and if the client is register then the client complete access information is recorded by the web server log. Web log server creates the log file and necessary information is then recorded in the relational database. (Fig. 5 shows) the initial entry while the client is registering for the distance education system. Entry for the respective user is made for each subject and the value of SubjectCount is assigned zero.

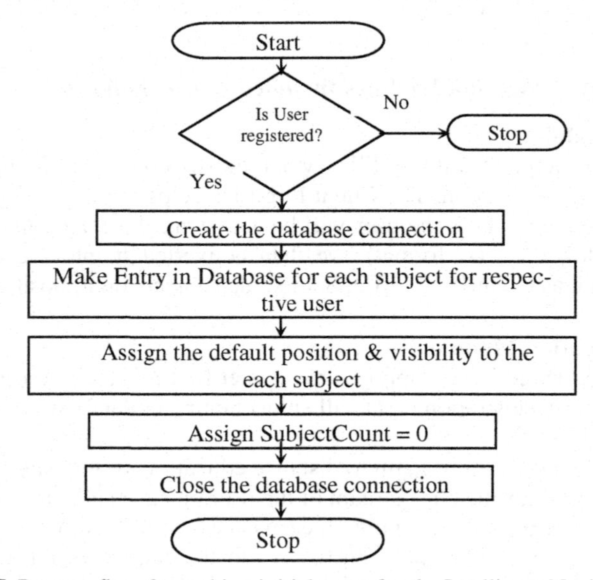

Fig. 5. Process flow for making initial entry for the Intelligent Navigation

(In Fig. 6) we have first recorded the number of attempt the respective user have login and if login attempt>3 then we provide a user with intelligent navigation on the user accessing browser and if login attempt<=3 then we keep on updating the SubjectCount.

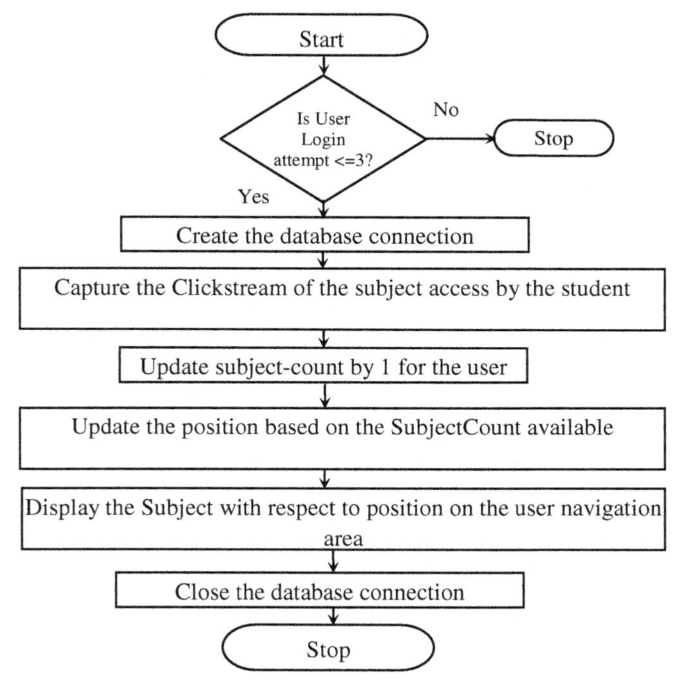

Fig. 6. Process flow for acquiring intelligent navigation on the User browser

6 Result

6.1 Comparison

We have captured a log for a month '19-07-2012' to '22-08-2012' of distance education system and depending upon the log captured we had following analysis.

(Table 4) shows the comparison between log table and summarized log table.

Table 4. Comparison between Log Table (LT) and Summarized Log Table (SLT)

Attributes	**LT**	**SLT**
File Size (KB)	2154	519
Number of Records	3432 54	12519

(Table 5 shows) the Number of user that have uniquely access the distance education system. It is used to record the distinct Client IP address and No. of times the user has log in the Distance Education system.

Table 5. Number of Access of the Unique Users

Client IP address	No. of Access
172.16.11.101	34
172.16.11.102	40
172.16.11.103	19
172.16.11.104	26
172.16.11.105	53
172.16.11.106	13
...	...
...	...

(Table 6 shows) summarized statistical report for the Errors pages count, Css files that were accessed and total number of hits on the Distance Education Site.

Table 6. Summary Statistical Report

Status	Records
Errors	1289
Css	868
Jpeg	2548
Hits	1524785

The Following Figures are showing total no. of records in Log table and summarized log table (Fig. 7), Number of access of the unique users (Fig. 8)

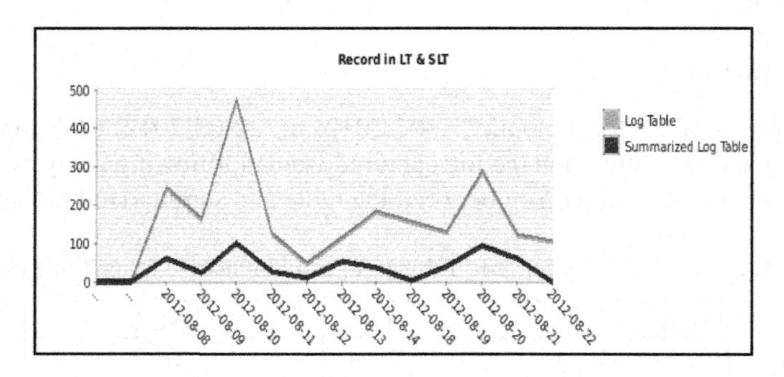

Fig. 7. Total no. of records in Log table and Summarized log table (Date-wise)

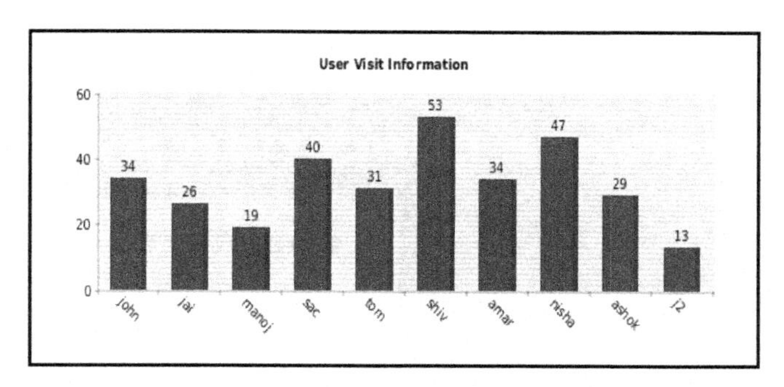

Fig. 8. No. of accesses of unique users

The environment required to perform this experiment is any the operating system with any of the browser but the user need to be a registered user.

(In Table 7) we have allowed numerous of user to register in distance education site and recorded their entry in the database in the form of log capture and kept on update the record based on their access i.e. Log entry of all user to provide intelligent navigation.

Table 7. Log Entry to Provide Navigation

UserID	Subject ID	Subject Name	Subject Count	Position	Visible
101	1	Java	10	1	1
101	2	CS	7	2	1
101	3	DBMS	6	3	1
101	4	CP	3	4	1
102	1	Java	4	4	1
102	2	CS	8	2	1
102	3	DBMS	15	1	1
102	4	CP	6	3	1
...	...	...	...	...	...
...	...	...	...	...	...
...	...	...	...	...	...

(Fig. 9. shows) The initial intelligent navigation representation for the respective registered user, it is used to display the user with the default navigation pattern on the browser.

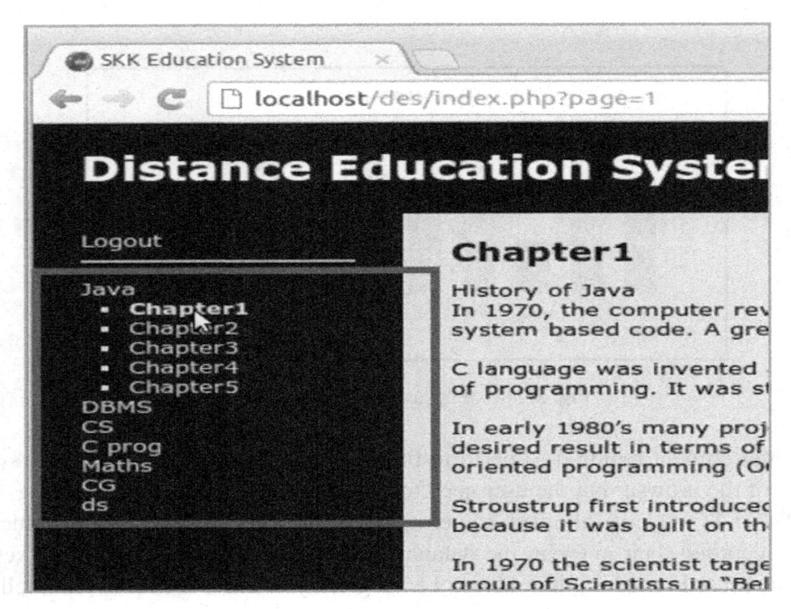

Fig. 9. Initial representation when User login for first time

(Fig. 10. shows)The Intelligent navigation representation of the user when the respective user access "DBMS Subject" frequently.

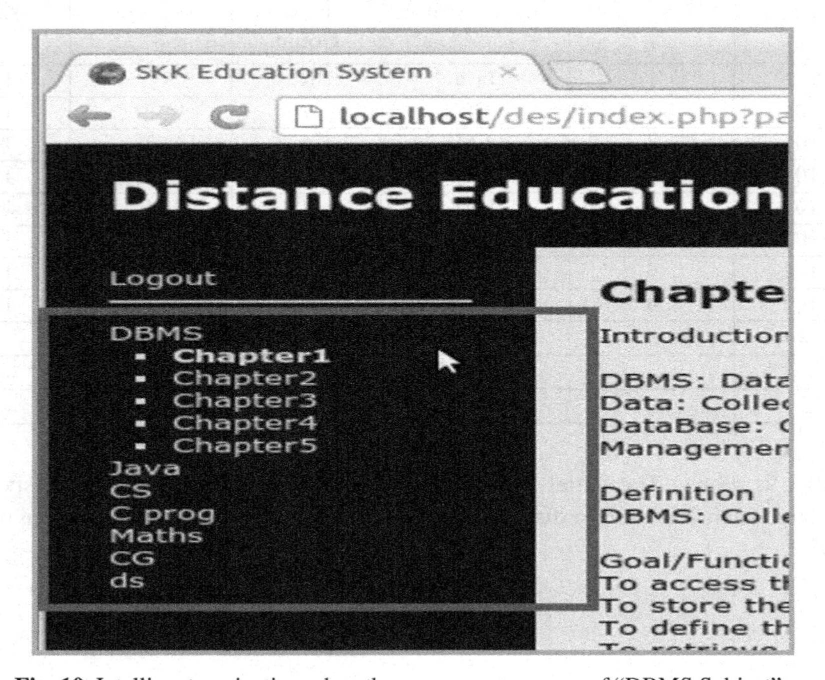

Fig. 10. Intelligent navigation when the user accesses more of "DBMS Subject"

7 Conclusion

Data Preprocessing helps in removing all the irrelevant information which speeds up the execution time and provides with valuable information to the users. It also allows the user to have different access pattern due to which better accessibility is achieved. We can also record the information in the cookies and when the user ends its session we can then transfer the complete information from the cookies to the database.

Web Usage pattern analysis is used to provide different fields that is used to discover hidden and interesting navigation patterns. We have found a solution to problem of Web Site improvement & recommendation on the user behavior by using intelligent navigation module. Intelligent navigation module allows faster access of the subject and allows to have numerous list of subject on the same page i.e. there is no need to open a new browser or tab of the browser every time whenever user request for a new subject. It also provides user with better learnability along with better operability. Intelligent navigation module can be further improved by using cookies & session for temporary storing of information, while the user is accessing the site and later on transfers the final result to the database.

References

1. Aye, T.T.: Web log Cleaning for Mining of Web Usage Patterns. IEEE, 978-1-61284-840-2/11
2. Hussian, T., Asghar, S., Masood, H.: Web Usage Mining: A Survey on Preprocessing of Web log File, 978-1-4244080306/10
3. Li, X., Zhang, S.: Application of Web usage mining in e-learning platform. In: International Conference on E-Business and E-Government (2010)
4. Cooley, R., Mobasher, B., Srivastava, J.: Data preparation for mining World Wide Web browsing patterns. Knowledge and Information Systems 1(1), 5–32 (1999)
5. Yuan, F., Wang, L.-J., et al.: Study on Data Preprocessing Algorithm in Web Log Mining. In: Proceedings of the Second International Conference on Machine Learning and Cybernetics, Wan, November 2-5 (2003)
6. Dong, Y., Li, J.: Personalized Distance Education System Based on Web Mining. In: International Conference on Educational and Information Technology (2010)
7. Cooley, R., Mobasher, B., Srivastava, J.: Data Preparation for Mining World Wide Web Browsing Patterns. Knowledge and Information Systems I(I) (February 1999)

Web Content Mining Using Genetic Algorithm

Faustina Johnson and Santosh Kumar

Department of Computer Science & Engineering
Krishna Institute of Engineering & Technology
Ghaziabad-201206, India
johnson.faustina@yahoo.in, santosh_k25@rediffmail.com

Abstract. Web mining is the application of data mining techniques on the web data to solve the problem of extracting useful information. As the information in the internet increases, the search engines lack the efficiency of providing relevant and required information. This paper proposes an approach for web content mining using Genetic Algorithm. Genetic Algorithm is being used for wide range of optimization problems. Evolutionary computing methods help in developing web mining tools which extract relevant and required information. It has been shown experimentally that the proposed approach is able to select good quality web pages as compared to the other existing algorithms proposed in the literature. The proposed approach considers several parameters like time website existed, backward link, forwards links and others for selecting good quality web pages.

Keywords: Genetic Algorithm, Web Mining, Backward Links, Forward Links.

1 Introduction

The World Wide Web is enormous and growing exponentially day by day. Finding the relevant and required information is tedious task. Information is so vast that it cannot be directly used for business purposes. Web mining is an approach in which data mining techniques are applied on the web data. Web mining approaches can be used in extracting the relevant information from the huge internet database. The data on the web is heterogeneous varying from structured to almost unstructured data like images, audios, and videos. There is enormous amount of redundant information available on the web resulting in multiple pages containing almost same or similar information differing in words and/or formats. A significant amount of information on the web is linked. Hyperlinks exist among web pages within a site and across different sites. Based on the nature of data in the web mining is categorized into three main areas: Web Content Mining, Web Structure Mining and Web Usage Mining. Web content mining search automatically and retrieves information from a huge collection of websites and online database using search engine. The data for content mining lies in various formats text, image, audio, video metadata and hyperlinks. Web Structure Mining is discovering the model underlying link structures (topology) on the web e.g. discovering authorities and hubs. Web Usage Mining mines the log file and data associated with a particular website to discover and analyze the user access patterns.

S. Unnikrishnan, S. Surve, and D. Bhoir (Eds.): ICAC3 2013, CCIS 361, pp. 82–93, 2013.

The data used for web content mining includes both text and graphical data. Based on the searching content mining is divided into two types. These are Web Page Content Mining and Search Result Mining [22, 23]. Web page content mining is the technique of searching the web via content. Search result content mining further searches the pages from a previous search. Evolutionary approaches have been used for web content mining [27]. The proposed approach uses different technique using Genetic Algorithm (GA) for web content mining. GA is a branch of Artificial Intelligence which was inspired by Darwin's theory of living organisms in which successful organisms were produced as a result of evolution [8, 13]. So GA is search algorithm based on the natural selection and natural genetics. The main significance of GA is the survival of the fittest which is also known as natural selection. It is different from other search methods in that it searches among population of points and works with coding of parameter set rather than parameter values themselves. There are problems which we cannot determine a priority to the sequence of steps leading to a solution. Search is a best method for such problems. There are two methods to perform search. These are blind strategies and heuristic strategies. Blind strategies do not use information about the problem domain. Heuristic strategies use additional information to guide the search. Two main issues in search strategies are exploring the search space and exploiting the best solution. Exploration is the method of searching new sources and exploitation is the method of using known sources. Hill Climbing is an example of exploitation and random search is an example for exploration. GA makes a remarkable balance between exploration and exploitation of the search space. The major steps in GA include generation of a population of solution, finding the objective function and the application of genetic operators such as reproduction, crossover and mutation. GA implementation starts with a population of chromosomes which are randomly generated. According to the fitness function the chromosomes are evaluated. The chromosomes with better solution are given more chance to reproduce than the chromosomes with poorer solution [13]. This paper proposes a GA based approach for web content mining to get the Top-T web links and the proposed algorithm has been compared with the algorithm proposed in [27] hereafter known as MA algorithm. It has been shown experimentally that the proposed approach performs better than the MA algorithm.

In Section 1.1 Literature review is done. Section 2 discusses GA based approach. Section 2.2 describes the proposed algorithm. An example based on the proposed approach is given in Section 3. Section 4 & 5 are experimentation and conclusion respectively.

1.1 Related Work

Web content mining is the most challenging area in the field of web mining. A lot of work has been done still the search engines lack in their efficiency and accuracy in responding the user queries. Evolutionary approaches can play a critical role in the mining of web content data. In [27], the authors proposed an algorithm (MA algorithm) for content mining. They have considered web search as a general problem of function optimization. Using the fact that the web is a graph in which nodes are web pages and edges are the links between these web pages. The search space in the optimization problem is a set of web pages. Evaluation or fitness function is done on a set of web pages. In the beginning individuals are generated with a heuristic creation operator by querying standard search engine to obtain pages. Individuals are selected or deleted based on fitness function. And then it gives birth to offsprings after crossover is performed. Evaluation function is based on the link quality and page quality. They are obtained using the number of keywords given by the user and mean number of occurrences in link. In [2] a genetic relation algorithm (GRA) was performed for additional searching of documents according to user interest. Evolutionary GRA optimizes the relationship between hyperlinks in web pages. GRA provided the search strategy with minimal user intervention.

Genetic Algorithm is an adaptive heuristic search algorithm which was inspired by Darwin's survival of the fittest. It is based on the evolutionary ideas of natural selection and genetics. GA can be used to solve the optimization problem. GA mimics the process of the nature such as Selection, Crossover, Mutation and Accepting. The most commonly used methods for selecting chromosomes for crossover are Roulette Wheel selection, Boltzmann selection, Tournament selection, Rank selection, Steady state selection. One Point, Two Point, Uniform, Arithmetic and Heuristic crossovers can be applied on selected chromosomes [8]. For data mining optimization in educational web based system, GA utilizes the data from educational web based system and predicts the final grade and classifies them according to their grades. So here GA is useful to optimize prediction accuracy [12]. GA can be applied to different types of mining such as Content Mining, Structure Mining and Usage Mining. There are several research trends and techniques in the field of web content mining [4, 6].

Web Content Mining can be done on structured and semi-structured data. It explains how web content mining helps in extraction of data in a simple way. Web consists of data such as audio, text, video etc. HTML document is semi-structured data and data in the form of tables is structured data. Techniques used for extraction of structured data are Web Crawler, Wrapper Generation and Page Content Mining. Techniques used for semi-structured data are top-down extraction using OEM (Object Exchange Model), WICCAP [5]. Web Mining helps in resource discovery, information selection and preprocessing, generalization, analysis and visualization. The technique of discovering usage pattern from web data is known as web usage mining. It consists of three phases such as pre-processing, pattern discovery and pattern analysis [3, 7]. Pre processing consist of usage pre-processing, content pre-processing, structure pre-processing. Pattern discovery consists of statistical analysis, association rules, clustering, classification, sequential patterns, dependency modeling. According to the user's interest additional searching is done using a genetic relation algorithm. The main feature of this algorithm is that it can optimize the relationships between the web hyperlinks using evolution theory. It satisfies the user by giving more quality pages than search engines. It uses only minimum user interaction and it provides similar hyperlinks according to the users need [2]. The technique of finding data objects which differ significantly from the rest of the data are known as outlier mining. This helps in identification of competitors which in turn help in the development of electronic commerce. WCOND - Mine using n-grams without a domain dictionary algorithm is proposed for mining web content outliers. N-grams means n contiguous character slice of a string which is divided into smaller substrings each of size n .The result show that this algorithm is capable of detecting web content outliers from web datasets [9].

Information retrieval, Information extraction and machine learning were some of the techniques used to extract knowledge from the web. In [3] these techniques were compared with web mining. In [24] useful information was selected by information retrieval by indexing text. Relevant document is selected by Information retrieval whereas information extraction focuses on extracting relevant facts. Information retrieval system and information extraction system is a part of web mining. Preprocessing phase is supported by information extraction before web mining. It also helps in indexing which further helps in retrieval. Machine learning indirectly supports web mining by improving text classification process better than traditional information retrieval process [25]. Information in the web is structured to facilitate effective web mining. Web mining is decomposed into resource discovery, information extraction, and generalization [10]. A multi object combination optimization model is constructed which handles GA for negotiation problem then considers global convergence, concurrency features. A new GA based approach is proposed to hybridize local meta-heuristic mechanism which helps to speed up the search process [11]. Web design patterns are useful tool for web data mining. Web pages are analyzed to find out which useful information is included in the web page. To find out useful information two methods were used. One is information retrieval and the second one is information extraction. Information retrieval is used to extract useful information from large collection of web pages. Information extraction is used to find structure information. It uses patterns for web page description and this description is used for another task. It detects pattern instances in web pages and then it is compared with information extraction [22]. After extracting information, an algorithm is used to match large number of schema in databases. In [1] an algorithm is explained which matches correlated attributes with less cost. The positive and negative correlated attributes are distinguished by Jaccard measure.

2 The Approach

2.1 Genetic Algorithm

John Holland invented Genetic Algorithms in 1960 and at the University of Michigan. The algorithm got developed by Holland and his students in the 1960 and 1970. GA is an evolutionary algorithm. Evolution is a method of searching a solution from an enormous number of possibilities. In Biology, enormous set of possibility means it is a set of possible genetic sequences. It is a solution of finding out fittest organisms. Evolution can also be explained as a method for designing innovative solution for complex problems [25]. GA was inspired by Darwin's theory of evolution. GA is used to solve optimization problems such as numerical optimization and combinatorial optimization [8]. Optimization is the problem of finding best solution. Best Solution implies that there is more than one solution and the solutions are of not equal value [26]. A GA mimics the process of natural evolution. The five phases are Initial Population, Fitness Function, Selection, Crossover and Mutation. The most beneficial part of GA is the crossover.

Biological background of GA is that all living beings are made up of cells. And cells in turn are made up of chromosomes and each cell contains the same set of one or more chromosomes. They are made up of strings of DNA which acts as a blueprint for the organism. Chromosomes are further divided into genes. The genes are responsible for exhibiting particular trait such as eye color. Alleles are different possible setting of a trait. Brown, blue, hazel colors of eye are examples for alleles. The position of each gene in the chromosome is known as locus. Complete set of Genetic material is called organism's genome. Genotype is the particular set of genes in a genome. The physical expression of genotype is called the phenotype.

The major steps in GA are the generation of initial population of solutions, finding the objective function, and applying genetic operators [13]. These are shown in Fig. 1.

Generate initial population of chromosomes randomly
Repeat
Evaluate each chromosome against objective function
Apply genetic operators
  Reproduction
  Crossover
  Mutation
Until stopping criteria

Fig. 1. Basic Genetic Algorithm

Crossover is the main operator in GA which results in generation of new offsprings. Crossover exchanges genes between parents and produces new offsprings. Mutation is another tool in GA. It is a process of changing the gene in a chromosome which results in generation of a new offspring. The offspring which survive the most are considered to be more fit. So fitness value is calculated on the basis of survival of the fittest [8, 25].

Here in the web content mining problem, the Chromosomes are a set of web pages. And each web page is the gene of the chromosome and locus is the position of web page connected to a Chromosome. GA is used to find the Top-T web links according to the need of the user.

Chromosome1 = (l_{11}, l_{12}, l_{13}, l_{14}, l_{15}, l_{16}, l_{17}, l_{18}, l_{19}, l_{20})

Where $l_{11,}$ l_{12}.... are web links.

2.1.1 Chromosome Representation

Initial population is the set of candidate solutions. Each candidate solution is represented by a chromosome. The structure of a chromosome is set of Top-T web links as given below.

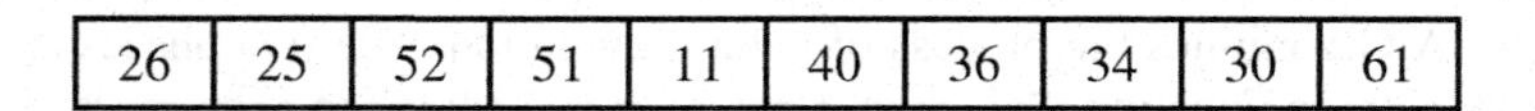

26	25	52	51	11	40	36	34	30	61

Fig. 2. Top-10 Web Links

2.1.2 Fitness Function

Fitness function is a quality measurement derived from a gene. Fitness function quantifies the optimality of a solution (chromosome), so that the particular solution may be ranked against all other solution. Function depicts the closeness of a given solution to the desired result [8].

1. No of Keywords such as $K_1, K_{2,\ldots\ldots} K_n$,
   Frequency of each keyword in a document d is $f_1, f_{2,}, \ldots f_n$ respectively.
2. The time the website existed in the internet
3. Number of backward links
4. Number of forward links

$Cost_{Keywords} = \sum_{j=1}^{Top\ T} \sum_{i=1}^{n} fi$

Where n=10 and f_i is the total frequency of keywords in documents in Chromo some Cj.

$Cost_{time} = \sum_{l=1}^{n} Tl$

Where n =10 andT_l is the time the web page existed in the net.

$Cost_{backward\ link} = \sum_{m=1}^{n} Bm$

Where n=10 and B_m is the number of backward links.

$Cost_{forward\ link} = \sum_{p=1}^{n} Fp$

Where n=10 and F_p is the number of forward link.

Cost Function

$$F(x) = C1.\ \sum_{j=1}^{TopT} \sum_{i=1}^{n} fi + C2.\ \sum_{l=1}^{n} Tl + C3.\ \sum_{m=1}^{n} Bm + C4.\ \sum_{p=1}^{n} Fp$$

Where C1, C2, C3 and C4 are Constants to adjust the different parameter.

Fig. 3. Parameters for the cost function

2.1.3 Selection

The Selection is a method of choosing chromosomes for performing crossover. The proposed algorithm is using binary tournament selection. Two individuals are randomly selected from the initial population. One out of these two randomly selected chromosomes is selected for crossover. The chances of selection of fitter individual is more because of the selection of predefined parameter k as 0.75.The process is described in Fig.4.

```
Choose a parameter 'k' (say= 0.75)
Choose two individuals randomly from population
Choose a random number 'r' between 0 and 1
if r < k then
Select the fitter among the two individuals
else
Select the less fitter individual
end if
```

Fig. 4. Algorithm for Tournament Selection

2.1.4 Crossover

Crossover is the genetic operator which combines two chromosomes to produce new offsprings. Crossover is used to exchange the genetic material of two chromosomes. Crossover is used to explore the search space. There are several types of crossover like One Point, Two Point, Uniform, Arithmetic, Heuristic and cyclic cross over. Here cyclic crossover has been used.

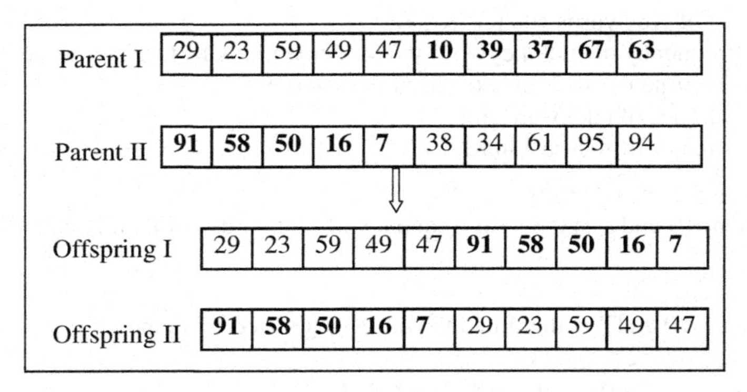

Fig. 5. Crossover

So we get two new chromosomes having different web links from their parents [8].

2.1.5 Mutation

Mutation maintains genetic diversity from one generation to the next. Mutation alters one or more gene value in a chromosome from its initial state. This results in entirely new gene. As a result GA arrives into a better solution. Mutation is used to exploit the search space. Different types of Mutation operators are Flip gene, Boundary, Uniform, Non uniform and Gaussian [8].

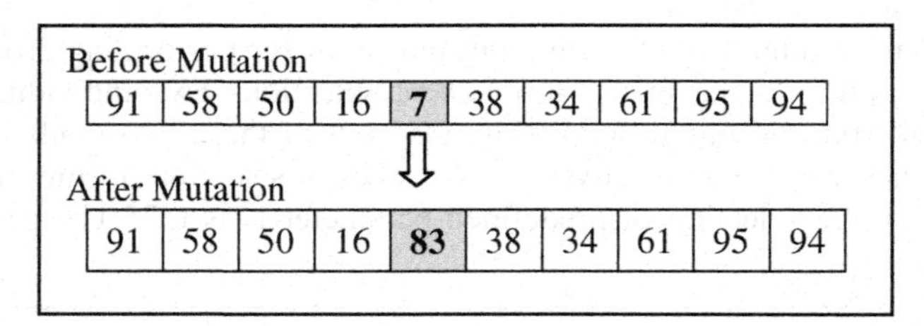

Fig. 6. Mutation

An example based on flip gene mutation is given in Fig. 6.

2.2 Proposed Algorithm

The proposed algorithm takes number of Top-T web links, Initial population size, number of generations and mutation rate as input parameter and generates Top-T web links as an output. The algorithm starts by generating initial population randomly. The binary tournament selection is used for selecting two chromosomes from the population for generating population for crossover. The algorithm proceeds by performing crossover followed by mutation to generate new population. The process is repeated for pre-specified number of generations. The whole process is shown in Fig. 7.

Input:

- Number of top-T Web links, TopT
- Initial Population Size, InitialPopulationSize
- Number of generations, N
- Mutation Rate, MR

Output:

Set of Top-T Web links

Notations:

- Tournament Size, TS
- Crossover Rate, CR
- Crossover Point, CP
- Cost Function

  $F(x) = C1.\ \sum_{j=1}^{TopT} \sum_{i=1}^{n} fi + C2.\ \sum_{i=1}^{p} Ti + C3.\ \sum_{m=1}^{n} Bm + C4.\ \sum_{p=1}^{n} Fp$

Method:

```
Randomly Generate Initial Population, InitialPopulation[InitialPopulationSize]
For Generation=1 to N
a) //Applying Tournament Selection
  For I=1 to CR
        Select TS number of chromosomes randomly from InitialPopulation[]
        Select a random variable, rand_var, between 0 and 1
        k=CR/100
        If rand_var < k Then
        CrossoverPopulation[I]= fitter(TS randomly selected chromosomes)
        Else
        CrossoverPopulation[I]=less_fitter(TS randomly selected chromosomes)
  End For
b) //Performing Crossover
  For J=1 To InitialPopulationSize
        Parent1=ChooseRandom(CrossoverPopulation[])
        Parent2= ChooseRandom(CrossoverPopulation [])
        (offspring1, offspring2)=Crossover(Parent1, Parent2)
 //Copy offsprings to new generation
        NewGeneration[J]=offspring1
        NewGeneration [J+1]=offspring2
        J = J +2
        End For
c) Apply mutation with mutation rate, MR
d) //Copy new generation to initial population
        InitialPopulation [] = NewGeneration[]
End For
Return InitialPopulation [0]
```

Fig. 7. Proposed Algorithm (PA) for Selecting Top-T Web Links

3 An Example

The proposed algorithm has been implemented using JDK1.7. The program was run for 50 generations with different crossover rates. Fig. 8 shows quality of PA for generation 1, 2, 49 and 50 for crossover rate of 75% obtained from the execution. The initial population with their quality values is shown against each chromosome. Ci represents chromosomes. Each Chromosome consists of 5 pages. The initial population consisting 10 chromosomes are represented in the tabular form. The initial population as in Generation 1 has different quality values and as the generation increases the quality values get converged.

Generation 1

	Chromosome					Quality
C1	3	2	1	9	8	1539.56
C2	3	2	1	5	4	1666.82
C3	1	5	4	9	8	1534.38
C4	1	7	6	4	9	1812.96
C5	2	7	6	5	4	1782.52
C6	3	2	7	6	9	1704.07
C7	3	1	7	4	9	1680.09
C8	2	7	6	5	8	1708.62
C9	3	1	6	5	8	1717.14
C10	3	2	6	4	9	1653.88

Generation 2

	Chromosome					Quality
C1	2	7	6	5	8	1708.62
C2	1	7	6	4	9	1812.96
C3	3	2	7	6	9	1704.07
C4	3	2	7	6	9	1704.07
C5	3	2	7	6	9	1704.07
C6	3	2	7	6	9	1704.07
C7	3	2	7	6	9	1704.07
C8	3	2	1	9	8	1539.56
C9	3	2	7	6	9	1704.07
C10	3	2	1	9	8	1539.56

Generation 49

	Chromosome					Quality
C1	3	2	7	6	9	1704.07
C2	3	2	7	6	9	1704.07
C3	3	2	7	6	9	1704.07
C4	3	2	7	6	9	1704.07
C5	3	2	7	6	9	1704.07
C6	3	2	7	6	9	1704.07
C7	3	2	7	6	9	1704.07
C8	3	2	7	6	9	1704.07
C9	3	2	7	6	9	1704.07
C10	3	2	7	6	9	1704.07

Generation 50

	Chromosome					Quality
C1	3	2	7	6	9	1704.07
C2	3	2	7	6	9	1704.07
C3	3	2	7	6	9	1704.07
C4	3	2	7	6	9	1704.07
C5	3	2	7	6	9	1704.07
C6	3	2	7	6	9	1704.07
C7	3	2	7	6	9	1704.07
C8	3	2	7	6	9	1704.07
C9	3	2	7	6	9	1704.07
C10	3	2	7	6	9	1704.07

Fig. 8. Computation of Quality for Generation 1,2,49,50

The cost function of the proposed Genetic Algorithm is:
F(x) = C1.CostKeywords+C2.Costtime+C3.Cost backward link+C4.Cost forward link
Where C1, C2, C3 and C4 are Constants to normalize the different parameters according to significance.

C1=0 .09 C2= 0.05 C3=0.08 C4=0.07

In the above example, CostKeywords=2453.0, Costtime=25834.0, Cost backward link=1198.0, Cost forward link=1368.0

Using the Cost Function

$$F(x) = 0.09*2453.0 + 0.05*25834.0 + 0.08*1198.0 + 0.07*1368.0 = 1704.07$$

4 Experimentation

Fig. 9 shows the comparison of the existing GA based approach with the proposed GA based approach for selecting Top-T web pages. Both the algorithms were implemented using JDK 1.7 in Windows 7 environment. The two algorithms were compared by conducting experiments on an Intel based 2 GHz PC having 3 GB RAM. The comparisons were carried out on Quality of web pages selected by the two algorithms. The experiments were performed for selecting the top-5 to top-10 web-pages over 500 generations using different crossover rate. The graphs are plotted with Top-T pages on the X-axis and Quality of the web pages on the Y-axis for different crossover rate which

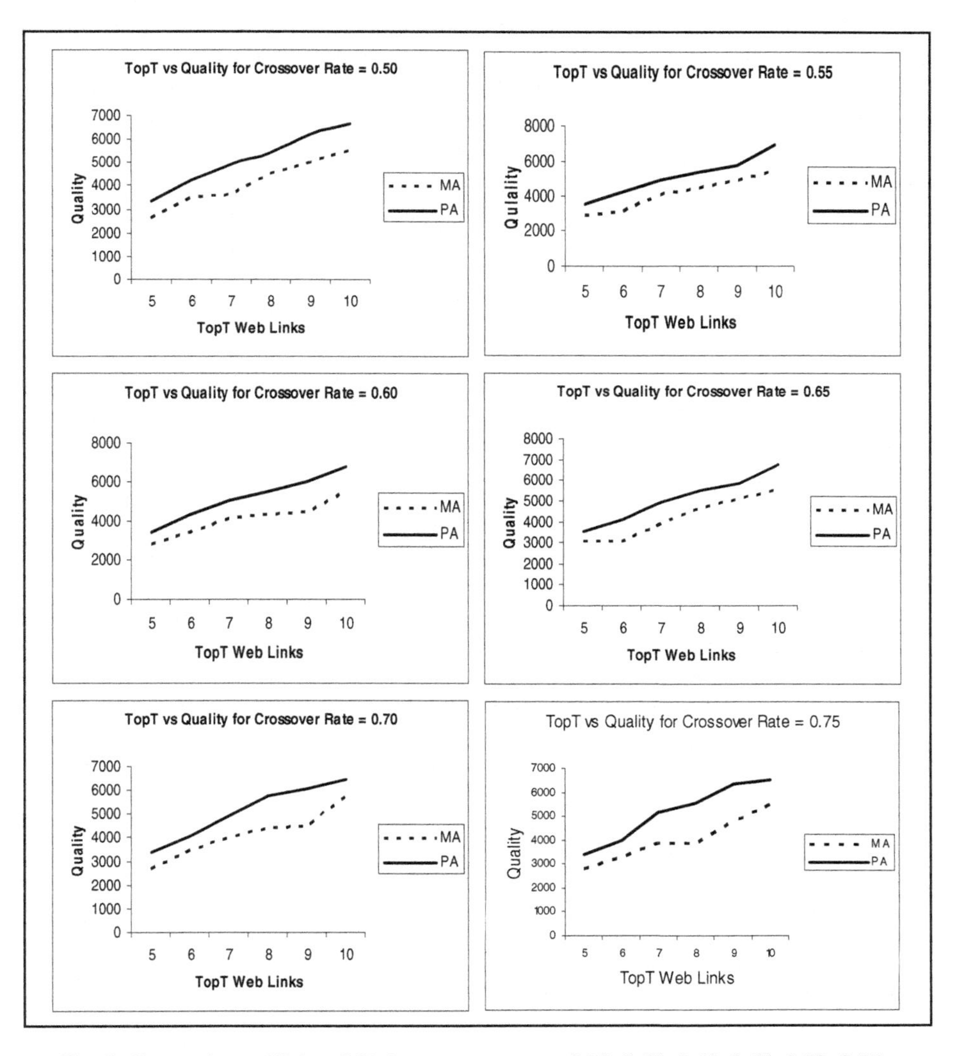

Fig. 9. Comparison of PA vs MA for crossover rate=0.50, 0.55, 0.60, 0.65, 0.70, 0.75

ranges from 0.50 to 0.75. The quality of the existing GA based approach was calculated based on the frequency of the keywords and their mean quality on the other hand the quality of PA was calculated based on frequency of keywords, time the website existed on the internet, number of backward links and forward links. Based on the significance of the factors, multiplicative constants have been used. The experimental results show that proposed algorithm performs better than the existing algorithm.

5 Conclusions

The proposed GA based algorithm is a new approach to select Top-T web links considering several important parameters like number of forward links, number of backward links, keywords and the time website existed. It helps to get relevant and required web pages. As the factors in the cost function increases, the GA provides better quality web links. Further, it has been shown experimentally that the web pages selected by the proposed algorithm are better than the existing algorithm MA. As the Top-T pages increases the quality of the proposed algorithm also increases.

References

1. Ajoudanian, S., Jazi, M.D.: Deep Web Content Mining. World Academy of Science, Engineering and Technology 49 (2009)
2. Gonzales, E., Mabu, S., Taboada, K., Hirasawa, K.: Web Mining using Genetic Relation Algorithm. In: SICE Annual Conference, pp. 1622–1627 (2010)
3. Kosla, R., Blockeel, H.: Web Mining Research: A Survey. SIGKDD Explorations 2, 1–15 (2000)
4. Liu, B., Chiang, K.C.: Editorial Special Issue on Web Content Mining. ACM Journal of Machine Learning Research 4, 177–210 (2004)
5. Nimgaonkar, S., Duppala, S.: A Survey on Web Content Mining and extraction of Structured and Semi structured data. IJCA Journal (2012)
6. Singh, B., Singh, H.K.: Web Data Mining Research: A Survey. In: IEEE International Conference on Computational Intelligence and Computing Research (ICCIC), pp. 1–10 (2010)
7. Srivastava, J., Cooley, R., Deshpande, M., Tan, P.N.: Web Usage Mining: Discovery and Applications of Usage Patterns from Web Data (2000)
8. Chakraborty, R.C.: Fundamentals of Gentetic Algorithms. Artificial Intelligence (2010)
9. Agyemang, M., Barker, K., Alhajj, R.S.: WCOND-Mine: Algorithm for detecting Web Content Outliers from Web Documents. In: 10th IEEE Symposium on Computers and Communication, pp. 885–890 (2005)
10. Etzioni, O.: The World Wide Web: Quagmire or Gold Mine? Communications of the ACM 39(11), 65–68 (1996)
11. Zhi, Z., Jun, J., Fujun, Z., Qiangang, D.: A New Genetic Algorithm for Web-based Negotiation Support system. In: IEEE International Conference on Natural Language Processing and Knowledge Engineering, pp. 209–214 (2003)

12. Bidgoli, B.M., Punch, W.F.: Using Genetic Algorithms for Data Mining Optimization in an Educational Web Based System. In: Cantú-Paz, E., Foster, J.A., Deb, K., Davis, L., Roy, R., O'Reilly, U.-M., Beyer, H.-G., Kendall, G., Wilson, S.W., Harman, M., Wegener, J., Dasgupta, D., Potter, M.A., Schultz, A., Dowsland, K.A., Jonoska, N., Miller, J., Standish, R.K. (eds.) GECCO 2003. LNCS, vol. 2724, pp. 2252–2263. Springer, Heidelberg (2003)
13. Mathew, T.V.: Genetic Algorithm. pp. 1–15 (2005)
14. Khalessizadeh, S.M., Zaefarian, R., Nasseri, S.H., Ardil, E.: Genetic Mining: Using Genetic Algorithm for Topic Based on Concept Distribution. World Academy of Science, Engineering and Technology (2006)
15. Juang, C.F.: A Hybrid of Genetic Algorithm and Particle Swarm Optimization for Recurrent Network Design. IEEE Transactions on System, Man and Cybernetics, 997–1006 (2004)
16. Dallal, A.A., Shaker, R.: Genetic Algorithm in Web Search Using Inverted Index Representation. In: 5th IEEE GCC Conference & Exhibition, pp. 1–5 (2009)
17. Nasaaroui, O., Dasgupta, D., Pavuluri, M.: S2GA: a soft structured Genetic Algorithm and its application in Web Mining. In: Fuzzy Information Processing Society. IEEE Proceedings, pp. 87–92 (2002)
18. Toth, P.: Applying Web-Mining Methods for Analysis in Virtual Learning Environment (2006)
19. Liu, B.: Web Content Mining. In: The 14th International World Wide Web Conference, Japan, May 10-14 (2005)
20. Nick, Z.Z., Themis, P.: Web Search using a Genetic Algorithm. IEEE Internet Computing 5(2), 18–26 (2001)
21. Kudelka, M., Snasel, V., Lehecka, O., Qawasmeh, E.E.: Web Content Mining Using Web Design Patterns (2008)
22. Dunham, M.H.: Data Mining Introductory and Advanced Topics. Pearson Education, India (2006)
23. Van, C.J.: Information Retrieval. Butterworths (1979)
24. MitChell, T.: Machine Learning, ch. 1-9. McGraw Hill (1997)
25. Mitchell, M.: An Introduction to Genetic Algorithms, ch. 1-6. MIT Press, pp. 1–203 (1998)
26. Haupt, R. L.: Practical Genetic Algorithms, ch. 1-7. John Wiley & Sons Inc., pp. 1-251 (2004)
27. Marghny, M.H., Ali, A.F.: Web Mining Based on Genetic Algorithm. In: AIML Conference, pp. 82–87 (2005)

Improved Implementation and Performance Analysis of Association Rule Mining in Large Databases

H.R. Nagesh, M. Bharath Kumar, and B. Ravinarayana

Dept. of Computer Science & Engineering, MITE, Mangalore, India
{hodcse,bharath,ravinarayana}@mite.ac.in

Abstract. Data mining is the process of extracting interesting and previously unknown patterns and correlations from data stored in Data Base Management Systems (DBMSs). Association Rule Mining is the process of discovering items, which tend to occur together in transactions. Efficient algorithms to mine frequent patterns are crucial to many tasks in data mining. The task of mining association rules consists of two main steps. The first involves finding the set of all frequent itemsets. The second step involves testing and generating all high confidence rules among itemsets. Our paper deals with obtaining both the frequent itemsets as well as generating association rules among them.

In this paper we implement the FORC (Fully Organized Candidate Generation) algorithm, which is a constituent of the Viper algorithm for generating our candidates and subsequently our frequent itemsets. Our implementation is an improvement over Apriori, the most common algorithm used for frequent item set mining.

1 Introduction

The quantity of data to be handled by the file systems and the databases is growing at an exponential rate. Thus, users who need to access them need sophisticated information from them. The task of data mining satisfies them. To use a simple analogy, it's finding the proverbial needle in the haystack. In this case, the needle is that single piece of intelligence your business needs and the haystack is the large data warehouse you've built up over a long period of time. It is simply defined as finding the hidden information in a database. It can also be called exploratory data analysis or data driven analysis.

In essence, data mining is distinguished by the fact that it is aimed at the discovery of information, without a previously formulated hypothesis. The information discovered must be previously unknown. The new information has to be valid. Most critical, the information has to be actionable, that is, it must be possible to translate it into some business advantage and take action towards that advantage. It is "a decision support process in which we search for patterns of information in data". This search may be done just by the user, i.e. just by performing queries, or by sophisticated statistical analysis and modeling techniques, which uncover patterns and relationships hidden in organizational databases. Once found, the information needs to be presented in a suitable form, with graphs, reports, etc.

S. Unnikrishnan, S. Surve, and D. Bhoir (Eds.): ICAC3 2013, CCIS 361, pp. 94–104, 2013.

One of the reasons behind maintaining any database is to enable the user to find interesting patterns and trends in the data. The goal of database mining is to automate this process of finding interesting patterns and trends. Once this information is available, we can perhaps get rid of the original database. This goal is difficult to achieve due to the vagueness associated with the term `interesting'. The solution is to define various types of trends and to look for only those trends in the database. One such type constitutes the association rule.

Association rules attempts to predict the class given a set of conditions on the LHS. The conditions are typically attribute value pairs, also referred to as item-sets. The prediction associated with the RHS of the rule is not confined to a single class attribute; instead it can be associated with one or more attribute combinations.

Association rule induction is a powerful method for so-called *market basket analysis*, which aims at finding regularities in the shopping behavior of customers of supermarkets, mail-order companies and the like. With the induction of association rules one tries to find sets of products that are frequently bought together, so that from the presence of certain products in a shopping cart one can infer (with a high probability) that certain other products present. Such information, expressed in the form of rules, can often be used to increase the number of items sold, for instance, by appropriately arranging the products in the shelves of a supermarket (they may, for example, be placed adjacent to each other in order to invite even more customers to buy them together) or by directly suggesting items to a customer, which may be of interest for him/her.

1.1 Problem Statement

Given a database D of T transactions and a minimum support ‘s’, the problem is to find all the frequent itemsets (i.e. those itemsets whose support is greater than ‘s’). Once the frequent itemsets are found, association rules can be derived from them using confidence ‘c’.

2 Basic Concepts and Terminologies

Association rules are extracted from the database, which can be viewed as a collection of **transactions,** keeping in mind the grocery store cash register. For example, a transaction could be {Pencil, Sharpener, Eraser}, each element in it is termed as an **item.** We are not interested in quantity or cost, but only in their purchase. A collection of items is called as the **itemset**.

Here, we provide some basic concepts and terminologies used in the association rule mining.

1. Given a set of items $I = \{I_1, I_2, \dots, I_m\}$ and a database of transactions $D = \{t_1, t_2, \dots, t_n\}$ where each t_i contains different combinations of I, an **association rule** is an implication of the form $X \Rightarrow Y$ where $X, Y \in I$ are sets of items called itemsets and $X \cap Y = \phi$.

2. The **support** (s) for an association rule X=>Y is the percentage of transactions in the database that contain $X \cup Y$.

   Let T be the set of all transactions under consideration, e.g., let T be the set of all "baskets" or "carts" of products bought by the customers of a supermarket - on a given day if you like. The support of an item set S is the percentage of those transactions in T which contain S. In the supermarket example this is the number of "baskets" that contain a given set S of products, for example S = {bread, wine, cheese}. If U is the set of all transactions that contain all items in S, then

$$\text{support}(S) = (|U| / |T|) * 100\%,$$

where |U| and |T| are the number of elements in U and T, respectively. For example, if a customer buys the set X = {milk, bread, apples, wine, sausages, cheese, onions, potatoes } then S is obviously a subset of X, hence S is in U. If there are 318 customers and 242 of them buy such a set U or a similar one that contains S, then support(S) = 76.1%.

3. The **confidence** or **strength** for an association rule X=>Y is the ratio of the number of transactions that contain $X \cup Y$ to the number of transactions that contain X.

   The confidence of a rule R="A and B->C" is the support of the set of all items that appear in the rule divided by the support of the antecedent of the rule, i.e. confidence(R) = (support({A, B, C})/support({A,B})) *100%.

   More intuitively, the confidence of a rule is the number of cases in which the rule is correct relative to the number of cases in which it is applicable. For example, let R="wine and bread -> cheese". If a customer buys wine and bread, then the rule is applicable and it says that he/she can be expected to buy cheese. If he/she does not buy wine or does not buy bread or buys neither, than the rule is not applicable and thus (obviously) does not say anything about this customer.

   If the rule is applicable, it says that the customer can be expected to buy cheese. But he/she may or may not buy cheese, that is, the rule may or may not be correct. We are interested in how good the rule is, i.e., how often its prediction that the customer buys cheese is correct. The rule confidence measures this: It states the percentage of cases in which the rule is correct. It computes the percentage relative to the number of cases in which the antecedent holds, since these are the cases in which the rule makes a prediction that can be true or false. If the antecedent does not hold, then the rule does not make a prediction, so these cases are excluded.

   With this measure a rule is selected if its confidence exceeds or is equal to a given lower limit. That is, we look for rules that have a high probability of being true, i.e., we look for "good" rules, which make correct (or very often correct) predictions.

4. If the support of an itemset is greater than or equal to a given support threshold, then it is termed a **frequent itemset,** otherwise it is infrequent.
5. Given a set of items $I = \{I_1, I_2, \dots, I_m\}$and a database of transactions $D = \{t_1, t_2, \dots, t_n\}$ where each t_i contains different combinations of I, the **association rule problem** is to identify all association rules X => Y with a minimum support and confidence. These values are fed as inputs to the problem.

2.1 Data Layout Alternatives

Conceptually, a market basket database is a two dimensional matrix where the rows represent individual customer purchase transactions and the columns represent the items on sale. The implementation of this matrix is as shown in the Fig 1.

Horizontal Item Vector(HIV). The database is organized as a set of rows with each row storing a transaction identifier (tid) and a bit vector of 1's and 0's to represent for each of the items on sale, its presence or absence, respectively in the transaction is as shown in Fig. 1 a [1].

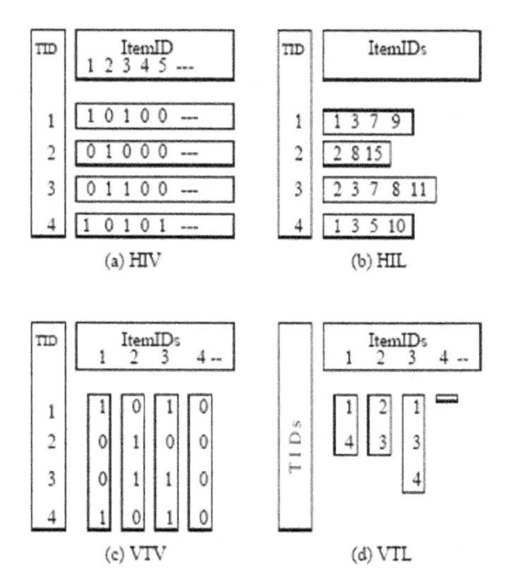

Fig. 1. Comparison of Data Layouts [1]

Horizontal Item List (HIL). This is similar to HIV, except that each row stores an ordered list of item identifiers(iid), representing only the items actually purchased in the transaction is as shown in Fig.1 b [1].

Vertical Tid Vector (VTV). The database is organized as a set of columns with each column storing an IID and a bit vector of 1's and 0's to represent the presence or absence, respectively of the item in the customer transactions is as shown in Fig 1 c [1].

Vertical Tid List (VTL). This is similar to VTV, except that each column stores an ordered list of only the tid's of the transactions in which the item was purchased is as shown in Fig 1 d [1].

2.2 The Viper Algorithm

VIPER (Vertical Itemset Partitioning for Efficient Rule Extraction) uses the vertical tid vector (VTV) format for representing an items' occurrence in the tuples of the database. The bit vector is stored in a compressed form called the 'snake'. It is a multipass algorithm, wherein data in the form of snakes is read from and written to the disk in each pass [1]. It proceed in a bottom up manner and at the end of the data mining, the supports of all frequent itemsets are available. Each pass involves the simultaneous counting of several levels of candidates via intersections of the input snakes. To minimize the computational costs, VIPER implements a new candidate generation scheme called **FORC** (Fully organized candidate generation), partly based on the technique of equivalence class clustering. VIPER also incorporates a novel snake intersection and counting scheme called FANGS (Fast Anding Graph for Snakes). The FANGS scheme is based on a simple DAG structure that has a small footprint and efficiently supports concurrent intersection of multiple snake pairs by using a pipelined strategy [1].

Our implementation is a hybrid version wherein we generate the candidate itemsets using the FORC procedure of the VIPER algorithm, the frequent itemsets using the conventional Apriori algorithm and the association rules using the Association rule generation algorithm described before.

3 Experiments and Results

3.1 Data Generation and Formatting

Here, we generate the sample data that would be used later for testing the execution of the algorithm. The data generated is stored in a file in the form of bit vectors, i.e. 0/1 indicating the presence/absence respectively of an item in the given transaction. We have the columns depicting the items and the rows depict the transactions. This is called as a Transaction Vector, as specified earlier in the literature. If required, this data can be compressed with the help of Run Length Encoding.

3.2 The FORC Candidate Generation Algorithm

We use the algorithm called FORC (Fully Organized Candidate Generation) for efficiently generating candidate itemsets. FORC is based on the powerful technique of equivalence class clustering, but adds important new optimizations of its own.

The FORC algorithm operates as follows: Given a set S_k (which can be either a set of frequent itemsets or a set of candidate itemsets) from which to generate $C_{k+1,}$ the itemsets in S_k are first grouped into clusters called "**equivalence classes**". The grouping criterion is that all itemsets in an equivalence class should share a common prefix of length k-1. For each class, its prefix is stored in a hash table and the last element of every itemset in the class is stored in a lexicographically ordered list, called the *extList.* With this framework, the following is a straightforward method of generating candidates: For each prefix in the hash table, take the union of the prefix with all ordered pairs of items from the extList of the class (the order ensures that duplicates are not generated). For each of the potential candidates, check whether all its k-subsets are present in S_k, the necessary condition for an itemset to be a candidate. This searching is simple since the k-1 prefix of the subset that is being searched for indicates which extlist is to be searched. Finally, include those, which survive the test in C_{k+1} [1].

The FORC Algorithm under discussion is as listed below.

```
SetOfItemsets FORC(S_k) {
   Input: Set of k-itemsets(S_k)
   Output: Set of candidate k+1-itemsets(C_k+1)
   for each itemset i in S_k do
         insert (i.prefix) into hashTable;
         insert(i.lastelement) into i.prefix→extList;
         C_k+1 = ∅;
         for each prefix P in the hashTable do {
         E = P→extList;
         for each element t in E do {
                  newP = P ∪ t ;
                  remList = {i|i ∪ E and i > t };
                  for each (k -1) subset subP of newP do
                  remList = remList ∩ (subP → extList);
         for each element q in remList do {
         newCand = newP ∪ {q};
         C_k+1 = C_k+1 ∪ newCand;
         }
         }
   }
   return C_k+1;
}
```

Fig. 2. FORC Algorithm [1]

3.3 Simultaneous Search Optimization

We can optimize the above-mentioned process by recognizing the fact that since the unions are taken with ordered pairs, the prefix of the subsets of the candidates thus formed will not depend on the second extension item, which in turn means that all these subsets are shared and the same for each element in the extlist. Hence, repeated searches for the same subsets can be avoided and they can be searched for simultaneously, as shown in the following example. Consider a set S3 in which we have the items ABC, ABD, ABE with prefix AB. These are grouped into the same equivalence class g, with prefix AB. The associated extlist is C, D and E. We now need to find all the candidates associated with each of the itemsets in g. We can illustrate this process by showing it for ABC. We first find the items that are lexicographically greater than C in the extlist, namely D and E. Now, the potential candidates are ABCD and ABCE and we need to check if all their 3-subsets are also present in S3. That means that we have to search for ABi, ACi, and BCi where i may be either D or E. We don't have to search for ABC, as it is already present. To search for ABD, for example, we access its group, say h, with prefix AB and then check if in the extlist D is present, which mean that ABD is present in S3. The important point is that, having come so far to check if ABD is present, we can as well check if ABE is present, as it would also be present in the same equivalence class with prefix AB. We can do this by going further down in the extlist to check if E is present in it.

Generalizing the above example, we can overlap subset status determination of multiple candidates by ensuring that all subsets across these candidates that belong to a common group are checked for with only one access of the associated extlist. This is in marked contrast to the standard practice of subset status determination on a sequential basis, resulting in high computational cost.

3.4 Calculating Support and Frequent Items

Once the candidates are ready in the array "candidates", we process each of them to find out their support in the database of transactions. As defined earlier, support(Item) is the following:

support (item) = no occurrences (item)/ total no of transactions.

So, for each item, which is a candidate, we count from the database the number of times it is present in the set of transactions. Many optimizations can be used in this method. Some methods like FP-Tree can be used to calculate and find support without even generating candidates. However, we do the simple method of counting the existence of 1's in the transaction list. This turns out inefficient and time consuming for very large databases, of size greater than a few gigabytes. So, better methods can be used here.

Once the support is calculated for each candidate item, it is checked against the threshold given by the user. If the threshold is crossed by the candidate, it becomes a "frequent item".

We repeat this process for each candidate at the given level. The items whose support crosses the threshold are stored in the frequent item set. These items are then made the candidates of the next level. This set of candidates is subject to similar process and this process is repeated till no frequent items are found at a given level. This means that we have reached the maximum possible combination of items; having support more than the threshold, and that we cannot proceed further to the next level to find more frequent items. This process of finding maximum possible frequent itemset stops here.

3.5 Rule Generation

At the end of the previous step, we have a set of items that form the maximum possible frequent itemset. Now, we have to mine rules out of this frequent itemset. This proceeds as follows in our implementation:

The user enters a minimum threshold confidence level. Confidence, as defined before is

$$\text{Confidence } (a \text{ -> } b) = \frac{(\text{no of transactions having } a \text{ and } b)}{(\text{no of transactions having } a)}$$

This provides the assurance required to publish a rule. We generate rules as follows:

```
Generate_rules_algorithm()
for (each item in frequent itemset)
{
  find all subsets of the frequent itemset and store it;
  for(each subset S of frequent itemset under consideration)
 {
   find out the support for subset S from the storage   structure;
  find out the ratio of support(frequent item) : support(S);
  if(ratio > threshold_confidence)
  {
    publish rule subset S → (frequent itemset – S) with confidence = ratio;
   }
 }
}
```

As seen above, the algorithm takes each frequent itemset and generates all its subsets. This is stored and one by one, for each subset, we find the confidence as specified. If it crosses the threshold set, it qualifies as a strong association rule. This rule is published and this process is repeated for all subsets of all frequent itemsets.

This ends the program execution with strong association rules being published, along with their confidences.

4 Results

From the literature, we obtained the following performance statistics of Apriori Algorithm. This is compared here under similar configurations with the execution of our algorithm, a hybrid between Apriori and VIPER algorithm.

We noted the performance of our algorithm in terms of the execution time for a number of transactions as shown in Table 1. We executed the algorithm for the following configuration data.

Total Number of Items = 5
Minimum Confidence = 45%

We have then varied the total number of transactions and found the total execution time for four different levels of minimum support. Fig. 3 shows the comparison of our implementation with Apriori.

It can be seen from the above table and chart that our implementation clearly out performs the traditional Apriori implementation. Moreover, the difference in performance is more clearly pronounced as the size of the database increases.

Table 1. No. of transactions V/S execution time

No. of transactions	Execution Time (sec)	
	Apriori	Our Implementation
10,000	11	2.2
50,000	51	8.12
1,00,000	103	14.57
2,00,000	209	28.34
5,00,000	512	95.05
10,00,000	1046	137.9

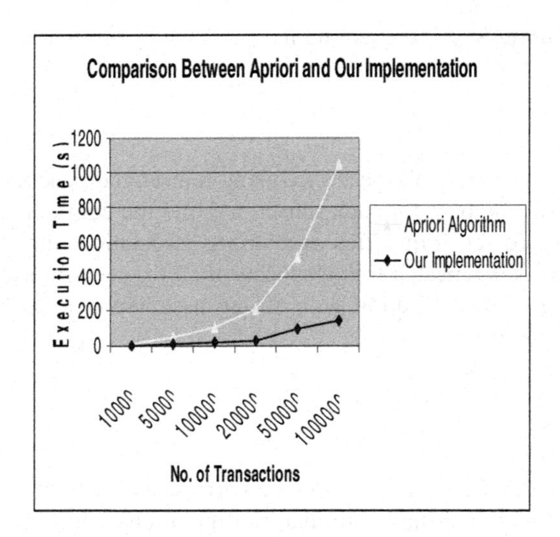

Fig. 3. Comparison of Apriori Algorithm and Our implementation

5 Conclusions

Our work "Association Rule Mining in Large Databases" has been successfully developed as a hybrid of the Viper and Apriori implementations. Based on the results of our development work, the execution time using our method is better compared to the Apriori algorithm. This is proved ever at higher number of transactions. This algorithm scales well compared to the original Apriori implementation as expected. The performance scales well with the size of the input database. Moreover, it is designed for a stand-alone system, as we have not focused on optimizing CPU cycles, which also becomes a concern in distributed databases. Some of the inherent drawbacks in the algorithm developed by us are as follows:

The algorithm reads directly from the database of transactions rather than storing it in a buffer and maintaining the buffer in main memory by periodically reading in blocks of data. This is not required though, if the database is not read repeatedly, as in the case of prefix tree methods of mining association rules. Here, the database is read only once and a prefix tree is constructed from it, which is used for further calculations of support and confidence.

The other drawback is that we make use of a synthetic data generator for generating our test case inputs. This is a drawback of the random number generator not being "random enough" over long series of data. Data is generated by the default rand() function and as it is programmed to follow a specific probability distribution, the output also follows a trend. The data generated follows similar patterns at regular intervals, thus making the output i.e. the frequent itemsets, their support and also the association rules quite predictable when the database size is scaled. In other words, the rules obtained when the number of transactions is say 10,000 is very similar to the ones obtained when no. of transactions is say 1, 00,000, except for the scaled levels of support and confidence.

The actual efficiency of our program will be better appreciated when our program is run on real time data.

6 Future Extensions

6.1 Minimizing I/O Costs

Routines to read data in blocks from the disk into main memory have to be incorporated in our program. More efficient mechanisms to find the support and confidence from the given database should be explored into and incorporated to make the number of I/O's lesser compared to the current implementation.

6.2 Compression of Data Stored

We can use compression techniques like "Run Length Encoding" to store data in the form of compressed bit vectors called snakes in the main memory thus gaining space. This requires conversion on-the-fly when data is required for counting etc.

6.3 FANGS Implementation

We can implement the FANGS (Fast Anding of Graphs) procedure of the VIPER algorithm to generate the frequent itemsets and subsequently the association rules as it involves fewer scans of the database.

6.4 Prefix-Tree Methods

Instead of generating candidates and then checking if they are frequent, we can directly use the method of prefix trees to mine rules.

References

[1] Shenoy, P., Bawa, M., Shah, D.: Turbo-charging Vertical Mining of large databases. In: Proceedings of ACM SIGMOD Intl. Conference on Mgmt of Data (2000)
[2] Mobasher, B., Cooley, R., Srivastava, J.: Web mining: Information and pattern discovery on the World Wide Web. In: 9th IEEE International Conference on Tools with Artificial Intelligence, ICTAI 1997 (1997)
[3] Agrawal, R., Imielinski, T., Swamy, A.: Mining association rules between sets of items in large databases. In: Proceedings of ACM SIGMOD Intl. Conference on Management of Data (1993)
[4] Bodon, F.: A Fast Apriori Implementation. Computer and Automation Research Institute. Hungarian Academy of Sciences, Budapest (2012)
[5] Agrawal, R., Srikant, R.: Fast algorithms for mining association rules. In: Proc. of 20th Intl. Conf. Very Large Databases, VLDB (1994)
[6] Golomb, S.W.: Run Length Encoding. IEEE Transactions on Information Theory 12(3) (1966)
[7] Savasere, A., Omiecinski, E., Navathe, S.: An efficient algorithm for mining association rules in large databases. In: Proc. of 21st Intl. Conf. on Very Large Databases, VLDB (1995)
[8] Chen, M., Han, J., Yu, P.S.: Data Mining: An Overview from a Database Perspective. TKDE 8(6) (December 1996)
[9] Adriaans, P., Zantinge, D.: Data Mining. Addison-Wesley (1996)
[10] Agrawal, R., Imielinski, T., Swami, A.: Database Mining: A Performance Perspective. TKDE 5(6) (December 1993)
[11] Han, J., Kamber, M.: Data Mining: Concepts and Techniques. Morgan Kaufmann (2001)
[12] Han, J., Pei, J., Yin, Y.: Mining Frequent Patterns without Candidate Generation. SIGMOD (2000)

An Optimized Formulation of Decision Tree Classifier

Fahim Irfan Alam[1], Fateha Khanam Bappee[2], Md. Reza Rabbani[1], and Md. Mohaiminul Islam[1]

[1] Department of Computer Science & Engineering
University of Chittagong, Chittagong, Bangladesh
{fahim1678,md.reza.rabbani,mohaiminul2810}@gmail.com
[2] Department of Mathematics, Statistics and Computer Science
St. Francis Xavier University, Antigonish, Canada
bappeenstu@gmail.com

Abstract. An effective input dataset, valid pattern-spotting ability, good discovered pattern evaluation is required in order to analyze, predict and discover previously unknown knowledge from a large data set. The criteria of significance, novelty and usefulness need to be fulfilled in order to evaluate the performance of the prediction and classification of data. Thankfully data mining, an important step in this process of knowledge discovery extract hidden and non-trivial information from raw data through useful methods such as decision tree classification. But due to the enormous size, high-dimensionality and heterogeneous nature of the data sets, the traditional decision tree classification algorithms sometimes do not perform well in terms of computation time. This paper proposes a framework that uses a parallel strategy to optimize the performance of decision tree induction and cross-validation in order to classify data. Moreover, an existing pruning method is incorporated with our framework to overcome the overfitting problem and enhancing generalization ability along with reducing cost and structural complexity. Experiments on ten benchmark data sets suggest significant improvement in computation time and better classification accuracy by optimizing the classification framework.

Keywords: Data Mining, Decision Tree, Knowledge Discovery, Classification, Parallel Strategy.

1 Introduction

Recent times have seen a significant growth in the availability of computers, sensors and information distribution channels which has resulted in an increasing flood of data. However, these huge data is of little use unless it is properly analyzed, exploited and useful information are extracted. Prediction is a fascinating feature that can be easily adopted in data-driven techniques of extracting useful knowledge tasks. Because, if we can not predict a useful pattern from an

S. Unnikrishnan, S. Surve, and D. Bhoir (Eds.): ICAC3 2013, CCIS 361, pp. 105–118, 2013.

enormous data set, there is little point of gathering these massive sets of data unless they are recognized and acted upon in advance.

Data mining [23], a relatively recently developed methodology and technology aims to identify valid, novel, non-trivial, potentially useful, and understandable correlations and patterns in data [6]. It does this by analyzing the datasets in order to extract patterns that are too subtle or complex for humans to detect [11]. There are number of applications regarding prediction e.g. predicting diseases [10], stock exchange crash [9], drug discovery [17] etc where we see an unprecedented opportunity to develop automated data mining techniques of extracting concealed knowledge. Extraction of hidden, non-trivial data from a large data set leads us to obtain useful information and knowledge to classify the data according to the pattern-spotting criteria [5].

The study of data mining builds upon the ideas and methods from various versatile fields such as statistical and machine learning, database systems, and data visualization. Decision tree classification method [4] [19] is one of the most widely used technique for inductive inference in data mining applications. A decision tree which is a predictive model is a set of conditions organized in a hierarchical structure [19]. In this structure, an instance of data is classified by following the path of satisfied conditions from the root node of the tree until reaching a leaf, which will correspond to a class value. Some of the most well-known decision tree algorithms are C4.5 [19] and CART [4].

The research in decision tree algorithms is greatly influenced by the necessity of developing algorithms for data-sets arising in various business, information retrieval, medical and financial applications. For example, various business organizations are using decision tree to analyze the buying patterns of customers and their needs as well, medical experts are using them for discovering exciting pattern from the data in order to facilitate the cure process, and credit card industry is using them for fraud detection.

But due to the enormous size, high-dimensionality, and heterogeneous nature of the data sets, the traditional decision tree classification algorithms sometimes fail to perform well in applications that require computationally intensive tasks and fast computation of classification rules. Also, computation and analysis of massive data sets in decision trees are becoming difficult and almost impossible as well. For example, in medical domain for disease prediction tasks that require learning the properties of atleast thousands of cases for a safe and accurate prediction and classification, it is now merely possible for a human analyst to analyze and discover useful information from these data-sets. An optimized formulation of decision trees hold great promises for developing new sets of tools that can be used to automatically analyze the massive data-sets resulting from such simulations.

However, the large data sets and their high-dimensionality make data mining applications computationally very demanding and in this regard, high-performance parallel computing is fast becoming an essential part of the solution. We can also improve the performance of the discovered knowledge and classification rules by utilizing available computing resources which mostly remain unused in an environment. This has motivated us to develop a parallel strategy for the existing decision tree classification algorithms in order to save computation time and a better utilization of resources. As the nature of the decision tree induction shows a natural concurrency, the parallel formulation is undoubtedly a suitable option and solution for an optimized performance.

However, designing such parallel strategy is challenging and require different issues to look into. Computation and communication costs are two such issues that most parallel processing algorithms consider while the formulation take place. As multiple processors work together in order to optimize the performance but at the same time the internal exchange of information (if any) between them increases the communication cost to some extent which in turn affects the optimization performance.

SLIQ [16] is a fast, scalable version decision tree algorithm which achieves better classification accuracy with small execution time. But the performance of SLIQ is limited by its use of a centralized, memory-resident data-structure - the class list which puts a limit on the size of the datasets SLIQ can deal with. SPRINT [21] is the parallel implementation of SPRINT which solves the problem of SLIQ regarding memory by splitting the attribute lists evenly among processors and find the split point for a node in the decision tree in parallel. However in order to do that the entire hash table is required on all the processors. In order to construct the hash table, an all- to- all broadcast [12] is performed which in turn makes this algorithm unscalable with respect to runtime and memory requirements. Because each processor requires $O(N)$ memory to store the hash table and O(N) communication cost for all- to- all broadcast , where N is the size of the dataset.

ScalParcC [8] is an improvised version of the SPRINT in the sense that it uses a distributed hash table to efficiently implement the splitting phase of the SPRINT. Here the overall communication overhead of the phase does not exceed above O(N) and the memory does not exceed O(N/p) for each processor. This ensures the scalable property of this algorithm in terms of both runtime and memory requirement. Another optimized formulation of decision trees is a concatenated parallelism strategy of divide and conquer problems [7]. In this method, the combination of data parallelism and task parallelism is used as a solution to the parallel divide and conquer algorithm. However, in the problem of classification decision tree, the workload cannot be determined based on the size of data at a particular node of the tree. Therefore, one time load balancing used in this method is not desired.

In this paper, we propose a strategy that puts the computation and communication cost to minimal. Moreover, our algorithm particularly considers the issue of load-balancing so that every processor handle roughly equal portion of the task and there is no underutilized resources left in the cluster. Another exciting feature of our algorithm is that we propose to work with both discrete and continuous attributes.

Section 2 discusses a sequential decision tree algorithm that we want to optimize in this paper. The parallel formulation of the algorithm is explained in section 3. Experimental results are shown in section 4. Finally, concluding remarks are stated in section 5.

2 Sequential Decision Tree Classification Algorithm

Most of the existing inductionbased decision tree classification algorithms e.g. C4.5 [19], ID3 [18] and CART [4] use Hunts method [19] as the basic algorithm. Those algorithms mostly fail to optimize for applications that require analysis of large data sets in a short time. The recursive description of Hunts method for constructing a decision tree is explained in Algorithm 1.

Algorithm 1. Hunt's Method [19]

Inputs: Training Set T of n examples $\{T_1, T_2, \ldots\ T_n\}$
with classes $\{c_1, c_2, \ldots\ c_k\}$ and attributes
$\{A_1, A_2, \ldots\ A_m\}$ that have one or more mutually
exclusive outcomes $\{O_1, O_2, \ldots\ O_p\}$
Output: A decision tree D with nodes $N_1, N_2, \ldots$
Case 1:
if $\{T_1, T_2, \ldots\ T_n\}$ belong to a single class c_j **then**
$D \leftarrow$ leaf identifying class c_j
Case 2:
if $\{T_1, T_2, \ldots\ T_n\}$ belong to a mixture of classes **then**
Split T into attribute-class Table S_i
for each $i = 1$ to m **do**
for each $j = 1$ to p **do**
Separate S_i for each value of A_i
Compute degree of impurity using either entropy, gini index or classification error.
Compute Information gain for each A_i
$D \leftarrow$ node N_i with largest information gain
Case 3:
if T is an empty set **then**
$D \leftarrow N_i$ labeled by the default class c_k

3 Optimized Algorithm

Decision tree is an important method for classification problem. A data-set called the training set is given as input to the tree first, which consists of a number of examples each having a number of attributes. The attributes are either continuous, when the attribute values are ordered, or categorical, when the attribute values are unordered. One of the categorical attributes is called the class value or the classifying attribute. The objective of inducing the decision tree is to use the training set to build a model of the class value based on the other attributes such that the model can be used to classify new data not from the training data-set.

In 3.1, we give a parallel formulation for the classification decision tree construction using partition and integration strategy followed by measuring predictive accuracy of the tree using cross-validation approach [3]. We focus our presentation for discrete attributes only. The handling of continuous attributes is discussed in Section 3-2. Later, a pruning will be done in order to optimize the size of the original decision tree and reduce its structural complexity, as explained in Section 3-3. With an artificially created training set, we will describe our parallel algorithm by the following steps:

3.1 Partition and Integration Strategy

Let us consider an artificial training set T_r with n examples, each having m attributes as shown in Table 1.

Table 1. Artificial Training Set

Gender	Car Ownership	Travel Cost	Income Level	Class
male	zero	cheap	low	bus
male	one	cheap	medium	bus
female	zero	cheap	low	bus
male	one	cheap	medium	bus
female	one	expensive	high	car
male	two	expensive	medium	car
female	two	expensive	high	car
female	one	cheap	medium	car
male	zero	standard	medium	train
female	one	standard	medium	train
female	one	expensive	medium	bus
male	two	expensive	medium	car
male	one	standard	high	car
female	two	standard	low	bus
male	one	standard	low	train

1. A root processor M will calculate the degree of impurity using either entropy, gini index or classification error for T_r. Then it will divide T_r into a set of subtables $T_{sub} = \{T_{r_1}, T_{r_2}, \ldots\ T_{r_m}\}$ for each m according to attribute-class combination and will send the subtables to the set of child processors $C = \{C_{p_1}, C_{p_2}, \ldots\}$ by following the cases below-

Case 1: If $|C| < |T_{sub}|$, M will assign the subtables to C in such a way that the number of subtables to be handled by each child processor C_{p_i} where $i = 1, 2, \ldots\ |C|$ is roughly equal.

Case 2: If $|C| > |T_{sub}|$, M will assign the subtables to C in such a way that each C_{p_i} handles exactly one subtable.

2. Each C_{p_i} will simultaneously calculate the information gain of respective attributes and will return the calculated information gain to M.

3. M will compare the information gain found from each C_{p_i} and will find the optimum attribute as the root node that produces the maximum information gain. Our decision tree now consists of a single root node as shown in Fig.1 for our training data in Table 1 and will now expand it.

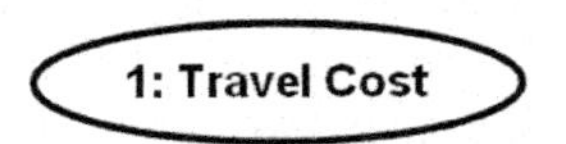

Fig. 1. Root node of the decision tree

4. After obtaining the root node, M will again split T_r into subtables according to the values of the optimum attribute found in Step 3. Then it will send the subtables for which impurities are found to the child processors by following the cases below. Pure class is assigned into leaf node of the decision tree.

Case 1: If $|C| < |T_{sub}|$, the case 1 explained in Step 1 will be applicable and each C_{p_i} will iterate in the same way followed by above steps.

Case 2: If $|C| > |T_{sub}|$, the case 2 explained in Step 1 will be applicable and each C_{p_i} will partition the subtables into a set of sub-subtables $T_{subsub} = \{T_{sub_1}, T_{sub_2}, \ldots T_{sub_m}\}$ according to attribute-class combination. The active child processors will balance the load among the idle processors in such a way that the total number of sub-subtables to be handled by all the child processors is roughly equal. After computing the information gain, the child processors will synchronize to find the optimum attribute and send it to M.

5. Upon receiving, M will add those optimum attributes as child nodes in the decision tree. According to the training data in Table 1, the current form of the decision tree is shown in Fig.2.

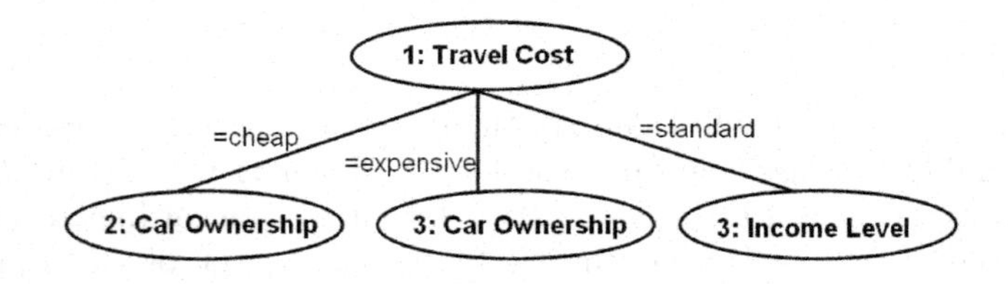

Fig. 2. Decision Tree after First Iteration

6. This process continues until no more nodes are available for the expansion. The final decision tree for the training data is given in Fig.3. The entire process of our optimization is depicted in Algorithm 2.

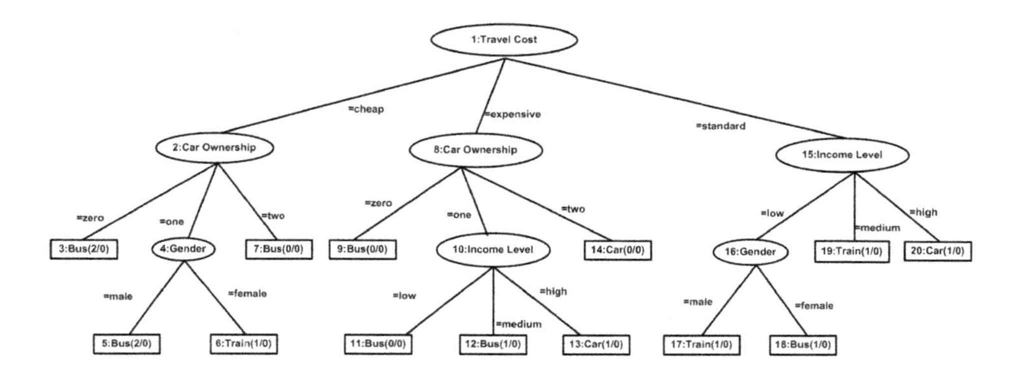

Fig. 3. Final Decision Tree

Next, we focus on determining predictive accuracy of the generated hypothesis on our dataset. The hypothesis will produce highest accuracy on the training data but will not work well in case of unseen, new data. This overfitting problem restricts the determination of predictive accuracy of a model. To prevent this problem, we carry out cross-validation [3] which is a generally applicable and very useful technique for tasks such as accuracy estimation. It consists of partitioning a data set T_r into n subsets T_{r_i} and then running the tree generation algorithm n times, each time using a different training set T_r-Tr_i and validating the results on T_{r_i}. An obvious disadvantage of cross-validation is its computation intensive nature. For example, an n-fold cross-validation is implemented by running the algorithm n times. To reduce the computation overhead, again a parallel strategy is carried out to generate the n number of trees for an n-fold cross-validation which is explain below:

1. A root processor will divide the original dataset into n folds of which *n-1* folds are considered as training set and the remaining fold is considered as validation or test set. The root will continue dividing the dataset until all of the examples in the dataset is used as validation example exactly once.
2. The root processor will send the divided datasets (consisting of both training and test) to the child processors in such a way that the assignment of datasets to the child processors is roughly equal.
3. The respective child processors will act as roots and will form n temporary decision trees by following algorithm 2.
4. Next, the child processors will calculate the predictive accuracy of the temporary decision trees by running the validation sets and send the results to the root processor.
5. Finally, the root processor will average the results and determine the actual predictive accuracy of the original decision tree.

Algorithm 2. Partition & Integration Algorithm

```
Inputs: Training Set T_r of n examples {T_r1, T_r2, ... T_rn}
        with classes {c_1, c_2, ... c_k} and attributes
        {A_1, A_2, ... A_m} that have one or more mutually
        exclusive values {V_1, V_2, ... V_p}, root processor
        M and child processors C = {C_p1, C_p2, ...}
Output: A decision tree D with nodes N_1, N_2, ...
  Processor M: Compute degree of impurity for T_r
        Divide T_r into T_sub={T_r1, T_r2, ... T_rm} for each m
        Send T_sub to C
  //SubTables Assignment Process to C
  if |C| < |T_sub| then
    j=1
    while (j ≠ |T_sub|) do
      for each i = 1 to |C| do
        C_pi ← T_rj
        j++
        if i == |C| then
          i=1
  if |C| >= |T_sub| then
    j=1
    while (j ≠ T_sub) do
      for each i = 1 to |C| do
        C_pi ← T_rj
        j++
  Child Processors C_pi; i = 1, ... |C|:
        Compute Information Gain for each T_rj
        Send computed Information gain to M
  Processor M: Find Optimum Attribute A_opt
        N_ROOT ← A_opt
        D ← N_ROOT
        Divide T_r into T_sub according to A_opt
        Send T_sub to C
  Repeat actions for SubTables Assignment Process to C
  Child Processors C_pi:
     Partition T_rj into T_subsub = {T_sub1, T_sub2, ... T_subm} for each m
  Repeat Actions for First Iteration
        Compute Information Gain and send A_opt's to M for each V_x of A_y; x =
  1, 2, ... p, y = 1, 2, ... m
  Processor M:
  if Entropy==0 then
        D ← C for V_x
  else
        N ← A_opt
        D ← N
```

3.2 Computing Continuous Attributes

Handling continuous attributes in decision trees is different than that of discrete attributes. They require a special attention and can not fit into the learning scheme if we try to deal with them exactly the same way as discrete attributes. However, the use of continuous attributes are very common in practical tasks and we can not ignore the use of these. One way of handling continuous attributes is to discretize them which is a process of converting or partitioning continuous attributes to discretized attributes by some interval [2]. One possible way of finding the interval would be selecting a splitting criteria for dividing the set of continuous values in two sets [20]. This again involves a critical issue of unsorted values which makes it difficult to find the splitting point. For large data sets, the sorting of values and then selecting interval require significant time to do.

In this regard, we chose to do the sorting and selecting splitting point in parallel approach in order to avoid the additional time needed for discretizing large set of continuous values. The root processor will divide the continuous attribute set into N/P cases where N = number of training cases for that attribute and P = number of processors. Then it will send the divided cases to each child processor. Therefore, if each processor contains N/P training cases and subsequently will do the sorting. After the individual sorting done by each child processor, another round of sorting phase will be carried out among the child processors by using merge sort [13]. The final sorted values will be send over to the root and the splitting criteria will be decided accordingly. The overall process is depicted in Algorithm 3.

Algorithm 3. Discretize Continuous Attributes

Inputs: Training Set T_r of n examples, Continuous Attributes $\{A_1, A_2, \ldots A_m\}$ from the training set , Root processor M and child processors $C = \{C_{p_1}, C_{p_2}, \ldots\}$
Output: Discretized Continuous Attribute Values

Processor M: **Send** $n/|C|$ of $A_i; i = 1, \ldots m$ to each C
Child Processors C_{p_i}: **Sort** values of A_i using [13]
Perform Merge Sort over the sorted individual groups from each C_{p_i}
Send the final sorted value to M where $a=$ lowest value; and $b=$highest value
Processor M: **Compute** Split Point = $\frac{a+b}{2}$

3.3 Pruning Decision Tree

Pruning is usually carried out to avoid overfitting the training data, and eliminates those parts that are not descriptive enough to predict future data. A pruned decision tree has a simpler structure and good generalization ability which comes at the expense of classification accuracy. A pruning method, called Reduced Error Pruning (REP) is simple in structure and provides a reduced tree in good speed but the classification accuracy is effected. For optimization purpose, we should produce a simpler structure but at the same time achieving classification accuracy is also our one of the main concerns. To minimize the trade-off, a novel method for pruning decision tree, proposed in [24], is used in this paper. This method Cost and Structural Complexity (CSC) takes into account both the classification ability and structural complexity which evaluate the structural complexity in terms of the number of all kinds of nodes, the depth of the tree, the number of conditional attributes and the possible class types. The cost and structural complexity of a subtree in the decision tree T(to be pruned) is defined as

$$CSC(Subtree(T)) = 1 - r(v) + Sc(v)$$

where $r(v)$ is the explicit degree of the conditional attribute v and $Sc(v)$ is the structural complexity of v.

Here, the explicit degree of a conditional attribute is measured in order to evaluate the explicitness before or after pruning. This measurement is absolutely desirable for maintaining explicitness as much as possible for achieving high classification accuracy.

A fascinating feature of the pruning method is its post-pruning action which deals with the problem of 'horizon effect' which causes inconsistency in the pre-pruning process [24]. The pruning pays attention to overcome the problem of over-learning of the details of data that takes place when the subtree has many leaf nodes against the number of classes. Along with this, the method also handles the number of conditional attributes of a subtree against the set of all possible conditional attributes. The final pruned tree is simple in structure which is a very desirable feature of an optimization algorithm.

This pruned tree is again considered as the originally induced decision tree which we again use to predict classification accuracy by using cross-validation as directed in section 3.1. It is to be noted that the time spent on pruning for a large dataset is a small fraction, less than 1% of the initial tree generation [22]. Therefore, the pruning process is not adding much overhead toward the computational complexity. Experimental results on classification accuracy based on both pruned and non-pruned decision trees are given in the following section.

Next, we perform a theoretical comparison between our proposed framework and other existing parallel strategies of decision tree algorithm. Other parallel strategies defined in [14] such as dynamic data fragment, static (both horizontal and vertical division) data fragments are considered to make a comparison in terms of load balancing and communication cost. Our framework proposes an optimization algorithm that particularly pays attention toward uniform load balancing among the child processors which we explained in section 3.1. Moreover, the communication cost is also reduced in our proposed framework. The comparison is given in Table 2.

Table 2. Comparison of Parallel Strategies

Strategies	Communication Cost	Load Balancing
Dynamic Data Fragment	High	Low
Vertical Data Fragment	Low	High
Horizontal Data Fragment	Low	Medium
Our Framework	Low	Excellent

4 Experimental Results

We used MPI (message passing interface)[15] as the communications protocol for the implementation purposes in order to facilitate processes to communicate with one another by sending and receiving messages. The reason for choosing MPI is its point-to-point and collective communication supported features. These features are very significant to fit into high performance computing today. That makes MPI a very dominant and powerful model today for doing computationally intensive works.

We worked with mpicc compiler for the compilation purpose. For testing purposes, we implemented and tested our proposed formulation with classification benchmark dataset from UCI machine learning repository [1]. We tested with ten datasets from this benchmark dataset collection. Table 3 summarizes the important parameters of these benchmark datasets.

Table 3. Benchmark Datasets

Dataset	# Examples	# Classes	# Conditional Attributes
Adult	32561	2	14
Australian	460	2	14
Breast	466	2	10
Cleve	202	2	13
Diabetes	512	2	8
Heart	180	2	13
Pima	512	2	8
Satimage	4485	6	36
Vehicle	564	4	18
Wine	118	3	13

The comparison with serial implementation in terms of computation time is given in Table 4.

The effect of pruning with REP and CSC in our framework is depicted in terms of the reduced tree size which we can compare in the following Table 5.

The classification accuracy is one of the major parts of a decision tree algorithm to measure its performance. In this proposed framework, the prediction was done by taking cross-validation approach into consideration. The average accuracy for all the ten datasets were calculated in case of both without and

Table 4. Comparison of Execution Times

Dataset	Serial	Parallel
Adult	115.2	17.6
Australian	1.09	0.0027
Breast	1.08	0.0032
Cleve	0.55	0.0015
Diabetes	1.09	0.004
Heart	0.31	0.0008
Pima	1.01	0.0022
Satimage	7.56	1.42
Vehicle	1.02	0.0018
Wine	0.31	0.00003

Table 5. Tree Size before (Hunt's Method) and after pruning (REP and CSC)

Dataset	Hunt's Method	REP	CSC
Adult	7743	1855	246
Australian	121	50	30
Breast	33	29	15
Cleve	56	42	29
Diabetes	51	47	27
Heart	41	41	22
Pima	55	53	32
Satimage	507	466	370
Vehicle	132	128	112
Wine	9	9	9

Table 6. Classification Accuracy on datasets (in %)

Dataset	Without Pruning	REP	CSC
Adult	84	85	87.2
Australian	85.5	87	87
Breast	94.4	93.1	95
Cleve	75.4	77	77.2
Diabetes	70.2	69.9	70.8
Heart	82.7	82.7	84.1
Pima	73.1	73.2	74.9
Satimage	85	84.2	86.1
Vehicle	68.4	67.7	69.2
Wine	86	86	86

with pruning effect. The following table 6 depicts the effect of before and after pruning in classification accuracy for test sets which we believe is a standard way of measuring the predictive accuracy of our hypothesis.

From the table we noticed that the inclusion of pruning using method in [24] achieves better accuracy.

5 Conclusion

In this paper, we proposed an optimized formulation of decision tree in terms of a parallel strategy as a inductive-classification learning algorithm. We designed an algorithm in a partition and integration manner which particularly reduces the work load from a single processor and distributes it among a number of child processors. Instead of performing computation for the entire table, each child processor computes a particular portion of the training set and upon receiving the results from the respective processor, the root processors forms the decision tree. An existing pruning method that particularly draws a balance between structural complexity and classification accuracy is incorporated in our framework which produces a simple structured tree that generalizes well for new, unseen data. Our experimental results on benchmark datasets indicate that the inclusion of parallel algorithm along with pruning optimizes the performance of the classifier as indicated by the classification accuracy.

In future, we will attempt to concentrate on several issues regarding the performance of the decision tree. Firstly, we will experiment our algorithm on image dataset in order to facilitate different computer vision applications. Also, we will focus on selecting multiple splitting criteria for discretization of continuous attributes.

References

1. http://archive.ics.uci.edu/ml/datasets.html/
2. An, A., Cercone, N.J.: Discretization of Continuous Attributes for Learning Classification Rules. In: Zhong, N., Zhou, L. (eds.) PAKDD 1999. LNCS (LNAI), vol. 1574, pp. 509–514. Springer, Heidelberg (1999)
3. Blockeel, H., Struyf, J.: Efficient algorithms for decision tree cross-validation. In: Proceedings of ICML 2001- Eighteenth International Conference on Machine Learning, pp. 11–18. Morgan Kaufmann (2001)
4. Breiman, L., Friedman, J.H., Olshen, R.A., Stone, C.J.: Classification and Regression Trees. Statistics/Probability Series. Wadsworth Publishing Company, Belmont (1984)
5. Chen, M.-S., Hun, J., Yu, P.S., Ibm, T.J., Ctr, W.R.: Data mining: An overview from database perspective. IEEE Transactions on Knowledge and Data Engineering 8, 866–883 (1996)
6. Chung, H., Gray, P.: Data mining. Journal of Management Information Systems 16(1), 11–16 (1999)
7. Goil, S., Aluru, S., Ranka, S.: Concatenated parallelism: A technique for efficient parallel divide and conquer. In: Proceedings of the 8th IEEE Symposium on Parallel and Distributed Processing (SPDP 1996), pp. 488–495. IEEE Computer Society, Washington, DC (1996)
8. Joshi, M.V., Karypis, G., Kumar, V.: Scalparc: A new scalable and efficient parallel classification algorithm for mining large datasets. In: Proceedings of the 12th International Parallel Processing Symposium on International Parallel Processing Symposium, IPPS 1998, pp. 573–579. IEEE Computer Society, Washington, DC (1998)

9. Senthamarai Kannan, K., Sailapathi Sekar, P., Mohamed Sathik, M., Arumugam, P.: Financial stock market forecast using data mining techniques. In: Proceedings of the International MultiConference of Engineers and Computer Scientists (IMECS 1996), Hong Kong, March 17-19, pp. 555–559 (2010)
10. Koh, H.C., Tan, G.: Data mining applications in healthcare. Journal of Healthcare Information Management 19(2), 64–72 (2005)
11. Kreuze, D.: Debugging hospitals. Technology Review 104(2), 32 (2001)
12. Kumar, V., Grama, A., Gupta, A., Karypis, G.: Introduction to parallel computing: design and analysis of algorithms. Benjamin-Cummings Publishing Co., Inc., Redwood City (1994)
13. Lipschutz, S.: Schaum's Outline of Theory and Problems of Data Structures. McGraw-Hill, Redwood City (1986)
14. Liu, X., Wang, G., Qiao, B., Han, D.: Parallel strategies for training decision tree. Computer Science J. 31, 129–135 (2004)
15. Madai, B., AlShaikh, R.: Performance modeling and mpi evaluation using westmere-based infiniband hpc cluster. In: Proceedings of the 2010 Fourth UKSim European Symposium on Computer Modeling and Simulation, Washington, DC, USA, pp. 363–368 (2010)
16. Mehta, M., Agrawal, R., Rissanen, J.: SLIQ: A Fast Scalable Classifier for Data Mining. In: Apers, P.M.G., Bouzeghoub, M., Gardarin, G. (eds.) EDBT 1996. LNCS, vol. 1057, pp. 18–32. Springer, Heidelberg (1996)
17. Milley, A.: Healthcare and data mining. Health Management Technology 21(8), 44–47 (2000)
18. Quinlan, J.R.: Induction of decision trees. Machine Learning 1(1), 81–106 (1986)
19. Quinlan, J.R.: C4.5: Programs for Machine Learning, 1st edn. Morgan Kaufmann, San Mateo (1992)
20. Quinlan, J.R.: Improved use of continuous attributes in c4.5. Journal of Artificial Intelligence Research 4, 77–90 (1996)
21. Shafer, J., Agrawal, R., Mehta, M.: SPRINT: A scalable parallel classifier for data mining. In: VLDB, pp. 544–555 (1996)
22. Srivastava, A., Han, E., Kumar, V., Singh, V.: Parallel formulations of decision-tree classification algorithms. Data Mining and Knowledge Discovery: An International Journal, 237–261 (1998)
23. Trybula, W.J.: Data mining and knowledge discovery. Annual Review of Information Science and Technology (ARIST) 32, 197–229 (1997)
24. Wei, J.M., Wang, S.Q., Yu, G., Gu, L., Wang, G.Y., Yuan, X.J.: A novel method for pruning decision trees. In: Proceedings of 8th International Conference on Machine Learning and Cybernetics, July 12-15, pp. 339–343 (2009)

MapReduce Frame Work: Investigating Suitability for Faster Data Analytics

Monali Mavani[1] and Leena Ragha[2]

[1] SIES College of Management Studies
Navi Mumbai, India
monamavani@gmail.com
[2] Ramrao Adik Institute of Technology
Navi Mumbai, India
leena.ragha@gmail.com

Abstract. Faster data analytics is the ability to generate the desired report in near real time. Any application that looks at an aggregated view of a stream of data can be considered as an analytic application. The demand to process vast amounts of data to produce various market trends, user behavior, fraud behavior etc. becomes not just useful, but critical to the success of the business. In the past few years, fast data, i.e., high-speed data streams, has also exploded in volume and availability. Prime examples include sensor data streams, real-time stock market data, and social-media feeds such as Twitter, Facebook etc. New models for distributed stream processing have been evolved over a time. This research investigates the suitability of Google's MapReduce (MR) parallel programming frame work for faster data processing. Originally MapReduce systems are geared towards batch processing. This paper proposes some optimizations to original MR framework for faster distributed data processing applications using distributed shared memory to store intermediate data and use of Remote Direct Access (RDMA) technology for faster data transfer across network.

Keywords: Map Reduce, faster data analytics, distributed shared memory, Remote Direct Memory Access.

1 Introduction

Today, with the growing use of mobile devices constantly connected to the Internet, the nature of User-generated data has changed, it has become more real-time. The New York Stock Exchange generates about one terabyte of new trade date per day, Facebook Hosts approximately 10 billion photos, taking up one petabyte of storage, The Internet Archives stores Around 2 petabyte of data, and is growing at a rate of 20 terabytes per month [1]. Processing data in batches is too slow to produce faster reports. Accumulated data can lose its importance in several hours or, even, minutes. Faster data processing requires stream processing and online aggregation. The process of mapping a request from the originator to the data source is called "Map"; and the

S. Unnikrishnan, S. Surve, and D. Bhoir (Eds.): ICAC3 2013, CCIS 361, pp. 119–130, 2013.

process of aggregating the results into a consolidated result is called "Reduce". This research investigates the suitability of Google's MapReduce (MR) parallel programming frame work for faster data processing. In the MR the user of the MR library expresses the computation as two functions: Map and Reduce. Map, written by the user, takes an input pair and produces a set of intermediate key/value pairs. The Map Reduce library groups together all intermediate values associated with the same intermediate key and passes them to the Reduce function. The Reduce function, also written by the user, accepts an intermediate key and a set of values for that key. It merges together these values to form a possibly smaller set of values. Typically, just zero or one output value is produced per Reduce invocation. The key benefits of this model are that it harnesses compute and I/O parallelism on commodity hardware and can easily scale as the datasets grow in size [2]. Hadoop[3] is an open source MR framework used in distributed computing environment. This paper investigates the suitability of MR framework to be used in faster data analytics and takes Hadoop as a platform for study.

Apache is managing an open-source based Hadoop project, derived from the Nutch project [4]. The Hadoop project releases the Hadoop Distributed File System (HDFS) and Map Reduce framework. Following section explain the detail execution environment of Hadoop. Each map function is given a part of input data called input split which is coming from HDFS. The map function buffers the output in memory. Map output is intermediate output: it's processed by reduce tasks to produce the final output, and once the job is complete the map output can be thrown away. Before it writes to disk, the thread first divides the data into partitions corresponding to the reducers that they will ultimately be sent to. Within each partition, the background thread performs an in-memory sort by key. Hadoop allows the user to specify a combiner function to be run on the sorted map output—the combiner function's output forms the input to the reduce function. Running the combiner function makes for a more compact map output, so there is less data to write to local disk and to transfer to the reducer. Each time the memory buffer reaches the spill threshold, a new spill file is created, so after the map task has written its last output record there could be several spill files. Before the task is finished, the spill files are merged into a single partitioned and sorted output file. The map output is compressed as it is written to disk. The data flow between map and reduce tasks is colloquially known as "the shuffle," as each reduce task is fed by many map tasks. The output file's partitions are made available to the reducers over HTTP. So shuffle stage involves disk I/O and network I/O. Hadoop uses the concept of barrier to synchronize map and reduce operations. Map functions run in parallel, creating different intermediate values from different input data sets. All worker processes executing map functions reach to a barrier and then wait for reduce operation assignment to them by master process. So Bottleneck is that reduce phase cannot start until map phase completes. The output of the reduce is normally stored in HDFS for reliability. The map tasks may finish at different times, so the reduce task starts copying their outputs as soon as each completes. This is known as the copy phase of the reduce task. The map outputs are copied to the reduce task JVM's memory if they are small enough otherwise, they are copied to disk. When the in-memory buffer reaches a threshold size or reaches a threshold number of map outputs, it is merged and spilled to disk. When all the map outputs have been copied, the reduce task moves into the sort phase which merges the map outputs, maintaining their sort ordering and applies reduce function to sorted data. It finally writes output to HDFS and informs the master about its location. In this execution environment heavy disk I/O and network I/O is observed.

Basic MR programming model gives simple yet powerful model for distributed data processing. MR model has got few limitations to be suitable for the execution environment for performing faster data analytics. Following section highlights its limitations.

The rest of the paper is organized as follows. Section 2 provides limitation of MR to be used directly in faster analytics. Section 3 presents literature survey. Section 4 proposes the optimization to MR frame work and Section 5 concludes the paper.

2 Faster Data Analytics

Faster data Analytics is the process of manipulating raw data and the ability to generate the desired report in near real time. Any application can be categorized that looks at an aggregated view of a stream of data, as an analytic application. Faster data Analytics involves stream processing and online aggregation. Many of the data intensive applications must combine new data with data derived from previous batches or iterate to produce results, and state is a fundamental requirement for doing so efficiently. For example, incremental analytics re-use prior computations, allowing outputs to be updated, not recomputed, when new data arrives [5]. To perform faster analytics, query results should be outputted as soon as input to the query is available rather than waiting for the whole input to get loaded. Reducers begin processing data as soon as it is produced by mappers, they can generate and refine an approximation of their final answer during the course of execution. This technique is known as online aggregation [6]. Faster data analytics requires fast in-memory processing of a MapReduce query program for all (or most) of the data. There are certain limitations of MR framework. Following subsection illustrate the same.

2.1 Limitations of Map Reduce for Faster Analytics

Various authors have identified limitation of MapReduce model for faster analytics following are some of the issues highlighted in various research.

- Group by implementation of SQL is implemented using sort-merge algorithm in MapReduce model which is CPU intensive.
- For huge datasets lot of intermediate data is generated so merge phase of sort merge step performs disk I/O due to memory overflow which is blocking.
- MapReduce systems wait until the entire data set is loaded to begin query processing and further employ batch processing of query programs, they are ill-suited for faster processing.
- Reduce phase cannot be started until map phase is over and all the data for the reduce function is available before starting the execution of reduce function. This further introduces latency.

- In shuffle phase a reduce task fetches data from each map task's local file system by issuing HTTP request. This involves network I/O as well as disk I/O which adds to latency.
- Traditional MapReduce implementations provide a poor interface for interactive data analysis, because they do not emit any output until the job has been executed to completion [7].
- Non suitability of MR for stream processing as mentioned by Lam et al. [8] is as follows. MapReduce runs on a static snapshot of a data set, while stream computations proceed over an evolving data stream. In MapReduce, the input data set does not (and cannot) change between the start of the computation and its finish. In stream computations, the data is changing all the time; there is no such thing as working with a "snapshot" of a stream.

Various researchers have worked on different solutions. They have addressed the possibility of faster and incremental data processing using MR framework in general and Hadoop in particular. Following section presents work done by different authors found in literature.

3 Literature Review

Faster data processing and online data aggregation is needed which becomes difficult by traditional MR framework due to above mentioned reasons. Various authors have suggested variant in MR framework to address above mentioned limitations.

- Condie et al. have proposed modified MR architecture -MapReduce Online, which implements Hadoop Online Prototype (HOP) with pipelining of data where map output is pushed eagerly to reducers [7]. Due to pipelining early returns on long-running jobs via online aggregation and continuous query over streaming data becomes possible. HOP pushes map output in finer granularity and hence increases network I/O cost and reduces CPU utilization. This moves some of the sorting work to reducers but still due to merge-sort implementation incurs disk I/O and n/w I/O.
- Elteir et al. have proposed enhancement in MR via asynchronous data processing namely incremental processing. They have tried to address drawback of synchronized communication between map and reduce of existing Hadoop implementations using two different approaches. The first approach, hierarchical reduction, starts a reduce task as soon as a predefined number of map tasks completes. The second approach, incremental reduction, starts a predefined number of reduce tasks from the beginning and has each reduce task incrementally reduce records collected from map tasks [9]. So they have tried to change run time system of original hadoop implementation for recursively reducible jobs. They evaluated different reducing approaches with two real applications on a 32-node cluster and found that incremental reduction can speed-up the original Hadoop implementation by up to 35.33% for the word count application and 57.98% for the grep application. But for hierarchical reduction network I/O cost will be high.
- Mazur et al. have addressed the drawback due to sort-merge phase of MR Runtime sequence which is CPU intensive affecting overall running time [2]. They have re-implemented sort –merge implementation of group by with hash based technique which can support fast in-memory processing whenever computation states of the

reduce function for all groups fit in memory. They have experimented the overhead due to sorting in the click stream analysis application and showed that up to 48% of CPU cycles, and up to 53% of running time can be saved. They also have experimented change at architecture level by using separate fast and small storage device to hold intermediate data for reducing disk contention and separate storage and computing nodes. Extra storage devices help reduces the total running time; from 76 minutes to 43 minutes for sessionization. Distributed storage system reduces the running time of sessionization from 76 minutes to 55 minutes. But the issues of blocking and intensive I/O remain unaffected.

- Bu et al. have proposed HaLoop, a runtime based on Hadoop which supports iterative data analysis applications especially for large-scale data [10]. By caching the related invariant data and reducers' local outputs for one job, it can execute recursively. It offers a programming interface to express iterative data analysis applications. HaLoop uses a new task scheduler for iterative applications that leverages data locality in the applications. From the original hadoop task scheduler and task tracker modules are modified, and the loop control, caching, and indexing modules are newly added modules. They evaluated HaLoop on real queries and real datasets. Compared with Hadoop, on average, HaLoop reduces query runtimes by 1.85, and shuffles only 4% of the data between mappers and reducers.
- Verma et al. [11] have presented techniques for supporting general purpose applications in a barrier-less MapReduce framework. They have modified Hadoop implementation. They have bypass the sorting mechanism, modified the invocation of the Reduce function so that it can be called with a single record. Reducer must maintain partial results for every key it has received. They used the Java implementation of Red-Black trees called TreeMap. They experimented seven different categories of MapReduce algorithms, and converted into barrier less version. Their results showed an average improvement of 25% (and 87% in the best case) in the job completion times, with minimal additional programmer effort. For large datasets huge intermediate results are generated which causes memory overflow and disk I/O. In order to address these memory overflow problems, they explored two possible memory management solutions: a disk spill and merge scheme and an off-the-shelf disk-spilling key-value store. In a disk spill and merge scheme when the memory usage reaches a predefined memory threshold, the structure moves the partial results to a newly created local spill file on the disk. Instead of flushing the entire contents of the memory to a file on the disk, the partial results can be maintained in a key/value store that has the capability of spilling to disk. They experimented BerkeleyDB (Java Edition) as key/value store. The disk spill and merge approach performed better than original hadoop. And disk-spillable key/value store has performed poorly on word count problem and they concluded that off-the-shelf key/-value store may not be suitable option for map reduce workload.
- Yan et al. have proposed IncMR framework which incrementally processing new data of a large data set, which takes state as implicit input and combines it with new data [12]. They have added additional modules and extended existing hadoop API. Map tasks are created according to new splits instead of entire splits while reduce tasks fetch their inputs including the state and the intermediate results of new map tasks from designate nodes or local nodes. Their work stresses on job scheduling based on data locality but lacks in state management issues like size and

location of state which incurs huge network I/O. In their work they highlighted the main overhead of IncMR in the storage and transmission of many intermediate results. Job scheduler will try to allocate reduce tasks to the nodes that have performed reduce tasks recently because they have cached related state.

- Lam et al. have describe Muppet - MapUpdate, a framework like MapReduce, but specifically developed for fast data. MapUpdate operates on data streams, so map and update functions must be defined with respect to streams [8]. Streams may never end, so updaters use storage called slates to summarize the data they have seen so far. In MapUpdate, slates are "memories" of updaters, distributed across multiple map/update machines as well as persisted in a key value store for later processing. It is continuously updated an in-memory data structure that stores all important information that the update function U must keep about all the events with key k that U has seen so far. Muppet uses Cassandra key-value store to store slates. They have run Cassandra key-value store on solid-state flash-memory storage (SSDs). In further optimization the workers pass events directly to one another without going through any master. The master in Muppet is used for handling failures. Muppet uses hashing techniques in shuffle stage.
- Yocum et al. have implemented MapReduce over a distributed stream processor. They have used in-network aggregate computation that avoids data reprocessing. They have used window based stream processing input is in streams rather than batches. Their system shows efficiency for huge unstructured data. They have used the distributed stream processing platform, Mortar to create and manage the physical MapReduce data flows. Mortar is a platform for instrumenting end hosts with user-defined stream processing operators. The platform manages the creation and removal of operators, and orchestrates the flow of data between them.
- Logothetis et al. have proposed a generalized architecture for continuous bulk processing (CBP) where prior results are reused to incrementally incorporate the changes in the input [5]. They have proposed group wise processing operator, translate, and dataflow primitives to maintain state during continuous bulk data processing. The translate operator is run repeatedly, allowing users to easily store and retrieve state as new data inputs arrive. They have modified hadoop to support full CBP model and optimize treatment of state. They have evaluated their CBP model on incremental crawl queue, clustering coefficients and page rank. In Page Rank application data movement is reduced by46 % and running time is cut by 53% reducing both network and CPU usage.
- Bhatotia et al. have proposed the architecture- Incoop based on incremental HDFS (content based input splitting), incremental map (memorization aware scheduler), incremental reduce (combiners for each reduce task) to maximize the reuse of results of previous computations [22]. Work and time speedups vary between 3-fold and 1000-fold for incremental modifications ranging from 0% to 25% of data. Higher speedups are observed for computation-intensive applications (K-Means, KNN) than for data-intensive applications (Word Count, Co-Matrix, and BiCount). Both work and time speedups decrease as the size of the incremental change increases, because larger changes allow fewer computation results from previous runs to be reused. Their evaluation shows that Incoop can improve efficiency in incremental runs, at a modest cost in the initial, first run where no computations can be reused.

Various methods have been proposed for faster data processing which involves either modifying the Hadoop API or extending the API by adding certain modules. Some of them have proposed changes in the runtime system or suggested architectural level changes. But disk I/O and network I/O still remains as the bottleneck. This paper proposes optimizations in original framework which can help in reducing latency due to disk I/O and network I/O.

4 Proposed Optimization

This research tries to address some of the limitations of MR model for fast data analytics. It proposes

- Use of distributed shared memory (shared tuple space) to store intermediate results.
- Use of Remote Direct Memory Access (RDMA) to address the latencies introduced due to MR execution environment.

This work is also in favor of using stateful map reduce framework. Figure 1 shows Architecture based on RDMA and tuple space.

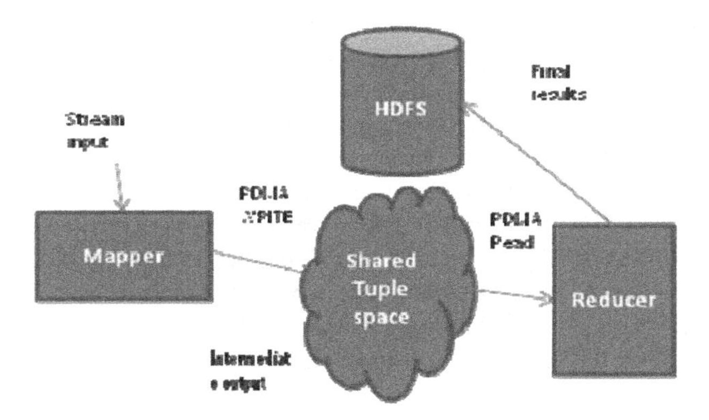

Fig. 1. Architecture based on RDMA and tuple space

1. Use of stateful approach during the execution of one MR Job helps in reduce the latency in producing results. Use of stateful approach in consecutive MR jobs helps iterative algorithms to be processed faster and more efficient way with MR framework. Stateful reducer allows partial results to be saved as state which can be reused with the new input and emitted as intermediate results for interactive query processing. In case of stream processing input continuously arrives. Mapper can store partially produce results (key, value pair) as a state and with each new input from the stream updates the state with corresponding key.
2. Mappers output can be written in shared tuple space. Reducer will start fetching data from shared tuple space and starts early execution. Inputs are fetched by reducer as it is emitted by mapper, so periodically slot in tuple space becomes available which can be reused to store another map output. Map outputs are deleted in shared space as soon as it is taken by reducer, so to have fault tolerance; map

outputs are also written to local disk of mapper's nodes. This induces disk I/O but does not affect execution time. A tuple space is an implementation of the associative memory paradigm for parallel/distributed computing. Producers post their data as tuples in the space, and the consumers then retrieve data from the space that match a certain pattern.[14] Linda [15] language was the first implementation of Tuple Spaces. Later on, the JavaSpaces specification was defined as part of Sun's Jini specification. The JavaSpaces technology [16] is a simple and powerful high-level tool for building distributed and collaborative applications based on the concept of shared network-based space that serves as both object storage and exchange area. The space has a built-in clustering model, which allows it to share data and messages with other spaces on the network. [17]

3. Network I/O latency can be reduced with use of Remote Direct Memory Access (RDMA), next generation network stack for high speed-high performance computing. Modern RDMA capable networks such as InfiniBand and Quadrics provide low latency of a few microseconds and high bandwidth of up to 10 Gbps. This has significantly reduced the latency gap between access to local memory and remote memory in modern clusters. [18]. RDMA provides read and write services directly to applications and enables data to be transferred directly into Upper Layer Protocol (ULP). It also enables a kernel bypass implementation. [19] InfiniBand [20] is an I/O architecture designed to increase the communication speed between CPUs and subsystems located throughout a network. One of the most important features of IB is RDMA. It provides applications with an easy-to-use messaging service. Instead of making a request to the operating system for access to one of the server's communication resources, an application accesses the InfiniBand messaging service directly. InfiniBand provides two transfer semantics; a channel semantic sometimes called SEND/RECEIVE and a pair of memory semantics called RDMA READ and RDMA WRITE. The "initiator" (mapper) places the block of data in a buffer in its virtual address space and uses a SEND operation to send a storage request to the "target" (e.g. the storage device –shared tuple space). The target, in turn, uses RDMA READ operations to fetch the block of data from the initiator's (mappers) virtual buffer. Once it has completed the operation, the target uses a SEND operation to return ending status to the initiator. The initiator (mapper), having requested service from the target, was free to go about its other business while the target asynchronouslv completed the storage operation, notifying the initiator (mapper) on completion. Similarly when reducer reads the data from shared space it uses receive operation to send a fetch request to the target. The target, in turn, uses RDMA write operation to write the data to reducer's internal virtual buffer. Once it has completed the operation, the target uses a SEND operation to return ending status to the reducer. The reducer having requested service from the target (shared space), was free to execute its code while the target asynchronously completed the storage operation, notifying the reducer on completion. This prefetch of input overlapped with reducer execution improves the performance and reduces the latency. The Sockets Direct Protocol (SDP) is a networking protocol developed to support stream connections over InfiniBand fabric. It acts as an interface between high-level application and IB fabric [21].

So with the use of RDMA aware communication channel and distributed shared memory to store intermediate data delay in execution can be reduced. These are few optimizations suggested to improve MR model in order to be suitable execution environment for faster data analytics. In this research we have tried to verify our proposal using performance model developed by Krevat et al. Next section describes performance modeling in terms of overall execution speed.

5 Performance Modeling

This section verifies the suggested optimization in improving speed of data analytics operation using distributed shared memory unit tuple space. For the reference performance model developed at Parallel Data Laboratory, Carnegie Mellon University is taken [23].

5.1 Assumptions

- data-intensive workloads
- computation time is negligible in comparison to I/O speeds
- assumption holds for grep- and sort-like jobs
- pipelined parallelism can allow non-I/O operations to execute entirely in parallel with I/O operations
- input data is evenly distributed across all participating nodes in the cluster
- nodes are homogeneous
- each node retrieves its initial input from local storage
- a single job has full access to the cluster at a time,with no competing jobs or other activities

Table 1 identifies the I/O operations in each map-reduce phase for two variants of the sort operator.

Table 1. Configuration parameters for performance modeling

Symbol	Definition
n	no.of nodes in the cluster
Dw	The aggregate d disk write throughput of a single node
Dr	The aggregate d disk read throughput of a single node
eM	The ratio between Map operator's output and input
eR	The ratio between reduce operator's output and input
N	The network throughput of a node
r	Replication factor used for the job's output data if no replication then r=1
I	Total amount of I/p data for a given computation

When intermediate data does not fit in memory then in traditional map reduce operation, an external sort is used where data is written on disk in multiple files and sorting phase is applied i.e. in disk sorting is used. Overall execution time for external sort bearing above assumptions can be calculated by following equation derived and verified by Krevat et al.:

$$i/n*\{\max[(1 \div Dr)+(Em \div Dw),(n-1)*Em \div n*N]+\max[(Em \div Dr)+r*Em*Er \div Dw, Em*Er*(r-1)\div N])\} \quad (1)$$

Overall execution time is coming out to be 462.0288 seconds for 1TB of i/p data, 25 node clusters, disk throughput is 780 MB/s, Network throughput is 110MB/s and replication factor for output is 1.

Now with the use of tuple space even with larger intermediate data that does not fit into memory of the node, external sort can be avoided and in memory sorting can be used where now memory is in the distributed form. Following equation for in memory sorting is used:

$$i/n*\{\max(1 \div Dr,(n-1)\div n)Em \div N)+\max(r*Em*Er \div Dw,[Em*Er*(r-1)\div N])\} \quad (2)$$

Overall execution time is reduced to 409.6 seconds for the same configuration. This shows improvement due to bypassing disk operations.

Authors of this model have verified their developed model using Parallel Data Series (PDA) -data analysis tool. They have also used this model for modeling various published benchmarks. Then they have demonstrated inefficiency of Hadoop for optimal configurations. Motivation to use this model to verify our suggested optimization is that this model accurately takes different system parameters in to account. As second equation calculates total execution time when disk operations are not needed and they have used this when intermediate data are fitted in single node's memory . However even if data is more than the capacity of single node's memory, it can be distributed across different node's memory and can be managed by layer above it using tuple space API. So second equation still holds true for distributed shared memory.

With the use of tuplespace disk I/O can be avoided but network I/O still remains as the speed limiting factor. As proposed in the previous section next generation interconnection technology based on Remote Direct Memory Access can reduce latency by bypassing Tcp stack which is totally controlled by kernel for network data transfer.

6 Conclusion

This paper investigates the suitability of MR frame work to use in faster data analytics and suggests optimization in traditional MR framework. Use of stateful execution engine helps mappers and reducers to execute faster by updating the state rather than generating the output as inputs come in. Asynchronous communication between mapper and reducer helps to start early execution of reduce function which lowers overall the execution time and less intermediate data to be handled and stored by MR runtime engine. Due to this, shared tuple space, a distributed shared memory unit, can be used to store intermediate data instead of mapper's local disk. Further optimization can be achieved through the use of Remote Direct Memory Access (RDMA) to reduce latency due to network I/O. As a part of future work, suggested optimizations can be evaluated in real experimental setup.

References

1. White, T.: Hadoop The Definitive Guide. Yahoo Press (January 2012)
2. Li, B., Mazur, E., Diao, Y., McGregor, A., Shenoy, P.: A Platform for Scalable One-Pass Analytics using MapReduce. In: International Conference on Management of Data, pp. 985–996 (2011)
3. http://hadoop.apache.org
4. Khare, R., Sitaker, D.C.K., Rifkin, A.: Nutch: A Flexible and Scalable Open-Source Web Search Engine, Oregon State University, Commerce Net Labs Technical Report, pp. 1–10 (2004)
5. Logothetis, D., Olston, C., Reed, B.: Stateful Bulk Processing for Incremental Analytics. In: Proceedings of the 1st ACM Symposium on Cloud Computing, pp. 51–62 (2010)
6. Hellerstein, J.M., Haas, P.J., Wang, H.J.: Online Aggregation. In: SIGMOD Conference, pp. 171–182 (1997)
7. Condie, T., Conway, N., Alvaro, P., Hellerstein, J.M., Elmeleegy, K., Sears, R.: Map reduce online. In: Proceedings of the 7th USENIX Conference on Networked Systems Design and Implementation, pp. 1–14 (2010)
8. Lam, W., Liu, L., Prasad, S.T.S., Rajaraman, A., Vacheri, Z., Doan, A.H.: Muppet: MapReduceStyle Processing of Fast Data. Proceedings of the VLDB Endowment 5(12), 1814–1825 (2012)
9. Elteir, M., Lin, H., Feng, W.-C.: Enhancing mapreduce via asynchronous data processing. In: IEEE 16th International Conference on Parallel and Distributed Systems, pp. 397–405 (2010)
10. Bu, Y., Howe, B., Balazinska, M., Ernst, M.: Haloop: efficient iterative data processing on large clusters. In: 34th International Conference on Very Large Data Bases, pp. 285–296 (2010)
11. Verma, A., Zea, N., Cho, B., Gupta, I., Campbell, R.H.: Breaking the MapReduce Stage Barrier. Journal on Cluster Computing, 1–16 (2011)
12. Yan, C., Yang, X., Yu, Z., Li, M., Li, X.: IncMR: Incremental Data Processing based on MapReduce. In: 5th IEEE International Conference on Cloud Computing, pp. 534–541 (2012)
13. Logothetis, D., Yocum, K.: AdHoc Data Processing in the Cloud. Proceedings of the VLDB Endowment 1(2), 1472–1475 (2008)
14. http://en.wikipedia.org/wiki/Tuple_space
15. http://en.wikipedia.org/wiki/Linda_coordination_language
16. http://java.sun.com/developer/technicalArticles/tools/JavaSpaces
17. http://wiki.gigaspaces.com/wiki
18. Liang, H., Noronha, R., Panda, D.K.: Swapping to Remote Memory over InfiniBand: An S Approach using a High Performance Network Block Device. In: IEEE International Conference on Cluster Computing, pp. 1–10 (2005)

19. Recio, R., Metzler, B., Culley, P., Hilland, J., Garcia, D.: A Remote Direct Memory Access Protocol Specification. RFC 5040
20. http://members.infinibandta.org/kwspub/Intro_to_IB_for_End_Users.pdf
21. http://docs.oracle.com/javase/tutorial/sdp
22. Bhatotia, P., Wieder, A., Rodrigues, R., Acar, U.A., Pasquini, R.: Incoop: MapReduce for Incremental Computations. In: Proceedings of the 2nd ACM Symposium on Cloud Computing, Article No. 7, pp. 1–14 (2011)
23. Krevat, E., Shiran, T., Anderson, E., Tucek, J., Wylie, J.J., Ganger, G.R.: Applying performance models to understand data-intensive computing efficiency. Technical Report Carnegie Mellon University, HP Labs, pp. 1–18 (2010)

Quality Factor Assessment and Text Summarization of Unambiguous Natural Language Requirements

R. Subha[1] and S. Palaniswami[2]

[1] Department of Computer Science and Engineering,
Sri Krishna College of Technology, Coimbatore – 641042, India
kris.subha@gmail.com
[2] Government College of Engineering, Bargur -635104, India

Abstract. The software requirements are documented in natural language to make it easy for the users to understand the document. This flexibility of natural language comes with the risk of introducing unwanted ambiguities in the requirements thus leading to poor quality. In this paper, we propose and evaluate a framework that automatically analyses the ambiguities in a requirements document, summarizes the document and assess its quality. We analyse the ambiguities that can occur in natural language and present a method to automate ambiguity analysis and consistency and completeness verification that are usually carried out by human reviewers which is time consuming and ineffective. The Open Text Summarizer based system summarizes the document and provides an extract of it. We use a decision tree based quality evaluator that identifies the quality indicators in the requirements document and evaluates it.

Keywords: Software Requirements Document, Natural language Processing, Ambiguity, Text summarization, Quality factors.

1 Introduction

Requirements collection plays a significant role during the development of a project. Requirements engineering is the process of discovering, by identifying stakeholders and their needs, and documenting these in a form that is amenable to analysis, communication and subsequent implementation. As a software development life cycle progresses, if the requirements retrieved are not formally documented or analyzed or updated, the quality of the software will degrade. Requirements consistency checking, Requirements tracing etc are done to improve the quality of the software. Most of these activities are done automatically and the rest are constituted by the human analyst. Therefore, to improve the quality of the software researchers are immensely involved in discovering better solutions. Ambiguity is an essential feature to be considered as it affects the Natural language requirements document and thereby affects the quality of the software. Many pre-processing activities are involved to carry out the ambiguity detection and classification. Earlier, the requirements gathering team was equipped with a handbook in order to remove ambiguity while preparing the requirements documents. When it comes to reading the requirements documents, they

S. Unnikrishnan, S. Surve, and D. Bhoir (Eds.): ICAC3 2013, CCIS 361, pp. 131–146, 2013.

are large in amount and time consuming. The Natural Language requirements documents are error-prone. Before analyzing the requirements documents for the levels of ambiguity the evaluation of requirements documents needs to be considered as it involves a significant role in analyzing the document characteristics. A Software Requirements document (SRS) is unambiguous if, and only if, every requirement stated therein has only one interpretation", as stated in IEEE Recommended Practice for Software Requirements Specifications. SRSs are usually written in natural language, often augmented or enhanced by information in other notations, such as formulae, and diagrams. Natural Language is preferred to write every initial conceptual document and every request to proposal virtually. A recent online survey of businesses requiring software, conducted at University of Trento in Italy shows that a majority of documents available for requirements analysis are provided by the user or are obtained by interviews [2]. Moreover,

- 71.8% of these documents are written in common natural language,
- 15.9% of these documents are written in structured natural language, and
- Only 5.3% of these documents are written in formalized language.

1.1 Natural Language Processing

Natural Language Processing (NLP) is a theoretically motivated range of computational techniques for analyzing and representing naturally occurring texts at one or more levels of linguistic analysis for the purpose of achieving human-like language processing for a range of tasks or applications. Automatic summarization, Co reference resolution, Morphological segmentation, Named entity recognition, Natural language generation, Natural language understanding, Part-of-speech tagging, Parsing, Relationship extraction, Sentence breaking, Sentiment analysis, Word segmentation, Word sense disambiguation are the tasks involved in NLP.

1.2 Text Summarization

A summary can be defined as a text that is produced from one or more texts, that contain a significant portion of the information in the original text(s), and that is no longer than half of the original text(s) [3][4]. Text summarization is the process of distilling the most important information from a source (or sources) to produce an abridged version for a particular user (or users) and task (or tasks).When the summary replaces the original document; the output may be extract or abstract. A differentiation can be made between the Generic summaries and user-focused summaries (query-driven). Based on the output, a detailed differentiation is made between indicative summaries that indicate what topics are addressed in the source text and the details of what the original text is about, and the informative summaries, which are intended to cover the topics in the source text.

1.3 Quality Indicators

Quality Indicators [5] are syntactic aspects of the requirements specifications that can be automatically calculated and that provide information on a particular quality property of the requirements specifications themselves. The Quality indicators incorporated in the Quality model are classified into indicators related to Requirement Sentences Quality (RSQ) and Requirement Document Quality (RDQ). Requirement Sentences Quality (RSQ) related indicators are Implicit Subject Sentences, Multiple Sentences, Optional Sentences, Subjective Sentences, Underspecified Sentences, Vague Sentences and Weak Sentences. Requirements Document Quality (RDQ) related indicators are Comment frequency, Readability Index, under referenced Sentences and Unexplained Sentences.

2 Related Work

Wee Meng Soon et.al, proposed "A machine learning approach to co reference resolution of noun phrases" [12]. Here, a prerequisite for co reference resolution is to obtain the majority, if not all, of the possible markables[11][12] in a raw input text. To determine the markables, a pipeline of natural language processing (NLP) modules is used. It consists of tokenization, sentence segmentation, morphological processing, part-of-speech tagging, and noun phrase identification, named entity recognition. As far as co reference resolution is concerned, the goal of these NLP modules is to determine the periphery of the markables, and to endow with the indispensable information about each markable for subsequent generation of features in the training examples.

Chinatsu Aone and Scott William Bennett proposed "Applying Machine Learning to Anaphora Resolution"[1].This system uses feature vectors for pairs of an anaphor[6][9] and its possible antecedent. A total of 66 features are used, and they include lexical (e.g. category), syntactic (e.g. grammatical role), semantic (e.g. semantic class), and positional (e.g. distance between anaphor and antecedent) features. Those features can be either unary features (i.e. features of either an anaphor or an antecedent such as syntactic number values) or binary features (i.e. features concerning relations between the pairs such as the positional relation between an anaphor and an antecedent).

Elena Lloret proposed "Text Summarization: An overview"[4] which gives an overall idea about text summarization.Traditionally, summarization has been decomposed into three main stages[3][8] .

According to the Sparck Jones[8] approach, the stages are:

- Interpretation of the source text to obtain a text representation,
- Transformation of the text representation into a summary representation, and,
- Finally, generation of the summary text from the summary representation

Sparck Jones[8] distinguishes three classes of context factors:

- **Input factors.** The features of the text to be summarized crucially determine the way a summary can be obtained. These falls into three groups, which are: text form (e.g. document structure); subject type (ordinary, specialized or restricted) and unit (single or multiple) documents as input.

- **Purpose factors.** These are the most important factors. They fall under three categories: situation refers to the context within the summary is to be used; audience (i.e. summary readers) and use (what is the summary for?).
- **Output factors.** In this class, material (i.e. content), format and style, are grouped.

Hui Yang et.al, [6] developed an architecture of an automated system to support requirements writing, by incorporating nocuous ambiguity detection into the requirements workflow. The core of such architecture comprises a classifier that automatically determines whether an instance of anaphoric ambiguity is nocuous or innocuous. The classifier is developed using instances of anaphoric ambiguity extracted from a collection of requirements documents. For each instance, a set of human judgments are used to classify. A classifier is then trained on the linguistic features of the text and the distribution of judgments to identify instances of nocuous ambiguity in new cases. Several approaches can be followed to ensure a good quality requirements document. Another approach is the linguistic analysis of a NL requirements document intended to confiscate most of the issues related to readability and ambiguity. A lot of studies dealing with the evaluation and the achievement of quality in NL requirement documents can be found in the literature and Natural Language Processing (NLP) tools have been recently applied to NL requirements documents for inspecting the consistency and completeness.

3 Methodology

In this paper, we propose a method for summarizing quality requirements which includes an open NLP based system using the MaxEnt models for the detection of sentence end words, tokens, parts-of-speech, named entities, anaphors and co referring phrases. The ambiguity detection module of the system is built on the model proposed by Hui Yang et.al [6] and is refined to reduce human intervention. The detection of anaphoric ambiguity and identification of the co-referring noun phrases based on the anaphors and their relationship with other words are done automatically by the system. The unambiguous requirements are summarized to the number of sentences specified by the user. The system has the quality evaluation module to evaluate the quality of the NL Requirements document. The proposed system takes requirements document [15] as input and reads the content. The sentence boundaries are detected by the sentence detector by reading the contents of the file. The tokenize module gets the input which is in the form of sentences and identifies the tokens in the sentences. These tokens are subsequently used by the POS tagger to mark the Parts-of-Speech tags. The construction of parse tree is done by the parser by utilizing the POS tag details. The parse tree is used by the ambiguity detection module which uses a classifier both to identify the co referring Noun Phrases (NP) present in the sentences as well as to classify the sentences as ambiguous. The significant sentences are extracted by the text summarizer from the document. The contents of the requirements document are utilized in order to discover the quality indicators and evaluate them by the quality evaluator.

3.1 Text Preprocessing

The critical part of any NLP system is the Text pre-processing since the characters, words, and sentences recognized at this stage constitute the elemental units passed to all advanced processing stages, from analysis and tagging components, such as morphological analyzers and part-of-speech taggers, through applications, such as information retrieval and machine translation systems. A "pipeline" of text processing components is used in order to provide value from text by the NLP applications, such as customizable information extraction or question answering application. By means of these systems, the performance of each successive system depends on the performance of each of the components that preceded it in the pipeline. In this way, errors made by an "upstream" component (like a part-of-speech tagging system) can cause a negative impact on the performance of each "downstream" system (such as a named entity recognizer or co reference resolution system).This is shown in Fig 1.

Fig. 1. Pipeline of Natural language processing modules

The modules of natural language processing can be described as follows.

Sentence Segmentation [14]. The documents are split into sentences by the Sentence Segmentation system that are later processed and annotated by "downstream" components. When scanned through the input text, one of these characters ('.', '!', '?') is encountered and a way for deciding whether or not it marks the end of a sentence is to be decided. Here maximum entropy model (MaxEnt) is used. A set of predicates related to the possible end-of-sentence positions is generated. Various features, relating to the characters before and after the possible end-of-sentence markers, are used to generate this set of predicates. This set of predicates is then evaluated against the MaxEnt model. The characters including the position of the end-of-sentence marker are separated off into a new sentence if the best outcome indicates a sentence break. The indication whether a punctuation character denotes the end of a sentence or not is detected by the Sentence Detector. This shows that a sentence is defined as the longest white space trimmed character sequence between two punctuation marks. An exception to this rule would be the first and last sentence. The first non whitespace character is assumed to be the beginning of a sentence, and the last non whitespace character is assumed to be an ending of a sentence. It is also possible to perform tokenization first and let the Sentence Detector process the already tokenized text but usually Sentence Detection is done before the text is tokenized thereby using the pre-trained models.

Tokenization [14]. Tokenization systems break sentences into sets of word-like objects which represent the smallest unit of linguistic meaning considered by a natural language processing system. This tokenize module will split words that comprises of contractions: for example, it will split "don't" into "do" and "n't", because it is designed to pass these tokens on to the other NLP tools, where "do" is recognized as a verb, and "n't" as a contraction of "not", an adverb modifying the preceding verb "do".

The input character sequences are split into tokens by the Tokenize module. Tokens are usually words, punctuation, numbers, etc. It is essential to ensure that tokenization produces tokens of the type expected by later text processing components.

Various tokenizer implementations are

- Whitespace Tokenizer - A whitespace tokenizer, non whitespace sequences are identified as tokens
- Simple Tokenizer - A character class tokenizer, sequences of the same character class are tokens
- Learnable Tokenizer - A maximum entropy tokenizer, detects token boundaries based on probability model

Parts-of-Speech tagging [14]. Part-of-speech tagging is the act of assigning a Part of Speech (POS) to each word in a sentence. The POS tags consist of coded abbreviations conforming to the scheme of the Penn Treebank, the linguistic corpus developed by the University of Pennsylvania. Based on the token itself and the context of the token the Part of Speech Tagger marks tokens with its consequent word type. A token can have multiple POS tags depending on the token and the context. The POS Tagger uses a probability model to identify the correct POS tag out of the tag set. A tag dictionary is used to increase the tagging and runtime performance of the tagger in order to restrict the possible tags for a token.

Named Entity Recognition[14]. Named Entity Recognition systems categorize phrases (referred to as entities) found in text with respect to a potentially large number of semantic categories, such as person, organization, or geopolitical location."Name finding" is the term used by the OpenNLP library to refer to the identification of classes of entities within the sentence - for example, people's names, locations, dates, and so on. Seven different types of entities, symbolized by the seven maximum entropy model files in the NameFind subfolder - date, location, money, organization, percentage, person, and time are established by the name finder .Other classes of entities are set up by utilizing the SharpEntropy library to train the new models. The algorithm is far from foolproof as it is dependent on the use of training data and there are many, many tokens that might come into a category such as "person" or "location".

NLP Models. OpenNLP[14] models are trained models developed for use in the NLP. The list of models used in the system is shown in Table 1.

Table 1. List of Models

Component	Description
Tokenizer	Trained on OPENNLP training data
Sentence Detector	Trained on OPENNLP training data
POS Tagger	MAXENT model with tag dictionary
Name Finder	Date name finder model
Name Finder	Location name finder model
Name Finder	Money name finder model
Name Finder	Organization name finder model
Name Finder	Percentage name finder model
Name Finder	Person name finder model
Name Finder	Time name finder model
Chunker	Trained on conll2000 shared task data

3.2 Ambiguity Detection

The ambigituy detection module includes coreference resolution and ambiguity classification.

Co Reference Resolution. Co reference Resolution is the process of identifying the linguistic expressions which make reference to the same entity or individual within a single document or across a collection of documents. Co reference occurs when multiple expressions in a sentence or document refer to the same entity in the world. Initially all possible references need to be extracted from the document before determining the co reference for a document. Every reference is a possible anaphor, and every reference before the anaphor in document order is a possible antecedent of the anaphor, except when the anaphor is nested. If the anaphor is a child or nested reference, then the possible antecedents must not be any reference with the same root reference as the current anaphor [11][12]. Still, the possible antecedents can be other root references and their children that are before the anaphor in document order. The new ambiguous instance, potential pairs of co referring NPs are offered to the classifier to resolve whether the two NPs co refer or not in order to estimate the co reference relations among the possible NPs antecedent candidates. In this system, heuristics-based methods are built-in to exploit the factors that influence co reference determination. The heuristics are incorporated in terms of feature vectors and are modeled based on the Table 2.

Table 2. Feature vector description for coreference resolution heuristics

Feature type	Feature	Description
String matching	Full string matching	Y if both NPs contain the same string aftere the removal of non-informative words,else N
	Head word matching	Y if both NPs contain the same Headword,else N
	Modifier matching	Y if both NPs share the same modifier substring, else N
	Alias name	Y if one NP is the alias name of the other NP, else N
Grammatical	NP type (NP_i)	Y if NP_i is either definite NP or demonstrative NP, else N
	NP type (NP_j)	Y if NP_j is either definite NP or demonstrative NP, else N
	Proper name	Y if both NPs are proper names, else N
	Number agreement	Y if NP_i and NP_j agree in number, else N
Syntactic	PP attachment	Y if one NP is the PP attachment of the other NP, else N
	Appostive	Y if one NP is in appostive to the other NP, else N
	Syntactic role	Y if both NPs have the same syntactic role in the sentence, else N

Each instance of an anaphor is associated with a set of candidate antecedents. A pair wise comparison of the NPs is accomplished by the classifier to identify potential co reference relations among the candidate antecedents. Likewise, each NP pair is tested for co reference, and sets of co referent candidates are identified.

Ambiguity Classification. Anaphoric ambiguity [7] occurs when the text offers two or more potential antecedent candidates either in the same sentence or in a preceding one, as in, 'The function shall build the parse tree, and then display it in a new window'. The expression to which an anaphor [9] refers is called its antecedent. Antecedents for personal pronoun [10] anaphora are nouns or noun phrases (NPs) found elsewhere in the text, usually preceding the anaphor itself. Based on multiple human judgments of the suitable NP antecedent candidate in terms of an anaphoric ambiguity instance [6], the antecedent can be classified. A number of preference heuristics are also included to model the factors that may favor a particular interpretation. A machine learning algorithm is implemented with a set of training instances to construct a classifier. Given an anaphor and a set of possible NP antecedents, the classifier then predicts how strong the preference for each NP is, and from there, whether the ambiguity is nocuous or innocuous. The Naive Bayes classifier is used to classify the antecedents.

Naive Bayes Classification

The Naive Bayes Classifier technique is based on the Bayesian theorem and is particularly suited when the dimensionality of the inputs is high.

Algorithm

D : Set of tuples

Each Tuple is an 'n' dimensional attribute vector

X : (x1,x2,x3,.... xn)

Let there be 'm' Classes : C1,C2,C3...Cm

Naive Bayes classifier predicts X belongs to Class Ci iff

$P(C_i/X) > P(C_j/X)$ for $1 <= j <= m$, $j <> i$

Maximum Posteriori Hypothesis

$P(C_i/X) = P(X/C_i)\ P(C_i) / P(X)$

Maximize $P(X/C_i)\ P(C_i)$ as $P(X)$ is constant

with many attributes, it is computationally expensive to evaluate $P(X/C_i)$.

Naive's Assumption of "class conditional independence"

$P(X/C_i) = P(x_1/C_i) * P(x_2/C_i) *...* P(x_n/ C_i)$

The Naive bayes classifier uses the feature vectors in Table 3 to classify the antecedent and the anaphoric ambiguity.

Table 3. Feature vector description for Antecedent classification heuristics

Feature Type	Feature	Description
Linguistics	Number agreement	Y if NP agree in number; N_P if NP does not agree in number but it has a person property; N if NP doesn't agree in number;UNKNOWN if the number information cannot be determined
	Definiteness	Y if NP is a definite NP; else N Non-prepositional NP Y if NP is a non-prepositional NP; else N
	Syntactic constraint	Y if NP satisfies syntactic constraint; else N
	Syntactic parallelism	Y if NP satisfies syntactic parallelism; else N
	Coordination pattern	Y if NP satisfies coordination pattern; else N
	Non-associated NP	Y if NP is a non-associated NP; else N
	Indicating verb	Y if NP follows one of the indicating verbs; else N
	Semantic constraint	Y if NP satisfies semantic constraint; else N
	Semantic parallelism	Y if NP satisfies semantic parallelism; else N
	Domain-specific term	Y if NP is contained in the domain-specific term list; else N
Context	Centering	Y if NP occurs in the paragraph more than twice; else N
	Section heading	Y if NP occurs in the heading of the section; else N
	Sentence recency	INTRA_S if NP occurs in the same sentence as the anaphor; else INTER_S
	Proximal	Integral value n, where n means that NP is the nth NP to the anaphor in the right-to-left order
Statistics	Local-based collocation frequency	Integral value n, where n refers to the occurrence number of the matched co-occurrence pattern containing NP in local requirements document
	BNC-based collocation frequency	Y if the matched co-occurrence pattern containing NP appears in the word list returned by the sketch engine; else N

The machine learning algorithm allots a weighted antecedent tag to the NP candidate while they are presented with a pronoun and a candidate. Likewise, the antecedent tag information is used by the system to predict whether the anaphora instance displays nocuous ambiguity and then they are disambiguated.

3.3 Text Summarization

Summarization can be exemplified as approaching the problem at the surface, entity, or discourse levels in the conventional system. In this paper, system surface level approach is used. Surface level approach inclines to represent information taking shallow features and then selectively combining them together in order to obtain a salience function that can be used to extract information.Among these features, are:

Thematic features rely on word (significant words) occurence statistics, so that sentences containing words that occur frequently in a text have higher weight than the rest which illustrates the fact that these sentences are the vital ones and they are hence extracted. Before doing term frequency, altering task must be done using a stop-list words which contains words such as pronouns, prepositions and articles. This is the classical statistical approach. Nevertheless, from a point of view of a corpus-based approach td*idf measure (commonly used in information retrieval) is extremely useful to determine keywords in text.

Location refers to the position in text, paragraph or any other particular section which exhibits the point that they contain the target sentences to be included in the summary. This is usually genre-dependent, but there are two fundamental wide-ranging methods, namely leadmethod and the title-based method with cue-word method .

Background assumes that the importance of meaning units is detemined by the presence of terms from the title or headings, initial part of the text or a user's query.

Cue words and phrases, such as "in conclusion", "important", "in this paper",etc. can be very useful to determine signals of relevance or irrelevance and such units are detected both automatically and manually.The text summarization is implemented based on Open Text Summarizer.

3.4 Quality Factor Assessment

Quality indicators are syntactic aspects of the requirements specifications that can be automatically calculated and that provide information on a particular quality property of the requirements specifications themselves. The system uses decision tree to evaluate the quality of the NL requirement document.Table 4 shows the different quality indicators and their description.

Table 4. Quality indicators and their description

Quality indicator	Description	Notes
Optionality	An Optionality Indicator reveals a requirement sentence containing an optional part (i.e. a part that can or cannot considered)	Optionality- revealing words: possibly, eventually, if case,if possible, if appropriate, if needed
Subjectivity	A Subjectivity Indicator is pointed out if sentence refers to personal opinions or feeling	Subjectivity-revealing wordings: similar, better, similarly, worse, having in mind, take into account, take into consideration, as[adjective] as possible
Vagueness	A Vagueness Indicator is pointed out if the sentence includes words holding inherent vagueness, i.e. words having a non uniquely quantifiable meaning	Vagueness-revealing words:clear, easy, strong, good,bad, efficient, useful,significant, adequate, fast,recent, far, close, in front
Weakness	A Weakness Indicator is pointed out in a sentence then it contains a weak main verb	Weak verbs: can, could,may.

Table 4. *(Continued)*

Implicity	An Implicity Indicator is pointed out in a sentence when the subject is generic rather than specific.	Subject expressed by: Demonstrative adjective (this, these,that, those) or Pronouns (it, they, ..). Subject specified by:Adjective previous,next, following, last,..)or Preposition
Readability	It is the value of ARI (Automated Readability Index) [ARI=WS + 9*SW where WS is the average words per sentence, SW is the average letters per word]	

4 Experiment and Results

The system is implemented using C#,.Net as front end and SQLite as the back end.

4.1 Text Preprocessing

Text preprocessing involves steps to be performed before the original text is being given as input to the next stage of the system. The requirements document written in natural language is given as the input and parsed sentences are the output. The Sentence Detector detects whether the punctuation character marks the end of a sentence or not.(i.e) Given a chunk of text, find the sentence boundaries. The input character sequence are broken into tokens.(i.e) Separate a chunk of continuous text into separate words by the Tokenizer. Tokens are usually words, punctuation, numbers, etc. Fig 2 shows the output of tokenizer.

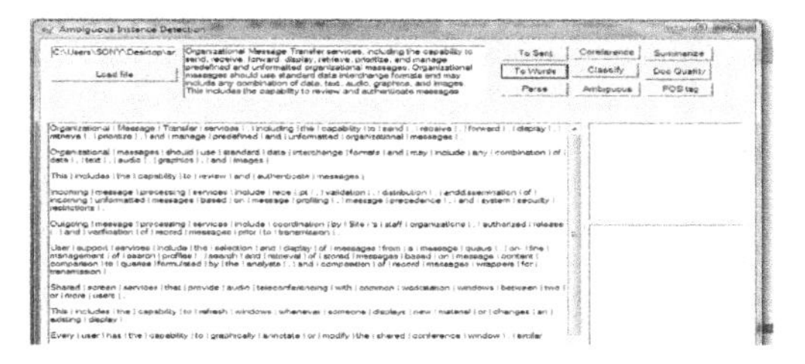

Fig. 2. Tokenization

The Part of Speech Tagger in scripts tokens with their equivalent word type based on the token itself and the context of the token. A token might have multiple POS tags depending on the token and the context. A tag dictionary is employed to increase the tagging and runtime performance of the tagger in order to restrict the possible tags for a token. Parsing determines the parse tree (grammatical analysis) of a given sentence. The output of parsing is shown in Fig 3.

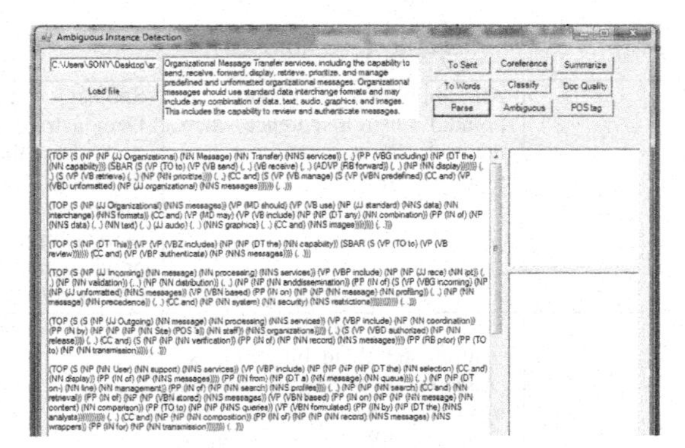

Fig. 3. Parser

4.2 Ambiguity Detection

The parsed sentences from requirements document are given as the input and Co referring NPs are the output. Based on multiple human judgments of the most likely NP antecedent candidate in terms of an anaphoric ambiguity instance a classifier is implemented. For every pronoun and associated NP antecedent candidates, the antecedent classifier allots one of the antecedent preference labels, Positive (Y), Questionable (Q), and Negative (N), to each candidate. Later the calculation is done to verify whether the anaphoric ambiguity is nocuous or innocuous. This information is then used to Fig 4 shows the output of the ambiguity detection module. The requirements statements are classified based on the threshold value.

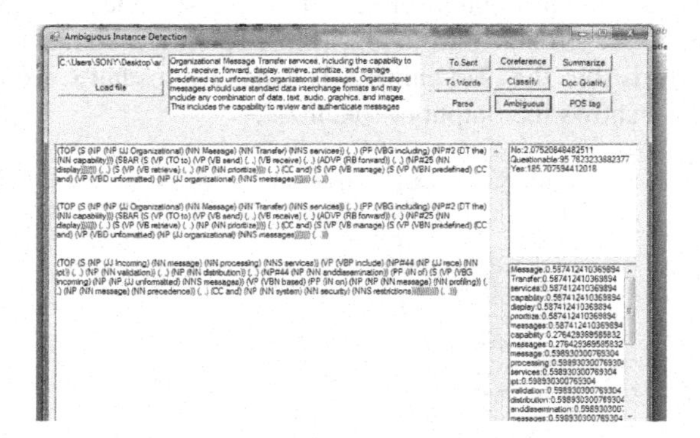

Fig. 4. Ambiguity Classification

4.3 Text Summarization

Text summarization involves reducing a text document or a larger corpus of multiple documents into a short set of words or paragraph that conveys the main meaning of the text. Sentence detection and tokenization,token ranking and finally prioritizing sentences based on the accumulated ranking are performed in Text Summarization.

4.4 Quality Factor Evaluation

The evaluation of the quality factors of the requirements document are carried out by involving qualtiy indicators.The document is verified and the quality values are assigned based on the quality indicators.This is shown in Fig 5.

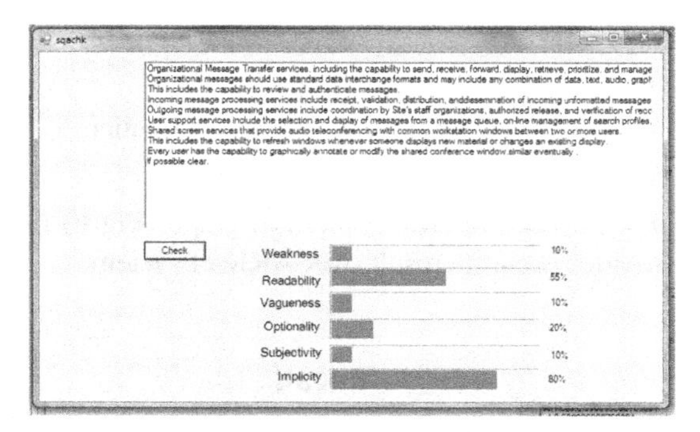

Fig. 5. Quality Factor Evaluation

5 Performance Evaluation

The output of the Ambiguity detection module is compared with "ARKref NP coreference system"[16]. The ARKref coreference resolution system is implemented in java and is available in the web. The ARKref resolution system uses the BNC corpora, web corpora and Wordnet to identify the NP coreferences among the NPs in the sentences. The result of the comparison is shown in Table 5. The input for the evaluation are taken from the requirements dataset collected from RE@UTS website [15].

Table 5. Results of Coreference resolution

S.No	Number of input sentences	Actual NP coreference	NP Coreferences detected	
			Proposed system	ARKref system
1	11	3	3	2
2	10	8	7	7
3	10	6	6	4

Table 6. Precision, Recall, F Measure values

S.No	Proposed system			ARKref system		
	Precision	Recall	F Measure	Precision	Recall	F Measure
1	1	0.2727	0.4284	0.666	0.1818	0.2848
2	0.8757	0.7	0.777	0.8757	0.7	0.777
3	1	0.6	0.75	0.666	0.4	0.4998

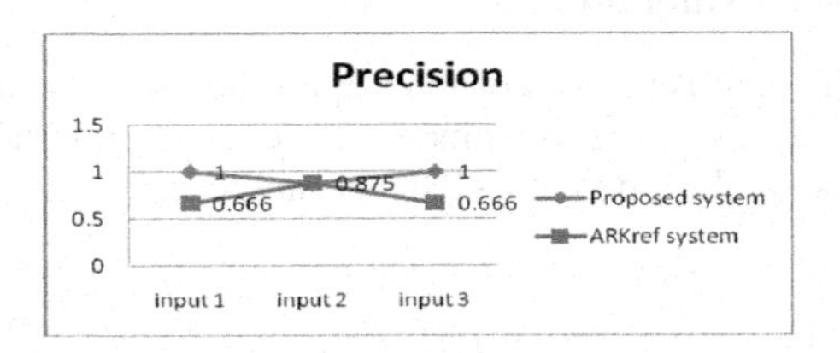

Fig. 6. Precision for co reference resolution

From the precision for co reference resolution graph (Fig 6) it is inferred that the proposed system provides accurate result over ARKref system.

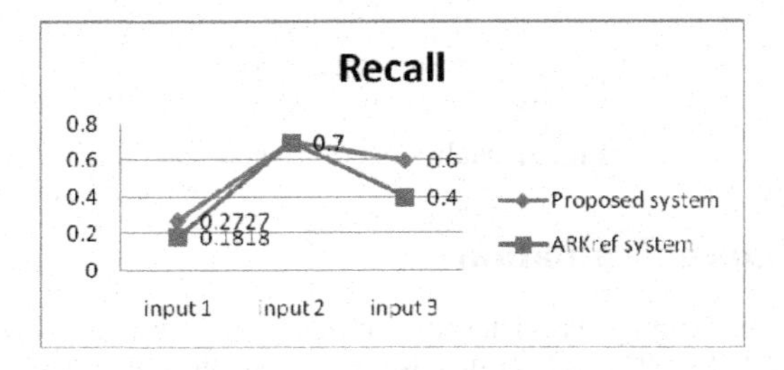

Fig. 7. Recall for co reference resolution

The Recall values in Fig 7 show that the proposed system provides false positives less than that of the ARKref system.

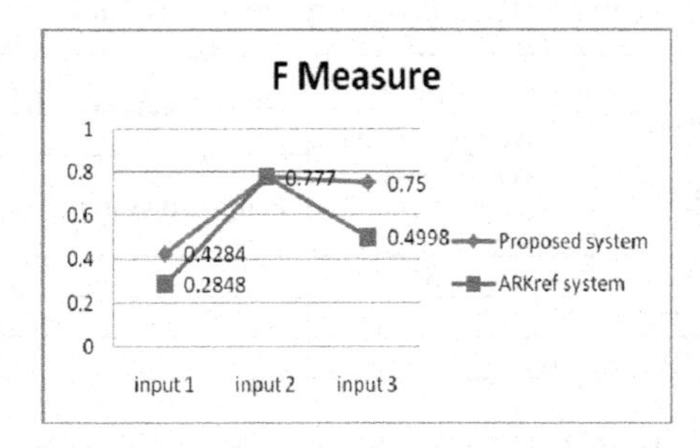

Fig. 8. F Measure for coreference resolution

Table 7 shows ambiguity judgement variation based on the threshold value set by the requirement analyst.

Table 7. Results of Ambiguity detection

Threshold value	Actual ambiguity	Detected nocuous ambiguity
0.5	3	1
0.6	3	2
0.65	3	3
0.7	3	3
0.75	3	3
0.8	3	3
0.85	3	6
0.9	3	7
1	3	8

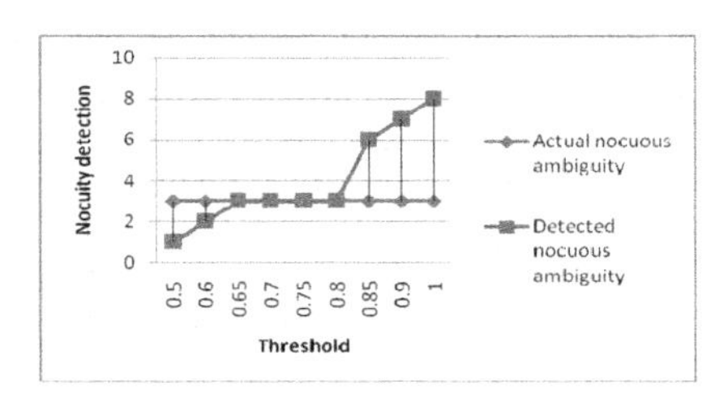

Fig. 9. Threshold vs Nocuous ambiguity detection

The graph in Fig 9 shows that the nocuity detection varies as the threshold value increases. The accuracy of nocuity detection drops when the threshold value is set above 0.8. Thus better accuracy is achieved in the proposed system when the threshold is set between 0.65 and 0.8.

6 Conclusion

Thus a framework is proposed that can automatically analyse the ambiguities for a given natural language requirement document. Initially the corefering NPs were utilised to classify the pronouns and further the ambiguity detection module was used to detect the nocuous statements from the requirments document by the antecedent classifier.Based on the threshold value the nocuous anaphoric ambiguity was differentiated from that of the innocuous ambiguities. The system used OpenNLP models developed by OpenNLP Apache software foundation. The accuracy of the proposed system was enhanced by Brown corpora and Wordnet dictionary. The Open Text Summarizer based system summarized the document and provided an extract of it. The decision tree based quality evaluator identified the quality indicators effectively in the requirements document and evaluated the quality of the requirement document provided for analysis.

References

1. Aone, C., Bennett, S.W.: Applying machine learning to anaphora resolution. In: Connectionist, Statistical and Symbolic Approaches to Learning for Natural Language Processing, pp. 302–314 (1996)
2. Berry, D.M., Kamsties, E., Krieger, M.M.: From Contract Drafting to Software Specification: Linguistic Sources of Ambiguity, University of Waterloo, Ontario, Canada (2003)
3. Hovy, E.: Automated Text Summarization. In: Mitkov, R. (ed.) The Oxford Handbook of Computational Linguistics (2005)
4. Lloret, E., Palomar, M.: Text summarization in progress: a literature review. Artificial Intelligence Review 37 (2012)
5. Lami, G.: QuARS: A Tool for Analyzing Requirement, CMU/SEI-2005-TR-014 (2005)
6. Yang, H., de Roeck, A., Gervasi, V., Willis, A., Nuseibeh, B.: Analyzing anaphoric ambiguity in natural language requirement. Requirements Engineering Journal 16, 163–189 (2011)
7. Dagan, I., Itai, A.: Automatic processing of large corpora for the resolution of anaphora references. In: Proceedings of the 13th International Conference on Computational Linguistics, pp. 1–3 (1990)
8. Jones, K.S.: Automatic summarizing: factors and directions. Advances in Automatic Text Summarization. MIT Press (1999)
9. Denber, M.: Automatic resolution of anaphora in English. Technical report. Eastman Kodak Co. (1998)
10. Brennan, S.E., Froedman, M.W., Pollard, C.J.: A centering approach to pronouns. In: Proceedings of the 25th Annual Meeting of the Association for Computational Linguistics (ACL), pp. 155–162 (1987)
11. Ng, V., Cardie, C.: Improving machine learning approaches to co reference resolution. In: Proceedings of the 40th Annual Meeting of the Association for Computational Linguistics, pp. 104–111 (2002)
12. Soon, W.M., Ng, H.T., Lim, D.C.Y.: A machine learning approach to co reference resolution of noun phrases. Computational Linguistics - Special Issue on Computational Anaphora Resolution Archive 27, 521–544 (2001)
13. http://incubator.apache.org/opennlp/
14. http://research.it.uts.edu.au/re/
15. http://www.ark.cs.cmu.edu/ARKref/

A Comparative Study of Spam and PrefixSpan Sequential Pattern Mining Algorithm for Protein Sequences

Rashmi V. Mane

Department of Technology, Shivaji University,
Kolhapur, Maharashtra, India
rvm_tech@unishivaji.ac.in

Abstract. Sequential Pattern Mining is an efficient technique for discovering recurring structures or patterns from very large dataset. Many algorithms are proposed for mining. Broadly data mining algorithms are classified into two categories as Pattern-Growth approach or candidate generation and Apriori – Based. By introducing constraints such as user defined threshold, user specified data, minimum gap or time, algorithms outperforms better. In this paper we have used dataset of protein sequences and comparison in between PrefixSpan from pattern growth approach and SPAM from Apriori-Based algorithm. This comparative study is carried out with respect to space and time consumption of an algorithm. The study shows that SPAM with constraints outperforms better than PrefixSpan for very large dataset but for smaller data PrefixSpan works better than SPAM.

Keywords: Data Mining, Sequential Pattern Mining, SPAM, PrefixSpan.

1 Introduction

Discovery of sequential patterns from large dataset is first introduced by Agrawal[1]. Sequential pattern mining is one of the important fields in data mining because of its variety of applications in web access pattern analysis, market basket analysis, fault detection in network, DNA sequences etc. It plays vital role different areas.

Many algorithms were proposed for sequential pattern mining. These mining methods are broadly classified into two approaches as Apriori-based candidate generation method [2] and pattern growth method [3]. Apriori based approach uses the Apriori principle presented with association mining [4], which states that any of the super sequence of non frequent pattern can not be frequent. These candidate generation tests are carried out by GSP [2], SPADE [5], and SPAM [6].These algorithms perform well but scans database frequently and requires large search space. To increase the performance of these algorithms constraint driven discovery can be carried out. With constraint driven approach system should concentrate only on user specific or user interested patterns or user specified constraints such as

S. Unnikrishnan, S. Surve, and D. Bhoir (Eds.): ICAC3 2013, CCIS 361, pp. 147–155, 2013.

minimum support, minimum gap or time interval etc. With regular expression these constraints are proposed in SPIRIT [7].

Pattern growth approach is used by algorithm PrefixSpan[3], FreeSpan[8].These algorithms uses a method which examine only prefix subsequence and project only prefix subsequences. With these respective postfix subsequences are projected into the projected database. These work faster than GSP [2] Apriori algorithm for smaller database without using any constraint.

Sequential pattern mining algorithms efficiently discovers the recurring structures present in the protein sequences [11] [12]. In this paper we have used large protein sequences as the input for algorithm. These protein databases are easily available with the protein databank [9].The comparative study on these large protein sequences shows that Apriori–all SPAM algorithms works better than that of pattern growth PrefixSpan.

2 Related Work

For mining sequential patterns an algorithm was proposed for very large database called SPAM using bitmap representation. This uses the depth first search strategy with effective pruning mechanism. The algorithm uses bitmap representation of database for efficient support counting. It first generates lexicographic sequence tree and traverses the tree in depth first search manner. To improve performance it uses pruning techniques, S-Step and I-step pruning, which are also Apriori based. SPAM is composed with SPADE and PrefixSpan for the dataset used by IBM Assoc Gen Program [1].The comparison shows that SPAM performs well on large dataset due to its bitmap representation of data for efficient counting. SPAM outperforms SPADE by a factor of 2.4 on small dataset. With respect to space SPAM is less space efficient than SPADE [6].

When SPAM used with incorporating constraints to it as gap and regular expression then runtime performs get increased .It also achieves reduction in output size by adding different level of constraints to it. Experimental study show that at high minimum support greater than 30 % there is not much difference in runtime performance but with low minimum support, runtime performance increases. By introducing gap constraint into SPAM output space can be reduced. By increasing mingap from 0 to 5 the output size is significantly reduced. This shows that adding constraints like gap and regular expression increases efficiency of SPAM sequential pattern mining algorithm [10].

3 Problem Definition

A sequence is a collection of an ordered list of item set where item set is a collection of unordered, non empty set of items. For a given set of sequence, to find a set of frequently occurring sub sequences. Any subsequence is said to be frequent if its support value greater or equal to minimum support or threshold value. Support is with which occurrences that item is present in the given number of sequences.

For example

Table 1. Protein Sequences

Sequence No.	Sequence
1	MKKV
2	KVM
3	MKKVM

If minimum support value is 50%, then those sequence with support greater than or equal to minsup are called frequent subsequences. So in the given number of sequence MK, MKK, MKKV, KV, KKV, VM and KVM etc. are some of the frequent subsequences.

4 SPAM Algorithm

SPAM is one of Apriori–all type of sequential pattern mining algorithm. It assumes that entire database used for algorithm completely fit into memory. For the given sequences it generates a lexicographic tree [10]. The root of tree always starts with empty string. The children node of it are formed by sequence-extended or an item set-extended sequence. Let for sequence containing M, K, V, we have to generate a lexicographic analysis of tree then it starts with root node as { }.

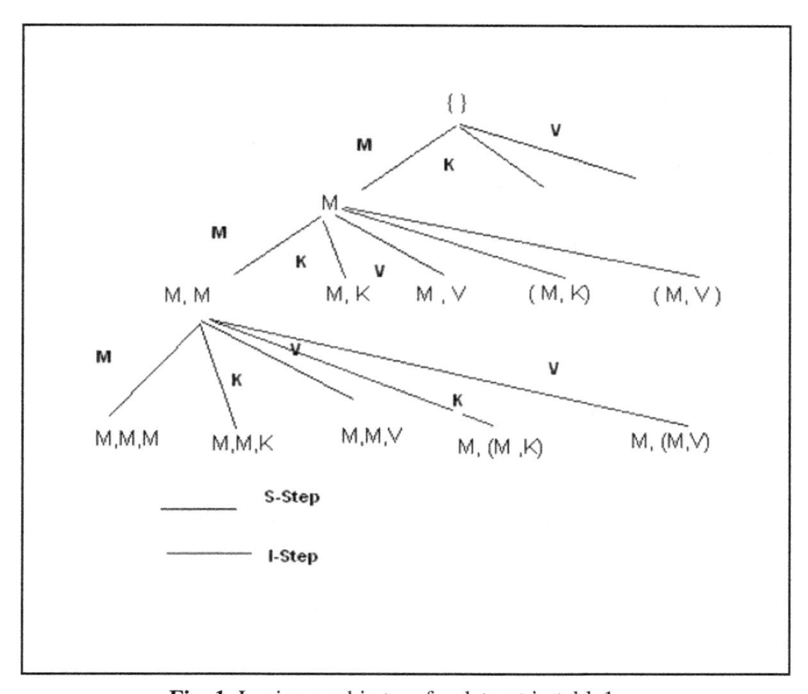

Fig. 1. Lexicographic tree for dataset in table1

At level 1 considers items M, K, V separately. Let us consider M item first, for candidate generation it will take sequence-extension step and generate different sequence as {M, M} {M, K} {M, V}.Similarly generates item set extended sequence by Item set extension step or I-Step as (M, K),(M,V) at level 2. Similarly with different items at level 3 it will generate a sequence of length 3.

Once the whole tree is generated for discovering or searching user specified set of subsequence algorithm traverses depth first search strategy. At each node, support of sequence is tested. Those nodes having support value greater than or equal to minimum support are stored and repeats DFS recursively on those nodes. Otherwise the node is not considered by the principle of Apriori.

For the support counting algorithm uses a vertical bitmap structure. For table 1 bitmap representation of dataset can be given as below. As bitmap representation of sequence length maximum is 5 so vertical bitmap consists of 5 bit and as the number of sequences are 3 in table 1 bitmaps with 3 slots.

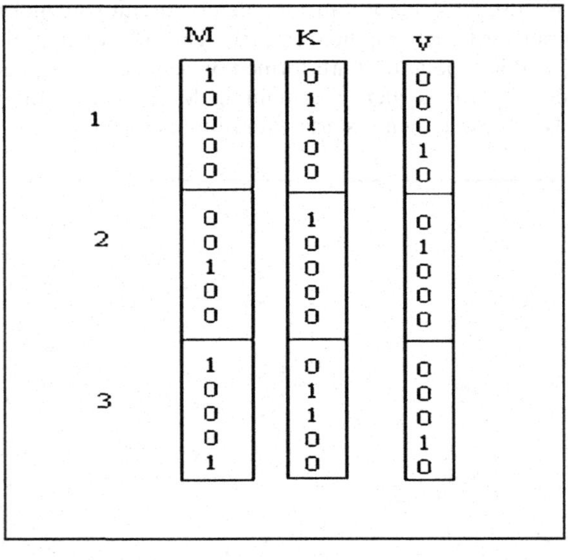

	M	K	V
1	1	0	0
	0	1	0
	0	1	0
	0	0	1
	0	0	0
2	0	1	0
	0	0	1
	1	0	0
	0	0	0
	0	0	0
3	1	0	0
	0	1	0
	0	1	0
	0	0	1
	1	0	0

Fig. 2. Vertical bitmap representation for items in dataset

For the bitmap presentation of S-Step, every bit after the first index of one is set to zero and every bit after that index position set to be one.

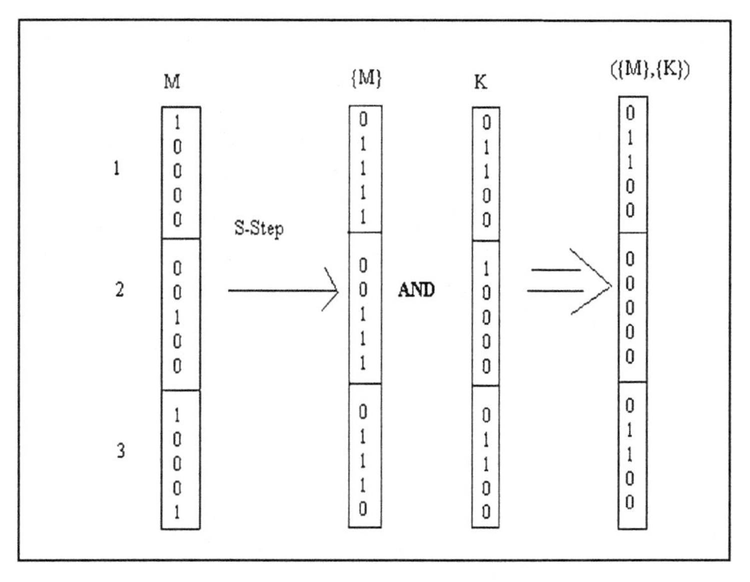

Fig. 3. Vertical bitmap representation for S-step with no gap constraint

For the bitmap representation of I-Step, newly added item set's bitmap are logically ANDed with the sequence generated from S-Step.

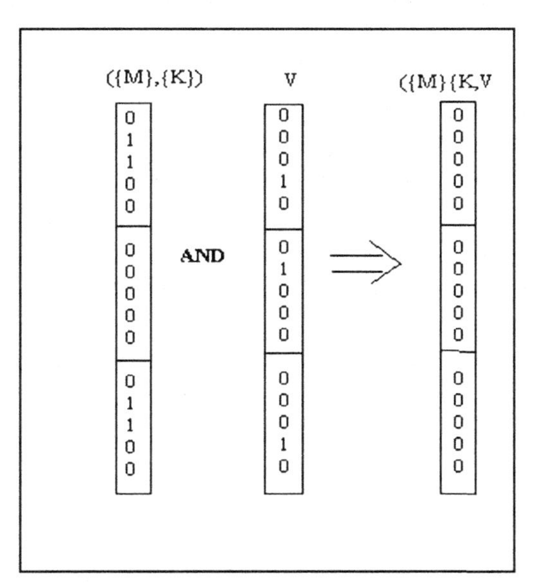

Fig. 4. Vertical bitmap representation for I-step

To improve the performance of an algorithm pruning techniques are used with S-extension and I-extension of a node. S-step pruning technique prunes S-step children. For pruning it applies Apriori principle i.e. Let for the sequence

({M}{M}),({M},{K}) and ({K}{V}) are given and if ({M}{K} is not frequent then ({M}{M}{K}),({M}{V}{K}) or ({M},{M,K}) or ({M},{V,K} are ignored. Similarly I-Step pruning technique prunes I-step children. For pruning at I-Step it applies similarly Apriori principle for item set i.e. Let for the item set sequences ({M,K}) and ({M,V}) if ({M,V}) is not frequent then ({M,K,V}) is also not frequent.

Constraints can be added as minimum and maximum gap in between the two items. With mingap and maxgap constraint, transformation step is modified to restrict number of position that next item can appear after the first item. If first item is {M} and next is {K} and constraint of mingap=1 and maxgap=1 then all the sequences {MVK},{MKK} etc are some of the sequences for above mentioned dataset.

Regular expressions are also can be used to limit the number of interesting patterns. Let M+K is given then it is considered that all those sequences containing M or K can be obtained.

5 PrefixSpan Algorithm

This algorithm uses the pattern growth method for mining sequence patterns. Here idea is that, instead of projecting sequence database by considering all possible occurrences of frequent subsequences, the projection is based only on frequent prefixes. Let {MKKV} be the sequence then prefixes of sequence may be <M>, <MK>, <MKK>.Let <M> be prefix then <_KKV> be the prefix of it.

The algorithm first finds out all length of sequences i.e. <M> <K> <V> etc .with respect to its support. Depending on the number of items present in given sequence, complete set of sequence patterns can be partitioned into those numbers of prefixes. For the above example all sets are partitioned into three different prefixes as <M> <K> and <V>.

Let <M> be prefix chosen then all {<KKV>}, {<KKVM>} be the projected postfix from the given postfixes all length-2 sequence patterns having prefix <M> can be found out like <MK> with its support. Similarly for length -3 pattern the process repeats.

Table 2. Number of Sequences Pattern from Given Prefix and Postfix

No.	Prefix	Postfix	Sequence Pattern
1	<M>	{-KKV},{-KKVM}	<M>,<MK>,<MKK>,<MKKV>,<MKKVM>
2	<K>	{-KV}{-V}{-VM}{-KVM}	<K><KK><KV><KKV><KVM><KKVM>
3	<V>	{-M}	<V> <VM>

The projected database keeps on shrinking for every step. But the major cost of PrefixSpan is during construction of projected database. If number of sequence patterns are more the cost will increased a lot as PrefixSpan constructs projected database for every sequence patterns.

6 Experimental Results

To test the efficiency of sequential pattern mining algorithms like SPAM and PrefixSpan we have taken a protein sequence as input dataset. This implementation is carried out on JAVA Standalone application. All experiment were performed on 1.46GHz Intel Pentium Dual CPU machine with 512MB of main memory Microsoft XP and J2SE Runtime Environment 1.5.

First experiment is performed to find the time taken by two algorithms with changing minimum support. Minimum support value varied from 10 to 50 and time measured for execution of an algorithm. Fig. 5 shows that time required for SPAM is too less then PreixSpan. Here the text files containing 528 numbers of sequences chosen.

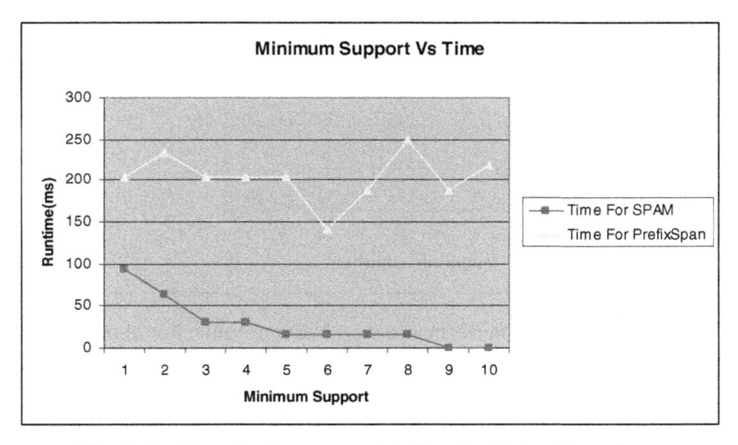

Fig. 5. Runtime Performance with Varying Value For minsup

Second experiment was performed for checking the runtime by varying number of sequence. Here number of sequence containing in the text file are only 2.

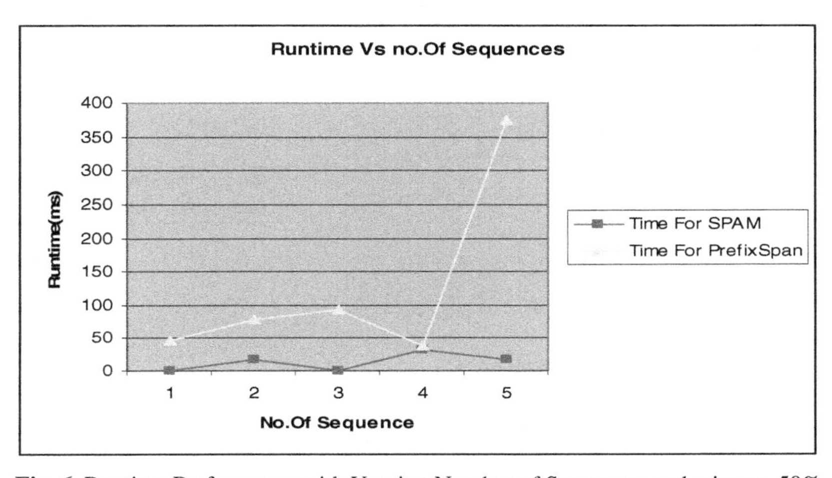

Fig. 6. Runtime Performance with Varying Number of Sequences and minsup=50%

If number of sequences is too less then it shows that PrefixSpan works better than SPAM.

Third experiment was performed by changing minimum support and memory utilization by two algorithms. Again minimum support value varied from 10 to 50.Memory utilization taken by both of algorithm is measured in bytes. As minimum support value gets increased memory utilization also increased.

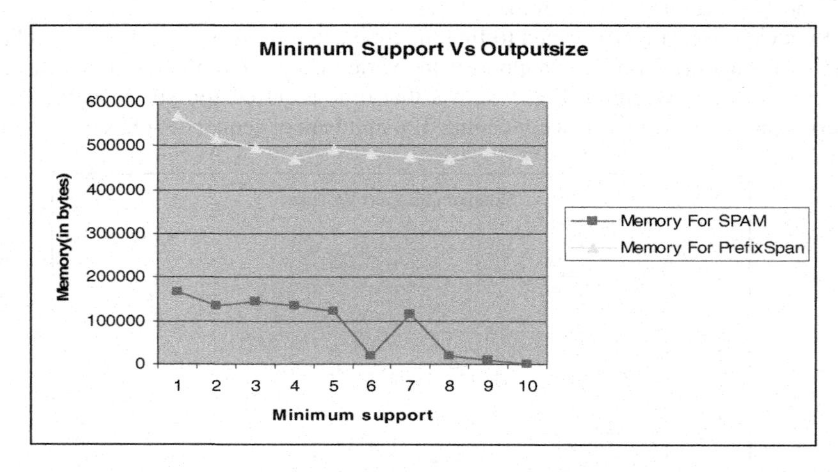

Fig. 7. Output Size Performance with Varying Value For minsup

Fourth experiment is performed for measuring output size by keeping minimum support constant as 50% and by changing number of sequence. As number of sequence are increased algorithm requires more memory. The major drawback of PrefixSpan is that lot of memory needed during performing postfix database for each frequently occurring prefix sequence.

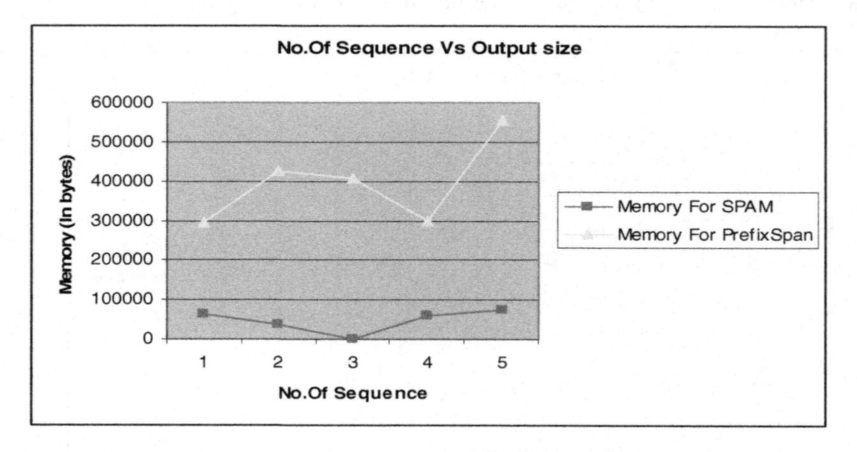

Fig. 8. Output Size Performance with Varying no. Of Sequences

7 Conclusion

For discovering frequently occurring sequential patterns from a large dataset can be possible with two approaches of mining algorithm as Apriori-all or Pattern growth. Experimental study shows that PrefixSpan performs better for smaller number of sequences. This is due to PrefixSpan does not have to waste time during creation of vertical bitmaps of items with in the sequences. It takes more time for SPAM though the number of sequence is less.

SPAM outperforms better than PrefixSpan for larger dataset. This is because of depth first search traversal of tree, vertical bitmap representation for support counting and pruning technique used during generation of candidate sequence. Though memory utilization by SPAM is higher but it is more time efficient than any of the pattern growth approach like PrefixSpan.

References

1. Agrawal, R., Srikant: Mining Sequential Pattern. In: Yu, P.S., Chen (ed.) Eleventh International Conference on Data Engineering (ICDE 1995), Taipei, Taiwan, pp. 3–14. IEEE Computer Society Press (1995)
2. Srikant, R., Agrawal, R.: Mining Sequential patterns: Generalization and performance improvements. In: Apers, P.M.G., Bouzeghoub, M., Gardarin, G. (eds.) EDBT 1996. LNCS, vol. 1057, pp. 3–17. Springer, Heidelberg (1996)
3. Pei, J., Han, J., Mortazavi-Asl, B., et al.: PrefixSpan: Mining Sequential Patterns efficiently by prefix projected pattern growth. In: ICDE 2001, Heidelberg, Germany, pp. 215–224 (2001)
4. Agrawal, R., Srikant, R.: Fast algorithm for mining association rules. In: Proceedings of International Conference on Very Large Data Bases (VLDB 1994), Santiago, Chile, pp. 487–499 (1994)
5. Zaki, M.: An efficient algorithm for mining frequent sequence. Machine Learning 40, 31–60 (2000)
6. Ayres, J., Gehrke, J., Yiu, T., Flannick, J.: Sequential Pattern Mining using Bitmap representation. In: Proceedings of ACM SIGKDD 2002, pp. 429–435 (2002)
7. Zaki, M.: Sequential mining in categorical domains-Incorporating constraints. In: Proceeding of CIKM 2000, pp. 422–429 (2000)
8. Han, H., Pei, J., Mortazavi-Asl, B., Chen, Q., Dayal, U., Hsu, M.-C.: FreeSpan: Frequent Pattern projected Sequential Pattern mining. In: Proceedings of 2000 International Conference on Knowledge Discovery and Data Mining, pp. 355–359 (2000)
9. `http://www.pdb.org`, `http://www.ebi.ac.uk/pdbc`, `http://www.rcsb.org`, `http://www.pdbj.org`
10. Ho, J., Lukov, L., Chawla, S.: Sequential Pattern mining with constraints on large protein databases. In: ICMD (2005)
11. Tao, T., Zhai, C.X., Lu, X., Fang, H.: A study of stastical methods for function prediction of protein motifs
12. Wang, M., Shang, X.-Q., Li, Z.-H.: Sequential Pattern Mining for Protein Function Prediction. In: Tang, C., Ling, C.X., Zhou, X., Cercone, N.J., Li, X. (eds.) ADMA 2008. LNCS (LNAI), vol. 5139, pp. 652–658. Springer, Heidelberg (2008)

Business Oriented Web Service Processing and Discovery Framework

Debajyoti Mukhopadhyay, Falguni Chathly, and Nagesh Jadhav

Department of Information Technology
Maharashtra Institute of Technology
Pune 411038, India
{debajyoti.mukhopadhyay,chathly.falguni,nagesh10}@gmail.com

Abstract. Information Technology business needs are refining virtually for meeting Client's requirement, since the beginning of the computer age. Importance of asking for service and time to get serviced is same. Most credible is how fast and accurate service reaches to Client. In order to provide fast and reliable service to the Client Service Discovery need to be superlative. Various researches are made using different computing methods and algorithms for discovering the required Service from the repository. In customary scenario, Client tries to search the repository of Web Service based on the keyword matching but the problem arises when no such service is available. With this proposed work, we introduce Business Oriented Web Service Processing and Discovery Framework to deal with the problem of undefined Web Service. As a start, request from Client is semantically analyzed and modeled to follow Hierarchical Task Network (HTN) ontology for such request is generated and if no match is found, relevant services are fetched based on relevance ranking and resulted to Client.

Keywords: Web Services, Web Service Discovery, Hierarchical Task Network, Ontology, Semantic Analysis.

1 Introduction

Concepts of interoperability and distributed technologies rule the world as best innovations till date in the field of Web Computing. Web service is introduced as the refinement of knowledge and after understanding human perceptions to search for his requirement. Orientation of Web Services, presented as reusable components has become relatively a great support to IT Market. Client's requirement and to-the-date fulfillment of requirement by Providers is what gets important and also Mainer approaches have been introduced to deal with it. Providing Service, as what actually present with Service Agents is their task, but dealing with what does not exist and providing relatively best match becomes superior these days. Effectively, Web-services allow businesses to share the information they master in and also the ones stored in their computer applications with other applications in the company or with those run by clients, providers and partners. By linking these intermediaries online,

S. Unnikrishnan, S. Surve, and D. Bhoir (Eds.): ICAC3 2013, CCIS 361, pp. 156–166, 2013.

Organization can significantly increase the efficiency--and thus recommends lower cost and enterprise benefits. Applications can be Software applications, which are web services that can be invoked remotely by users or programs, or Programs and scientific computations which are important resources in the context of emerging collaborative business processes, eventually even more vital than data.

Business oriented Analyzing and specifying Client's requirement of a Web Service, processing it, ranking and finally servicing with most appropriate outcome, is need of business oriented ontology. Dealing with technically complex business needs is vitally tuff and leads to online concept to fail sometimes. Business oriented ontology, is the derivation of business requirement, from the user's specified requirement to build Semantic tree formation, where business required Web Services lies on leaf node, which is then relatively matched and ranked for terming the output well. Web Services are available as reusable components, with in Service Registry, sometimes when well matched, they are provided as the ones.

Services can be defined as a technology for offering software services or general-purpose architecture that will trigger an essential shift in the way that all distributed systems are created. Services provides strong interface for collection of operations being accessed on network [1].

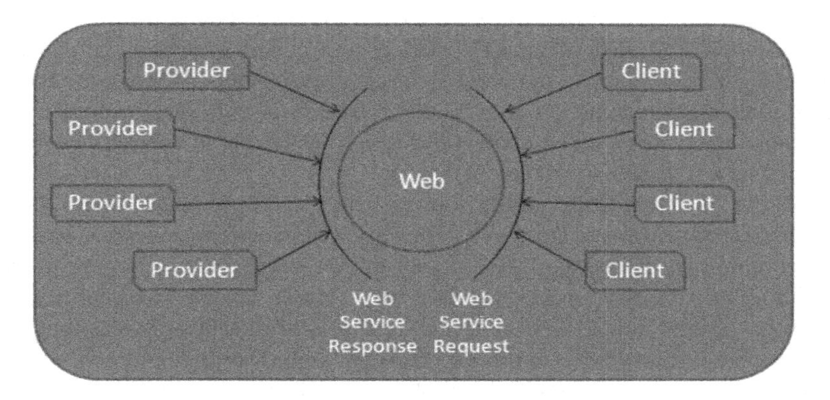

Fig. 1. Web Service Framework

Semantics of any language can be defined as the meaning of each included word or symbols in sentence an also the relation that exist between those. Semantics play an important role in the complete lifecycle of Web services as it is able to help service development, improve service reuse and discovery, significantly facilitate composition of Web services and enable integration of legacy applications as part of automatic business process integration [2]. Hierarchical Task Networks (HTN) deals with planning to perform tasks rather than to attain results. It acts as a helping mechanism to build a modeled way to provide good and accurate service to the client. HTN gives detailed understanding of the existing relationships among the lower decomposed matches.

In this paper we introduce, the method of retrieving most suitable Web Service to serve client. Our research in this area, relates the human behavior of asking for Service to the way it actually gets processed. Processing Web Service is to be meant for best suitable output. At end, we conclude, with the most suitable tools that are used for creating this ontological idea.

2 Related Work

Predicting Web services play a significant role in e-business, ecommerce applications. Web service applications are platform independent and interoperable. Finding most suitable web service from large collection of web services is very crucial job. Web services are published in UDDI and search operations are performed on the UDDI. Generally it is a two-step process, first step involves web service matching i.e. looking for required by user and second step involves web service selection i.e. choosing the best and right service. Many researchers have provided different approaches for searching a web services. [3]

Wenge Rong and Kecheng Liu have proposed context aware web service discovery approach. This approach helps in request optimization, result optimization and personalization. The web service discovery context is defined as any information that explicitly and implicitly affects the user's web service request generation. It is divided into two categories as Explicit and implicit. User is responsible for providing Explicit context during matchmaking process. Implicit context is collected in automatic or semi-automatic manner. Implicit context plays significant role in web service discovery as users are not directly involved. [4]

Falak Nawz, Kamram Qadir and H. Farooq Ahmad proposed push model for web service discovery using semantic web. They used semantic web service matching using Web Ontology Language (OWL-S). In this model service requesters are provided with service notification prior to discovery. It also involves concept matching for ranking published web services. It is a two phase process which involves subscription phase, which starts when a subscriber registers himself onto registry for notification of required services and notification phase, which starts when a new service is published on registry. [5]

Web service discovery based on Keyword clustering and concept expansion is suggested by J. Zhou, T. Zhang, H. Meng, L. Xiao, G. Chen and D. Li. They used Pareto principle to calculate similarity matrix of words. That is in turn used for semantic reasoning to find requested web service.Bipartite graphs are used to calculate degree of matching between the service requests and available services.[6]

A. Paliwal, N. Adam and C. Bornhovd suggested an enhanced web service discovery approach. For web service discovery they proposed concept combining ontologies and latent semantic indexing. They used domain ontology to build the service request vector, the training set of the LSI classifier are built by extracting features from selected WSDL files. They used vector space model and the cosine similarity measure to find similarity between the web services and to retrieve relevant service descriptions from WSDL file. [7]

Guo Wen-yue, Qu Hai-cheng and Chen Hong have proposed layered architecture for web service discovery. In this architecture each layer performs a search and applies a filter so as to minimize search area. They have applied this approach on intelligent automotive manufacturing system. Three layers for service matching include service category matching, service functionality matching and quality of service matching.[8]

Georgios Meditskos and Nick Bassiliades provided semantic web service discovery framework using OWL-S. Based on object based matching technique authors suggested a web service matchmaking algorithm which extends object-based matching techniques. It allows retrieval of web services not only based on subsumption relationships, but also using the structural information of OWL ontologies. Structural case based reasoning done on web service profiles provide classification of web services, which allows domain dependent discovery. Service matchmaking is performed on Profile instances which are represented as objects considering domain ontologies. [9]

In order to overcome the drawback of centralised web service discovery F. Emekci, O. Sahin, D. Agrawal and A. Abbadi proposed a peer-to-peer framework for web service discovery which is based on process behaviour. Framework considers how service functionality is served. All available web services are represented using finite automaton. [10]

U. Chukmol, A. Benharkat and Y. Amghar have proposed feedback based web service discovery. In this method they used the feedback from the users who already have used web service. They proposed collaborative tagging system for web service discovery. Tags are labels that a user can associate to a specific web service. Web services are tagged by different keywords provided by different users. For each tag, tag weight is assigned. [11]

3 Proposed Business Oriented Web Service Processing and Discovery Framework

Based on our hand-on experiences in developing a Service-based query-processing environment for retrieving best match Service, we present a framework, which processes the higher level business oriented entries of requirement in to decomposable lower levels and processes it to get best match. Refer Fig.2.We consider the example of train booking (elaborate). Proposed Framework moves as follows.

3.1 Procedure for Creation of Business Oriented Framework

With the growing dependence on the Web as a medium of business progressions, there is an increasing need to dynamically match the requirements. A procedure that can be followed to present the Web Service needs by Client till the appropriate Web Service gets facilitated.

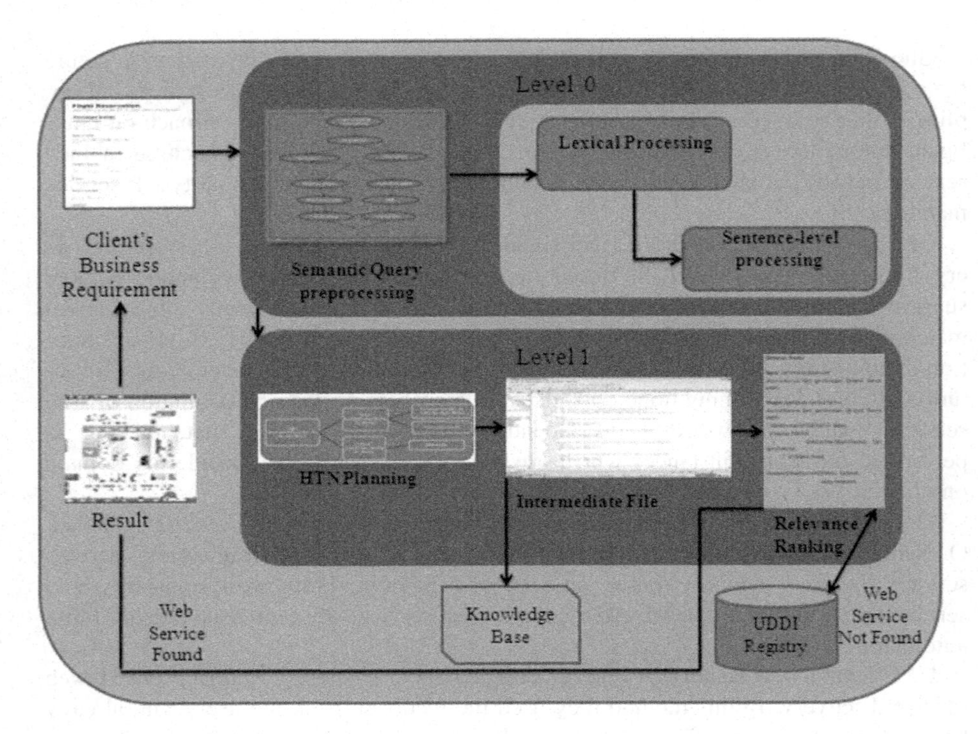

Fig. 2. Business Oriented Web Service Processing and Discovery Framework

3.1.1 Business Oriented GUI Interaction

Client requirement for Service to be facilitated matters a lot. For Semantic Web Services the terms used by Client to search for Service becomes the base to make search forward. The input defined by Client is in the higher level form and the same values become intermediate input to next phase. Client defines the input in the form of query defined in the form. Here in example, the Passenger Details and Reservation Details entries of the form Flight Reservation, provides the business dominated GUI format. The provided information in the form gets forwarded as input to ontology formation. According to the Web Interface binding contract agreements the details of Web Service is taken and processed.

3.1.2 Creating Business Oriented Ontology

The Ontological approach gives detailed view of tasks to be done and existing relationships between them. Each service is concomitant with a service domain and has an interface and theoretically many implementations. Controlling complex domains like robot movements, browsing Internet, etc., require step-wise and graphical representation of the levels and steps to be dealt with.

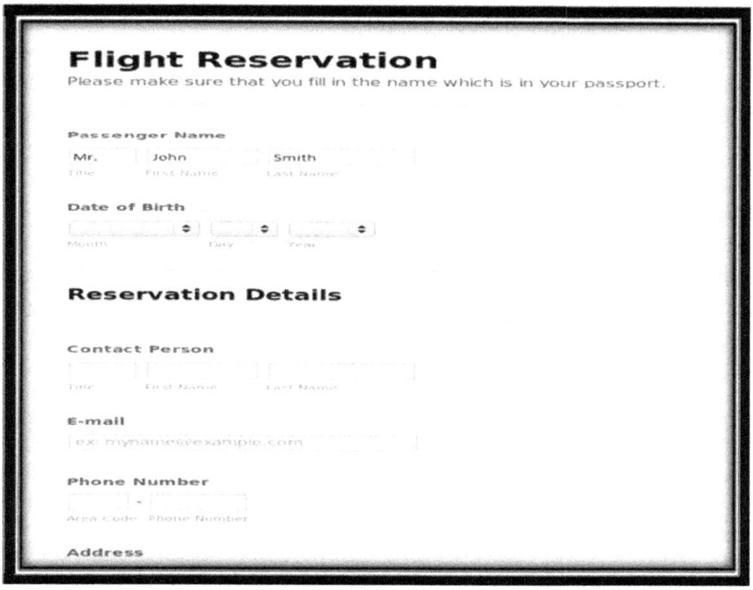

Flight Reservation
Please make sure that you fill in the name which is in your passport.
Passenger Name
Mr.
John
Smith
Title
First Name
Last Name
Date of Birth
Month
Day
Year
Reservation Details
Contact Person
Title
First Name
Last Name
E-mail
ex: myname@example.com
Phone Number
Area Code
Phone Number
Address

Fig. 3. Snapshot of Flight Reservation Form

3.1.3 Semantic Query Preprocessing

Semantic analysis must follow following steps: 1. It must map individual words into appropriate objects in the knowledge base. 2. It must create the correct structures to correspond to the way the meanings of the individual words combine with each other.

Semantic Processing
Lexical Processing
Sentence Level Processing

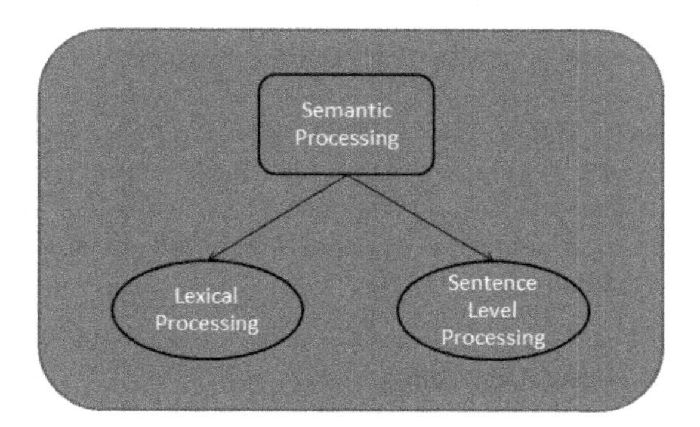

Fig. 4. Semantic Query Preprocessing

- Lexical Processing

The first step in the semantic processing system is to look up the individual words in a dictionary and extract their meaning. Based on the business input lexical processing phase will try to find meaning of the word by referring knowledge base. This further can be used to generate the graph.

- Sentence-level processing

Sentence level processing includes semantic grammars, case grammars, conceptual parsing. Semantic grammars combine syntactic, semantic and pragmatic knowledge into a single set of rules in the form of a grammar. The result of parsing with such a grammar is a semantic.

3.1.4 Forming HTN Planning and Searching Paradigm

For understanding the input made by Client for Web Service search is semantically bifurcated and split. Branch of Artificial Intelligence called HTN (Hierarchical Task Network), an analytical approach to automated planning in which the dependency among actions can be given in the form of networks. HTN is suitable for domains where tasks are naturally organized in a hierarchy.

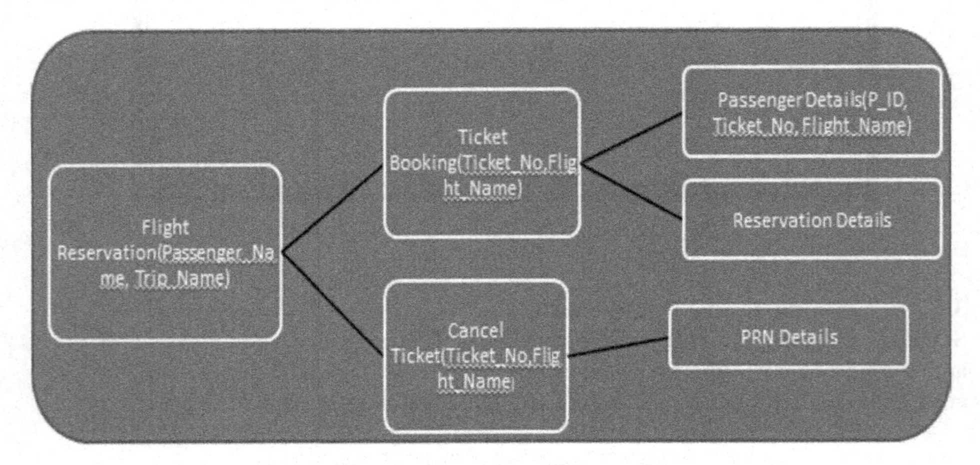

Fig. 5. HTN for Flight Booking Example

Planning of difficulties is identified in the HTN approach by providing a set of tasks that can be: abstract operators to start a plan (Flight Reservation), primitive tasks, that can be executed (Passenger Details); compound tasks, which can be complex task composed of a sequence of actions; goal tasks, which task of satisfying a condition (Flight Booked). Refer Figure 5.

3.1.5 Relevance Ranking Algorithm

Following Table 1 shows the relevance ranking algorithm for discovering the web service.

Table 1. Relevance Ranking Algorithm

```
Relevance Ranking
InputWeb Service to be responsed

Rank_Service(Service name)

Output Appropriate matched Service

Rank_Service( Service name)
{ Matches=searchUDDI(Service name);
 if Matches FOUND
  {
   If Matches are Exact
    return Matches;
     else
     Relavent_Match=Relevance_Calculation(Matches);
      {
        Finalmatch=RankServices(Relevant_Match);
                return finalmatch;
       }
     }
  }
```

4 Step for Business Oriented Web Service Processing and Discovery

Table 2. Step for Business Oriented Web Service Processing and Discovery

Step 1. Service Client forwards business request using Web Interface

Step 2. Client Request gets forwarded to Query Preprocessing phase (level 0).
  a. It maps individual words into appropriate objects in the knowledge base.
  b. It creates the correct structures according to the meaning of words.

Step 3. Query preprocessing phase result is forwarded to Level-1block for further processing
  a. Appropriate graph is generated using Graph-viz
  b. Based on leaf node values UDDI is searched for Web Service matching
  c. If exact matches found result is returned back to Client
  d. If not then Result of HTNgets stored into intermediate file for further processing
  e. Intermediate file is processed using Relevance ranking algorithm to generate final result.
  f. Generated results are ranked based on similarity measure.

Step 4. Final result is returned back to Client

5 Implementation

Business oriented web service processing and discovery framework can be implemented using following tools

1. Eclipse

Eclipse is used as front end for the development of client application. The user is supposed to enter the business details for discovery. Appropriate Soap message generated by client sent to UDDI for web service discovery. We used Eclipse Version: 3.4.2 with Axis2-1.4.1 and OWL-S plug-in.

2. OWL-S

OWL-S is a web service description language that develops web service descriptions based on the WSDL using semantic information from OWL ontology's. OWL-S plug-in is configured with Eclipse. We used Code-Lib-Feature-1.1 and OWL-SEditor-1.1 for implementation purpose. Figure 6

3. Graph-viz

Graph-viz is open source graph visualization software for representing structural information in the form of diagrams. We have used Graph-viz for generating a graph based on the results generated from semantic preprocessing phase.

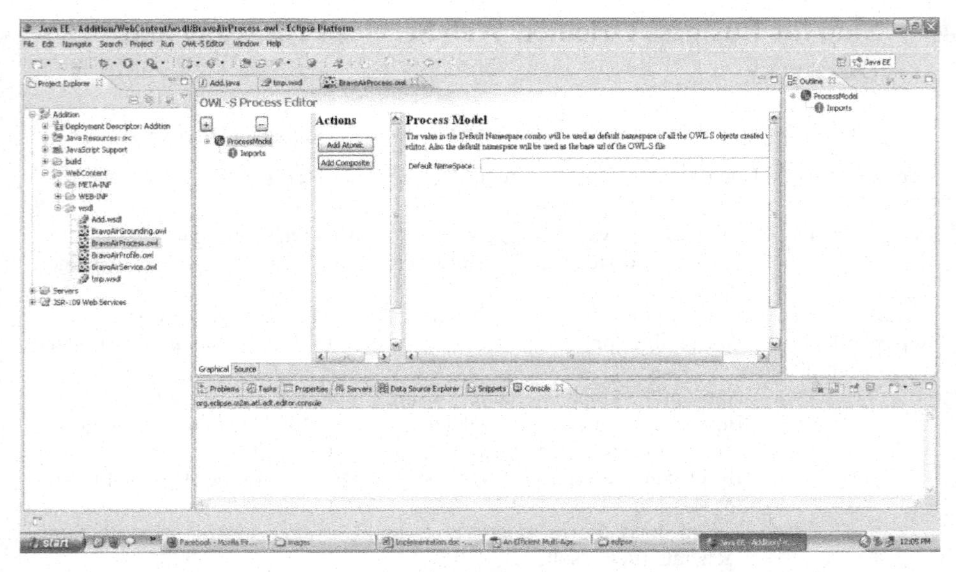

Fig. 6. OWL-S Editor

4. Apache Tomcat and jUDDI

For implementation we used Apache Tomcat as server and jUDDI as web service repository. We have configured jUDDI with MySql 5 as a backend.

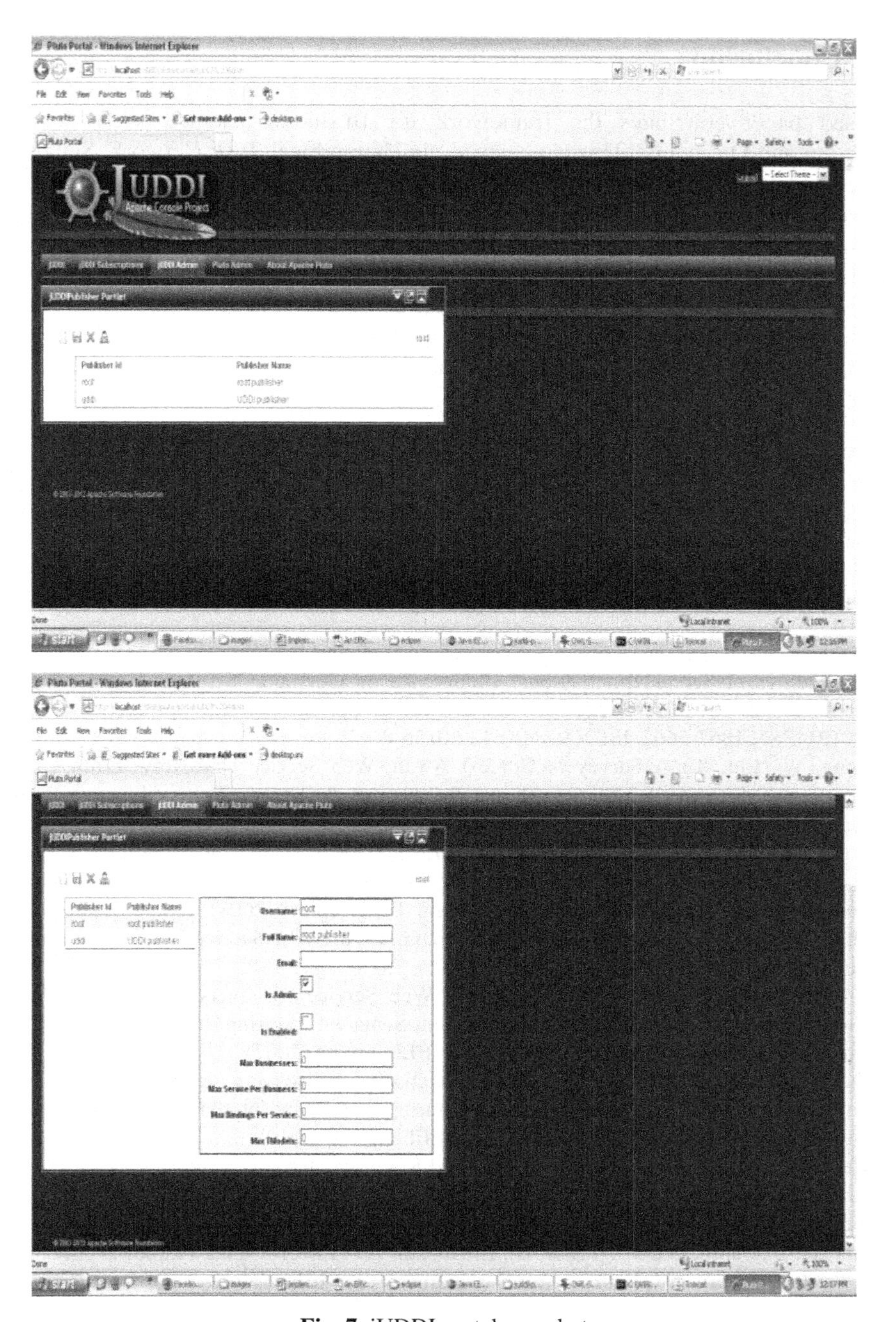

Fig. 7. jUDDI portal snapshot

6 Conclusion

Proposed paper concludes the framework for Business Oriented Web Service Processing and Discovery. This is achieved via Hierarchical Task Network, Semantically analysis and using algorithm of Relevance Ranking. It is highly suitable when on urgent bases, organization require searching for whether Web Service is actually available or what are the relevant available Service in repository for making deals well. It uses ontological modeling along with HTN planning network. Semantic analysis phase strengthens the searching process if exact match is not found. As a future work, more advanced modeling approach or more advance tools can be used and also can provide effectively higher measures of increasing QoS.

References

1. Wu, C., Chang, E., Aitken, A.: An empirical approach for semantic Web services discovery. In: 19th Australian Conference on Software Engineering (2008)
2. Mukhopadhyay, D., Chathly, F.J., Jadhav, N.N.: Framework for Effective Web Service Cloud (WS-Cloud) Computing. A Journal of Software Engineering and Applications (2012)
3. Mukhopadhyay, D., Chougule, A.: A Survey on Web Service Discovery Approachesp. In: Proceedings of Second International Conference on Computer Science, Engineering & Applications, ICCSEA 2012, May 25-27, vol. 8, pp. 1001–1012. Springer, Germany (2012) ISSN 1867-5662, ISBN 978-3-642-30156-8
4. Rong, W., Liu, K.: A Survey of Context Aware Web Service Discovery: From User's Perspective. In: Fifth IEEE International Symposium on Service Oriented System Engineering (2010)
5. Nawaz, F., Qadir, K., Farooq Ahmad, H.: SEMREG-Pro: A Semantic based Registry for Proactive Web Service Discovery using Publish Subscribe Model. In: Fourth International Conference on Semantics, Knowledge and Grid. IEEE Xplore (2008)
6. Zhou, J., Zhang, T., Meng, H., Xiao, L., Chen, G., Li, D.: Web Service Discovery based on Keyword clustering and ontology
7. Paliwal, A.V., Adam, N.R., Bornhovd, C.: Web Service Discovery: Adding Semantics through Service Request Expansion and Latent Semantic Indexing. In: International Conference on Services Computing (SCC 2007). IEEE Xplore (2007)
8. Guo, W.-Y., Qu, H.-C., Chen, H.: Semantic web service discovery algorithm and its application on the intelligent automotive manufacturing system. In: International Conference on Information Management and Engineering. IEEE Xplore (2010)
9. Meditskos, G., Bassiliades, N.: Structural and Role-Oriented Web Service Discovery with Taxonomies in OWL-S. IEEE Transactions on Knowledge and Data Engineering 22(2) (February 2010)
10. Emekci, F., Sahin, O.D., Agrawal, D., El Abbadi, A.: A Peer-to-Peer Framework for Web Service Discovery with Ranking. In: Proceedings of the IEEE International Conference on Web Services, ICWS 2004 (2004)
11. Chukmol, U., Benharkat, A., Amghar, Y.: Enhancing Web Service Discovery by using Collaborative Tagging System. In: 4th International Conference on Next Generation Web Services Practices. IEEE Xplore (2008)

Constraint Driven Stratification of RDF with Hypergraph Graph (HG(2)) Data Structure

Shiladitya Munshi[1,4], Ayan Chakraborty[2,4], and Debajyoti Mukhopadhyay[3,4]

[1] Meghnad Saha Institute of Technology, Kolkata, India
shiladitya.munshi@yahoo.com
[2] Techno India College of Technology, Kolkata, India
achakraborty.tict@gmail.com
[3] Maharastra Institute of Technology, Pune, India
debajyoti.mukhopadhyay@gmail.com
[4] Web Intelligence and Distributed Computing Research Lab, Golfgreen, Kolkata: 700095, India

Abstract. Current paper introduces a Constraint Driven Stratificatified Model for RDF with Hypergraph-Graph (HG(2)) model of data storage which can be represented as a hybrid data structure based on Hypergraph and Graph. In general, the schema component of RDF i,e RDFS is unstratified in nature. In order to establish a sustained information integration of RDF and Topic Map or other stratified semantic metamodel, imposing some constraints on RDF (without any semantic loss) evolves as an imperative pre condition. While HG(2) data structure is claimed to realize complex combinatorial structures, the current investigation reports a novel HG(2) based model for constraint driven stratification of RDF without any loss of semantic expressiveness.

Keywords: RDF, RDFS, Stratification, HG(2).

1 Introduction

As proposed by Tim Berners-Lee et al [1] semantic web is viewed as the web of information for machines to understand. With the evolution of many semantic web standards like RDF, Topic Map etc, the disagreement of common information framework has always hindered the advancement of Semantic Web as a whole. The accessibility, readability and searchability of web information have been considerably limited due to the coexistence of these two types of information frameworks.

Since last few years, many efforts have been found [2 , 3] which address this problem. Many inter-operable translation mechanisms for semantic web information models have been designed and developed keeping RDF and Topic Map in focus, but none of them were able to overcome the intricacies of information exchange among the said two frameworks completely. Recent researches [4, 5] investigates this issue from a different perspective where a direct translation mechanism from one format to other and vice-versa is not considered to be a

S. Unnikrishnan, S. Surve, and D. Bhoir (Eds.): ICAC3 2013, CCIS 361, pp. 167–180, 2013.

proper solution to the problem of RDF-Topic Map interoperability, rather integration of information or semantic expressions from both the framework is believed to be the best possible and meaningful way of go beyond the walled garden of RDF and Topic Map framework.

Inherently, RDF follows a unstratified model where as Topic Map is always a stratified one. The difference of nature of stratification becomes a costly affair, while porting the semantic elements of one model to the other. Hence stratification of RDF with some constraints is essential in order to restore semantic equivalence between RDF and Topic Map.

Motivated by this requirement, the objective of the current paper has been set to model the mechanism of constratnt driven stratification of RDF based on some suitable data structure capable of representing combinatorial and complex relationships.

In the current paper, a data structure "Hypergraph - Graph" (HG(2)) has been reported. HG(2) takes the advantages of both the data structures of Graph and Hypergraph. This data structure can model a problem space into two logical partitions with different levels of complexities. The more complex level can be represented by Hypergraph in order to take the leverages of its power to designate the relationships among a group of objects. The other level which is characterized by comparatively lesser complexity and better orderedness, could be represented by Graph in order to take the advantage of its naturalness and simplicity.

The present research focuses the potential of HG(2) in constraint driven stratification of RDF. With a high objective of modeling the semantic lossless stratification of RDF through HG(2), the present paper has been organized as follows -

Section 2 reports the notions of Stratification in model architectures in general with a specific discussion on stratification nature of RDF. Next, Section 3 proposes HG(2) and related theoretical aspects formally with definitions and illustrative examples. Section 4 reports the constraint driven HG(2) based stratification model of RDF which is followed next by conclusion of the current investigation in section 5.

2 Notion of Stratification in Model Architectures

In some model frameworks, the objects are divided into strata based on classification. Each object belongs to exactly one stratum, and all objects in a stratum must be in-stances of classes in the next higher stratum. In principle, a metamodel could be stratified according to any other intransitive relation, but in practice there appears to be little use for such variations. At most, strata are defined based on specializations of the classification relation, as in [6].

Those classes, considered as objects in their own right, are in turn instances of metaclasses in the next stratum, and so on. If the meta-tower is finite, top stratum is eventually reached where objects are not instances of any classes; they just are. No relationships beside classification are allowed across strata. The Unified Modeling Language (UML), for example, employs a four-layer stratified

Layer / Stratum	Description	Examples
M_3: meta-metamodel	The infrastructure of the metamodeling architecture.	MetaClass, MetaAttribute, MetaOperation
M_2: metamodel	Objects that specify the language for defining the model.	Class, Attribute, Operation, Artifact
M_1: model	Objects that specify the structure of the information domain; user-defined classes.	Person, name, writeThesis, Dissertation
M_0: user objects	Objects that represent individuals; most of the actual data resides here.	<com.ideanest.Person 14218>, "Peter", void writeThesis(), <Dissertation final copy>

Fig. 1. Stratified model of UML

metamodel. The definition of the layers is detailed in Fig 1. Objects in M_n are instances of classes in M_{n+1}, for $0 \leq n \leq 2$. Objects in M_3, the top layer, are not classified; they are completely self-defined.

The alternative to a stratified metamodel is an unstratified one, where classes and their instances are intermixed and can be directly related. Classification, though it may be a privileged relationship, does not partition the model. A trademark characteristic of unstratified metamodels is that they do not arbitrarily cap the classification hierarchy. Instead, the most meta class is an instance of itself, forming a classification loop that provides a finite representation of an infinite classification tower.

The arguments advanced against non-fixed layer metamodeling architectures include following:
(a) Unstratified metamodels are non-standard and difficult to understand.
(b) If classes are considered as sets, then having a set be a member of itself may break the Foundation Axiom of set theory and is suspiciously close to the Russell Paradox.
(c) The semantics of unstratified metamodels are difficult to formalize and lend themselves to the layer mistake problem.
Based on the above discussions, following subsection reports the unstratified nature that is being exhibited by RDF.

2.1 RDF as a Unstratified Model

RDF Schema or RDFS governs the RDF infrastructures. It exhibits RDFs simplistic notion of *typing* by introducing the *Class* concept. The type of every *resource* must be an instance of *Class* (that is, a *resource* of type *Class*). For example, since *type* is of type *Property*, *Property* is of type *Class*. *Class* is itself of type *Class*, giving RDF an unstratified type model. RDF Schema also insists that every *resource* must be the instance of at least one *Class*, so it introduces *Resource*, the class of all resources. Finally, the class *Literal* is the class of all literals.

It is noteworthy that predefined properties can be used to model instance of and subclass of relationships as well as domain restrictions and range restrictions of attributes. A specialty of RDFS is that properties are defined globally and are not encapsulated as attributes in class definitions. Therefore, a frame or object-oriented ontology can only be expressed in RDFS by reifying the property names with class name suffixes.

RDF distinguishes between a *class* and the set of its instances. Associated with each *class* is a set, called the *class extension* of the *class*, which is the set of the instances of the *class*. Two classes may have the same set of instances but still they remain as different classes. It is possible for these classes to have exactly the same instances, yet to have different properties. Furthermore, a class may be a member of its own class extension and may be an instance of itself, clearly establishing the unstratified nature. For example, the group of *resources* that are RDF Schema classes is itself a *class* called $rdfs : Class$.

On the basis of previous discussions, the unstratified nature of RDF is justified. In order to impose stratification on RDF, Inter stratum constraints are to introduced. The introduction of constraints readily forms a complex structure within RDF framework. The complex combinatorial structure of RDF with stratifying constraints has been investigated (in this paper) to be realized with HG(2) data structure which is described next.

3 Theories of Hypergraph-Graph (HG(2)) Data Structure

Hypergraph - Graph data structure denoted as HG(2) is conceptualized as a model to represent a complex problem space based on certain criteria. The criteria could be formalized as follows -

The problem space (PS) must logically be divided into two levels with different complexities, one (PSG) with relatively lesser complexities, better orderdness and bounded by formalized set of rules, and another (PSH) which could be characterized by greater complexities and absence of ordered rule sets.

The some or all interrelationship between objects of PSH must be dictated by the objects of PSG and even the behavior of some or all objects of PSH must be defined by the objects of PSG. Here the term "object" is being used informally and must not necessarily indicate any Object Oriented paradigm.

The type of problem space mentioned above could easily be identified in real life scenarios. For example, the organizational behavior within an establishment supports the criteria of identifying PSG and PSH. The PSG layer could successfully represent the rule based relationships between any two capacities, the scope of responsibilities, range of visibilities etc, while PSH layer could represent the individual employees of that organizations, their group relationships etc. where any rule book hardly exists beyond the scope of human emotions and reactions. Here the behavior of individual employees and the group relationships are dependent on the PSG objects.

As the complex real life combinatorial structures are not rare at all, proposed HG(2) has an intrinsic objective to represent PSH with Hypergraphs and PSG with Graphs. The inter dependencies of PSH and PSG form the basis of evolution of the behavior of HG(2) as a whole.

On this background, following subsection revisits (originally presented in [18]) some essential concepts of Hypergraphs which will be exploited in designing HG(2). The discussion on Graph concepts that are useful for HG(2) are not studied (only mentioned) here due to its obviousness and generality.

3.1 Hypergraph Concepts

A Hypergraph [7 - 18] is a pair $H = (V, E)$, where $V = v_1, v_2, \cdots, v_n$ is the set of vertices (or nodes) and $E = E_1, E_2, \cdots, E_m$, with $E_i \subseteq V$ for $i = 1, \cdots, m$, is the set of Hyperedges. Clearly, when $|E_i| = 2$, $i = 1, \cdots, m$, the Hypergraph is a standard graph.

While the size of a standard graph is uniquely defined by n and m, the size of a Hypergraph depends also on the cardinality of its Hyperedges; hence the size of H could be defined as the sum of the cardinalities of its Hyperedges: $Size(H) = \sum_{E_i \in E} |E_i|$.

A directed Hyperedge or Hyperarc is an ordered pair, $E = (X, Y)$, of (possibly empty) disjoint subsets of vertices; X is the *tail* of E while Y is its *head*. In all the subsequent discussions, the *tail* and the *head* of Hyperarc E will be denoted by $T(E)$ and $H(E)$, respectively.

A directed Hypergraph is a Hypergraph with directed Hyperedges. In the following discussions, directed Hypergraphs will simply be called Hypergraphs. An example of Hypergraph is illustrated in Fig. 2. It is to be noted that Hyperarc E_5 has an empty head, whereas for edge E_1, nodes with values A and B are heads and nodes with values D, E and F are tails.

As for directed graphs, the incidence matrix of a Hypergraph H is a nxm matrix $[a_{ij}]$ defined as follows (see Fig. 2):

$$a_{ij} = \begin{cases} -1 & \text{if} \quad v_i \in T(E_j), \\ 1 & \text{if} \quad v_i \in H(E_j), \\ 0 \ otherwise & \end{cases}$$

Clearly, there is a one-to-one correspondence between Hypergraphs and (-1, 0, 1) matrices.

Paths and Connection in the Context of Hypergraphs. A path P_{st}, of length q, in Hypergraph $H = (V, E)$ is a sequence of nodes and Hyperarcs $P_{st} {=} (v_1 = s, E_{i_1}, v_2, E_{i_2}, \cdots, E_{i_q}, v_{q+1} = t)$, where:
$s \in T(E_{i_1}), t \in H(E_{i_q})$, and $v_j \in H(E_{i_{j-1}}) \cap T(E_{ij}), j = 2, \cdots, q$.

Nodes s and t are the origin and the destination of P_{st}, respectively, and it is said that t is connected to s. If $t \in T(E_{i_1})$, then P_{st} is said to be a cycle; this is in particular true when $t{=}s$. In a simple path all Hyperarcs are distinct, and

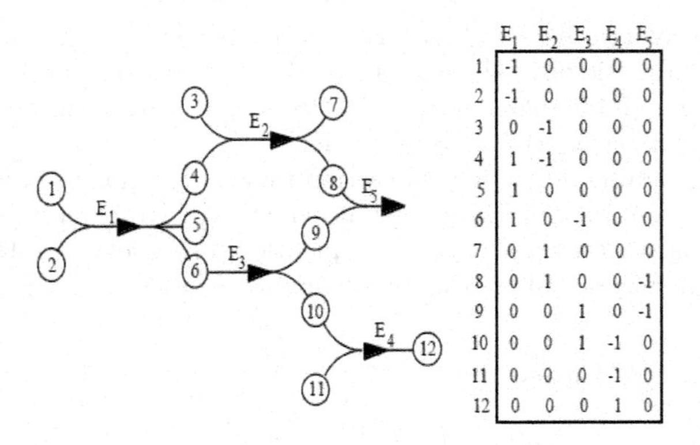

	E_1	E_2	E_3	E_4	E_5
1	-1	0	0	0	0
2	-1	0	0	0	0
3	0	-1	0	0	0
4	1	-1	0	0	0
5	1	0	0	0	0
6	1	0	-1	0	0
7	0	1	0	0	0
8	0	1	0	0	-1
9	0	0	1	0	-1
10	0	0	1	-1	0
11	0	0	0	-1	0
12	0	0	0	1	0

Fig. 2. An illustration of Hypergraph and its corresponding Adjacency Matrix

a simple path is elementary if all nodes $v_1, v_2, \cdots, v_{q+1}$ are distinct. Similarly simple and elementary cycles may be defined. A path is said to be cycle-free if it does not contain any subpath which is a cycle.

In Fig. 3, node with value 8 is connected to node with value 1, while node with value 9 is not. The elementary path connecting node with value 8 to node with value 1 is drawn in thick line.

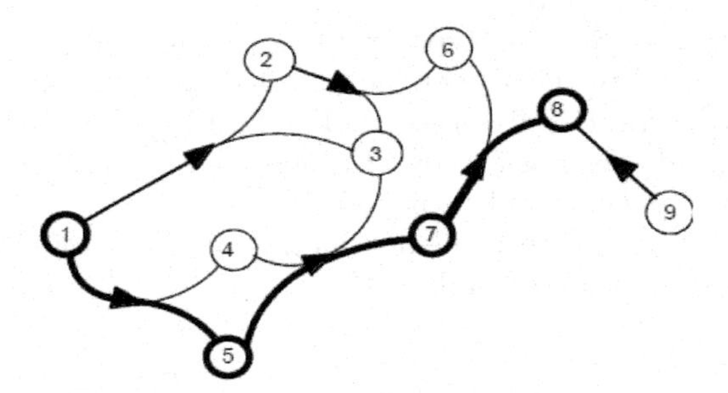

Fig. 3. A Hypergraph having a path P_{18}

Above discussion sets the right ambiance for introducing HG(2) which follows in next subsection.

3.2 Introducing HG(2)

A Hypergraph-graph data structure HG(2) is a triple denoted as $HG(2) = (H, G, C)$ where H is a Hypergraph, G is a graph and C is a set of *connectors*.

H is a Hypergraph defined as $H = (V^h, E^h)$, where $V^h = v_1^h, v_2^h, \cdots, v_n^h$, $n = 2, 3, \cdots$; and $E^h = E_1^h, E_2^h, \cdots, E_m^h$, $m = 2, 3, \cdots$ where each $E_i^h \subseteq V^h$

G is a Graph defined as $G = (V^g, E^g)$, where $V^g = v_1^g, v_2^g \cdots v_p^g$, $p = 2, 3, \cdots$; and $E^g = e_1^g, e_2^g \cdots e_q^g$, $q = 2, 3, \cdots$ where each e_i^g could be expressed in the form of e_{xy}^g which connects v_y^g from v_x^g.

C is a set of *connectors*, which could be conceptualized as a set of all the dependencies between PSH and PSG (as described earlier) which are characterized by H and G respectively.

here we define two types of connectors:
(a) c_{xy}^v which connects a node in the Graph v_y^g from a node in the Hypergraph v_x^h. It is to note that the behavioral dependency of an object of PSH on an object of PSG gets realized through c_{xy}^v; and
(b) c_{xy}^e which connects a node in the Graph v_y^g from a Hyperedge E_x^h. Here c_{xy}^e realizes the dependencies of collective behavior (bound with a specific relation) of the objects of PSH on the objects of PSG.

The set of all c_{xy}^vs is noted as C^v while C^e represents all c_{xy}^es. Hence on the basis of ongoing discussion, it could be concluded that $C = (C^v, C^e)$. For the rest of the paper, it is assumes for simplicity that the dependency flows from PSH to PSG. That is the Hypergraph layer is dependent on the Graph layer. No dependency flows through a connector backward from the Graph layer to the Hypergraph layer. Above discussion can be illustrated with the example as shown in Fig 4. In this figure, an HG(2) is shown to have a Hypergraph H and a Graph G. The H and the G are connected with *connectors* C. The Hypergraph H is composed of (V^h, E^h) where V^h consists nodes 1, 2, 3, 4, 5, 6 & 7 and E^h consists E_1, E_2, E_3 & E_4. Further, the Graph G is composed of nodes a, b, c, d, e & f. For simplicity the edges of the G are not highlighted.

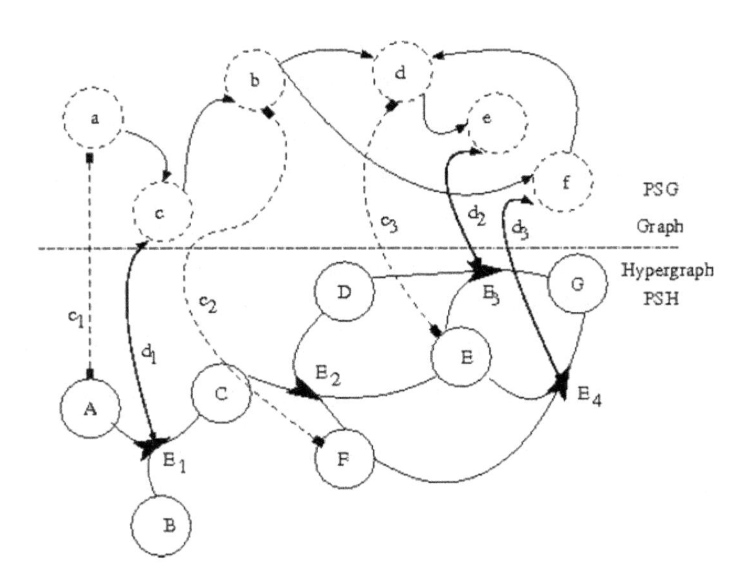

Fig. 4. Illustration of an HG(2) data structure

It is to be noted that E_1 is composed of nodes 1, 2, 3, E_2 is composed of nodes 3, 4, 5, 6, E_3 is composed of nodes 4, 5, 7 and finally E_4 is composed of nodes 5, 6 and 7. Following is the identification of head and tail nodes for each Hyperedge:

$H(E_1) = 1, 2$ and $T(E_1) = 3$
$H(E_2) = 3, 4$ and $T(E_2) = 5, 6$
$H(E_3) = 4, 5$ and $T(E_3) = 7$
$H(E_4) = 5, 6$ and $T(E_4) = 7$

The Fig 4 further illustrates that there are three connectors (marked with dashed square ended bi-directional arrow) that connect Hyperedge nodes and Graph nodes. They are $c^v_{1a} = c_1$, $c^v_{6b} = c_2$ and $c^v_{2d} = c_3$. Moreover there are three connectors (marked with bold bidirectional arrow) that connect a Hyperedge with a Graph node. They are identified as $c^e_{1c} = d_1$, $c^e_{3e} = d_2$ and $c^e_{4f} = d_3$. Hence formally C which is a pair (C^v, C^e) holds the following:
$C^v = c_1$, c_2 and c_3; and $C^e = d_1$, d_2 and d_3.

Though the diagram shows the individual connectors with bidirectional arrow, the earlier assumption still holds that the direction of the dependency is from Hypergraph to Graph layer. The bidirectional arrows have been given for easy diagrammatic identification of connectors out of many different edges.

In this context, the size of HG(2) could be defined as the cardinality of the Hyperedges present in C^e added with the cardinality of edges within the subgraph of G composed of the graph nodes present in C^e. Hence the formal definition of size of HG(2) is as follows:

S{HG(2)} = $|\acute{E}^h| + |\acute{E}^g|$ where

(i) $\acute{E}^h$ is a set of E^h_i, where each such $E^h_i \in C^e$; and

(ii)$\acute{E}^g$ is a set of e^g_i, where each such $e^g_i \in \acute{G}$; $\acute{G}$ is the subgraph composed with $v^g_i \in C^e$.

In the example shown in Fig(4), C^e contains three Hyperedges E_1, E_3 and E_4, hence $|\acute{E}^h| = |E_1| + |E_3| + |E_4| = 7$. Moreover, C^e contains three graph nodes c, e and f; the subgraph constituted by these nodes when investigated, yields no edges hence $\acute{E}^g = 0$. Finally the size of HG(2) comes out to be 7 + 0 = 7.

Next we define *Order* of HG(2) which is denoted as $O\{HG(2)\}$. The order of HG(2) can be computed from the number of nodes (both on H and G) involved in C^v. Each of the connectors present in C^v incorporates two nodes (each from H and G respectively) and hence $O\{HG(2)\}$ could be given as folllowing

$O\{HG(2)\} = 2 * |C^v|$

For the HG(2) shown in Fig 4, $C^v = \{c_1, c_2, c_3\}$. Hence $O\{HG(2)\}$ - 2* 3 = 6

However, Fig 4 shows the entire problem space to be logically divided (by a dashed horizontal divider) into *PSG* and *PSH* which are represented by the Graph and Hypergraph respectively.

On the basis of current discussion, next subsection reports critical issues related with concept of Path in a HG(2).

3.3 Concept of Path in HG(2)

The study of Paths in HG(2) and the related algorithms demand formalization of certain essential terminologies which are presented next.

A *node dependent pair* or simply the *node pair* is defined as the pair of Hypergraph and graph nodes which share a dependency relation through c_{xy}^v and is denoted as $v_x^h(v_y^g)$. In Fig 3, $1(a)$, $6(b)$ and $5(d)$ could be identified as three *node pairs*. It is to be noted that, in context to a Hypergraph node not involved in C^v, the *node pair* is symbolized as $v_x^h(\ )$. The *node pair* of a Hypergraph node n is designated as NP^n.

Similarly a *edge dependent pair* or simple *edge pair* is defined as the pair of Hyperedges and graph nodes which share a dependency relation through c_{xy}^e and is denoted as $E_x^h(v_y^g)$. Any *edge pair* in context to a Hyperedge not involved in C^e is symbolized as $E_x^h(\ )$. In Fig 3, three *edge pairs* could be identified as $E_1(c)$, $E_3(e)$ and $E_4(f)$. However, the *edge pair* of a Hyperedge n is designated as EP^n.

Based on the terminologies discussed above, A path $P_{st}^{HG(2)}$ in HG(2) (known as *HG(2) path* of length q could be defined as a sequence of node pair and edge pair. That is $P_{st}^{HG(2)} = \{NP^{v_1^h=s}, EP^{E_1^h}, NP^{v_2^h}, EP^{E_2^h} \cdots EP^{E_q^h}, NP^{v_{q+1}^h=t}\}$; where following conditions are all true;

i) $P_{st}^{HG(2)}$ is a valid Hyperpath considering only the Hyperedges in H
ii) each $NP^{v_i^h} = v_x^h(v_y^g)$ and each $EP^{E_i^h} = E_z^h(v_w^g)$; i and x $= 2, 3, \cdots$
iii) for any $NP^{v_i^h}, EP^{E_i^h}$ pair, there exists a valid path from v_y^g to v_w^g in G
iv) for any $EP^{E_i^h}, NP^{v_i^h}$ pair, there exists a valid path from v_w^g to v_y^g in G

A path in Graph G traced by the Hyperpath in Hypergraph H is defined as *Graph Path* or simple as *GPath*. A *GPath* is formed by the graph nodes involved in constituent *node pairs* and *edge pairs* (in sequence) of a *HG(2) path*.

An *HG(2) path* $P_{st}^{HG(2)}$ is said to contain a *Graph Loop* or simply *GLoop* if there exists a cycle in *GPath*

An *HG(2) path* $P_{st}^{HG(2)}$ is said to contain a *Hypergraph Loop* or simply *HLoops* if there exists a cycle in Hyperpath considering the Hypergraph only. The condition of identification of loop in Hyperpath has been discussed in Section 3.1.

An *HG(2) path* $P_{st}^{HG(2)}$ containing *GLoop* must not necessarily contain *HLoop* and the vice-versa is also true. Even the *Gloop* and *Hloop* may co-exist in a single $P_{st}^{HG(2)}$.

An *HG(2) path* $P_{st}^{HG(2)}$ is qualified as *Elementary* if all the nodes in the Hyperpath are distinct, and the same is termed as *Simple* if all the Hyperedges are distinct.

However, the idea regarding HG(2) path, that evolved up from the ongoing study, could be illustrated with the example shown in Fig 4.

Let us consider an *HG(2) path* $P_{17}^{HG(2)}$. This path starts from Hypergraph node 1 ($V_1^h = 1$) and reaches another Hypergraph node 7 ($V_7^h = G$). $P_{17}^{HG(2)}$ follows two Hyperpaths The nodes and edges of the first and second Hyperpaths result sequences of *node pair* and *edge pair* as shown below:

$\{1(a), E_1(c), 3(\ ), E_2(\ ), 5(d), E_3(e), 7(\ )\}$ and;
$\{1(a), E_1(c), 3(\ ), E_2(\ ), 6(b), E_4(f), 7(\ )\}$

Both these *HG(2) paths* ensure that the every following graph node in the sequence are connected to its previous node. For example in $\{a, c, d, e\}$, c is connected to a, d is connected to c (through b) and e is connected to d That is $\{a, c, b, d, e\}$ and $\{a, c, b, f\}$ form two valid paths in Graph G. Hence these two graph paths can be termed as *GPath.*

There exists another Hyperpath $\{1, E_1, 3, E_2, 5, E_4, 7\}$ that can be followed for $P_{17}^{HG(2)}$. The corresponding *node pair* and *edge pair* sequence is generated as $\{1(a), E_1(c), 3(\ ), E_2(\ ), 5(d), E_4(f), 7(\ )\}$ and the sequence of Graph nodes corresponding the *node pair* and *edge pair* sequence is generated as $\{a, c, d, f\}$. It is to be noted here that node f is not connected to node d and hence this sequence can not form a valid path in the Graph G.

Hence $\{1(a), E_1(c), 3(\ ), E_2(\ ), 5(d), E_4(f), 7(\ )\}$ is not a valid *HG(2) path.*

A path in a HG(2), thus, as seen in the above discussion, a valid Hyperpath with some additional constraints. These additional constraints can be attributed to the *GPath.* Hence it is to be considered that an HG(2) path can have greater potential than a Hypergraph to model sequence of complex relationships.

On this background, the next subsection discusses the notions of Weighted HG(2) and consequently the cost involved in HG(2) paths.

3.4 Cost Analysis of Path on a Weighted HG(2)

A Weighted HG(2) is composed of a weighted Hypergraph, a weighted Graph and weighted Connectors. In this investigation, the weight simply means the weight of edges that means the cost of connecting nodes within individual Graph, Hypergraph, nodes of Graph an Hypergraph and nodes of Graph and edges of Hypergraph. The weight of graph edges that represents the cost of connecting two Graph nodes v_i^g and v_j^g is denoted by $\mathcal{C}_{e_{ij}^g}$ or $\mathcal{C}_{e_k^g}$ where e_k^g connects node i and node j. Similarly, the weight of hyperedge that represents the cost of connecting $v_a^h, v_b^h, v_c^h \cdots$ related by hyperedge E_i^h is denoted by $\mathcal{C}_{E_i^h}$. The weight of node to node connector c_{ij}^v is the cost of connecting a node of Hypergraph v_i^h and a node of graph v_j^g, and it is denoted as $\mathcal{C}_{c_{ij}^v}$. Hence $\mathcal{C}_{c_{ij}^e}$ will determine the weight of edge to node connector c_{ij}^e which connects E_i^h to c_j^v. Based on this discussion, the cost associated with a path can be viewed as follows.

Before proceeding any further, it is to be noted that ant *HG(2) Path* may give rise to multiple valid Hyperpath routes. For example, *HG(2) Path* $P_{17}^{HG(2)}$ in the

HG(2) shown in Fig 3 gives rise three valid Hyperpath routes $\{1, E_1, 3, E_2, 5, E_3, 7\}$, $\{1, E_1, 3, E_2, 6, E_4, 7\}$ and $\{1, E_1, 3, E_2, 5, E_4, 7\}$. Consequently we get two valid HG(2) route out of these three Hyperpath routes as :

$\{1(a), E_1(c), 3(\,), E_2(\,), 5(d), E_3(e), 7(\,)\}$ and;
$\{1(a), E_1(c), 3(\,), E_2(\,), 6(b), E_4(f), 7(\,)\}$

Each of these valid HG(2) routes will be denoted by $R^1_{P^{HG(2)}_{17}}$ and $R^2_{P^{HG(2)}_{17}}$. Hence in general, $R^k_{P^{HG(2)}_{st}}$ denotes the kth route for *HG(2) path* $P^{HG(2)}_{st}$.

It is to be further noted that for each individual valid HG(2) route, there may exist multiple *GPaths*. Each of these *GPaths* are the valid connectors among the subgraph of G which has been induced by corresponding c^v_{ij} and c^e_{ij} of the chosen HG(2) route $R^k_{P^{HG(2)}_{st}}$. Hence the notation $\acute{G}_{R^k_{P^{HG(2)}_{st}}}$ will denote the subgraph of G induced by the valid HG(2) route $R^k_{P^{HG(2)}_{st}}$ of *HG(2) Path* $P^{HG(2)}_{st}$. An ith *GPath* GP_{pq} on $\acute{G}_{R^k_{P^{HG(2)}_{st}}}$ denotes a path from graph node p to graph node q where p and q are the first and last graph nodes encountered tracing a *HG(2) Path* $P^{HG(2)}_{st}$. The ith *GPath* for graph node pair p and q is thus denoted by GP^i_{pq}.

The simplified concept of *HG(2) Path* presented in the previous section could be represented by single or multiple *HG(2) routes* where each such *HG(2) routes* can be traced with single or multiple *GPaths*. The cost of *HG(2) Path* hence will be dependent on cost of *HG(2) routes* and cost of *GPath.*

However Cost of *HG(2) route* is the summation of cost of Hyperedges that is $\mathcal{C}_{R^k_{P^{HG(2)}_{st}}} = \sum_{i=0}^{n} \mathcal{C}_{E^h_i}$ where E^h_0 to E^h_n are the Hyperedges that constitute the Hyperpath of the *HG(2) route.* Similarly the cost of GPath $\mathcal{C}_{GP^i_{pq}} = \sum_{i=0}^{n} \mathcal{C}_{e^g_i}$ where e^g_i to e^g_n constitute the *GPath.*

This paragraph will investigate the effects of cost of connectors into the computation of cost of *HG(2) Path.* While considering GP^i_{pq}, there might be some nodes which are connected with Hypernodes and Hyperedges for the Hyperpath selected, and at the same time ther might exist some graph nodes v^g_is who are not connected with the any of the Hypernodes or Hyperedges, rather they got traced in order to establishing a valid path between two consecutive graph nodes of *node pair* and *edge pair* (discussed in previous section).

Let us consider the case as shown in Fig 3. For the *Hg(2) Path* $P^{HG(2)}_{st}$ and for *G(2)* route $R^1_{P^{HG(2)}_{17}}$, node b got selected to establish the path between node c and node d. Here node c and node d are directly connected to the Hyperedge E_1 and Hypernode 5 respectively, but node b has no connection with the selected Hyperpath. While the nodes connected with the selected Hyperpath are referred to as *Participating Nodes*, the other types of graph nodes are referred to as *Auxiliary Nodes.*

The characterization of *Auxiliary Nodes* results that there might be three cases as follows:

(i) A particular *Auxiliary Node* has no connection at all. Neither it participates in any c^v_{ij} nor in any c^e_{ij}
(ii) A particular *Auxiliary Node* has connections out of the scope of the selected Hyperpath; and
(ii) A particular *Auxiliary Node* has connections within the scope of the selected Hyperpath.

The example shown just above represents the case (ii).

The engagement of Case (i) clearly has no effect in computing Cost of *HG(2) Path*, while engagement of Case (ii) and case (iii) induces some dependency of Graph layer bacj to Hypergraph layer. But as stated earlier in Section 3, for simplicity, there exists no dependency from Graph layer to Hypergraph layer, so even the existence of Case (ii) and case (iii) does not make any difference in computation of Costs of *HG(2) Path*. Had there been a scope of mutual dependency between Graph and Hypergraph layer, the existence of Case (ii) and Case (iii) had have severe effect in *HG(2) Path* cost, but in present research, this is beyond the scope of any further discussions.

On this background, for all the *Participating Nodes*, there are wighted connectors and only the connectors connected to the *Participating Nodes* will contribute the cost in computing *HG(2) Path*. Hence the cost of connectors for a set of specific $R^x_{P^{HG(2)}_{st}}$ and specific GP^y_{pq}, which is denoted as $\mathcal{C}_{xy}$, could be given by the relation:

$\mathcal{C}_{xy} = \sum_{r=0}^{n} \mathcal{C}_{c^v_r} + \sum_{r=0}^{n} \mathcal{C}_{c^e_r}$ where each $c^v_r = c^v_{ij}$ and each $c^e_r = c^e_{ij}$ denote node to node and edge to node connector respectively which are connected to each of the participating node j for a specific selected GP^w_{pq} $(w = 1, 2 \cdots)$

All the preceding discussions identify the cost expression of a *HG(2) Path* $P^{HG(2)}_{st}$ for a specific *HG(2) route* $R^u_{P^{HG(2)}_{st}}$ and specific *GPath* GP^z_{pq} as

$$\mathcal{C}_{P^{HG(2)}_{st}}(u, z) = \mathcal{C}_{R^u_{P^{HG(2)}_{st}}} + \mathcal{C}_{GP^z_{pq}} + \mathcal{C}_{uz}.$$

For a given $P^{HG(2)}_{st}$ there will be multiple set of (u, z) pair and the $min(\mathcal{C}_P^{HG(2)_{st}}(u, z))$ will denote the least cost.

4 HG(2) Based Model for Constraint Driven Stratification of RDF

This section reports a HG(2) based model for Constraint Driven Stratification of RDF. The constraints are imposed as an inter stratum cost. For example, if the constraint (in cost form) between two classes C_1 and C_2 be c, then to be C_1 as an instance of C_2, cost c is incurred, and vice versa. For example the un-stratification nature of rdfs:Class could be attributed for a constraint with zero (0) cost. That means, incurring zero cost, rdfs:Class could be an instance of $rdfs : Class$ (itself).

The notion of constraint just described could be mapped to a HG(2) data structure. The mapping of constraints to any HG(2) element needs a formal point to point mapping between RDFS and HG(2) elements.

The nodes of the Graph within HG(2), i,e V^gs are mapped with the Classes of RDFS hierarchies, similarly, the edges of the Graph E^grepresent the $rdfs : subClassOf$. Hence, the classes which are not the part of the RDFS Class Hierarchy, should not be treated as valid V^g.

The Nodes of the Hypergraph are considered to be the RDFS property literals or URIs, i,e the literal or URI values of $rdfs : label$, $rdfs : isDefinedBy$ etc should be considered as V^h and the corresponding properties are the Hyperedges. All the literal and URI values of $rdfs : seeAlso$ give rise to the Hyperedge $rdfs : seeAlso$.

The two types of connectors represent the constraints, The node to node connectors that connect a URI or literal value of a RDFS *property* to a RDFS *class*, represents the possibility of a V^h to be included as the instance of V^g. The weight of the C^v elements quantifies the possibility as a cost. Specific Cost function could be derived case to case basis.

On the other hand, C^e elements that is the edge to node connectors that connect a hyperedge to a graph node, represent the $rdfs : type$ which could be treated as a binary constraint. A specific RDFS *property*, if is a type of RDFS *classes*, then the constraint that is the connector or the element of C^e will appear, otherwise not.

The constraint driven HG(2) based model just presented has a potential to represent the entire RDFS ontology with added information. The constraints are conceived as the Connectors of HG(2) and the weight of the connectors reveals the cost of that constraint. The cost functions must be introduced in order to evaluate or quantify the constraint. The stratification of RDF thus obtained will have all the semantic expressiveness just due to the fact that the entire RDFS ontology has been mapped to HG(2).

5 Conclusion

The present research has introduced a HG(2) based model of constraint driven and semantic lossless RDF stratification. This proposed model has a great potential in representing constraints as an HG(2) element which opens up new avenues in complex characterizations of constraints in stratification of RDFs.

Within the current discussion, th Ce future scope of studies on HG(2) based startification of RDF has aptly been identified to focus on utilizing HG(2) Path based algorithms to gain insights in stratification of RDF.

References

1. Bernees-Lee, T., Hendler, J., Lasilla, O.: The Sumentic Web. Scientific American (2001)
2. Graham, M.: RDF and Topic Maps: An exercise in convergence. In: Proceedings of XML Europe (2001)
3. Lacher, M.S., Stefan, D.: On the Integration of Topic Maps and RDF Data. In: Extreme Markup Languages Conference (2001)
4. Munshi, S., Chakraborty, A., Mukhopadhyay, D.: A Hybrid Graph based Framework for Integrating Information from RDF and Topic Map: A Proposal. In: Cube Conference (2012)
5. Hayes, J.: A Graph Model for RDF. Diploma Thesis, Technische Universität Darmstadt, Universidad de Chile (2004)
6. Auillans, P., Ossona de Mendez, P., Rosenstiehl, P., Vatant, B.: A Formal Model for Topic Maps. In: Horrocks, I., Hendler, J. (eds.) ISWC 2002. LNCS, vol. 2342, pp. 69–83. Springer, Heidelberg (2002)
7. Berge, C.: Graphs and Hypergraphs. North-Holland, Amsterdam (1973)
8. Boullie, F.: Un modèle universel de banque de données, simultanément partageable, portable et répartie. Thèse de Doctorat d'Etat es Sciences Mathématiques (mention Informatique), Université Pierre et Marie Curie (1977)
9. Berge, C.: Minimax theorems for normal hypergraphs and balanced hypergraphs - a survey. Annals of Discrete Mathematics 21, 3–19 (1984)
10. Berge, C.: Hypergraphs: Combinatorics of Finite Sets. North-Holland, Amsterdam (1989)
11. Nilsson, N.J.: Problem Solving Methods in Artificial Intelligence. McGraw-Hill, New York (1971)
12. Nilsson, N.J.: Principles of Artificial Intelligence. Morgan Kaufmann, Los Altos (1980)
13. Nguyen, S., Pallottino, S.: Hyperpaths and shortest hyperpaths. In: Simeone, B. (ed.) Combinatorial Optimization. Lecture Notes in Mathematics, vol. 1403, pp. 258–271. Springer, Berlin (1989)
14. Alpert, C.J., Caldwell, A.E., Kahng, A.B., Markov, I.L.: Hypergraph Partitioning with Fixed Vertices. IEEE Trans. on CAD 19(2), 267–272
15. Boros, E., Elbassioni, K., Gurvich, V., Khachiyan, L.: An efficient incremental algorithm for generating all maximal independent sets in hypergraphs of bounded dimension. Parallel Processing Letters 10, 253-266
16. Boros, E., Elbassioni, K.M., Gurvich, V., Khachiyan, L.: Generating Maximal Independent Sets for Hypergraphs with Bounded Edge-Intersections. In: Farach-Colton, M. (ed.) LATIN 2004. LNCS, vol. 2976, pp. 488–498. Springer, Heidelberg (2004)
17. Trifunovic, A.: Parallel Algorithms for Hypergraph Partitioning. PhD Thesis, University of London (February 2006) (Abstract)
18. Gallo, G., Longo, G., Nguyen, S., Pallottino, S.: Directed Hypergraphs and Applications, http://www.cis.upenn.edu/~lhuang3/wpe2/papers/gallo92directed.pdf

Ontology Based Zone Indexing Using Information Retrieval Systems

Rajeswari Mukesh[1], Sathish Kumar Penchala[2], and Anupama K. Ingale[2]

[1] Hindustan University, Chennai, 600016 India
rajeswarim@hindustanuniv.ac.in
[2] Rajarambapu Institute of Tecnology, Islampur, 415414 India
{p.sathishkumar,anupama.ingale}@ritindia.edu

Abstract. Information Retrieval [IR] is a technique used for searching documents, information within documents, and for metadata about documents, and also the searching relational databases and the World Wide Web. There is overlap in the usage of the terms data retrieval, document retrieval, information retrieval, and text retrieval, but each also has its own body of literature, theory, and technologies. In this paper the collection of documents is matched through ontological concepts instead of keywords and zone based indexing is used to sustain retrieval status value (RSV). Zones are similar to fields, except the contents of a zone can be arbitrary free text. A field will take on a relatively small set of values, a zone can be thought of as an arbitrary, unbounded amount of text. For instance, document titles and abstracts are generally treated as zones. We have built a separate inverted index for each zone of a document. The proposed system aims for better recall and precision.

Keywords: Information Retrieval, Ontological Concepts, Zone based Indexing (ZBI).

1 Introduction

In Information retrieval systems challenging blend of science and engineering in many interesting unsolved problems can span many areas of CS: architecture, distributed systems, algorithms, compression, information retrieval, machine learning, UI, etc. Scale far larger than most other systems; small teams can create information systems used by hundreds of millions. Engineering difficulty roughly is equal to the product of these parameters called as document indexing, queries / sec, index freshness/update rate, query latency, information kept about each document, complexity/cost of scoring/retrieval algorithms. With the present growing World Wide Web the use of metadata will become necessary.

In conceptual querying a set of concepts are examined. The general idea is to restrict general world knowledge ontology to the given set of concepts, extending this with relations and related concepts, and thereby providing a structure called instantiated ontology, for navigation and further investigation of the concepts [2].

S. Unnikrishnan, S. Surve, and D. Bhoir (Eds.): ICAC3 2013, CCIS 361, pp. 181–186, 2013.

Conceptual investigation of a set of documents can be performed by extracting the set of concepts appearing in the documents and by providing means for navigation and retrieval within the set of extracted concepts.

The languages which represent ontology are Resource development framework Schema (RDFs), Ontology Interface Layer (OIL), Web Ontology Language (OWL) [4]. The Resource Description Framework (RDF) [5] is an infrastructure that enables the encoding, exchange and reuse of structured metadata.

Design objectives and goals of RDF are:

(i) Independence: It should be possible for anyone to define its own properties or classes and use and/or reuse them in a specific semantic way. Moreover, RDF based applications should be future-proofing, regardless on the evolution of the schemas.
(ii) Interchange: It should be easy to transport and storage the metadata described by RDF.
(iii) Scalability: Even for a huge set of metadata it should be easy to handle and process them.

The notions of controlled vocabularies and Ontologies, their formal notations, and how they should be implemented is controversial. However, the ontology definition is very similar to the prevalent RDF recommendation, and to Stumme's definition of core ontology [8].

Different approaches on word sense disambiguation are reviewed in [3]. Word sense disambiguation is a very active field of research, with potential applications in language translation, information extraction and in search engines.

2 Importing and Mapping of Ontologies

There are many different formats for storing Ontologies; a common mechanism for exchanging them is needed. The RDF recommendation was recently released by the W3 consortium. RDF is a data format based on XML that is well suited as an interchange format of Ontologies and controlled vocabularies. RDF is widely accepted for data interchange and RDF versions of most important Ontologies are available.

In order to parse and import Ontologies to the database, Jacob kohler et al [2006][7] used the Jena API, v1.6.0 (McBride). It was developed by McBride and is promoted by HP under BSD-style license for free. Importing ontology consists of mainly three steps:

1. *Convert RDF-file to JENA-model:*

In this method RDF-file document is taken and is converted to JENA-model [8].

2. *Read configuration file for ontology import*: configuration file stores individual properties of ontology and maps the different notations for equivalent characteristics (fig.1). Ontology has a set of characterizing attributes. Beside name and language the relation types of an entry are essential. The Unique attribute configures the import to use complete URIs or concept names as concept IDs. New Ontologies that are available in RDF can be imported simply by adding a new line to the configuration file.

3. Write concepts and relations into the database:
The concepts, relations and synonyms are written to the relational backend via the Java JDBC interface (Java Database Connectivity).

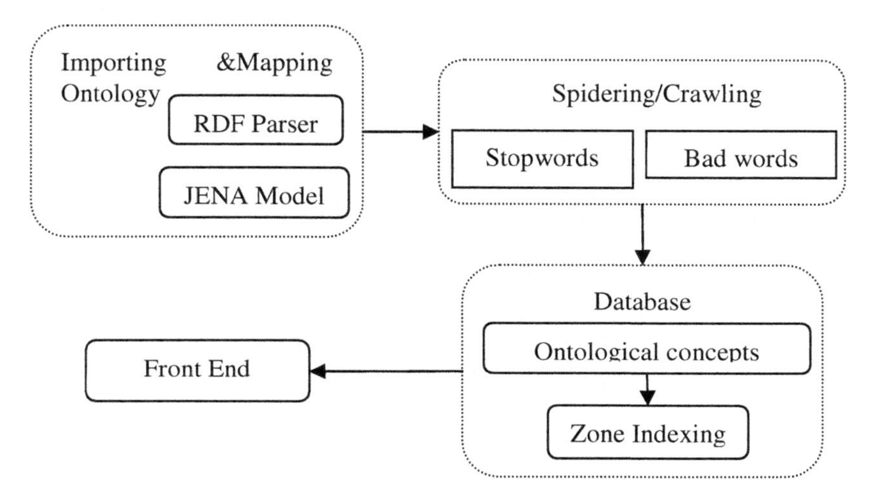

Fig. 1. Ontological concepts using Zone Indexing

3 Spidering and Indexing

Spidering and indexing denotes for downloading web pages from the Internet and mirror it in the local file system (see Fig.1). Apart from the ontological indexing, also a simple keyword based index is generated [7]. This allows the comparison of the ontology based indexing method against standard keyword based indexing. This involves several steps:

(i) Removal of special characters.
(ii) Drop stopwords and badwords: stopwords [6] are common words that are not included in the index since they occur on almost all web pages (Fig.2). Badwords filter can be used in order to exclude the indexing of vulgar language.

3.1 Situating Descriptors in Ontologies

In this approach [9] discusses that the taxonomic ontology is enriched with descriptors attached to the nodes in the ontology. Thus noun phrases are each mapped onto a node in the so called generative ontology, whose potentially infinite structure is generated by means of a set of rules constituting an ontological grammar. Augmentation of inclusion structures with descriptors may be achieved by means of (typed) features structures well-known from unification-based grammar systems, where features are used to express and combine information from the syntactic, semantic and pragmatic dimensions of linguistic analysis. Another option is to adopt (basic) description logic for representation of and comparison of–descriptors.

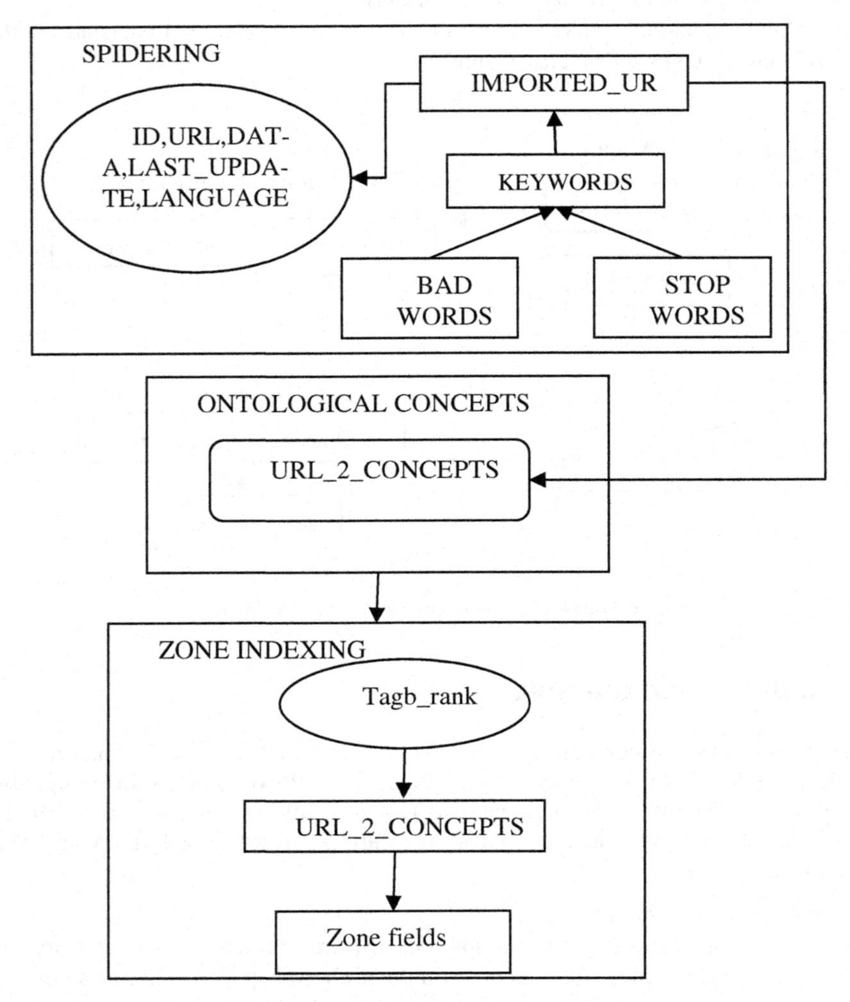

Fig. 2. Tag based ranking in Zone Indexing

4 Zone Indexing

Zones are similar to fields (Fig.2), except the contents of a zone can be arbitrary free text. Whereas a field may take on a relatively small set of values, a zone can be thought of as an arbitrary, unbounded amount of text [1]. For instance, document titles and abstracts are generally treated as zones (fig.3). We may build a separate inverted index for each zone of a document, to support queries.

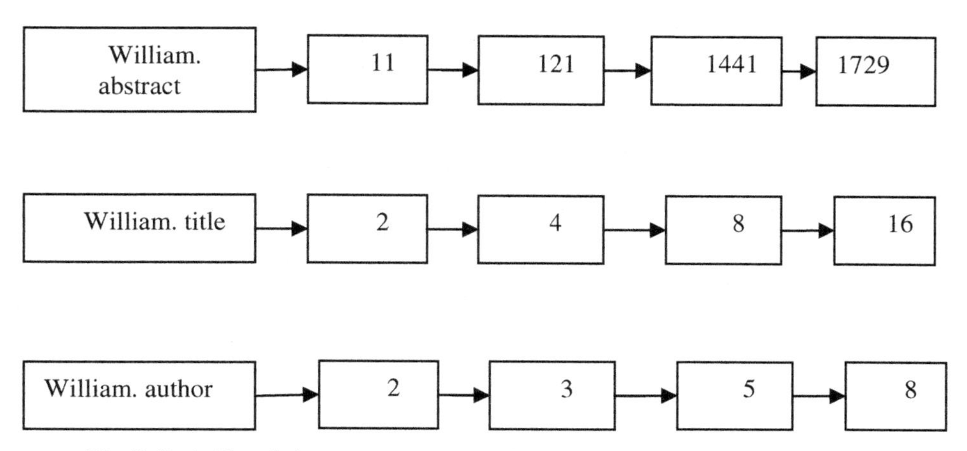

Fig. 3. Basic Zone index; zones are encoded as the extensions of dictionary entries

Algorithm:
(i) Calculate tag based rank tbr

$$tbr= \sum Weighting\ (w_i, t_i) \tag{1}$$

for each occurrence word wi. The position of a word between certain XML tags as title, meta keywords, description, headlines (e.g. (h1), (h2), . . .), links (<a>) and the accentuation (strong, underlined, italic) was used as a primitive method for weighting the relevance of a webpage w.r.t. a keyword by assigning words in these tags a weighting factor.

(ii) Drop XML tags.
(iii) Write words and their ranking to the zone indexing in the database.

Consider a set of documents, each of which has *zones*. Let g1,... g ∈ [0, 1] such that $\sum_{i=1}^{l}(g_i = 1)$ for 1<i<l , let t_i be the tag based rank which gives the ranking position of the document depending on the space between each word.
Finally weighted zone indexing is termed as:

$$\sum_{i=1}^{l}(g_i\ t_i) \tag{2}$$

Consider the query Shakespeare in a collection in which each document has three zones: *author, title,* and *body*. The tag base rank function for a zone takes on the value 1 if the query term Shakespeare is present in the zone and 0 otherwise. Weighted zone scoring in such a collection requires three weights $g1$, $g2$, and $g3$, respectively corresponding to the *author, title,* and *body* zones (fig.4). Suppose we set $g1 = 0.2$, $g2 = 0.3$, and $g3 = 0.5$ (so that the three weights add up to 1); this corresponds to an application in which a match in the *author* zone is least important to the overall score, the *title* zone somewhat more, and the *body* contributes even more.

Fig. 4. Zone index in which the zone is encoded in the postings rather than the dictionary

Thus, if the term Shakespeare were to appear in the *title* and *body* zones but not the *author* zone of a document, the score of this document would be 0.8.

Displaying the same results for different searches will be reduced. Ontological descriptors along with zone indexing will give us accurate results and saves time. Time complexity will be eradicated.

5 Conclusion

This paper illustrates about the zone indexing and situating tbr (tag based rank) in zone indexing. The collection of documents is matched through ontological concepts instead of keywords and zone based indexing is used to sustain retrieval status value (RSV). To achieve the above mentioned task we have to preprocess the documents, import Ontologies and map ontological concepts to metadata. We are working on ontological descriptors to enhance the effectiveness of semantic web.

References

1. Saruladha, K., Aghila, G., Penchala, S.K.: Design of New Indexing Techniques Based on Ontology for Information Retrieval Systems. In: Das, V.V., Vijaykumar, R. (eds.) ICT 2010. CCIS, vol. 101, pp. 287–291. Springer, Heidelberg (2010)
2. Andreasen, T., Bulskov, H.: Conceptual querying through Ontologies. Science Direct, 2159-2172 (2009)
3. Pinheiro, V., Furtado, V., Freire, L.M., Ferreira, C.: Knowledge-Intensive Word Disambiguation via Common-Sense and Wikipedia. In: Barros, L.N., Finger, M., Pozo, A.T., Gimenénez-Lugo, G.A., Castilho, M. (eds.) SBIA 2012. LNCS, vol. 7589, pp. 182–191. Springer, Heidelberg (2012)
4. He, G., An, L.: Ontology Language OWL Research Study. In: International Conference on Study Management and Service Science (MASS), pp. 1–4 (2011)
5. Almendros-Jimenez, J.M.: An RDF Query Language based on Logic Programming. In: 3rd Int. Workshop on Automated Specification and Verification of Web Systems, vol. 200, pp. 67–85. ScienceDirect (2008)
6. Dragut, E., Fang, F., Sistla, P., Yu, C., Meng, W.: Stop word and related problems in web interface integration. ACM Journal, Proceedings of the VLDB Endowment 2(1), 349–360 (2009)
7. Kohler, J., Philippi, S., Specht, M., Ruegg, A.: Ontology based text indexing and querying for the semantic web. Knowledge-Based Systems 19(8), 744–754 (2006)
8. Stumme, G., Maedche, A.: FCA-Merge: A Bottom-Up Approach for Merging Ontologies. In: Proceedings of the International Joint Conference on Artificial Intelligence, Seattle, Washington, USA, pp. 225–234 (2001)
9. Fischer Nilsson, J.: Concept descriptions for text search. In: Information Modeling and Knowledge Bases XIII, pp. 296–300. IOS Press (2002)

Framework for Blog Software in Web Application

Karan Gupta[1] and Anita Goel[2]

[1] Department of Computer Science, University of Delhi, New Delhi, India
guptkaran@gmail.com
[2] Department of Computer Science, Dyal Singh College, University of Delhi, New Delhi, India
agoel@dsc.du.ac.in

Abstract. Web applications integrate blog software to facilitate effective communication with the users of web application. There is no mention of a design or requirement specification document for blog software, even academically, which makes integrating the blog software in a web application a tedious task. In this paper, we present a framework for integrating blog software into a web application. The framework consists of different components corresponding to the different stages of the software development life cycle. The requirement component of the framework aids the web application in deciding features of blog software, to be included in a web application. For this, a weighted requirement checklist is presented. The design component provides a structural design of blog software which facilitates the developer in understanding the design of existing blog software as well as modifying it or developing a new one. The framework also aids the user during validation and verification of blog software integrated in a web application.

Keywords: Web Application, Blogging Software, Integration, Framework.

1 Introduction

Nowadays, web application use blog software to interact with their users. Blog software allows web application to provide different kinds of content to its users like regular updates on their products, pros and cons, latest news and events and sometimes, simple discussion opportunity. The users can comment, share or like content presented to them by the web application. The integration of blog software in a web application allows communication to occur effectively between the web application and the users.

Since there is no formal design document or requirement specification available for integration of blog software in web application, even academically, the integration is carried out on-the-go, depending on whims and fancy of the developer and stated requirements of a web application. Web application can use either existing freely available blog software or may write completely new code. Usually, the needs, and, look and feel of a web application determines the modifications that are required to be made to the blog software, both functionally and cosmetically.

Here, in this paper, the focus is to define a framework for blog software which eases the integration of the blog software into a web application. The framework aids in

S. Unnikrishnan, S. Surve, and D. Bhoir (Eds.): ICAC3 2013, CCIS 361, pp. 187–198, 2013.

the specification of requirement and their elicitation during the requirement phase. Also, the framework provides a structural design for understanding structure of the blog software, so as to ease the adaption and development of the blog software depending on the user's requirements.

In this paper, we present a framework of blog software to aid the integration of the web application. The structural design and the weighted requirement checklist of our framework are based on our earlier work [1]. The framework consists of four components - (1) Blog_Requirement, (2) Blog_Design, (3) Blog_Development, and (4) Blog_Test. The Blog_Requirement performs the task of eliciting and specifying requirements. The Blog_Design component aids in generation of the design. The Blog_Development and Blog_Test components help during developing and testing the blog software. The web application and the developer interact with these components for integration of the blog software in a web application. Each component consists of sub-components to simplify the completion of associated tasks.

Both the web application owner and the developer are benefitted by the framework presented here. During integration of the blog software into a web application, the developer uses the framework to understand structure of the blog software. Also, by using design, the developer can understand the various interactions that occur between the different components of the blog software. The framework assists the web application owner to understand various requirements that are provided by the blog software and also to select them.

The framework can be used during the various phases of the software development process. During the requirement phase, the framework helps in defining the requirements of the blog software. The framework helps in the design phase by providing an outline of design of the blog software. The framework is also used during the testing phase for verification and validation of the integrated blog software.

In this paper, the background of our framework is explained in section 2. Section 3 describes the framework. The limitations and future work are described in section 4. Section 5 contains related work. Conclusion is stated in section 6.

2 Background

Web applications are being used in different domains of our lives like - hotels, news, sports, medicine, business or education. To survive in today's competitive world, web application provides the user with regular updates and information using various techniques like writing articles, providing product updates. Web applications integrate blog software to facilitate promotion of products and services.

The content presented in the blog is maintained and controlled by the owner of the blog, i.e. web application and administrator. Blog software allows visitors to a web application to comment on the content provided in the blog. They can also share and like this content. Moreover, the visitors can communicate among themselves and the web application by replying to the comment already made.

During our previous work [2], we have identified the different type of users that can access the blog software in a web application. Moreover, the requirement of the blog software have identified and presented as weighted requirement checklists in [1]

The weighted requirement checklists can be used by the web application owner to identify the requirements. Also, a structural design has been identified in [1] which aids the developer in understanding the structure of blog software. In the following subsection, an overview of our earlier work is presented for the type of users, weighted requirement checklists and the structural design, which form the basis for this framework.

2.1 Users of the Blog Software

The users of the blog software in a web application have been identified in our earlier work in [2]. The identification of the users is based on their interaction with the blog software in a web application. Three kinds of users identified based on the use-case based approach are as follows –

- *Web Application* is the software in which the blog software is integrated.
- *Administrator* is any person performing task of maintaining the blog functionality.
- *Visitor* is a user who uses the blog software in a web application.

2.2 Weighted Requirement Checklists

Web applications incorporate blog software to serve different kinds of purposes. Sometimes, web applications might use blog software to provide periodic updates to their users. Other web applications may allow their users to interact among themselves and web application by allowing them to post comments on the posts. Since web applications can have differing requirements, weighted requirement checklists have been generated in [1] to meet those requirements. The generated weighted requirement checklists can be used for selecting operations of the blog software in a web application. To denote the importance of a feature in the weighted requirement checklists, weights have been used. The importance of features has been defined into three categories as follows -

- *Required-Basic* category is assigns highest level of importance and is denoted by weight '3'.
- *Optional-Basic* category is for those features that may be helpful in the software but are not a necessary requirement. It is denoted by weight '2'.
- *Optional-Advanced* specify the lowest level of importance, and is denoted by weight '1'.

The weighted requirement checklists for blog software consist of three checklists, namely, (1) Blog Home, (2) Blog Dashboard and (3) Blog Parameters. Blog Home lists operations provided to a visitor of the blog software. Using this checklist, the operations for a visitor to the blog software can be selected. In the Blog Parameter checklist, operations available to the web application are listed. During the requirement phase, this checklist is used for identifying parameters and settings that are to be included into a web application. Blog Dashboard checklist contains tasks performed by the administrator in the blog software in a web application. The checklist is used to identify the functionality to be included in the dashboard. Table 1 shows a small portion of Blog Home Weighted Requirement Checklist.

Table 1. Portion of Weighted Checklist for Blog Home

Entities		Sub-entities		Operations	
Name	W	Name	W	Name	W
Post & Page Use	3	Options Present	3	Previous item; Next item; Like; No. of comments; Share; Report Abuse; View likes; View shares; Categories; Comment	3
				Rate; Subscribe; View ratings; Tags	2
		Comment	3	Comment Box; View Likes; View Shares; View; Like comment; Share comment; Report Abuse	3
				View Ratings; Reply; Rate comment	2
				Subscribe	1
		Other	3	View	3
Blog Main	3	Options Present	3	Same as Post (Options Present)	3
					2
		Others	3	Same as Post (Other)	3

During the requirement phase, the three weighted requirement checklists can be used to select features of the blog software. In the next sub-section, an overview of the structural design of blog software is presented.

2.3 Structural Design

During our study of blog software in [1], it was found that the blog software has the following building blocks – content, content catalogue, appearance, setting and maintenance. Here, 'content' represents content stored in post, pages & media of the blog and 'content catalogue' is the index associated with the content. 'Appearance' part of the blog software is used to match the look and feel of web application. 'Setting' is used for manipulate various features of the software, and 'maintenance' is used for maintaining the blog software in the web application.

The structural design presented in [1], is based on these building blocks. The design displays interaction of users and the blog software in a web application. In the structural design, entities for the building blocks are identified as follows –

- **Content** - Post & Page Update, Post & Page Use, Post & Page Setting, Media Update, Media Use and Media Setting.
- **Content Catalogue** - Blog Main, Blog Title and Blog Main/Title Setting.
- **Appearance** - Appearance Use, Appearance Theme, Appearance Widget and Appearance Menu.
- **Setting** - Setting Blog.
- **Maintenance** - Maintenance Blog.

The structural design is based on the entities identified above. The identified entities are extensible in nature and can accommodate any new feature or functionality. The identified entities are further divided into a group of sub-entities where each sub-entity provides a set of correlated tasks within a particular entity. Each sub-entity contains a group of operations performed by the sub-entity.

3 Blog Software Integration Framework

Blog software is integrated into a web application on-the-go, due to unavailability of a formal document for blog software. The blog software integration framework presents the structural design and weighted requirement checklists which aids during integration of blog software in a web application. The framework is designed to be used during different phases of the software development life cycle. The four components of the framework to be used during the different phases of the software development life cycle are as follows -

1. Blog_Requirement
2. Blog_Design
3. Blog_Development
4. Blog_Test

The correlation between the software development lifecycle and our framework is depicted in Figure 1.

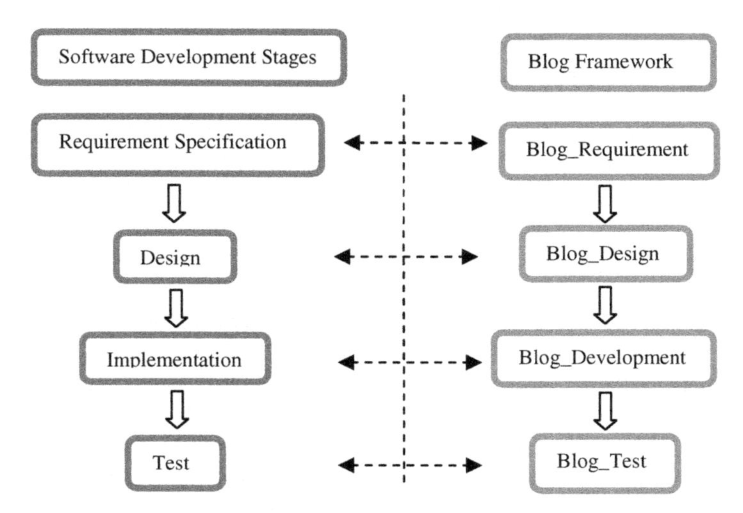

Fig. 1. Correlation between Software Development Lifecycle and our Framework

Figure1 shows the relation of the components of framework to the different phases of software development lifecycle. We see that the Blog_Requirement component is used during the requirement specification phase and the Blog_Development component is used during the implementation phase. During software development process, the web application interacts with Blog_Requirement and Blog_Test components. On the other hand, the developer interacts with Blog_Design, Blog_Development and Blog_Test components. In the following sub-sections, we discuss the different components.

3.1 Blog_Requirement

Blog_Requirement component aids in requirement specification of the blog software in a web application during the requirement elicitation phase of software development. The component consists of three sub-components, namely, *Blog Purpose*, *Requirement Gatherer* and *Requirement Estimator*. Figure 2 shows the composition of the Blog_Requirement component.

Blog Purpose sub-component requires defining the purpose that blog will serve in a web application. The identification of purpose helps in identifying the scope of blog software in a web application, like, restrictions to be applied, key role of blog software in a web application etc.

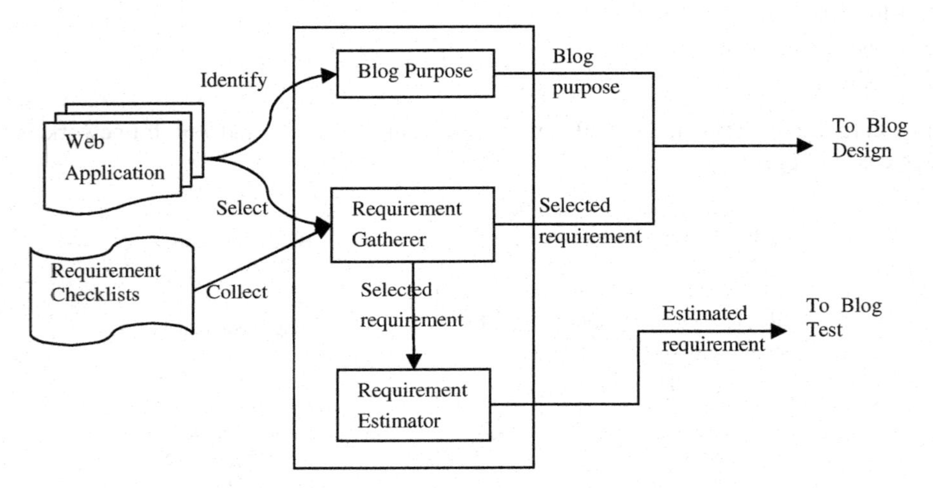

Fig. 2. Composition of Blog_Requirement Component

Requirement Gatherer sub-component accepts the requirement checklists defined in [1], as an input to identify features that must be available in the blog software integrated in a web application. Each checklist has three parts- required-basic, optional-basic and optional-advanced, based on the weights assigned to them. This sub-component allows the web application owner to interact with the checklists to determine the features and functionality that is to be included in the blog software. The web application owner may select some, all or none features from these checklists.

Requirement Estimator sub-component uses output of the requirement gatherer to quantify number of features selected by the web application owner. For this, the requirement estimator sub-component uses the estimation mechanism, presented in [1]. Using the mechanism, the weighted percentage for each checklist is calculated. The output of the requirement estimator forms an input to the Blog_Test component, to be used for validation and verification purposes.

3.2 Blog_Design

The Blog_Design component is used for the design of blog software for integration in a web application, during the design phase of software development. The Blog_Design component consists of two sub-components – (1) Splitter, and (2) Design Creator. Figure 3 show working of the Blog_Design component.

Splitter performs the task of splitting incoming requirements into the basic buildings blocks of blog software, namely, Content, Content Catalogue, Appearance, Setting and Maintenance. This sub-component takes its input from the Requirement Gatherer sub-component of Blog_Requirement component to produce split requirements.

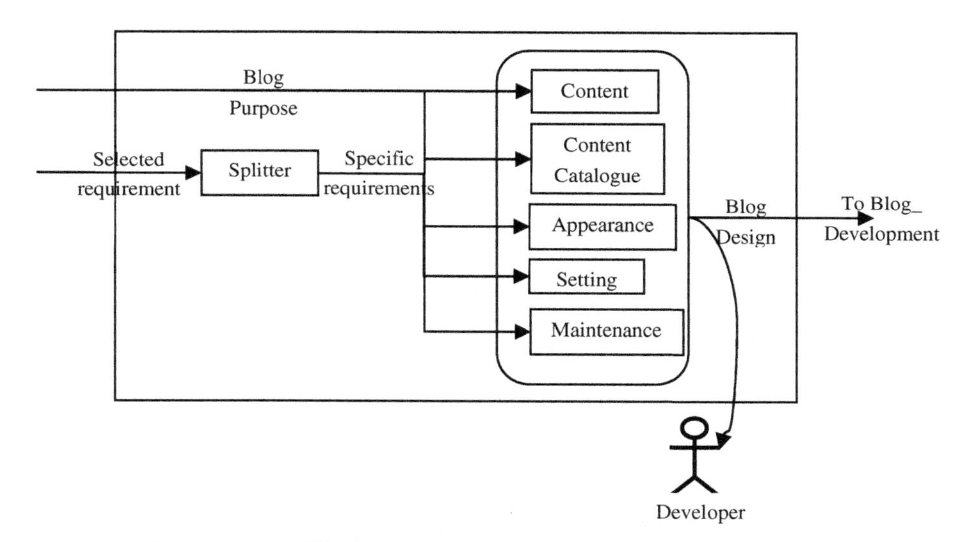

Fig. 3. Composition of Blog_Design

Design Creator accepts the requirements split based on the building blocks and creates design of the blog software. The requirements are taken as input from its peer sub-component, Splitter. The design created by this sub-component is based on the structural design presented in [1]. This sub-component consists of five parts, Content, Content Catalogue, Appearance, Setting and Maintenance. Each part of this sub-component performs the task creating the design for its name equivalent building block, like Content part create the structural design for Content.

3.3 Blog_Development

Blog_Development is used during the implementation phase of the software development cycle to create and integrate blog software into a web application. The component is further divided into two sub-components, namely, *Creator* and *Integrator*. The component is displayed in figure 4.

Creator carries out the task of creating the blog software. For creating blog software, firstly, the developer chooses from among the following three options -

- Use already available blog software
- Use freely available code
- Write new code

The developer can either select some already available blog software like Wordpress [3] or can use freely available code for the creation of blog software. However, the developer needs to adapt the selected blog software or code to match the web application owner's stated requirements as well as the specified design. Also, the selected software or code has to be matched according to the look and feel of the web application. In case the developer chooses to write new code, no adaptation is required.

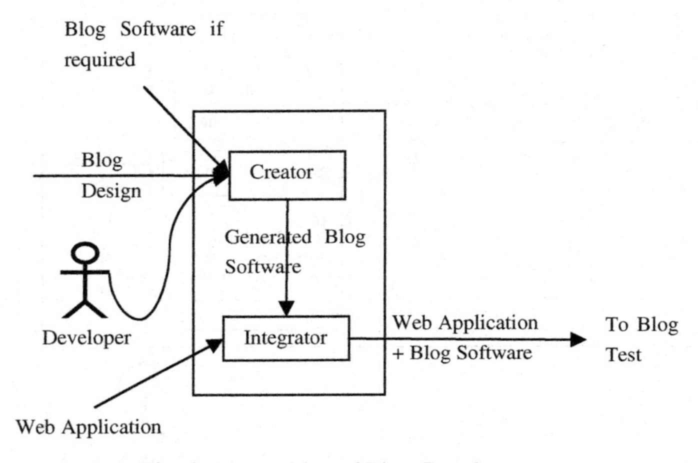

Fig. 4. Composition of Blog_Development

Integrator carries out the task of integrating the developed blog software in the web application. This sub-component provides guidelines for integrating blog software in a web application. Table 2 lists guidelines for integration based on structural design of the blog software.

Table 2. Guidelines for integration of Blog Software

Part of Structural Design	User Type	Specific Instruction
Content	Visitor	Public Domain/ Login-based
	Administrator	Private Domain
Content Catalogue	Visitor	Public Domain/ Login-based
	Administrator	Private Domain
Appearance	Visitor	Public Domain/ Login-based
	Administrator	Private Domain
Setting & Maintenance	Administrator	Private Domain

As shown in Table 2, the visitor part of structural design of content is placed in public domain of the web application or login-based access of the web application. On the other hand, the part of structural design belonging to the administrator is not available in public domain but only in private domain.

3.4 Blog_Test

Blog_Test is used during the testing of integrated blog software. There are two sub-components in this component, namely, verification and validation. The composition of Blog_Test component is displayed in figure 5.

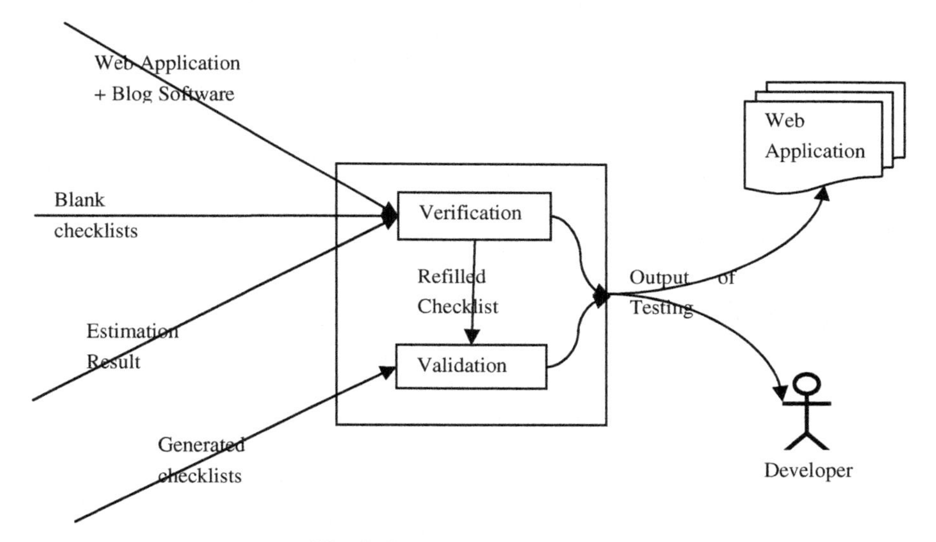

Fig. 5. Composition of Blog_Test

Verification sub-component uses the integrated blog software to refill the weighted requirement checklists, present in Requirement Gatherer sub-component of Blog_Requirement. The checklists are filled by marking the features that are present in the integrated blog software. Then, estimation of these refilled checklists is done using the Requirement Estimator sub-component of Blog_Requirement. The result of this calculation is compared with the result calculated previously during the Blog_Requirement. In the ideal world, there would be no difference between these two sets of checklists, and it can be assumed that "the blog software is built right".

Validation sub-component uses the checklists refilled by Verification, and the checklists originally filled by the web application owner. These two sets of checklists are then provided to the web application owner for validation purposes. The web application can validate that "the blog software built is the right thing" by checking whether the blog software meets the needs of the web application.

The blog software integration framework explained here, eases the task of integration of blog software in a web application. However, there are limitations of this framework, which we discuss in the next section. We also discuss the future work.

4 Limitation and Future Work

The framework presented here for blog software is for integration purpose only. Nevertheless, the framework nor structural design nor weighted requirement checklists can be used for developing standalone blog software, since standalone software would have a different set of requirements and a different set of users. Standalone blog software requires much more functionality and features, not supported by this framework.

Moreover, the structural design and the weighted requirement checklist used here for the framework have been arrived at after analyzing the most popular and freely available blog software. The commercial versions of blog software may contain some functionality or feature which is not included in the freely available software. Also, other freely available blog software may contain some extra features. This is a limitation of our work. However, the structural design and the weighted requirement checklist presented here are extensible in nature and can be updated to accommodate any new feature.

In future, we are working towards developing a tool based on this framework. The tool will facilitate during integration of blog software in a web application. The tool will be able to help during the different stages of the software development life cycle by easing the requirement elicitation, design creation and testing.

5 Related Work

Researchers and scholars have carried out a lot of research on blog software. Work has been done in various fields related to blog distillation, style of writing blog post, using blog in various fields like education, corporation and for mining of blog logs.

Blog distillation is performed to find blogs that match user's recurring central interest. The initiative for blog distillation started in 2007 in TREC conference [4]. Different methods like ad-hoc search and resource selection are used for distillation. Ad-hoc search is the simplest method in which a complete blog is considered as a single document [5]. Some authors use expert search techniques in this field. Balog et al. [6] and Macdonald et al. [7] assume that each post in a blog is a separate entity, and each post is an evidence of blog having interest in the topic. Arguello et al. [8] use resource selection methods to achieve an aim of finding the desired blog.

Another area of research is corporate blogs; blogs provided by corporation to their employees. Studies have been conducted to investigate corporate blogging and its objectives with respect to fortune 500 companies [9]. Research has also been done on differences between blogging activities of different cultures [10] and impact of these cultures on various aspects of the blog.

Blogging has been used in generation of requirements as well. Seyff [11] present a requirements elicitation tool for the end-users to enhance service oriented architecture. The requirement gathering tool allows end-users to blog needs without analysts' facilitation. Maiden [12] argue that in future software will be commonplace and the users will uses different means which include blog, to tell about their requirements.

Different kind of frameworks has been developed for blog software. Ferreira et al. [13] creates a framework for the development of blog crawlers. For the framework, they use techniques both from information retrieval and information extraction fields to handle the dynamic nature of blog software. Costa et al. [14] proposes a software framework for building Web mining applications in the blog world. The architecture of the proposed framework combines the use of blog crawling and data mining algorithms, in order to provide a complete and flexible solution for building general-purpose Web mining applications.

Usually, freely available blog software like Wordpress or b2evolution [15] is adapted and integrated into the web application. Generally, the web application has little amount of knowledge about the number of features that blog software can provide. Similarly, the developer is at pains to update the blog software with no design document available to understand the structure. Thus, both the developer and the web application require some kind of formal specification to ease the integration of blog into web application. During our search for research papers related blog software, integrated into web application, no results were found.

6 Conclusion

In this paper, we present a framework for integration of blog software into web application. The framework can be used during the various phases of the software development process. During the requirement generation phase, the framework can be used for specifying of requirements. The framework also aids in outlining the design during the design phase. Moreover, in the testing phase, the framework eases the verification and validation of the integrated blog software. The components of the framework are extensible in nature and can be easily updated to add new features and functionality.

References

1. Gupta, K., Goel, A.: Requirement Estimation and Design of Blog Software in Web Application (manuscript submitted for publication, 2012)
2. Gupta, K., Goel, A.: Requirement checklist for blog in web application. International Journal of Systems Assurance Engineering and Management 3(2), 100–110 (2012)
3. Wordpress, http://wordpress.com/
4. Macdonald, C., Ounis, I., Soboroff, I.: Overview of the TREC-2007 Blog Track. In: Proceedings of the Sixteenth Text REtrieval Conference, TREC 2007 (2007)
5. Elsas, J.L., Arguello, J., Callan, J., Carbonell, J.G.: Retrieval and feedback models for blog feed search. In: Proceedings of the 31st Annual International ACM SIGIR Conference on Research and Development in Information Retrieval, Singapore, pp. 347–354. ACM (2008)
6. Balog, K., de Rijke, M., Weerkamp, W.: Bloggers as experts: feed distillation using expert retrieval models. In: Proceedings of the 31st Annual International ACM SIGIR Conference on Research and Development in Information Retrieval, Singapore, pp. 753–754. ACM (2008)
7. Macdonald, C., Ounis, I.: Key blog distillation: ranking aggregates. In: Proceeding of the 17th ACM Conference on Information and Knowledge Management, Napa Valley, California, USA (2008)

8. Arguello, J., Elsas, J., Callan, J., Carbonell, J.G.: Document representation and query expansion models for blog recommendation. In: Proceedings of the 2nd International Conference on Weblogs and Social Media, ICWSM (2008)
9. Lee, S., Hwang, T., Lee, H.-H.: Corporate blogging strategies of the Fortune 500 companies. Management Decision 44(3), 316–334 (2006)
10. Jeon, S., Yoon, S.N., Kim, J.: A cross cultural study of corporate blogs in the USA and Korea. Int. J. Inf. Technol. Manage. 7(2), 149–160 (2008)
11. Seyff, N., Graf, F., Maiden, N.: End-user requirements blogging with iRequire. In: Pro-of 2012, vol. 2, Cape Town, South Africa, pp. 285–288. ACM (2010)
12. Maiden, N., Wever, A.: Requirements Analysis: The Next Generation. IEEE Software 28(2), 22–23 (2011)
13. Ferreira, R., Lima, R., Melo, J., Costa, E., Freitas, F., Pacca, H.: RetriBlog: a framework for creating blog crawlers. In: Proceedings of the 27th Annual ACM Symposium on Applied Computing, Trento, Italy, pp. 696–701. ACM (2012)
14. Costa, E., Ferreira, R., Brito, P., Bittencourt, I.I., Holanda, O., Machado, A., Marinho, T.: A framework for building web mining applications in the world of blogs: A case study in product sentiment analysis. Expert Syst. Appl. 39(5), 4813–4834 (2012)
15. b2evolution: More than a blog! http://b2evolution.net/

Demonstration of Signature Analysis Using Intelligent Systems

Aishwarya Venkatesh, Alifiya Bhanpurawala, Subhashree Chowdhury, and D.R. Kalbande

Computer Engineering, University of Mumbai, India
{aishwaryavenkatesh227,alifiyabnw,subhi.smile4ever}@gmail.com, k_dhananjay@yahoo.com

Abstract. Signature is a form of self portrayal through which the author is publicly recognized. Personality traits of a person, like his emotional stability, sensitivity, creativity, thoughtfulness and his general attitude towards himself and others, can easily be estimated by analyzing his signature. This paper proposes a new technique to implement signature analysis and predict the dominant personality traits the author exhibited while signing. For implementing this technique, the image of the signature is processed and certain characteristics of the signature are measured. These quantified values are forwarded as the input to an intelligent system, which analyzes the inputs and generates the output as a group of appropriate adjectives. Thus, the current temperament of the author is described.

Keywords: Signature Analysis, Human Behavior, Personality Traits, Image Processing, Intelligent Systems.

1 Introduction

A signature is the trademark of its author. It is supposed to manifest the author's identity and reflect his personality. Consequently, signature analysis developed as a science, in which analysts interpreted the signature of an individual and described his personality. In this intricate and perceptive science, various characteristic parameters of the signature such as pressure, size of the letters, baseline slant, letter slant, zones, length of the strokes, size of the first letter in the signature, margins, length of curves, etc. are considered to assess the signature. Each of the values, indicating the different manifestations of the various parameters, corresponds to different personality traits. Therefore, a combined analysis of these parameters gives the overall personality of the author.

Thus, using signature analysis, the dominant qualities of a person can be drawn out and the potential of an individual in various fields can be estimated. This information can prove to be of great use in human resourcing, to psychologists for analyzing people, and to everyone, for understanding the people they deal with in their day-to-day lives.

This paper describes a technique that attempts to computerize the whole procedure of signature analysis. The proposed system focuses only on four parameters. They are: (i) Baseline Slant, (ii) Size of the signature, (iii) Pressure, (iv) Spacing between the letters.

S. Unnikrishnan, S. Surve, and D. Bhoir (Eds.): ICAC3 2013, CCIS 361, pp. 199–206, 2013.

The baseline is the imaginary line on which bottoms of the letters align [2]. Pressure is defined by how much force you apply to the writing surface with the writing instrument [1]. Letter spacing is the horizontal space between the characters in the signature [2]. Size of the signature indicates only the vertical expanse of the signature, because the horizontal expanse will depend on the number of letters that make up the signature.

For demonstrating signature analysis using these parameters, the first step in this technique is to take a digital copy of the signature in the form of an image. This image is processed using MATLAB. The processing results are stored in a MySQL database. This database is forwarded to an intelligent system, which is implemented in C#.net. This system is tested using a sample set consisting of 200 signatures. On successful analysis, the system reveals the corresponding personality traits exhibited by the author of the signature.

2 Proposed Methodology

Various existing systems have analyzed different combinations of handwriting characteristics like baseline slant [4], [6], [7], pen pressure [7], size [6]. The proposed system incorporates Signature Analysis by considering the significance of four parameters namely, baseline slant, pen pressure, size of the signature and spacing between the letters. The baseline slant determines the emotional stability of the person. The system considers the three most common baselines, which are ascending, descending and straight. It interprets a moderate upward slant as a person, who is an optimist, has faith in future and feels being loved; conversely, a moderate downward slant shows pessimism, fatigue, and depression. A straight baseline denotes reliability, even temper and control of emotions. According to the pressure exerted by the author while writing, this system classifies him as a light pressure writer, moderate pressure writer or heavy pressure writer. For drawing conclusions, the system uses the signature analysis principle, which states that emotional intensity of a person is proportional to the pressure applied by an author while signing. Heavy pressure indicates a deeply intense personality and strong determination whereas light pressure reveals sensitive nature and adaptability. The size of the signature indicates the author's desire to be noticed by the world. The system hence relates very large signatures to an author, who demands for attention, while small sized signatures show the author's shyness and his analytical nature. The spacing in the signature talks about the behavior of the author in society. Closed spaced letters show the resentful, inhibited and selfish nature of the individual whereas wide spaced letters indicates that the person is understanding, extrovert and sympathetic in nature [2], [3].

For incorporating such principles of signature analysis, different systems under the scanner have employed Artificial Neural Networks [7], Segmentation method [6], Apriori Algorithm [5] and even Neuro-Fuzzy Concepts [4] for handwriting analysis and classification.

The proposed methodology demonstrates the use of an Intelligent System designed specifically for implementing the above mentioned principles and mimicking the human thought process involved during signature analysis. For achieving this, first the system focuses on recognizing the various forms in which these parameters appear in the digital image of the signature sample. These quantified outputs are then fed into the Intelligent system for interpretation. Finally the system generates a comprehensive description of the author as the result.

3 Implementation and Result

Figure 1 represents the basic architecture of the system.

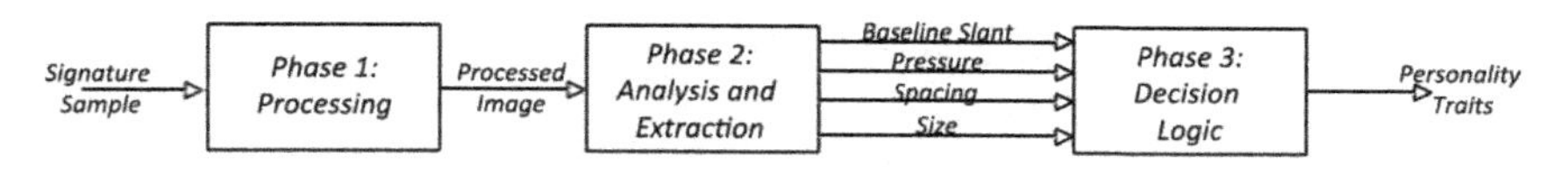

Fig. 1. Basic architecture

As the figure illustrates, the working of the system is broken down into three phases. The detailed proceedings of every phase are explained as follows.

3.1 Phase 1: Processing

To obtain the signature sample in a digital format, it is scanned and converted into an image in the Portable Network Graphics format. This image is subjected to some basic pre-processing, through which the unwanted pen marks and noise are removed. This processed image is forwarded to the next phase as the input.

3.2 Phase 2: Analysis and Parameter Extraction

Initially the image of the signature is pre-processed and subjected to standardization. Then it is subjected to the workings of four algorithms – for the extraction of each parameter. These algorithms are described as follows:

Algorithm 1: Estimation of Baseline Slant.
The baseline slant of the image is calculated by finding the centroid of every column in the image. The centroid of the column is calculated only with respect to the black pixels in the column. Then line connecting the centroid of all the columns is polygonalized. The slope of this line is calculated. This gives the baseline slant of the signature.

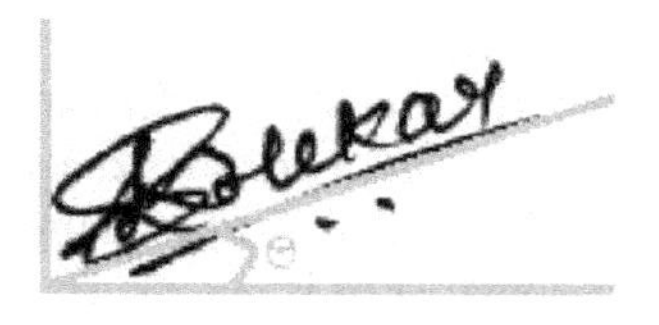

Fig. 2. Detection of baseline slant

Algorithm 2: Estimation of the Pressure Level.
A special version of the image is generated, in which the entire signature is defined on the basis of the pressure exerted while making it. This highlighting between the heavy and light pressure regions is done using a thresholding function. This version of the image is compared with the original image and the pressure level in the signature is determined.

Fig. 3. Original image

Fig. 4. Reference image after setting threshold

Algorithm 3: Estimation of the Spacing between the letters.
To measure the spacing, the breaks in the signature or the low density regions indicating the discontinuity are counted. A space counter is maintained. The number of pixels in every column is counted and stored in an array and then it is checked against a threshold. The threshold value is set to 0.08*total number of rows. If the value is less than the threshold for that particular column, then the space counter is incremented. The stretch of the signature is taken into consideration and the calculated count is used to determine the spacing percentage. The percentage ranges are then used to classify the writer's signature either as more spaced or less spaced.

Fig. 5. Spacing between the letters

Algorithm 4: Estimation of the Size of the Signature.
First, the signature image is rotated to bring it on a horizontal level (only if the baseline is ascending or descending). Of all the resulting horizontal strips of the image, the strip containing 90% of the signature is considered for further processing. This strip is called the *targetRegion*. The top and bottom boundaries of the *targetRegion* are recursively adjusted until the smallest horizontal strip containing 90% of the signature is obtained. The height of this horizontal strip is calculated and the size of the image is determined.

Fig. 6. Original image

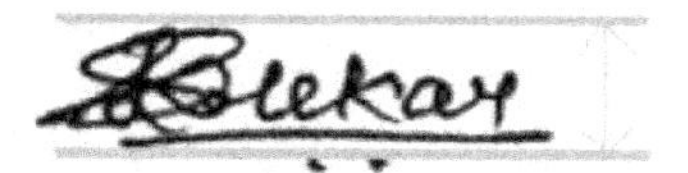

Fig. 7. Image after rotation

On completion of this phase numerical values are forwarded to the next phase as its input.

3.3 Phase 3: Decision Logic

The four inputs values indicate the current manifestation of the corresponding parameter. Using the tables and algorithms that are mentioned as follows, the system completes analyzing these inputs and provides the appropriate personality traits.

Table 1 gives the list of values corresponding to all the considered manifestations of every parameter.

Table 1. Range of values for every parameter

Parameter	Manifestation	Numerical Range
	Downward	<= -5
Baseline Slant	Upward	> 5
	Straight	>-5 and < 5
Pressure	Light	<= 20
	Heavy	> 20
Spacing	Less	<= 50
	More	> 50
Size	Small	<= 30
	Large	> 30

Different personality traits are mapped to the different ranges of these parameter values on the basis of signature analysis principles. *Table 2* lists all the chosen personality traits, along with their respective indexes, using which they are represented during the operation of this phase. The indexes are assigned randomly and hold no numeric value; instead they are pointers to the assigned personality trait, used for convenience of programming.

Table 2. Personality traits with their indexes

Index	Personality Trait	Index	Personality Trait
1	Optimistic	16	Extrovert
2	Pessimistic	17	Forgiving
3	Excited	18	Resentful
4	Happy	19	Meticulous
5	Depressed	20	Visionary
6	Dominating	21	Modest
7	Obedient	22	Boastful
8	Hostile	23	Proud
9	Passive	24	Fatigued
10	Understanding	25	Selfish
11	Adaptable	26	Focused
12	Resourceful	27	Distracted
13	Creative	28	Reliable
14	Emotionally Strong	29	Reasonable
15	Sensitive	30	Analytical

For every input *i*, value of a parameter an array of four elements, denoted by X_i[1], is generated. The first three elements of the array are indexes corresponding to the personality traits (as given in *Table 2*) that are implied by the current value of the parameter. The fourth member is called the *Deciding Factor* (*DF*) which indicates the level of influence of each parameter in the given sample. The value of *DF* is calculated as a multiple of a default value, '*n*', after comparing the deviation of the parameter's value from its closest range limit, in terms of percentage. The value of *n* is set to 100. The calculation of *DF* is done as follows:

- For the parameter with the highest percentage deviation, and hence the highest influence, $X_i[4] = 4n$
- For the parameter with lower percentage deviation, and hence lower influence, $X_i[4] = 3n$

[1] X[j] represents the member of Array 'X' at position j.

- For the parameter with still lower percentage deviation, and hence still lower influence, $X_i[4] = 2n$
- For the parameter with the lowest percentage deviation, and hence the lowest influence, $X_i[4] = 1n$

A new 2D array, Z, is generated to hold, in the first column - the list of all the personality trait indexes that appear in even one of the X_i arrays, and in the second column the priority assigned to the respective personality trait. For computing these priorities, the system considers Signature Analysis principles and the DF value in the X_i in which that particular trait appears. It also takes into account the influence that the contradictory personality traits have on each other. *Table 3* enlists all the pairs of contradictory personality traits.

Table 3. Opposing Personality Traits

Sr. No.	Pair of Opposing Personality Traits	
1	Optimistic	Pessimist
2	Excited	Fatigued
3	Joyful	Depressed
4	Dominating	Obedient
5	Hostile	Passive
6	Forgiving	Resentful
7	Boastful	Modest
8	Proud	Modest
9	Focussed	Distracted

The following algorithm is incorporated in this process:

Step 1: For every entry, set the priority to the default value i.e. 0

Step 2: For every entry, add $\Sigma X_i[4]$ to update the current values[2]

Step 3: For every entry corresponding to a personality trait that belongs in an active pair of opposing personality traits[3], subtract $\Sigma X_i[4]$ to reduce the priorities of all the opposing personality traits

---

[2] $\Sigma Xi[4]$ represents the sum obtained by adding the DF values corresponding to the parameters to which a particular personality trait is mapped.

[3] A pair of opposing personality traits is active if both the members of the pair appear in any of the input arrays *Xi*.

After the priorities for every entry is computed, the entries with the four highest priorities are selected. The personality traits corresponding to the indexes in these entries are displayed as the output of the system. Thus, the system incorporated signature analysis to determine the dominant personality traits of an individual.

4 Conclusion

This paper proposes an off-line, writer independent signature analysis system with some revised algorithms for parameter extraction. This system provides a non-subjective implementation of signature analysis which is based on principles that are generally accepted in the field of Graphology. This system disregards the concept of personal interpretation that is involved in the procedure of signature analysis. Hence it proves to be a supporting tool in this process and can be used as a reference to verify the interpretation of experts.

One improvement can involve the consideration of other characteristics of the signature like its margins, its curves and formation of specific letters, its embellishments, etc. Further improvements on this system can also involve incorporating features that aim to translate the techniques specific to an expert who is using this system to aid his analysis of signature.

References

1. Handwriting Analysis, Graphology, Personality Profile, http://www.handwritingpro.com/theory.html
2. Gardner, R.: Instant Hand Writing Analysis – A Key to Personal Success. Llewellyn Publications, St. Paul, Minnesota 55164-0383, USA
3. McNichol, A., Nelson, J.A.: Handwriting Analysis – Putting It To Work For You. Jaico Publishing House (2007)
4. Mogharreban, N., Rahimi, S., Sabharwal, M.: A combined crisp and fuzzy approach for handwriting analysis. In: IEEE Annual Meeting of the Fuzzy Information Processing, NAFIPS 2004, vol. 1. IEEE (2004)
5. Cha, S.-H., Srihari, S.N.: A priori algorithm for sub-category classification analysis of handwriting. In: Proceedings of the Sixth International Conference on Document Analysis and Recognition. IEEE (2001)
6. Prasad, S., Singh, V.K., Sapre, A.: Handwriting Analysis based on Segmentation Method for Prediction of Human Personality using Support Vector Machine. International Journal of Computer Applications, IJCA 8(12), 24–28 (2010)
7. Champa, H.N., AnandaKumar, K.R.: Artificial Neural Network for Human Behavior Prediction through Handwriting Analysis. International Journal of Computer Applications, IJCA 2(2), 36–41 (2010)

A Question Answering System Supporting Temporal Queries

Suresh Kumar Sanampudi[1] and Vanitha Guda[2]

[1] Department of IT, JNTUH College of Engg. Jagityal, Nachupally, Karimnagar Dist., AP
sureshsanampudi@gmail.com
[2] Department of CSE, Chaitanya Bharathi Institute of Technology, Gandipet, Hyderabad, AP
vanitha_cse@cbit.ac.in

Abstract. A Question Answering (Q.A) system supporting temporal queries must be able to deduce which qualitative or quantitative temporal aspect holds for a particular event or between a pair of events. As the data changes in the dynamic world it becomes important to access the information and facts which are true at current point of time. This paper presents a temporal Question Answering architecture suitable for enhancing current Q.A. capabilities with the possibility of answering temporal queries. That is, to be able to answer questions that involves events occurrence at a point or interval or time periods. Specifically, we have designed a model which is integrated with search system for specific time relevant questions. This model defines functionality process into two major steps with respect to the process of adequate response. The question is processed by decomposition unit to identify important components in the question. Answers are extracted and integrated to eruct the final answer for a given question.. Furthermore, proposed system has been evaluated with TREC 2007 question corpus which consist of 200 questions out of which 100 are temporal questions. The series of experiments are conducted for different class of questions and results are evaluated to measure the performance of our system. When compared with that of general Q.A systems our model has showed improved performance in obtaining accurate answers for temporal queries.

Keywords: Temporal Question Answering, Information Extraction, Question classification.

1 Introduction

Question Answering systems (Q.A Systems) are defined as the process of computer retrieving short and correct answers to a natural language query formulated by the user. Research in current state-of-art question answering systems [1][2][3][4][5] are unable to extract the exact one word or one sentence answers for the temporal queries such as “who is Prime Minister (PM) for India during 2010 ?”.

S. Unnikrishnan, S. Surve, and D. Bhoir (Eds.): ICAC3 2013, CCIS 361, pp. 207–214, 2013.

Temporal question answering requires the extraction of temporal information encoded in natural language text. Temporal information may exist either in explicit form such as dates and time expressions, or in implicit form such as "after", "before", "during" etc,. A temporal question answering system needs to extract times of the events occurring in the text. For e.g. answering temporal questions such as "what happened before the attack on Taj Hotel?" require the issues of temporal properties or ordering of events, which remained still as an open research problem in current Q.A Systems. Answering such temporal based questions becomes essential in the present Q.A systems, for instance the questions like "Who is the president of India during 1950-52?", needs temporal event detection, temporal context reasoning, as well as temporal context indexing to improve correct candidate answer selection.

The current Q.A systems[5] accepts user's natural language question as input, the system will submit the question to a search engine, then extract all plausible answers from the search results according to the question type identified by the question classification module, finally select the most plausible answers to return as an output.But it is unable to obtain the answers for temporal queries.

In this paper we are presenting a model which aims to enhance the current Q.A capabilities with the possibility of processing temporal questions whose answer needs to be gathered from pieces of factual information that is scattered in different documents and by increasing the precision for the answers obtained.

The remaining part of the paper is organized as follows, Section 2 explains the classification of temporal questions. Section 3 discuss about the Layered Architecture of the temporal Q.A system. In section 4, Experiments and results are discussed which exhibits the performance of our model. Section 5 concludes the paper.

2 Classification of Temporal Queries

Answering of temporal queries requires handling of either time-denoting expressions or event-denoting expressions. Time-denoting expressions comprises of temporal information that can be stated with reference to a calendar or clock system (e.g. on Friday or today, or the fourth quarter). In Time-denoting expressions temporal, reference can be expressed in three different ways:

Explicit Reference: Date expressions such as 08.04.2001 refer directly to the entries of a calendar system. Also, time expressions such as 3 p.m. or Midnight denote a precise moment in our temporal representation system.

Indexical Reference: All temporal expressions that can only be evaluated via a given index time are called indexical. Expressions such as today, by last week or next Saturday need to be evaluated with respect to the article's time stamp.

Vague Reference: Some temporal expressions express only vague temporal information and it is rather difficult to precisely place the information expressed on a time line. Expressions such as in several weeks, in the evening or by Saturday the latest cannot be represented by points or exact intervals in time.

Table 1. Classification of temporal Queries

Types in Queries				
	Simple Queries		*Complex Queries*	
Types	***Type-1***	***Type-2***	***Type-3***	***Type-4***
Type Description	Single event questions without a Temporal Expression	Single event questions with a Temporal Expression	Multiple event temporal questions with a Temporal Expression	Multiple event temporal questions without a Temporal Expression
Example query for the types	"When did Jordan close the port of Aqaba to Kuwait"? TE : -nil-	"Who won the Bharat Ratna award in 1954 ? TE: 1954	"What did Gandhiji do after the law to handle the British government in May 1893?" TE: May 1893 and multiple events	"What happened to world oil prices after the Iraqi annexation of Kuwait?". TE : Multiple Events without TE
Procedure	Without pre or post processing is required to answer this type	Temporal event reorganization is required to answer this type	Preprocessing Decomposition and temporal signal identification is must for this type	Preprocessing and Decomposition is required for this type
Complexity	Easy to handle because only single event	Easy to handle time expression is present with single event	Difficult to handle because decomposition is required for multiple events	Difficult to handle because decomposition is required for multiple events

Event denoting expressions refers to pair of events which are related with implicit temporal information for e.g. election in the phrase after the election - that serve as temporal reference pointers in building the temporal structure. For these expressions, there is no direct or indirect link to the calendar or clock system. Temporal queries are categorized as simple or complex based on number of events occurred in a question. Table-1 illustrates the classification in detail.

3 Layered Architecture of Temporal QA System

In this paper we present a layered architecture that can increases the functionality of current Q.A systems by enhancing the feature to crack complex temporal question. Temporal layer is added as an supplementary component to the existing Q.A systems [5]. Divide and conquer strategy is applied in the construction of temporal layer. The layer mainly decomposes a complex question into several simple questions. This task identifies a variety of components in the query for which answers can be extracted based on the type to which it belong to. Answers obtained to these simple questions are integrated to erect a final answer for the given complex temporal query. The tasks performed in the architecture are independent of each other but collectively provide answer for a given temporal query.

The architecture of the proposed question answering system that can answer temporal queries is revealed in figure 1. In this method a complex queries s given as an input to the temporal layer. Decomposition Phase in this layer splits the given complex question into simple questions, based on the temporal relationships [1] conveyed in the original question. These simple questions are then processed independently by a question answering engine. These simple queries are given to information processing engine for answers extraction. The answers obtained are sent to answer recomposition phase of the temporal layer. The respective answers are filtered and are integrated based on temporal relations extracted from the original complex question. Thus obtained outcome constitutes the final desired answer. These components are elaborated with the help of an example.

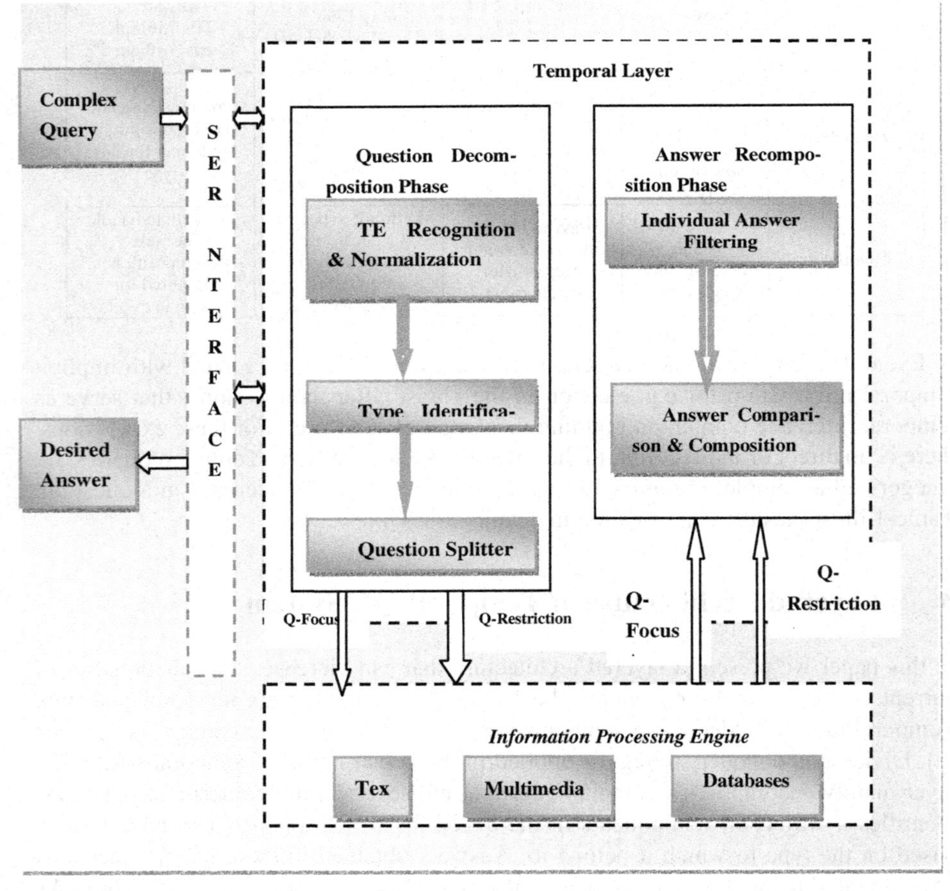

Fig. 1. Architecture of temporal Q.A system

- **Question Decomposition**

In this unit input query is traced for temporal expressions (TE). Depending on number of events and temporal information, the given query is categorized to any one of the four types as shown in table 1. If the question is found to be complex in nature then the query is split into simple questions. The decomposition of a complex question is based on the identification of temporal signals, which link simple events to form complex questions.

The task of TE Recognition and Normalization detects temporal expression in the query and are normalized to standard form. The temporal tags (TE tag with a value attributes) obtained from the questions forms the output of this phase and they are used in the Answer Recomposition Unit in order to filter the individual answers obtained by the Q.A system. The TE tag is necessary in order to determine the temporal compatibility between the answers to the Q-Focus and the answer to the Q-Restriction. Type Identification activity involves the classification of temporal queries depending on the features as described in table 1. This task is necessary for complex questions to be divided into simple ones.

Using the temporal signal as a referent, the two events related by it will be transformed into two simple questions: Question-Focus (Q-Focus) and Question-Restriction (Q-Restriction). The first sub-question is defined as the question focus (Q-Focus) and it specifies the type of information the user needs to find. Q-Focus is processed by a Q.A system, the system will return a list of possible answers. The second sub-question is called the question restriction (Q-Restriction) and the answer to this sub-question establishes the temporal restrictions on the list of answers to the Q-Focus. The Q-Restriction is constructed using the part of the complex question that follows the temporal signal. This question is always transformed to a "When" question using a set of lexical and syntactic patterns defined in the layer. When the Q-Restriction is processed by a Q.A system, only one appropriate answer is expected. For example, the question "Where did Mahatma Gandhi studied before going to London University?", is divided into two sub-questions that are related by the temporal signal "before".

Q-Focus: "Where did Mahatma Gandhi study?" and
Q-Restriction:"When did Mahatma Gandhi go to London University?".

The two simple sub-questions (Q-Focus and Q-Restriction) generated are processed by search engine, In order to obtain an answer for each of them. The simple questions generated are processed by an information processing engine. For E.g processing of Q-Focus an Q-Restriction by search engine returns the following answers:

Q-Focus Answers:
Georgetown University (1964-68)
University College London(UCL) (1888-91)
Yale Law School (1970-73)
Q-Restriction Answer: 1888

- **Answer Recomposition Unit**

Once the complex questions have been split by the Decomposition Unit into the Q-Focus and the Q-Restriction and the answers to these questions have been obtained by information processing engine. The main task of the Answer recomposition Unit is to obtain the final answer to the complex question using all the available outputs of the Decomposition Unit and the answers obtained from the search engine. The Recomposition Unit carries out Individual Answer Filtering and Answer Comparison based recomposition activities to obtain final answer for a given query.

The list of possible answers to the Q-Focus and the answer to the Q-Restriction given by the information processing engine forms the input of the Individual Answer Filtering task. It selects only those answers that satisfy the temporal constraints obtained by the TE Identification and Normalization Unit. The date of the answer should be temporally compatible [7] with the temporal tag [8] i.e the date of the answer must lie within the date values of the tag otherwise it will be rejected. Only the answers that fulfill the constraints go to the Answer Re-Composition module.

Once the answers have been filtered using the signals and the ordering key, the results for the Q-Focus are compared with the answer to the Q-Restriction in order to determine if they are temporally compatible. The temporal signal establishes the appropriate order between the answers of Q-Focus and the Q-Restriction. Analyzing the temporal compatibility between the list of possible answers of Q-Focus and Q-Restriction answer, it constructs the appropriate answer to the complex question. For e.g the processing of answers to the questions obtained in query decomposition unit is shown in figure 2.

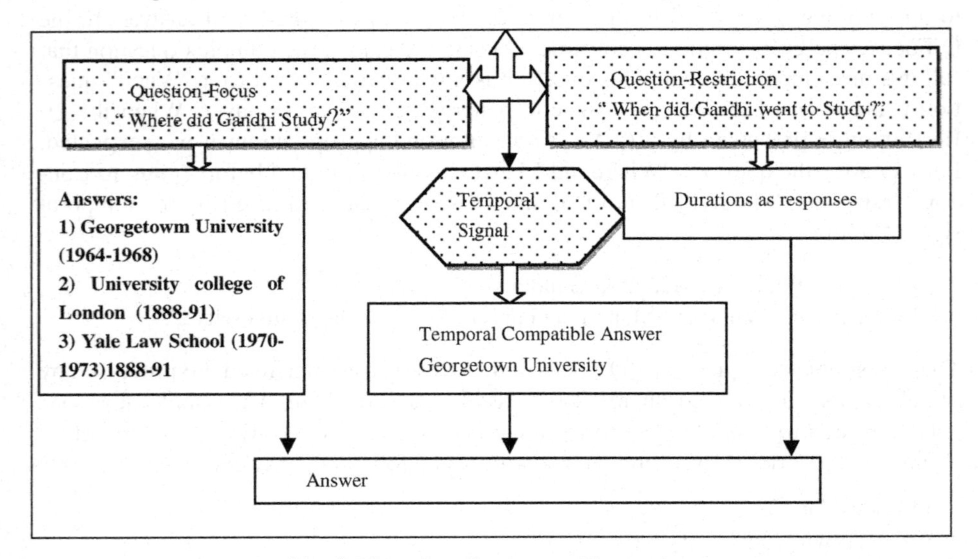

Fig. 2. Flow chart for Answer Exteraction

4 Evaluations

To evaluate the performance of our question answering system we have selected the corpus used in TREC 2007 [6] QA track. It consists of 200 questions that belong to different types. The evaluation is done manually by four annotators by considering four different aspects such as temporal expression resolution, type identification, signal identification and answer type. Precision and recall obtained for different questions are used as a measure to evaluate the performance of our system. The attributes used for evaluation are as follows:

- ACT: Number of items treated by the system POS : total number of items
- CORR: Number of items properly treated (Correct)
- PREC: Precision (CORRACT) percentage of items in the output of the system that are properly treated
- REC: Recall (CORRPOS) percentage of items treated by the system (CLEF Accuracy)
- F: ((1+β2)(P ∗R)) Combination of precision and recall in a single value. β = 1

Table 2. Evaluation of our Q.A Sytem

	POS	ACT	CORR	PREC	REC	F
Type1	**50**	**50**	**35**	**70.00%**	**70.00%**	**70.00%**
Type 2	**50**	**45**	**23**	**51.11%**	**46.00%**	**48.42%**
Type 3	**50**	**8**	**1**	**12.50%**	**2.00%**	**3.45%**
Type 4	**50**	**18**	**2**	**11.11%**	**4.00%**	**5.88%**
GLOBAL	**200**	**121**	**32**	**50.41%**	**30.50%**	**38.01%**

The results, in terms of precision and recall are shown in Table 2. In this evaluation, 50 queries are posed for each type of category, out of which Type1 and Type2 queries are properly recognized by our system. In Type3 and Type 4 over all 8 and 18 questions are recognized remaining are not treated due to wrong detection of temporal signal. Globally from 200 queries 121 are recognized out of which 32 are correctly evaluated resulting with the precision of 50.41% and recall of 30.50%. The results of evaluation are compared with general Q. A's like START and Answer Bus and it is observed that their systems had shown lowest precision and recall values when compared with our system.

5 Conclusions

This paper extended a temporal layer to present existing Q.A systems. Insertion of this layer augmented the performance of system by answering diverse class of temporal queries. The system involves two major tasks like question decomposition unit and answer re-composition unit for correctly answering a given question. The evaluation of the proposed Q.A system is carried out on TREC 2007 data set. Precision and recall are calculated to measure the performance of our Q,A system. The system has found to be answering most of the simple questions and a few complex questions. As a conclusion, it could be said that the results of our Q,A system are found to be encouraging when compared with that of other general Q,A systems. In future, our work it is directed to refine the proposed system in order to be able to effectively answer more kinds of complex questions.

References

[1] Allen, J.F.: Maintaining Knowledge about Temporal Intervals. Communications of the ACM 26(1), 832–843 (1983)

[2] Saquete, E., Martínez-Barco, P., Muñoz, R., Vicedo, J.L.: Splitting Complex Temporal Questions for Question Answering systems. In: Proceedings of the 42nd Annual Meeting on Association for Computational Linguistics, pp. 566–573 (2004)

[3] Katz, B., Lin, J.: START and Beyond. In: Proceedings of the 6th World Multi conference Systemics, Cybernetics and Informatics (2002)

[4] Katz, B.: Annotating the World Wide Web using Natural Language. In: Proceedings of the 5th Conference on Computer Assisted Information Searching on the Internet (1997)

[5] Zhang, D., Lee, W.S.: A Web-based Question Answering System (2002)

[6] TREC. Text retrieval Conference, `http://trec.nist.gov/`

[7] Saquete, E., Muñoz, R., Martínez-Barco, P.: TERSEO: Temporal Expression Resolution System Applied to Event Ordering. In: Matoušek, V., Mautner, P. (eds.) TSD 2003. LNCS (LNAI), vol. 2807, pp. 220–228. Springer, Heidelberg (2003)

[8] Schilder, F., Habel, C.: From Temporal Expressions to Temporal Information: Semantic Tagging of News Messages. In: Proceedings of the ACL 2001 Workshop on Temporal and Spatial Information Processing, ACL 2001, Toulouse, France, July 6-11, pp. 65–72 (2001)

Domain Identification and Classification of Web Pages Using Artificial Neural Network

Sonali Gupta and Komal Kumar Bhatia

Department of Computer Engineering, YMCA University of Science & Technology
Faridabad, India
Sonali.goyal@yahoo.com, Komal_bhatia1@rediffmail.com

Abstract. A huge amount of data has been made available on the WWW [3] lately most of which remains inaccessible to the usual Web crawlers as those web pages are generated dynamically in response to users queries through Web based search form interfaces [5, 6, 9]. A Hidden Web crawler must be able to automatically annotate such Hidden Web data. The goal can only be accomplished if the crawler has been provided with some knowledge or data that pertains to a domain similar to that of the search form interface. The paper seems to provide a solution in this regard by exploiting the information present in the HTML structure of the Web pages, efficiently obtaining domain specific data to facilitate the crawler's access to the dynamic web pages through automatic processing of these search form interfaces. Finding the domain of the webpage further eases the process of organization and understanding of the web content.

Keywords: World Wide Web, Hidden Web, Crawlers, domain identification, classification, HTML structure.

1 Introduction

Over the years, the web has been rapidly "deepened" with the growth of countless searchable databases that provide dynamic query-based data access through Web based search form interfaces, the so-called "hidden web" [5, 6]. The Hidden Web has been estimated to be significantly large and of higher quality [5, 9] than the so-called "Surface Web" that is indexed by search engines [3, 6]. However, the usual crawlers are unable to acquire and search material from the Hidden Web as these search interfaces have been designed primarily for human understanding. Accessing the contents of the Hidden web may prove useful for extending the content visibility on traditional search engines index [3, 5, 6, 8, 9] or forming a concept hierarchy for a domain [7, 8]. This has necessitated the need of suggesting innovative measures that would effectively help in accessing the Hidden Web. A critical step involves facilitating the crawler's act of automatically processing these Web based search form interfaces [5,9] thus making these crawlers intelligent enough to behave likewise and automatically enter legitimate values in the form fields to extract the hidden data. To address this problem, a possible strategy has been presented by the creation of numerous domain specific data repositories with each one containing data from a different domain of

S. Unnikrishnan, S. Surve, and D. Bhoir (Eds.): ICAC3 2013, CCIS 361, pp. 215–226, 2013.

consideration. The process of the creation of repositories has been based on the domain identification of the Web pages that can be reached by following hyperlinks through them. In this paper, a novel approach that facilitates extracting data from the Hidden Web based on some learned knowledge from the pages of the Surface Web has been proposed.

Existing algorithms [16, 17, 18, 19] have been relying over the usage of text content of a web page for the purpose. But, web pages contain a lot more information apart from their text content [12], to be utilized for their efficient understanding: like determining its keywords, predicting its domain information etc. The approach predicts the domain information of the web page based on its HTML elements like <META> tag and the <TITLE> tags [14, 18, 19]. Therefore, by combining and exploiting the information present in the META tags and the TITLE tag, an efficient way for finding domain specific data has been provided. As has been found, HTML-meta tags for keywords and descriptions happen to be a rich source of information indicating the domain of a web page and are used by several major search engines for ranking [21] and display of search results, they haven't been used widely enough to help organize and classify domain specific data [7, 8].

Another distinguishing feature of our approach is the use of an ANN [2, 4, 10, 13] that has been first trained by using a set of exemplary documents and later on used to generate domain specific knowledge from the whole new set of documents with the help of the learnt data. A critical look at the available literature indicates that most of the work done in the field of domain identification has been carried out using clustering approaches, Bayesian Networks, Decision Trees, Support Vector Machines etc. [16, 20].

2 Artificial Neural Networks

Artificial Neural Networks (ANNs) are relatively crude electronic models based on the neural structure of the human brain. NNs are powerful techniques for representing complex relationships between inputs and outputs [4, 21].

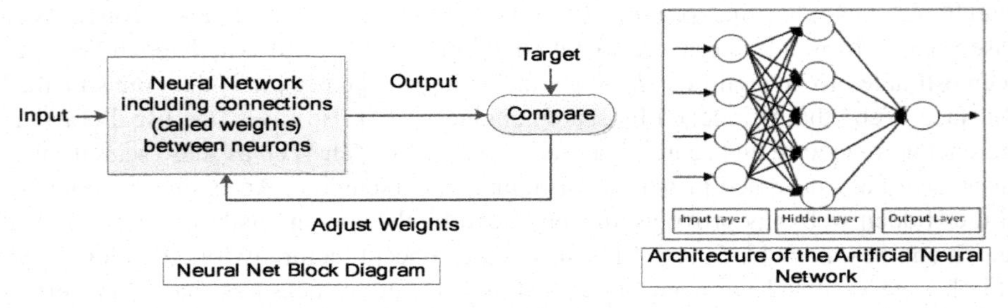

Fig. 1. Block Diagram of a basic Neural Network (back propagation) [21] and the abstract architecture of the same as used by the proposed system

A Neural networks is typically composed of simple elements operating in parallel that can be trained to perform a particular function by adjusting the values of the connections (weights) between elements, so that a particular input leads to a specific target output [4, 16]. Such a situation is shown in Fig 1. There, the network is adjusted, based on a comparison of the output and the target, the network output matches the target. Typically many such input/target pairs are used, in the learning process to train a network.

The proposed system uses a back propagation neural network model as a basic network that has been designed with three layers, one input layer, one output layer and one hidden layer as shown in the architecture of Fig 1. Table 1 illustrates the biases used at each of the three layers and illustrates the various activation functions that have been used by our network model.

Table 1. Biases and Activation Functions Used

Biases	
Number of Inputs	80 (20*4)
Number of Neurons in Input layer	4
Number of Neurons in Hidden layer	5
Number of Neurons in Output layer	1

Activation Functions	
For Input layer	Piece-wise linear
For Hidden layer	Sigmoid
For Output layer	Sigmoid
Error criteria used	Mean Squared Error
Target accuracy	0.00000015

3 Proposed Work

In this paper, a novel approach for organizing domain specific information into domain specific data repositories has been proposed and implemented using the HTML elements [1, 14, 15, 18, 19] of a webpage and a neural network. Our work involves creating four different domain specific repositories for four domains, i.e., entertainment, food, medicine and sports that can be extended to include any number of such domain specific data repositories as per the system requirement.

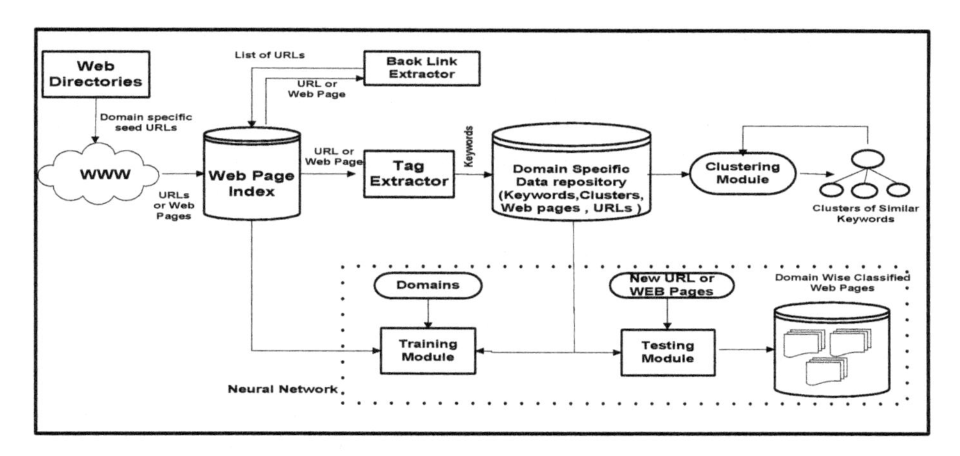

Fig. 2. The proposed domain identification system for the creation of Domain Specific data repositories

Fig 2 shown next illustrates the architecture of the proposed system which consists of the following components and modules:

- Tag Extractor,
- Back-link Extractor,
- A Web page index to store webpages and their URLs,
- A Clustering Module,
- A domain-specific data repository that contains keywords, clusters, web pages and URLs,
- A repository of web pages classified as belonging to domains as used by our system, and
- A Neural Network model that constitutes a training module and a testing module.

Initially, a list of seed URLs classified as belonging to a particular domain is taken (either from a search engine or from the various web directories) and the corresponding web pages are downloaded from the World Wide Web. The next step prepares the index to store the downloaded web pages. Also, the taken initial set of URLs or seeds have been directly stored in an index in our system.

Every time a URL or web page is taken from this index, it is fed to the meta-tag extractor. The meta-tag extractor extracts meta-tag and title keywords associated with the URL or web page up to a pre-specified depth. However, if no meta-tags and title keywords are found, back-links are extracted for that URL or web page by a back-link extractor. The extracted back-links are further added to the index and it is updated accordingly for extracting more new keywords of a domain. Further, the extracted keywords may be grouped together under a cluster based on various similarity metrics that are already known in the art, by using a clustering module. The clusters of similar keywords that belong to the domain under consideration are saved in its associated domain-specific repository of keywords. Thereafter, the keywords and their clusters that have been stored in the domain-specific repositories are assigned weights. The process of calculation of weights and their assignment to every keyword/cluster is discussed in detail later in section 3.2.

As a next step, the neural network that have been used by our system for the purpose of classification purpose is trained based on the assigned weights to clusters/ keywords associated with a web site whose domain is already identified. Next, the neural net is tested on a web site that needs to be classified into its topical domain by fetching the weights from the created domain-specific repositories. The step is discussed in detail in section 3.3. The classified web sites, thereafter, are stored in a repository as per their respective domains as elaborated in section 3.4.

3.1 Keyword Extraction Using META and TITLE Tags

The system starts by considering a set of web pages whose domain is already identified. These web pages have been utilized for the creation of training set required in the later stages of our system when the Neural Network has to be trained and tested. In our approach, the URLs of web pages that have already been identified for their domains are fed to a tag extractor which extracts meta-tags and title keywords from those URLs by traversing the URL up to a depth as specified by the system. However, if a URL or a web page does not contain meta-tags and title keywords, the back-links for the corresponding URL are extracted by a back-link extractor (as shown in Fig 2). The back-links are, further, fed to the meta-tag extractor which extracts the meta-tags and title keywords for those URLs. Thereafter, the extracted keywords are saved for future reference. The whole process of meta-tag and title extraction is repeated for the extracted back-links.

The Fig 3 below represent a snapshot of the system when a URL has been provided to the tag extractor for extracting keywords from the Meta and Title tags. Also the required depth up to which the keywords must be extracted has been specified.

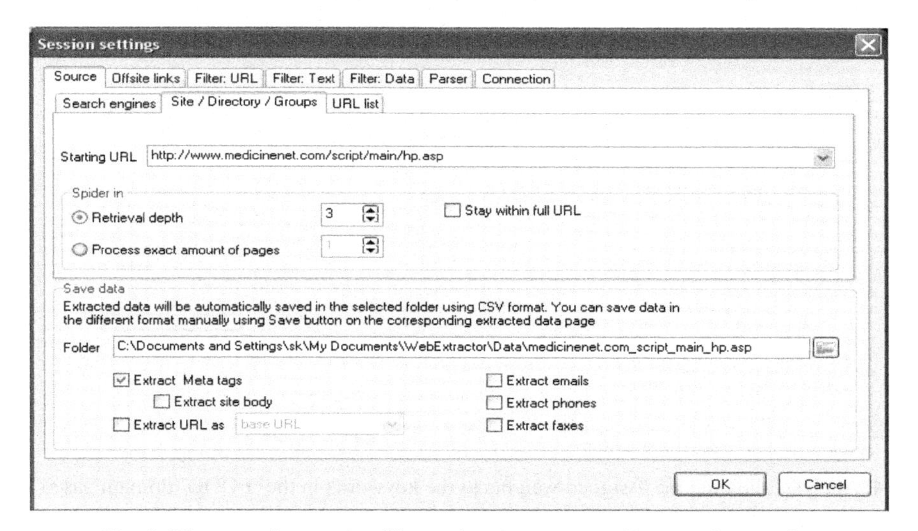

Fig. 3. The sytem's snapshot illustrating the process of keyword extraction

3.2 Assigning Weights to the Various Keywords Extracted in Step 1

After meta-tags and title keywords have been extracted, they are stored in a repository where the keywords are categorically saved (i.e., saved domain-wise). In order to assign weights to the keywords based on their no. of occurrences, the extracted keywords have been analyzed.

Fig 4 shows a tabular representation of the analysis where column1 in the table lists the keywords that belong to 'food' domain, column 2 stores the name of the domain to which the keywords belong, column 3 stores the number of occurrences of each keyword or cluster of keywords across all the web pages that have been found as belonging to the specified domain, column 4 stores the weight that has been calculated using the formula mentioned in equation 1 below and column 5 stores no. of web sites traversed for a particular domain.

$$Weight\ (x) = \frac{No.of\ occurrences\ of\ a\ keyword\ belonging\ to\ a\ domain}{Total\ no.of\ web\ pages\ traversed\ for\ that\ domain} \quad (1)$$

However, before assigning weights, keywords with similar context may be grouped together in order to form small clusters of keywords. For example, the following keywords with similar context may be clustered together:

- carbohydrates, fats, proteins, minerals, vitamins
- hills, mountains, hill stations
- sex, sexual, sexual health
- swim, swimmer, swimming, swimming pool, swimsuit

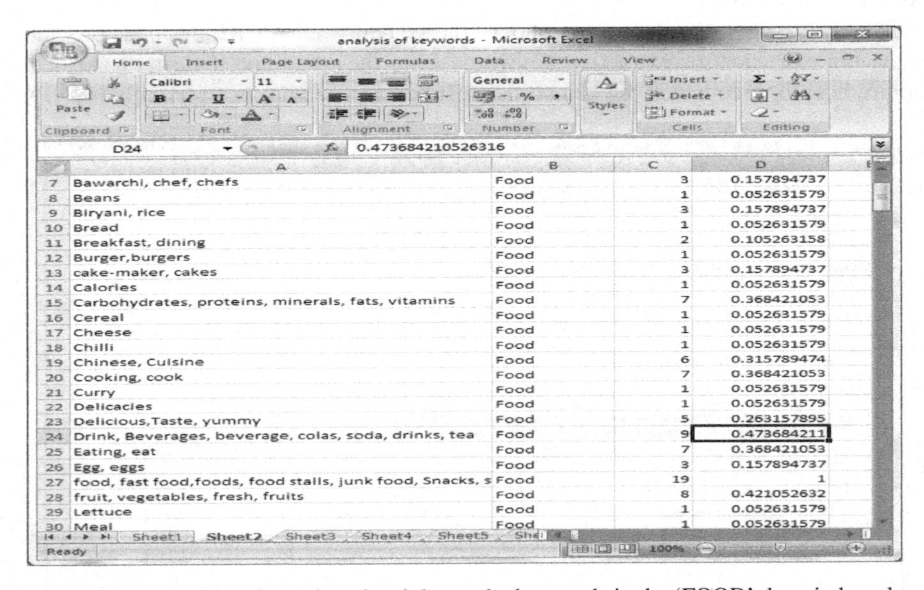
analysis of keywords - Microsoft Excel

D24 0.473684210526316

	A	B	C	D
7	Bawarchi, chef, chefs	Food	3	0.157894737
8	Beans	Food	1	0.052631579
9	Biryani, rice	Food	3	0.157894737
10	Bread	Food	1	0.052631579
11	Breakfast, dining	Food	2	0.105263158
12	Burger,burgers	Food	1	0.052631579
13	cake-maker, cakes	Food	3	0.157894737
14	Calories	Food	1	0.052631579
15	Carbohydrates, proteins, minerals, fats, vitamins	Food	7	0.368421053
16	Cereal	Food	1	0.052631579
17	Cheese	Food	1	0.052631579
18	Chilli	Food	1	0.052631579
19	Chinese, Cuisine	Food	6	0.315789474
20	Cooking, cook	Food	7	0.368421053
21	Curry	Food	1	0.052631579
22	Delicacies	Food	1	0.052631579
23	Delicious,Taste, yummy	Food	5	0.263157895
24	Drink, Beverages, beverage, colas, soda, drinks, tea	Food	9	0.473684211
25	Eating, eat	Food	7	0.368421053
26	Egg, eggs	Food	3	0.157894737
27	food, fast food,foods, food stalls, junk food, Snacks, s	Food	19	1
28	fruit, vegetables, fresh, fruits	Food	8	0.421052632
29	Lettuce	Food	1	0.052631579
30	Meal	Food	1	0.052631579

Fig. 4. Analysis showing the assigned weights to the keywords in the 'FOOD' domain based on the number of occurrences

Herein, similar context may include keywords with same base forms, keywords that are frequently found together, keywords with similar meanings and the like. A total of 258 clusters of keywords have been prepared to be used by our proposed system. Further, it is assumed that every domain has a unique set of keywords, i.e., no two domains have a common keyword. These keywords and their corresponding clusters are stored in a 'domain-specific repository of keywords and clusters' as shown in Fig 2, along with their computed weights.

3.3 Training the Neural Network

After the keywords have been found and assigned weights for every web page, they are used for training the neural network so as learn what kinds of keywords belong to which domain. For example, the keywords such as:

- 'fun, humor, jokes, travelling, tourism' etc. belong to entertainment domain,
- 'burger, snacks, restaurants, cuisine' etc. seems to belong to food domain,

- 'doctor, patients, nurses, treatment, illness' etc. are expected to belong to medicine domain, and
- 'cricket, stadium, team, matches, score, and rank' etc. seem to belong to the domain of sports.

For every web site that is traversed for a domain, an input and output matrix is prepared that contains the keywords and their associated weights where the weights are fetched from the domain specific data repository that also contains the keywords and clusters. These input and output matrices are fed to the neural network during the training process. The domains have been represented in our system by assigning integral values to each like Entertainment=0; Food=1; Medicine=2 and Sports=3and the proposed algorithm that has been used by our system for training the neural network for web page domain identification and classification is depicted in Fig 5 below.

Table 2. Training data: Input & Output Matrices

Training Data (Input Matrix)																					
Domains \ Keyword Weights	K1	K2	K3	K4	K5	K6	K7	K8	K9	K10	K11	K12	K13	K14	K15	K16	K17	K18	K19	K20	Sum
Entertainment (0)	0.2	0.27	0	0	0	0	0	0	0	0	0	0	0	0	0	0	0	0	0	0	0.47
Food (1)	0.36	1	0.36	0.42	0.36	0.1	0	0	0	0	0	0	0	0	0	0	0	0	0	0	2.6
Medicine (2)	0.04	0.18	0.54	0.27	0.09	0.04	0.59	0.13	0.09	0.63	0.13	0.95	0.04	1.04	0.04	0.04	0.04	0.13	0.59	0.04	**5.64**
Sports (3)	0.16	0.05	0.05	0.33	0.16	0.05	0.11	0.22	0.27	0.16	0	0	0	0	0	0	0	0	0	0	1.56
Testing Data (Output Matrix)																					
Result (Domain-Wise)	1	1	2	1	1	1	2	3	3	2	2	2	2	2	2	2	2	2	2	2	**2**

Input Matrix: The input matrix used for training (as shown above in table II) is a 21x4 matrix that is prepared for every web site where we provide an input of 20 keyword weights, along with their row-wise 'sum' in the 21st column for the domains used by our system (i.e. entertainment, food, medicine and sports). For example, the weight in the first column for entertainment domain is 0.2. It has been calculated by using the afore-mentioned formula for a 1-keyword cluster 'beach'. Similarly, all other keyword clusters are also assigned weights using the afore-mentioned formula. The last column depicts the row-wise 'sum' of all the assigned weights. Herein, the row-wise sum of entertainment domain is 0.47. However, if a web site consists of keywords less than 20 for a particular domain, those entries are provided an input of 0, as can be depicted from the table below.

Output Matrix: The output matrix used for training is also 21x1 matrixes similar to the input matrix which reflects the domain of the web site to be classified based on entry in the last cell. For every entry in the input matrix, the maximum value is found in every column. For example, the maximum entry in first column of input matrix as shown in table II above is 0.36 which belongs to food domain. The code for food domain is 1, which is reflected in the first column of output matrix. Similarly, the whole output matrix is prepared.

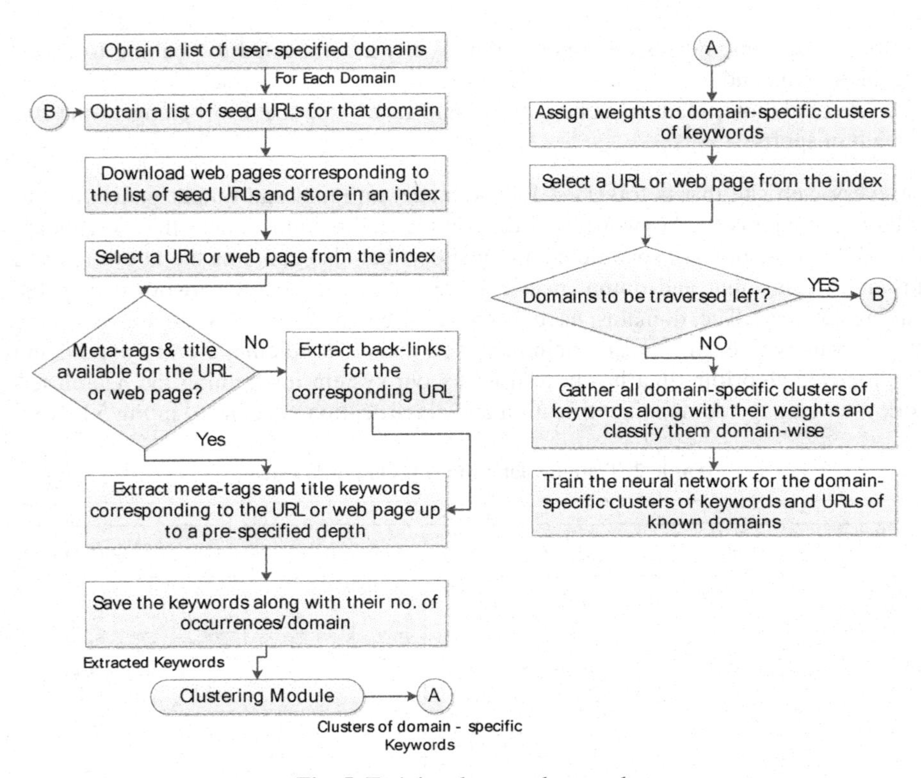

Fig. 5. Training the neural network

The overall output (i.e., the domain of the web site) is reflected by the entry in last cell i.e. '2' in the output matrix which represents the code for medicine. The last cell represents row-wise 'sum' and column-wise 'maximum' value amongst the sums of all the 4 domains. The output can also be supported by counting the no. of occurrences of each domain in the output matrix. For example, here, the no. of occurrences for entertainment domain is zero, for food domain is five, for medicine domain is thirteen and for sports domain is two. This implies that the maximum no. of occurrences is for medicine domain. The domain, whose sum and no. of occurrences would be 'maximum', will be the predicted domain for the corresponding web page. For example, the following data belongs to a web site of medicine domain.

3.4 Creation of Domain Specific Data Repositories (Testing the Neural Network)

After the neural network has been trained with web pages whose domain is already identified, the trained neural network is used for predicting the domain of web pages whose domain needs to be identified, in order to classify them. Again, here input matrices that have been prepared for various web sites are fed to the neural network. The network predicts the domain of the web site based on the input matrix fed to it.

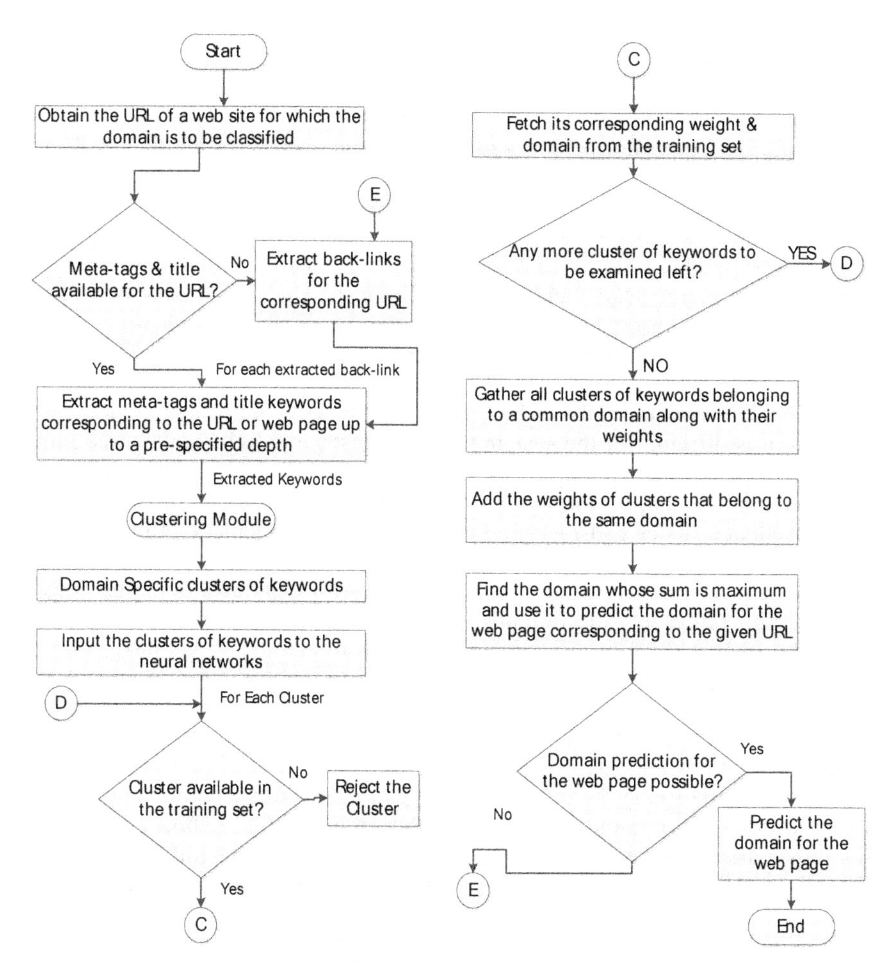

Fig. 6. Flowchart for testing the neural network

Further, for every cluster that has been obtained, it is checked whether its keywords (or a subset of keywords of the cluster) exist in the domain-specific repository or not. If the keywords of the cluster (or a subset of keywords of the cluster) are present in the domain-specific repository, weight associated with the corresponding cluster and the associated domain is fetched from the repository. However, if none of the keywords of the cluster exist in the repository, the system simply discards the cluster. After the weights and domains of all clusters have been fetched, they all are gathered together and are separated domain-wise for the corresponding web page or URL. Thereafter, the weights are input to the testing module in order to predict the domain of the URL or web page and classify it accordingly, as illustrated by the flowchart in Fig 6 above.

4 Result Analysis

Various URLs have been examined by our proposed system in order to create the domain specific data repositories that will facilitate the process of automatically filling the search form interfaces by any Hidden Web Crawler. Domain prediction has been done by our system on the basis of the cell having 'maximum-sum' under the column labeled sum of table 2. It can be depicted from the graph in Fig 7.that our system has correctly classified 89.75% of the whole set of URLS that have been considered whereas the percentage stands to just 7.69 for the URLS that have been incorrectly classified. The reason behind this inaccuracy must have been the fact that though web pages contain useful features as discussed above but, these features are sometimes missing, misleading or unrecognizable in some particular web pages. For example, web pages containing large images or flash objects but little textual content. In such cases, it is difficult for the system to make reasonable judgments based on the features on the page. Our system deals this problem to some extent by extracting hints from neighboring pages (through a back link extractor) that are related in some way to the webpage under consideration and supply supplementary information necessary for prediction thereby diminishing the value to just 7.69%.

The graph in Fig 8 represents the systems accuracy for candidate domains. The X-axis in the graph represents the domain of web pages i.e. Entertainment has been represented by integer 0 while the domains food, medicine and sports have been assigned the values 1, 2 and 3 respectively. The Y-axis in the graph represents the systems performance and accuracy of domain prediction corresponding to the various candidate web pages from those domains. As can be depicted from the graph in Fig 8, the system shows the worst performance for the web pages that actually belong to the entertainment domain giving the interpretation that page from such a domain is likely to have keywords from a number of various other domains like hobbies of cooking, listening music, travelling, playing (sports) or current news in the similar domains etc. For example: a URL, http://www.cookingchanneltv.- com that belongs to entertainment domain in reality, the system predicts to be belonging to food domain, which is incorrect due to the numerous words present from the food domain. For web pages in most of the other domains, the system achieves nearly 95 % of identification accuracy while raising the percentage to 100 for some domains. Thus, the system outperforms many of the existing [11] systems for the purpose.

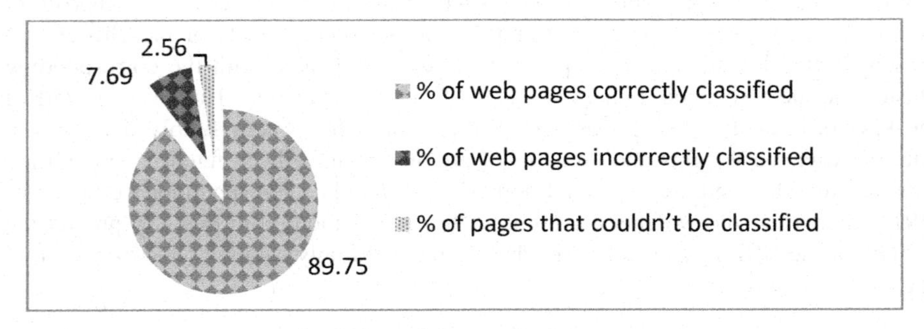

Fig. 7. Overall System Performance

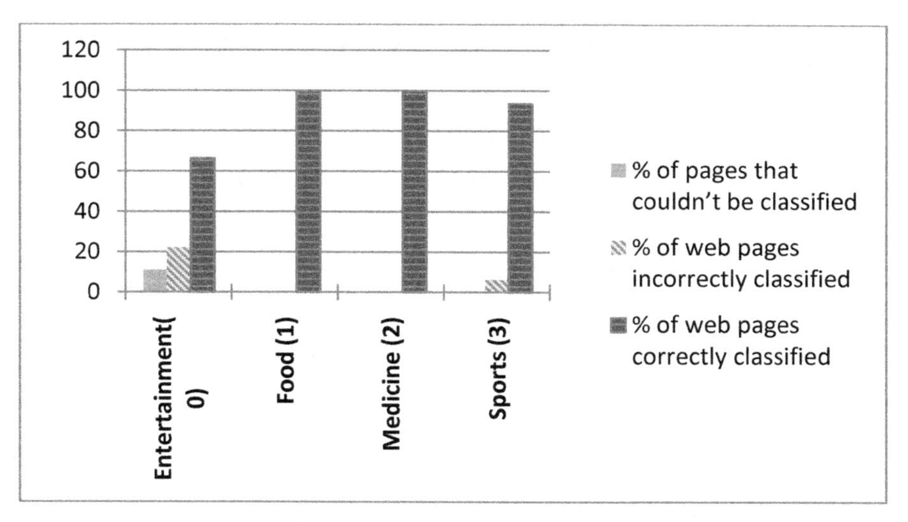

Fig. 8. Accuracy of domain prediction for various candidate domains

5 Conclusion

In this proposal, both meta-tags and title tag keywords have been used for the purpose of domain prediction. The approach seems of help to any Hidden Web crawler which can utilize the domain information from the pages that lie on the Surface of the Web to create numerous domain specific data repositories and can comprehend the process of automatically filling search form interfaces to extract the data hidden in databases behind them. These data repositories being domain specific will act as a source of information and knowledge to the crawler.

As good quality document summarization can accurately represent the major topic of a web page, so can be used to increase the accuracy of domain identification.

However, the proposed system suffers from the drawback that since most HTML tags are oriented toward representation rather than semantics, web page authors may generate different but conceptually equivalent tag structures. Therefore, using HTML tagging information in web classification may suffer from the inconsistent formation of HTML documents.

References

1. Meta tags, Frontware International
2. The Maths Work, `http://www.mathworks.com/products/neuralnet/`
3. Lawrence, S., Giles, C.L.: Searching the World Wide Web. Science 280, 98–100 (1998), `http://www.sciencemag.org`
4. Russell, S., Norvig, P.: Artificial Intelligence: A Modern Approach. Prentice Hall, London (2003)

5. Raghavan, S., Garcia-Molina, H.: Crawling the Hidden Web. In: 27th International Conference on Very Large Databases, VLDB 2001, Rome, Italy, September 11-14, pp. 129–138. Morgan Kaufmann Publishers Inc., San Francisco (2001)
6. Bergman, M.K.: The deep web: Surfacing hidden value. The Journal of Electronic Publishing 7(1) (2001)
7. Diligenti, M., Coetzee, F., Lawrence, S., Giles, L., Gori, M.: Focused crawling using context graphs. In: Proc. of 26th International Conference on Very Large Data Bases, Cairo, Egypt, Chakrabarti, pp. 527–534 (2000)
8. Chakrabarti, S., van den Berg, M., Dom, B.: Focused crawling: a new approach to topic-specific web resource discovery. Computer Networks 31(11-16), 1623–1640 (1999)
9. Gupta, S., Bhatia, K.K.: Exploring 'hidden' parts of the Web:The Hidden Web. In: Proceedings of the International Conference on ArtCom 2012. LNEE, pp. 508–515. Springer, Heidelberg (2012)
10. Manchanda, P., Gupta, S., Bhatia, K.K.: On the automated Classification of Web Pages Using Artificial Neural Network. IOSR Intl J. of Computer Engg. 4(1), 20–25 (2012)
11. Yi, J., Sudershesan, N.: A classifier for semi structured documents. In: KDD 2000, Boston, USA (2000)
12. Pierre, J.M.: Practical Issues for Automated Categorization of web pages (September 2000)
13. Kamruzzaman, S.M.: Web Page Categorization Using Artificial Neural Networks. In: Proc. of the 4th Intl Conf. on Electrical Engg. & 2nd Annual Paper Meet, January 26-28 (2006)
14. Qi, X., Davison, B.D.: Knowing a web page by the company it keeps. In: International Conference on Information and Knowledge Management (CIKM), pp. 228–237 (2006)
15. Qi, X., Davison, B.D.: Web Page Classification: Features and Algorithms, Department of Computer Science & Engineering, Lehigh University (June 2007)
16. Xhemali, D., Hinde, C.J., Stone, R.G.: Naïve Bayes vs. Decision Trees vs. Neural Networks in the Classification of Training Web Pages. Intl J. of Computer Sc. IJCSI 4(1) (2009)
17. Yu, H., Chang, K.C.C., Han, J.: Heterogeneous learner for web page categorization, University of Illinois at Urbana-Champaign
18. An, A., Huang, X.: Feature selection with rough sets for web page categorization. York University, Toronto
19. Asirvhatam, A.P., Kranti, K.R.: Web Page Categorization based on Document Structure. International Institute of Information Technology, Hyderabad, India 500019
20. Riboni, D.: Feature Selection for Web Page Classification, Universita' di Milano, Italy
21. Attardi, G., Gulli, A., Sebastiani, F.: Automatic Web Page Categorization by Link and Context Analysis

Artificial Intelligence Based Integrated Cricket Coach

Chirag Mandot and Ritika Chawla

Dwarkadas J. Sanghvi College of Engineering, Vile Parle (West), Mumbai-400056, India
chirag_mandot2000@yahoo.com, ritschawla@yahoo.co.in

Abstract. In the 21st century, the population is increasing exponentially with time. Majority of the people in the developing as well as underdeveloped countries are subdued by poverty. This deeply hampers their progress in the field of their interest. Many aspiring cricketers cannot be a part of good training and expert coaching due to lack of resources. For this, we are developing a system called artificial intelligence based integrated cricket coach which will be under the government authorities. The upcoming cricketers can use it to practice and gain expertise in cricket. The system will be initially trained by expert coaches by giving various inputs and after the complete learning process it can be used to train individuals. It is an integrated system which has expertise in all aspects of cricket like batting, bowling and fielding.

Keywords: Sports Coach, Intelligent Game Tutor, Cricket Coach, Artificial Intelligence, Neural Networks.

1 Introduction

Cricket is an internationally recognized game and with advancement in technology, this game has witnessed major changes in itself. New technology has made the observation of the game intricate and minute details of the game can be easily observed. The technology has great impact on the training of the players to make them more efficient by providing them the game environment. Majority of individuals, approximately 85-90%, play cricket and rate this game as their best choice for entertainment in India. With the country having 70 percent of its population below poverty line and living in rural areas[1], it is not possible for upcoming players to receive cricket coaching which is at par with international standards. In a survey conducted by us in few nearby villages, it has been revealed that people are not receiving desired training, and few training institutes which are providing rudimentary training are not affordable and moreover the training is far below the expected standards.

The previous research work focused on development of equipments like bowling machine, analysis of the game through video inspection and other equipments which has led to easy observation of game. There are other systems which compare different batting styles using fuzzy logic. The use of Virtual Reality to train players with a virtual environment of playground has also made advances to the training purpose. Therefore, a need for a system arises which provides actual comprehensive playing experience and where the need of actual coach is avoided. Hence the system can be used in places where the presence of excellent human coach is difficult.

S. Unnikrishnan, S. Surve, and D. Bhoir (Eds.): ICAC3 2013, CCIS 361, pp. 227–236, 2013.

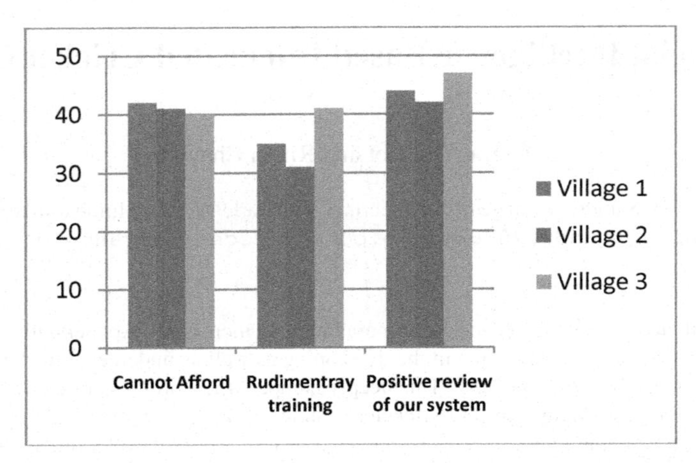

Fig. 1. Survey Report based on questionnaire regarding acceptance of the system

We have planned to develop an integrated system based on the previous research done, which will provide training considering all the aspects of game, which includes batting, bowling and fielding. The system is being planned to be kept with the government authorities so that people with different interests like batting, bowling or fielding, can use it. This makes it imperative for the system to have all the features. The technologies used by the system are Image Processing and Artificial Intelligence. The system will be programmed and implemented in Matlab software language. The system takes inputs based on Image Processing and the computation is carried out using Neural Network which helps in training the system. The system aims at devising its own hardware components which includes camera and bowling machine assembly, bat with installed gyroscope sensor and a camera installed fielding training machine. The system will be based on user login and after the login; the session of the player will begin. The system presents rich user interface which will display all the vital details to the player after the training session ends. Along with that, an analysis where details related to area of improvements will be provided as well as stored in the machine. Further, next time when the user logs in, the machine will adapt according to the user. The system will develop training algorithm based on the user performance, for example if the user is weak in playing spin bowling, faces difficulty in playing off-side balls, and usually gets bowled, then the system will decide the series of bowling process accordingly and will try to improve the weak areas by providing detailed graphical report and improvements to be implemented. But there are a few limitations as for precise results huge computation is required and the system requires initial coaching to give efficient output at a later stage. Hence, we are building a system which will train cricket enthusiast in a comprehensive way and help them to achieve acme quality of training.

2 Methodology

2.1 Image Processing

Image Processing plays a very crucial role on which the system is dependent. It provides majority of the input data. Every move of the bowler and batsman along with the movement of the ball and bat needs to be recorded and various parameters need to be calculated. For this image processing is employed. The bowler's arm is at what angle when he throws the ball and the bowler runs at what speed, are the parameters calculated with the help of image processing. Similarly, the body parts of the batsman, that is, the arm up to the elbow, the knee and the head should be detected. For this we use a specific color, for example, red, which will be the color of the batsman's costume. The parameters calculated will be the angle of the bat, the position of the batsman's leg (for deciding LBW) and the point where the ball hits the bat. The path of the ball can also be reviewed and whether it hits the wicket or not, can be determined.

2.2 Neural Network[2]

Neural Network will help to make the system self learning and adaptive. The system will be self-learning and improving unit, in which after gaining experience, the system as a whole develops itself and extrapolates new information based on its own intellect. The system will be encountered with huge amount of input data related to various configurations of cricket. We plan to use neural network technique called Radial Basis Function network which uses radial basis functions as activation function. It will assist us in classification, approximation and forecasting of the output data. The data which will be used for training is classified as Training Set, Generalization Set and validation Set. The training data is used to feed a large amount of input in the system with various probabilities. In training set, after the input is given to the system, error is computed and accordingly the classification is done. The generalization data manages to handle the unseen data after the training set has been fed to the system. The validation data will be run through the system when the training has finished and we get final validation errors which are improved in the next cycle. In the learning procedure, the input is given and an expected output is also fed, therefore, the system will compare the expected output and resultant output and progressively will try to improve the efficiency by using backtracking technique and will try to come close the expected output and hence improving itself. The output data obtained is then matched with the expected output data and hence using the backtracking technique, the system error is continuously reduced and finally making the system with least error. The radial basis function has the parameters as centre, distance measure and shape. Therefore, while simulation when the outputs falls into the radius of the shape, then the output will be matched to the maximum within that specified area.

2.3 Fuzzy Logic[3]

Fuzzy logic is implemented in our system to tell what should be the output for different values of inputs. The inputs are taken from image processing system. This is then used to calculate the output accordingly using a Mamdani filter and the feedback is given to the player informing him if he played good or bad and what improvements should be made.

For example, following are the inputs

- Expected shot deviation: {small, medium, large}
- Strength of shot: {weak, medium, strong}
- Gap: {small, medium, large}

Rules:

- ExpectedShotDeviation(large)^Strength(weak)^Gap(small)=bad
- ExpectedShotDeviation(medium)^Strength(weak)^Gap(small)= below average
- ExpectedShotDeviation(medium)^Strength(medium)^Gap(medium)=average
- ExpectedShotDeviation(small)^Strength(medium)^Gap(large)=good
- ExpectedShotDeviation(small)^Strength(strong)^Gap(large)=excellent

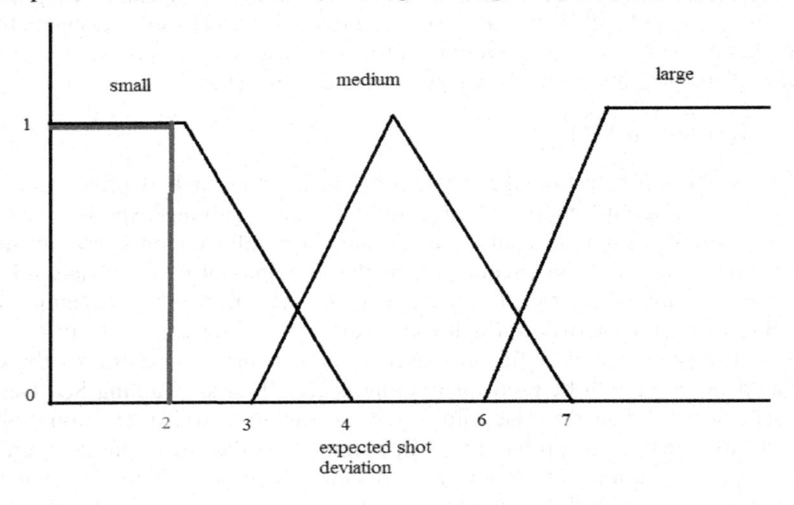

Fig. 2. Input graph of expected shot deviation based on Mamdani filter

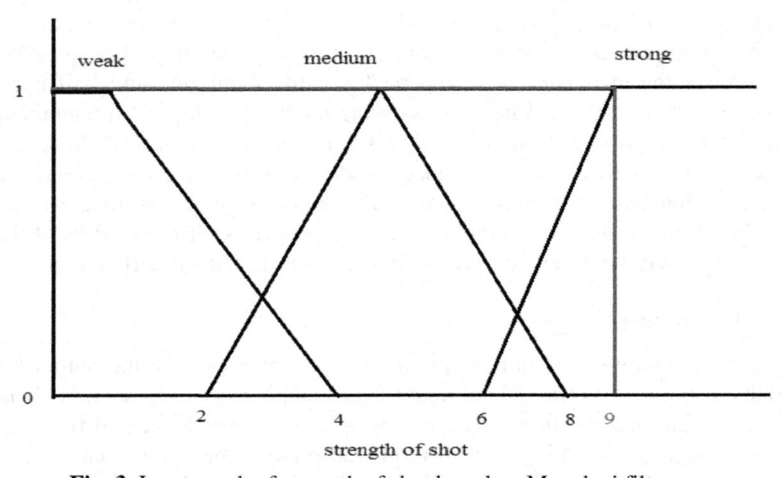

Fig. 3. Input graph of strength of shot based on Mamdani filter

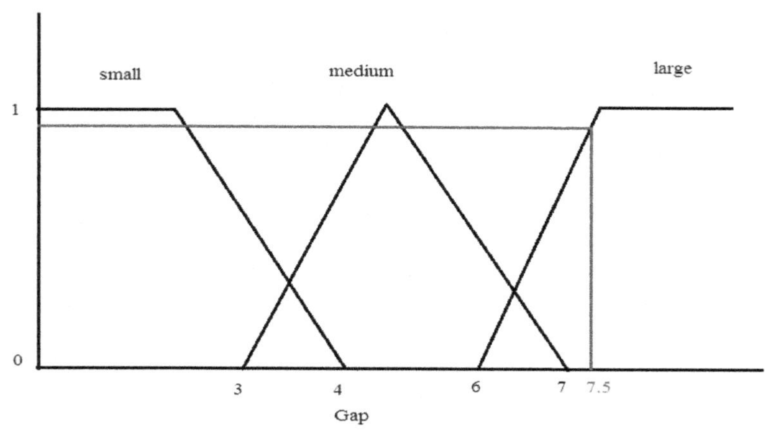

Fig. 4. Input graph of gap on field based on Mamdani filter

2.4 Simulation of a Trial System

This section provides what processes will be carried out for the actual computation and analysis of the person who is practicing. After the training of system, when it is in position to classify major data related to training, the actual coaching procedure begins. In this every ball will be assigned a type, for example spin or swing and based on the balling style the system will have virtual fielding arrangement. Based on the speed and direction of the ball obtained from bowling machine and the resultant force and angle of the shot played by bat, the final direction and force on the ball can be computed. If the player misses the ball, the shot is qualified bad and added to the database. If the player plays the shot and the ball ends up in the area of virtual fielder, then the shot is qualified average.

In the figure below, the grey parts are the gaps in the fielding. If the ball is hit such that it falls in the gap then it adds points for the shot to be classified as a good shot. Similarly, the speed of the shot and other factors are computed.

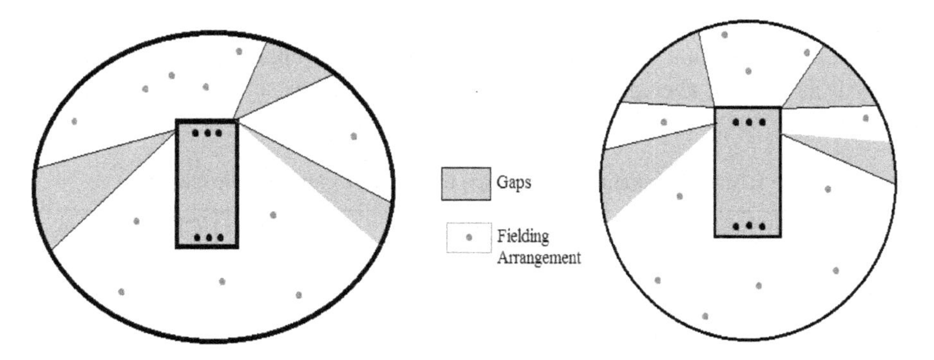

Fig. 5. Fielding arrangement as per the bowling type

After the evaluation of various shot the machine will generate a report where it will mention that the weak areas of player like the player is facing difficulty in playing bouncer, and the strong areas and average areas are mentioned. Along with that, the machine will give strategy to improve like tips regarding the timing of the shot, the way the bat is held, display of various recorded videos of how the shot is played successfully and next time when player will be trained on the system, the machine will test user on the weak point and try to improve in the best way possible. The bowling part will be purely based on image processing. The bowling technique will take input parameters as the speed of the approaching bowler, the angle of the bowler shoulder, the spin, bounce or swing of the ball, the line of ball with respect to wickets and perfection of the bowler to throw same ball number of times repeatedly. All these values will be computed by Image Processing. In Bowling, the improvement in bowling angle, running speed will be modified to achieve a successful ball. Further, the spin and swing deviation will be calculated, and if it needs to improve the bowler will be given various types of finger movements. The finger movements are fed into the system in prior and mathematically computed movement of wrists and fingers will be simulated and required suggestions will be provided to the bowler. All these statistics will help the bowler for a deeper evaluation of the strategy and self improvements can be made accordingly. In fielding part, there will be ball throwing machine, which is slight modification of the Bowling Machine. This machine will throw ball in many possible ways and the system will record the way the fielder handles the bowl. The system will therefore tell the player to improve parameters like the speed with which the player is approaching the ball, the body movements and quickness of the fielder. After computing all these parameters, a report of the fielder will be generated.

All the three systems will have computation table where values of various parameters will be given out of 10 and holistic grade will be generated with a graphical analysis of weak and strong points.

Batting Table

The parameters in the table can be explained as: Ball played is 1 if bat touches the ball, 0 if bat doesn't touch the ball; Expected Shot indicates the deviation of the shot played from the expected shot; Strength of the shot indicates how good the shot was on a rating of 10; If the batsman hits the ball successfully in the gap then the value will be 10. If not, then accordingly the value decreases; if the wicket fall then the value will be 0 for batsman. If not, then it will be 1. Then all values are taken into consideration to classify it as a good, average or bad shot and various other subtle classifications are done. Similarly, the other tables are computed considering the same logic.

Table 1. Various parameters on which the player's game will be classified

Ball Played*	Deviation from Expected Shot	Strength of the shot	Gap	Wicket*	Classification
1	7.5	1.5	3.1	0	Below Average
1	1.1	9.1	6.9	0	Good
0	8.8	0	0	1	Bad

Table 2. Various parameters on which the player's bowling will be classified

Speed of the delivery	Expected Delivery	Swing/Spin	Desired angle	Wicket line	Consistency of delivery	Classification
8.8	9.4	8.8	8.2	9.7	8.0	Very good
6.5	7.2	7.2	6.5	5.1	6.6	Average
9.1	6.6	6.9	5.2	3.2	7.6	Average

Table 3. Various parameters on which the player's fielding will be classified

Agility	Speed	Expected location of player	Grip	Classification
9.5	9.2	8.2	6.2	Good
9.9	9.5	9.1	2.2	Good
6.5	8.0	5.5	3.3	Average.

*Boolean

The system design and implementation will be explained in the following section.

3 Hardware

3.1 Batting

In this model, we have designed a hardware model of the bat, where we have a gyro sensor, accelerometer and an x-bee module. The gyro will measure the roll, yaw and pitch value of the position of the bat before and until the shot is played. The accelerometer will provide with acceleration of the bat movement. The x-bee will transfer data to the computing machine wirelessly. Based on the values the angular velocities are calculated and a vector addition is computed to calculate the force and direction of the impact and to predict the direction and force of the ball after the impact. We have computed a series of values of the gyro and classified them as respective shot. The shots are classified as square cut, long on, long off and similarly other shots. The gyro sensors are used to get precise values for calculating exact impact of force and direction of the bat movement. The other ergonomics of batsman posture are determined using Image Processing. The simulation calculates the way the shot is played and precise bat movement and force of the shot. Rotation is calculated by the mathematical equation [4] as:

$$R(\alpha,\beta,\gamma)= R_x(\alpha)R_y(\beta)R_z(\gamma)=$$

$$\begin{pmatrix} cos\alpha\, cos\beta & cos\alpha sin\beta sin\gamma - sin\alpha cos\gamma & cos\alpha sin\beta cos\gamma + sin\alpha sin\gamma \\ sin\alpha cos\beta & sin\alpha sin\beta sin\gamma + cos\alpha cos\gamma & sin\alpha sin\beta cos\gamma - cos\alpha sin\gamma \\ -sin\beta & cos\beta sin\gamma & cos\beta cos\gamma \end{pmatrix}$$

Using the above equation the position of the bat before the motion of impact (Ri) and till the impact happens (Rj) is calculated. The transition from Ri to Rj will help us to calculate the movement of bat along with acceleration. We are also aware of speed and direction of ball which will be obtained from the Bowling Machine. Hence by using Collision Theory [5], we can compute the ball's strength of impact and direction of impact.

On the other hand, the computation performed by Image Processing provides the precise body movements of the player. The correctness of the movement of the body, leg and bat movement are calculated. These parameters are taken as input and are compared with the standard and hence, final output is generated. The images will be compared using Hausdorff Distance Algorithm [6]. Various modification required will include increasing the comparing pixel. The comparison will have to scale to 800*800 pixels to maintain maximum accuracy.

3.2 Bowling

The Bowling System [7] will be implemented with the basic concept of differential speed of the rotating wheels. There is another system with three wheels [8] placed at 120 degree. The upper two wheels serve as two fingers and the lower one serves as thumb. Thus, it provides additional variety of balls like Googly, Wrong-un, In-swing seam out, Out-swing seam in, Top Spin, Flipper. This feature enables machine to replicate any kind of ball assisting player to get trained on all the possible outcomes. The existing bowling System will be integrated with image processing module like a camera and a computation machine like computer which will carry out computation and operate the machine smartly. Varieties of ball are achieved by manipulating speed of the wheels. In our system, considering the weak area of player the intelligent system will automatically generate different balls for the player and the weak and strong forte of player will be computed.

The actual bowling body movements [9] are as displayed in the system. The terms are explicitly associated with the figure and are self explanatory.

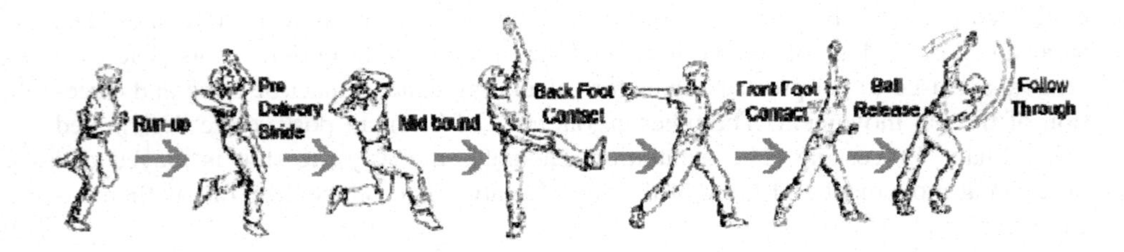

Fig. 6. The bowling process depicted with body movements

Every ball has specified perfect body movements associated with it. By setting up standards and then making comparison will help us to evaluate how much the bowler complies with the system and where exactly the errors are happening.

The neural network will plot values of all the movements on the graph. Then a line will be passed which will pass through maximum points and that should be the line which the bowler should try to achieve. The table will explain the concept as:

Table 4. Different balls as per different parameters

Type of ball	Run–up (km/hr)	Pre-delivery stride (vertical) In feet	Mid-bound (horizontal) In feet	Back foot contact (sec)	Front foot contact (sec)	Ball release (configuration)	Release angle (degree) W.r.t. vertic-al	Follow through (distance) in meters
Off-swing	30	0.7	1.1	0.2	0.15	Holding seam on one side and thrown	32	5
Leg-Spin	15	0.3	0.5	0.3	0.25	Holding the seam and ro-tate hand anti clockwise	30	2
Bounc e	35	0.8	0.9	0.2	0.18	Holding the seam and thrown	40	5.5

The data in the table is calculated by trying and computing approximate values based on the configuration provided in the picture. Hence, by taking all these parameters we can compute the bowling style and to coach players.

3.3 Fielding

The fielding is a crucial part of the game and many systems overlook this aspect. Our system provides fielding training using the same machine used in the bowling mechanism. The machine can be used by changing the parameters like throwing direction, speed and swing. The benefit of using bowling machine is that we can have different types of throws, with varying speed and the best part is the swing in the ball. The training process is carried out under high quality camera which captures all the body movements done to capture the ball. The machine has various perfect videos based on which the system expects user to accomplish the task. If the task is accomplished with violating the other parameter like the body movement was ahead of time, but luckily the the ball was caught, then system here will inspect the move. This fielding technique will check the agility of player, the speed with which the player moves the grip on the ball and various other reflexes which includes diving. The practice session will improve the timing of the player, by computing at exactly what time the player should dive and similarly such parameters will be calculated.

4 Conclusion and Future Work

This paper presented an integrated system which will assist the existing coaching program and to make the overall system self dependent, hence can be used in absence of any guide. The system is based on Artificial Intelligence and improves itself every time it is used. This system can produce deliveries like all the different bowlers, suggest the best stroke as well as bat movements and can train fielders considering every crucial aspect of training. The advantage of this system as compared to human coach is that the system can learn endlessly, knowledge base is not confined as the system can expand with technology, along with that, after trained by experts in respective fields, this will be a comprehensive system. The system then can be replicated; hence everybody can use the excellent system and attain superlative training. Considering the system to be so huge and comprehensive, there can be numerous modifications and many features can be added to it. For example: During most of the matches the performance is degraded with high pressure and tension, so the players can be trained in pressure simulated situation by using neuroscience after a good amount of research. Further, all these (AI based coach) machines can be connected to each other via some communication channel like a cloud where information can be exchanged and the system can be trained taking input from everywhere the machine is used. Hence this would lead to fast growth in the development of the system.

References

1. Rural-Urban distribution Census of India: Census Data 2001: India at a glance > > Rural-Urban Distribution. Office of the Registrar General and Census Commissioner, India (retrieved on November 26, 2008)
2. `http://takinginitiative.net/2008/04/23/basic-neural-network-tutorial-c-implementation-and-source-code/`
3. Shivanandam, Deepa: Principles of Soft Computing. Wiley; Ross, T.J.: Fuzzy Logic with Engineering Application. MacGraw-Hill
4. Hibbeler, R.C.: Engineering Mechanics - Statics and Dynamics, 3rd edn., ISBN 0-02-354140-7
5. Tolman, R.C.: The Principles of Statistical Mechanics. Clarendon Press, Oxford (1938); Reissued. Dover, New York (1979) ISBN 0-486-63896-0 (collision)
6. Huttenlicher, D.P., Klanderman, G.A., Rucklidge, W.J.: Comparing Images using the Hausdorff distance. IEEE Transaction on Pattern Analysis and Machine Intelligence 15(9) (September 1993)
7. Design and development of indigenous cricket bowling machine. Journal of Scientific & Industrial Research; Roy, S.S., Karmakar, S., Mukherjee, N.P., Nandy, U., Datta, U.: Central Mechanical Engineering Research Institute, Durgapur 713 209, Vol. 65, pp. 148–152 (February 2006) (received February 01, 2005; revised October 20, 2005; accepted December 12, 2005)
8. `http://www.bowlingmachine.co.in/threewheelcricketbowlingmachine.html`
9. `http://thames2thayer.com/wiki/index.php?title=Cricket`

AHP Driven GIS Based Emergency Routing in Disaster Management

Varsha Mali[1], Madhuri Rao[1], and S.S. Mantha[2]

[1] Thadomal Shahani Engineering College, Bandra, Mumbai, India
varsha_mali@yahoo.co.in, my_rao@yahoo.com
[2] AICTE, India
ssmantha@vjti.org.in

Abstract. In emergency situations, finding suitable route to reach destination is critical issue now a days. Traditional shortest path cannot always travel in minimum time because of several factors such as traffic conditions, road width, mass density etc. Hence there is need to find another path to reach destination in minimum travel time. The objective of current research is to find a method which considers major travel time delay factors and calculates optimal path for given traffic conditions. Methodology includes identifying travel time delay factors and obtaining overall road weight depicting real traffic conditions at that time. Route with least travel time is obtained by applying dijkstra's algorithm on weighted road network. Analytical hierarchical processing (AHP) is used to compare seven delay factors and to obtain overall weight of each factor. Actual distance is modified according to weight obtained for road length. Results obtained from emergency routing system showed that optimal path obtained by modified routing method takes less time to reach destination than traditional shortest path routing.

Keywords: Geographical Information System (GIS), Emergency Routing, Path planning, Analytical hierarchical processing (AHP), Dijkstra's Algorithm, Optimal Path, Delay factors, High Traffic Comparison Matrix (HTCM), Low Traffic Comparison Matrix (LTCM), Medium Traffic Comparison Matrix (MTCM).

1 Introduction

Natural disasters such as Floods, droughts, cyclones, earthquakes, landslides are frequent in India. Recent manmade disasters like Bombing, terror attacks in India have once again raised questions on Disaster management and traffic handling capacity of our country. Thus better disaster management system needs to be implemented to handle emergency situations. Post disaster situation can be handled effectively if relief aid and helping man power reaches rapidly in affected regions. The most critical issue is how to reach affected region in minimum time. One can take one of the two paths to reach the destination, first is shortest path from given source to destination and another is optimal path which takes minimum travel time. Shortest path does not always take minimum time because of lot of factors like traffic volume, mass density,

S. Unnikrishnan, S. Surve, and D. Bhoir (Eds.): ICAC3 2013, CCIS 361, pp. 237–248, 2013.

junction delay, road width etc. These factors affect overall travel time to reach destination and if these factors are not considered in obtaining route to the destination then it can result in delay in traveling. This paper presents a routing method based on GIS for emergency situations as well as for day to day traveling. Aim of this system is to overcome the drawbacks of existing routing techniques and to develop a system which not only considers delay factors to find out optimal path but also consider different traffic conditions of a day to give more dynamic and appropriate travel time estimation.

Geographical Information System's are designed to capture, analyze, represent spatial data in a way that user can easily understand. Current research in the field of GIS based routing system focuses on obtaining minimum travel time path as traffic is overwhelming problem now a days. GIS based dynamic traffic management system was developed to improve traffic management [1]. Recently Dijkstra's algorithm is modified to combine the impedance factors to give optimal path [2]. Model for disaster management system using web services and AJAX (Asynchronous JavaScript and XML) is developed so that spatial data can be loaded efficiently into the client browser and loading time can be minimized [3]. Few efforts have been taken till date to match real time traffic conditions and post emergency evacuation methods.

2 Architecture of the System

Universal purpose of any disaster management system is to deliver emergency services such as police, fire brigade and medical service as quickly as possible in affected areas. Disasters such as flood, earthquakes, terrorist attacks need huge mass evacuation to prevent mortality and reduce economical damage. Emergency routing system presented in this research paper, helps to find the optimal route from a given source to destination based on seven delay factors such as distance, road width, road type, traffic count, population density, junction delay and velocity. Out of these factors traffic count is the crucial and dynamic factor. The salient feature of this system is to compare these seven factors and to determine factors which play greater role in minimizing travel time. Analytical hierarchical processing (AHP) is used to compare seven delay factors and compute overall contribution weight of each factor by building comparison matrix. Actual distance between two nodes is modified with respect to its contribution weight obtained and Dijkstra's algorithm is applied to get optimal path. Traffic is a dynamic factor and it does not remain constant throughout a day. For some roads which show great deviation in its peak hour traffic count and non peak hour traffic count, it is desirable to have different comparison matrix for realistic output. Two types of comparison matrix are made for such roads. In High traffic matrix, traffic count factor is given lowest contribution weight and should be used when user requests for route in peak hour. Similarly, in Low traffic matrix, traffic count factor is given highest contribution weight and should be used when user request for route in non peak hour. Architecture of emergency routing system is shown in the Figure 1.

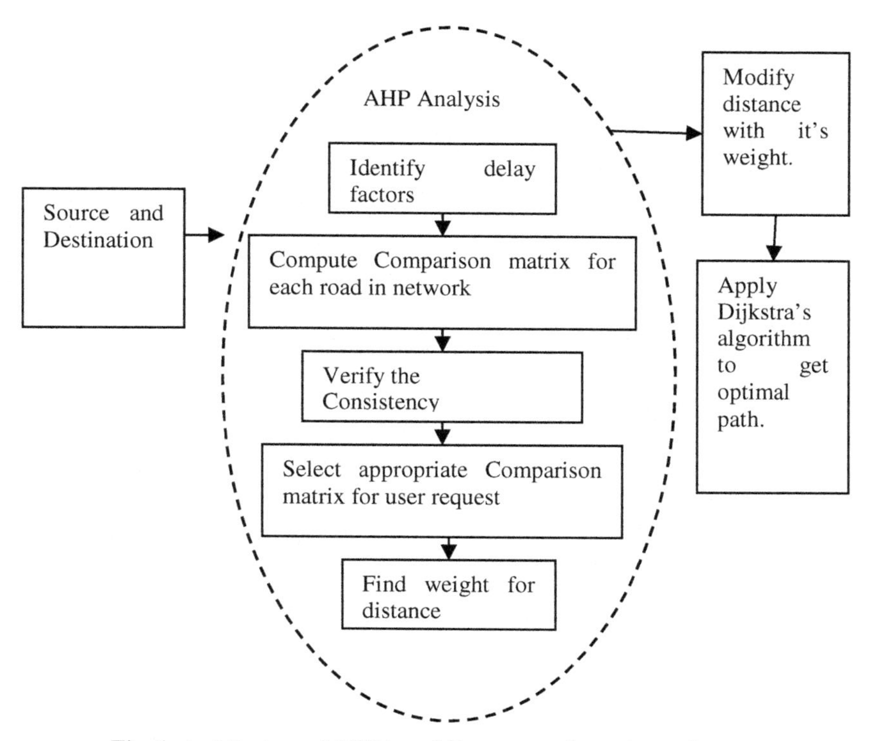

Fig. 1. Architecture of AHP based Emergency dynamic routing system

3 Scheme of Implementation

3.1 Delay Factor Selection

There are many techniques to evaluate factors which delays travel time. Delay factors are selected by studying real time traffic conditions and surveying traffic. Travel time on particular road depends on some or all of these factors. Deciding delay factors which contributes more to increase travel time is purely subjective issue. Selected seven factors are enlisted in Table 1. Next subsection explains how AHP is used to compare these seven factors.

Table 1. Delay factors's list

Delay Factors						
Distance	Road width	Road type	Traffic count	Population	Junction delay	Velocity

3.2 AHP Processing

The Analytic Hierarchy Process (AHP) developed by Saaty (1980) [4], is a multi-criteria decision-making approach which deals with complex decision making problems. The simpler mathematical structure and easy inputs are strong points of this method. It uses a multi-level hierarchical structure of final goal and its alternative to reach its goal. A set of pairwise comparisons are made to compare these alternatives. These comparisons are used to obtain weights of decision criteria, and relative performance measures of the alternatives in terms of each individual decision criterion. If the comparisons are not perfectly consistent, then AHP provides a mechanism for improving consistency. Detailed steps of AHP processing are explained below.

Represent Decision Making Problem as Hierarchy Containing Overall Objective and Alternative. The first step in AHP processing is representing the problem as hierarchy in which overall objective of decision making is considered as a root and number of alternatives are considered as branches from that root. All delay factors towards obtaining minimum travel time are considered here in the form of alternative options and hierarchy is created as shown in the Fig. 2.

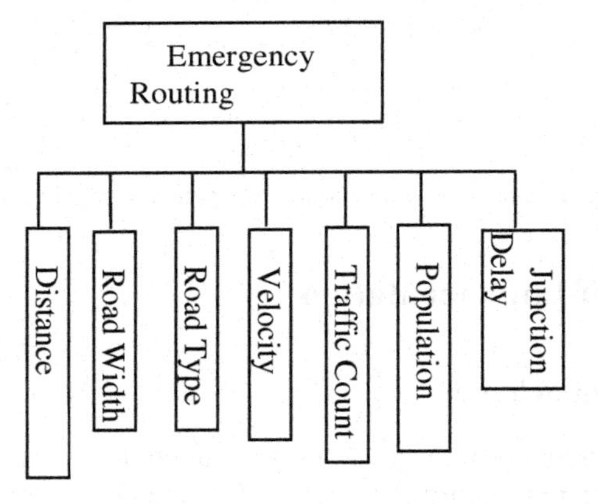

Fig. 2. The Hierarchy model of Emerency Routing for AHP

Build Comparison Matrix by Evaluating Contribution of Each Factor towards Final Goal i.e. Obtaining Minimum Travel Time. In AHP process, Pair wise comparisons are made to judge importance of one alternative over another and to obtain contribution vector for all alternatives. Judging the importance is purely subjective issue and depends on one's perception. Saaty has developed quantitative scales (Refer Table 2) for pair wise comparisons [4].

Evaluate Judgments to Yield Overall Contribution Values for Hierarchy. Contribution of each alternative is calculated by first calculating normalized comparison matrix by geometric mean method [2].

$$r_i = \prod_{j=1}^{n} (a_{ij})^{1/n} \tag{1}$$

Then Contributions vector is calculated as,

$$w_i = \frac{r_i}{\sum_j r_j} \tag{2}$$

Where a_{ij} (i,j=1...n) denotes comparison values in the pair wise comparison matrix and n is number of alternatives.

Table 2. Scales for pair wise comparions(according to saaty(1980))

Value of Importance	Definition
1	Equally Importance
3	Moderate Importance
5	Strong Importance
7	Extreme Importance
9	Extremely More Importance
2,4,6,8	Intermediate values between Adjacent Scale Values
Reciprocals of above	If alternative i has one of the above non-zero numbers assigned to it when compared with alternative j, then j has the reciprocal value when compared with i
1.1–1.9	If the alternatives are very Close

Verify Consistency of Comparison Matrix. When many pair wise comparisons using several alternatives are performed, some inconsistencies may typically arise. Consistency of judgment means if decision maker evaluates 1st alternative slightly more important than 2nd alternative and 2nd alternative as slightly more important than 3rd one, then it is inconsistent to say that 1st alternative is equally or less important to 3rd alternative. Consistency of comparison matrix holds transitivity condition, $a_{ij} * a_{jk} = a_{ik}$ where a_{ij} is the ijth element of the matrix. The AHP incorporates an effective technique for checking the consistency by calculating Consistency Ratio (C.R.) [5][6][7].

$$C.R = C.I / R.I. \tag{3}$$

$$C.I. = (\lambda_{max} - n) / (n-1) \tag{4}$$

In this λ_{max} is maximum eigen value of comparison matrix, n is number of alternatives and R.I is Random Index as shown in Table 3.

If C.R. < 0.1 then consistency is tolerable otherwise comparison matrix should be modified again until it reaches consistency criterion [4].

Table 3. Random Index for n = 1…7

n	1	2	3	4	5	6	7
R.I.	0.00	0.00	0.58	0.90	1.12	1.24	1.32

Contribution weight vector of n alternatives indicates contribution of each alternative toward final objective i.e. obtaining minimum travel time. Large value indicates that a particular criterion is having greater contribution in achieving final goal.

Combining AHP Weights with Dijkstra's Algorithm. This can be done by combining weight of Road length factor obtained from AHP processing, with its real value to give modified distance in between two nodes. Traffic Count on particular road varies throughout a day, resulting in variation of contribution of traffic count in getting minimum travel time. To achieve this dynamicity of traffic count, two types of comparison matrix (for high traffic and low traffic) and their respective Contribution weight vectors are constructed. Depending on time at which user requests for route, high or low contribution weight vector is selected.

Apply Modified Dijkstra's Algorithm. Dijkstra's algorithm is applied on street network with modified road length to get minimum travel time path.

4 Experimental Analysis

For testing this system under real time GIS usage, experiments are carried out to ensure algorithm works properly.

4.1 Input

Study area: City of Somerville, in Middlesex County, Massachusetts, United States, located just north of Boston [5]. Data collected is in ESRI shape files of streets, hospitals, police stations etc from different sources. For implementing this system street network of 87 streets is taken. Comparison matrixes are made for each of these streets to define contribution weight by each delay factor.

4.2 AHP Processing

AHP Processing starts with building Comparison matrix and defining Contribution vector for all delay factors. For some streets whose average daily traffic is more than usual, they typically show huge deviation in between their peak hour and non-peak hour traffic. Thus their contribution differs with respect to traffic count. Two different matrices are built by assigning different contribution for extreme traffic conditions on some streets. Now if user requests for a route in peak hours then comparison matrix for high traffic (HTCM) will be used and if user requests for a route in non-peak hours then comparison matrix for low traffic (LTCM) will be used. Weights in HTCM are assigned in a way to get low weight for traffic, because traffic now has low contribution in obtaining final goal i.e. minimal travel time. An example of such street is given in Table 4.

Table 4. Comparison Matrix for High Traffic (HTCM)

	Distance	Width	Type	Velocity	Traffic Count	Popula-tion	Junction delay
Distance	1.00	4.00	8.00	4.00	4.00	2.00	2.00
Width	0.25	1.00	2.00	1.00	1.00	0.50	0.50
Type	0.13	0.50	1.00	0.50	0.50	0.25	0.25
Velocity	0.25	1.00	2.00	1.00	1.00	0.50	0.50
Traffic Count	0.25	1.00	2.00	1.00	1.00	0.50	0.50
Population	0.50	2.00	4.00	2.00	2.00	1.00	1.00
Junction delay	0.50	2.00	4.00	2.00	2.00	1.00	1.00

The contribution weight vector obtained for this HTCM clearly indicates that street with high traffic gets lower contribution weight value to traffic volume and as a consequence of it, Road Length gets relative higher contribution weight value as shown in Table 5.

Table 5. Contribution weight vector for High Traffic

Distance	**Width**	**Type**	**Velocity**	**Traffic Count**	**Population**	**Junction delay**
0.35	0.09	0.04	0.09	0.09	0.17	0.17

In case of low traffic street, contribution weight value assigned for Distance factor is smaller than Traffic Count. When Road length contribution weight value of each Road is combined with its actual distance, high traffic streets gets higher weight than low traffic streets and Dijkstra's algorithm includes streets with minimum weight for finding optimal path from given source to destination avoiding high traffic streets.

4.3 Apply Modified Dijkstra's Algorithm

Fig. 3 shows shortest route from some source to destination obtained by simple Dijkstra's algorithm. When Modified Dijkstra's algorithm is applied on same network to find route between same source to destination, it alters the route to give optimal path i.e. minimum travel time path.

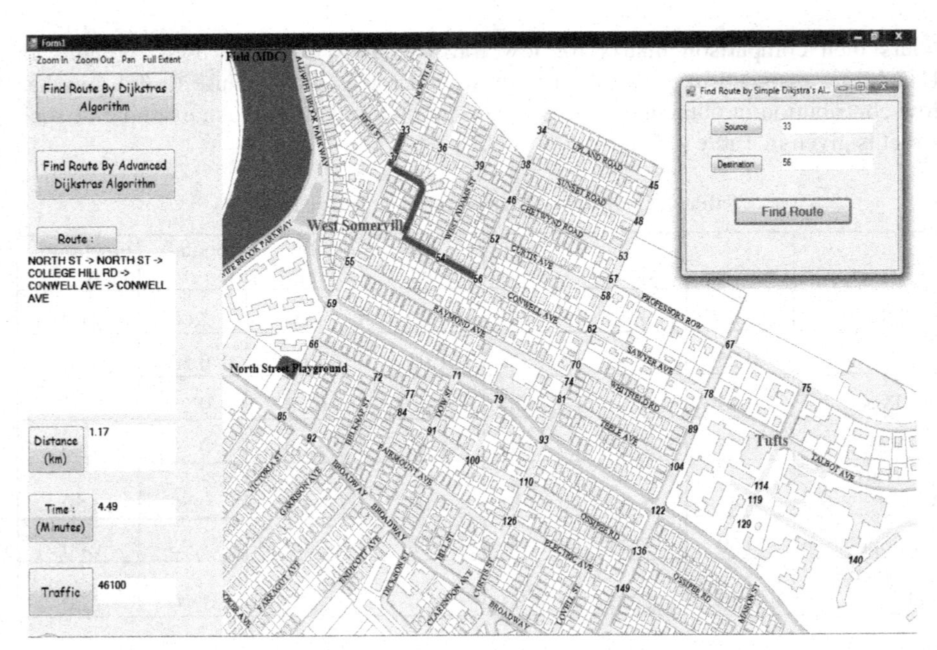

Fig. 3. Result obtained by simple Dijkstra's algorithm

Route displayed by modified Dijkstra's algorithm in peak hour and non-peak hour in shown in Fig. 4 and Fig.5 respectively.

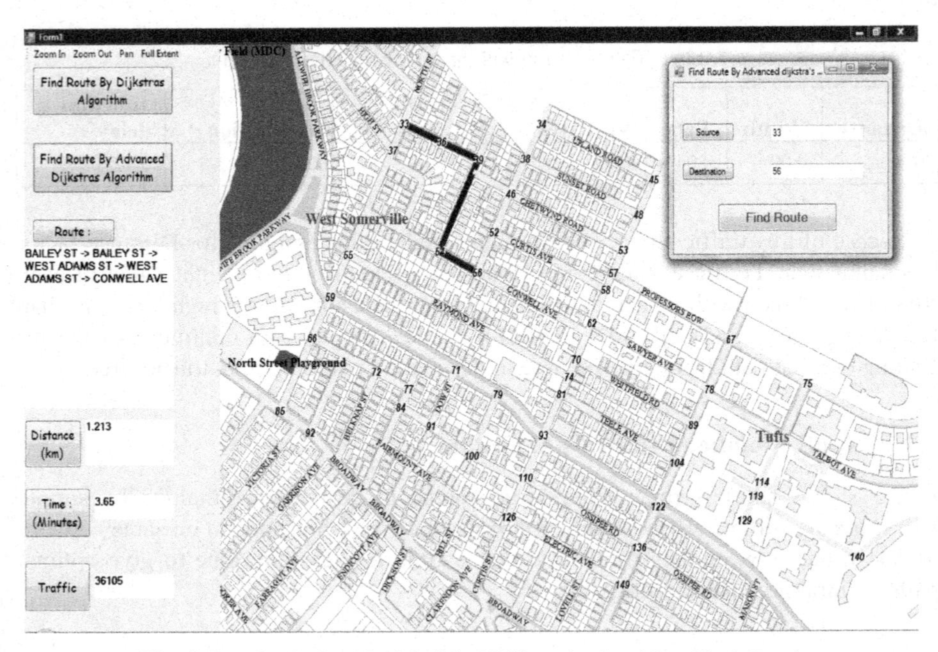

Fig. 4. Result obtained by Modified Dijkstra's algorithm (Peak hour)

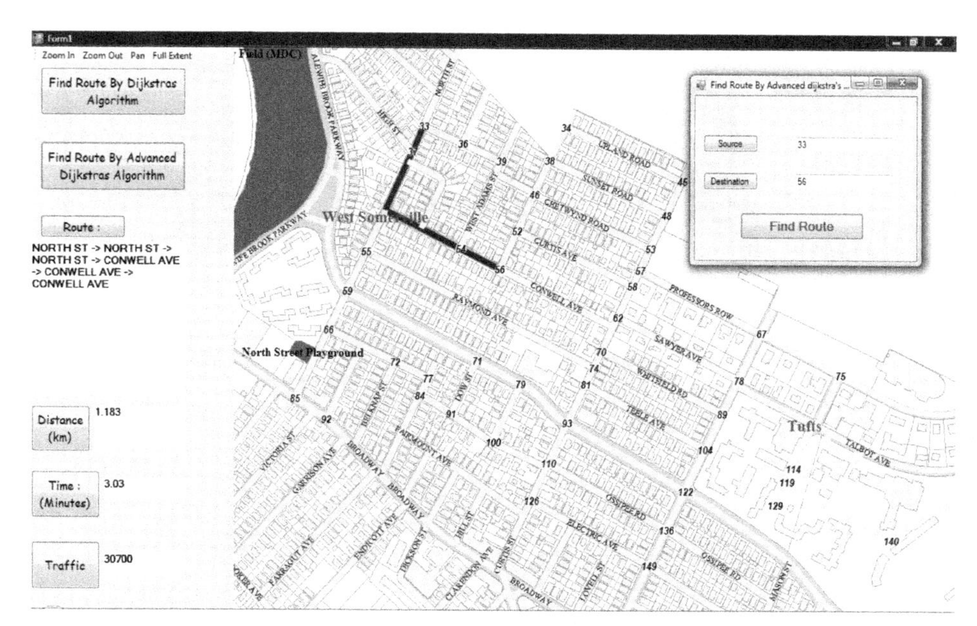

Fig. 5. Result obtained by Modified Dijkstra's algorithm (Non-Peak hour)

Result obtained by Modified Dijkstra's algorithm is compared with result obtained by simple Dijkstra's algorithm and it is found that although a path found by Modified method is slightly longer than former but it has less traffic congestion, resulting in taking less travel time. Considering different traffic conditions at different time in a day, more robust, emergency routing system is implemented. Result comparison indicates that this method is most suitable in any case of emergency or daily routing where traffic is a travel time determining factor (refer Table 6).

Table 6. Results comparison

Node 33 to 56	Shortest route by simple Dijkstra's algorithm	Optimal path by Emergency Routing Method	
		Peak hour	Non-peak hour
Distance (km)	1.17	1.21	1.183
Time (minutes)	4.49	3.65	3.03
Traffic	46100	36105	30700

In case of emergency situations, streets in affected region are explicitly assigned high traffic contribution weight values and it is found that these streets are omitted by Emergency routing method. It helps to divert traffic away from affected region and implements a better disaster management system. Assume that some emergency condition has arisen on streets from street id 42 to 46 and from street id 46 to 53 which are circled in fig. 6. In case of emergency, simple Dijkstra's algorithm displayed a path without considering real time traffic conditions as shown in fig. 6.

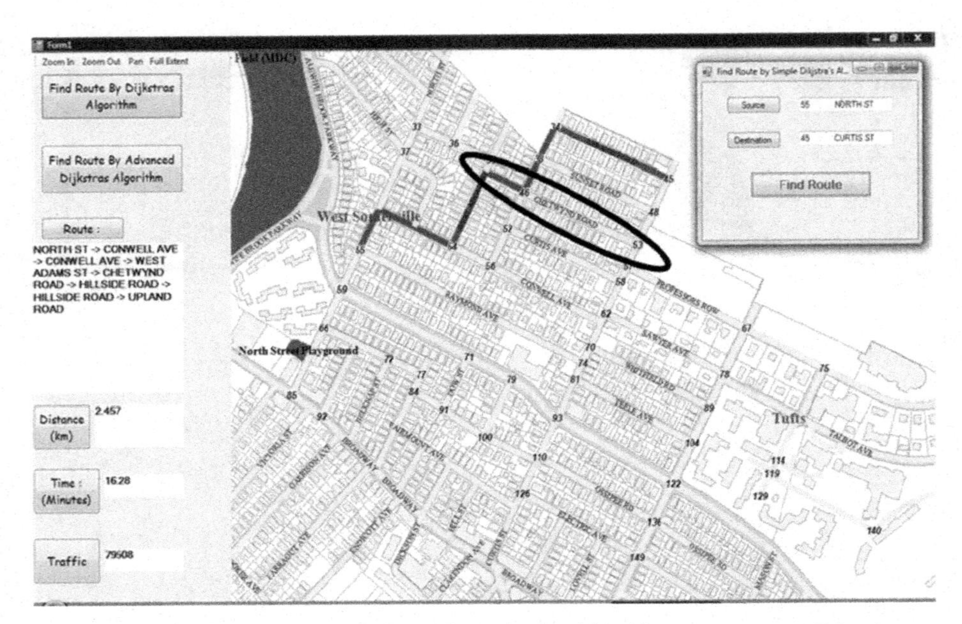

Fig. 6. Result obtained by Simple Dijkstra's algorithm (in emergency conditions)

When High traffic status assigned to circled streets as shown in Fig. 7, it is seen that modified Dikjstra's algorithm alters the path to avoid those conjusted streets as shown in Fig. 8 while simple Dijkstra's output remained unaltered as shown in Fig. 6.

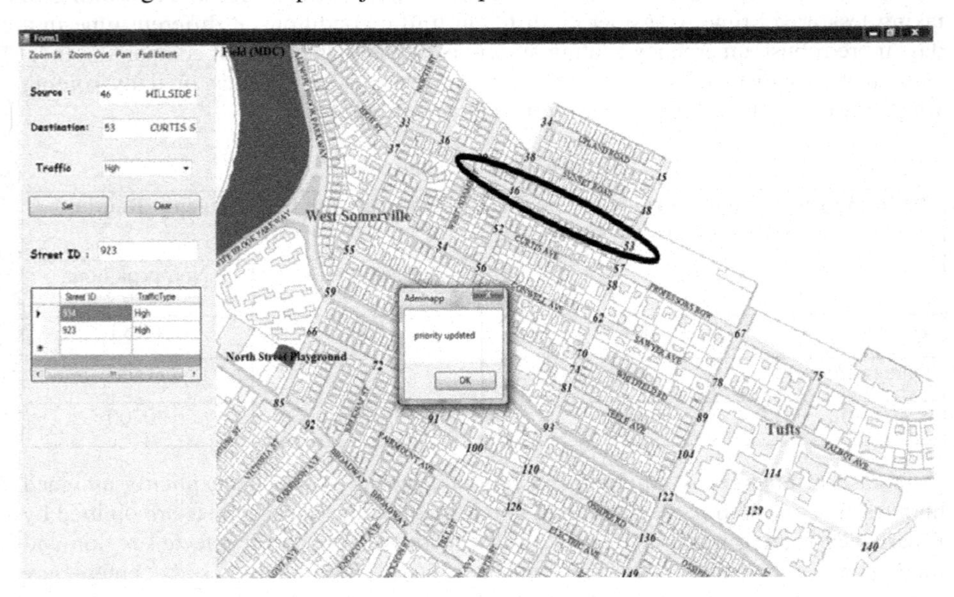

Fig. 7. High Traffic priority is assigned to the street from 42 to 46 and from 46 to 53

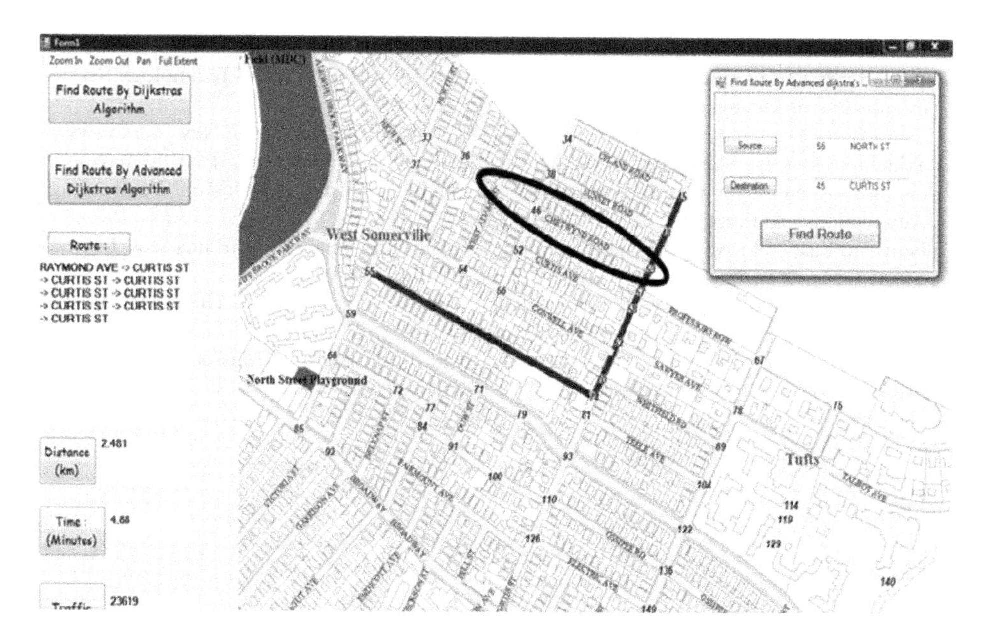

Fig. 8. Optimal path obtained by Modified Routing System in emergency conditions

5 Conclusion

In emergency conditions, finding a path which takes minimum time to reach destination is a big problem. Finding shortest path is not a solution all the time because there are several factors affecting travel time. Using AHP, these delay factors are easily compared to find optimal path. With AHP based Emergency routing method, a better emergency routing scheme is achieved which plays significant role in any disaster management system. Paths obtained in AHP based emergency routing method takes less travel time than traditional shortest path and hence it is most practical method for daily routing and emergency conditions as well. Further research is focused on integrating this system with real time on-road traffic count to display more dynamic, reliable and accurate routes to end user.

References

1. Shashikiran, V., Sampath Kumar, T.T., Sathish Kumar, N., Venkateswaran, V., Balaji, S.: Dynamic Road Traffic Management based on Krushkal's Algorithm. In: IEEE International Conference on Recent Trends in Information Technology (ICRTIT), pp. 200–204 (2011)
2. Yang, S., Li, C.: An Enhanced Routing Method with Dijkstra Algorithm and AHP Analysis in GIS-based Emergency Plan. In: 18th International Conference on Geoinformatics, Beijing, pp. 1–6 (2010)

3. Jeberson Retna Raj, R., Sasipraba, T.: Disaster Management system based on GIS Web services. In: IEEE International Conference on Recent Advances in Space Technology Services and Climate Changes (RSTCC), pp. 256–261 (2010)
4. Saaty, T.L.: The Analytic Hierarchy Process. McGraw-Hill International, New York (1980)
5. Triantaphyllou, E., Mann, S.H.: Using the analytic hierarchy process for decision making in engineering applications: some challenges. International Journal of Industrial Engineering: Applications and Practice 2(1), 35–44 (1995)
6. Ishizaka, Lusti, M.: An expert module to improve the consistency of AHP matrices. International Transactions in Operational Research 11, 97–105 (2004)
7. Coyle, G.: Practical Strategy. Open Access Material. The Analytic Hierarchy Process (Ahp). Pearson Education Limited (2004)
8. The Official Website of the Office of Geographic Information of Massachusetts country (MassGIS), `http://www.mass.gov/mgis/database.htm`

An Efficient Genetic Based Algorithm for an Irregular Low Density Parity Check Code with Low Computational Complexity and Low Error Floor

D.P. Rathod and R.N. Awale

Department of Electronics Engineering
Veermata Jijabai Technological Institute, Mumbai, India
{dprathod,rnawale}@vjti.org.in

Abstract. Very low error rate is necessary for several coding applications like satellite communications, Ethernet transmission, and data storage applications. Therefore, construction of practical tools for anticipating error floors and assessing the success of LDPC codes in the low frame error rate region is necessary. Completely randomly generated codes are good with high probability. The problem that will arise is encoding complexity of such codes is usually rather high as the length of codes increases. We propose an efficient Genetic base algorithm for construction of irregular LDPC codes, which reduces error floor and computational complexity in code design process. Our proposed method is compared with an algorithm in which the codes are generated randomly,. Bit Error Rate (BER), Frame Error Rate (FER) and Computational Complexity are calculated using Matlab platform and compared using both algorithms. Simulations results shows that the codes constructed with proposed method produce low error floors and reduced computational complexity. LDPC Codes based on this design suits for the short block length.

Keywords: Irregular LDPC codes, randomly generated code, genetic algorithm, belief propagation algorithm, bpsk modulation.

1 Introduction

A category of codes possessing features and performance identical to that of turbo codes has been rediscovered following the unparalleled success of the latter. Codes of this category are termed as low-density parity-check (LDPC) codes. One of the capacity approaching error correction code that has emerged as an eminent competitor over several vital channels is the Low-density parity-check (LDPC) code. They are appropriate for use in extremely efficient parallel decoding algorithms and have a good performance. Protographs, also known as base graphs or projected graphs is used as the basis for a well-known construction of LDPC codes. Efficient encoding of LDPC codes that have a block length up to some thousand bits a may be difficult if the code does not have any algebraic structure .Error Correction Codes (ECC) is one of many tools made available for achieving consistent data transmission. The

S. Unnikrishnan, S. Surve, and D. Bhoir (Eds.): ICAC3 2013, CCIS 361, pp. 249–258, 2013.

error-correcting performance of low-density parity check (LDPC) codes, when decoded using practical iterative decoding algorithms, is known to be close to Shannon limits for codes with suitably large block lengths. Girth of Quasi-Cyclic LDPC codes is an important issue and several current researches are going on this topic [8]. It has been shown that increasing the girth or average girth of a code increases its decoding performance. The girth also determines the number of iterations before a message propagates back to its original node. Performance of structured codes could therefore be improved by increasing their girths. Optimized irregular codes are capable of excellent performance with reasonable decoding complexity; one of the main hurdles in the implementation of LDPC codes is the computational complexity of the encoding algorithm. Encoding is, in general, performed by matrix multiplication and so complexity is quadratic in the code length. One option for efficient encoding is to use algebraic code constructions and exploit the subsequent code structure.

2 Related Work

Thomas J. Richardson and Rüdiger L. Urbane [1] present a general method for determining the capacity of low-density parity-check (LDPC) codes under message-passing decoding when used over any binary-input memoryless channel with discrete or continuous output alphabets. Transmitting at rates below this capacity, a randomly chosen element of the given ensemble will achieve an arbitrarily small target probability of error with a probability that approaches one exponentially fast in the length of the code. (By concatenating with an appropriate outer code one can achieve a probability of error that approaches zero exponentially fast in the length of the code with arbitrarily small loss in rate.) Conversely, transmitting at rates above this capacity the probability of error is bounded away from zero by a strictly positive constant which is independent of the length of the code and of the number of iterations performed. Results are based on the observation that the concentration of the performance of the decoder around its average performance, as observed by Luby et al in the case of a binary-symmetric channel and a binary message-passing algorithm, is a general phenomenon. For the particularly important case of belief-propagation decoders, we provide an effective algorithm to determine the corresponding capacity to any desired degree of accuracy. The ideas presented in this paper are broadly applicable and extensions of the general method to low-density parity-check codes over larger alphabets, turbo codes, and other concatenated coding schemes are outlined.

Jaehong Kim, and Steven W. McLaughlin† [4] present a new class of irregular low-density parity-check (LDPC) codes for finite block length (up to a few thousand symbols). The proposed codes are efficiently encodable and have a simple rate-compatible puncturing structure which is suitable for incremental redundancy hybrid automatic repeat request (IR-HARQ) systems. The codes outperform optimized irregular LDPC codes and (extended) irregular repeat accumulate codes for rates 0.67~0.94, and are particularly good at high puncturing rates where good puncturing performance has been previously difficult to achieve. These characteristics result in good throughput performance over time-varying channels in IR-HARQ systems.

Fatma A. Newagy, and Magdi M. S. El-Soudani [7] This paper introduces different construction techniques of parity-check matrix H for irregular Low-Density Parity-Check (LDPC) codes. The first one is the proposed Accurate Random Construction Technique (ARCT) which is an improvement of the Random Construction Technique (RCT) to satisfy an accurate profile. The second technique, Speed Up Technique (SUT), improves the performance of irregular LDPC codes by growing H from proposed initial construction but not from empty matrix as usual. The third and fourth techniques are further improvements of the SUT that insure simpler decoding. In Double Speed up Technique (DSUT), the decoder size of SUT matrix is fixed and the size of H is doubled. In Partitioned Speed Up Technique (PSUT), the H size is fixed and the decoder size decreases by using small size of SUT matrices to grow H. Simulations show that the performance of LDPC codes formed using SUT outperforms ARCT at block length N = 1000 with 0.342dB at BER = 10^{-5} and LDPC codes created by DSUT outperforms SUT with 0.194dB at BER = 10^{-5} . Simulations illustrate that the partitioning of H to small SUT sub matrices not only simplifies the decoding process, it also simplifies the implementation and improves the performance. The improvement, in case of half, is 0.139dB at BER=10^{-5} however as partitioning increases the performance degrades. It is about 0.322dB at BER=10^{-5} in case of one-fourth.

Chad A. Cole Eric. K. Hall [10] this paper outlines a three-step procedure for determining the low bit error rate performance curve of a wide class of LDPC codes of moderate length. Traditional method to estimate code performance in the higher SNR region is to use a sum of the contributions of the most dominant error events to the probability of error. For even moderate length codes, it is not feasible to find all of these dominant error events with a brute force search. The proposed method provides a convenient way to evaluate very low bit error rate performance of an LDPC code without requiring knowledge of complete error event weight spectrum or resorting to a Monte Carlo simulation. This new method can be applied to various types of decoding such as the full belief propagation version of message passing algorithm or commonly used min-sum approximation to belief propagation. This result will provide a solid foundation for analysis and design of LDPC codes & decoders that are required to provide a guaranteed very low bit error rate performance at certain SNRs.

Benjamin Smith *et al.* [22] have proposed a numerical method for decreasing the decoding complexity of long-block-length irregular low-density parity-check (LDPC) codes. In binary-input memory less symmetric channels and iterative message-passing decoding algorithms with a parallel update schedule, this design methodology could be employed. The number of operations necessary to perform a single decoding iteration and the number of iterations necessary for convergence that have been incorporated in a new complexity measure has been considered as a significant feature of the proposed optimization method. They have demonstrated that density-evolution and extrinsic information transfer chart analysis of the code could be used to precisely assess the complexity measure. They also presented a sufficient condition for the convexity of the complexity measure in the variable edge-degree distribution.

In case it is not satisfied the numerical experiments have indicated that still a unique local minimum exists. The results presented in the article demonstrated that when the decoding complexity is constrained, the threshold-optimized codes have been substantially outperformed by the complexity-optimized codes at long block lengths within the ensemble of irregular codes.

The research enlisted here could not produced efficient LDPC codes with low error floors and irregular LDPC code complexity has not been efficiently reduced. For example, the recent PEG-ACSE algorithm has been used to avoid the small trapping sets and detrimental ETS in the connected sub graph. This algorithm calculates the CSE (Cycle-Set-Extrinsic-message-degree) and ACSE (Approximate-minimum-Cycle-Set-Extrinsic-message-degree) values based on sub graph cycles. This algorithm has computational complexity because of the complexity in determining those values at each time. There is a need to propose an LDPC code construction system with reduced computational complexity as well as low error floors. Moreover, there is a little possibility that their graphical representations contain many short cycles, so the long codes with length greater than 10^6 are of asymptotically good performance. However, due to hardware complexity as well as the incurred time-delay problems, such long codes cannot be practically employed in several applications. In this paper, we propose a genetical algorithm for irregular LDPC code design to reduce the computational complexity and error floors. The paper is organized as - Section 3 describes the construction of irregular LDPC codes , section 4 deals with simulation results and section 5 concludes the paper.

3 Matrix for Irregular LDPC Codes

Two stages exist in the construction of irregular codes. The first stage selects a profile that represents the required number of columns and rows of each weight. The second stage constructs the parity-check matrix that realizes the specified profile [7]. The parity-check matrix of a code can be regarded as describing a bipartite graph with "variable" vertices representing the columns and "check" vertices representing the rows. Design of irregular codes described on graphs and computation of the convergence threshold of a specified collection of codes for diverse decoding algorithms have been given much importance. There are two different ways to represent LDPC codes; one matrix representation and other graphical representation. In the graph point of view, Tanner graph is an efficient graphical representation of LDPC codes. There are m check nodes (c-nodes; number of parity bits) and n variable nodes (v-nodes; number of bits in a codeword). LDPC codes are said to be regular if Wc is constant for every column, and Wr=Wc. (n/m) is constant for each row. If the parity matrix H is low density but the number of "1" in each row or column are not constant, the code is said to be an irregular. One matrix representation and graphical representation is shown below. Let, the parity check matrix 'H' with 16-columns and 8 rows represented as,

$$H = \begin{bmatrix} 0101011101\ 101011 \\ 1000000000\ 000000 \\ 0110011001\ 101010 \\ 0000000010\ 000000 \\ 1101000100\ 101001 \\ 0000010001\ 000000 \\ 0100000100\ 001101 \\ 0010111001\ 100010 \end{bmatrix}$$

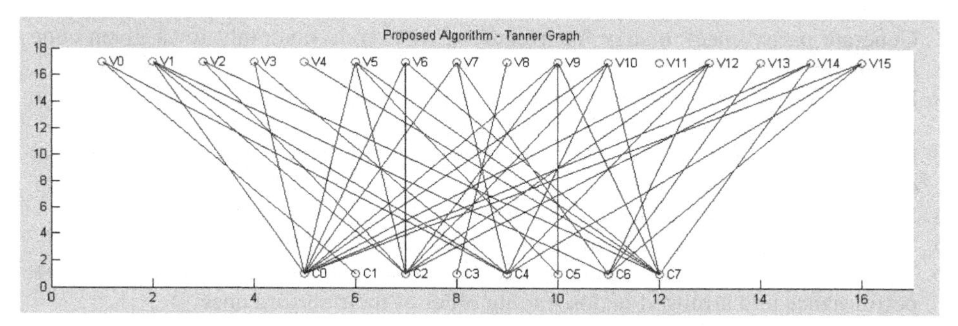

Fig. 1. Tanner graph for parity check matrix (16, 8)

3.1 Tanner Graph

The upper v-nodes are called variable nodes or message nodes and the bottom c-nodes are called parity nodes or check nodes. The messages are exchanged between v-nodes and c-nodes with the edges of the graph which act as the information pathways. The tanner graph is also called as bipartite graph.

3.2 Mutation

Mutation operation is the process in which the few elements of matrix are changed randomly to get new matrix whose performance is compared with previous and best is retained, like obtaining best fit chromosome. This process is repeated for several numbers of times and final matrix is called as best chromosome.

3.3 Discussion

We consider an Additive White Gaussian Noise Binary Symmetric (AWGNBSC) Channel, whose input is a random variable. We assume that the transmitter uses an LDPC code and the receiver decodes using the Belief Propagation algorithm. We also assume that the information bit “0” is mapped to -1 and information bit “1” is mapped to 1 on the channel, and that the all-zero codeword is transmitted. Notice that since the channel is symmetric and Belief Propagation decoding satisfied the check-node symmetry and variable-node symmetry conditions, the decoder’s performance is independent of the transmitted codeword. This algorithm passes messages between variable and check nodes, representing the probability that the variable nodes are '1' or '0' and whether the check nodes are satisfied. The Bit Error Rate (BER), Frame Error Rate (FER) and computational complexity is calculated. MATLAB Platform is used. Results obtained using our proposed algorithm is compared with method in which matrix generated randomly without mutation operation. Rate one half codes are considered.

3.4 Genetic Algorithm

1. Generate parity check matrix for irregular LDPC code randomly for a given code length and code rate.
2. Each matrix is evaluated over Additive White Gaussian Noise binary symmetric memory less Channel (AWGNC) for its performance.
3. Evaluate each matrix for BER, FER against Eb/N0(dB) and computational complexity against code length
4. Matrices are arranged in descending order according to their performance values
5. Select first few matrices for mutation operation to get new matrices.
6. This new generated matrix after mutation and old few matrices are evaluated for performance and arranged in descending order of their performance.
7. After certain number of iteration first matrix is the best of all which achieves low error floor and computational complexity.
8. Results of proposed method are compared with code constructed completely randomly

4 Simulation Results

The performance of the proposed method for irregular LDPC code design is implemented and evaluated on MATLAB platform. Simulation results shows that the proposed approach performs efficiently with low computation complexity and low bit error rate (BER), frame error rate (FER).

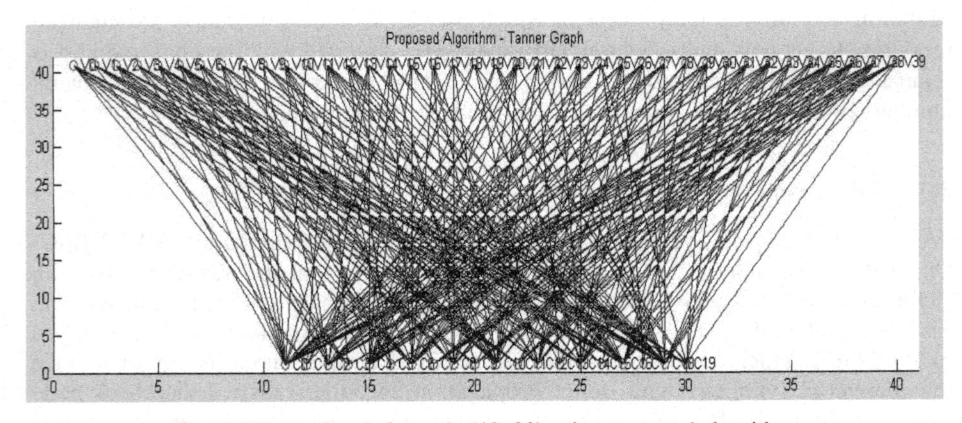

Fig. 2. Tanner Graph for code (40, 20) using proposed algorithm

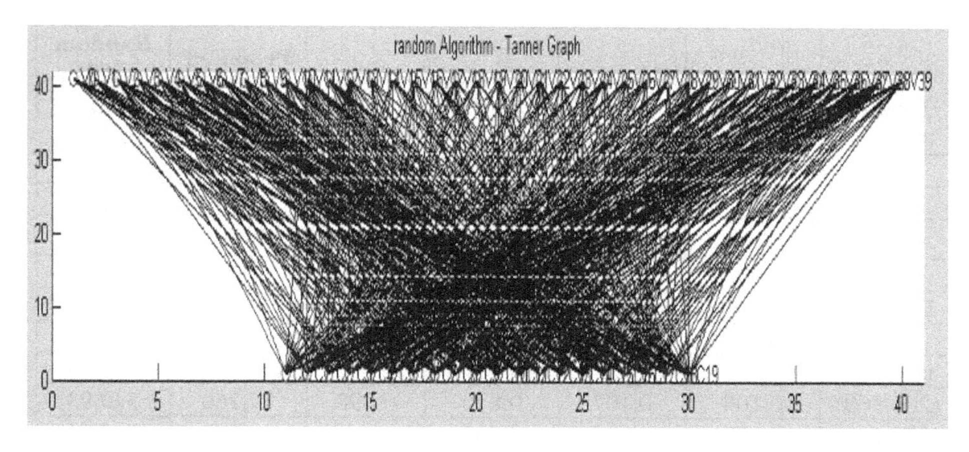

Fig. 3. Tanner Graph for code (40, 20) using completely random Generation algorithm

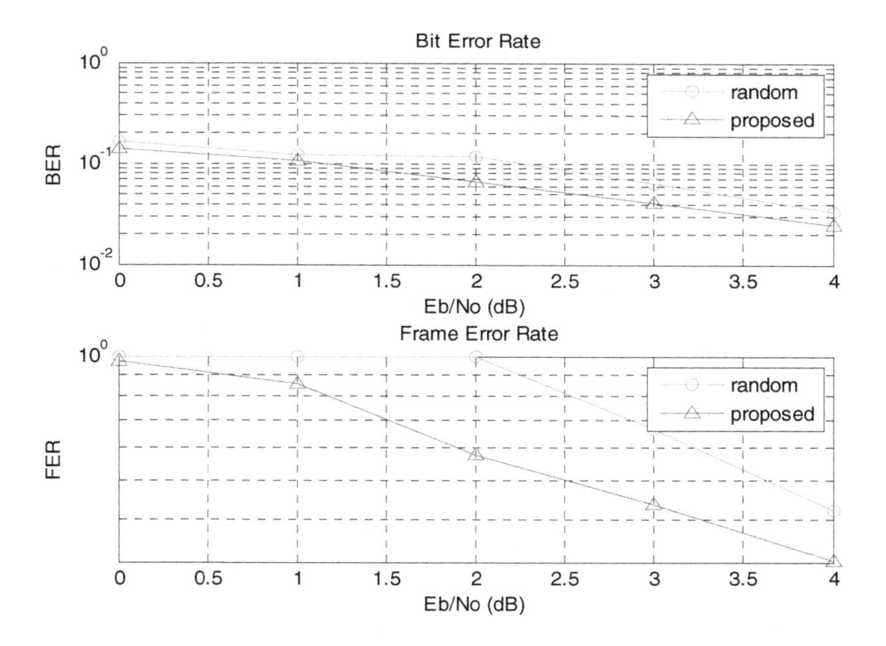

Fig. 4. Comparison of BER, FER against Eb/N0 (dB) for code (40,20)

Table 1. Performance of parity check matrix for different code length with the Shannon capacity of 1 Mbits/sec

Code Length	Proposed BER	Random BER	proposed FER	random FER	Proposed time (in sec)	Random CPU time (in sec)
(16,8)	0.061	0.069	0.2540	0.400	0.35297	0.5077
(20,10)	0.064	0.080	0.2448	0.4635	0.9902	0.9831
(40,20)	0.074	0.102	0.630	0.813	1.128	1.673
(80,40)	0.099	0.103	0.862	0.949	1.880	4.500
(100,50)	0.101	0.106	0.931	0.987	3.011	7.849
(128,64)	0.106	0.107	0.981	1	5.275	15.553
(150,50)	0.073	0.075	0.911	0.987	6.203	16.423
(200,100)	0.106	0.107	0.870	0.898	17.760	54.693
(300,100)	0.077	0.078	0.993	1	36.887	115.16
(320,160)	0.0848	0.085	0.9499	0.9910	124.27	406.39

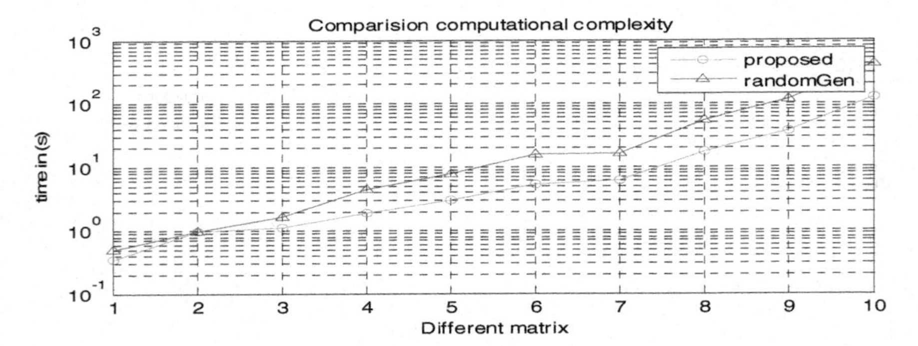

Fig. 5. Computational complexity using both methods with different matrices shown in table 1

5 Conclusion

We have designed an irregular LDPC codes for binary-input (AWGNC) channels. Graph of Bit Error Rate (BER), Frame Error Rate (FER), versus signal to noise ratio (SNR in dB) is plotted. Tanner graph which helps describing decoding operation is also plotted. Simulation results show that our proposed genetic algorithm lowers Bit Error Rate (BER), Frame Error Rate (FER) and computational complexity drastically as compared with algorithm in which codes are generated randomly. Results show that this method is suitable for short code length in the order of few hundred, where error floor is lowered. If the length of code increases computational complexity decreases drastically as compared to other method, where as there is no significant difference in BER and FER for code length more than 300 and rate one half code. Further BER and FER can be reduced drastically if code rate 1/3, 1/4 and so on is used.

References

1. Richardson, T.J., Urbanke, R.L.: The Capacity of Low-Density Parity-Check Codes Under Message-Passing Decoding. IEEE Transactions on Information Theory 47(2), 599–618 (2001)
2. Asvadi, R., Banihashemi, A.H., Ahmadian-Attari, M.: Lowering the Error Floor of LDPC Codes Using Cyclic Liftings. In: Proceedings of IEEE International Symposium on Information Theory, Austin, TX, pp. 74–728 (June 2010)
3. Eckford, A.W., Kschischang, F.R., Pasupathy, S.: On Designing Good LDPC Codes for Markov Channels. IEEE Transactions on Information Theory 53(1), 5–21 (2007)
4. Kim, J., Hur, W., Ramamoorthy, A., McLaughlin, S.W.: Design of Rate-Compatible Irregular LDPC Codes for Incremental Redundancy Hybrid ARQ Systems. In: Proceedings of IEEE International Symposium on Information Theory, Seattle, WA, pp. 1139–1143 (July 2006)
5. Yadav, K., Srivastava: Improved performance of Irregular LDPC Codes by attaining regularity for AWGN channel. In: Proceedings of SPIT-IEEE Colloquium and International Conference, Mumbai, India, vol. 3, pp. 7–11 (February 2008)
6. Dolecek, L., Lee, P., Zhang, Z., Anantharam, V., Nikolic, B., Wainwright, M.J.: Predicting Error Floors of Structured LDPC Codes: Deterministic Bounds and Estimates. IEEE International Journal on Selected Areas in Communications 27(6), 908–917 (2009)
7. Newagy, F.A., Fahmy, Y.A., El-Soudani, M.M.S.: Novel Construction of Short Length LDPC Codes For Simple Decoding. Journal of Theoretical and Applied Information Technology 3(3), 64–69 (2007)
8. Islam, M.R., Kim, J.: Quasi Cyclic Low Density Parity Check Code for High SNR Data Transfer. Journal of Radio Engineering 19(2), 356–362 (2010)
9. Johnson, S.J., Weller, S.R.: A Family of Irregular LDPC Codes With Low Encoding Complexity. IEEE Communications Letters 7(2), 79–81 (2003)
10. Cole, C.A., Hall, E.K., Wilson, S.G., Giallorenzi, T.R.: A General Method for Finding Low Error Rates of LDPC Codes. IEEE Transcations on Information Theory, 1–30 (2006)
11. Ardakani, M., Kschischang, F.R.: A More Accurate One-Dimensional Analysis and Design of Irregular LDPC Codes. IEEE Transactions on Communications 52(12), 2106–2114 (2004)
12. Prasartkaew, C., Choomchuay, S.: A Design of Parity Check Matrix for Irregular LDPC Codes. In: International Symposium on Communication and Information Technology, Incheon, Korea (September 2009)
13. Gunnam, K.K., Choi, G.S., Yeary, M.B., Atiquzzaman, M.: VLSI Architectures for Layered Decoding for Irregular LDPC Codes of WiMax. In: Proceedings of IEEE International Conference on Communications, Glasgow, pp. 4542–4547 (June 2007)
14. Bhardwaj, V., Pathak, Kumar, A.: Structured LDPC Codes with Linear Complexity Encoding. In: Proceedings of WRI International Conference on Communications and Mobile Computing, vol. 1, pp. 200–203 (2009)
15. Eckford, A.W., Kschischang, F.R., Pasupathy, S.: Designing Very Good Low-Density Parity-Check Codes for the Gilbert-Elliott Channel. In: Proceedings of 8th Canadian Workshop on Information Theory, Waterloo, pp. 18–21 (2003)
16. Landner, S., Hehn, T., Milenkovic, O., Huber, J.B.: Two Methods for Reducing the Error-Floor of LDPC Codes. IEEE Transactions on Information Theory 1, 1–50 (2007)

17. Cole, C.A., Hall, E.K., Wilson, S.G., Giallorenzi, T.R.: Analysis and Design of Moderate Length Regular LDPC Codes with Low error Floors. In: Proceedings of 40th Annual Conference on Information Sciences and Systems, Princeton, NJ, pp. 823–828 (March 2006)
18. Iliev, T.B., Hristov, G.V., Zahariev, P.Z., Iliev, M.P.: Application and evaluation of the LDPC codes for the next generation communication systems. In: Novel Algorithms and Techniques in Telecommunications, Automation and Industrial Electronics, pp. 532–536. Springer (2008)
19. Milenkovic, O., Soljanin, E., Whiting, P.: Asymptotic Spectra of Trapping Sets in Regular and Irregular LDPC Code Ensembles. IEEE Transactions on Information Theory 53(1), 39–55 (2007)
20. Chen, J., Michael Tanner, R., Zhang, J., Fossorier, M.P.C.: Construction of Irregular LDPC Codes by Quasi-Cyclic Extension. IEEE Transactions on Information Theory 53(4), 1479–1483 (2007)
21. Zheng, X., Lau, F.C.M., Tse, C.K.: Constructing Short-Length Irregular LDPC Codes with Low Error Floor. IEEE Transactions on Communications 58(10), 2823–2834 (2010)

Channel Allocation for Cellular Networks Using Genetic Algorithm

Jish Elizabeth Joy and Reena Kumbhare

Sardar Patel Institute of Technology
Bhavan's College, Munshi Nagar
Andheri (W), Mumbai – 400053
{jishelizabeth,rk.extc}@gmail.com

Abstract. With the limited frequency spectrum and increasing demand for cellular communication services, the problem of channel assignment becomes increasingly important. The three main constraints impeding the efficient usage of frequency spectrum are co-channel constraint (CCC), the adjacent channel constraint (ACC) and the co-site constraint (CSC). The objective is to obtain a conflict-free channel allocation scheme using modified immune genetic algorithm which satisfies both the electromagnetic compatibility (EMC) constraint and traffic demand requirement. The minimum separation encoding scheme is introduced to meet the co-site constraint. The genetic operators, viz. crossover and mutation are proposed to ensure the traffic demand through the iterative process. To increase the efficiency and velocity of convergence, immune operators are given, such as immune clone, vaccination, immune selection etc.

Keywords: cellular networks, channel assignment problem, genetic algorithm, electromagnetic compatibility constraints.

1 Introduction

The number of users of cellular communication have increased manifold. The increase has been driven by the ease of use, convenience and technological advancement. The recent technological advancement has enabled mobile communication to support and provide high quality voice communication and high speed data services. The new generations of wireless access system offers wideband services such as multimedia. As larger number of cellular users increases, the available frequency spectrum has to be shared between an ever increasing user-base. As frequency spectrum is a limited resources, availability and efficient usage of frequency becomes critical. In cellular communication, the coverage area is divided into polygons call cells and each of these cells is allocated a portion of the total available frequency spectrum. The spectrum allocated to a cellular service provider is divided into number of channels and allocated to a cell depending on the number of users and service requirements. However, unlike wired communication, the users of mobile communication are exposed to the possibility of interference if frequency channel used are in close proximity to each other. In order to minimize communication interference and enable reuse of

S. Unnikrishnan, S. Surve, and D. Bhoir (Eds.): ICAC3 2013, CCIS 361, pp. 259–265, 2013.

limited resource while satisfying the large demands cellular communication services, channels need to be assigned in an optimum way. This is referred to as the Channel Allocation Problem (CAP).

The important principles of channel assignment are should meet the overall user / traffic demand, should meet the electromagnetic compatibility constraint and minimize cost through minimum frequency reuse. The problem of finding a conflict-free channel assignment with minimum channel span is a NP hard [2]. What it means is that as the size of the problem increases, the time required to solve the problem increases in an exponential manner instead of in a polynomial manner. CAP can be classified into two categories: 1) fixed channel assignment (FCA), where channels are permanently allocated to each cell and 2) dynamic channel assignment (DCA), where all channels, which are available for every, are allocated dynamically upon request. While DCA gives better performance, FCA outperforms when cellular traffic condition increases, which is expected going forward. Many approaches have been proposed in solving CAP. Most of these methods are based on a heuristic ranking of cells according to the difficulty of meeting the electromagnetic constraints. In recent years, some approaches based on the Hopfield neural network, simulated annealing and genetic algorithm have been proposed [1].

Biologically motivated computing activity forms the backbone of Genetic Algorithm (GA). GA is a prominent example of evolutionary computation which was introduced in 1950s and 1960s but gained momentum in early 1980s. It is based on the principle of evolution, operations such as crossover and mutation, and the concept of fitness. This project looks at a modified genetic algorithm for the channel allocation problem. The modified genetic algorithm is based on the minimum-separation encoding scheme. In order to increase the efficiency and velocity of convergence, immune operators such as immune clone, vaccination, immune selection etc. are used [1].

2 Genetic Algorithm

Genetic algorithms (GAs) were invented by John Holland in the 1960s and were developed by Holland and his students and colleagues at the University of Michigan in the 1960s and the 1970s. In contrast with evolution strategies and evolutionary programming, Holland's original goal was not to design algorithms to solve specific problems, but rather to formally study the phenomenon of adaptation as it occurs in nature and to develop ways in which the mechanisms of natural adaptation might be imported into computer systems. Holland's introduction of a population–based algorithm with crossover, inversion, and mutation was a major innovation [4].

GA is a search algorithm that is based on the mechanics of natural selection, genetics and evolution. Evolution is, in effect, a method of searching among an enormous number of possibilities for solutions. Hence, this mechanism of evolution is well suited for some of the most pressing computational problems in many fields. Many computational problems require searching through a huge number of possibilities for solutions. Hence, GA has been used in a variety of applications.

One of the common applications of GA is function optimization, where the goal is find a set of parameter values that maximize, say, a complex multi-parameter function.

For example, $y = f(x_1, x_2, x_3, x_4)$

Here the solution y with different values of x_1, x_2, x_3 and x_4 forms the population.

GA has following elements: population of chromosomes, selection according to fitness, crossover to produce new offspring and random mutation of new offsprings. Each solution is represented as an individual. A collection of individual forms a population. Just as in genetics, an individual contains several genes. As an individual is represented in binary numbers, each binary bit corresponds to a gene. From the population of solutions, GA selects the best possible solution on the basis of a threshold or fitness function. This fitness function is unique for every optimization problem. The fitness of each chromosome in the population is measured, and the best chromosome is selected. GA ensures a quicker convergence to the near-optimal solution. Any problem that can be represented as an optimization problem can be solved using GA. The process keeps repeating in an iterative manner till a particular termination criterion has reached.

2.1 Initialization

The population size can be designed by the user. This population contains many chromosomes, and should be present in a binary form. The number of chromosomes in a population forms the population size. This parameter plays an important role in the performance of GA.

In channel assignment problem, the CSC constraint is required therefore the literature [2] presents a minimum separation encoding scheme.

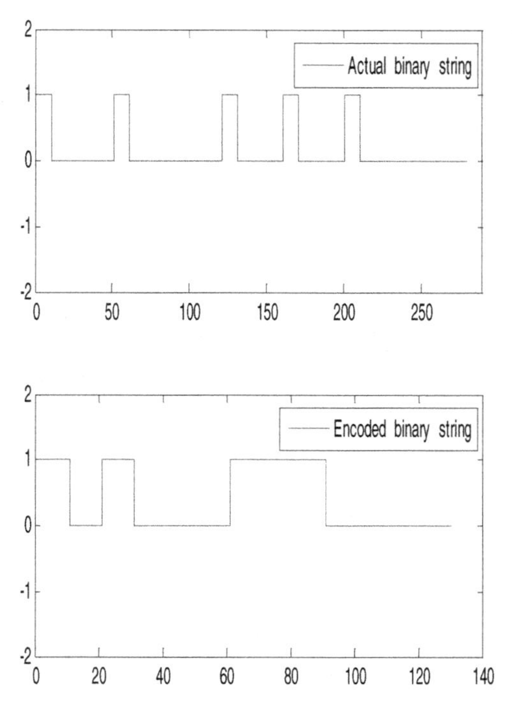

Fig. 1. A typical encoded binary string

2.2 Minimum Separation Encoding Scheme

Let an individual be represented by **p**-bit binary string with **q** fixed element and let dmin be the minimum separation between consecutive elements. The idea of the encoding scheme is to represent the solution in such a way that a ‘1’ is followed by (d_{min} – 1) zeroes encoded as a new ‘1’ denoted as 1’. Hence, the length of representation can be substantially reduced. However, a problem still remains if “one” is at a position within $(d_{min} - 1)$ from the end of the string. To cope with this shortcoming, we need to first augment the original individual with $(d_{min} - 1)$ “zero” before performing the encoding scheme. Thus, the total bit length is $(p + d_{min} - 1)$. Using the minimum separation encoding scheme, a bit binary string with elements can be encoded by a binary string of $[p - (q - 1)(d_{min} - 1)]$ bits only.

In the literature [1], the solution to the problem is given by N x M binary matrix F. The minimum separation encoding scheme is used to ensure CSC, where $d_{min} = c_{ii}$. The special genetic operator to meet the traffic requirement, namely, $\sum f_{ii} = d_i$

3 Evaluation

Before starting with the GA, a fitness function has to be formulated first. This fitness function is the most crucial part of the algorithm, and varies depending upon the application GA is used in. The fitness function must be designed such that the best chromosome corresponds to the one with the least fitness value. In the evaluation phase, the fitness functions of all the chromosomes present in the initial population are calculated. The fitness function should be formulated such that the individual chromosome be its variable input parameter. In certain applications, real numbers have to be converted to binary form, to be applied in GA.

Basic requirements of a cellular network are the ability to serve the expected and avoidance of interference. A total of d_i channels are required for cell i. This implies that the total number of ones in row i of **F** must be d_i.

Mathematically,

$$\sum_{q=1}^{m} f_{iq} = d_i \tag{1}$$

The second requirement is CSC. If channel is within distance from an already assigned channel in cell, then channel must not be assigned to cell.

Mathematically, it means that if the assignment of channel to cell violates CSC, then

$$\sum_{\substack{q=p-(c_{ij}-1) \\ q \neq p \\ 1 \leq q \leq m}}^{p+(c_{ij}-1)} f_{iq} > 0 \tag{2}$$

The third requirement is CCC and ACC. If channel $\boldsymbol{p}$ in cell $\boldsymbol{i}$ is within distance c_{ij} from an already assigned channel $\boldsymbol{q}$ in cell $\boldsymbol{j}$, where $c_{ij} > 0$ and $i \neq j$, then channel must not be assigned to cell $\boldsymbol{i}$.

Mathematically, it means that if the assignment of channel $\boldsymbol{p}$ to cell i violates CCC and/or ACC, then

$$\sum_{\substack{j=1 \\ j\neq 1 \\ c_{ij}}}^{n} \sum_{q=p-(c_{ij}-1)}^{p+(c_{ij}-1)} f_{jq} > 0 \tag{3}$$

Equations (1), (2) and (3) are taken from [1].

Our objective is to find the optimum allocation based on iterative process of algorithm through crossover and mutation such that the cost function attains minimum value with all the constraints being satisfied.

$$C(F) = \sum_{i=1}^{n} \sum_{p=1}^{m} \left| \sum_{\substack{j=1 \\ j\neq i \\ c_{ij}>0}}^{n} \sum_{\substack{q=p-(c_{ij}-1) \\ 1\leq q\leq m}}^{p+(c_{ij}-1)} f_{jq} \right| f_{ip} \tag{4}$$

3.1 Creation of Initial Population

By using the minimum separation encoding scheme, a row of **F** is encoded as a [m – (d_i – 1) (c_{ii} – 1)] binary strings. In order to meet the traffic demand requirement, we set d_i "one" in row of F randomly. Because there are n rows binary strings in **F**, we need to add them together in sequence to construct individual.

3.2 Crossover

Once a portion of the population has been selected, the number of chromosomes in the initial population decrements. But the population size must be maintained throughout. For this purpose, new chromosomes have to be generated from the existing ones. This is done with the help of two functions: crossover and mutation.

Crossover is the process where two chromosomes are combined to form two new chromosomes. The strings that are selected from the selected population for this purpose are called parent chromosomes. The offsprings produced are called child chromosomes. Crossover can be either one-point or two-point crossover.

Given two points A and B, we create a first-in last-out (FILO) stack to store the bit position k corresponding to the opposite bit pair ($\mathbf{A}_k$, $\mathbf{B}_k$). $\mathbf{A}_k$ and $\mathbf{B}_k$ are set to be opposite if $\mathbf{A}_k \oplus \mathbf{B}_k = 1$. The crossover is performed by first generating two crossover points c_1 and c_2 at random along the string length such that $c_1 < c_2$ and then moving right from c_1 until a i is found such that $\mathbf{A}_i \oplus \mathbf{B}_i = 1$. We push i into the FILO stack and continue the process until we find a j such that $\mathbf{A}_j \oplus \mathbf{B}_j = 1$. Then we compare $\mathbf{A}_j$ with $\mathbf{A}_{s1}$, where $s1$ is the top element in the stack. If they are the same, we push j into the stack, otherwise we swap the pair indexed by j with the pair indexed by $s1$ and pop $s1$ from the stack. The process continues until c_2 is reached.

	Before Crossover	After Crossover
A	1 0 **1 1 0 0 0 1** 1 0	0 1 **0 1 1 0 1 0** 1 0
B	0 1 **0 1 1 1 1 0** 0 0	0 1 **1 1 0 1 0 1** 0 0

3.3 Mutation

Mutation is the process where only one parent is involved to form a new chromosome. Some random genes are selected for mutation or change. Usually the probability of mutation is chosen to be less than the probability of crossover. In order to balance the number of one in an individual, the mutation operation must always be done in pairs of opposite bits. This can be implemented as follows:

Let b_i be the ith bit position of an individual. To mutate b_i, we need to find the random b_j such that b_i XOR $b_j = 1$. Then, we swap b_i with b_j. In case of binary array representation, both b_i with b_j must be in the same row.

3.4 Termination

These processes take place in an iterative manner. But certain terminating criterions are provided. It could either be the number of iterations, or a particular threshold value that has to be attained, or else the time taken for implementation. When any of these criterion reaches, the iteration automatically stops, and the first chromosome in the current population is selected as the best individual, or as the optimum solution to the problem. Hence stop criterion is given by

- A given generation is reached
- The optimal solutions are not improved in a given generation

4 Conclusion

The problem of conflict-free frequency channel allocation in cellular radio network is studied here. We propose an approach based on modified genetic algorithm. Using the minimum-separation encoding scheme, the required number of bits for representing the solutions decreased considerably. The simulation of the genetic algorithm optimizes the computation time and hence the costly process of implementing it can be greatly reduced thereby speeding up search.

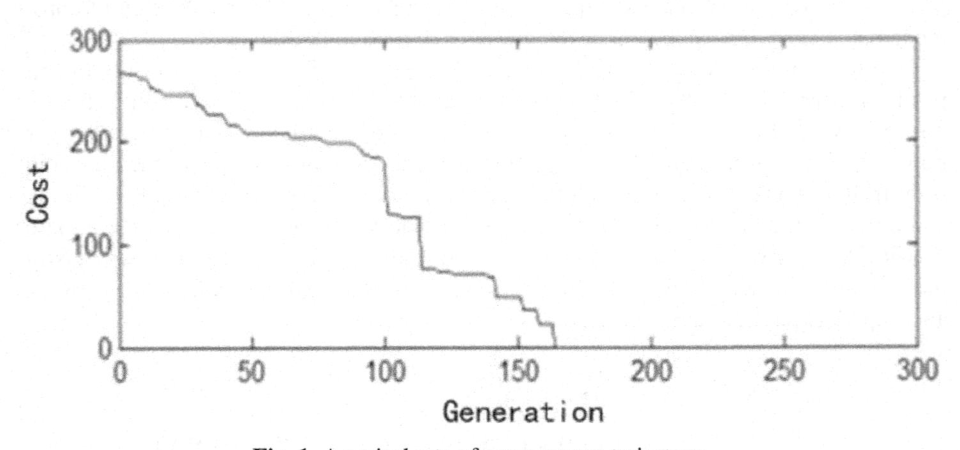

Fig. 1. A typical rate of convergence trajectory

References

[1] Yu, Z., Lv, S., Yan, G., Zhang, Q.: Modified Immune Genetic Algorithm for Channel Assignment Problems in Cellular Radio Networks. In: International Conference on Intelligent System Design and Engineering Application, pp. 823–826. IEEE (2010)
[2] Ngo, C.Y., Li, V.O.K.: Fixed Channel Assignment in Cellular Radio Networks Using a Modified Genetic Algorithm. IEEE Transactions on Vehicular Technology 47(1), 163–172 (1998)
[3] Pinagapany, S., Kulkarni, A.V.: Solving Channel Allocation Problem in Cellular Radio Networks Using Genetic Algorithm. In: IEEE Proc. COMSW App., pp. 239–244 (2008)
[4] Funabiki, N., Takefuji, Y.: A Neural Network Parallel Algorithm for Channel Assignment Problems in Cellular Radio Networks. IEEE Transactions on Vehicular Technology 41(4), 430–437 (1992)
[5] Beckmann, D., Killat, U.: A new strategy for the application of genetic algorithms to the channel assignment. IEEE Trans. Vehi. Technol. 48(44), 1261–1269 (1999)
[6] Mitchell, M.: An Introduction to Genetic Algorithm, 1st edn. MIT Press (1999)

NACPred: Computational Prediction of NAC Proteins in Rice Implemented Using SMO Algorithm

N. Hemalatha[1,*], M.K. Rajesh[2], and N.K. Narayanan[3]

[1] AIMIT, St. Aloysius College, Mangalore, Karnataka, India
hemasree71@gmail.com
[2] Division of Crop Improvement, Central Plantation Crops Research Institute, Kasaragod 671124, Kerala, India
mkraju_cpcri@yahoo.com
[3] School of Information Science and Technology, Kannur University, Kannur, India
csirc@rediffmail.com

Abstract. The impact of abiotic stresses, such as drought, on plant growth and development severely hampers crop production worldwide. The development of stress-tolerant crops will greatly benefit agricultural systems in areas prone to abiotic stresses. Recent advances in molecular and genomic technologies have resulted in a greater understanding of the mechanisms underlying the genetic control of the abiotic stress response in plants. NAC (NAM, ATAF1/2 and CUC2) domain proteins are plant-specific transcriptional factors which has diversified roles in various plant developmental processes and stress responses. More than 100 NAC genes have been identified in rice. In the proposed method, NACPred, an attempt has been made in the direction of computational prediction of NAC proteins. The well-known sequential minimum optimization (SMO) algorithm, which is most commonly used algorithm for numerical solutions of the support vector learning problems, has been used for the development of various modules in this tool. Modules were first developed using amino acid, traditional dipeptide (i+1), tripeptide (i+2) and an overall accuracy of 76%, 90%, and 97% respectively was achieved. To gain further insight, a hybrid module (hybrid1 and hybrid2) was also developed based on amino acid composition and dipeptide composition, which achieved an overall accuracy of 90% and 97%. To evaluate the prediction performance of NACPred, cross validation, leave one out validation and independent data test validation were carried out. It was also compared with algorithms namely RBF and Random Forest. The different statistical analyses worked out revealed that the proposed algorithm is useful for rice genome annotation, specifically predicting NAC proteins.

Keywords: Rice, gene prediction, NAC, SMO.

1 Introduction

Rice (*Oryza sativa* L.), a source of staple food, has a major influence on human nutrition and food security. Billions of people world-wide depend on rice-based production systems for their main source of employment and development. Rice production

S. Unnikrishnan, S. Surve, and D. Bhoir (Eds.): ICAC3 2013, CCIS 361, pp. 266–275, 2013.

continuously faces the challenge of keeping pace with rapid rise in human population and declining natural resource base, two of the critical resources being land and water. In addition, abiotic stresses, such as drought, adversely affect the growth and productivity of rice-based farming systems.

The development of stress-tolerant crops will be of immense advantage in modern agriculture, especially in areas that are prone to such stresses. In recent years, several advances have been made towards identifying potential stress related genes which are capable of increasing the tolerance of plants to abiotic stress. NAC transcription factors have major functions in plant development as well as in abiotic stress responses. NAC (NAM, ATAF1/2 and CUC2) domain proteins comprise of one of the largest plant-specific transcriptional factors which is represented by approximately 140 genes in rice [1]. These transcription factors (TFs) regulate gene expression by binding to specific cis-acting promoter elements, thereby activating or repressing the transcriptional rates of their target genes [2, 3]. Thus, for the reconstruction of transcriptional regulatory networks, the identification and functional characterization of these transcription factors is essential [4].

Computational prediction methods, compared with the experimental methods are fast, automatic and more accurate especially for high-throughput analysis of large-scale genome sequences. Therefore, a fully automatic prediction system for NAC transcription factors in rice is a systematic attempt in this direction. The SMO module for the prediction of NAC proteins in genome of indica rice (*Oryza sativa* L. ssp. indica) was developed using various features of a protein sequence and the performance of these models was evaluated using cross-validation techniques.

2 Materials and Methods

2.1 Datasets

The selection of dataset is the most important concern during development of a prediction method. The data set used in the present study, consisted of 95 NAC proteins of indica rice taken from Uniprot Knowledgebase. These 95 proteins were screened strictly in order to develop a high quality data set for the prediction tool. Fifteen NAC proteins were randomly selected from the main dataset for the creation of test set and remaining 80 proteins were used for positive dataset / training set. Non-NAC protein sequences were used as the negative data set. For training and testing, independent datasets were used which means training set and test set were entirely different.

2.2 Performance Evaluation and Parameters

Three methods often used for examining the effectiveness of a predictor, in statistical prediction are single independent dataset test, cross-validation test and jackknife test. Out of these, the jackknife test is considered to be most rigorous and objective one, as illustrated by a comprehensive review [5]. However, since the size of the dataset in the present study was large and jackknife test method takes much longer time to train a predictor based on SMO, cross-validation (5-fold, 8-fold) and independent dataset test were adopted for performance measurement. In n-fold cross validation, all the positive and negative datasets were combined and then divided equally into n parts, keeping the same distribution of positive and negative datasets in each part. Then n-1 parts were merged into a training data set with the one part left out taken as a test data set and the average accuracy of n-fold cross validation was used to estimate the performance. In the independent dataset test, although none of the data to be tested occurs in the training dataset used to train the predictor, the selection of data for the testing dataset could be quite arbitrary. In "leave-one-out" cross-validation (LOO), each sample in the dataset is separated out in turn as an independent test sample, and all the remaining samples are used as training data. This process is repeated until every sample is used as test sample one time with no repetition. All models were implemented in the WEKA software package [6].

2.3 The Machine Learning Algorithms

Sequential Minimal Optimization(SMO) is a support vector machine learning algorithm (SVM) that is conceptually simple, easy to implement, generally faster, and has better scaling properties for difficult SVM problems than the standard SVM training algorithm [7]. Training a support vector machine requires the solution of a very large quadratic programming (QP) optimization problem which is quite time consuming. In the case of SMO, it breaks this large QP problem into a series of smallest possible QP problems which are solved analytically and avoids using a time-consuming numerical QP optimization as an inner loop. The amount of memory required for SMO is linear in the training set size allowing SMO to handle very large training sets. Because matrix computation is avoided, SMO scales somewhere between linear and quadratic in the training set size for various test problems, where as SVM algorithm scales somewhere between linear and cubic in the training set size. SMO's computation time is dominated by SVM evaluation which makes SMO faster for linear SVMs and sparse data sets.

The Radial Basis Function (RBF) network is as a variant of artificial neural network [8]. An RBF is embedded in three layers, *viz.*, the input layer, the hidden layer, and the output layer. The input layer broadcasts the coordinates of the input vector to each of the nodes in the hidden layer and contains one neuron in the input layer for each predictor variable. Each node in the hidden layer then produces an activation based on the associated radial basis function and this layer has a variable number of neurons based on the training process. Finally in the output layer each node computes a linear combination of the activations of the hidden nodes. The result of an RBF network to a given input stimulus is completely determined by the activation functions associated with the hidden nodes and the weights associated with the links between the hidden layer and the output layer.

Random Forests (RF) grows many classification (decision) trees. To classify a new object from an input vector, RF puts the input vector down into each of the trees in the forest. Each tree gives a classification, and it is said that the tree "votes" for that class. The forest chooses that classification which has the most votes (over all the trees in the forest).The forest error rate depends on the correlation between any two trees in the forest and the strength of each individual tree in the forest. Decreasing forest error rate increases the strength of the individual trees.

2.4 Features and Modules

Amino-acid composition: Amino-acid composition is the fraction of each amino acid occurring in a protein sequence. This representation completely misses the order of amino acids. To calculate the fraction of all 20 natural amino acids following equation was used:

$$\text{Fraction of amino acid} = \frac{\text{Total number of amino acid i}}{\text{Total number of amino acids in protein}} \quad (1)$$

Traditional dipeptide composition: Traditional dipeptide composition gives information about each protein sequence giving fixed pattern length of 400 (20x20). This composition encompasses the information of the amino-acid composition along with the local order of amino acids. The fraction of each dipeptide was calculated according to the equation:

$$\text{Fraction of dep}(i + 1) = \frac{\text{Total number of dep (i+1)}}{\text{Total number of all possible dipeptides}} \quad (2)$$

In addition, to observe the interaction of the ith residue with the 3rd residue in the sequence, tripeptide (i + 2) was generated using Equation 3,

$$\text{Fraction of tripep}(i + 2) = \frac{\text{Total number of (i+2) tripep}}{\text{Total number of all possible tripeptides}} \quad (3)$$

where tripep (i + 2) is one of 8000 tripeptides.

Hybrid SMO module(s): The prediction accuracy was further enhanced with various hybrid approaches by combining different features of a protein sequence.

Hybrid 1: In this approach, we developed a hybrid module by combining amino acid composition and dipeptide composition features of a protein sequence as calculated by using Eqs. (1) and (2), respectively. This module was provided with a WEKA input vector pattern of 420 (20 for amino acid and 400 for dipeptide composition).

Hybrid 2: In the second approach we developed another hybrid module by combining amino acid composition and tripeptide composition as calculated using Eq. (1) and (3), respectively. The WEKA input vector pattern thus formed was 8020-dimensional [20 for amino acid and 8000 for tripeptide)].

2.5 Sequence Similarity Search

In this study, a query sequence was searched against the existing non-redundant database of NAC proteins (95 sequences used in training set) using PSI-BLAST

(Position-Specific Iterative Basic Local Alignment Search Tool). Here PSI-BLAST was used instead of normal BLAST because it has the capacity to detect remote homologies. Position-Specific Iterated BLAST (PSI-BLAST), a variant of BLAST is used for the discovery of weak but relevant protein sequence matches. This carries out an iterative search in which sequences found in one round was used to build score model for next round. Thus, PSI-BLAST provides a method of detecting distant relationships between proteins.

2.6 Evaluation Parameters

We adopted five frequently considered measurements for evaluation, *viz.*, accuracy (Ac), sensitivity (Sn), specificity (Sp), precision (Pr) and Mathew's Correlation Coefficient (MCC). Accuracy (Ac) defines the correct ratio between both positive (+) and negative (-) data sets. The sensitivity (Sn) and specificity (Sp) represent the correct prediction ratios of positive (+) and negative data (-) sets of NAC proteins respectively. Precision is the proportion of the predicted positive cases that were correct. However, when the number of positive data and negative data differ too much from each other, MCC should be included to evaluate the prediction performance of the developed tool. MCC is considered to be the most robust parameter of any class prediction method. The value of MCC ranges from -1 to 1, and a positive MCC value stands for better prediction performance. Among the data with positive hits by NACPred, the real positives are defined as true positives (TP), while the others are defined as false positives (FP).

$$\text{Sensitivity} = \frac{TP}{FN+TP} x\ 100 \tag{4}$$

$$\text{Specificity} = \frac{TN}{FP+TN} x\ 100 \tag{5}$$

$$\text{Accuracy} = \frac{TP+TN}{TP+TN+FP+FN}\ \text{x}\ 100 \tag{6}$$

$$\text{Precision} = \frac{TP}{TP+FP}\ x\ 100 \tag{7}$$

$$\text{MCC} = \frac{(TP\ X\ TN) - (FP\ X\ FN)}{\sqrt{(TP+FP)(TP+FN)(TN+FP)(TN+FN}} \tag{8}$$

where TP and TN are truly or correctly predicted positive NAC protein and negative (non- NAC protein), respectively. FP and FN are falsely or wrongly predicted NAC and non-NAC proteins, respectively.

2.7 ROC Curves

To compare the performance of different algorithms and performance of different composition methods on best algorithm in detail, ROC curves were used for intuitively visualizing prediction performance. ROC curves plots the true positive rate (TPR) as function of the false positive rate (FPR) which is equal to 1-specificity. The area under the ROC curve is the average sensitivity over all possible specificity values which can be used as a measure of prediction performance at different thresholds. ROC curves of random predictors will be around the diagonal line from bottom left to top right of the graph with scores of about 0.5, while a perfect predictor will produce a curve along the left and top boundary of the square and will receive a score of one.

3 Results and Discussion

The prediction accuracy was assessed by two different validation techniques namely cross-validation and independent data set tests. In order to achieve maximum accuracy, five different feature extraction techniques, including three composition-based and two hybrid-based, were used and models were developed with three different algorithms namely SMO, RBF and RF. Performance accuracy of SMO algorithm was found to be the best compared to other algorithms. A graphical representation of the accuracy values of the different feature extraction methods using SMO is shown in Figure 1.

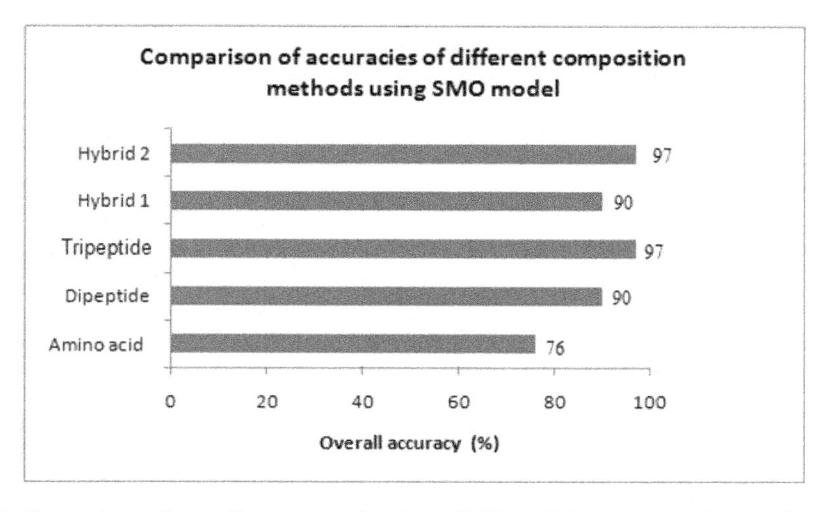

Fig. 1. Comparison of overall accuracy of various SMO modules constructed using five composition methods

3.1 Composition-Based Modules

The amino-acid composition-based module, with RBF algorithm, achieved an accuracy of 76% with different validation techniques applied in this study. The module implemented based on traditional dipeptide composition (i+1) gave more information about frequency and local order of residues. This module could achieve a maximum accuracy of 90% with RBF algorithm. Tripeptide (i+2) composition-based module was also developed to obtain more comprehensive information on the sequence order effects. This could achieve an accuracy of 97% with sequential minimum optimization algorithm (SMO) with various validation techniques. It could be observed that traditional dipeptide composition-based modules achieved higher accuracy compared to other independent compositions (Tables 1 and 2). This may be because dipeptide composition uses the actual order of sequence while calculating the composition where as the tripeptide is based on the pseudo sequence order. The detailed performance of amino acid, traditional dipeptide and tripeptide based modules with different validation techniques are presented in Tables 1 and 2.

3.2 Sequence Similarity Search

PSI-BLAST was used to compare a protein sequence with a created database to generate the homology of the given sequence with other related sequences [9]. This provided a broad range of information about each functional encoded protein. A 10-fold cross-validation was conducted with no significant hits and an accuracy of only 50% was obtained. This result suggests that similarity-based search tools alone cannot be efficient and reliable as compared to different composition-based modules.

Table 1. Validation of independent data test results of NAC proteins with SMO

Approaches	Algorithm	Sn(%)	Sp(%)	Acc(%)	Pr(%)	MCC
Aminoacid	RBF	100	79	90	83	0.81
	SMO	93	57	76	70	0.55
	RF	93	64	79	74	0.61
Dipeptide	RBF	93	93	93	93	0.86
	SMO	93	86	90	88	0.79
	RF	87	93	90	93	0.80
Tripeptide	RBF	33	86	59	71	0.22
	SMO	100	93	97	94	0.93
	RF	93	57	76	70	0.55
Hybrid 1	RBF	100	86	93	88	0.87
	SMO	93	86	90	88	0.79
	RF	93	64	79	74	0.61
Hybrid 2	RBF	33	86	59	71	0.22
	SMO	100	93	97	94	0.93
	RF	93	57	76	70	0.55

Table 2. Comparison of the prediction performance of three machine learning algorithms with different compositions

Approach	Algorithm	5-fold cross validation					8-fold cross validation					Leave one out cross validation				
		Sn	Sp	Ac	Pr	MCC	Sn	Sp	Ac	Pr	MCC	Sn	Sp	Ac	Pr	MCC
Amino acid	RBF	100	79	90	83	0.81	100	79	90	83	0.81	100	79	90	83	0.81
	SMO	93	57	76	70	0.55	93	57	76	70	0.55	93	57	76	70	0.55
	RF	93	64	79	74	0.61	93	64	79	74	0.61	93	64	79	74	0.61
Dipeptide	RBF	93	93	93	93	0.86	93	93	93	93	0.86	93	93	93	93	0.86
	SMO	93	86	90	88	0.79	93	86	90	88	0.79	93	86	90	88	0.79
	RF	87	93	90	93	0.80	87	93	90	93	0.80	87	93	90	93	0.80
Tripeptide	RBF	33	86	59	71	0.22	33	86	59	71	0.22	33	86	59	71	0.22
	SMO	100	93	97	94	0.93	100	93	97	94	0.93	100	93	97	94	0.93
	RF	93	57	76	70	0.55	93	57	76	70	0.55	93	57	76	70	0.55
Hybrid1	RBF	100	86	93	88	0.87	100	86	93	88	0.87	100	86	93	88	0.87
	SMO	93	86	90	88	0.79	93	86	90	88	0.79	93	86	90	88	0.79
	RF	93	64	79	74	0.61	93	64	79	74	0.61	93	64	79	74	0.61
Hybrid 2	RBF	33	86	59	71	0.22	33	86	59	71	0.22	33	86	59	71	0.22
	SMO	100	93	97	94	0.93	100	93	97	94	0.93	100	93	97	94	0.93
	RF	93	57	76	70	0.55	93	57	76	70	0.55	93	57	76	70	0.55

3.3 Hybrid Approach

In addition to the different composition methods, hybrid methodologies were also developed and used by combining various features of a protein sequence. Firstly, hybrid 1 was developed by combining amino acid composition and dipeptide composition. This obtained an accuracy of 90% with SMO algorithm. Secondly, hybrid 2 was developed by combining amino acid and tripeptide composition which also had a higher accuracy of 97% with SMO algorithm. Comparison of both of these hybrid approaches revealed that hybrid 2 composition method achieves an accuracy rate equivalent to tripeptide (i+2) composition method (Fig. 1).

3.4 ROC Curves

A ROC curve is a measure which shows the relationship between sensitivity and specificity of a given class. To evaluate the best classifier obtained, we plotted ROC curves based on the results of independent data test and cross validation (results obtained were similar). Figure 2 shows the ROC curve for SMO algorithm for five different compositional methods and it can be observed from the figure that all the curves result in a straight horizontal line. This is a desirable property of ROC curves and such models have high probability of correct prediction, with a minor chance of negative prediction. This is also reflected by area under the curve values of all compositions of SMO models. Figure 3 shows the best results of each algorithm.

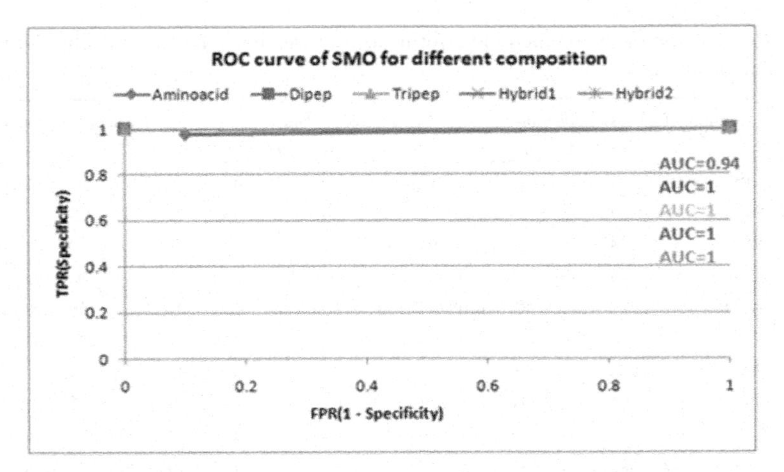

Fig. 2. ROC curve of SMO algorithm with various composition methods

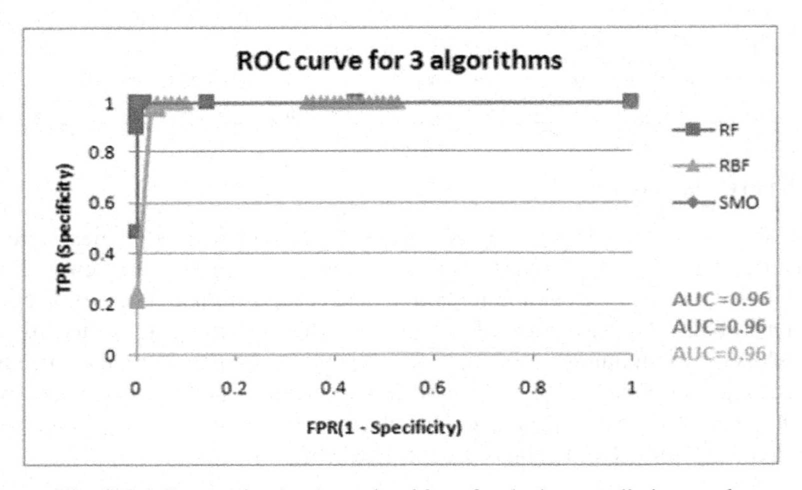

Fig. 3. ROC curve for the three algorithms for the best prediction results

4 Conclusions

Tools and resources are being developed to maximally construe the rice genome sequence. A major difficulty with rice annotation is the lack of accurate gene prediction programs. Rice has a substantial number of genes that are hypothetical in that they are predicted solely on the basis of gene prediction programs, making it vital that the quality of gene prediction programs for rice be improved further. Moreover rice, which is a model species, is the plant in which the function of most cereal genes will be discovered. Thus, the availability of systems/tools that can predict characteristics from sequence is essential to the full characterization of expressed proteins. Computational tools provide faster and accurate access to predictions for any organism and plants.

Identification of NAC proteins from sequence databases is difficult due to poor sequence similarity. In this work, we present a new method for NAC prediction based on SMO implemented in WEKA. The performance was found to be highly satisfactory. Comparison between different machine learning algorithms viz. RBF Network and Random forest was also carried out. Very high prediction accuracies for the validation tests show that NACPred is a potentially useful tool for the prediction of NAC proteins from genome of indica rice.

References

1. Fang, Y., You, J., Xie, K., Xie, W., Xiong, L.: Systematic Sequence Analysis and Identification of Tissue-specific or Stress-responsive Genes of NAC Transcription Factor Family in Rice. Mol. Genet. Genomics 280, 547–563 (2008)
2. Riechmann, J.L., Heard, J., Martin, G., Reuber, L., Jiang, C., Keddie, J., Adam, L., Pineda, O., Ratcliffe, O.J., Samaha, R.R., Creelman, R., Pilgrim, M., Broun, P., Zhang, J.Z., Ghandehari, D., Sherman, B.K., Yu, G.: Arabidopsis Transcription Factors: Genome-wide Comparative Analysis Among Eukaryotes. Science 290, 2105–2110 (2000)
3. Wray, G.A., Hahn, M.W., Abouheif, E., Balhoff, J.P., Pizer, M., Rockman, M.V., Romano, L.A.: The Evolution of Transcriptional Regulation in Eukaryotes. Mol. Biol. Evol. 20, 1377–1419 (2003)
4. Riano-Pachon, D.M., Ruzicic, S., Dreyer, I., Mueller-Roeber, B.: PlnTFDB: An Integrative Plant Transcription Factor Database. BMC Bioinformatics 8, 42 (2007)
5. Chou, K.C., Zhang, C.T.: Prediction of Protein Structural Classes. Crit. Rev. Biochem. Mol. Biol. 30, 275–349 (1995)
6. Witten, I.H., Frank, E.: Data Mining: Practical Machine Learning Tools and Techniques, 2nd edn. Morgan Kaufmann, San Francisco (2005)
7. Platt, J.C.: Fast Training of Support Vector Machines using Sequential Minimal Optimization. In: Schölkopf, B., Burges, C., Smola, A. (eds.) Advances in Kernel Methods - Support Vector Learning, pp. 185–208. MIT Press, Cambridge (1999)
8. Mitchell, T.: Machine Learning. McGraw Hill, New York (1997)
9. Altschul, S.F., Madden, T.L., Schaffer, A.A., Zhang, J., Zhang, Z., Miller, W., Lipman, D.J.: Gapped Blast and PSI-Blast: A New Generation of Protein Database Search Programs. Nucleic Acids Res. 25, 3389–3402 (1997)

Transmission and Reception Based Relay Deployment for Database Transactions in MANETs

B. Diwan and M.R. Sumalatha

Department of Information Technology, MIT Campus,
Anna University, Chennai, India
diwandiwan@gmail.com, sumalatha@annauniv.edu

Abstract. Traditionally in mobile ad-hoc network (MANET), mobile nodes can move freely and communicate with each other using limited energy resources. Database transactions in MANET should be done by considering the issues such as transmissions and receptions energy limitation. In order to manage energy limitation special nodes with high energy resources called relay nodes are deployed whose mobility can be controlled by the network. The proposed network presents two cases of the relay deployment problem to minimize the energy consumption by the traditional mobile nodes. In this paper, Min-Tx aims to minimize the energy consumption by traditional mobile nodes during data transmission and Min-Rx aims to minimize the energy consumption by traditional mobile nodes during data reception. The simulation study of proposed solution is experimented in terms of energy consumption with the effect of relay nodes reducing energy consumption.

Keywords: MANETs, relay nodes, transmission, reception, energy efficiency.

1 Introduction

In MANETs traditional mobile node (TMN) mobility leads to frequent and dynamic changes in network connectivity. Topology should be able to locate the dynamic changes in such type of network where traditional mobile nodes move freely while performing the database transactions. Because of the traditional mobile node mobility, mobile ad hoc networks face frequent disconnections that are a single connected network topology partitioned into two or more network topologies. Traditional mobile nodes within each partition can communicate with each other but cannot communicate with traditional mobile nodes in other partitions. Thus the network partitioned results in failed/aborted transactions. Important issue in mobile ad hoc networks is energy constraints on traditional mobile nodes because traditional mobile nodes will run on small and low power devices. Energy awareness becomes important factor in the algorithm design [1]. Energy consumption at traditional mobile node is more when it is transmitting data packets as compared to the packet reception and idle time. In order to reduce energy consumption further, each TMN will operate in different energy modes namely active mode, doze mode and sleep mode.

Many applications in mobile database transaction management environment need to be executed within their deadlines. More energy is consumed at the TMN with more data flows. Thus TMN with more data transmissions/receptions play vital role in executing transactions within their deadlines. Energy consumption should be taken care in

S. Unnikrishnan, S. Surve, and D. Bhoir (Eds.): ICAC3 2013, CCIS 361, pp. 276–283, 2013.

particularly TMNs with more data flows. As wireless systems continue to evolve, there has also been a growing interest to study and build mobility-aided systems where the mobility of certain nodes, often referred to as relay nodes, can be controlled by the underlying network protocol [10]. We can optimize the network performance by using these relay nodes. For example, special infrastructure nodes called relay nodes can be used to improve network connectivity. The spatial diversity and mitigate the effects of path loss [14] are exploited through relay nodes. According to IEEE specifications of the network interface card (NIC) with 2 Mbps the energy consumption at receiving node varies from 240 mA and the energy consumption at transmitting node varies from 280 mA using 0.5V energy. In this paper, two cases are proposed namely Min-Tx and Min-Rx for minimizing energy consumption by the traditional mobile nodes using relay nodes during more than one data transmissions and more than one data receptions.

2 Related Work

In MANETs, more power is needed for the transmission over long distance compared to the transmission over shorter distance. Source node transmits data to the destination node through an intermediate node namely relay node for reducing the power consumption [10].As the number of nodes in an ad hoc network increase then the wireless channel capacity will decrease. Developing scheduling algorithms with channel and topology uncertainty is a challenging one [5]. A transmission pair should use the same channel [7]. Relay nodes used for minimizing power consumption in ad hoc and sensor networks [15]. Low power and busy intermediate nodes are avoided to maintain energy efficiency [8]. Mobile node's power constraints influence the real time applications [2]. Different issues about nodes and energy constraints are proposed [3],[4],[13]. Network links cannot receive or transmit at a same time if it shares a common node [9]. Transmission powers are fixed [11]. In order to adapt specified transmission range, transmission power of each node is controlled [6]. Many researchers published number of research papers, concentrating on deploying relay nodes at transmission side. More energy is consumed even when the node receives more than one transaction. As the number of nodes increase then the wireless channel capacity will decrease. This paper aims to deploy relay nodes for maintaining effective transactions in Mobile ad hoc networks.

3 Deploying Relay Nodes

The importance of deploying relay nodes is explained here through the mathematical equations. The energy consumption E(p) to transmit a packet p is given by

$$E(p)=i*v*tp \quad (1)$$

Where i is the current, v is the voltage and tp is the time taken to transmit the packet p. Therefore the required energy Etx(p) to transmit a packet p is calculated as

$$Etx(p)=280mA*v*tp \quad (2)$$

If the node transmits more than one packet then the required energy to transmit the packets is calculated as

$$Etx(p)= \sum_{p=1}^{n} Etx(p) \quad (3)$$

Where n refers the number of packets transmitted by the node. The required energy Erx(p) to receive a packet p is calculated as

$$\text{Erx(p)=240mA*v*tp} \tag{4}$$

If the node receives more than one packet then the required energy to receive the packets is calculated as

$$\text{Erx(p)}=\sum_{p=1}^{n} Erx(p) \tag{5}$$

where n refers the number of packets received by the node. The node consumes more energy when it transmits or receives more than one data packets. So relay nodes are deployed where the nodes need to transmit or receive more than one data packets in order to reduce the energy consumption in the network.

3.1 Minimize Transmission Energy: Min- Tx

The objective of this case is to deploy a relay node in such a way as to minimize the overall transmission energy across the traditional mobile nodes in the network.

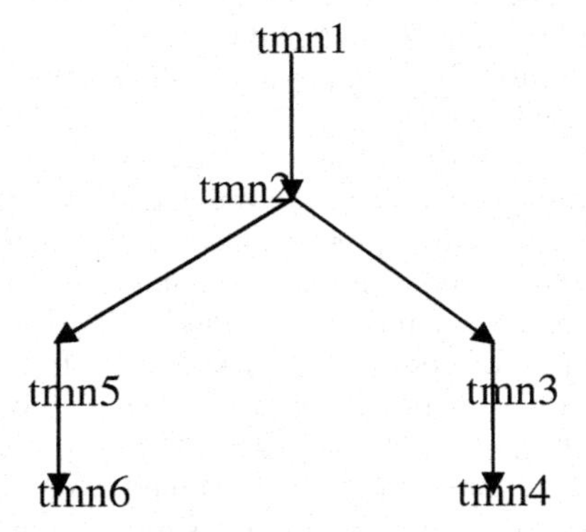

Fig. 1a. Network without relay node

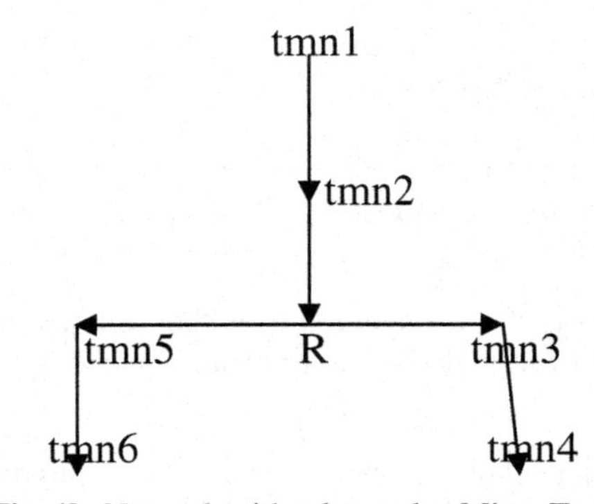

Fig. 1b. Network with relay node : Min – Tx

Where tmn – traditional mobile node , R – relay node.

From Fig 1.a, the transmission energy at the traditional mobile node tmn2 is calculated as the summation of required energy to transmit packets from tmn2 to tmn5 and from tmn2 to tmn3. That is, Etx(tmn2) = Etx(tmn2-5) + Etx(tmn2-3). From Fig 1.b, the transmission energy at the traditional mobile node tmn2 is calculated as the required energy to transmit packet from tmn2 to relay node. That is, Etx(tmn2) = Etx(tmn2-R).

3.2 Minimize Reception Energy: Min- Rx

The objective of this case is to deploy a relay node in such a way as to minimize the overall reception energy across the traditional mobile nodes in the network.

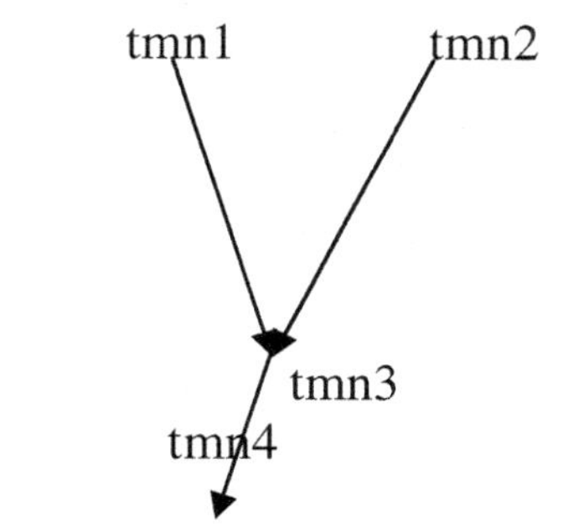

Fig. 2a. Network without relay node

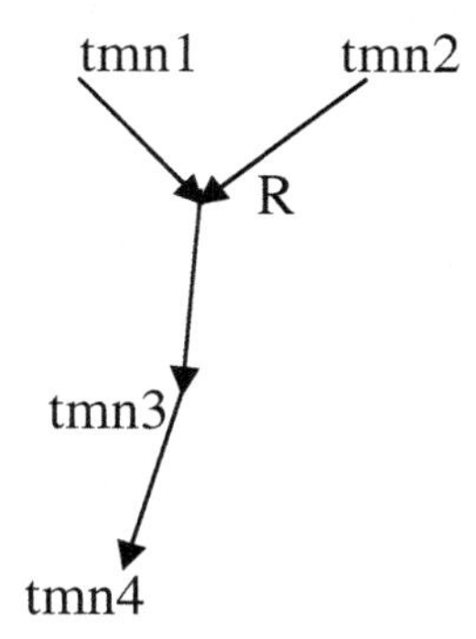

Fig. 2b. Network with relay node: Min – Rx

From Fig 2.a, the reception energy at the traditional mobile node tmn3 is calculated as the summation of required energy to receive packets from tmn1 to tmn3 and from tmn2 to tmn3. That is, Erx(tmn3) = Erx(tmn1-3) + Erx(tmn2-3). From Fig 2.b, the reception energy at the traditional mobile node tmn3 is calculated as the required energy to receive a packet from relay node to tmn3. That is, Erx(tmn3) = Erx(R-tmn3).

4 Mobility Epochs

The mobile network is dynamic due to the mobility of the traditional mobile nodes. Because of this dynamic network, it is necessary to find the position of traditional mobile node over fixed time interval length called mobility epochs before submitting transactions. Each traditional mobile node transmits 'hello' message in the network. All neighboring nodes receive the 'hello' message and update their neighboring nodes details. The proposed network is updating the speed and distance of every traditional mobile node in the every epoch time so that the network can predict the position of traditional mobile node.

5 Transaction Management

Data manager consists of two servers such as server id1 and server id2. The client sends the request to the data manager, data manager forwards the request to either server id1 or server id2 based on the data availability in the database, server will execute the transaction and sends the result to the data manager, data manager sends the result to the client. If the database in the local site is not enough to process the data then the request is sent to the global data manager, global data manager forwards the request to the server, server will execute the transaction and sends result back to the global data manager, global data manager sends result to respective local data manager, local data manager sends result to respective client.

6 Algorithm for Relay Deployment Problem: Min-Tx / Min-Rx

```
Execute Transaction (transaction T_ID, relay position
R_pos, epoch time, T_deadline,
node_with_transmission_flow TMN-Tx,
node_with_reception_flow TMN-Rx)
Begin
  Set epoch time=value
  Set timer=deadline
  While deadline • 0 do
Find TMN_Pos  // find position of TMN in every epoch time
If TMN-Tx || TMN-Rx > 1  // TMN serving more than one
Tx/Rx flow
      Find R_pos  // deploy relay node
      Execute T with relay node
  else
      execute T without relay node
    End if
  End while
```

7 Experimental Evaluations

The proposed algorithm described in "Transmission and Reception based Relay deployment for data base transactions in MANETs" is implemented in network simulator ns2. The simulator version used here is 2.32. The network consisting of 50 traditional mobile nodes is simulated. The results shown in this paper use aodv routing as routing protocol. Each simulation run was executed and the results showed are average of ten simulation runs.

7.1 Transmission Energy Results

The Fig 3 shows the comparison between the average transmission energy consumption in the network without relay nodes and with relay nodes. The graph is plotted between time (seconds) and average transmission energy consumption*10-3 (Joules). The mathematical equation (3) shows that more transmission energy is required when the node transmits more number of transactions. Thus the network without relay node consumes more energy as shown in the Fig 3. The Fig 3 shows the transmission energy consumption is reduced when the network uses relay nodes.

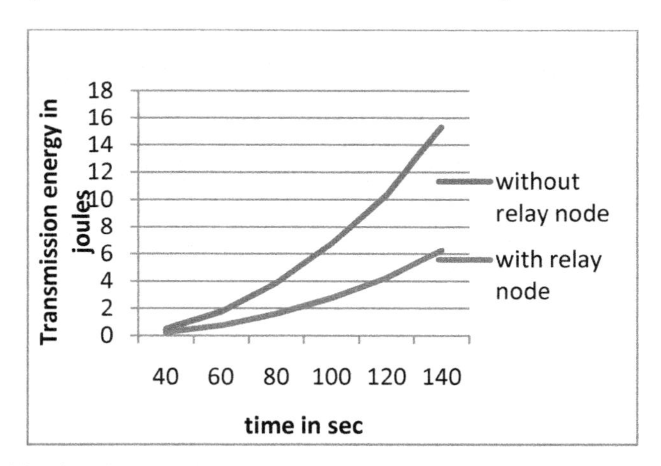

Fig. 3. Effects of relay nodes on transmission Energy consumption

7.2 Reception Energy Results

The Fig 4 shows the comparison between the average reception energy consumption in the network without relay nodes and with relay nodes. The graph is plotted between time (seconds) and average reception energy consumption*10^{-3} (Joules). The mathematical equation (5) shows that more receptive energy is required when the node receives more number of transactions. Thus the network without relay node consumes more energy as shown in the Fig 4. The Fig 4 shows the reception energy consumption is reduced when the network uses relay nodes.

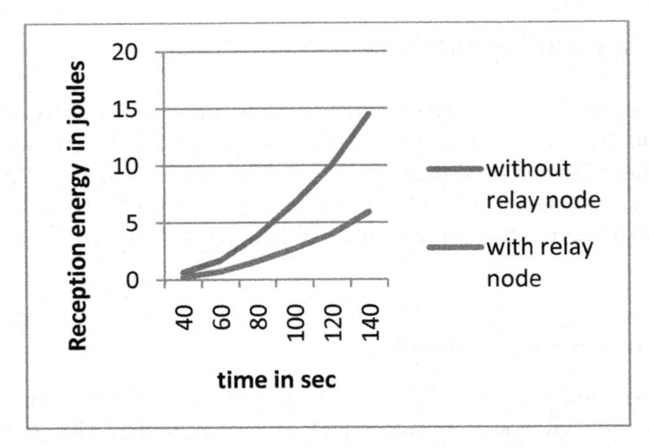

Fig. 4. Effects of relay nodes on reception Energy Consumption

7.3 Effect of Varying Number of Relay Nodes

The Fig 5 shows the performance of the network by varying number of relay nodes. From the Fig 5, $2.65*10^{-3}$ Joules energy consumption is saved when the network uses 2 relay nodes, $3.75*10^{-3}$ Joules energy consumption is saved when the network uses 10 relay nodes. Thus the Fig 5 depicts the varying the number of relay nodes influence the energy consumption in the network. The number of relay nodes required can be optimized based on the application. For example, if the application considers energy as the most important factor then the number of relay nodes can be increased. If the application considers channel capacity as the most important factor then the number of relay nodes should be limited.

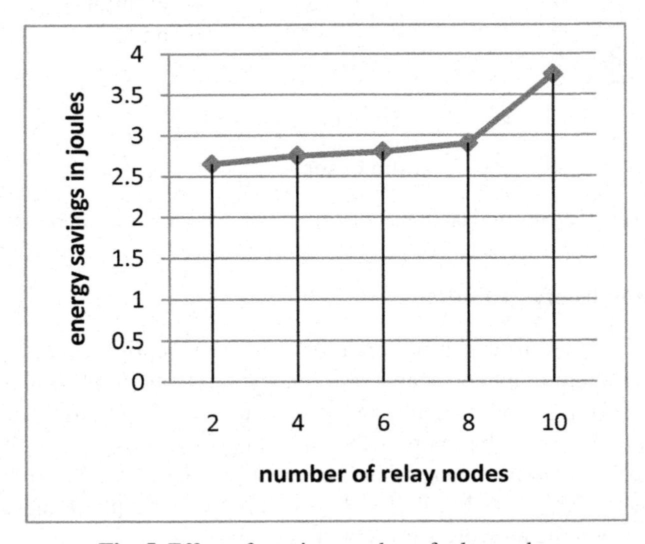

Fig. 5. Effect of varying number of relay nodes

8 Conclusion

In this paper, the relay nodes are deployed for minimizing the transmission and reception energy consumption by the traditional mobile nodes when serving more than one transactions. The proposed Min-Tx and Min-Rx algorithm uses mobility prediction with MANET routing algorithm to predict the position of all traditional mobile nodes. The simulation results indicate the significant energy savings in the network. As part of ongoing research, the effect of selfish nodes in the database transactions in MANETs is being studied.

References

1. De Rango, F., Guerriero, F., Fazio, P.: Link-Stability and Energy Aware Routing Protocol in Distributed Wireless Networks. IEEE Transactions on Parallel and Distributed Systems 23(4) (April 2012)
2. Rekik, J.D., Baccouche, L., Ghezala, H.B.: An energy efficiency and delay guarantee service selection protocol in MANET. IEEE (2012)
3. Liu, J., Jiang, X., Nishiyama, H., Kato, N.: Generalized Two-Hop Relay for Flexible Delay Control in MANETs. IEEE/ACM Transactions on Networking (2012)
4. Abdelmoumen, M., Arfaoui, I., Frikha, M., Chahed, T.: On the performance of MANETs under different mobility patterns and routing protocols and its improvement based on fixed relay nodes. IEEE (2012)
5. Ying, L., Shakkottai, S.: Scheduling in Mobile Ad Hoc Networks with Topology and Channel-State Uncertainty. IEEE Transactions on Automatic Control (2012)
6. Liu, J., Jiang, X., Nishiyama, H., Kato, N.: Exact Throughput Capacity under Power Control in Mobile Ad Hoc Networks. IEEE (2012)
7. Chao, C.-M., Tsai, H.-C., Huang, C.-Y.: Load-Aware Channel Hopping Protocol Design for Mobile Ad Hoc Networks. IEEE (2012)
8. Rekik, J.D., Baccouche, L., Ghezala, H.B.: Load-balancing and Energy Aware Routing Protocol for real-time flows in Mobile Ad-Hoc Networks. IEEE (2011)
9. Miao, X.-N., Zhou, X.-W., Du, L.-P., Lin, L.: Cross-layer Rate Control and Dynamic Scheduling in Mobile ad hoc Networks. IEEE (2009)
10. Venkateswaran, A., Sarangan, V., La Porta, T.F., Acharya, R.: A Mobility-Prediction-Based Relay Deployment Framework for Conserving Power in MANETs. IEEE Trans. Mobile Computing 8(6) (June 2009)
11. Akyol, U., Andrews, M., Gupta, P., Hobby, J., Saniee, I., Stolyar, A.: Joint Scheduling and Congestion Control in Mobile Ad-Hoc Networks. IEEE (2008)
12. Gruenwald, L., Banik, S.M., Lau, C.N.: Managing real-time database transactions in mobile ad-hoc networks. Springer (2007)
13. Brahma, M., Kim, K.W., El Hachimi, M., Abouaissa, A., Lorenz, P.: A Buffer and Energy Based Scheduling in Mobile Ad hoc Networks over Link Layer. IEEE (2006)
14. Serbetli, S., Yener, A.: Relay Assisted F/TDMA Ad hoc Networks: Node Classification, power Allocation and Relaying Strategies. IEEE Trans. Comm. 56(6) (November 2006)
15. Tang, C., McKinley, P.K.: Energy optimization under informed mobility. IEEE Trans. Parallel Distributed Systems 17(9), 947–962 (2006)

Optimization of Data Allocation for Mobile Value Added Services

Ayesha Kazi[1], Sonal Shah[1], Zinnia Katib[1], Meera Narvekar[1], and S.S. Mantha[2]

[1] Department of Computer Engineering, D.J. Sanghvi College of Engineering, Mumbai, India
{kaziayesha,zinniak_91}@hotmail.com
{meera.narvekar,sonalshah8591}@gmail.com
[2] SNDT Women's University, Mumbai, India
ssmantha@vjti.org

Abstract. The advancement of wireless mobile services, decline in Average Revenue per User (ARPU) due to decrease in call tariffs and the 'pull effect' from consumers demanding additional services has led to increased usage of Mobile Value Added Services (MVAS) on a daily basis. The need of the hour is now to improve the efficiency of Mobile Value Added Services. Thus, an attempt is made to optimize the allocation of MVAS data in the server databases by utilizing the knowledge of user moving patterns. Making use of this information, four algorithms have been devised to globally allocate data in a mobile system that will cater to the characteristic of service offered. By employing these algorithms, the access cost, response time and accuracy can be minimized thus improving the performance.

Keywords: Mobile Valued Added Services, Mobile Computing, MVAS Data Allocation.

1 Introduction

Rapid growth of usage of mobile telephone, various satellite services and now the wireless internet are generating tremendous changes in the mobile computing field. Mobile computing technology enables the realization of global marketplace with scattered information, where people can seamlessly access any information from anywhere through any device while stationary or mobile. The information is in the form of graphics/wallpaper downloads, ring tones, weather, news, stock prices, games, etc and is collectively termed as Mobile Value Added Services (MVAS).

Over the last few years, the telecom industry has realized the importance of MVAS. Given the decline in the ARPU, the reduction in the industry revenue due to decreasing call tariffs and increasing competition among the service providers, the challenge is to retain customers and develop alternative revenue streams [1]. Thus the changing industry looks at MVAS as a major contributor of the revenue with increasing personalization of content and cheaper handsets.

A mobile computing system is similar to distributed server architecture [2]. Mobile device uses a wireless communication medium. The device is partially connected to the network due to high cost and unavailability of the wireless cellular network. Also

S. Unnikrishnan, S. Surve, and D. Bhoir (Eds.): ICAC3 2013, CCIS 361, pp. 284–300, 2013.

limited availability of bandwidth of the wireless medium leads to increase in communication cost for the users. Another issue in the cellular network is the determining the current location of the user. The user might change location frequently resulting in delay of requested service as the whole service delivery process is started all over again for the user's current location. Keeping these in mind, we have developed four algorithms, each of which address a certain issues, improving the overall efficiency of data allocation. Each of these algorithms focus on different categories of MVAS:

1. Entertainment Value Added Services e.g.:Jokes, Ring tones, Wallpapers, Newspaper, Caller Ring Back Tone, and Games where communication cost must be minimized.
2. Infotainment Value Added Services e.g. News Alerts and updates, Stock Market Prices, and Internet where response time is crucial.
3. Location Based Value Added services such as maps, weather, shopping alerts etc where location specific data availability is necessary.
4. M-Commerce Value Added Services e.g.: Mobile Banking, Movie Ticket Booking, and Stock Trading [1] where good response time is essential.

The work in this paper explains how to optimize data allocation for MVAS. Section 2 gives an overview of the literature survey. In Section 3 gives a brief description of the system model. The algorithms for allocation of MVAS data are presented in Section 4. In Section 5 how the system is implemented along with performance evaluation is discussed. In Section 6 conclusion of the work is presented.

2 Background and Literature Review

Various attempts to optimize delivery of services to the user based on prediction and personalization have been made. An adaptive algorithm which limits the reservation and configuration procedure to a subset of cells around the user in WAN was proposed [1] but it does not incorporate user behavioural patterns, wireless link characteristics and the handover decision-making mechanism which are essential for accurate prediction of user behavioural patterns. Also technological barriers resulting from the diversity and cost of approaches to location determination which create a complex set of choices for operators and potential interoperability problems that have not been addressed. Certain temporal and spatial methods are used for data replication and allocation [3]. But additional techniques such as user moving pattern, personalization of data according to user preference could be used to replicate and allocate data which could provide more accuracy and result in efficient use of memory too have not been considered. Methods to learn user profile data through a combination of machine learning and human factors directed user interaction have been carried out that combine network observable data with specific queries to the user ;but incorporation of human behavioural aspects into the query sequencing framework have not been addressed [4]. Hidden Markov model which can be adjusted to reflect the latest user behaviour without re-scanning the original dataset is also used but needs further research [5].[6] discusses user movement behaviour patterns where a PC-based experimental evaluation under various simulation conditions, using hypothetical data. Although prediction process consumes more time and no precise prediction of mobile user behaviour is observed.

This paper utilizes user behaviours for allocating mobile value added service data globally in mobile computing environments. Its aim is to allocate right data at the right location for whole users in the environment. The user behaviour includes user moving patterns and user request patterns. By utilizing the information, four algorithms are developed which focus on different views, respectively, such as hit ratios, communication costs, and response time.

3 System Model

As shown in Fig. 1 a mobile environment consists of BS (Base Stations) and BSC (Base Station Controller) [7]. The coverage of a particular area is taken care by the BS. When a user enters a particular area with MS (Mobile Station), it communicates with the BS. Also the MS can request MVAS from the BS. A supervisory BSC can manage several BSCs and all the information of the services is stored in a supervisory BSC not in BS. The BSCs are essentially connected with each other via a core network. Thus the MSs in the different BSCs can communicate with each other. The MVAS data request by a MS to its BS is forwarded to the corresponding supervisory BSC. If the requested MVAS data is found it would return the result to the MS immediately however, if not found then the BSC would forward the request through the wired core network to the BSC storing the required data. This forwarding involves overhead of transmission time and communication cost. To minimize these overheads we need to know the location of the required MVAS data so that we can get the required data from local BSC instead of accessing remote BSCs.

Thus following issues are important for optimized delivery of services:

- Minimize access time of frequently accessed data.
- Minimize overhead of communication cost.
- Minimize response time of MVAS data.

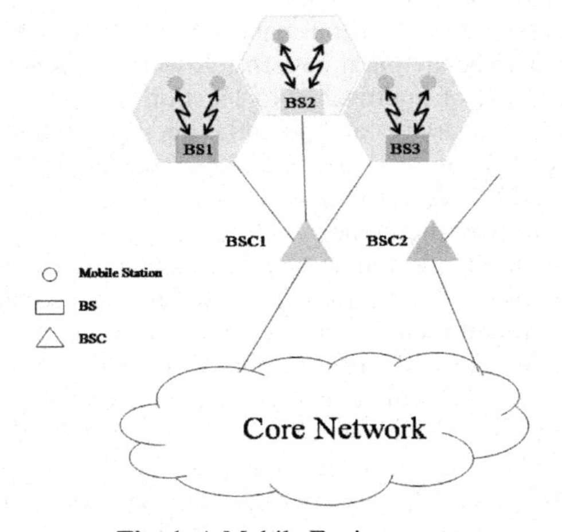

Fig. 1. A Mobile Environment

4 Algorithms for Global Allocation of MVAS Data

4.1 Generation of User Records

The user's mobile usage behaviour pattern is obtained by mining data from the network service provider using SS7 signalling toolbox. In a mobile network each node is viewed as a VLR (Visitor Location Register) and the links between them can be seen as connections between the VLRs [7]. A location log will be generated for the movement pattern of users. An activity log will be generated whenever the user requests for a Mobile Value Added Service. Both these logs will help us generate the user record required.

The MVAS data allocation algorithms are developed depending on patterns of user movement and request of services by the user. According to the subordinate behaviour between the BSCs and BSs; the visited BSs and visited BSCs are mapped as the MVAS data requested is stored in the BSCs and the users roam around the BSs in the mobile network.

Table 1 shows mapping between the BS to the corresponding BSC. Table 2 gives the movement pattern of the user, whereas Table 3 maps the patterns of Table 2 into BSC Frequency Matrix (BSCF matrix). Table 4 called as User MVAS Request Matrix (UM matrix) shows the requested MVAS by the user. Table 5 gives the BSC Communication Cost Matrix (BSC CC matrix). In Table 6 the BSC Transmission Time Matrix is shown (BSC TT matrix) [5].

Table 1. Subordinancy between BSCs and base stations

BSC	Managed BS
BSC1	(BS1, BS2, BS4)
BSC2	(BS3, BS5, BS9, BS10)
BSC3	(BS6, BS7, BS8)

Table 2. UserMovement Pattern

User ID	User Pattern
User1	<BS1, BS3, BS7, BS8><BS4, BS6, BS7>
User2	< BS1, BS5, BS9. BS10>
User3	<BS5, BS9, BS10><BS1, BS2, BS3><BS6, BS9>
User4	<BS2, BS3, BS8><BS1, BS2, BS4, BS5, BS7>

Table 3. BSCFMatrix

	User1	User2	User3	User4
BSC1	2	1	2	4
BSC2	1	3	5	2
BSC3	4	0	1	2

Table 4. UM Matrix

	M1	M2	M3	M4	M5	M6	M7	M8
User1	0	1	1	0	0	1	0	1
User2	1	0	1	1	1	0	0	1
User3	1	1	1	0	0	0	1	0
User4	0	0	0	0	1	1	1	0

Table 5. BSC CC Matrix

	BSC1	BSC2	BSC3
BSC1	0	8	2
BSC2	8	0	4
BSC3	2	4	0

Table 6. BSC TT Matrix

	BSC1	BSC2	BSC3
BSC1	0	7	3
BSC2	7	0	2
BSC3	3	2	0

Four algorithms are presented. First two focus on frequently accessed MVAS data. Out of which the first is from user point of view and second is from BSC point of view. Third algorithm focuses on minimizing overhead of communication costs and fourth on minimization of response time of service to user.

4.2 Frequent Data Algorithm

The frequent data algorithm is based on access frequency of MVAS data in the corresponding BSC. Here UM matrix is first multiply by the BSCF matrix to produce BSC-MVAS Data matrix (BMD matrix).

	M1	**M2**	M3	M4	M5	M6	M7	M8
BSC1	3	**4**	5	1	5	6	6	3
BSC2	8	**6**	9	3	5	3	7	4
BSC3	1	**5**	5	0	2	6	3	4

In the next step, the threshold value is calculated for each Mobile Value Added Service i.e. for each column of M1, M2…M8 to get the frequently accessed BSCs for each data. For example for column M2 the threshold is calculated as (4+6+5)/3 = 15/3 = 5. The elements less than the threshold such as BSC1 is set to 0. By this the more frequently accessed services are found in the BMD matrix and hence called as 'Frequent Data Algorithm'.

	M1	M2	M3	M4	M5	M6	M7	M8
BSC1	0	0	0	0	5	6	6	0
BSC2	8	6	9	3	5	0	7	4
BSC3	0	5	5	0	0	6	0	4

Finally, the MVAS data allocation pattern is obtained where each non zero element in the BMD matrix is attached to its corresponding BSC row by row. The MVAS data allocation pattern for the above BMD matrix would be as follows:

	Allocation pattern
BSC1	{M5, M6, M7}
BSC2	{M1, M2,M3, M4, M5, M7, M8}
BSC3	{M2, M6, M8}

Frequent Data Algorithm.

```
Input:
BSCF matrix (m x n);
UM matrix (n x k);
Output:
MVAS data allocation pattern;
Begin
    Step 1:
      Input: BSCF matrix, UM matrix
      Output: BMD matrix
      /* Multiply UM matrix with BSCF matrix to produce BMD ma-
      trix */
      Begin
    For (i = 1; i• m; i++)
       {
    For (j = 1; j• k; j++)
         {
    BMD[i][j] = 0;
  For (l=1; l • n; l++)
           {
              BMD[i][j]=BDF[i][l]*UM[l][j]+
              BMD[i][j];
           }
        }
    }
      End
    Step 2:
    /* Select the frequent elements column by column */
      For every column j in BMD matrix do
        Let s= sum of the elements in column j;
```

```
        Let n= number of non-zero elements in column j;
        threshold = s/n;
      For every element i in column j do
        If element i< threshold
        then Set element i to 0;
    Step 3:
    /* Produce the MVAS data allocation pattern */
      For every row i in BMD matrix do
      Append every non-zero element to BSCi;
End
```

4.3 BSC Oriented Algorithm

BSC algorithm is very similar to the Frequent Data Algorithm. In the first step, like the frequent data algorithm, the UM matrix is multiplied by BSCF matrix to produce BSC-MVAS Data matrix (BMD matrix). In the second step more frequent elements from the BMD matrix are selected row by row instead of selecting them column by column, thus calling it 'BSC Oriented Algorithm'. The more frequently accessed MVAS data is found within a BSC. The threshold for row BSC2 is calculated as (8+6+9+3+5+3+7+4)/8 = 5.625. Elements less than this value such as M4, M5, M6, and M8 are set to 0.

	M1	M2	M3	M4	M5	M6	M7	M8
BSC1	0	0	5	0	5	6	6	0
BSC2	8	6	9	0	0	0	7	0
BSC3	0	5	5	0	0	6	0	4

In the third step, the MVAS data allocation pattern is produced where each non zero element in the BMD matrix is attached to its corresponding BSC row by row. The MVAS data allocation pattern for the above BMD matrix would be as follows:

	Allocation pattern
BSC1	{M3,M5, M6, M7}
BSC2	{M1, M2,M3, M7}
BSC3	{M2, , M3, M6, M8}

As seen in the above table, some data such as M4 is not allocated to any of the BSCs as it is less frequently accessed in all the BSCs, but will be incorporated in at least one of the BSCs in the final step. All columns from the original BMD matrix except columns for unallocated data are left out thus paying attention only on the unallocated data. Step 2 and 3 are reapplied and step would be repeated till there is no unallocated data. Finally the MVAS data allocation pattern would be as follows:

	M4			M4
BSC1	1	⇨	BSC1	1
BSC2	3		BSC2	3
BSC3	0		BSC3	0

	Allocation pattern
BSC1	{M3, M4, M5, M6, M7}
BSC2	{M1, M2,M3, M4, M7 }
BSC3	{M2, , M3, M6, M8}

BSC Oriented Algorithm

```
Input:
  BSCF matrix (m x n);
  UM matrix (n x k);
Output:
  MVAS data allocation pattern;
Begin
  Step 1:
    Input: BSCF matrix, UM matrix
    Output: BMD matrix
    /* Multiply UM matrix with BSCF matrix to produce BMD ma-
    trix */
    Begin
  For (i = 1; i• m; i++)
      {
    For (j = 1; j• k; j++)
       {
    BMD[i][j] = 0;
  For (l=1; l • n; l++)
  {
              BMD[i][j]=BDF[i][l]*UM[l][j]+BMD[i][j];
          }
  }
    }
    End
  Step 2:
    /* Select the frequent elements row by row */
    For every row j in BMD matrix do
      Let s= sum of the elements in row j;
      Let n= number of non-zero elements in row j;
        threshold = s/n;
    For every element i in column j do
      If element i< threshold
      then Set element i to 0;
```

```
  Step 3:
    /* Produce the MVAS data allocation pattern */
    For every row i in BMD matrix do
    Append every non-zero element to BSCi;
  Step 4:
    /*If there is any unallocated data execute this step*/
    While (unallocated data)
      Leave out all columns from original BMD matrix except the
      columns for unallocated data;
      Apply Step 2 and Step 3;
End
```

4.4 Minimizing Communication Cost Algorithm

As this algorithm considers user communication cost it is very different from the other two algorithms. It is based on the logic that higher hit ratios do not necessary translate into lower communication cost therefore this algorithm focuses on minimizing communication cost and is named 'Minimizing Communication Cost Algorithm'. The first step is to multiply the UM matrix by BSCF matrix to produce BMD matrix. In the second step, Dijkstra's algorithm is applied to BSC CC matrix to obtain the least communication cost between any two BSCs, giving us LCC matrix. For the third step the BMD matrix is multiplied by the LCC matrix to give LCCBMD matrix.

An element (i, j) in the LCCBMD matrix i.e. data Mj allocated at BSCi gives the total communication cost for other BSCs. For example, for M1 allocated at BSC1 the total communication required for BSC2 and BSC3 is 50.

	BSC1	BSC2	BSC3
BSC1	0	6	2
BSC2	6	0	4
BSC3	2	4	0

	M1	M2	M3	M4	M5	M6	M7	M8
BSC1	50	46	64	18	34	30	48	32
BSC2	22	44	50	6	38	60	48	34
BSC3	38	32	46	14	30	24	40	22

In the last step the smallest element from each column is selected as shown in the table above, and the MVAS data is allocated to the corresponding BSC as follows:

	Allocation pattern
BSC1	{}
BSC2	{M1, M4}
BSC3	{M2, , M3, M5, M6, M7, M8}

Unless all the requested copies of each MVAS data are allocated, the algorithm cannot be concluded. In order to minimize the communication cost only copies of data up to half number of BSCs in the system are allowed.

Minimizing Communication Cost Algorithm

```
Input:
  BSCF matrix (m x n);
  UM matrix (n x k);
  BSC CC matrix (m x m);
Output:
  MVAS data allocation pattern;
Begin
  Step 1:
    Input: BSCF matrix, UM matrix
    Output: BMD matrix
    /* Multiply UM matrix with BSCF matrix to produce BMD ma-
    trix */
    Begin
      For (i = 1; i• m; i++)
      {
        For (j = 1; j• k; j++)
        {
          BMD[i][j] = 0;
          For (l=1; l • n; l++)
          {
            BMDij = BSCF[i][l] * UM[l][j] + BMD[i][j];
          }
        }
      }
     End
  Step 2:
    Input: BSCC matrix
    Output: LCC matrix
    LCC matrix= Function Dijikstras(BSCCC matrix);
  Step 3:
    Input: LCC matrix, BMD matrix;
    Output: LCCBMD matrix;
    /* Multiply BMD matrix with LCC matrix to produce LCCBMD
    matrix */
    Begin
  For (i = 1; i• m; i++)
  {
        For (j = 1; j• k; j++)
        {
          LCCBMD[i][j] = 0;
```

```
            For (l=1; l • n; l++)
            {
               LCCBMD[i][j] = LCC[i][l] * BMD[l][j] +
               LCCBMD[i][j];
            }
          }
      }
    End
  Step 4:
     /* Produce the MVAS data allocation pattern */
     Backup BSC CC matrix and BMD matrix;
     For every column j in LCCBMD matrix do
       For (k = 1; k • BSC/2; k++)
          Select the minimum element (say in row i) in column j;
       Append data Mj to the corresponding BSC;
       Delete row i and column i in BSC CC matrix;
       LCC matrix = Function Dijkstra (BSC CC matrix);
       Delete row i in BMD matrix;
       Multiply only column j in BMD matrix by LCC matrix;
       Recover BSC CC matrix and BMD matrix;
End
```

4.5 Minimizing Response Time Algorithm

The Minimizing Response Time algorithm is similar to Minimizing Communication Cost Algorithm but considers user response time. The first step is to multiply the UM matrix by BSCF matrix to produce BMD matrix. In the second step, Dijkstra's algorithm is applied to BSC TT matrix to obtain the least response time between any two BSCs, giving us LRT matrix. For example given, the LRT matrix would be as follows:

	BSC1	BSC2	BSC3
BSC1	0	5	3
BSC2	5	0	2
BSC3	3	2	0

The last step is to multiply each column in BMD matrix, column-wise with LRT matrix to give LRTBMD matrix. For M2, Column j in LRTBMD matrix (i.e. data allocated at BSC_j) shows the response time for other BSCs. For example, MVAS data M1 allocated at BSC2 the response time required for BSC1 and BSC3 is 15 and 2 respectively. The maximum element is selected from each of column in the LRTBMD matrix as user response time depends on longest response time. Then the minimum element from all the maximum elements is selected, and the MVAS data is assigned to the corresponding BSC. According to LRTBMD matrix, M2 is allocated at BSC3.

	BSC1	BSC2	BSC3		M2
BSC1	0	5	3	BSC1	4
BSC2	5	0	2	BSC2	6
BSC3	3	2	0	BSC3	5

	BSC1	BSC2	BSC3
BSC1	0(0 x 4)	**20(5 x 4)**	12 (3 x 4)
BSC2	**30 (5 x 6)**	0 (0 x 6)	**12 (2 x 6)**
BSC3	15(3 x 5)	10 (2 x 5)	0 (0 x 5)

Similar to the Minimum Communication Cost Algorithm, unless all the requested copies of each MVAS data are allocated, the algorithm cannot be concluded. In order to minimize the response time only copies of data up to half number of BSCs in the system are allowed.

Minimizing Response Time Algorithm

```
Input:
  BSCF matrix (m x n);
  UM matrix (n x k);
  BSC TT matrix (m x m);
Output:
  MVAS data allocation pattern;
Begin
  Step 1:
    Input: BSCF matrix, UM matrix
    Output: BMD matrix
    /* Multiplycoloumn j of UM matrix with BSCF matrix to pro-
    duce BMD matrix */
    Begin
  For (l = 1; l• m; l++)
    {
    For (i= 1; i• m; i++)
       {
    BMD[i][l] = BSCF[i][l] * UM[i][j];
       }
     }
     End
  Step 2:
    Input: BSC TT matrix;
    Output: LRT matrix;
    LRT matrix= Function Dijkstra (BSC TT matrix);
  Step 3:
    /* Produce the MVAS data allocation pattern */
    Backup BSC TT matrix, LRT matrix and BMD matrix;
```

```
    For every column j in BMD matrix do
      For (k = 1; k • BSC/2; k++)
      /* Multiplycoloumn j of BMD matrix with LRT matrix to
      produce LRTBMD matrix */
        Begin
  For (l = 1; l• m; l++)
    {
      For (i= 1; i• m; i++)
      {
              LRTBMD[i][l] = LRT[i][l] * BMD[i][j];
    }
        }
    Select the maximum element of each column into Max{};
    Select the minimum element (say in coloumni) in Max{};
    Append data Mj to the corresponding BSC;
    Delete row i and column i in BSC TT matrix;
    LRT matrix = Function Dijkstra (BSC TT matrix);
    Delete row i in BMD matrix;
    Recover BSC TT matrix, LRT matrix and BMD matrix;
End
```

5 Implementation and Performance Evaluation

The algorithms have been implemented using the following setup:

- Operating System: Ubuntu 11.04 Natty Version
- Processor: Pentium I5 2nd Gen. with 2.93 GHz
- RAM: 6 GB
- Hard Disk: 650 GB

Several experiments were conducted to compare these algorithms using three parameters such as number of BSCs, number of users and number of MVAS data. For the performance experiments 10 BSCs, 1000 users and 100 MVAS data were consider. Additionally three more parameters such as hit ratio cost of communication and response time are used for comparison.

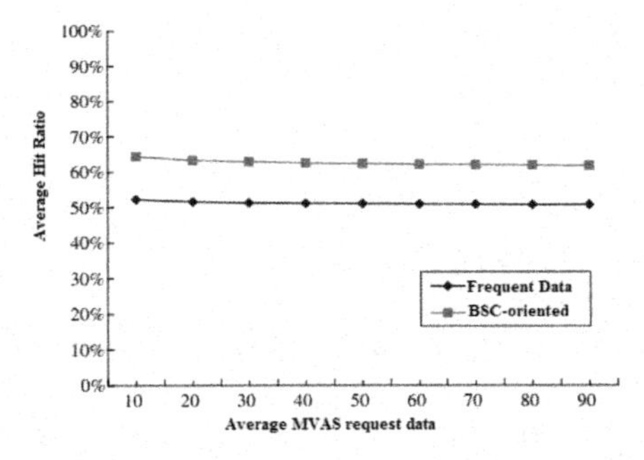

Fig. 2. Average hit ratios with average MVAS request data varied

Test 1: By Changing The Average Requested MVAS Data of Individual User.
In this experiment, average requested MVAS data is varied to check for its performance. The hit ratio of both the algorithms does not change and remain constant even though there are changes in the requested MVAS data as shown in Fig. 2. The average hit ratios of both algorithms keep a significant percentage in steady no matter how the average request MVAS data is varied, but the BSC-oriented algorithm is better than the frequent data algorithm. For the measure of communication cost, though BSC oriented is better for average hit ratio but its average communication cost is worse than the other two algorithms viz. frequent data algorithm and the minimize communication cost algorithm as shown in Fig. 3. This makes minimizing communication cost algorithm a better algorithm for communication cost measure. Since the high hit ratios do not directly imply short response time, we also conducted another experiment to evaluate the average response time measure which resulted in BSC-oriented algorithm being the worst, and the minimize response time algorithm being the best where the average response time would decrease when the average request data increases as shown in Fig. 3 and 4.

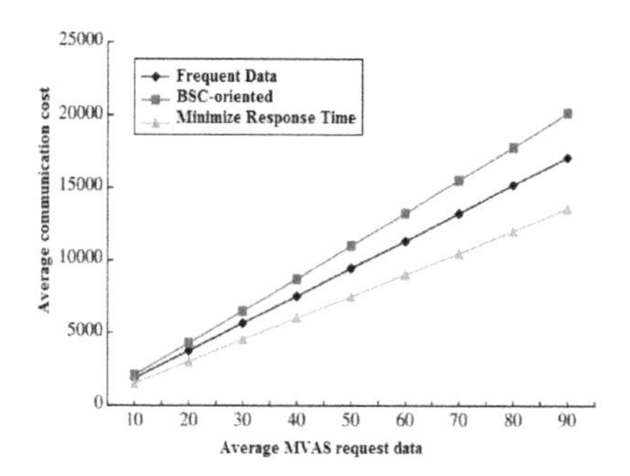

Fig. 3. Average communication costs with the average MVAS request data varied

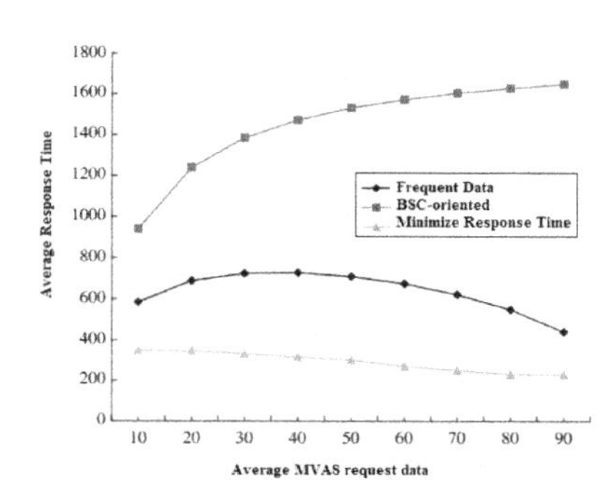

Fig. 4. Average response time with average MVAS request data varied

Test 2: By Changing the Number of Frequent MVAS Users.
In this experiment, the number of frequent MVAS users is changed to check for its performance. The frequent users are the ones who visit BSCs more number of times than other users i.e., the frequencies of frequent users in BSCF matrix are much more than other users. Experimentally it is observed that the BSC-oriented algorithm

performs better than the frequent data algorithm, and the hit ratio increases as frequent users increase as seen in Fig. 5. Thus, it was found that the BSC-oriented algorithm is more adaptive than the frequent data algorithm. As shown in Fig. 6 and 7 for average communication cost and average response time measures, it was observed that both the BSC-oriented algorithm and the minimize communication cost algorithm perform better than the frequent data algorithm. However, as the number of frequent users increases to around 300, the BSC-oriented algorithm performs even better than the minimize communication cost algorithm. Thus, it was found again that the BSC-oriented algorithm is more adaptive than the other two algorithms. Besides, for the average response time, it was found once again that the BSC-oriented algorithm was the worst, and the minimize response algorithm was the best.

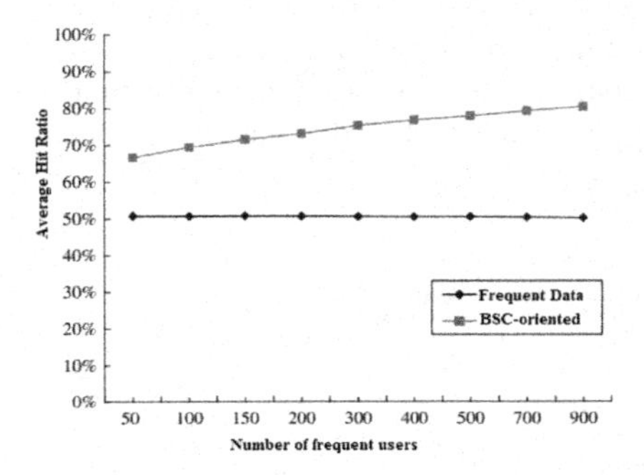

Fig. 5. Average hit ratios with number of frequent users varied

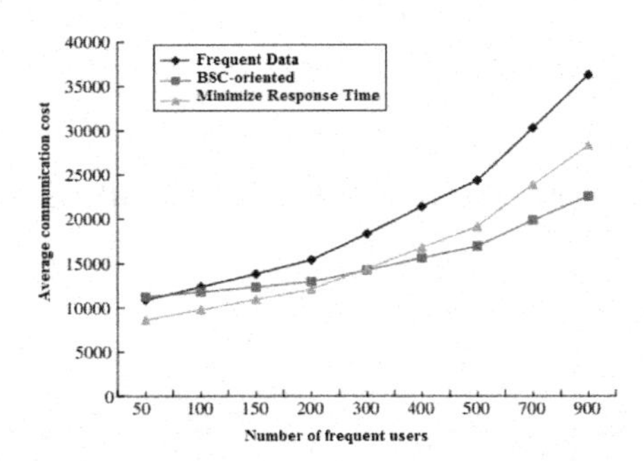

Fig. 6. Average communication costs with numbers of frequent users varied

Test 3: By Biasing BSCF Matrix.

In this experiment, BSCF matrix is biased from 10% to 90%; for e.g., 10% bias means that the MVAS users usually visit 10% of the same BSCs as they visited before. The average hit ratio of the BSC-oriented algorithm is better than the frequent data algorithm when the bias is below 30% but it changes after the bias increases up to 30% as can been seen in Fig. 8. This is because that the biased BSCF matrix makes the frequent data algorithm allocate much more data than the BSC-oriented algorithm. In the frequent data algorithm, the bias makes all the larger elements in RD matrix be selected and more the bias, the more the larger elements are selected. But, the BSC-oriented algorithm is not influenced by the bias. For average communication cost, it was observed that the frequent data algorithm is the worst and the minimize communication cost algorithm is the best when the bias is below 30%, but, as the bias increases up to 40%, the frequent data algorithm performs even better than the minimize communication cost algorithm. For the average response time, when the bias is below 30%, the BSC-oriented algorithm was found to be the worst, and the minimize response time algorithm was the best. However, as the bias increases up to 40%, the frequent data algorithm performs even better than the minimize response time algorithm. Thus, the frequent data algorithm is more sensitive than the other two algorithms i.e. BSC-oriented and minimize communication cost algorithms.

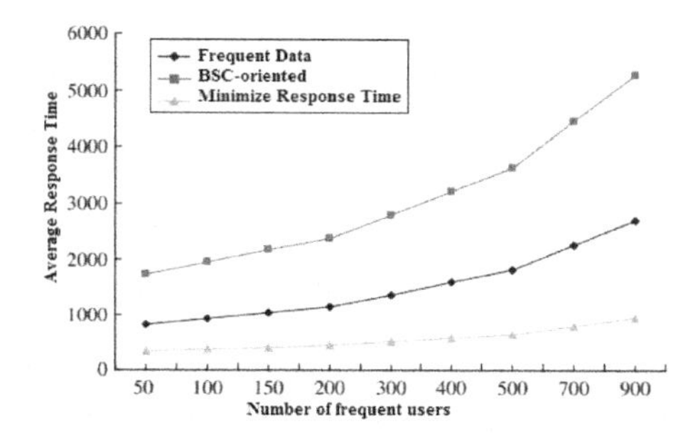

Fig. 7. Average response time with number of frequent users varied

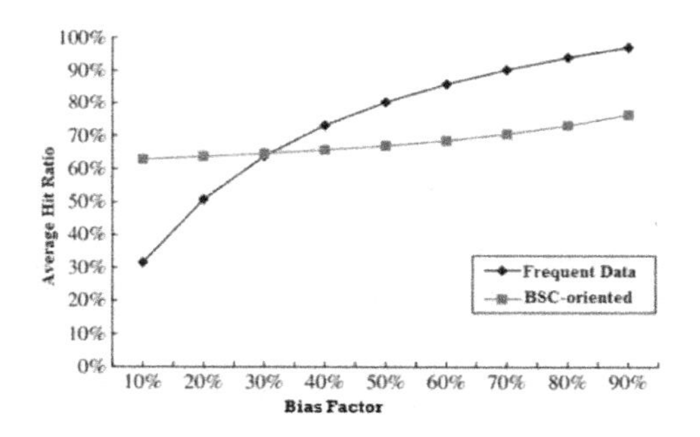

Fig. 8. Average hit ratio with biased BSCF matrix

6 Conclusion

In this paper, a global data allocation scheme for MVAS based on location logs and activity logs generated from user behaviour has been developed. Four different algorithms were developed which focused on frequent data, communication costs, and response time. The Frequent Data Algorithm and the BSC Oriented Algorithm focused on the frequently accessed data. The Minimizing Communication Cost Algorithm and the Minimizing Response Time Algorithm focus on communication costs between BSCs and response time of resulting service to reach to the user, respectively. Where to allocate the MVAS data and what MVAS data should be allocated depends on information obtained from user behavior. The table given below proposes the most optimum algorithms based on the type of MVAS service:

MVAS	Recommended Algorithms
Ring tones, Wallpapers, Jokes	Frequent Data Algorithm, Minimizing Communication Cost Algorithm
Online Games	Frequent Data Algorithm, Minimizing Response Time Algorithm
Media	Minimizing Communication Cost Algorithm
Stock Market updates, Stock Trading, Auctions	Minimizing Response Time Algorithm
Internet	Minimizing Response Time Algorithm
Mobile Banking	Minimizing Response Time Algorithm
Movie Ticket Booking	Minimizing Response Time Algorithm
News & Sport Score Updates	Minimizing Response Time Algorithm
Maps & Restaurants	BSC Oriented Algorithm

Note: Minimizing Communication Cost Algorithm can be used for delivering all MVAS.

Several experiments are conducted to compare these algorithms using three parameters The Service providers can use the above algorithms to store data related to their Mobile Value added Services to enhance user experience.

References

1. Chan, J., Seneviratne, A.: A Practical User Mobility Prediction Algorithm for Supporting Adaptive QoS in Wireless Network. In: IEEE International Conference on Networks, September 28-October 1 (1999)
2. Adjusted Replica Allocation in Ad Hoc Networks for Improving Data Accessibility (2007)
3. Cao, J., Dinoff, R., Ho, T.K., Hull, R., Kumar, B., Santos, P.: Mobile User Profile Acquisition through Network Observables and Explicit User Queries (2008)
4. Mobile VAS in India, A Report by IAMAI & eTechnology Group, Internet and Mobile Association of India (August 2008)
5. Novel approach for efficient and effective mining for mobile user behaviour (2010)
6. Huang, Y.-F., Lin, K.-H.: Global data allocation based on user behaviours in mobile computing environments (2008)
7. Schiller, J.: Mobile Communications, 2nd edn. (2000)
8. Lu, K.-C., Hsu, C.-W., Yang, D.-L.: A Novel Approach for Efficient and Effective Mining of Mobile User Behaviors (2010)
9. Huang, Y.-F., Chen, J.-H.: Fragment allocation in distributed database design. Journal of Information Science and Engineering 17(3) (2001)
10. Barbara, D.: Mobile computing and databases – a survey. IEEE Transactions on Knowledge and Data Engineering 11(1), 108–117 (1999)
11. Bruegge, B., Bennington, B.: Applications of mobile computing and communication. IEEE Personal Communications 3(1), 64–71 (1996)
12. Dunham, M.H., Helal, A.: Mobile computing and databases: anything new? ACM SIGMOD Record 24(4), 5–9 (1995)

Context Aware, Mobile Phone Based Sink Model of Wireless Sensor Network for Indian Agriculture

Prashant Ingole[1], Jayant Pawar[2], and Hrishikesh Vhatkar[2]

[1] G.H. Raisoni College of Engineering and Management, Amravati (M.S.) India
[2] Atharva College of Engineering Mumbai (M.S.) India
pvingole@gmail.com, jayantppawar@yahoo.co.in

Abstract. Agriculture is the fundamental area of human society in which research and development has still a wide scope. Water is an important parameter in agricultural practices. In the limited water how to increase the irrigation area with out compromising the productivity is the important area of research in agriculture. Context aware computing means sensing the context and other input channels, taking smart decisions and feedback tracking from the context. Agriculture is a very rich context aware domain. Context aware solution in agriculture is the next step of the precision agriculture.

Wireless sensor network [WSN] is the promising technology for instrumentation and control. The sink is the major node in WSN and can be implemented in smart phone [2]. The integration of WSN and cellular networks generate the many useful application. Here, we have suggested the model in which the mobile phone based sink controls the water irrigation using simple pyramid based context aware model.

Keywords: pyramid structure, Sensor-mobile architecture.

1 Introdcution

Agriculture may be defined as an integrated system of techniques to control the growth and harvesting of crops. In the Indian agriculture scenario two main drawbacks are less availability of water and low productivity i.e. low yield per area. How to improve the productivity and irrigation area in the available water is a good tread off.

Also the smart decision in irrigation depends upon the ambient temperature, soil temperature, water holding capacity of the soil, humidity etc. Types of crops, growth and productive status of that crop also are the deciding factors in irrigation. Due to individual agriculture practice, the type of soil of each field in the vicinity of same village is different.

This paper proposes a system for automation of agricultural methods by integrating context aware computing, mobile communication and sensor-actuator network so as to implement more efficient agricultural methods.

S. Unnikrishnan, S. Surve, and D. Bhoir (Eds.): ICAC3 2013, CCIS 361, pp. 301–308, 2013.

The goal of context computing is to acquire and utilize information about the contexts.

The rest of the paper is organized as follows. In section 2, Pyramid context aware model is illustrated. Integration of WSN and cellular networks is explained in section 3, while context aware sensor network using mobile phone has been described in section 4. Finally Conclusion is drawn and future works are listed in section 5.

2 Context Aware Model

Let the agriculture information space be represented as a concept pyramid with the vertices representing the various contexts of information retrieval.

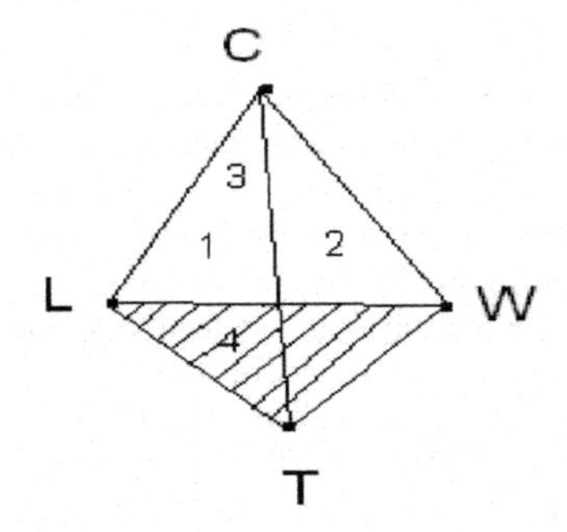

Planes:-1, 2, 3, 4

Edges :-

1-CL = LC	4-LT =TL
2-CT =TC	5-TW =WT
3-CW =WC	6-LW =WL

Fig. 1. Concept Pyramid

The simple pyramid context aware model for water irrigation is shown in figure 1. In this information space the nature of query may be single dimension query, edge level query and space level query.

The query is like "What is the soil type in this location?. " so the client is referring only the land context and the context point present exactly at the land vertex is the single dimension query. In the edge level the client is requesting information regarding two contexts at a time. E.g.- the query which crops are feasible for this particular water availability? So here the client refers to two contexts, crop and water and hence the context point is present somewhere on the edge connecting the crop and the water context.

When the client needs information regarding three contexts simultaneously, it lies in space level. For example the query how much irrigation should be done for this particular crop in this particular soil?, has a context point present in the concept space between land, crop and water contexts.

The context migration is shown in table 1. In some cases where land is not a major deciding factor or posses uniform properties, or crop may be same, then we can replace this context by humidity.

Table 1. Context Migration Table

Context point	Context edge	Context plane
Crop	CL,CT,CW (1,2,3)	1,2,3
Land	LT,LW,LC (4,6,1)	1,3,4
Water	WC,WT,WL (3,5,6)	2,3,4
temp	TC,TL,TW (2,4,6)	1,2,4

3 Integration of WSN and Cellular Networks

Improvements in wireless network technology interfacing with emerging sensor-based technology is allowing sophisticated inexpensive sensing, storage, processing, and communication capabilities to be unobtrusively embedded into our everyday physical world. Proactive WSNs are well suited for applications requiring periodic monitoring of data. In our application proactive WSN is useful.

The international standardization bodies are working toward integration of different wireless technologies with cellular networks [3].

In homogeneous network, all network entities run the same protocol stack, where each protocol layer has a particular goal and provides services to the upper layers. In a heterogeneous environment different mobile devices can execute different protocols for a given layer[3]. The protocol stack is given in figure. 2.

This protocol stack consists of multiple physical, data link and MAC layers, and network, transport and application layers. Therefore, it is critical to select most appropriate combination of lower layers (link, MAC and physical) that could provide the best service to the upper layers, which also implies the best communication interface. It is expected that the future advances in software defined radio(SDR) technology[SDRFORUM] will make multi-interface, multi-mode and multi-band communication device a commonplace. Mobile device equipped with multiple network interfaces can access multiple networks simultaneously. Even though SDR based mobile device is not fully capable of simultaneously accessing multiple wireless systems, discovering access networks available for given connection need to be performed.

In our application, several questions like what technology(or communication interface) to select to start a connection for this application or when to switch an on going connection from one interface to another (i.e. vertical handoff) are not complex because of fixed sensor nodes and also the fixed location of mobile phone sink in the WSN.

Also other important issues include transmission power selection for a given communication interface, co-channel interference, topology discovery and route creation, mobility and handoff management are less complex because of fixed WSN and interference free atmosphere in agriculture land.

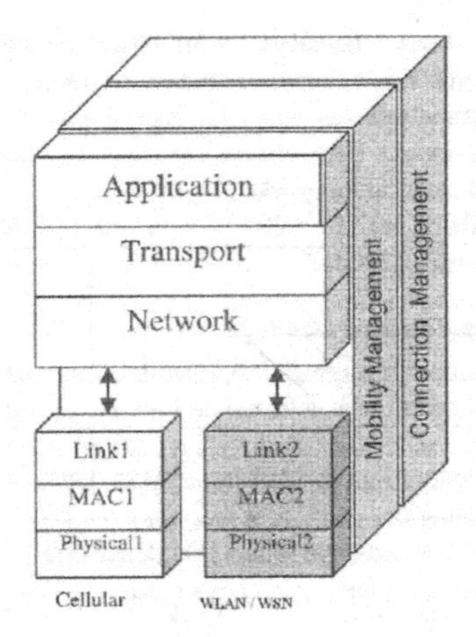

Fig. 2. Protocol stack of a dual- mode mobile terminal in heterogeneous network environment [9]

In this application horizontal handoffs are not required because mobile phone will have fixed location in the farm land. When mobile phone based sink collect the data from farm land periodically and wants to control the water irrigation system based on the context aware model, it will execute the vertical handoffs using switching the networks.

3.1 Algorithm for Switching from WSN to Cellular Network

Following is the proposed algorithm for vertical handoff

1. Read input parameters from sensor nodes periodically (soil moisture, humidity, temperature etc.)
2. Compare it with previous data.
3. If the total percentage change is less than or greater than precision value, then hand over the data to server through vertical handoff.
4. Collect the instruction from server.
5. Compare it with present condition (e.g. irrigation is on or off).
6. Take the decision (i.e. if irrigation is on and still need irrigation, keep irrigation on otherwise off).

4 Context Aware Sensor Network Using Mobile Phone

4.1 Mobile Phone Based Sink

In order to empower a smart-phone to work as a mobile data sink, two gaps have to be filled. In one, is the way that smart-phones communicate with wireless sensor nodes, and the other one is a software on smart phones which performs the mobile sink's functionalities.

Additionally, because there are many kinds of mobile platforms and smart-phones, cross-platform is also an important requirement for such a system [2].

While designing the mobile phone based sink, four issues are important. Energy efficiency, computational complexity, extensibility and cross-platform decide the flexibility and efficiency of the sink. Mobile phone's energy comes from battery which has a limited capacity, the middleware is required to push received sensor data to applications, and applications don't need to repetitively poll such data.

Current smart phones are being equipped with power full processor and large memory, still their capabilities are much lower than personal computer. So some algorithms which can be applied on fixed sinks may not smoothly run on a mobile phone. In a result, low complexity algorithm should be explored in the implementation of mobile phone based sink.

Because WSNs are a kind of application-specific networks, sensor nodes may have different sensors and produce data in different formats. In order to handle the heterogeneity in sensor nodes and sensor data, the solution must have good extensibility to support as many sensor networks as possible.

Currently, there are many mobile platforms for smart phone. It is very important feature for a mobile solution to support as many platforms and smart phones as possible.

4.2 Contex Aware Sensor-Mobile Architecture

Context aware sensor network makes energy efficient network which can help in arriving at a smart decision in active agriculture process Sensor network are useful to sense the physical attributes of surrounding for variation and current situation in the area of irrigation, cattle monitoring land disease monitoring, data generation for research in agriculture etc. the physical attributes of the surrounding like air temperature humidity ambient light, soil moisture and temperature are helpful.

A context aware computing is an application ability to detect and offer response to environmental variation. Context awareness can be achieved by analyzing the user request and based upon that the data from the distributed data-bases stored in the servers can queried. The relevant information can be retrieve and delivered to the users.

Let us say if any client needs information about crops feasible in his area he will just need to load the requirement and the server will respond as per the context [4].

Context awareness appears to be a fruitful idea for increasing the usability of web based services with mobile routers so that the client can access dynamic data as per requirement routed through the interface designed for users.

For example Agricultural Universities as a service provider are connected to the network, serving needs of the client with variety of information from various sources. This is a client sever base service where the client places the requested and the distributed data- base present in the server responds to the client's request to provide services that are appropriate to the specific people, place, time and event. Context awareness helps to get and utilize information about the physical situation in which user and their wireless network is embedded in order to provide improvised services that are suitable to the specific users in the given interaction platform. Hence the information retrieved from the distributed database has to be achieved with minimum interactions.

The internetworking of cellular systems with different wireless network technology has also been investigated by standardization bodies [03].The fusion of mobile phone network, wireless sensor and actuators (WSAN) network is very useful in vast crop land and its wire free option which suitable for precious instrumentation and control with hassle free operation and maintenance.

The smart phone based mobile sink is possible for WSN [02]. The integration of WSN and cellular network increases the application many fold. Furthermore, when all these technologies are integrated with internet, the possibilities are countless.

The rest of the paper is organized as follows. In section 2, Pyramid context aware model is illustrated. Integration of WSN and cellular networks is explained in section 3, while context aware sensor network using mobile phone has been described in section 4. Finally Conclusion is drawn and future works are listed in section 5.

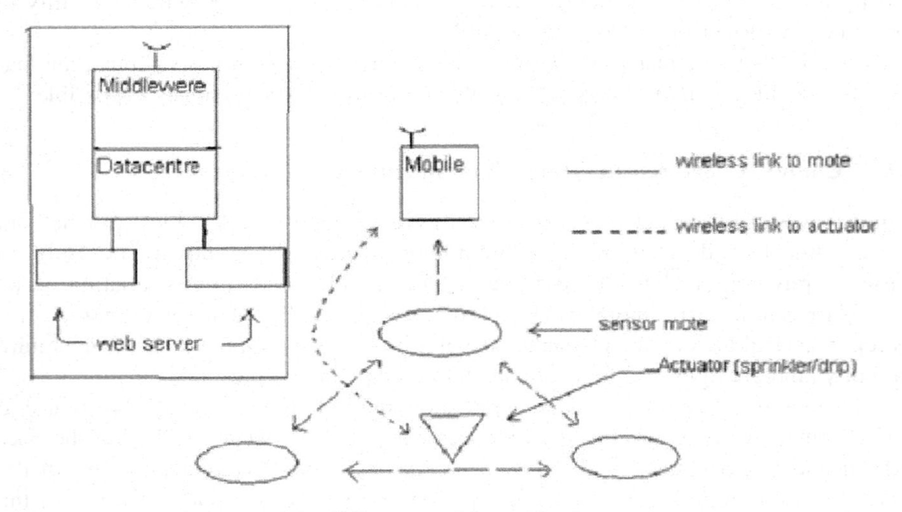

Fig. 3. Sensor-mobile architechture

Mobile service provider can update the data center and web servers and provide the feedback for control action. The computation ability of the sys em is divided into two locations. The functions of sink to monitor and control are locally available on mobile phone, while context awareness monitor & control functions are executed remotely at network domain.

There are two existing solutions to expand mobile devices with new hardware modules in low cost and convenient ways ,including the well known SDIO-based solution [6] [7] and utilizing SD interface as general purpose IO port (GPIO) [8].

This model is excellent where water availability is very less. Location awareness in mobile sink is added advantage because the central database contains the information regarding land soil, and land is also a vertex in pyramid context model.

In the Indian agriculture, use of computers and grid computing are costly techniques. Instead of computer, the mobile devices are easy to use and installation of such a model is cheap, because of falling mobile prices.

The moving mobile phone sink with farmer is also the another model. Here the mobile sink increases the coverage area of WSN. Also single mobile phone can be the sink for different WSNs. Here the routing protocols get changed. The mass implementation and co-operative practices in agriculture also reduces the cost of implementation. Government agricultural universities are the perfect choice for such type of service provider.

5 Conclusion and Future Work

Integration of wireless sensor networks with cellular technology where mobile phone works as a sink is an effective technology in water irrigation. In context aware feedback model, we have presented Pyramid model for simple but effective solution. We have also suggested the algorithm for vertical handoff for irrigation application. In the future, our research will be supporting above model for the benefit of farmers in developing countries.

References

[1] Ingole, P.V., Pawar, J.P.: Role of Information Technology in Global Agro Business. In: Second International Conference at Nirma Institude of Technology, January 1-4 (2000)

[2] Zhang, J.-F., Chaen, C.F., Ma, J., He, N., Ren, Y.: uSink: smart phone-based mobile sink for wireless sensor networks. In: 5th IEEE Workshop on Personalized Networks, Pernets 2011 (2011)

[3] de Morais Cordeiro, C., Agarwal, D.P.: Adhoc Sensor Networks: Theory and Applications. World Scientific Publication

[4] Chandrasekaran, S., Dipesh Dugar, M., Jitendra Kumar Jain , D., Kamlesh Jain, S., Dinesh Kumar Jain, N.: Context Aware Mobile Service Deployment Model of Agricultural Information System for Indian Farmers. 2010 International Journal of Computer Applications (0975-8887) 1(29) (2010)

[5] Lim, H.B., Teo, Y.M., Mukharjee, P.: Sensor Grid: Integration of Wireless Sensor Networks and the Grid. In: Proceeding of the IEEE Conference on Local Computer Networks 30th Anniversary

[6] SD Card Association, SDIO Simplified Specification, ver. 2 (2007)

[7] SDIO-Zigbee Card, `http://www.spectec.com.tw`

[8] Pering, T., Zhang, P., Chaudhri, R., Anokwa, Y., Want, R.: The PSI Board: Realizing a Phone-Centric Body Sensor Network. In: Proc. of 4th BSN 2007, Germany (2007)

[9] Cavalcanti, D., Cordeir, C., Agarwal, D., Xie, B., Kumar, A.: Issues in Integrating cellular Networks, WLAN and MANETs: A Furturistic Heterogeneous wireless networks. IEEE Wireless Communication 12(3), 30–41 (2005)

[10] Aqeel-ur-Rehman, Shaikh, Z.A.: Towards Design of Context- Aware Sensor Grid Framework for Agriculture. To appear in the Fifth International Conference on Information Technology (ICIT 2008), XXVIII- WASET Conference, Rome, Italy, April 25-27 (2008)

[11] Zhang, W., Kantor, G., Singh, S.: Integrated Wireless Sensor/Actuator Networks in an Agricultural Application. In: Proc. 2nd ACM Intl Conf. Embedded Networked Sensor Systems (SENSYS 2004), p. 317.8. ACM Press (2004)

[12] Basu, T., Thool, V.R., Thool, R.C., Birajdar, A.C.: Computer Based Drip Irrigation Control System with Remote Data Acquisition System. In: Proc. 4th World Congress Conf. Computers in Agriculture and Natural Resources, USA (July 2006)
[13] Kim, Y., Evans, R.G., Iversen, W.: Remote Sensing and Control of Irrigation System using a Distributed Wireless Sensor Network. IEEE Trans. Instrumentation and Measurement (2007)
[14] Baggio, A.: Wireless Sensor Networks in Precision Agriculture. In: Proc. ACM Workshop Real-World Wireless Sensor Networks (2005)
[15] Wark, T., Corke, P., Sikka, P., Klingbeil, L., Guo, Y., Crossman, C., Valencia, P., Swain, D.: Transforming Agriculture through Pervasive Wireless Sensor Networks. IEEE Pervasive Computing, 50–57 (April-June 2007)

TruVAL: Trusted Vehicle Authentication Logic for VANET

Suparna DasGupta[1], Rituparna Chaki[2], and Sankhayan Choudhury[3]

[1] Department of Information Technology,
JIS College of Engineering, West Bengal, India
suparnadasguptait@gmail.com
[2] Department of Computer Science & Engineering,
West Bengal University of Technology, West Bengal, India
rituchaki@gmail.com
[3] Dept. of. Computer Sc. & Engg. University of Calcutta, West Bengal, India
sankhayan@gmail.com

Abstract. Vehicular ad hoc networks are characterized by nodes with relatively high mobility and comprise of vehicle-to-vehicle and vehicle-to-infrastructure communications based on wireless network technologies. The deployment of vehicular communication systems is strongly dependent upon their underlying security and privacy features. As vehicular ad hoc networks (VANETs) are vulnerable to malicious attacks, the security in VANETs is receiving a significant amount of attention in the field of wireless mobile networking. It has been observed that trust establishment in VANET is a challenging task due to the lack of infrastructure, and the high speed mobile nodes causing frequent changes to the network topology. In this paper we have proposed a Trusted Vehicle Authentication scheme for secured communication in VANET. The use of a layered framework for assigning trust values to a vehicle helps in detecting a node taking part in malicious activity. Conducted simulation experiments on different scenarios show the performance and effectiveness of our new proposed vehicle authentication logic for vehicular ad hoc networks.

Keywords: Vehicular ad hoc networks, Security, Privacy, Trust Value.

1 Introduction

Vehicular ad hoc networks have become a popular area for both the academic research community and automobile industry, with specific attention to improving driving experience and road safety. Prerequisite to communicate within VANETs is an efficient route between network nodes which must be adaptive to the rapidly changing topology of VANET. The VANET consist of vehicles and road side units (RSUs) as network nodes and enables inter-vehicle communications (IVC) along with the road side to vehicle communications (RVC). Each vehicle that is part of a VANET contains an onboard wireless computing unit, commonly known as the onboard unit (OBU). Vehicles can communicate with nearby vehicles known as a vehicle-to-vehicle (V2V) communication and also with road side infrastructure also

S. Unnikrishnan, S. Surve, and D. Bhoir (Eds.): ICAC3 2013, CCIS 361, pp. 309–322, 2013.

known as vehicle-to-Infrastructure (V2I). IVC and RVC can be divided into two categories; safety-related application and information-related application.

The security of VANETs is crucial as their existence relates to critical life threatening situations. The network has to be secure without compromising current transmission quality. Recent studies on VANETs identify several issues, including those in security and privacy, which need to be addressed for widespread adoption. The security algorithm which is to be implemented in VANET, aims for lesser computational cost and should utilize limited resource effectively. Identifying malicious node and preserving privacy are also one of the major aspects of the VANETs security. The security of vehicular ad hoc networks can be affected mostly due to illegal access and impersonation attacks. For avoiding illegal access a vehicle must have a desired trust value and be registered under an authorized registration authority for accessing the network. An attacker can perform much different type of attacks; they can disturb the network topology and attack other node's dataflow by identity spoofing.

Authentication is an important security requirement for VANET. Large number of high speed vehicles and dynamic topology of the vehicular network are some of the key factors which makes authentication task difficult. Many emerging vehicular ad hoc networks use a trust value based mechanism which is dynamic and context dependent.

In this paper we have introduced a new trust based security mechanism to track malicious and misbehavioral vehicles and present new trusted vehicle authentication logic for vehicular ad hoc networks. This proposed scheme consists of two different steps. (i) Registration procedure has been introduced for new vehicles and trust value has been assigned to each registered vehicles. (ii) Trust value updating mechanism has been presented for existing vehicles.

The rest of the paper is organized as follows. Comprehensive surveys of related works of different secure routing protocols for VANETs are discussed in section 2. In section 3 we have presented new trusted vehicle authentication logic for VANET. Intensive performance analysis of our proposed scheme is presented in section 4. We conclude our paper with final remarks in section 5.

2 Related Works

For full deployment of VANETs two paramount issues should be resolved, namely security and privacy. The information communicated by vehicles should be secured. Many researchers have been already published number of research papers, addressing the security issue of vehicular ad hoc networks. In this section we have discussed some of the security related research challenges of VANET.

In [7] B. Aslam et al. proposed a certificate based distributed approach for VANET. The focus of this approach is to achieve the desired security attributes (Privacy, authentication, confidentiality and non-repudiation) during the initial deployment phase of VANET. Architecture has been proposed to achieve desired security requirements and enables service providers to offer incrementally various VANET services with minimal investment thus encourages both service providers and users to try/adopt VANET. According to this protocol when a user wants to participate in a VANET, he/she purchases a payment-processing-device, consists of an associated certificate. At the time of initialization, the device was registered with the user's account. The vehicle's information is maintained by the provider. In this case they need to introduce some roaming or cross certification mechanisms between the service areas.

S. Biswas et al. [17] proposed an ID-based message authentication mechanism for VANETs. This is a safety message authentication scheme for vehicular ad hoc networks using an ID-based signature and verification mechanism. This offers a certificate-less public key verification, while a proxy signature provides flexibilities in message authentication and trust management. An ID-based proxy signature framework with the standard ECDSA for VANET's road-side unit (RSU) has been incorporated in this scheme. RSU originates safety application message. Forwarding of signed messages is specially handled to ensure the trust and authentication of RSU's application messages. In this mechanism the current location information of a signer has been used as the signer's identity in order to sign and verify the proxy signature. An emergency/road-safety application message has been generated by a trusted central authority (e.g. department of transportation), while the issued message is signed and delivered to end users (OBUs) by corresponding road-side units (RSUs) on behalf of the originator of the message. If any OBU is the outside communication range of an RSU, it may receive the broadcast through an intermediate "message-forwarder" OBU. The receiver vehicle checks the signature contents for the verification of the message. One easy way to accomplish for forwarding exactly the same signature materials as received from RSU. The receiving OBU verifies the signature as if it has received the message from the corresponding RSU. This approach is appropriate for authentication and trust management in highly dynamic and untrustworthy vehicular network environment as it is resilient to potential security threats. It is also compatible to the VANET's standard specifications.

In [8] Terence Chen et al. proposed a distributed routing framework for authentication of messages, nodes and routes. The architecture is distributed and uses limited assistance from a Certificate Authority (CA). By using digital signature control messages message integrity and originality can be ensured. A secure neighbor discovery method is included in the node-to-node authentication module. The link status evidence mechanism included in the cumulative routability verification module regulates the behavior of internal nodes. This proposed mechanism is applied to the OLSR [3] routing protocol, resulting in an OLSR extension which guarantees trusted routing using only the routes with verified nodes. The routability verification module adds a substantial amount of overhead, which may result in scalability problems.

Y. Sun et al. [9] Proposed ECHO, an efficient certificate updating scheme by Vehicle-to-Vehicle communications. They had used GPSR [1] for transmitting the message from the source node to the certain location. In case an improved route is discovered, the next relay node prefers the OBUs on the shortest path from source to the destination. The shortest path is calculated based on the selected roads with high traffic density. In this scheme the OBU not only sends the certificate request to an immobile RSU but also inform the RSU where the response message should be sent back. The RSU also issues a new certificate for the legitimate OBU. The OBU receives the response message at the expected callback address if the whole process is success.

In [6] Charles Harsch et al. presented a scheme that secures geographic position-based routing for VC. They design mechanisms to safeguard the functionality of PBR

[4]. A public key infrastructure with a Certification Authority (CA) has been introduced in this routing scheme. CA issues public/private key pairs and certificates containing public key, attribute list, the CA identifier, the certificate lifetime and the CA signature. Each received packet is first submitted to a sequence of plausibility checks using the packet's time and location fields as inputs. If at least one test fails, the packet is discarded. Otherwise, if all checks succeed, the packet is validated cryptographically. First, the certificate is validated, unless it was previously validated and cached. Then, the signature(s) on the packet are validated and, if failed, discarded. Otherwise, the packet is processed further. We have discussed s in more detail the security mechanisms hereafter.

Ming Chin Chuang et al. [18] proposed a decentralized lightweight authentication scheme called Trust Extended Authentication Mechanism (TEAM) for vehicle-to-vehicle (V2V) communication networks. This scheme adopts the concept of transitive trust relationships to improve the performance of the authentication procedure. It also claim to satisfy requirements like anonymity, location privacy, mutual authentication to prevent spoofing attacks, forgery attacks, modification attacks and reply attacks, as well as no clock synchronization problem, no verification table, fast error detection, and session key agreement. The amount of cryptographic calculation under TEAM is substantially less than in existing schemes because it only uses an XOR operation and a hash function.

V. Paruchuri et al. proposed [10] a protocol for anonymous authentication in vehicular Networks (PAAVE) to address the privacy preservation issue with authority traceability in vehicular ad hoc networks (VANETs). This protocol is based on smart cards to generate on-the-fly anonymous keys between vehicles and Roadside units (RSUs). PAAVE [10] is lightweight and provides fast anonymous authentication and location privacy while requiring a vehicle to store one cryptographic key.

In [5] J. Serna et al. proposed a geo-location based trust for VANET's privacy. This paper used as an authorization paradigm based on a Mandatory Access model and a novel scheme which propagates trust information based on a vehicle's geo-location. In this proposed scheme different levels of authorization is defined; Such as; personal (i.e., driver's "consent" is required), emergency (i.e., in this case driver's "consent" is not needed; however the system will require the credential of the entity (i.e. police) accessing the information) and public (i.e., no authorization checks are required). For trustworthiness certifying authority has been presented in this scheme.

Asif Ali Wagan et al. [12] presented a hardware based security framework which uses both standard asymmetric PKI and symmetric cryptography for secure and faster safety message exchange. The paper proposed to develop trust relationship among the neighboring node, finally leading to the formation of trust groups. The trust has been established via Trusted Platform Module (TPM) and group communication. In [12] Wagan et al. proposed an extended version of above discussed security framework. Within these two schemes, one is for efficient group formation for improving life time of a group leader role. They also presented hybrid (symmetric /asymmetric) message dissemination scheme for faster and secure communication.

A trust based privacy preserving model for VANETs has been presented by Ayman Tajeddine et al. [13], which is unique in its ability to protect privacy while maintaining accurate reputation based trust. VANET users are anonymous within their groups and yet identifiable and accountable to their group managers. In this proposed scheme each vehicle is a part of a static group assigned offline and should generate a group signature on each of its outgoing messages and includes with it an authentic group ID. Group Managers (GMs) have the responsibility to admit new vehicles and to evict attacker/malicious vehicles. Every group has a reputation level that accounts for the trustworthiness of the messages signed in the name of this particular group. In this proposed mechanism no scheme has been described for estimation of this reputation level.

In [14] a reputation based trust model has been presented by Qing Ding et al. This is an event based reputation model to filter bogus warning messages. A dynamic role dependent reputation evaluation mechanism has been presented to determine whether an incoming traffic message is significant and trustworthy to the driver. In this proposed scheme initially different roles played by vehicles are defined for reputation estimation. These roles are Event Reporter (ER), Event Observer (EO) and Event Participant (EP). After encountering a traffic event at first ER calculate reputation value of that event. If this value is over the redefined threshold, the event message will be sent to the traffic safety application in the vehicle and to all neighbors in one hop, namely EO. When an EO receives a traffic warning message from an ER, it will first store this message into the event table if there is no the same ID record in the table. Depending on the message truthfulness send by an ER an EO can calculate ER's reputation. EP receives message only form EO's and other EP's. Complex formula for calculating reputation values are given in this paper. Reputations are estimated based on the behavior of vehicles. The behavioral characteristics are not clearly identified.

G. Kavitha et al. [15] introduced a grid based approach for providing a quantitative trust value, based on the past interactions and present environment characteristics. This quantitative trust value has been used to select a suitable resource for a job and eliminates run time failures arising from incompatible user resource pairs. This act as a tool to calculate the trust values of the various components of the grid and there by improves the success rate of the jobs submitted to the resource on the grid. The access to a resource not only depend on the identity and behavior of the resource but also upon its context of transaction, time of transaction, connectivity bandwidth, availability of the resource and load on the resource. The quality of the recommender has been evaluated based on the accuracy of the feedback provided about a resource. The jobs are submitted for execution to the selected resource after finding the overall trust value of the resource. The parameters depend on which trust value of a node has been calculated are not clearly defined in the paper.

Sumit kumar Singh et al. [16] proposed two levels of security based on Signcryption and node trust. Signcryption is less cost effective than Signthenencryption and it also conserved confidentiality and integrity of the message. In [16] authors proposed a network model consisting VANET server, Roadside Unit, On-Board Unit, Source node and Destination node. The VANET server is a trusted entity by all nodes participating in the network. The shared key used between nodes and VANET server cannot be compromised in any conditions. The public and private keys sent from VANET server to nodes are through secure channel and cannot be compromised by any means. The trust value assigned by the destination node is appropriate and depending on this trust value, trust level has been assigned. This proposed mechanism does not focus on the estimation of trust value and the trust value calculation parameters have not mentioned in this research.

The above discussions lead to the conclusion that mainly cryptographic and certificate based techniques are being preferred by the researchers for securing communication within VANETs. The metrics influencing the certificate based technique have not been clearly identified. The cryptographic approaches such as, encryption, digital signature, Signcryption etc. has been introduced additional complexities for key management. Some researchers have chosen trust value based authentication, but the parameters influential in trust value assignment of a vehicle are not properly identified. This paper aims to identify the parameters in a trust based authentication scheme as to ensure authenticity of communication in a VANET.

3 Proposed New Routing Protocol: TruVAL: Trusted Vehicle Authentication Logic for VANET

The previous section leads to the observation of different security related research challenges of VANET. In this section we are going to propose TruVAL: Trusted Vehicle Authentication Logic for VANET. In our proposed solution we have distributed VANET in a layered architecture.

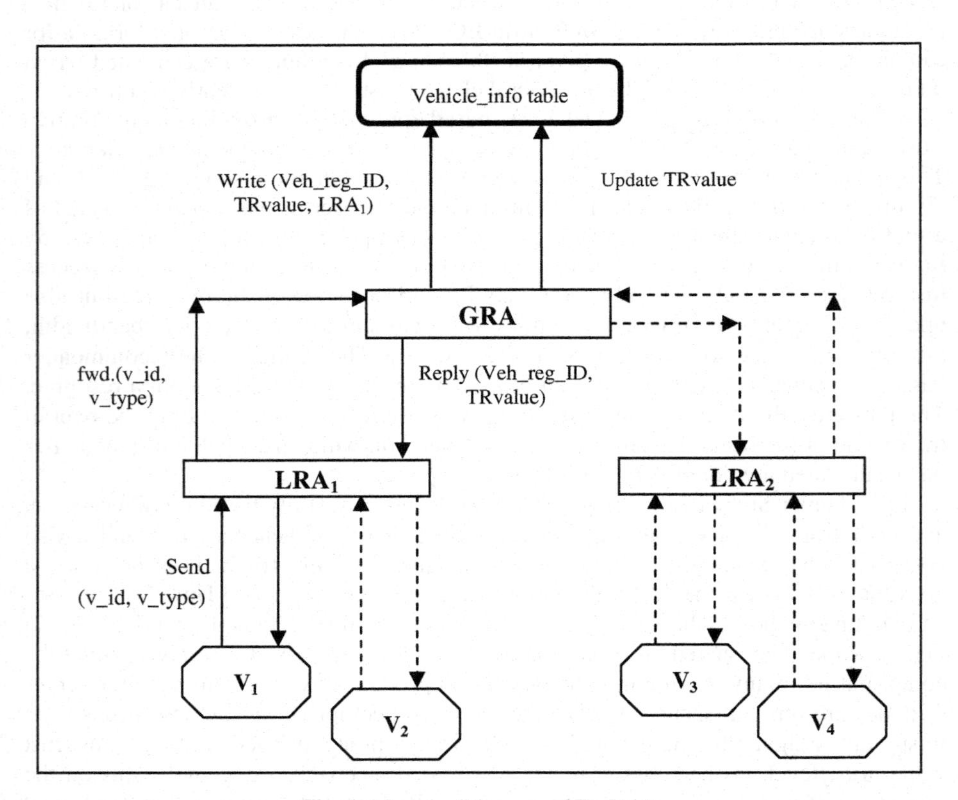

Fig. 1. Modular Diagram of TruVAL

In the lowest layer, all nodes (i.e., vehicles) present in the system. LRA_i (Local Registration Authority) implies a road-side unit that acts as a middle layer element within the framework. LRA_i is responsible for estimating an initial trust value and also updated the trust value based on some pre-defined parameters for that vehicle. The highest layer component, called Global Registration Authority (GRA), is nothing but a repository having all lower layer information. TruVAL consists of two phases: registration procedure and trust value updating mechanism.

I. Registration Procedure

Here we have assumed that all nodes in a VANET are distributed according the proposed layered architecture. On entry of a new vehicle in the system, it requests the LRA_i for a registration certificate. This request is termed as Registered to Communicate (RTC). In this request, new vehicle sends it vehicle number and other details to LRA_i. Depending on that information trust value of that vehicle is estimated. LRA_i forwards vehicle number and trust value to GRA. GRA generates a unique sequence number, i.e., USN for that particular vehicle. This USN acts as Veh_reg_ID for the corresponding vehicle. At the registration time LRA_i have no idea about the behavior of the vehicle. For this reason at this time an initial trust value is given to the vehicle.

❖ Trust Value Initialization

In this subsection we are going to present an algorithm for new vehicle entering in the system. Every new vehicle has to register under its local LRA_i. For this reason it sends a registration request to LRA_i. In this request each vehicle has to send their types and the unique features of it. After receiving the request LRA_i assigns a unique number and a trust value to the requesting vehicle. After initialization of trust value it will forward to GRA and GRA keeps all this information in Vehicle_info table and the corresponding vehicle is registered under the communicating LRA_i.

Table 1. Vehicle_info table

Veh_reg_ID	LRA_i	TR_{value}

Algorithm 1. Registration_proc

New vehicle sends (vehicle_id, v_type) to LRA_i

LRA_i call Trust_init_func(vehicle_id, v_type)

LRA_i forward that vehicle_id and TRvalue to GRA

GRA generates a Veh_reg_ID

GRA write Veh_reg_ID, TRvalue and LRA_i in Vehicle_info table.

New vehicle is registered under LRA_i

END.

After completion of the above discussed procedure, the systems now have a new registered vehicle. Apart from registered a new vehicle and supply information about that registered information, LRA_i have another very important role. LRA_i have to update trust value of every registered vehicle after a certain interval. In the next section we are going to discuss the trust value updating procedure.

II. Trust Value Updating Procedure

After a certain time interval LRA_i updates trust value of every registered vehicle. This mechanism has been done based of some parameters defined by us. Those parameters are as follows:

✓ **Active Factor:** For each communication every sender vehicle maintains a communication counter after every interaction. Every receiving vehicle also maintains a list of name of every sender vehicle. In this way LRA_i can easily know how many times a vehicle have taken part into communication. This can measure by a metric called active factor.

Active factor α is defined as follows: Let cn be the total number of communications in which vehicle v have participated during time interval t, then,

$$\alpha = 1/n \sum_{n=0}^{n} (c_s + c_r) / c_n \quad (1)$$

Where c_s is the value of communication counter maintained by v and c_r denotes the number of times v appeared in the sending vehicle list of other vehicle present in the network. Using equation (i) LRA_i can easily calculate value of active factor for v.

✓ **Feedback Factor:** After completion of every communication every vehicle who was taken part in that communication sends feedback [R_S, R_F, R_{PS}, R_{PF}] to LRA_i about its neighbors taking part in that communication. LRAi calculates feedback value from those reports. Feedback value can be calculated using some metrics defined in feedback form.

- Request success ratio (R_S): This is defined as the request success ratio which is calculated based on number of neighboring nodes who have successfully received from the source node which has broadcasted it. Once a vehicle sends a request to its neighbor and gets the acknowledgement within the timeout period, the respective R_S value is incremented by one.
- Request failure ratio (R_F): This is defined as the request failure ratio which is calculated based on number of neighboring nodes which have not received the request.
- Reply success ratio (R_{PS}): This is defined as the reply success ratio which is calculated as successful replies received by source node which has sent the request.
- Reply failure ratio (R_{PF}): This is defined as the reply failure rate which is calculated based on the number of neighboring nodes which have not sent the replies for the request received.

Using these metrics feedback value can be defined. Let feedback value is denoted by f. Then,

$$f = 1/n \sum_{0}^{n} (t_p * \left(\frac{R_S - R_F}{T_{mr}} + \frac{R_{PS} - R_{PF}}{T_{mn}}\right)) \quad (2)$$

Where, t_p is time period, T_{mr} is the total number request sends in that time period and T_{mn} is total number of reply receives in that time period.

- ✓**Experience Factor:** After successful registration of each new vehicle every LRA_i assigns a timestamp value to them. This value indicates the entry time of the vehicle in the system. From this value LRA_i can easily estimate the duration of the instance of vehicle in the system. According to this time it calculates experience factor which is denoted by € symbol. Let, at time t_0 a vehicle enters in the system.

At time t_n the value of

$$€= (t_n - t_0) \tag{3}$$

Trust updation of a vehicle is done based on this active, feedback and experience factor. The updated trust value (TRvalue) can be estimated using the following equation.

$$TR_{value} = W_1 * \alpha + W_2 * f + W_3 * € \tag{4}$$

Where, W_1, W_2 and W_3 are weighting factors. The multiplying values with different factors actually indicating weight given to that factor. Active, authentic and reliable new vehicle is more desirable than inactive and less reliable vehicle. For this reason more weight is given to α and *f* than €. The main aim for calculating TRvalue is secure authenticity and reliability in message passing. For this reason highest weight is given to *f* factor.

Algorithm 2. Trust_val_updt

/*updating mechanism of trust value for vehicle v by respective LRA_i */

Calculate active factor, $\alpha = 1/n \sum_{n=0}^{n} (c_s + c_r)/ c_n$

Calculate feedback factor,

$f = 1/n \sum_{0}^{n} (t_p * (\frac{R_S - R_F}{T_{mr}} + \frac{R_{PS} - R_{PF}}{T_{mn}}))$

Estimate experience factor, €.

Update TRvalue = $W_1 * \alpha + W_2 * f + W_3 * €$.

END.

After calculation of trust value this value is attached with the vehicle. From the above discussion we can see the trust value is calculated based on some basic behavioral activities of a node. These activities are not immortal. For this reason assigned trust value should be updated by LRA_i, after a certain time interval.

Vehicles are all highly mobile in nature. In a very short time interval it can move from one LRA_i's region to another LRA_i's region. After become registered once all information about the corresponding vehicle is maintained by parent LRA_i. Information of every registered vehicle's is also stored in GRA. GRA actually acts as a global repository of all registered vehicles. LRA_i monitors all vehicles registered under it. When a vehicle moves out of its region it broadcast a message consisting information about that vehicle. In this way the new LRA_i in which region the vehicle enters can know information about it. The information sends by the following message format.

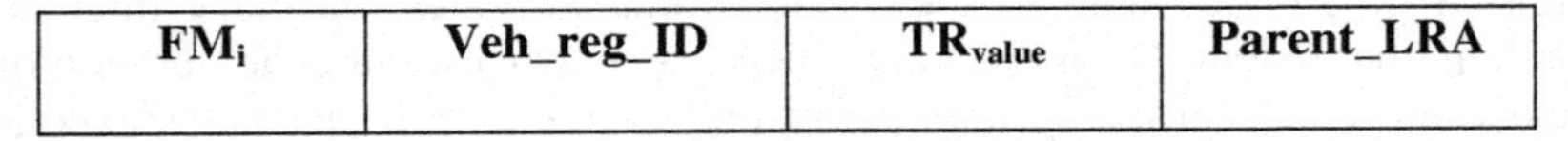

FM_i	Veh_reg_ID	TR_{value}	Parent_LRA

Fig. 2. Frequent message format

In the above message format FM_i denotes an identifier that uniquely identifies the message. Veh_reg_ID, TR_{value}, Parent_ LRA_i denotes registration identifier, trust value and initial LRA_i respectively for the corresponding vehicle. In this way when LRA_i found any new vehicle in its region it also have some essential information about that vehicle. If LRA_i needs more detail information then it can query to GRA and gets required information from it.

Table 2. Data Dictionary

Parameter	Details
LRA	Local Registration Authority
GRA	Global Registration Authority
RTC	Registered to Communicate
USN	Unique Security Number
Veh_reg_ID	Vehicle Registration Identifier
TR_{value}	Trust Value
Vehicle_id	Vehicle Identifier
V_type	Vehicle Type
FM_i	Frequent Message Identifier
α	Active Factor
C_s	Sending Message Counter
C_r	Receiving Message Counter
C_n	Total Number of Communication Maintained Counter
R_S	Request Success Ratio
R_F	Request Failure Ratio
R_{PS}	Reply Success Ratio
R_{PF}	Reply Failure Ratio
T_p	Total Time Period
f	Feedback Factor
€	Experience Factor
W_i	Weighted Factor
T_o	Time of A Vehicle Entering In System
T_{mr}	Total Number of Request Send Within Time Period
T_{mn}	Total Number of Reply Received Within Time Period

IV. Performance Analysis

In our proposed protocol we haven't used any cryptographic techniques like digital signature etc. Thus there is no overhead of maintaining public key and private key. Extra bandwidth is also not needed for transforming the data. The layered network structure helps us to overcome the problems of centralized approach like performance bottleneck, no scalability etc. Based on three clearly defined behavioral parameters of vehicle a trust value is assigned to all registered vehicles. Every node has to maintain this trust value which result to very less overhead. For performance analysis the above discussed trust estimation mechanism is integrated into existing AODV routing protocol. The basic aim is to restrict malicious vehicles to take part in communication.

We choose the NS2 simulator for this analysis because it realistically models arbitrary node mobility as well as physical radio propagation effects such as signal strength, interference, capture effect and wireless propagation delay. Our propagation model is based on the two-ray ground reflection model. The simulator also includes an accurate model of the IEEE 802.11 Distributed Coordination Function wireless MAC protocol.

Using NS2 we evaluate the performance of the proposed protocol TruVAL and present the following metrics for comparing the performance of traditional AODV with the TruVAL. The simulation model consists of a network model that has a number of wireless nodes, which represents the entire network to be simulated.

Table 3. Simulation Environment Parameters

Parameter	Value
Channel type	Wireless Channel
Radio-propagation model	Two Ray Ground
Antenna model	Omni Antenna
Network interface type	Wireless Phy
Mac type	802.11
Number of nodes	25

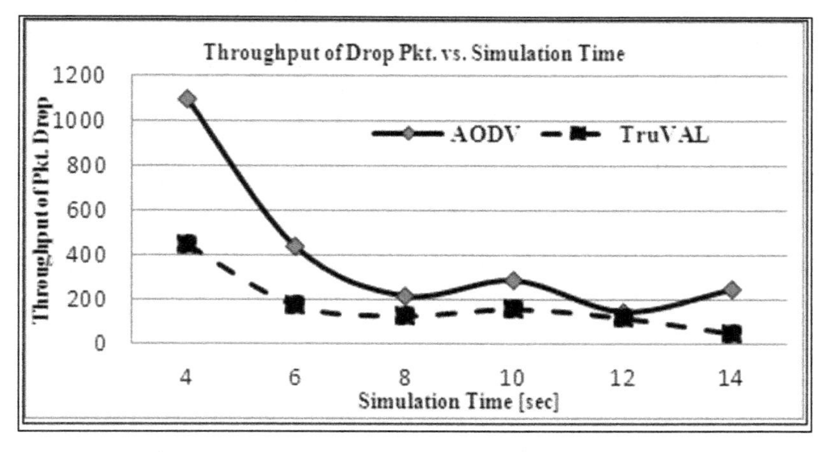

Fig. 3. Throughput of Drop Pkt. vs. Simulation Time

Figure 3 depicts a comparison of AODV and TruVAL with respect to packet dropping in presence of malicious nodes. The graph clearly shows that the number of packets dropped is less in case of TruVAL. Simulation result shows that in the same network scenario, traditional AODV dropped 4628 packets whereas TruVAL dropped 2444 packets.

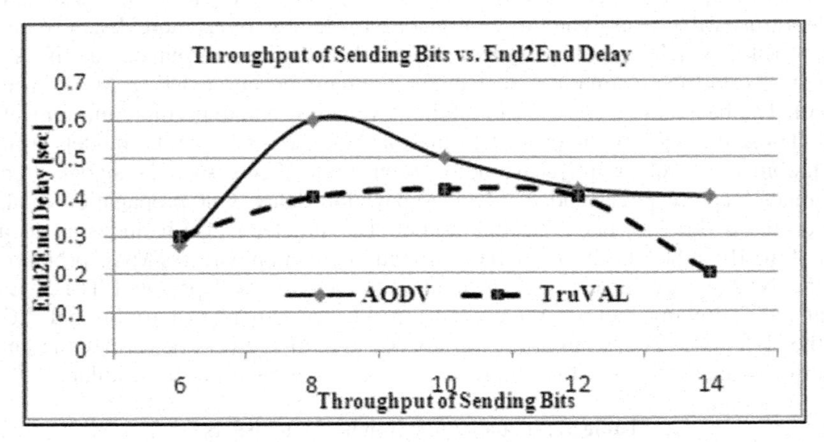

Fig. 4. Throughput of Sending Bits vs. End2End Delay

From figure 4, we have compared end2end delay for packet sending of AODV with TruVAL in same network condition. It is observed from the above graph that the delay for TruVAL is much lesser than AODV. Simulation result show average end2end delay for sending packet for AODV is 0.40226. For TruVAL it is 0.36542. In this case also TruVAL gives a better result than AODV.

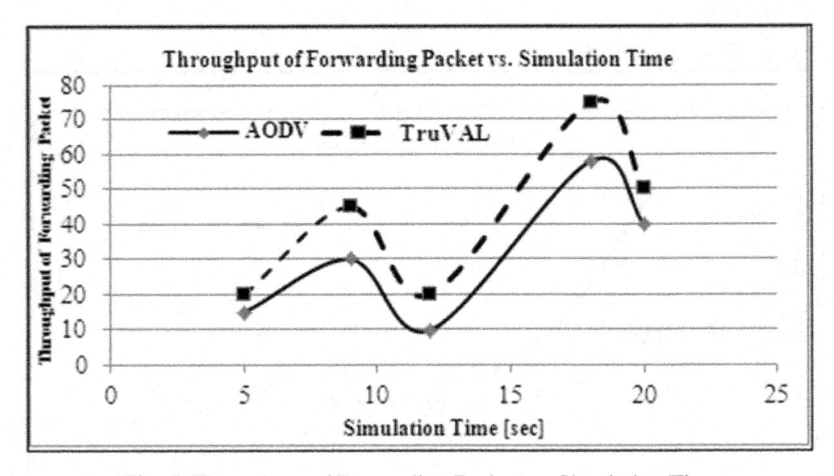

Fig. 5. Throughput of Forwarding Packet vs. Simulation Time

Figure 5 depicts a comparison of AODV and TruVAL with respect to number of forwarding packet in presence of malicious nodes. The graph clearly shows that the number is high in case of TruVAL. Simulation result shows, that in the same network scenario our proposed scheme forwards more bits with respect to traditional AODV.

From the above three comparisons we can see in present of malicious nodes our proposed scheme gives better result than traditional AODV.

4 Conclusions

In this paper we have summarized the generic characteristics of some well-known security approaches for VANETs and proposed a trust based authentication logic for VANET named TruVAL. We presented a layered structure for authenticating vehicles to communicate. A trust estimation mechanism is also proposed for trust calculation. Depending on this estimated trust malicious nodes are detected. In this way malicious nodes can be easily avoided during communication. For performance analysis of TruVAL, a simulation environment is created using NS2. The results show that TruVAL result in lesser number of packet drop and end2end delay as compared with traditional AODV. TruVAL also forward more bits than traditional AODV. In future this scheme shall be extended into a secure routing scheme.

References

1. Karp, B., Kung, H.T.: GPSR: greedy perimeter stateless routing for wireless networks. In: 6th Annual International Conference on Mobile Computing and Networking, pp. 243–254 (2000)
2. Perkins, C.E., Royer, E.M., Das, S.: Ad Hoc on Demand Distance Vector (AODV) Routing. IETF Internet draft, draft-ietf-manet-aodv-08.txt (March 2001)
3. Clausen, T., Jacquet, P.: Optimized Link State Routing Protocol (OLSR). RFC 3626, IETF Network Working Group (October 2003)
4. Namboodiri, V., Gao, L.: Prediction-Based Routing for Vehicular Ad Hoc Networks. IEEE Transaction on Vehicular Technology 56(4) (July 2007)
5. Serna, J., Luna, J., Medina, M.: Geo-location based Trust for Vanet's Privacy. In: Fourth International Conference on Information Assurance and Security, IAS 2008 (2008) 978-0-769-3324
6. Harsch, C., Festag, A., Papadimitratos, P.: Secure Position-Based Routing for VANETs. In: IEEE 66th Vehicular Technology Conference (VTC Fall), Baltimore, USA (2008)
7. Aslam, B., Zou, C.: Distributed Certificate and Application Architecture for VANETs. In: 28th IEEE Conference on Military Communications, MILCOM 2009 (2009) 978-1-4244-5238-5
8. Chen, T., Mehani, O., Boreli, R.: Trusted Routing for VANET. In: 9th International Conference on Intelligent Transport Systems Telecommunications, ITST 2009 (2009) 978-1-4244-5347-4
9. Sun, Y., Zhao, B., Su, J.: ECHO: Efficient Certificate Updating Scheme by Vehicle-to-Vehicle Communications. In: Fourth International Conference on Frontier of Computer Science and Technology, FCST 2009 (2009)

10. Puruchuri, V., Durresi, A.: PAAVE: Protocol for Anonymous Authentication in Vehicular Networks using Smart Cards. In: Proc. GLOBECOM (2010) 978-1-4244-5638-3
11. Wagan, A.A., Mughal, B.M., Hasbullah, H.: VANET Security Framework for Trusted Grouping using TPM Hardware. In: 2010 International Conference on Communication Software and Networks, ICCSN (2010) 978-0-7695-3961
12. Wagan, A.A., Mughal, B.M., Hasbullah, H.: VANET Security Framework for Trusted Grouping using TPM Hardware: Group Formation and Message Dissemination. In: 2010 International Symposium in Information Technology, ITSIM (2010) 978-1-4244-6716-711
13. Tajeddine, A., Kayssi, A., Chehab, A.: A Privacy-Preserving Trust Model for VANETs. In: 10th IEEE International Conference on Computer and Information Technology, CIT 2010 (2010) 978-0-7695-4108-2
14. Ding, Q., Jiang, M., Li, X., Zhou, X.H.: Reputation-based Trust Model in Vehicular AdHoc Networks. In: IEEE Conference on Wireless Communications and Signal Processing, WCSP (2010) 978-1-4244-7555
15. Kavitha, G., Sankaranarayann, V.: Secure Resource Selection in Computational Grid Based on Quantitative Execution Trust. International Journal of Computer and Information Engineering (2010)
16. Singh, S.K., Vijayan, R.: Enhanced Security for Information Flow in VANET using Signcryption and Trust level. International Journal of Computer Applications (0975-8887) 16(5) (February 2011)
17. Biswas, S., Misic, J., Misi, V.: ID-based Safety Message Authentication for Security and Trust in Vehicular Networks. In: 31st International Conference on Distributed Computing Systems Workshops (2011)
18. Chuang, M.C., Lee, J.F.: TEAM: Trust-Extended Authentication Mechanism for Vehicular Ad Hoc Networks. In: International Conference on Consumer Electronics, Communications and Networks, CECNet (2011) 978-1-61284-459

Deterministic Evaluation of Identity Based Random Key Distribution Using Deployment Knowledge in WSNs

P. Saritha Hepsibha and G. Sanjiv Rao

Dept. of IT, Sri Sai Aditya Institute of Science and Technology, Kakinada, India
{sarithahepsibha}@yahoo.co.in, {sanjiv_gsr}@yahoo.com

Abstract. With the recent developments in wireless communications, to achieve secure communications among sensor nodes in a Wireless Sensor Network (WSN), the research community has to resolve different technical issues like design and development of efficient node deployment and key distribution schemes. As the sensors are the resource constrained and vulnerable nodes, and are often deployed in hostile environments, an adversary may physically capture and compromise some of the nodes. Hence, deployment of nodes and key distribution are remained as challenging and basic issues to be addressed as they can influence the network performance metrics. Many schemes have been proposed in recent years for sensor node deployment and for key distribution. In this paper, we investigate a secure key distribution mechanism, such as Identity Based Random key Distribution scheme, for the two simple node deployment strategies, namely, Random and Grid. We first evaluate the key distribution protocol, and then we provide a comparative study that shows the improvements in the probabilistic prediction of the network performance resulting from inclusion of the two deployment schemes with the examined technique. We also show the simulation of deployment schemes of WSN using Network Simulator 2.34 with Mannasim.

Keywords: Wireless Sensor Network, Deployment Knowledge, Key distribution, Network Connectivity.

1 Introduction

A wireless sensor network (WSN) is a collection of small, low cost and battery powered devices called sensor nodes. The sensor nodes are also called as MICA2-DOT motes [3]. Each sensor node is capable of communicating with other sensor nodes wirelessly and monitoring an environment by sensing and gathering local data such as temperature, humidity, light or motion .These tiny sensor nodes with multi-functionalities make it possible to deploy wireless sensor networks (WSNs), which represent a significant improvement over traditional wired sensor networks [12].

A WSN is composed of densely deployed sensor nodes in a geographical area with a purpose of collecting and processing some useful data, and transmitting the data to nearby base stations for further processing. The transmission between the sensors is done by short range radio communications. The base station is assumed to be computationally well-equipped whereas the sensor nodes are resource- starved. A hierarchical wireless sensor network is explained in [1].

S. Unnikrishnan, S. Surve, and D. Bhoir (Eds.): ICAC3 2013, CCIS 361, pp. 323–336, 2013.

Usually the sensor networks are composed of one or more base stations and many sensor nodes scattered in a sensor field (i.e., deployment field or target field) as shown in Fig. 1. Each of these scattered sensor nodes has the capability to collect data and route data back to the base station. A base station or sink node serves as a gateway to another network, which is a powerful data processing unit, and also responsible for network management, collecting and performing costly operations on behalf of sensor nodes. As explained in a broadcast authentication scheme in WSNs [2], a base station acts like a commander, and broadcasts the commands to the sensors, and upon receiving those commands, sensors send results back to the base station. To route data back to the base station, sensor nodes use a broadcast or multi-hop communication paradigm instead of point-to-point communications. The base station may communicate with the task manager node via Internet or Satellite.

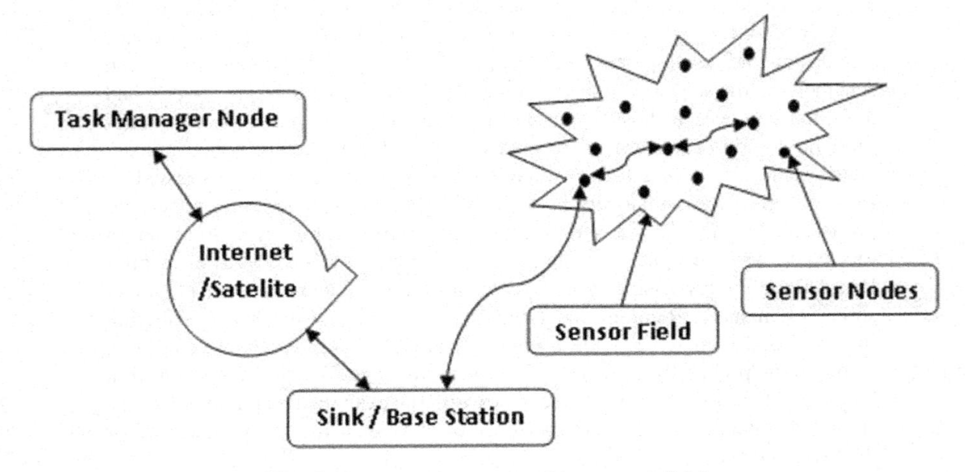

Fig. 1. Sensor nodes scattered in a target field

Using different sensors, WSNs can be implemented in many application areas including, security, entertainment, health care monitoring, automation, industrial monitoring, public utilities, and asset management. To support these applications, WSNs require to exhibit ad hoc networking properties. Ad hoc network properties can be found in [4]. However, the sensor networks need to have some unique features and application specific requirements, which made them different from ad hoc networks as listed in [12].

WSN devices have severe resource constraints in terms of energy, computation, memory and require large number of nodes. For many applications, sensors networks are usually deployed in sensing area to collect some information, and exhibits frequent changes in its topology. Therefore, deploying sensor nodes to provide complete area coverage is a basic issue to be addressed. Three alternative deployment schemes have been proposed in [4]. Depending upon the application requirements, node deployment can be either, application specific deterministic deployment, or random deployment or grid based deployment. Deterministic deployment is suitable for small-scale applications, where the sensors are placed directly in the required region. For large-scale applications, random deployment is suitable, where sensors are thrown randomly to form a network. Grid-based deployment is suitable for moderate to large-scale applications, where individual sensors are placed exactly at grid points. In this paper, we make a comparison on the amount of network coverage for two simple node deployment strategies namely, random and grid.

In typical applications like military monitoring, environmental monitoring, etc., sensor networks are deployed over the deployment region to collect data by sensing the surroundings. When sensor networks are being used for such kind of applications, security becomes another important factor to be addressed. Moreover, sensors are easily prone to attacks or failures. In order to provide secure communications among sensors, we required to set up secret keys for encryption and authentication.

Due to resource limitations of sensor nodes, a trusted-server scheme for authentication (such as Kerberos [15]) and traditional public key security protocols [16] are not suitable for sensor networks. Hence, the suitable solution is to distribute key information among all sensors prior to deployment. Previous research has explored a number of key pre-distribution schemes [9] [10] [11]. However, due to vulnerability of the sensor nodes, pair wise key establishment scheme is suitable for wireless sensor networks, where each sensor carry N-1 secret pair wise keys, each of which is known only to this sensor and one of the other N-1 sensors. Moreover, in this scheme, compromising of node does not affect the security of other nodes. A discussion on key pre-distribution schemes can be found in [14].

In order to establish pair wise secret keys among sensor nodes in large-scale sensor networks with better network connectivity, an improved version of identity-based random key establishment scheme proposed by Ashok Kumar Das is given in [1]. An over view of this scheme is provided in Section 3. In this paper, we attempt to use the sensor node deployment knowledge in key pre-distribution. We will show how this deployment knowledge can help to improve the efficiency of a identity-based key pre-distribution using a pseudo random function (IBPRF). In our study, we illustrate two node deployment strategies, random and grid and evaluate the identity based random key pre- distribution scheme, and then we provide a comparative analysis that show how the network connectivity probabilities can be improved for the key pre-distribution scheme with deployment knowledge.

To demonstrate our approach more analytically, we have simulated the node deployment strategies using the simulation tools (such as Network Simulator-2.34 [17] and Mannasim [18]). NS-2.34 is a discrete event network simulator that has been developed in 1989 as a variant of the REAL network simulator. Implementation and simulation of a network using NS-2 consists of 4 steps (1) Implementing the protocol by adding a combination of C++ and OTcl code to NS-2's source base (2) Describing the simulation in an OTcl script (3) Running the simulation and (4) Analyzing the generated trace files. NS-2 is a powerful tool for simulating ad-hoc networks. But to simulate WSNs in NS-2, we need to have an additional module namely mannasim to represent the protocols, that are specific to WSNs.

MANNASIM is a framework for WSNs simulation, and is based on NS-2. It introduces new modules for design, development and analysis of different WSN applications. An excellent survey on simulation tools is given in [20]. WSN simulations with mannasim for various deployment schemes have been explained in [19].

The rest of the paper is organized as follows. Section 2, discusses random and grid deployment schemes for WSNs. Section 3 analyses identity based random key pre-distribution scheme. Section 4, focuses on evaluation and comparative analysis of identity based random key pre-distribution scheme using deployment knowledge and Section 5 concludes our work.

2 Deployment Schemes for WSNs

In WSNs, knowledge about node deployment plays a very essential role, since an inappropriate node deployment may increases the complexity of communication, computations, energy consumption and reduces the network coverage in sensor networks. Deployment knowledge can be derived from the way the nodes are dropped on the required region. The sensors that are dropped next to each other have the highest probability of being close to each other. The importance of this deployment knowledge to improve the performance of key pre-distribution is explained in [14]. Hence, in many applications of WSNs, we consider the sensor node deployment to provide complete area coverage as an essential design problem. A study on various deployment schemes for WSNs can be found in [19]. In this work, we focus on two deployment schemes for sensor networks environments, Random Deployment and Grid Deployment.

2.1 Random Deployment

A proper node deployment scheme can be a very effective optimization means for achieving desired design problems in WSNs as, for example, routing, communication, etc. Furthermore, it can extend the network lifetime while minimizing energy consumption. An excellent survey of the current state of the research on optimized node placements in WSNs is provided in [21]. There are two ways to deploy a wireless sensor network: random and precise deployment. Random deployment means setting positions of wireless sensor nodes randomly and independently in the target deployment area. On the other hand, in precise deployment, nodes are set at exact positions one by one according to the communication range of the nodes.

Depending upon the application, type of sensors and the environment that the sensors will operate in, the decision can be made on the choice of the deployment scheme. Precise deployment is suitable for small-scale applications of range-finders, underwater acoustics, imaging and video sensors, etc. However, precise deployment method is time consuming even though costing the least number of sensor nodes. Random deployment is a feasible and practical method, and sometimes it is the only feasible strategy though, it costs a relatively larger number of nodes to achieve the deployment goal, especially for large-scale applications of WSNs. When practical application scenarios are considered, as, for example, in disaster recovery and forest fire detection etc., deterministic or precise deployment of sensors is very risky and/or infeasible. In such scenarios, randomized sensor placement often becomes the only option, where the sensors are dropped using some moving carriers (such as helicopter, grenade launchers or clustered bombs, etc.).

2.2 Grid Deployment

In a non-deterministic node deployment scheme, such as random deployment, excess redundancy is required to overcome the uncertainty in number of nodes needed to provide full network connectivity. Due to this, it could be very expensive. The number of nodes needed to provide full coverage and connectivity, when we have a second chance in random deployment is discussed in [5].

Previously many studies have explored the properties of grid deployment in the ideal circumstance where individual sensors are placed exactly at grid points. In [7], the authors have considered the situation where sensors are deployed in a battlefield in grid fashion. Grid deployment is conducted by dropping sensors row-by row using a moving carrier. The desired distance between consecutive droppings is achieved by controlling the time intervals. In WSNs, the area coverage is very important. Moreover in moderate to large-scale applications of WSNs, overall network coverage much depends on both the sensing ranges and the node deployment scheme. To achieve complete network coverage, grid deployment is considered as a good and attractive approach. The coverage performance of random and grid deployments is explained in [6]. There are three popular grid layouts namely, unit square, an equilateral triangle, a hexagon, etc. Among them, we simulated a square grid where nodes are placed at the corners and their sensing ranges intersections form a tessellation of the region as explained in [6]. However, in practice, it is often infeasible to guarantee exact placement due to various errors, including misalignments and random misplacement. The relation between the energy consumption and deployment strategy has been explained in [4].

3 Key Management

To provide security in WSNs, secret keys must be established among sensor nodes. Key management is a process of key setup, initial distribution of keys, and removal of a compromised key. Key distribution refers to distribution of multiple keys among the sensors. As sensors are resource constrained nodes, most deployed sensors may not be able to receive proper keys. Hence, key distribution is a difficult task in WSNs. A survey on key management and the three simplest keying models that are used to compare the different relationships between the WSN security and operational requirements are widely discussed in [22]. Moreover, due to the constraints of sensor node resources, traditional key management schemes are not applicable for wireless sensor networks. Hence, a novel approach has been proposed for WSNs, random key pre-distribution, where key information is distributed among all sensor nodes prior to deployment. A study on various random key pre-distribution schemes for WSNs is given in [8].

3.1 Random Key Pre-distribution

Many random key pre-distribution schemes were proposed for WSNs, as, for example Eschenauer and Gligor [9]; q-composite scheme [8]; Liu and Ning [10]; Zhu and Setia,[11]; where, for each sensor node, a subset of keys will be selected randomly from a large key pool. After being deployed, to perform secure communication, each node finds a common key with each of its neighbors. Though, the resilience of these schemes is perfect, these schemes are impractical for sensors having limited amount of memory. Moreover, once the network is deployed, addition of new nodes is a difficult task as the already deployed nodes do not have the keys for these new nodes. An overview on random key pre-distribution schemes for WSNs is provided in [1] [8]. In this section we study a novel random key pre-distribution scheme, proposed by Ashok Kumar Das, an identity based random key pre distribution for static sensor networks, which uses disjoint key pools approach and a pseudo random function. A detailed description for an identity based random key pre distribution scheme can be found in [1].

Motivation

In this section we first discuss the motivation behind choosing this key distribution scheme followed by the detailed analysis on various phases of the work. From the analysis of the direct key establishment phase of the basic random key distribution schemes (Eschenauer and Gligor [9] and q-composite scheme [8];), it has been identified that, to keep higher network connectivity, the key pool size should not be chosen larger. On the other hand, if the key pool size is small, an attacker could easily compromise all the nodes, there by the security degrades dramatically.

This problem can be solved with the identity based random key pre distribution scheme, where the direct key establishment phase can be completed within a short time duration. Hence, sensor nodes may be protected very well, so an attacker could compromise only a limited number of nodes during this short time period of the direct key establishment phase. Moreover, in the basic random pair wise keys scheme [8], addition of new nodes after initial deployment, introduces communication overhead. This problem can also be fixed with our current working scheme. A detailed description for the phases involved in this scheme is provided in [1].

The assumptions made by this scheme are summarized as below.

- Sensor nodes are static
- Base station will not be compromised
- Nodes can be protected during the network initialization period
- Uses an efficient pseudo-random function (PRF) (For example, as a PRF function proposed by Goldreich et al. in 1986 [13]).
- There exists a master key shared between each sensor node and the base station

This scheme further describes different phases such as.

- Key pre-distribution
- Direct key establishment
- Addition of nodes

3.2 Analysis of Identity Based Random Key Pre Distribution Scheme

In this section, we briefly describe and analyze the phases of identity based random key pre distribution scheme.

Key Pre Distribution

In this phase the key management server is responsible for generation of a large key pool K, of M size as shown in Fig. 2, where the key pool consists of 64 keys, i.e., from K_1 to K_{64}.

K_{40}	K_8	K_{48}	K_{16}	K_{56}	K_{24}	K_{64}	K_{32}
K_{39}	K_7	K_{47}	K_{15}	K_{55}	K_{23}	K_{63}	K_{31}
K_{38}	K_6	K_{46}	K_{14}	K_{54}	K_{22}	K_{62}	K_{30}
K_{37}	K_5	K_{45}	K_{13}	K_{53}	K_{21}	K_{61}	K_{29}
K_{36}	K_4	K_{44}	K_{12}	K_{52}	K_{20}	K_{60}	K_{28}
K_{35}	K_3	K_{43}	K_{11}	K_{51}	K_{19}	K_{59}	K_{27}
K_{34}	K_2	K_{42}	K_{10}	K_{50}	K_{18}	K_{58}	K_{26}
K_{33}	K_1	K_{41}	K_9	K_{49}	K_{17}	K_{57}	K_{25}

Fig. 2. Randomly generated key pool K

Once the generation of key pool is completed, the key management server selects two disjoint sub key pools (such as, K1 of size M1, and K2 of size M2), from the key pool K, as show in Fig. 3. (a) & (b) respectively. Since, K1 and K2 are disjoint i.e., $K1 \cap K2 = \Phi$, we can observe that, the sub key pools consists of 32 keys are also different from each other.

K_{56}	K_{24}	K_{64}	K_{32}
K_{55}	K_{23}	K_{63}	K_{31}
K_{54}	K_{22}	K_{62}	K_{30}
K_{53}	K_{21}	K_{61}	K_{29}
K_{52}	K_{20}	K_{60}	K_{28}
K_{51}	K_{19}	K_{59}	K_{27}
K_{50}	K_{18}	K_{58}	K_{26}
K_{49}	K_{17}	K_{57}	K_{25}

K_{56}	K_{24}	K_{64}	K_{32}
K_{55}	K_{23}	K_{63}	K_{31}
K_{54}	K_{22}	K_{62}	K_{30}
K_{53}	K_{21}	K_{61}	K_{29}
K_{52}	K_{20}	K_{60}	K_{28}
K_{51}	K_{19}	K_{59}	K_{27}
K_{50}	K_{18}	K_{58}	K_{26}
K_{49}	K_{17}	K_{57}	K_{25}

Fig. 3. a) Randomly chosen sub key pool K1; (b) Randomly chosen sub key pool K2

Now the key management server selects key sub sets from each sub key pools K1 and K2, for each node to be deployed in the target region, and assigns each node with an unique identifier. Thus, each node's memory consists of the following information in its memory as show in Table .1.

Table 1. Information resides in each node's memory

Node ID
Randomly chosen Key sub list of size m1 from Key Pool K1
Randomly chosen Key sub list of size m2 from Key Pool K2
PRF function

Here the m1 key units selected from the first key pool K1 will be used by the sensor nodes, deployed during the network initialization phase, in order to establish direct pair wise keys with maximum number of neighbor nodes. On the other hand m2 key units that are selected from the second key pool K2 are to be used by the newly deployed sensor nodes after the initial deployment in order to establish direct pair wise keys with already deployed nodes.

Direct Key Establishment

Let u and v be the two sensor nodes deployed in the required area. In this phase each sensor node's key ring (for example, let Ku1 be the key ring for sensor u), uses m1 key units, selected from the first key pool K1. After the sensor nodes have been deployed, each node broadcast its own ids, and also m1 key ids from their key rings to all its neighbors, in order to identify every neighbor node with which it shares at least one common key id. From these shared key ids, the corresponding keys will be considered as a set. Now, to generate a unique secret key for the communicating nodes, the key distribution scheme uses a secret one way function called pseudo random function (PRF) proposed by Goldreich et al. in 1986 [13]. For which, the inputs are, the two neighbor node's ids and the composite key formed by considering the XOR operation on the set of shared keys. Then, this secret key will be stored in the node's memory for its future communications. This secret key establishment phase explained in [1] can be depicted as in the below Fig.4.

To elaborate our analysis, let n0 and n1 be the two sensor nodes deployed in a target deployment area, which share at least one common key id. Now, by considering XOR operation, among these set of shared keys between the nodes n0 and n1, a composite key can be generated. Then, the two node's identifiers, and the generated composite key are given as inputs to the pseudo random function (PRF), to produce a unique pair wise secret key, which can be used by the two nodes for their future communications. As soon as each node establishes secret pair wise keys with its neighbor nodes, the m1 key units selected from first key pool K1 are deleted from the nodes key rings.

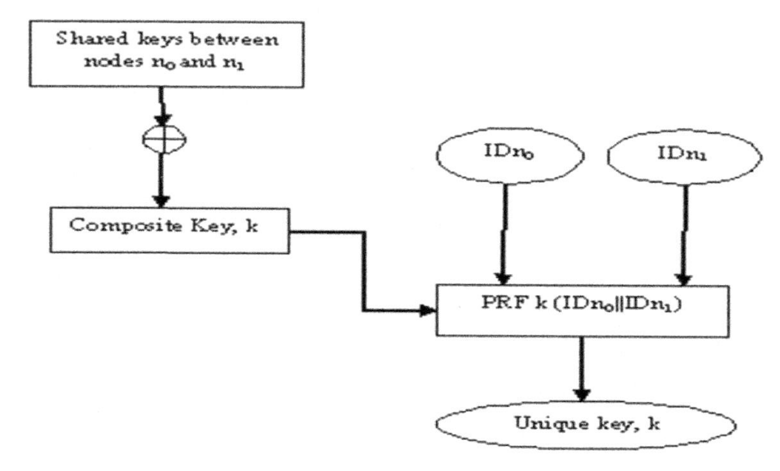

Fig. 4. Generation of secret pair wise key

Addition of Sensor Nodes

Sometimes it is necessary to redeploy some new sensor nodes to replace the faulty or compromised nodes. In such cases, the key management server selects a random subset key ring ku2 of size m2 from key pool K2 and also assigns a unique identifier IDu to this newly deployed node u. We note that the already deployed nodes have to use remaining m2 key units in their key rings in order to establish pair wise secret keys with newly deployed nodes.

4 Our Contribution

In this session, we present the deterministic evaluation of identity based random key pre-distribution mechanism described in the previous section for two basic node deployment strategies random and gird .Through this study we provide comparative analysis of these deployment schemes with respect to network performance in terms of connectivity.

4.1 Evaluation of Identity Based Random Key Pre-distribution Using Deployment Knowledge

In Fig. 5(a) & (b) we had simulated the deployment of 10 sensor nodes in random and 9 sensor nodes in grid fashion respectively. Here, for each sensor node ni to be deployed in the target deployment filed, let the key management server assigns a unique identifiers (for example, for the sensor nodes n0, n1, n2, n3 ….n9, the assigned identifiers are 0, 1, 2, 3,--9).

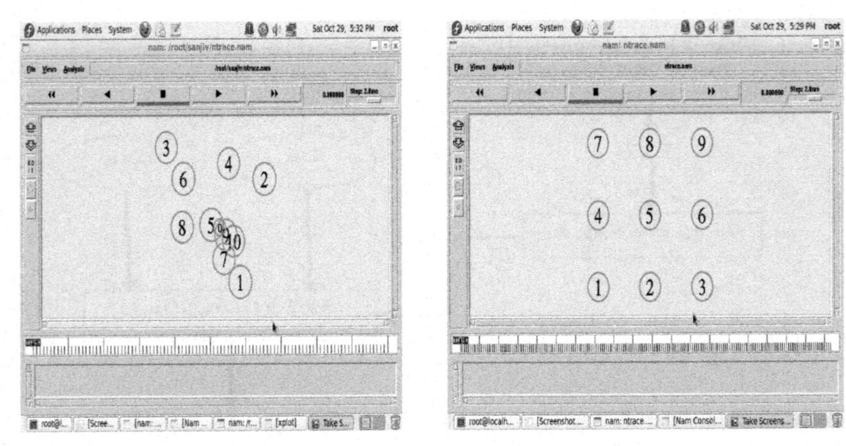

Fig. 5. (a) Simulation of WSN using Random Deployment ; (b) Simulation of WSN using Grid Deployment

Phase 1: Key Pre Distribution

Let n0 and n1 be the sensor nodes deployed in the target field.

For the node n0 the key management server assigns.

- A unique identifier, say, node 0.
- A random key subset say, $\{K_8, K_6, K_1, K_4, K_3, K_2, K_7\}$ from the first key pool K1.
- Another Random key subset say, $\{K_{45}, K_{48}, K_{51}, K_{53}, K_{58}, K_{42}, K_{40}, K_{46}\}$ from the second key pool K2.

Similarly, for node n1, the key management server assigns.

- A unique identifier, say, node 1.
- A random key subset says, $\{K_8, K_{21}, K_{40}, K_4, K_{33}, K_2, K_{11}. K_5\}$ from the first key pool K1.
- Another Random key subset say, $\{K_{40}, K_{60}, K_{62}, K_{53}, K_{51}, K_{64}, K_{59}, K_{56}\}$ from the second key pool K2.

Phase 2: Direct Key Establishment at Each Node n0 and n1

- For nodes n0 and n1, this phase uses first m1 key units say, $\{K_8, K_6, K_1, K_4, K_3, K_2, K_7\}$ and $\{K_8, K_{21}, K_{40}, K_4, K_{33}, K_2, K_{11}. K_5\}$ selected from the first key pool K1
- The shared keys identified from the above m1 key units are $\{K_8, K_4, K_2,\}$.
- Now the composite key can be calculated as follows: k= XOR $\{K_8, K_4, K_2\}$.
- Now secret pair wise key can be generated as : $K^1 = PRF_K(ID_0 || ID_1)$
- The nodes n0 and n1 uses the secret key for the future communication.

Phase 3: Addition of Sensors

- To establish direct key with newly added nodes in the network, this phase uses the second m2 key units selected from the second key pool of already deployed nodes ie $\{K_{45}, K_{48}, K_{51}, K_{53}, K_{58}, K_{42}, K_{40}, K_{46}\}$ and , $\{K_{40}, K_{60}, K_{62}, K_{53}, K_{51}, K_{64}, K_{59}, K_{56}\}$.

4.2 Comparative Analysis of Identity Based Random Key Pre-distribution Mechanism Using Deployment Knowledge

By considering the node deployment strategies along with their characteristics in large-scale wireless sensor networks [6], and by investigating the properties of identity based random key pre-distribution mechanism [1], in this section, we present a comparative analysis on the network connectivity, of identity based random key pre-distribution mechanism, using two deployment schemes random and grid.

We have observed the following interesting considerations during this process.

- In the basic identity based random key pre-distribution scheme [1], the researcher state that, if that if n1,n2 represent the deployed nodes and P1,P2 be the network connectivity probabilities during the network initialization and new nodes addition phases respectively, then the probability of overall network connectivity for the network of size n=(n1+n2) is given by P, Thus we have

$$P1 = 1 - \frac{\binom{M1 - m1}{m1}}{\frac{M1}{m2}}$$

$$P2 = 1 - \frac{\frac{M2 - m2}{m2}}{\frac{M2}{m2}}$$

$$P = \frac{n1\ P1 + n2\ P2}{n1 + n2}$$

- By investigating deployment schemes in [4][6] and basic random key pre-distribution schemes in [1][8], we note that the grid deployment scheme exhibits
- greater connectivity than the random deployment scheme, and the identity based random key pre-distribution scheme performs well than the remaining schemes.
- Moreover, in order to secure the communications among sensor nodes, knowing which sensors are close to each other is essential for key pre-distribution. Due to randomness of deployment, it is unrealistic to know the exact set of neighbors of each sensor.
- Therefore, if we use the identity based random key pre-distribution scheme, along with grid deployment, there is a chance of improvement in the probability of network connectivity, than using the key pre-distribution scheme all alone.

Thus based on our survey on deployment schemes and identity based random key pre-distribution, we present a comparative analysis with respect to network connectivity, in the following figures, Fig .6 and Fig. 7, by considering network initialization phase and new nodes addition phases respectively, as specified in [1].

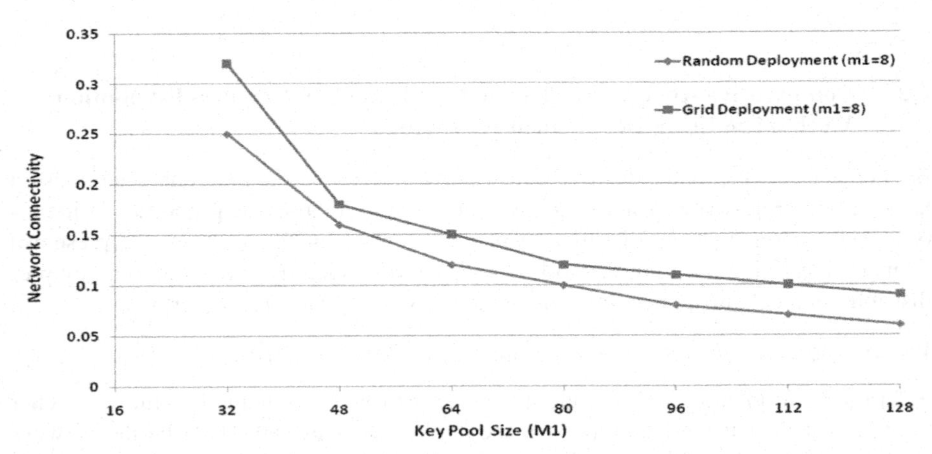

Fig. 6. Analysis vs. simulation results of network connectivity probabilities of Identity Based Random key pre distribution mechanism during the direct key establishment phase using grid and random deployments

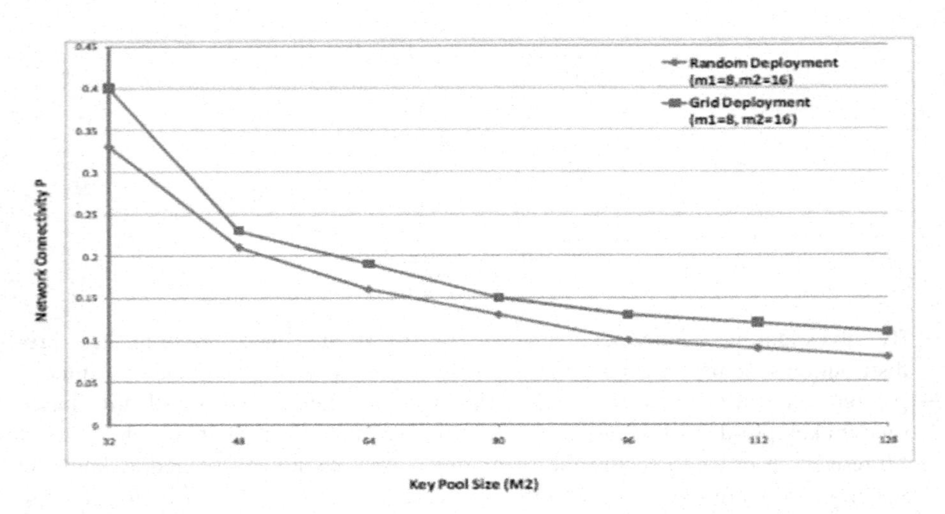

Fig. 7. Analysis vs. simulation results of overall network connectivity probabilities of identity based random key pre distribution mechanism using grid and random deployments

Here, we note that, during the direct key establishment phase with key ring size=8 and for various values of key pool of size M1 (for example, M1= 16, 32, 48, 64, 80, 96, 112, 128), the current working scheme exhibits greater network connectivity probabilities, when it is used with grid deployment than with random deployment. We also analyzed the overall network connectivity probabilities of the current scheme for various values of key pool M2, by assuming that, initially n1=10 nodes and later n2=5 nodes are deployed in the target field. In this case also, we can expect a better performance when the current working scheme is used with grid deployment knowledge than with random deployment.

5 Conclusion

In this paper, we have studied and analyzed an identity based random key pre distribution mechanism in static sensor networks using deployment knowledge. We observed that the analyzed scheme provides improved network connectivity than the existing random key pre-distribution schemes [8] and [9], as they exhibit security degradation problem. Moreover, we pointed out that, the grid deployment exhibits better network connectivity than random deployment. We evaluated the identity based random key pre-distribution mechanism for the two deployment schemes .Through the comparative and analytical results, we concluded that the probabilities of network connectivity in identity based random key pre-distribution mechanism can be expected to improve when the sensor nodes are deployed in grid fashion than in random deployment.

References

1. Das, A.K.: Improving Identity-based Random Key Establishment Scheme for Large-scale Hierarchical Wireless Sensor Networks. International Journal of Network Security 13(3), 181–201 (2011)
2. Wang, Du, Liu: ShortPK: A Short- Term Public Key Scheme for Broadcast Authentication in Sensor Networks. ACM Transactions on Sensor Networks 6(1), Article 9 (2009)
3. Crossbow Technology Inc., Wireless Sensor Networks(2010), `http://www.xbow.com`
4. Monica, Sharma, A.K.: Comparative Study of Energy Consumption for Wireless Networks based on Random and Grid Deployment Strategies. International Journal of Computer Applications (0975 – 8887) 6(1) (September 2010)
5. Yang, X., Wang, Y., Feng, W., Li, M., Liu, Y.H.: Random Deployment of wireless sensor Networks: Power of Second chance
6. Poe, W.Y., Schmitt, J.B.: Node Deployment in Large Wireless Sensor Networks: Coverage, Energy Consumption, and Worst-Case Delay. In: AINTEC 2009, Bangkok, Thailand, November 18-20 (2009)
7. Ruj, S., Roy, B.: Key predistribution using combinatorial designs for grid-group deployment scheme in wireless sensor networks. ACMTrans.Sensor Netw. 6(1), Article 4 (December 2009)
8. Chan, H., Perrig, A., Song, D.: Random Key Predistribution Schemes for Sensor Networks. In: IEEE Symposium on Security and Privacy, Berkeley, California, pp. 197–213 (2003)
9. Eschenauer, L., Gligor, V.D.: A Key Management Scheme for Distributed Sensor Networks. In: 9th ACM Conference on Computer and Communication Security, pp. 41–47 (November 2002)

10. Liu, D., Ning, P., Li, R.: Establishing Pairwise Keys in Distributed Sensor Networks. ACM Transactions on Information and System Security 8(1), 41–77 (2005)
11. Zhu, S., Setia, S., Jajodia, S.: LEAP+: Efficient Security Mechanisms for Large-Scale Distributed Sensor Networks. ACM Transactions on Sensor Networks 2(4), 500–528 (2006)
12. Akyildiz, I.F., Su, W., Sankarasubramaniam, Y.: A Survey on Sensor Networks IEEE Communications Magazine (August 2002)
13. Goldreich, O., Goldwasser, S., Micali, S.: How to construct random functions. Journal of the ACM 33(4), 792–807 (1986)
14. Du, W., Deng, J., Han, Y.S., Varshney, P.K.: A Key Predistribution Scheme for Sensor Networks Using deployment Knowledge. IEEE Transactions on Dependable and Secure Computing (1) (2006)
15. Neuman, B.C., Tso, T.: Kerberos: An Authentication Service for Computer Networks. IEEE Comm. 32(9), 33–38 (1994)
16. Perrig, R., Szewczyk, V., Wen, D.: Cullar, and J.D. Tygar, Spins: Security Protocols for Sensor Networks. In: Proc. Seventh Ann. ACM/ IEEE Int'l Conf. Mobile Computing and Networking MobiCom, pp. 189–199 (July 2001)
17. The Network Simulator, NS-2, http://www.isi.edu/nsnam/ns/
18. http://www.mannasim.dcc.ufmg.br/
19. Sanjiv Rao, G., Valli Kumari, V.: A Study on Various Deployment Schemes for Wireless Sensor Networks. In: Meghanathan, N., Chaki, N., Nagamalai, D. (eds.) CCSIT 2012, Part III. LNICST, vol. 86, pp. 495–505. Springer, Heidelberg (2012)
20. Lopez, Alonso, Sala, Marino, Haro: Simulation Tools for wireless sensor Networks
21. Younis, M., Akkaya, K.: Strtegies and techniques for node placement in wireless sensor networks: A survey. Ad Hoc Netw. 6(4), 621–655 (2008)
22. Lee, J.C., Leung, V.C.M., Wong, K.H., Cao, J., Chan, H.C.B.: Key management issues in wireless sensor networks: current proposals and future developments. IEEE Wireless Communications (October 2007)

Performance Evaluation of GMSK for Digital Mobile Radio Telephony

Gejo George[1,2] and Srija Unnikrishnan[2]

[1] Don Bosco Institute of Technology, Mumbai, India
[2] Father Conceicao Rodrigues College of Engineering, Mumbai, India
gejo84@gmail.com, srija@frcrce.ac.in

Abstract. GMSK is most notably used in the Global System for Mobile Communications (GSM). Performance of a cellular system is dependent on the efficiency of the modulation scheme in use. This paper evaluates the performance of GMSK for different channel conditions. The strong point of GMSK is its high spectral efficiency and power efficiency advantages which make it the preferred choice of modulation in fading mobile channels. Here, the GMSK system is implemented using SIMULINK and its performance under different channel models i.e. AWGN, Rayleigh and Rician fading channels has been evaluated using different analysis parameters such as BER (bit error rate), effect of change in BT, hard/soft decision decoding on system performance.

Keywords: BER, GMSK, AWGN, Fading channels, BT, Hard/Soft Decision Decoding.

1 Introduction

Mobile wireless systems operate under harsh and challenging channel conditions. Factors such as multipath and shadow fading, Doppler spread, etc. are all related to variability that is introduced by the mobility of the user and the wide range of environments that may be encountered as a result [1].

Modern communication system demands special attention to use primary resources like transmitted power and channel bandwidth efficiently [2][3]. Due to optimum requirement of bandwidth and power, the Gaussian Minimum Shift Keying (GMSK) is well suited for mobile communication [2][4].

GMSK implemented in GSM uses the concept of differential encoding which offers the advantage that information is carried in the phase changes, rather than in the phase itself [2][3]. Knowing the absolute phase becomes unnecessary. A GMSK signal can easily be detected by an orthogonal coherent detector which is exactly same as for classical MSK [3][4].

2 System Implementation

2.1 GMSK Model

GMSK has high spectral efficiency and provides a straightforward, spectrally efficient modulation method for wireless data transmission system [5]. GMSK is most notably

S. Unnikrishnan, S. Surve, and D. Bhoir (Eds.): ICAC3 2013, CCIS 361, pp. 337–345, 2013.

used in the Global System for Mobile Communications (GSM) [5]. Figure 1 shows the block diagram of implementation of GMSK system model in SIMULINK. This paper models a system that includes convolutional coding and GMSK modulation. The receiver in this model includes two parallel paths, one that uses soft decisions and another that uses hard decisions. The model uses the bit error rates for the two paths to illustrate that the soft-decision receiver performs better. This is what you would expect, because soft decisions enable the system to retain more information from the demodulation operation to use in the decoding operation.

The Error Rate Calculation block compares input data from the transmitter with input data from the receiver. It calculates the error rate as a running statistic, by dividing the total number of unequal pairs of data elements by the total number of input data elements from one source. Using BERTool from SIMULINK, we observe the performance of the GMSK system under different channel conditions.

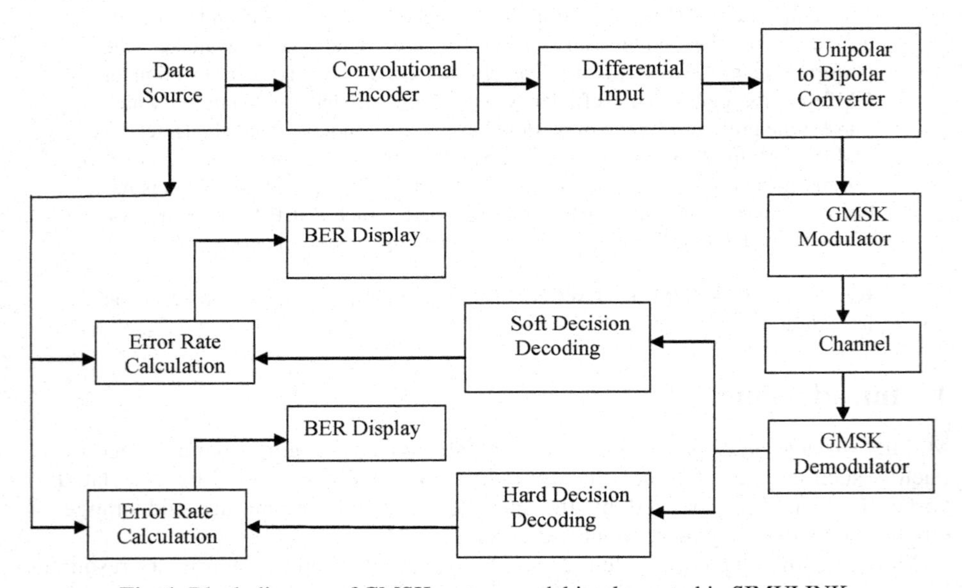

Fig. 1. Block diagram of GMSK system model implemented in SIMULINK

2.2 Key Components of the SIMULINK Model

2.2.1 Generating Binary Data Stream

By using Bernoulli binary generator block in the communication tool box, we can generate binary data stream [7]. By adjusting the parameters like initial seed, sample time and output data type, we can achieve fixed binary stream.

2.2.2 Convolutional Encoder

The Convolutional Encoder block encodes a sequence of binary input vectors to produce a sequence of binary output vectors. This block can process multiple symbols at a time. The convolutional encoder is defined by using the Trellis structure parameter. The operation mode is set to continuous mode where, the block retains the encoder states at the end of each frame, for use with the next frame.

2.2.3 Differential Input

The Differential Encoder block encodes the binary input signal. The output is the logical difference between successive bits.

2.2.4 Unipolar to Bipolar Converter

The Unipolar to Bipolar Converter block maps the unipolar input signal to a bipolar output signal. This block is set by adjusting the parameters such as M-ary number, polarity and output data type. When the parameter is set to its default setting, Inherit via internal rule, the block determines the output data type based on the input data type. If the input signal is floating-point (either single or double), the output data type is the same as the input data type [7]. The input to the GMSK modulator is of integer type. For integer input, the data type must be double or signed int. Therefore, the output data type of this block is set to double.

2.2.5 GMSK Modulator

The GMSK Modulator Baseband block modulates using the Gaussian minimum shift keying method. The output is a baseband representation of the modulated signal [7]. The BT product parameter represents bandwidth multiplied by time. This parameter is a nonnegative scalar. It is used to reduce the bandwidth at the expense of increased intersymbol interference. Here, we evaluate the performance of GMSK system for different values of BT (0.25, 0.3 and 0.5). The Pulse length parameter measures the length of the Gaussian pulse shape, in symbol intervals. The Phase offset parameter is the initial phase of the output waveform measured in radians and is set to 0 in our case. The input type parameter is set to Integer as the block accepts values of 1 and -1. This block can output an upsampled version of the modulated signal. The Samples per symbol parameter is the upsampling factor.

2.2.6 Channels

Communication channels introduce noise, fading, interference and other distortions into the signals that they transmit. Simulating a communication system involves modeling a channel based on mathematical descriptions of the channel. Different transmission media have different properties and are modeled differently. In a simulation, the channel model usually fits directly between the transmitter and receiver.

AWGN

The term additive means the noise is superimposed or added to the signal where it will limit the receiver's ability to make correct symbol decisions and limit the rate of information [6]. An AWGN channel adds white Gaussian noise to the signal that passes through it. This block uses the Signal Processing Blockset™ Random Source block to generate the noise. The Initial seed parameter in this block initializes the noise generator. The Initial seed parameter specifies the initial seed for the pseudorandom number generator. The generator produces an identical sequence of pseudorandom numbers each time it is executed with a particular initial seed. The port data types are inherited from the signals that drive the block.

Fading Channels

The Multipath Rayleigh Fading Channel block implements a baseband simulation of a multipath Rayleigh fading propagation channel. This block is useful for modeling mobile wireless communication systems. This block accepts only frame-based complex signals at its input. Relative motion between the transmitter and receiver causes Doppler shifts in the signal frequency. Since a multipath channel reflects signals at multiple places, a transmitted signal travels to the receiver along several paths that may have different lengths and hence different associated time delays. Fading occurs when signals traveling along different paths interfere with each other. The block multiplies the input signal by samples of a Rayleigh-distributed complex random process [7]. The scalar Initial seed parameter seeds the random number generator.

The Multipath Rician Fading Channel block implements a baseband simulation of a multipath Rician fading propagation channel. This block is useful for modeling mobile wireless communication systems when the transmitted signal can travel to the receiver along a dominant line-of-sight or direct path. Fading causes the signal to spread and become diffuse. The block multiplies the input signal by samples of a Rayleigh-distributed complex random process. The scalar Initial seed parameter seeds the random number generator.

2.2.7 GMSK Demodulator

Gaussian Filter block filters the input signal provided by the pulse generator using a Gaussian FIR filter. The block expects the input signal to be upsampled, so the Input samples per symbol parameter, N, is set to 8 (same as the GMSK modulator block). The block's icon shows the filter's impulse response [6]. From the point of view of performance evaluation, the parameter BT in the modulator and the demodulator is varied. The simulation results show the effect of change in BT on the system performance for the following three cases: BT = 0.25, 0.3 and 0.5. The complex phase shift block shifts the phase of the complex input signal which is fed to the digital filter alongwith the output of the Gaussian filter.

The Digital Filter block independently filters each channel of the input signal with a specified digital IIR or FIR filter. The output frame status and dimensions are always the same as those of the input signal that is filtered. When inputs are frame based, the block treats each column as an independent channel; the block filters each column. When inputs are sample based, the block treats each element of the input as an individual channel. The Complex to Real-Imaginary block accepts a complex-valued signal from the digital filter and outputs the real part of the input signal, when the output parameter is set to real. This output is then downsampled by a factor 8 (same as the upsampling factor).

2.2.8 Soft and Hard Decision Decoding

The section labeled Soft Decisions uses an eight-region partition in the Quantizing Encoder block to prepare for 3-bit soft-decision decoding using the Viterbi Decoder block. The section labeled Hard Decisions uses a two-region partition to prepare for hard-decision Viterbi decoding. Using a two-region partition here is equivalent to having the demodulator make hard decisions. In each decoding section, a Delay block aligns codeword boundaries with frame boundaries so that the Viterbi Decoder block can decode properly. This is necessary because the combined delay of other blocks in the system is not an integer multiple of the length of a codeword.

2.2.9 Error Rate Calculation

The Error Rate Calculation block compares input data from the transmitter with input data from the receiver. It calculates the error rate as a running statistic, by dividing the total number of unequal pairs of data elements by the total number of input data elements from one source. This block produces a vector of length three, whose entries correspond to: the error rate the total number of errors, that is, comparisons between unequal elements and the total number of comparisons that the block made [7]. The display blocks show BER of the system with each type of decision for different values of BT for the three channel models.

2.2.10 BERTool Environment

BERTool is an interactive GUI for analyzing communication systems' bit error rate (BER) performance [7]. Eb/No versus BER graph is plotted using Bit Error Rate Analysis Tool in MATLAB/SIMULINK. Monte Carlo simulation results for different channel conditions are shown in the results.

3 Results

3.1 Eb/No versus Bit Error Rate

Eb/No versus BER graph is plotted using Bit Error Rate Analysis Tool in MATLAB/SIMULINK for the following three values, i.e. BT=0.25, 0.3 and 0.5.

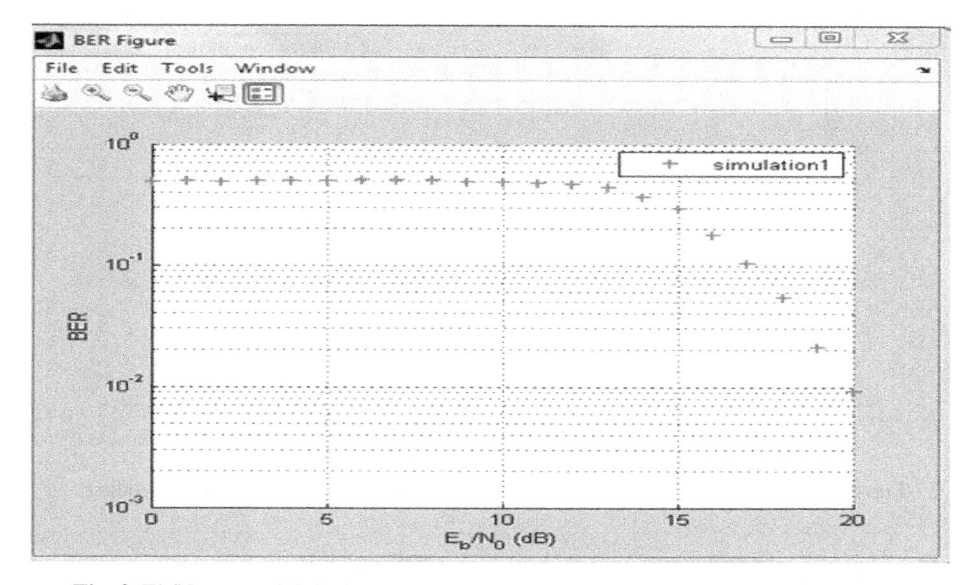

Fig. 2. Eb/No versus BER plot - AWGN channel (BT = 0.25) Hard Decision Decoding

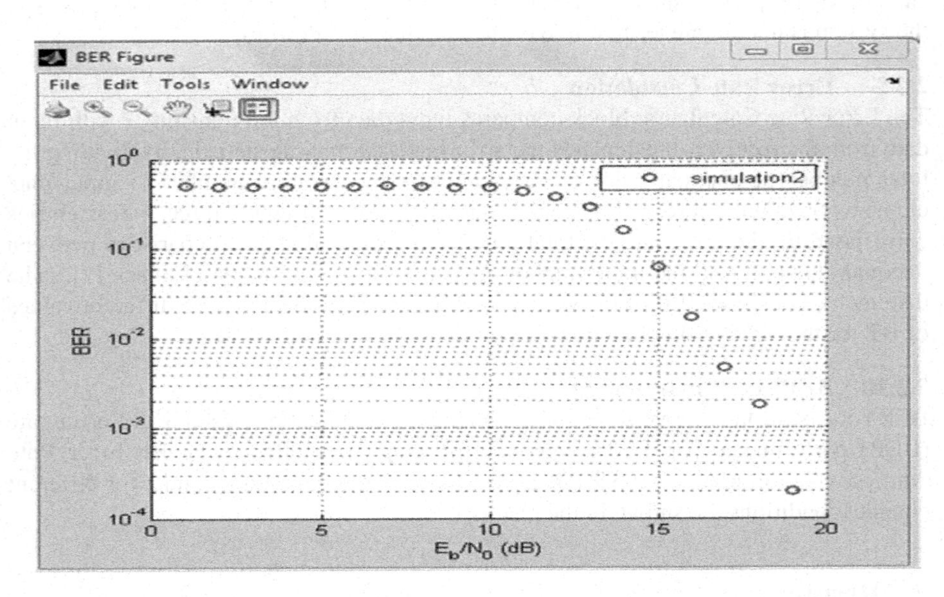

Fig. 3. Eb/No versus BER plot - AWGN channel (BT = 0.25) Soft Decision Decoding

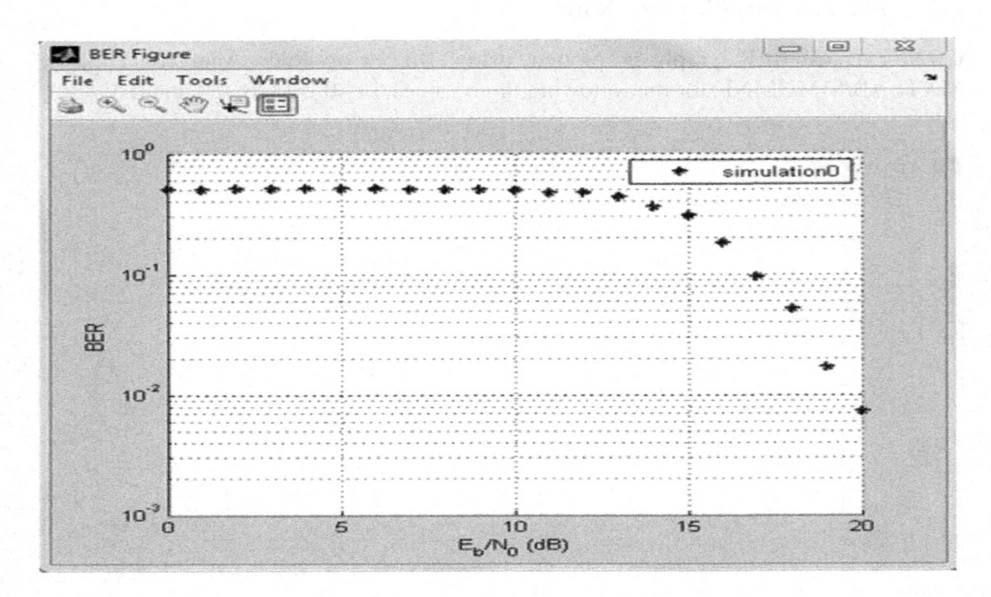

Fig. 4. Eb/No versus BER plot - AWGN channel (BT = 0.3) Hard Decision Decoding

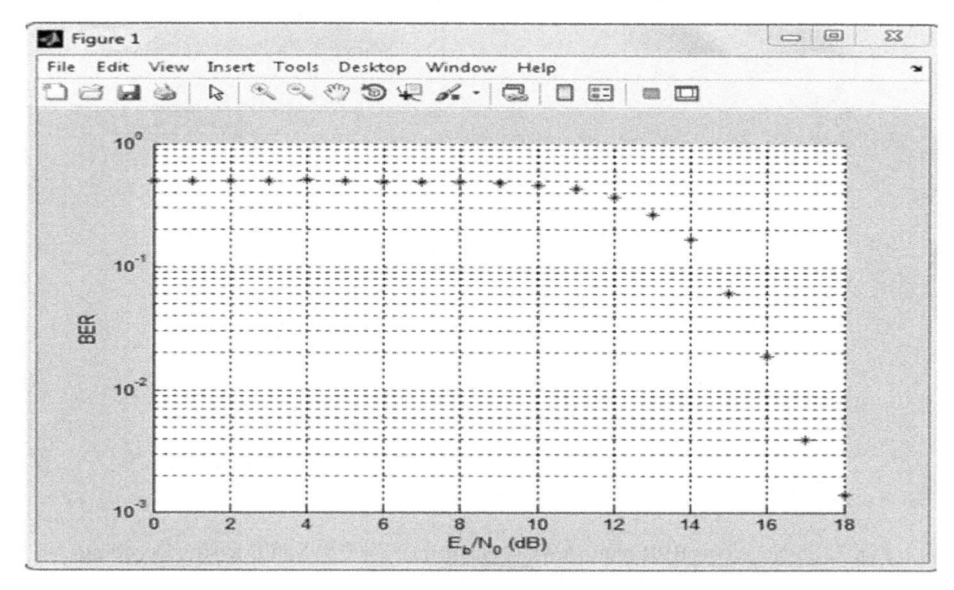

Fig. 5. Eb/No versus BER plot - AWGN channel (BT = 0.3) Soft Decision Decoding

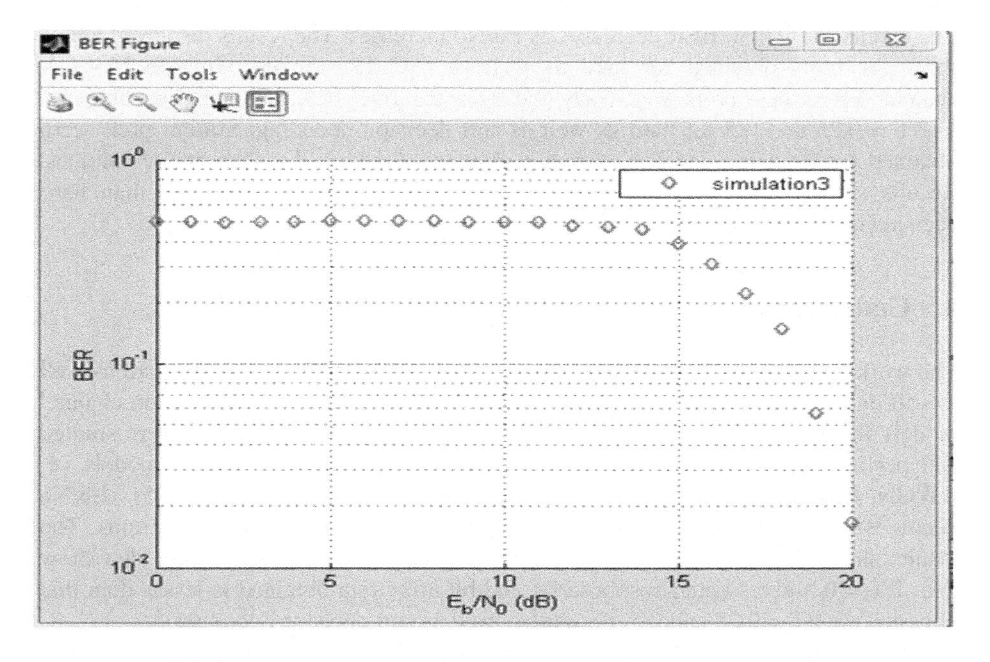

Fig. 6. Eb/No versus BER plot - AWGN channel (BT = 0.5) Hard Decision Decoding

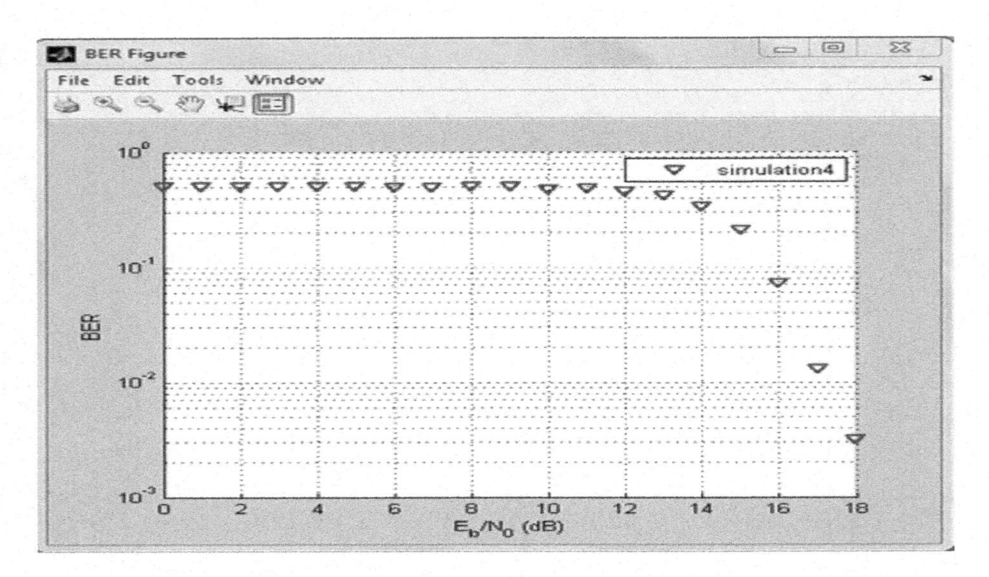

Fig. 7. Eb/No versus BER plot - AWGN channel (BT = 0.5) Soft Decision Decoding

3.2 Observations

The results show that, BER decreases as Eb/No increases. The results displayed above show the plots obtained for hard as well as soft decision decoding for AWGN channel. From these plots we can say that BT = 0.3 gives better response as compared to BT = 0.25 and 0.5 for hard as well as soft decision decoding. Similar plots were obtained for Rayleigh and Rician fading channels for both the decoding techniques. Results showed that, soft decision decoding gives lesser bit error rate than hard decision decoding for AWGN as well as fading mobile channels.

4 Conclusion

The work presented here, gives the implementation of GMSK system for hard as well as soft decision decoding, using MATLAB/SIMULINK. The communication channel models for ideal (AWGN) and worst case (multipath fading) channels were studied and performance of the system is evaluated for all the three channels models i.e. AWGN, Rayleigh and Rician, for different values of BT (0.25, 0.3 and 0.5). Eb/No versus Bit Error Rate (BER) plots are obtained using Monte Carlo simulations. The results show that, BER decreases as Eb/No increases. Simulation results also show that, BT = 0.3 gives better response i.e. the bit error rate obtained is lesser than that obtained for BT = 0.25 and 0.5, for hard as well as soft decision decoding.

The system gives better performance under AWGN channel as compared to multipath fading channels.

Acknowledgment. I wish to acknowledge with deep gratitude the valuable guidance received from ***Dr. Srija Unnikrishnan*** who has encouraged me throughout this venture. I am also thankful to the management of Father Conceicao Rodrigues College of Engineering. Mumbai. India for providing all the facilities and help required from time to time.

References

1. Paizi, W.F.B.: BER Performance Study of PSK Based Digital Modulation Schemes in Multipath Fading Environment. Faculty of Electrical Engineering. University Technology, Malaysia (June 2006)
2. Laster, J.D.: Robust GMSK Demodulation Using Demodulator Diversity and BER Estimation. Doctor of Philosophy in Electrical Engineering, Virginia Polytechnic Institute and State University, Blacksburg, Virginia (March 1997)
3. George, G., Unnikrishnan, S.: BER Performance Evaluation of GMSK for Fading Mobile Channels. International Journal of Engineering Research and Applications (IJERA) 2(3), 375–380 (2012)
4. Murota, K., Hirade, K.: Spectrum Efficiency of GMSK Land Mobile Radio. IEEE Transactions on Vehicular Technology vt-34(2) (May 1985)
5. Dalwadi, D.C., Soni, H.B.: Performance Evaluation of GMSK Modulated Signal Under Rician Channel Model. In: 2010 IEEE International Conference on Computational Intelligence and Computing Research, ICCIC, December 28-29 (2010)
6. Tahir, A.A., Zhao, F.: Performance Analysis on Modulation Techniques of W-CDMA in Multipath Fading Channel. Blekinge Institute of Technology (January 2009)
7. MathWorks India, `http://www.mathworks.in/`

Reducing Interference in Cluster Based Sensor Network Using FDMA-CDMA Technique

Sankar Mukherjee[1] and G.P. Biswas[2]

[1] Durgapur Institute of Advanced Tech. & Mgt., Durgapur, West Bengal, India
sankar_mukherjee2000@yahoo.co.in
[2] Indian School of Mines, Dhanbad, India
gpbiswas@gmail.com

Abstract. The performance of Code Division Multiple Accesses in sensor networks is limited by the collisions due to Multiple Access Interference (MAI). In this article, we propose a frequency division technique to reduce the MAI in a DS-CDMA sensor network. Our proposal also reduces the energy consumption of the network. In the model, first a new clustering technique is used over several numbers of randomly deployed sensor nodes in a square area. The whole area is further divided virtually into some small squares which actually form the clusters. Each cluster then uses FDMA and CDMA combinedly to reduce the MAI. Simulation is done for the proposed system and compared it with other systems, which do not use frequency division. The study found that, by using few number of frequency channels, the MAI can be reduced significantly. The system also has less channel contention, and lower energy consumption.

Keywords: FDMA, CDMA, MAI.

1 Introduction

A wireless sensor network consists of a number of sensor/actuator devices that can sense the environment, calculate/aggregate data and transmit the aggregated data to nodes within its transmission range. Every node has hardware for sensing, microprocessors for computation and low-power communication radios for transmission. The sensor nodes are normally battery operated. Hence, they have very limited energy. They also have limited memory to buffer data packets. The contention based medium access (MAC) protocols are obviously not a good choice for this network. The RTS/CTS control packets usually employed in contention based protocol produces significant overhead. According to Woo and Culler [2] this overhead is up to 40% in small packet size sensor network. Although IEEE 802.11 standard specifies that RTS/CTS can be avoided with small data packet transmission, this may not be a suitable choice for sensor networks. The IEEE 802.11 network has a data rate of 2 MB/s where as sensor network usually have data rate of around 20 KB/s. A packet will take much longer time in sensor network compared to IEEE 802.11 networks. Hence the probability of collision is quite high in sensor network. Hence the proposal made in IEEE 802.11 network that for small packet RTS-CTS packet can be avoided

S. Unnikrishnan, S. Surve, and D. Bhoir (Eds.): ICAC3 2013, CCIS 361, pp. 346–355, 2013.

is not suitable in sensor network. CDMA may come very handy in sensor network application.

CDMA is based on spread spectrum (SS) techniques, in which each user occupies the entire available bandwidth. At the transmitter, a digital signal of bandwidth, say B1 bits/s, is spread using (i.e., multiplied by) a pseudo-random noise (PN) code of bandwidth, say B2 bits/s (B2/B1>>1 is called the processing gain). The PN code is a binary sequence that statistically satisfies the requirement of a random sequence, but that can be exactly reproduced at the intended receiver. Using a locally generated PN code, the receiver de-spreads the received signal, recovering from it the original information. The enhancement in performance obtained from spreading the signal makes it possible for several, independently coded signals to occupy the same channel bandwidth, provided that each signal has a distinct PN code. This type of communication in which each transmitter-receiver pair has a distinct PN code for transmitting over a common channel is called code division multiple access [16]. Interference occurs in the CDMA based system basically due to two types of collisions, primary and secondary. For example, consider two non-neighboring nodes A and C that have two different codes. These nodes may have a common neighbor, say B, with its own code. A primary collision may occur if nodes A and C simultaneously attempt to transmit to node B using B's code. Primary collision can be avoided by allocating proper code to the nodes and we have done the same to overcome it. However, the nonzero cross-correlations between different CDMA codes can induce multi-access interference (MAI), resulting in secondary collisions at a receiver (collisions between two or more transmissions that use different CDMA codes).

The major problem in using CDMA is multiple access interference (MAI). In a cellular DS-CDMA network there is a central base station. This station controls the MAI by controlling the transmission power of the active nodes. The received power from all active nodes at the base station is the same. But in the case of sensor network there is no central base station. So it is really difficult to control the MAI. Consider the situation in Fig 1, where sensors are randomly deployed. Here Rx represents the communication range. Each node has a number of neighbors situated at different distances. For example, A has neighbors B, E, F, D, and G, with each having different distance to A. Assume that each node uses the minimum required power to communicate with each other. When A is transmitting to a neighbor, the interference power caused by this transmission at other neighbors can have different values. Considering two simultaneous transmissions from A to B and C to D, where distance $L_{AB} >> L_{AD}$, the interference power at D caused by the closer neighbor A is much higher than that of the desired power from C and this makes the desired signal difficult to be recovered. However, if instead of from A to B, the transmission is from A to E the interference caused to D's reception is negligible. The problem caused due to interference signal(s) makes desired signal go down at a receiver is an effect of MAI. So it is found that MAI can't be reduced by using the power control in a CDMA based sensor network. The MAI may cause significant degradation in network throughput and is considered the main problem prohibiting the usage of CDMA in sensor networks.

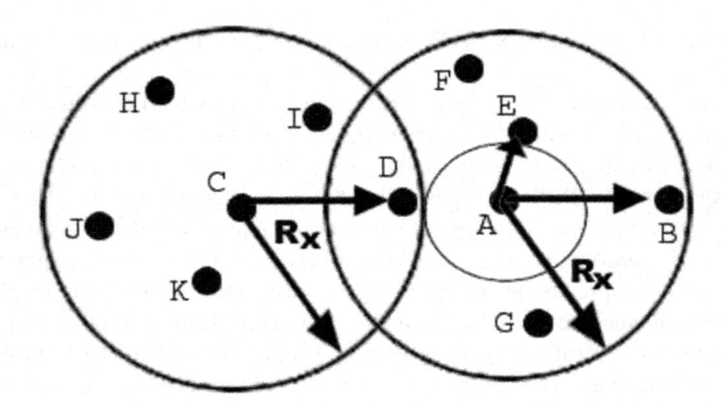

Fig. 1. Randomly deployed sensor network

In this paper FDMA and clustering are used both for reducing the MAI in CDMA based sensor network. The rest of the paper is organized as follows. Section 2 outlines proposals made by researchers for controlling MAI in CDMA-based sensor network. In section 3, we describe our system model and describe a technique for reducing the MAI in the system. Section 4 provides Experimental results and finally we sum up and conclude in section 5.

2 Related Works

Dow, Lin, and Fan [7] used DS-CDMA over cluster-based wireless networks to avoid the hidden node problem. It is well known that energy consumption is the crucial factor in sensor network design. This may lead to sensor network MAC protocols which prioritize energy savings over network throughput and packet latency. The multi-user, multiple access interference (MAI) environment of DS-CDMA introduces significant challenges on how interference can be properly controlled. Code assignment to the nodes in each cluster is done by the clusterhead in two phases. Muqattash and Krunz [1] proposed a CDMA-based MAC protocol for wireless ad hoc networks where out-of-band RTS/CTS are used to dynamically bind the transmission power of a node in the vicinity of a receiver. Both RTS and CTS are enlarged to accommodate MAI related information. However, our design goal is to reduce MAI by using FDMA and clustering technique. Liu, Chou, Lipman, Jha [6] proposed using frequency division to reduce the MAI in a DS-CDMA sensor network. They provide theoretical characterization of the mean MAI at a given node and show that a small number of frequency channels can reduce the MAI significantly. Each pair of nodes are assigned different frequency channel so that no interference occurs in the neighboring nodes. There is no clustering concept is used in this paper.

3 System Model

Our system consists of several identical sensor nodes deployed in a region along with two sink nodes that accumulates the sensed data. The sink nodes are powerful laptops or personal computers. All the sensor nodes are deployed randomly in a square shaped area ABCD (Fig 2) where the sink nodes are placed in two corners A and B of the ABCD square field. Here we have used a new position based clustering technique to form clusters.

3.1 Clustering Technique

Assume the length of the each side of the square is L and distance of the diagonal AC and BD is $L\sqrt{2}$. The maximum transmission range of the sink nodes is set to $L\sqrt{2}$ so that every sensor nodes can listen to all the sink nodes. The whole area has been divided into some virtual small squares and size of each side of the small squares is l. Each small square forms a cluster. So numbers of clusters are equal to the numbers of small squares in the whole field. Initially all the sink nodes will broadcast a hello signal only once with a maximum transmission range $L\sqrt{2}$ so that every sensor nodes can listen them. To avoid the collision the sink nodes will send the signal with a fixed

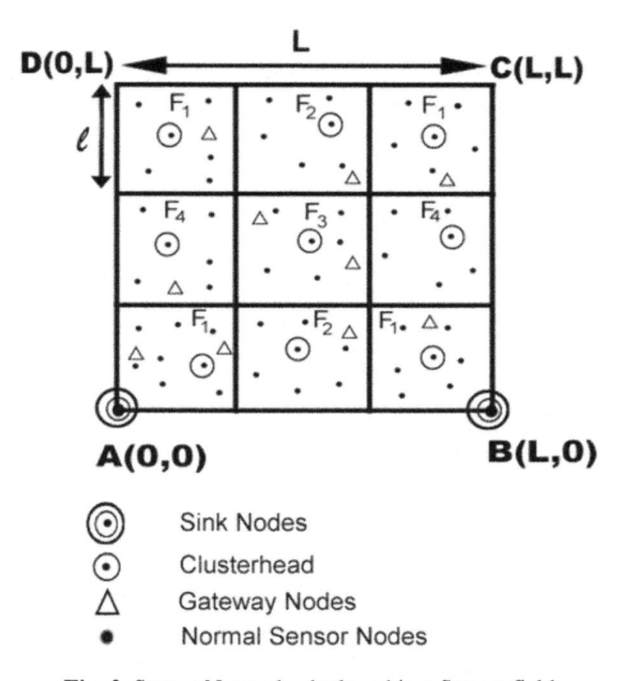

Fig. 2. Sensor Networks deployed in a Square field

time gap τ. The use of this broadcasting is to get the position of every sensor nodes. From the figure 3 it is clear that by considering the co-ordinate of A as (0,0) ,we can find out the position (x,y) of every sensor nodes. A sensor node received two hello signals from two sink nodes which are placed at A and B. From the received signal strength the node can estimate the distance d_1(AP) and d_2(BP) from the two respective sink nodes. Then the other parameter can be calculated accordingly.

$$\theta = \cos^{-1}\left[\frac{\left(d_1^2 + d_3^2 - d_2^2\right)}{2d_1 d_2}\right] \tag{1}$$

$$x = d_1 \cos\theta \tag{2}$$

$$y = d_1 \sin\theta \tag{3}$$

Where $d_3=L$, Once the sensor nodes calculate their position, they can identify the square they belong. As each square is a cluster so each node knows the cluster he belongs by using the position parameter(x, y). Node which belongs to the boundary of more than one squares, chooses any one as his own cluster. Every node in each cluster will broadcast their position parameter (x, y) with a maximum transmission range $l\sqrt{2}$.

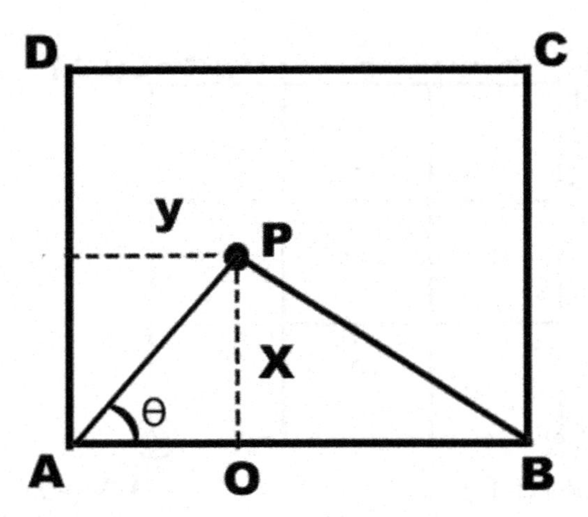

Fig. 3. Calculation of coordinate(x,y) of sensor node at P using the received signal strength from the sink nodes A and B

So each nodes in a cluster will get the position information of all other nodes in the same cluster. The node which is closest to the centre of the cluster will choose own-self as clusterhead. In case of tie, the node having lower id number will be the clusterhead. The clusterhead will broadcast this information to all members of the cluster.

3.2 FDMA-CDMA Technique

Each and every node is a member of one and only one cluster. Every normal member node communicates through their clusterheads. Intra-cluster communication of the sensor nodes are totally handled by the clusterheads. Inter-cluster communications involves two clusters and so are done through the gateway nodes. Every clusterhead aggregates the data received from his cluster members and send it to the sink node through different clusterheads. So to reach the sink node, conventional shortest path routing algorithm through clusterheads and gateways is chosen. A gateway is a node that directly connects to more than one clusterheads. Every clusterhead will assign different pseudorandom codes to its members and itself. These codes can be reused in other clusters. In the cluster based CDMA system two types of MAI can occur.

- Primary MAI: Inside clusters MAI caused at the clusterhead by the simultaneous transmission of its cluster members.
- Secondary MAI: MAI caused at the clusterhead, gateway node and the intermediate node due to simultaneous transmission of other nodes belonging to other neighbor clusters

We now elaborate on the performance implications of the MAI problem. Consider the reception of a packet at terminal i.. Let P_0^i be the average received power of the desired signal at the *i*th terminal. Suppose that there are K interfering transmissions with received powers P_j^i ,$j= 1, \ldots ,K$. The quality of the intended reception is adequately measured by the effective bit energy-to-noise spectral density ratio at the detector, denoted by μ^i. For an asynchronous direct-sequence BPSK system, μ^i is given by [1, 5, 6]

$$\mu^i = \frac{E_b}{N_{oeff}} = \left(\frac{2\sum_{j=1}^{k} P_j^i}{3WP_0^i} + \frac{1}{\mu_0} \right)^{-1} \tag{4}$$

Where W is the processing gain and μ_0 is the E_b/N_{0eff} ratio at the detector in the absence of interference. As the interfering power increases, μ^i decreases, and the bit error probability increases.

Primary MAI can be solved by the clusterhead itself by synchronizing all its member nodes. It is quite easy to do all member nodes are in directly connected to clusterhead. To mitigate the problem of secondary MAI, we have proposed a new technique which is a combination of FDMA and CDMA. Because MAI is caused by the non-perfect orthogonality of CDMA codes, the rationale of the design is to orthogonalize the reception in the vicinity of a sensor node by using frequency division. As most sensor network applications normally operate with low data rate, it is possible to use a narrow band CDMA system. Let's assume that data rate of the application is 20Kbps, and we use 50 chip/bit Pseudorandom codes (PN) to spread the baseband signal. The resulting bandwidth requirement is 1MHz. With 2.4GHz ISM band (2400-2483.5MHz) we can have more than 80 similar frequency channels. So if different frequency band is assigned to each cluster then there will not be any MAI due to transmission of other nodes belonging to its neighbor clusters.

3.3 Frequency and Code Assignment

According to the clusters formed only 4 frequency bands are sufficient. This is similar to graph coloring problem. The constraint is that the neighbor clusters will not use same frequency band. Assigning the frequency band to each cluster (each small square) in a proper sequence requires only four frequency bands (F1, F2, F3, F4). So the reuse factor is 4. If 80 MHz ISM band is divided into 4 equal sub bands then these can be assigned to the clusters and each will get 20 MHz band for communication. Gateway nodes will operate with more than one frequency band for intercluster communication.

In each cluster clusterhead assigns unique CDMA code to his cluster members and himself. Here the numbers of nodes in each cluster should not be more than the available codes. In that case we have to increase the number of chip in each code, which reduces the throughput. Every cluster will use the same code set. As the neighbor clusters use different frequency band and each clusterhead synchronizes his members so there will not be any MAI in the system.

Though the frequency band used by each cluster is reduced to 1/4th of the whole ISM band still it is sufficient for low data rate sensor node communication and the system will be totally collision free. On the other hand total energy consumption will be reduced a lot. If all clusters use the same frequency then due to MAI, in a cluster, data transmission will collide with transmission from nodes of other neighboring clusters. More collision and hence more retransmission means more energy consumption. Even by simulation it can be shown that if all clusters use same frequency of 80 MHz then as the load increases the successful transmission will decrease a lot and it is worse than the proposed model even at medium load.

4 Experimental Results

This section introduces a simulation environment used in the experiments. Simulation has been done on finite random CDMA based sensor networks with and without using the proposed FDMA scheme, by varying the load (packets generated in the system) and number of sensor nodes. The nodes are placed on a 1000 x 1000 area randomly. The network is fully connected one. Simulation is done using OMNet++. From equation (4), as the interfering power increases, μ decreases, and the bit error probability increases. As an example in equation (1), consider a CDMA system that uses BPSK modulation and a convolutional code with rate 1/2, constraint length 7, and soft decision Viterbi decoding. Let W = 100. To achieve a bit error probability of 10^{-6}, the required E_b/N_0eff is 5.0 dB. Ignoring the thermal noise and using (1), the total interference power must satisfy:

$$\frac{\sum_{i=1}^{k} P_i}{P_0} \leq 47.43 \tag{5}$$

Using this above equation in the simulation, it is found that, how much data will be successfully transmitted under varying load applied as well as varying numbers of nodes in the network. We assume that the transmission power attenuates with the distance d as k/d^n (k is a constant and n>>2 is the loss factor). Here we have considered the packet size is 500B and CDMA data rate is 1.6Mbps. From the Fig. 4 it is found that as the load (packets transmitted by different sensor nodes) increases for time duration, packet loss will be more due to MAI occurred in CDMA system. Same result (Fig.5) occurred when we increase the number of nodes in the network. As the density of node increases, MAI also increases. Here successful transmission means numbers of packets received by the receiver without any interference .As a result successful transmission will be reduced. In both the cases either nodes increases or load increases, interference will increase. But if we use FDMA with CDMA we can avoid the interference due to MAI, which is shown in both cases (Fig4 and Fig5). As the neighbor clusters use different frequency band so there will not be any interference due to other nodes in neighboring clusters. From Simulation result (Fig.6) it is observed that when CDMA is used without FDMA, network throughput will degrade rapidly. Here we have varied the number of packet generation λ (considering the packet generation followed in the system is Poisson distribution). When λ increases, the probability of numbers of simultaneous transmission will increase. So the MAI will be increased. But in the proposed FDMA-CDMA system, no MAI will occur and the throughput will reach its maximum gradually.

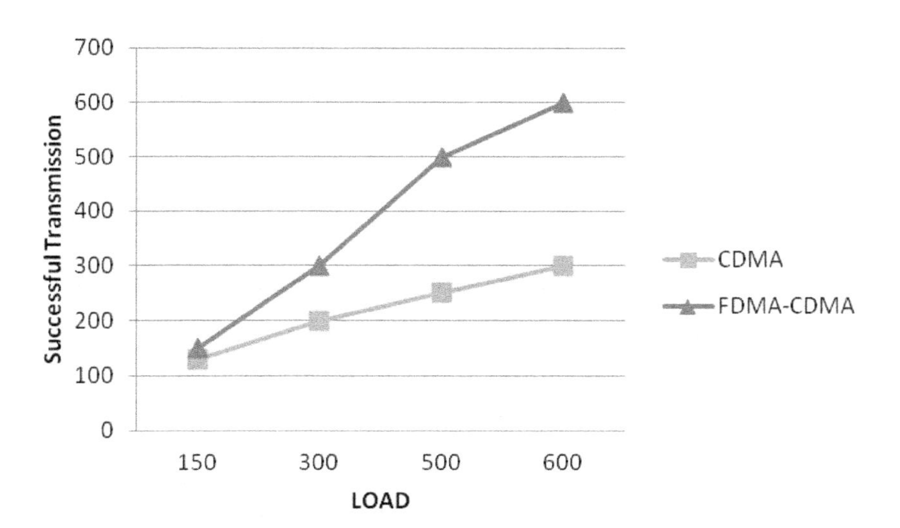

Fig. 4. Successful transmission with varying Load (packets)

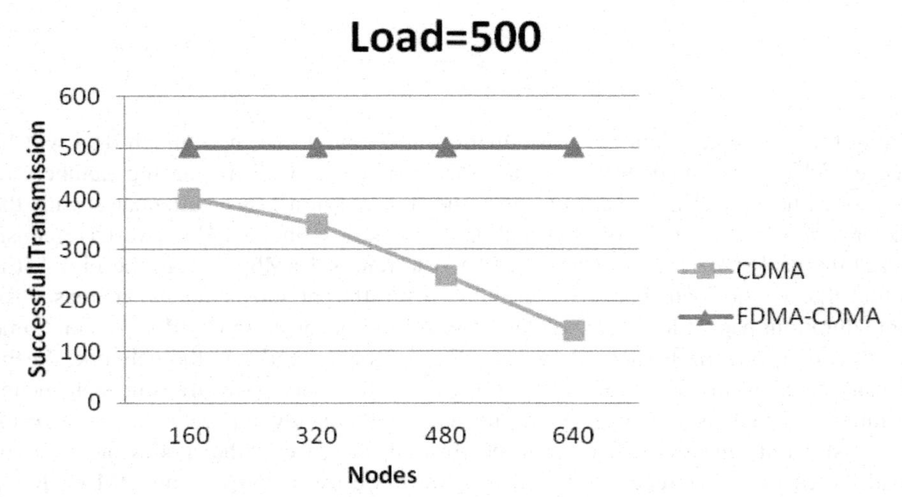

Fig. 5. Successful transmission with varying numbers of nodes (for Load=500)

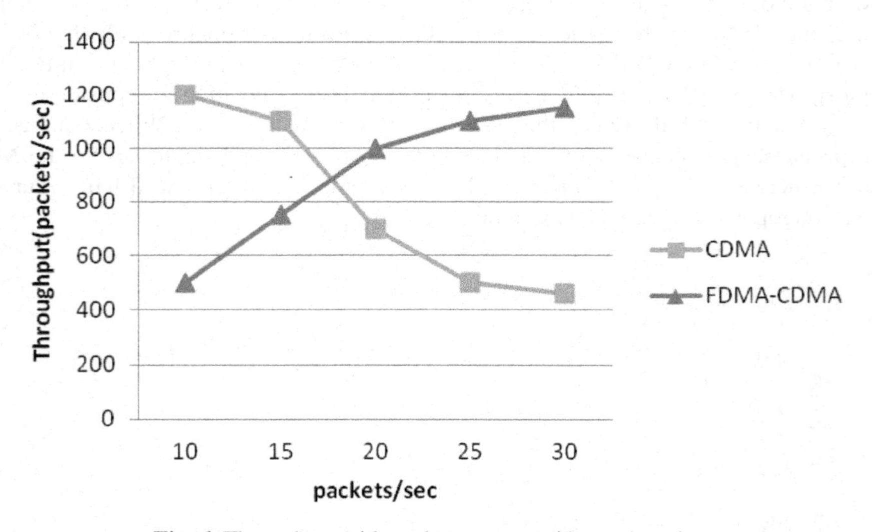

Fig. 6. Throughput with packets generated by each node

5 Conclusion

In this article, we discussed the effect of MAI and how it is reduced by FDMA-CDMA technique. We presented an effective technique to reduce the MAI in CDMA-based sensor network by dividing the available frequencies into several groups and then reusing them. We also proposed an algorithm for clustering the sensor network. Our protocol achieves less channel contention and high packet delivery ratio compared to normal CDMA-based system. By reducing the channel contention, we also achieved energy savings in energy scare sensor network.

References

1. Muqattash, A., Krunz, M.: CDMA-Based MAC Protocol for Wireless Ad Hoc Networks. In: MobiHoc 2003 (2003)
2. Woo, A., Culler, D.: A Transmission Control Scheme for Media Access in Sensor Networks. In: Proc. ACM MobiCom 2001, pp. 221–235 (2001)
3. Ye, W., Heidemann, J., Estrin, D.: An Energy-Efficient MAC Protocol for Wireless Sensor Networks. In: IEEE Proc. Infocom, pp. 1567–1576 (June 2002)
4. Hu, L.: Distributed Code Assignment for CDMA Packet Radio Networks. IEEE/ACM Trans. on Networking 1(6) (December 1993)
5. Rappaport, T.S.: Wireless Communications, Principles and Practice, 2nd edn. Prentice Hall (2002)
6. Liu, B.H., Chou, C.T., Lipman, J., Jha, S.: Using Frequency Division to Reduce MAI in DS-CDMAWireless Sensor Networks
7. Dow, Lin, Fan: Avoidance of Hidden Terminal Problems in Cluster-Based Wireless Networks Using Efficient Two-Level Code Assignment Schemes. IEICE Trans. Commun. E84-B(2) (February 2001)
8. Sohrabi, K., et al.: Protocols for Self-Organization of a Wireless Sensor Network. IEEE Pers. Commun., 16–27 (October 2000)
9. Blough, D.M., et al.: The k-Neigh Protocol for Symmetric Topology Control in Ad Hoc Networks. In: Proc. of IEEE MobiHoc 2003 (2003)
10. Liu, C.H., Asada, H.H.: A Source Coding and Modulation Method for Power Saving and Interference Reduction in DS-CDMA Sensor Networks. In: Proc. American Control Conf. (May 2002)
11. Ye, W., Heidemann, J., Estrin, D.: An Energy-Efficient MAC Protocol for Wireless Sensor Networks. In: IEEE Proc. Infocom, pp. 1567–1576 (June 2002)
12. Schurgers, C.: Optimizing Sensor Networks in the Energy-Latency-Density Design Space. IEEE Trans. on Mobile Computing 1(1) (January-March 2002)
13. Guo, C., Zhong, L.C., Rabaey, J.M.: Low Power Distributed MAC for Ad Hoc Sensor Radio Networks. In: IEEE Proc. GlobeCom 2001, San Antonio, November 25-29 (2001)
14. Sousa, E.S., et al.: Optimum Transmission Ranges in a Direct-Sequence Spread-Spectrum Multihop Packet Radio Network. IEEE Journal on Selected Areas in Communications 8(5) (1990)
15. Pursley, M.B.: Performance evaluation for phase-coded spread-spectrum multiple-access communications - Part I: System Analysis. IEEE Trans. Commun. COM-25, 795–799 (1977)
16. Proakis, J.G.: Digital Communications. McGraw-Hill, New York (2001)
17. Mukherjee, S., Singh, J.P.: Reducing MAI in Cluster Based Sensor Network Using FDMA-CDMA Technique. In: IEEE International Advance Computing Conference, March 6-7 (2009)

Priority Based Scheduler for Bluetooth Network

M.J. Sataraddi, Jayashree D. Mallapur,
S.C. Hiremath, and K. Nagarathna

E&CE Dept. Basaveshwar Engg., College Bagalkot, Karnataka, India
bdmallapur@yahoo.co.in,
kdmamata@rediffmail.com,
{shiremath837,rathnarajur}@gmail.com

Abstract. Bluetooth is a personal wireless communication technology and is being applied in many scenarios. It is an emerging standard for short range, low cost, low power wireless access technology. Current existing MAC (Medium Access Control) scheduling schemes only provide best-effort service for all master slave connections. In computer science, scheduling is the method by which threads, processes or data flows are given access to system resources (e.g. processor time, communications bandwidth). The work presented here in this scheme is aimed at efficient energy consumption by reducing the delay of each packet in case of single user as well as multiple users. The delay reduction is done by using priority scheduler. The scheduler has two types of services called best effort services and real time services. The real time services are considered as high priority applications where as best effort services are considered as low priority application. Hence real time services are served faster and better where as best effort is served after real time services are served. The delay for real time services are much reduced compared to best effort services. The simulation results show that the delay is much reduced for real time applications.

Keywords: Bluetooth, Priority, Scheduling.

1 Introduction

Bluetooth is a wireless technology that allows communication devices and accessories to interconnect using a short-range, low-power, inexpensive radio signals. Bluetooth was developed initially as a replacement for short-range cable linking portable consumer electronic products, but it can also be adopted for printers, keyboards etc. To date bluetooth has been seen as a promising category for ad-hoc wireless networking and wireless personal area network (WPAN). It has been expanded on wireless LANs as given in [1]. The smallest bluetooth unit is called a piconet, which consists of one master node and many slave nodes (up to seven active slaves). All the nodes in a same piconet should follow same frequency hopping pattern. Multiple piconets can also exist in the same area and can be connected via a bridge node, forming a scatternet. In a bluetooth system, full-duplex transmission is supported

S. Unnikrishnan, S. Surve, and D. Bhoir (Eds.): ICAC3 2013, CCIS 361, pp. 356–365, 2013.

using a master driven TDD (Time Division Duplex) scheme to divide the channel into 625 μs time slots. The time slots are alternatively switched between the master and the slaves. The master sends a poll or a data packet to a slave using the even numbered time slots, the slave sends a packet to the master in the immediate odd numbered slot. Thus, the Medium Access Control (MAC) scheduling in bluetooth is controlled by the master.

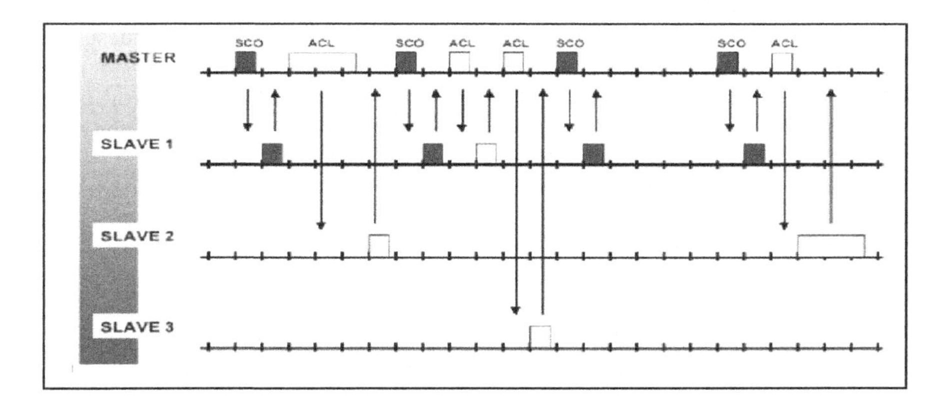

Fig. 1. Master driven TDD syste

As shown in Fig. 1 bluetooth system supports two types of data communication channels between the master and the slave: a Synchronous Connection Oriented (SCO) link and Asynchronous Connection Less (ACL) link. An SCO connection supports a circuit-oriented service with a constant bandwidth using a fixed and periodic allocation of slots. An SCO connection is suitable for delay-sensitive multimedia traffic like voice traffic, whereas an ACL connection supports a packet-oriented service between the master and slave. The ACL connection is suitable for various applications such as ftp, telnet, audio and video applications. Because these applications have various QOS requirements (such as delay and bandwidth), it is very important to provide different QOS for them. However, current bluetooth specification doesn't address how to meet these different QOS requirements, and current implementations only provide best-effort service to all applications.

The Round Robin (RR) scheme is a default MAC scheduling algorithm for bluetooth that uses a fixed cyclic order. In RR scheme, every slave has the same opportunity to send one data packet even when they have no packet to transmit. Once the master polls a slave, the next time slot is then assigned to the slave without considering whether the slave has data to transmit or not. Several bluetooth MAC scheduling algorithms have already been proposed to improve the system performance.

The need for a scheduling algorithm arises from the requirement for most modern systems to perform multitasking (execute more than one process at a time) and multiplexing (transmit multiple flows simultaneously). The scheduler is concerned mainly with:

- Throughput - The total number of processes that complete their execution per time unit.
- Latency, specifically:
  - Turnaround time - total time between submission of a process and its completion.
  - Response time - amount of time it takes from when a request was submitted until the first response is produced.
- Fairness / Waiting Time - Equal CPU time to each process (or more generally appropriate times according to each process' priority). It is the time for which the process remains in the ready queue.

Preference is given to any one of the above mentioned concerns depending upon the user's needs and objectives.

Types of Operating System Schedulers

Operating systems may feature up to three distinct types of scheduler, a long-term scheduler (also known as an admission scheduler or high-level scheduler), a mid-term or medium-term scheduler and a short-term scheduler. The names suggest the relative frequency with which these functions are performed. The scheduler is an operating system module that selects the next jobs to be admitted into the system and the next process to run. Scheduling disciplines are algorithms used for distributing resources among parties which simultaneously and asynchronously request them. The simplest best-effort scheduling algorithms are round-robin, fair queuing (a max-min fair scheduling algorithm), proportionally fair scheduling and maximum throughput. If differentiated or guaranteed quality of service is offered, as opposed to best-effort communication, weighted fair queuing may be utilized.

The RR scheme is a default MAC scheduling algorithm for bluetooth that uses a fixed cyclic order. The POLL packet does not have any information and just gives the polled slave the privilege of transmitting packet in the next slot. If the polled slave does not have any data to transmit, it replies to the master by sending a NULL packet which also does not have any information. As a result, numerous slots will be wasted with POLL and NULL packet exchanges in the case of no data to transmit.

To address this problem, it is proposed a new scheduling algorithm to enhance efficiency parameter as delay for wireless network using bluetooth. The connections that have no data to transmit in both directions belong to the lowest threshold range. Each real time connection is assigned a high threshold range and the best effort connection is given the middle threshold range.

The paper is organized as follows. In section 2 we present some of the related work on Bluetooth MAC scheduling algorithms for QoS parameters. In section 3 we explain the proposed priority based scheduling algorithm and in section 4 we compared the performance of our algorithm for real time applications and best effort applications using simulation. Section 5 gives the graphs for our results. Concluding remarks are summarized in section 6.

2 Related Works

The paper [2] presents priority gated round robin scheme for intra-piconet scheduling in bluetooth piconets. The algorithm is analyzed for different packet sizes and variable number of slaves and is compared with traditional gated round robin algorithm.

The issue of how to enhance QoS support and a Round Robin Scheduler (RRS) as possible solution for scheduling the transmissions in a bluetooth piconet is addressed in paper [3]. A back-off mechanism is then presented for best-effort slaves, used to increase bandwidth utilization.

The paper [4] proposes a polling policy that aims to achieve increased system throughput and reduced packet delays while providing reasonably good fairness among all traffic flows in a bluetooth piconet. The proposed scheduling algorithm outperforms the RRS algorithm by more than 40%.

The issue of supporting real-time service in a piconet of a bluetooth network was presented in [5]. It proposes a novel-Scheduling algorithm for the polling protocol at the MAC layer to support real-time traffic streams to meet their delay constraints.

The authors in [6] propose Blue-Park, a flexible networking structure that achieves energy efficiency and fairness, without requiring change to the bluetooth specification. Blue-Park utilizes the bluetooth park mode and can be described as a self-referencing structure, where each node parks all the other nodes as a master.

The work [7] addresses the question of providing throughput guarantees through distributed scheduling, which has remained an open problem for some time. It considers a simple distributed scheduling strategy, maximal scheduling, and prove that it attains a guaranteed fraction of the maximum throughput region in arbitrary wireless networks.

A novel-polling algorithm called Adaptive Share Polling (ASP) is introduced in [8] that is designed to perform well when the network consists of sources sending short data packets at constant rate that may fluctuate over time. In ASP, the scheduler at the master implicitly learns the share of the bandwidth that needs to be allocated to each of the slaves in order to meet the latency and/or power requirements of the application.

Paper [9] proposes a MAC scheduling policies for centrally controlled TDD wireless systems. The proposed policies reduce the power consumption of mobile devices by putting them into low power mode intelligently, using probabilistic estimates of inactivity based on the previous traffic arrival pattern. The policies ensure that the QoS parameters such as end-to-end packet delays are not violated.

Paper [10] proposes RR-FCFS (First Come First Serve), a simple MAC scheduling scheme that has the same advantages as RR(Round Robin) and ERR(Exhaustive Round-Robin). RR-FCFS acts as RR if the master's queue is empty and starts transmitting packets in first-come-first-serve order otherwise.

Marson et al. [11] investigates the optimal Bluetooth topology that minimizes energy consumption of the most congested node while the constraints posed by system specifications and the traffic requirements are met.

3 Proposed Work

Current bluetooth can provide voice support over SCO connections. In this section, the priority based scheduling algorithm for real time applications and best effort applications over ACL connections. In master driven TDD, slave can transmit a packet only after it has received a polling packet from the master. It has two issues.

1. The master cannot know whether a slave has data to transmit unless it sends polling packet to the slave, which leads to traditional time-stamp based algorithm, such as EDD (Earliest Due Deadline) cannot be used.
2. When one or more slaves have data to transmit, to poling them will decrease bandwidth utilization.

To overcome these issues priority based scheduling algorithm is proposed. The proposed work is as shown in fig.2.

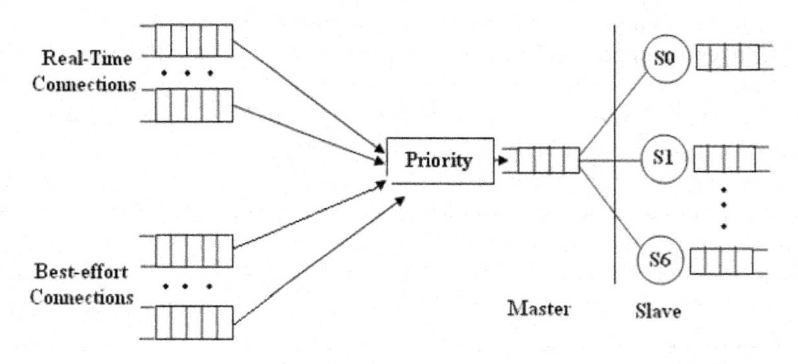

Fig. 2. Priority based scheduler

In our proposed algorithm, we assign a high priority to real time slaves and low priority to best effort slaves.The Master maintains a queue for each slave, which maintains its own queue. To provide different efficiency parameters for different slaves, categorize the master-slave connections in to two classes: One is real time connection whose packet should deliver as soon as possible, to meet their delay requirements. The other one is best-effort connection that has no efficiency parameters requirement.

For each real-time application R_i, define the following parameters: average bit rate A_i bits per second, maximum tolerable delay D_i seconds, packet length L_i bits, token counter C_i (same as polling counter), and token counter generation interval T_i seconds. Then the traffic of R_i can be estimated by increasing C_i by one per T_i seconds, where $T_i = L_i / A_i$ seconds, to assign a unique priority to R_i according to D_i and schedule real-time applications strictly based on their priorities. For every best effort application B_i the priority is given based on binary exponential back off mechanism. To schedule these applications the algorithm works as follows:

Begin

1. Each real time connection is assigned a unique priority based on its delay requirement.
2. If the data packet comes from real time applications schedule with its priority,
   else
   schedule with best effort priority applications.
3. If multiple packets of real time applications arrive
   then
   begin
   check for average delay
   If two real time connections have same delay requirement
   then
   earlier one get higher priority
   end
4. If packet arrives with multiple best effort applications
   then
   assign the priority within the threshold range.
5. Repeat step 2 to 4 until all the packets from all applications are over.

End.

4 Simulation

The simulation is to demonstrate the performance of priority based scheduling algorithm applied for real-time slaves is better than the algorithm applied for best-effort slaves in bluetooth network.

As depicted in Figure 3, the simulation configuration in a bluetooth piconet composed of a Master and seven Slaves. Each slave communicates with master based on priority. Each slave has fixed number of data byte per second and maximum delay. Each slave shows number of packets sent, number of packet receive, number of packets in queue with varying average delay in microsecond. The slaves 0-3 are served as real time services and 4-6 as best effort services.

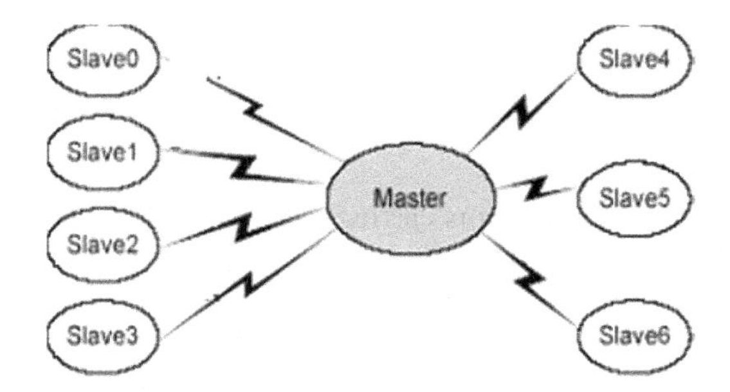

Fig. 3. Simulation configuration

Simulation Procedure

The proposed scheme is simulated with server maximum delay, server transmission rate, real time and best effort connection for given client-server IP address, client number and client transmission rate as simulation inputs. The simulation procedure is as follows:

Begin

1. Enter maximum delay, data byte per second, real time connection or best effort connection for corresponding client in server.
2. Enter server IP address, client number and data bytes per second in client.
3. Repeat above steps for other clients.
4. Note the average delay for each client.
5. Compare results of real time connection and best effort connection.

End

The average delays are tabulated in table 1 for real time connection and table 2 for best effort connection.

Table 1. Average Delays (µs) of Real-Time Connection

Slave number	**0**	**1**	**2**	**3**
Our algorithm for real time applications	**8522.2**	**8634.4**	**8428.0**	**8971.8**

Table 2. Average Delay (µs) of Best-Effort Connection

Slave number	**4**	**5**	**6**
Our algorithm for best effort applications	**45948.5**	**47296.1**	**48567.5**

These results show that the delay required to transmit real time applications is less than for best effort applications.

5 Results

The following are simulation results carried out on Intel Pentium IV processor using Java 1.4 programming language.The results measured are average delays in microsecond Vs packet sent, packet receive for single and multiple clients.

Figure 4 shows the results for Average Delay (µs) Vs Packet Sent for real time and best effort applications. This clearly shows real time performs better than best effort.

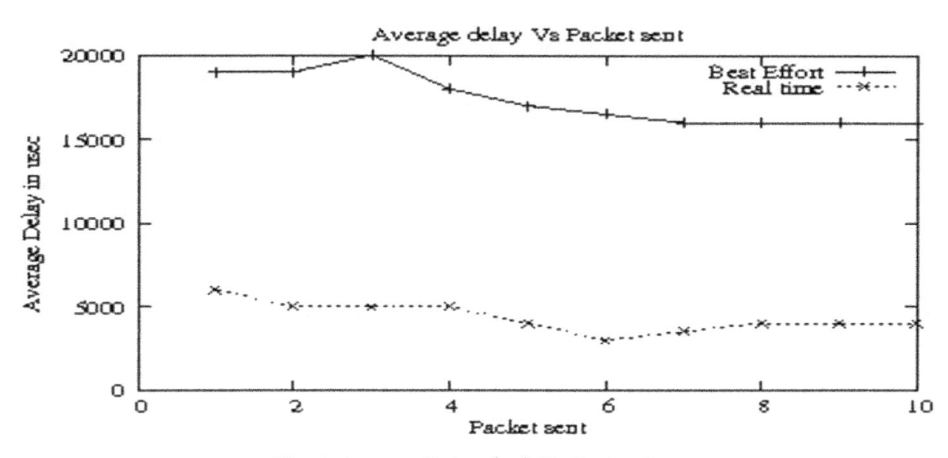

Fig. 4. Average Delay (μs) Vs Packet Sent

The Figure 5 shows result of Average Delay (μs) Vs Packet Receive.The receiver performance measure is done for the average delay. The receiver also receives the real time applications faster than best effort applications. Hence on both ways i ,e transmitter and receiver negotiate the real time applications.

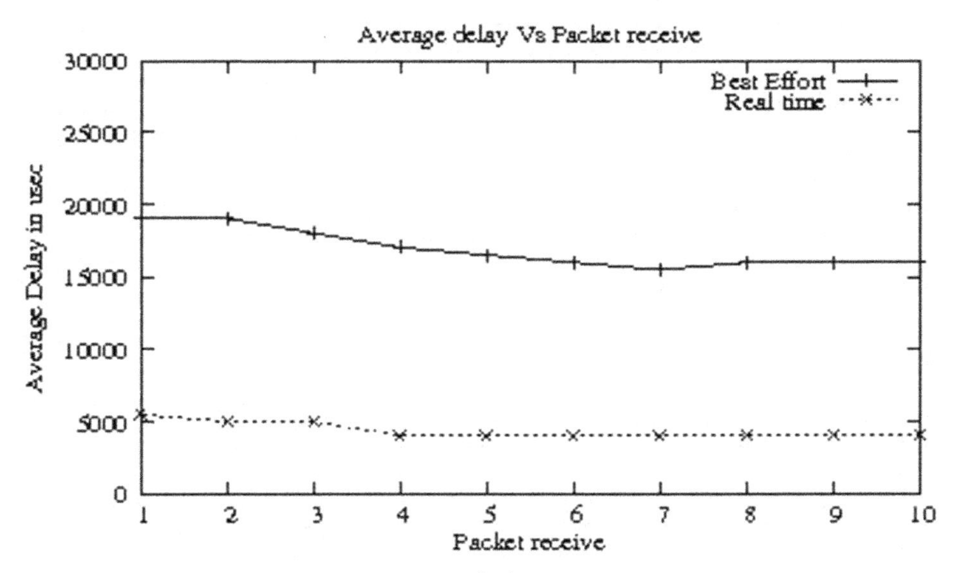

Fig. 5. Average Delay (μs) Vs Packet Receive

The Figure 6 shows result of Average Delay (μs) Vs Packet Sent for multiple clients. For different packet sent average delay of each real time is less than each best effort, which clearly indicates real time shows better performance than best effort.

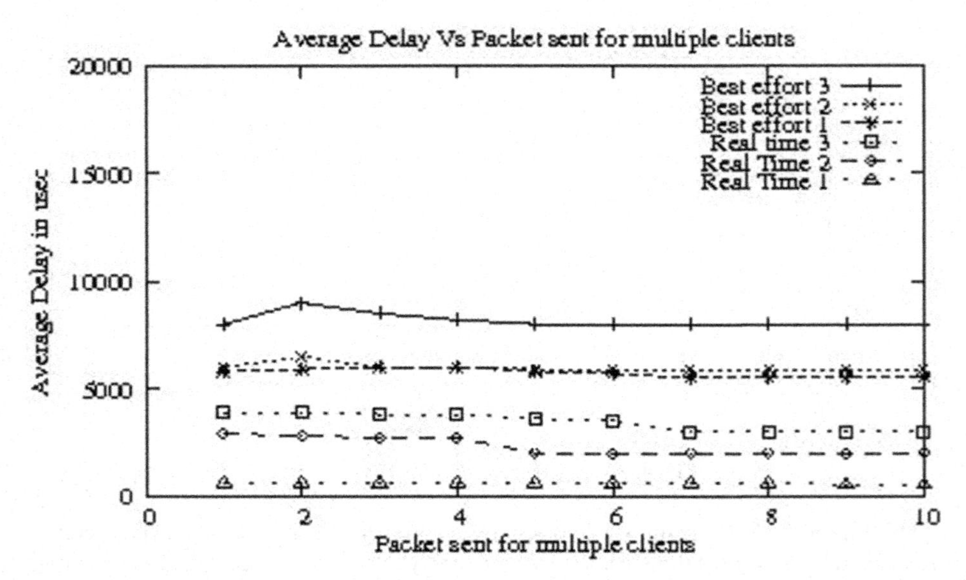

Fig. 6. Average Delay (µs) Vs Packet Sent for multiple clients

6 Conclusions

A priority based scheduling algorithm is presented in this paper to distinguish the priority between the real time applications and best effort applications. The multimedia applications are much useful nowadays than non real time one. The results shows that there is reduction in the transmission delay for real time applications which in turn increase s the efficiency of the system.

The scheme can be further enhanced taking other QoS parameters like jitter, bandwidth utilization etc.

References

1. Bluetooth Special Interest Group, `http://www.bluetooth.com/`
2. Ibrahim, A., Rizk, R., Mahmoud, G.: A priority gated round robin polling scheme for bluetooth piconets. Proceedings of the International Arab Journal of Information Technology 5(2), 176–182 (2008)
3. Chourishi "Maitraya", D., Seshadri, S.: Enhancing Performance of Bluetooth Piconets Using Priority Scheduling and Exponential Back-Off Mechanism. Proceedings of World Academy of Science, Engineering and Technology 50, 414–418 (2009)
4. Chaturvedi, V.P., Rakesh, V., Bhatnagar, S.: An Efficient and Optimized Bluetooth Scheduling Algorithm for Piconets. In: Janowski, T., Mohanty, H. (eds.) ICDCIT 2007. LNCS, vol. 4882, pp. 19–30. Springer, Heidelberg (2007)
5. Ma, M., Low, S.Y.: Adaptive polling interval scheduling to support real-time service in Bluetooth networks. International Journal of Wireless and Mobile Computing (IJWMC) 1(3/4), 204–210 (2006)

6. Popovski, P., Yomo, H., Kuijpers, G., Madsen, T.K., Prasad, R.: Blue-Park: Energy efficient operation of Bluetooth networks using park mode. Published in Journal Computer Communications 29(17), 3416–3424 (2006)
7. Chaporkar, P., Kar, K., Sarkar, S., Luo, X.: Throughput and fairness guarantees through maximal scheduling in wireless networks. IEEE Transactions on Information Theory (2008)
8. Perillo, M., Heinzelman, W.B.: ASP: An Adaptive Energy-Efficient Polling Algorithm for Bluetooth Piconets. In: Proceedings of the 36th Hawaii International Conference on System Sciences, pp. 1–10 (2003)
9. Chakraborty, I., Kashyap, A., Kumar, A., Rastogi, A., Saran, H., Shorey, R.: MAC Scheduling Policies with Reduced Power Consumption and Bounded Packet Delays for Centrally Controlled TDD Wireless Networks. In: Proceedings of IEEE International Conference on Communications (ICC), Helsinki, Finland (June 2001)
10. Yen, L.-H., Liao, C.-H.: Round-robin with FCFS preemption: a simple MAC scheduling scheme for bluetooth piconet. Published in Journal of Mobile Multimedia 1(3), 211–223 (2005)
11. Marsan, M.A., Chiasserini, C.-F., Nucci, A., Carello, G., Giovanni, L.D.: Optimizing the topology of Bluetooth Wireless Personal Area Networks. In: Proceedings of the IEEE INFOCOM 2002, New York (June 2002)
12. Marsan, M.A., Chiasserini, C.-F., Nucci, A., Carello, G., Giovanni, L.D.: Optimizing the topology of Bluetooth Wireless Personal Area Networks. In: Proceedings of the IEEE INFOCOM 2002, New York (June 2002)

QP - ALAH: QoS Provisioned - Application Layer Auxiliary Handover in IEEE 802.16e Networks

M. Deva Priya[1,*], M.L. Valarmathi[2,*], K. Jaya Bharathi[1], and V. Sundarameena[3]

[1] Department of Computer Science and Engineering,
Sri Krishna College of Technology, Coimbatore, Tamil Nadu, India
devapriyame@gmail.com

[2] Department of Computer Science and Engineering,
Government College of Technology, Coimbatore, Tamil Nadu, India
ml_valarmathi@rediffmail.com

[3] Department of Computer Science and Engineering,
Manakula Vinayagar Institute of Technology, Puducherry, India

Abstract. There is a growing demand for mobile multimedia services that urges a rapid growth in wireless mobile Internet access. IEEE 802.16e can provide a high data rate and better QoS compared to other mechanisms. However, vertical handover becomes hectic. A seamless handover mechanism is to be developed. This paper proposes an Application Layer Auxiliary Handover (ALAH) mechanism involving 802.21 Media Independent Handover (MIH). L2, L3 and L7 handovers are dealt with. An Auxiliary Server (AS) caches the media streams and offers to the Mobile Host (MH) until it remains in its service area. Here inter-Gateway vertical and horizontal handovers are taken into consideration and is implemented for WiMAX and WiFi networks. Selection of AS based on RSS and load improves the performance of the system in terms of handover latency and packet loss.

Keywords: WiMAX, Vertical handover, Handover latency, Auxiliary Server, MIH, L2 handover, L3 handover, L7 Handover, Anycast, MIPv6, Multicast.

1 Introduction

IEEE 802.16, also called Worldwide interoperability for Microwave Access (WiMAX) is an air Interface for Fixed BWA (Broadband Wireless Access systems). It supports all kinds of real-time application in wireless networks. WiMAX allows two modes of communication: Point-to-Multi-Point (PMP) and mesh mode. In PMP mode, communications are through the Base Station (BS) and the BS acts as the central entity. It decides the transmission and reception schedule of the Subscriber Stations (SSs). In Mesh mode, SSs communicate with each other without the BS. WiMAX provides speeds around 70 Mbps with coverage of 50 kms. The MAC and the PHYsical layers are defined in WiMAX. To overcome the disadvantage of the WiFi, designers developed WiMAX based on the Scalable Orthogonal Frequency-Division Multiple Access (SOFDMA). The SOFDMA offers cheaper, effective and

S. Unnikrishnan, S. Surve, and D. Bhoir (Eds.): ICAC3 2013, CCIS 361, pp. 366–380, 2013.

easily implementable services. The basic difference between WiMAX and WiFi are cost, speed, distance and so on. Range of WiMAX is about 30 miles while that of WiFi is very less. Other benefits of WiMAX technology include, VoIP calling, support for mobility, video making and high speed data transfer.

2 Handover

Mobile WiMAX, 802.16e is based on the same OFDM based technology. It is designed to deliver services across more sub-channels when compared to the OFDM 256-FFT. Mobile services will have to find some way to deal with multiple networks (3G core, LTE, Wi-Fi, 3G RN, and mobile WiMAX). Architecture to fulfill such requirements is indispensable. In any cellular network, when a user travels from one region of coverage of a BS to another within a call's duration, the call should be forwarded to the new BS. It is the process of maintaining a user's active sessions when a mobile terminal changes its connection point to the access network, a BS or an AP. If not retained, the call may be dropped because the link of the current BS may become too weak.

2.1 MIPv6

The WiMAX Forum proposed an architecture that defines how a WiMAX network can be connected to an IP based core network, which is typically chosen by operators that serve as Internet Service Providers (ISP); Nevertheless the WiMAX BS provides seamless integration capabilities with other types of architectures as with packet switched Mobile Networks. MIPv6 is a protocol developed as a subset of Internet Protocol version 6 (IPv6) to support mobile connections that allows a mobile node to transparently uphold connections, while moving from one subnet to another, retaining permanent IP address. It is an update of the IETF (Internet Engineering Task Force) Mobile IP standard (RFC 2002) designed to authenticate mobile devices using IPv6 addresses. Each device is identified by its Home Address (HA) although it may be connecting to another network. MIPv6 allows a mobile node to transparently maintain connections while moving from one subnet to another. When connecting through a foreign network, a mobile device sends its location information to a home agent which intercepts packets intended for the device and tunnels them to the current location [1].

2.2 Types of Handover

There are different types of handover based on various factors. The handovers based on Layers are:

- **Layer 2 (L2) Handover:** Layer 2 handover occurs when a Mobile Host (MH) changes its connection from one Access Point (AP) to another; hence termed as AP / BS handover. The AP sends a re-association message to the MH. Time taken by the AP to send the response to the MH after receiving the probe request is the layer 2 handover latency [2, 3]. L2 handover takes place whenever the signal strength becomes weak. The phases identified for L2 handover process are - Discovery phase, Re-authentication phase, Re-association phase [4].

- **Layer 3 (L3) Handover:** Layer3 handover known as IP handover occurs when a MH moves from one Gateway (GW) / Foreign Agent (FA) to another [5]. In other words, the MH moves from one subnet to another and both the subnets belong to different IP domains. Mobile IP provides solution to this [6, 7]. Mobile IP uses two IP addresses, i.e., HA and Care-of-Address (CoA). The HA of an MH is always the same irrespective of its location. The IP address of the MH changes when it moves to a new subnet. i.e there is a change in the CoA. The CoA address must be sent to the HA. In short, new address has to be created, verified and the MH should register.
- **Layer 7 (L7) Handover:** L7 handover can be divided into two categories - handover using the L7 method and handover behavior of applications. It is discussed in detail in the proposed system.

2.3 Media Independent Handover (MIH)

It is a standard being developed by IEEE 802.21 to enable the handover of IP sessions from one L2 access technology to another to achieve mobility of end user devices. The key functionality provided by MIH is communication among the various wireless layers and between each layer and the IP layer. The required messages are relayed by the MIH Function (MIHF) that is located in the protocol stack between the L2 wireless technologies and IP at L3. MIH may communicate with various IP protocols including Session Initiation Protocol (SIP) for signaling, Mobile IP for mobility management, and DiffServ and IntServ for QoS.

When a session is handed off from one AP to another using the same technology, the handover can usually be performed without involving MIHF or IP. For example, between 2 WiFi networks [8]. However, if the handover is from a Wi-Fi AP in a corporate network to a public Wi-Fi hotspot, then MIH is required, since the two APs cannot communicate with each other at the link layer, and are in general, on different IP subnets.

2.4 IPv6 Anycast and Multicast

It is a network addressing and routing methodology in which datagrams from a single sender are routed to the topologically nearest node in a group of potential receivers, all identified by the same destination address. With the growth of the Internet, network services increasingly have high-availability requirements. As a result, operation of anycast services (RFC 4786) has grown in popularity among network operators. In IPv4 to IPv6 transitioning, anycast addressing may be deployed to provide IPv6 compatibility to IPv4 hosts. Anycast addresses are a new, unique type of address that is new to IP in IPv6; the IPv6 implementation is based on the material in RFC 1546,

Host Anycasting Service. Anycast addresses can be considered a conceptual cross between unicast and multicast addressing [9, 10].

IPv6 is the latest revision of the Internet Protocol (IP), the primary communications protocol upon which the entire Internet is built. It is intended to replace the older IPv4, which is still employed for the vast majority of Internet traffic as of 2012. IPv6 was developed by the Internet Engineering Task Force (IETF) to deal with the long-anticipated problem of IPv4 running out of addresses. Using the application-layer anycasting technique, a packet sent to an Anycast Domain Name (ADN), which represents an anycast group, is delivered to the "appropriate" node, the node selected by the MH based on its capability. The ADN can be a set of unicast or multicast IP addresses. Clients interact with the members in an anycast group by sending a request to the ADN. Once the members of the corresponding anycast group receive the anycast request, they respond with their metrics.

3 Related Work

In [11], the handover coordinator utilizes a network-based mobility management protocol as a basis to effectively and efficiently handle vertical handovers in a localized heterogeneous wireless environment. Potential handover-related research issues in the existing and future WiMAX mobility framework is presented in [12]. IEEE 802.21 aims at providing a framework for media-independent handover (MIH) among heterogeneous networks is discussed. In [13], authors propose a handover Scheme with Geographic Mobility Awareness (HGMA), which considers the historical handover patterns of mobile devices. In [14], authors present a proxy mobile IP based L3 handover scheme for mobile WiMAX based wireless mesh networks. Seamless handovers within WiMAX and between WiMAX and WLAN is focused in [15]. In [16], authors propose fast handover techniques between WiMAX and WiFi networks to speed up handover process in vehicular communications.

If an MT is associated with the WLAN for the longest possible duration, the user throughput is improved and even during the transition period, the RSS oscillates around the receiver sensitivity level [17]. A general framework for the vertical handover process based on fuzzy logic and neural networks is presented in [18]. Policies considering different parameters like cost, power, available bandwidth and other parameters for different heterogeneous networks are analyzed in [19]. The available interfaces and the system resources to collect information is to be known. A score function is used that considers the usage expenses, link capacity, and power consumption for the available access technologies [20].

In [21], authors have focused on VHO decision that depends on the popular Signal to Noise Ratio (SNR) criteria or other performance metrics like bit error rate, delay, jitter etc or combination of these metrics. An efficient handover scheme within a PMIPv6 domain in WiMAX network is proposed in [22]. Mobility management techniques that support fast handover by enhancing currently available mobility management protocols are proposed in [23]. In [24], authors have devised a mechanism to reduce handover delay at L2 and L3 respectively. These solutions, however, do not discuss heterogeneous access scenarios. In [25], vertical handover between UMTS and WLAN is discussed. They did not take WiMAX networks into considerations. In [26], Application-Layer Proxy Handover (APH) enables applications to be executed smoothly when mobile clients move in the server-proxy-client architecture. APH utilizes IPv6 multicast to switch the session from the original proxy to the next proxy smoothly and to forward the unsent cache content in the original proxy to the next proxy for keeping the original session continuous. Vertical handovers are classified based on RSS [27, 28], bandwidth [29, 30] and monetary cost [31, 32].

4 Auxiliary Handover

In the architecture of ALAH, the ASs within the same service area form a Regional Auxiliary Group (RAG) and share their workload. These ASs provide different QoS for each MH. Once the MH moves to a new subnet/ gateway that involves a new RAG, it is the responsibility of the MH to select a suitable AS. The new ASs becomes candidate ASs for the MH. The selection process is performed using application layer anycasting.

5 Proposed System

In this work, vertical handover and application layer auxiliary handover in mobile heterogeneous networks are discussed. The two networks taken into consideration are WiMAX and WiFi. To improve QoS, L7 handover was discussed in [26, 33]. Handover using the L7 method [33] means resolving the IP connectivity of an MH, is within the application layer and not the network layer. For example, an MH that is getting service from the Internet moves from a wired network to a wireless network. Since these two networks are heterogeneous, the MH seems unable to get the service with the same quality. If this MH gets the network service via a proxy, the proxy can help the service be adaptive to the network environment, i.e., changing the quality of the requested data dynamically. Thus, this kind of L7 handover can improve the quality of service (QoS) of applications. Application-layer Auxiliary Handover (ALAH) is devised to solve this kind of L7 handover for multimedia streaming service in the mobile heterogeneous networks. Based on the observation from the past work, the vital issues are mobility management across heterogeneous networks, QoS Provisioning, selection of the next AS and session recovery. The above issues are dealt in the proposed system. First, Mobility management is achieved by performing vertical handover using MIH. Second, QoS provisioning is done by applying server auxiliary MH architecture to mobile heterogeneous networks. AS selection and session recovery are done by using anycast and multicast techniques.

5.1 Overall System Architecture

Application Layer Auxiliary Handover (ALAH) system structure involves the following members. Fig.1 depicts the system architecture of ALAH.

- The Mobile Host (MH) that moves from one location to another.
- Access Points (APs) associated with each subnet.
- A subnet with more than one AP and Auxiliary Server (AS).
- A router to connect two subnets.
- An Access Service Network Gateway (ASN GW) which acts as the Foreign Agent (FA) to connect two types of networks – WiFi and WiMAX.
- A Sever Cluster (SC) consisting of a collection of Media Servers (MSs) and a Supervisory Server (SS).
- A Connectivity Service Network (CSN) provides IP connectivity services and it contains the Home Agent of the MH.

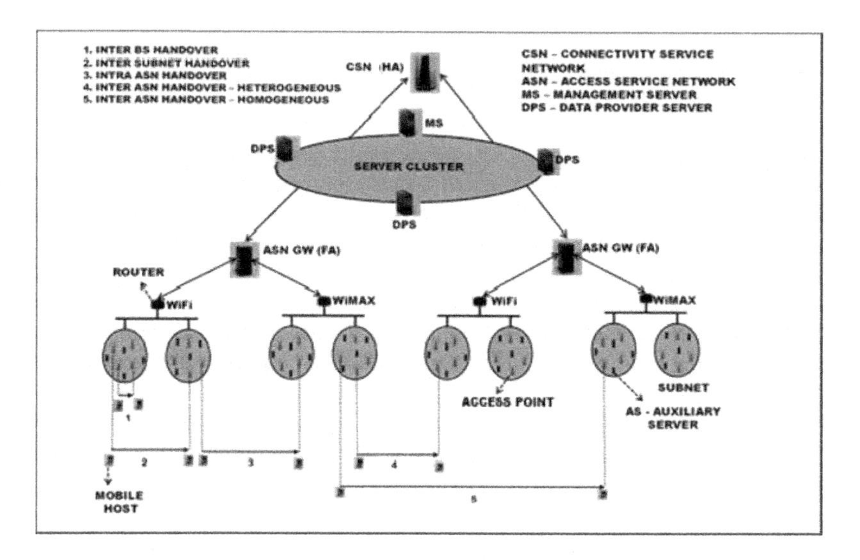

Fig. 1. 3 - Tier Handover Architecture of ALAH

5.2 Operational Procedure of APH

When MH moves out of the service area of the current LMP, L2 and/or L3 handover should be executed initially. After the L2/L3 handover, the MH reconnects to the Former AS (FAS) and receives the packets. However, if the MH goes out of the service area of the FAS, ALAH should be performed.

5.3 Server Cluster (SC)

The Server Cluster is composed of a number of Data Provider Clusters (DPCs). Each DPC provides services; i.e here data service is multimedia streaming service to the Internet. There are 2 kinds of servers in DPC. They are - Management Server (MS) and Data Provider Server (DPS). The MH should register with the MS. The MS verifies the MHs by the AAA mechanism; i.e Authentication, Authorization, and Accounting.

If a MH desires a multimedia stream from a DPS, it sends the request to the corresponding MS via an AS. The MH is granted service by a DPS of the DC only after verification by the MS. The DPS stores multimedia files and delivers the requested streams using multicast to corresponding AS. There may be many DPSs based on the number of MHS, but a single MS. When a multimedia stream is requested from a DPS, the multimedia stream is assigned a multicast group for transmission.

5.4 Auxiliary Server (AS)

An AS is responsible for

- Receiving multimedia streams from a DPS using multicast
- Caching the received streams
- Delivering the requested streams to MHs through unicast and
- Supporting the proposed ALAH.

Each AS acts as a cache, buffering parts of the multimedia streams so as to reduce the response time of similar requests. Virtually, it is assumed that each AS has its service area which is based on the response time and the wireless network environment concerned. A MH may get connected to an AS in the same service area or another. The QoS might differ based on many factors. i.e, once a MH moves from one AS, the throughput may decrease. In that case, the MH has the freedom to select another AS based on Signal to Noise Ratio (SNR), load, RSS, bandwidth available and so on. The service area may be a subnet. The ASs in a subnet having the same service area are combined into a RAG.

In the last two cases, a Care-of-Address in the Foreign network is to be generated by the HA. When a MH requests a multimedia stream via an AS, the required stream may be available with the AS. In that case the MH receives service from the AS instead of DPS. There are different possibilities.

- AS does not have a cache of the requested stream. Therefore the AS should join the corresponding multicast group to receive the stream.
- In another case, the AS may be serving a MH. Another MH sends a request for the same stream. In this situation, if the difference of the requesting times of both the MHs is less than the cache length of a stream, the AS can afford the cached copy of the stream to the new MH, immediately without requesting a new stream from the MS, i.e., without joining a new multicast group.
- In the third case, if the difference of the requesting times of both the MHs is longer than the cache length of a stream, the AS cannot afford the cached copy of the stream to the new MH. The AS must request the stream from the DPS by joining a new multicast group. If the transmission schedule of the new multicast group can go in line with that of the former, the AS serves two clients with the same multicast group and terminates the new stream.

5.5 Mobile Hosts (MHs)

A MH requests media streams and takes control of the process of ALAH. The whereabouts of MHs cannot be predicted. Based on their location, they are serviced dynamically by different ASs. The type of Handover should be identified and necessary action should be taken for transfer of media traffic to the host without interruption.

- **Inter-AP Handover** - When a MH moves from one AP to another, the associated AP has to be changed. This is known as L2 Handover. By persistently observing the signal qualities of the current wireless link, MH can query the surrounding APs about their capabilities and shift to an appropriate AP, in case the signal strength falls below the minimum defined value.
- **Inter-Subnet Handover** - When a MH moves from one subnet to another, within the same GW, an IP handover takes place. This is known as L3 Handover. IPv6 protocol stack in MMH is notified by a local event that triggers IP handover. The protocol stack uses the address allocation mechanism to get a new IP address. Finally, MH sends the updated IP to the corresponding RAG so that the session will not be interrupted.
- **Intra-ASN Handover** - This handover takes place, when a MH moves from one network to another. This takes place through the GW. Since the MH remains within the coverage of the GW, there is no need for a Care-of- Address (CoA).
- **Inter-ASN Handover - Homogenous** - This takes place when a MH moves from one GW to another and the source and destination networks are the same, say Wi-Fi and WiFi. The HA must be informed and a CoA should be obtained.
- **Inter-ASN Handover - Heterogeneous** - This takes place when a MH moves from one GW to another and the source and destination networks differ, say WiFi and WiMAX. CoA should be obtained as the previous case.

Steps involved in the proposed system are listed below.

1. The MH sends a selection intimation message to the entire set of ASs in the service area, querying about the details of the New ASs (NASs).
2. The NASs process the request and analyse the metrics like the load, SNR, RSS, bandwidth and so on.
3. The NAS sends the information about their capability to the MH. In the proposed system, the AS with less load and high RSS is selected.
4. If the MH is willing to connect to a NAS, it accepts and registers with the NAS.
5. The NAS now registers with the DPS.
6. NAS sends an acknowledgement to the MH stating that it will service the further requests.
7. The DPS forwards the stream to the NAS and not the Former AS (FAS).
8. The MH sends a termination request to the FAS to make it aware of its new selection of AS.
9. The FAS analyses and finds the part of the unsent stream.
10. The FAS accepts the termination.
11. The FAS sends termination intimation to the DPS to inform it to stop sending the stream for the corresponding MH.
12. It sends a HELLO message to the NAS.
13. The NAS responds with an acknowledgement.
14. The FAS forwards the portion of the stream that was left out.

15. The NAS forwards the media stream to the MH.
16. It sends intimation of completion to the FAS.
17. The NAS fetches the left out media stream from the DPS.
18. NAS forwards it to the MH.

A MH receives packets from an AS. Once it moves to another subnet belonging to another RAG, another AS that satisfies the requirements of the MH is selected. The new AS must be a member of the corresponding Multicast group to receive the packets without disruption. Else a new connection should be established. It is mandatory to keep track of the ASs joining or leaving the corresponding multicast group.

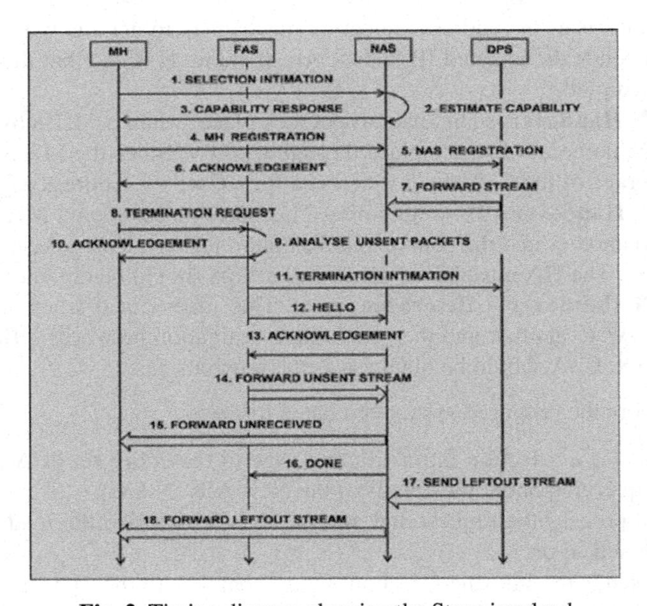

Fig. 2. Timing diagram showing the Steps involved

6 Implementation

The system was implemented using ns-2. Mobile nodes move from one location to another. Movement detection for L3 is to be performed. IP addresses of the MHs change when they move from one network to another. In this work, two vertical handovers are considered - Vertical handover from one subnet to another in the same GW and from one subnet to another in different GW.

In the first case, there is a change in the IP address alone. While in the second case, the CoA address is generated and the HA is intimated about the movement of the MH. A module, a part of the MIH packet is developed and is intended to support multiple interface types, such as Ethernet, WLAN, UMTS, and, in this case, Mobile WiMAX and WiFi.

The agent uses broadcast or unicast messages according to the technology in use. It is located in all nodes but the configuration in ns-2 should be done according to the type of the network and the role of the node. The router functionality consists of sending unsolicited Router Advertisements (RA) periodically to the hosts. The possible sending period is defined with parameters minRtrAdvInterval and maxRtrAdvInterval. In case a router receives a Router Solicitation (RS) from a host, it sends an RA, assuming that the time from previous sending is between the values of parameters described above. If a router receives an RA, it is discarded. The hosts can ask for an RA with RS messages. When an RA is received, the included prefix information is compared with the existing tables and new values are added.

6.1 The Media Independent Handover (MIH)

The Media Independent Handover (MIH) module is a part of the NIST Seamless Mobility project and was developed to control handovers with various technologies. The functionality is based on MIH Function (MIHF). It works on L3 and can communicate between local and remote interfaces. The remote interfaces can be contacted via another MIHF. Some modifications were included to support for multiple technologies and modification of default implementation intended for 802.11. Additionally, a special node suitable for multiple interfaces designed with support for subnet discovery and change of address was added. It is a virtual node that controls different technologies and interfaces.

The MIHF is implemented as an Agent and therefore can send L3 packets to remote MIHF. The MIHF contains the list of local interfaces to get their status and control their behavior. The MIH User is implemented as an Agent that registers with the MIHF to receive events from local and remote interfaces. The cross layer information exchange has been added by modifying the MAC layer and linking the MIHF to the MAC layers via TCL.

6.2 Parameters

Some of the parameters are taken as constants. The BS coverage area, transmit power as well as operating frequency are unchanged and they are the same for all BSs. Additionally, the velocity is taken as 12 m/s. Velocity of Mobile Node is varied for different speeds, by keeping all other parameters constant in order to measure handover latency and packet loss for various velocities. The simulations are done with speeds between 1 and 40 m/s. Handover delay and packet losses are calculated by varying the sizes of the packet, keeping velocity of the mobile node constant. In the simulation, each MH has two independent radio interfaces (Table 1).

One is for 802.11, and the other for 802.16e. The MH switches between network interfaces using the proposed handover procedure. A server is configured to deliver video traffic. An AS is configured to receive video traffic from the server and deliver the packets to MH through BSs 802.16e/802.11.

Table 1. Parameters

Packet Size			4960
Interval			0.1 ms
Modulation			OFDM_16QAM_3_4
Queue Length			50
Queue Type			Drop Tail 1000
MAC			802_11 / 802_16
Basic Rate			1Mbps
Bandwidth			11Mbps
Velocity			12 m/s
Speed			1 to 40 m/s
Frequency			2412e+6
802_11	RXThresh		6.12277e-09 for 20 m
	duplex-link	Bandwidth	100MBit
		Delay	30 ms
802_16	RXThresh		7.91016e-15 for 500 m
			1.26562e-13 for 100 m
	duplex-link	Bandwidth	100MBit
		Delay	15 ms

6.3 Observations during L2 and L3 Handover

Handover latency and packet loss are the two parameters that are to be considered to provide QoS provisioning. The observations are made on these parameters by varying the velocity of mobile node and the video traffic. Latency is the difference between the time in which the last packet was received from the FAS and the first received packet received from the NAS.

6.4 Velocity and Packet Loss

Velocity of the mobile node is varied from 2 ms to 40 ms and the packet loss rate is observed during L2 and L3 handover. Fig. 3 gives the details of packet loss for various velocities. From the figure, it is observed that packet loss decreases with increase in the velocity of the MH. This is due to the variation in RSS when the MH is moving towards NAS. The packet loss for inter-GW vertical handover is high when compared to that of inter-subnet handover. In both the cases, the packet loss is very much reduced when an AS is included.

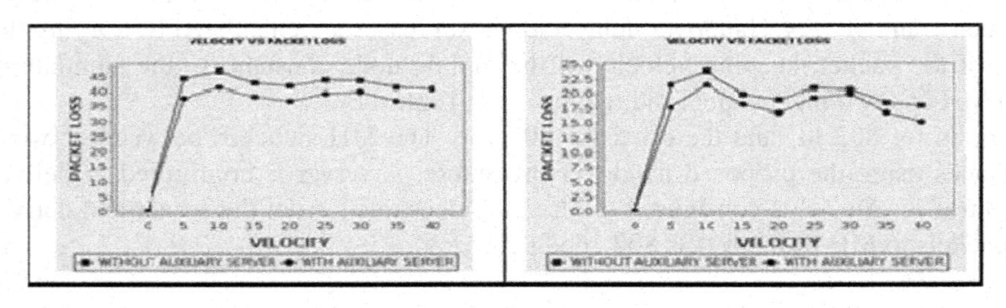

Fig. 3. Velocity and Packet Loss of Inter-GW Handover and Inter-Subnet Handover

6.5 Velocity and Handover Latency

Velocity of the mobile node is varied from 2 ms to 40 ms and the handover latency is observed during layer 2 and layer 3 handover. From the graph, it is observed that handover latency increases when the velocity of the mobile node is increased. Fig. 4 shows the latency for various velocities of the MH. The handover latency for inter subnet vertical handover is less when compared to that of inter-GW handover. Use of AS yields better results. Instead, if the MH contacts the DPS directly, the latency involved is high.

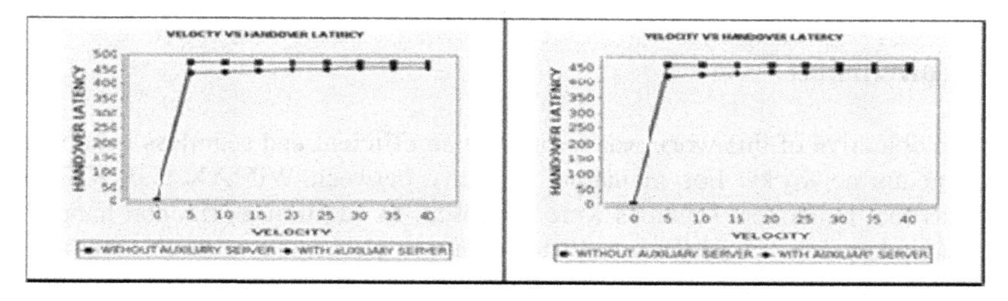

Fig. 4. Velocity and Handover Latency of Inter-GW Handover and Inter-Subnet Handover

6.6 Traffic and Packet Loss

By varying the video traffic, i.e. by varying the sizes of the packet, the number of packets lost during handover is observed. Packet loss for inter-GW vertical handover is high. From the graph, it is evident that packet loss increases when the traffic rate is increased. Packet losses are high, when AS is not involved. The traffic values plotted should be multiplied by 10^3 (Fig. 5).

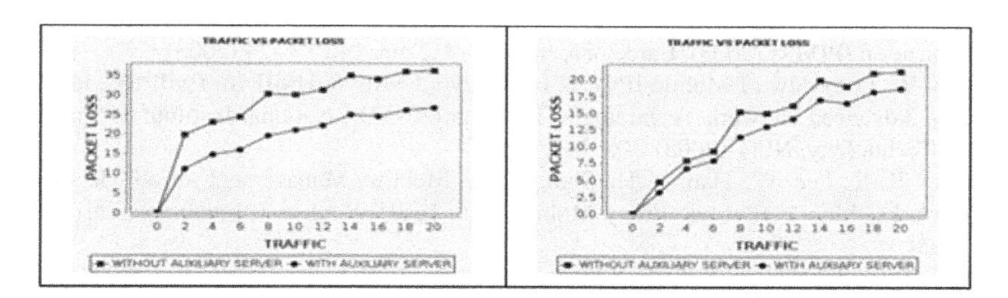

Fig. 5. Traffic and Packet Loss of Inter-GW Handover and Inter-Subnet Handover

6.7 Traffic and Handover Latency

By varying the video traffic i.e. by varying the sizes of the packet, latency during handover is observed. Figure gives the handover latency for various sizes of the packets. From the graphs, it is observed that, for vertical handovers, there is no much variation in handover latency when the traffic rate is varied. Latency increases if the DPS is communicated instead of AS. The traffic values plotted should be multiplied by 10^3 as the previous case (Fig. 6).

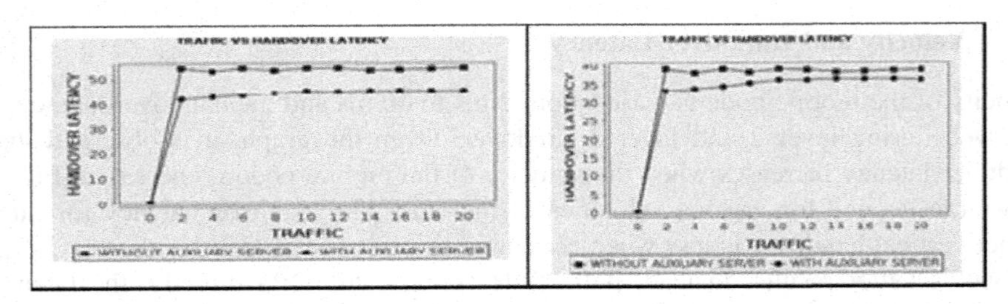

Fig. 6. Traffic and Handover Latency of Inter-GW Handover and Inter-Subnet Handover

7 Conclusion

The main objective of this work was to provide an efficient and seamless handover in heterogeneous networks. For managing mobility between WiMAX and WiFi networks, vertical handovers schemes were proposed. In addition, horizontal handovers were also taken into consideration. They were implemented with the support of 802.21 Media Independent Handover (MIH). For QoS provisioning, Auxiliary Server (AS) architecture was applied to mobile heterogeneous networks. From the results, it is evident that usage of an AS provides better performance for both inter and intra Gateway handovers. The handover latency and the packet loss are very much reduced.

References

1. Jang, H., Jee, J., Han, Y.H., Park, S.D., Cha, J.: Mobile IPv6 Fast Handovers over IEEE 802.16e Networks draft (2005)
2. Portoles, M., Zhong, Z., Choi, S., Chou, C.-T.: IEEE 802.11 Link - Layer Forwarding for Smooth Handoff. In: Proceeding of 14th IEEE Personal, Indoor and Mobile Radio Communication (PIMRC 2003), Barcelona, Spain, vol. 2, pp. 1420–1424 (2003)
3. Kim, Y.: Overview of Mobile IPv4, IPv6, MIPv6, FMIPv6, HMIPv6, PMIPv6 and 3GPP IMS. Advanced Network Technologies Division (ANTD), National Institute of Standards and Technology, NIST (2008)
4. Kong, K.-S., Lee, W., Han, Y.-H., Shin, M.-K.: Mobility Management for all - IP Mobile Networks: Mobile IPv6 vs. Proxy Mobile IPv6. IEEE Wireless Communications, 36–45 (2008)
5. Alvi, A.N., Babakhanyan, T.: Fast Layer-3 handover in Vehicular Networks (2009)
6. Johnson, D., Perkins, C., Arkko, J.: Mobility Support in IPv6. IETF RFC 3775 (2004)
7. McCann, P.: Mobile IPv6 Fast Handovers for 802.11 Networks, draft-ietf-mipshop-80211fh-03.txt (2004)
8. Neves, P., Soares, J., Sargento, S.: Media Independent Handovers: LAN, MAN and WAN Scenarios. In: IEEE GLOBECOM Workshops. Portugal Telecom Inovacao, Aveiro, Portugal, pp. 1–6 (2009)
9. Wang, X., Changshu: A secure communication model of solving anycast scalability in IPv6. In: Proceedings of Second International Conference on Networking and Digital Society, Changshu, China, vol. 1, pp. 171–174 (2010)

10. Wang, X., Changshu: Implementing Anycast Services in Ad Hoc networks connected to IPv6 networks. In: Proceedings of First International Conference on Future Information Networks, Changshu, China, pp. 1–6 (2009)
11. Magagula, L.A., Chan, H.A., Falowo, O.E.: PMIPv6-HC: Handover Mechanism for Reducing Handover Delay and Packet Loss in NGWN. In: Proceedings of Global Telecommunications Conference, Cape Town, South Africa, pp. 1–5 (2010)
12. Ray, S.K., Pawlikowski, K., Sirisena, H.: Handover in Mobile WiMAX Networks: The State of Art and Research Issues. Proceedings of IEEE Communications Surveys & Tutorials 12(3), 376–399 (2010)
13. Yang, W.-H., Wang, Y.-C., Tseng, Y.-C., Lin, B.-S.P.: An Energy-Efficient Handover Scheme with Geographic Mobility Awareness in WiMAX-WiFi Integrated Networks. In: Proceedings of IEEE Wireless Communications and Networking Conference, Hsinchu, pp. 1–6 (2009)
14. Min-Kim, Kim, J.-M., Kim, H.-S., Ra, I.-K.: A proxy mobile IP based layer-3 handover scheme for mobile WiMAX based wireless mesh networks. In: Proceedings of Second International Conference on Ubiquitous and Future Networks (ICUFN), Seoul, Korea, pp. 33–38 (2010)
15. Damhuis, J.R.: Seamless handover within WiMAX and between WiMAX and WLAN. In: Proceedings of 8th Twente Student Conference on IT, Enschede, Netherlands (2008)
16. Guo, J., Yim, R., Tsuboi, T., Zhang, J., Orlik, P.: Fast Handover Between WiMAX and WiFi Networks in Vehicular Environment. In: ITS World Congress, Cambridge, Massachusetts (2009)
17. Hatami, A., Krishnamurthy, P., Pahlavan, K., Ylianttila, M., Makela, J., Pichna, R.: Analytical framework for handoff in non-homogeneous mobile data networks. In: Proceedings of PIMRC 1999, Osaka, Japan, pp. 760–764 (1999)
18. Ylianttila, M., Pichna, R., Vallstram, J., Makela, J., Zahedi, A., Krishnamurthy, P., Pahlavan, K.: Handoff procedure for heterogeneous wireless networks. In: Proceedings of IEEE Global Telecommunications Conference, GLOBECOM 1999, Finland, vol. 5, pp. 2783–2787 (1999)
19. Pahlavan, K., Krishnamurthy, P., Hatami, A., Ylianttila, M., Makela, J.P., Pichna, R., Vallstron, J.: Handoff in hybrid mobile data networks. IEEE Communication Magazine 7(4), 34–47 (2000)
20. Chen, L., Sun, T., Chen, B., Rajendran, V., Gerla, M.: A smart decision model for vertical handoff. In: The Proceedings of 4th International Workshop on Wireless Internet and Reconfigurability (ANWIRE 2004), Athens, Greece (2004)
21. Kumar, R., Singh, B.: Comparison of vertical handover Mechanisms using generic QoS trigger for Next generation network. International Journal of Next-Generation Networks (IJNGN) 2(3) (2010)
22. Banerjee, K., Islam, S.M.R., Tahasin, Z.I., Uddin, R.: An Efficient Handover Scheme for PMIPv6 in IEEE 802.16/WiMAX Network. Int. J. Electrical & Computer Sciences, IJECS-IJENS 11(5) (2011)
23. Montavont, N., Noel, T.: Handover Management for Mobile Nodes in IPv6 Networks. IEEE Communication Magazine 40(8), 38–43 (2002)
24. Sharma, S., Zhu, N., Chiueh, T.-C.: Low-Latency Mobile IP Handoff for Infrastructure-Mode Wireless LANs. IEEE J. Selected Areas in Communication 22(4), 643–652 (2004)
25. Nkansah-Gyekye, Y., Agbinya, J.I.: Vertical Handoff Decision Algorithm for UMTS-WLAN. In: The Proceedings of Second International Conference on Wireless Broadband and Ultra Wideband Communications, Bellville, pp. 27–30 (2007)

26. Huang, C.-M., Lee, C.-H.: Layer 7 Multimedia Proxy Handoff Using Anycast/Multicast in Mobile Networks. IEEE Transactions on Mobile Computing 6(4), 411–422 (2007)
27. Zahran, H., Liang, B., Saleh, A.: Signal threshold adaptation for vertical handoff in heterogeneous wireless networks. Mobile Networks and Applications 11(4), 625–640 (2006)
28. Yan, X., Mani, N., Cekercioglu, Y.A.: A traveling distance prediction based method to minimize unnecessary handovers from cellular networks to WLANs. IEEE Communications Letters 12(1), 14–16 (2008)
29. Yang, K., Gondal, I., Qiu, B., Dooley, L.S.: Combined SINR based vertical handoff algorithm for next generation heterogeneous wireless networks. In: Proceedings of the 2007 IEEE Global Telecommunications Conference (GLOBECOM 2007), Washington, DC, USA, pp. 4483–4487 (2007)
30. Chi, Cai, X., Hao, R., Liu, F.: Modeling and analysis of handover algorithms. In: Proceedings of the 2007 IEEE Global Telecommunications Conference (GLOBECOM 2007), Washington, DC, USA, pp. 4473–4477 (2007)
31. Hasswa, Nasser, N., Hassanein, H.: Tramcar: A context aware cross-layer architecture for next generation heterogeneous wireless networks. In: Proceedings of the 2006 IEEE International Conference on Communications (ICC 2006), Istanbul, Turkey, pp. 240–245 (2006)
32. Tawil, R., Pujolle, G., Salazar, O.: A vertical handoff decision scheme in heterogeneous wireless systems. In: Proceedings of the Vehicular Technology Conference (VTC 2008-Spring), Paris, pp. 2626–2630 (2008)
33. Tourrilhes, J.: L7-Mobility: A Framework for Handling Mobility at the Application Level. In: Proceedings of Fifteenth IEEE International Symposium on Personal, Indoor and Mobile Radio Communication (PIMRC 2004), Palo Alto, CA, USA, vol. 2, pp. 1246–1251 (2004)

Improved Latency Handover in Fast MIPv6 for Streaming Video

Vidhate Amarsinh[1] and Satish Devane[2]

[1] Ramrao Adik Institute of Technology, Nerul, Navi Mumbai
vidhate.amarsinh@gmail.com
[2] Datta Meghe College of Engineering, Airoli, Navi Mumbai
srdevane@yahoo.com

Abstract. Uses of multimedia, video and audio oriented applications are increasing day by day on mobile devices. The growth of these applications is hampered because of limitation in MIPv6 during handover, as MIPv6 is not designed for continuous streaming. Limitations to support QoS parameters like variable jitter, delays in addition to loss of packets for streaming video during handoff. This paper makes an attempt to improve the latency in handover by modifying the signals related to handover, which has resulted in reducing the signaling cost and latency.

Keywords: Handover, MIPv6, Latency.

1 Introduction

In streaming applications, media streams have to be transmitted continuously in several application domain that span from traditional telecom services such as Voice over IP (VoIP), video conference–to entertainment and Video on Demand (VoD), overcoming the fluctuations of network resources. One of the challenging issues in supporting wireless internet continuous services is flow of data when clients roam from one locality to another. In order to provide uninterrupted services and maximum user-perceived quality, a successful video streaming solution needs to adapt good mobile handoff scenarios. While the mobile users switch between networks of different capacities and various service providers. Fast MIPv6 is a good protocol for normal handover operations with the non real time applications, but latency in seamless handover limits continues support to streaming media.

Mobile client selects the best connecting access point based on their service requirements and user's requirements. The process of changing point of connection in support of mobility is termed as handover. The main objective of Mobile IP[3] protocol is to ensure correct routing of packets to mobile node as mobile node changes its point of attachment with the internet. Many research proposals are available for streaming applications to reduce packet loss and delays but fails during handover [3][4][5].

Our work is motivated by a problem, to reduce the packet loss and handover delay during handoff by improving the MIPv6 signals and buffering. This paper provides

S. Unnikrishnan, S. Surve, and D. Bhoir (Eds.): ICAC3 2013, CCIS 361, pp. 381–392, 2013.

the solution by embedding MIPv6 signals to make it suitable for uninterrupted media streaming during handover.

2 Background

This section presents the main issues of handover, requirements of mobile multimedia in context of streaming continuity.

2.1 Handover

Handover is a process where mobile Node (MN) moves from one to other similar or dissimilar technology network by keeping connection active and seamless throughout. Handover Management is the process which allows MN to keep its connection active while transferring from one point of attachment area to other. The typical reason for handover is the signal degradation below a threshold value caused by the mobility or other issues [7].

Handover is classified into two types based on the movement of the MN from one region to other, if MN moved to region belonging to the same service technology called as horizontal handover or homogeneous handover and for other service technology region is called as vertical handover or heterogeneous handover [1, 7, 8].

Handover is also described as soft handover or hard handover based on how MN makes a new connection with the new Base Station (BS), before terminating the current connection if it makes connection to new BS, is referred as soft handover else hard handover.

Fast MIPv6 is the improved MIPv6 protocol to allow continuous streaming service but yet issues like handover latency, buffering mechanism to take care of lost packets during disconnections and sufficient prefetching of packets before its playout, are not addressed properly to improve streaming continuity.

The actual handover process takes major delay at layer 2 and layer 3. Layer 2 handover includes scan, authentication and association as the various phases. The link delay may be caused by either heavy traffic in an access router or the transmission latency, thus there is no guarantee for this delay and it is hardly unavoidable also, but can be minimized by using pre-authorization, and pre-distributes authentication information to the access points that are the potential targets of a future handover [17]. The scanning is required to search potential access points (Base Stations) in the near-by proximity area. Authentication is a single round trip message to verify the claimed identity. Association is a process where MN requests association or reassociation to a new base station with all the service capabilities. Once completed, MN is really connected and ready for higher layer operations [19].

Layer 3 handoff includes MD (Movement Detection) Latency, CoA Configuration latency which includes Duplicate Address Detection (DAD) and BU (Binding Update) Latency. By inspecting the current prefix with the previous one MN comes to know its movement to the other subnet. To insure that all configured addresses are likely to be unique on a given link, nodes run a "duplicate address detection" algorithm on addresses before assigning them to an interface. Binding Update will take a small latency which will be depends upon the distance between MN and HA/CN [3].

3 Related Work

There are various proposals like MIPv6[3], Fast MIPv6[4], Hierarchical MIPv6[5], Hierarchical Fast MIPv6[6], etc which claims to reduce the signaling delay and handover latency but they are not efficient to handle packet loss by optimizing signaling sequences, and even support about the streaming continuity. Many of the L3 events are taking place before L2 proactively, few takes care of buffering at the agents. The available proposals lack in two major solutions as lack of dynamic size buffering and coordination of playout buffer at the client side with the PAR/NAR's buffer [10] [11] [12] [13] [14].

A. Dutta et al. [11], have proposed a node near to PAR or NAR as Buffering node (BN) which will take care of buffering during handover. Even the size of the buffer goes on increasing or decreasing based on the need. They have provided a solution to this scenario by buffering packets for the Mobile Node at an access router or network node near the edge of the network where mobile may be moving away from or moving towards. The buffered packets are then forwarded to the Mobile Node once the handoff process completes. Ability to control the buffer dynamically provides a reasonable trade-off between delay and packet loss which is within the threshold limit for real-time communication.

Salim M. Zaki et al.[12], have proposed scheme based on the two tier buffering mechanism which claims that could reduced packet loss occurred during Mobile Node handover as the Correspondent Node sending data to Mobile Node.

P. Bellavista et al.[13], have proposed three contributions to enable handoff management to self-adapt to specific application requirements with minimum resource consumption. First, it proposes a simple way to specify handoff-related service-level objectives that are focused on quality metrics and tolerable delay. Second, it presents how to automatically derive from these objectives a set of parameters to guide system-level configuration about handoff strategies and dynamic buffer tuning. Third, it describes the design and implementation of a novel handoff management infrastructure for maximizing streaming quality while minimizing resource consumption.

Wei-Min Yao et.al.[14], have proposed an enhanced buffer management scheme for Fast handover. By means of the proposed scheme, we are able to improve the buffer utilization on routers.

V. Berlin Hency et al. [16], have proposed a enhancement for mobility by reducing the handoff latency along with reduced packet loss across the heterogeneous networks such as WLAN and UMTS by comparing the predicted RSS (received signal strength) against two thresholds calculated by the MN to make handoff decision. In the proposed approach the packet losses can be reduced by having jitter buffer with some extra delay added. The handoff speed can be improved by having pre-handoff notification using CoA (Care of Address) maintained in each router and hence the binding updates to home agent and correspondent node are performed once Previous Access Router knows a Mobile Node's new CoA. By this during handoff only the node movement happens hence reducing handoff latency.

András Bohák et. al.[17], have proposed an authentication scheme that is designed to reduce the authentication delay during a WiFi handover process. In proposed approach, in order to eliminate remote communications, it uses pre-authorization, and pre-distributes authentication information to the access points that are the potential targets of a future handover. This ensures that only local communications take place during the handover itself. The advantage of the proposed approach is that it can be applicable in real time applications such as telephony and video streaming for WiFi users.

Balaji Raman et. al.[18], have shown their results with significant reduction in output buffer size compared to deterministic frameworks. According to them, in future mobile devices, the playout buffer size is expected to increase, so, buffer dimensioning will remain as an important problem in system design.

Ruidong Li et. al.[19], have proposed a fast handover mechanism which provides rapid handover service for the delay-sensitive and real-time applications. They have proposed an enhanced fast handover scheme for Mobile IPv6. Here, each AR maintains a CoA table and generates the new CoA for the MN that will move to its domain. At the same time, the binding updates to home agent and correspondent node are to be performed from the time point when the new CoA for MN is known by PAR. Also the localized authentication procedure cooperated with the proposed scheme is provided.

Some of the main challenges faced in video streaming with respect to buffering are [20,21,22,23],

- Loss of important data in the network: Most video encoding schemes encode video into packets with different importance and the packets are dropped in the transmission randomly. So the major hindrance in effective multimedia streaming is the loss of important data in the network.
- Jitter: Video also suffers from jitter due to the variation in rate of the congestion control scheme.
- Burst losses in network that lead to losing a set of packets containing information about a single frame making estimation techniques at the receiver ineffective.
- Loss of synchronization between the encoder and decoder due to network losses.
- Loss of a significant amount of data with a loss of single packet that renders quality reconstruction almost impossible.

4 MIPv6 and Fast MIPv6

Following are the signaling description used in the sections

CN : Corresponding Node, **MN**: Mobile Node, **HA**: Home Agent, **PAR**: Previous Access Router, **NAR**: New Access Router , **PAP**: Previous Access Point, **NAP**: New Access Point, **RtSolPr**: Router Solicitation for Proxy, **RtSolPr [F-BU,P]** : modified router solicitation, **PrRtAdv**: Proxy Router Advertisement, **PrRtAdv [P]** : modified router Adv., **FBU**: Fast Binding Update, **HI** : Handover Initiate, **Hack**: Handover Acknowledgment, **F-NA**: Fast Neighbor Advertisement, **BU**: Binding Update, **Back**: Binding Acknowledgment, **HoTI**: Home Test Init, **CoTI**: Care-of Test Init, **HoT**: HoT from CN, **CoT**: CoT from CN, $\mathbf{D_{L2}}$: The layer 2 Handover latency or delay, **AR**: Association Request($_{ARt}$)/Response($_{ARs}$), **RR** : Reassociation Request ($_{RRt}$)/ Response ($_{RRs}$), $\mathbf{D_{BU}}$: Binding Update latency to update HA/CN, $\mathbf{D_{IP}} = D_{DAD} + D_{MD}$: Time required performing fast neighbor advertisement, $\mathbf{D_{MD}}$: Time required to detect movement by MN, $\mathbf{D_{DAD}}$: Time delay required for duplicate address detection, **m** : time required for a packet to be transferred from MN to PAR or NAR. The required time is assumed to be symmetric, **n**: time required for a packet to be transferred between intermediate routers and PAR and NAR, Δ : is the reduction in overall delay due to early binding updation with HA. **a**: cost of packet delivery between MN and HA, **b**: cost of packet delivery between HA and CN, **c**: cost of packet delivery between MN and CN.

$$D_{L2} = D_{AR_t} + D_{AR_s} + D_{RR_t} + D_{RR_s} \quad (1)$$

$$D_{L3} = D_{MD} + D_{DAD} + D_{BU} \quad (2)$$

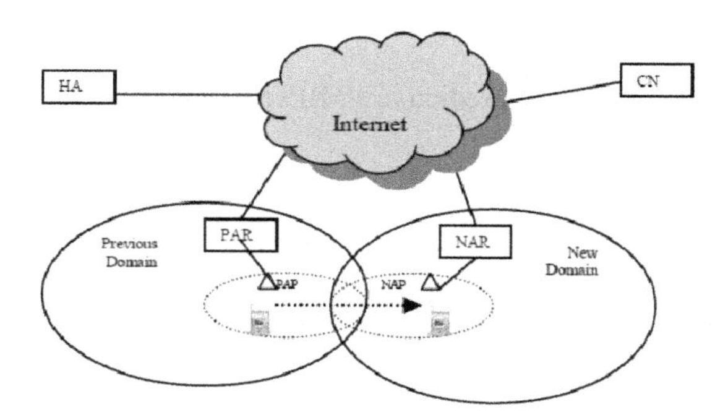

Fig. 1. System architecture: Basic Mobile IP

Fast MIPv6 (FMIPv6) is the improved version of the MIPv6 and of two types Reactive FMIPv6 and Predictive FMIPv6 differentiated based on the time to establish the tunnel between the PAR and NAR. Where in predictive handover, the tunnel is established before L2 handover, and reactive handover establishes tunnel directly after L2 handover. We focus on the predictive fast handover scheme in this paper because it has shorter latency than the reactive one and if the CoA formation and updating with PAR is not happening before the disconnection, FMIPv6 has forced to operate in reactive mode which will take longer HO delays[19].

FMIPv6 : Assumption : All the access points and routers are assumed that, they support Fast MIPV6.

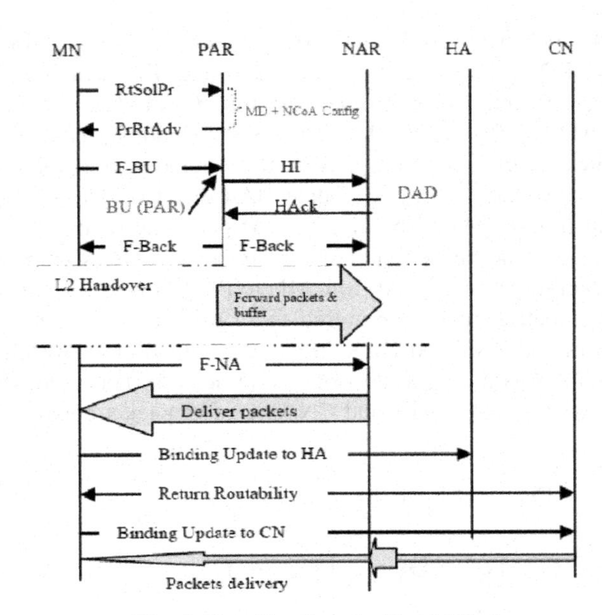

Fig. 2. Signaling flow for Fast MIPv6

There are 03 major phases in predictive FMIPv6 as handover initiation, tunnel establishment and packet forwarding.

MN can make use of layer 2 scanning techniques to identify other APs within its reach. By sending RtSolPr (Proxy router solicitation) message to its current AP, asking to get information about a certain AP identified by scanning i.e. (AP-ID) tuple. The trigger for sending RtSolPr can originate from a link-specific event, such as the promise of a better signal strength from another AR coupled with fading signal quality with the current access point.

In response PrRtAdv (Proxy Router Advertisement) message is sent, If the new access point is known and the PAR has information about it, then PAR must respond indicating that the new Access point is known and supplies the [AP-ID, AR-Info] tuple. Now MN can able to formulate NCoA, when it is still connected on PAR's link, so that latency due to prefix discovery can be eliminated. After a PrRtAdv message is processed, the MN sends an FBU which includes the proposed NCoA, so that it should be known to PAR so that binding between PCoA and NCoA should take place.

A tunnel will be formed between PCoA and NCoA, as a result, PAR begins tunneling packets arriving for PCoA to NCoA. Such a tunnel remains active until the MN completes the Binding Update with its correspondents.

The PAR and NAR can exchange messages to confirm that a proposed NCoA is acceptable.

Handover Initiation (HI) message will be sent by PAR to NAR to initiate the process of a MN's handover. It includes L2 address of MN, PCoA and NCoA. The NAR must verify if the NCoA present in HI is already in use. NAR must assign the proposed NCoA. The Handover Acknowledgment (HAck) message should be sent (typically by NAR to PAR) as a reply to the Handover Initiate message. If the NCoA is already in use, it should assign new CoA which should be included in HAck. HI and HAck are also useful to transfer the network resident contexts, such as access control, QoS, and header compression, in conjunction with handover. The Fast Binding Acknowledgment (FBack) message, which includes NCoA, is sent by the PAR to Acknowledge receipt of a Fast Binding Update message. The above four steps are important to find duplicate address on the links. Even if the probability of duplicate address is low but this step can not be ignored. If duplicate address exists in worst case, the address configuration process has to be repeated, which will take lot of extra delay. Even the PAR would be updated with the NCoA responded back.

When MN is preparing to attach to a new link (L2 handover), the packets reaching to PAR are buffered.

Return routability test is necessary to assure that the right MN is sending the message. It includes four messages like HoTI from (includes HoA to CN), CoTI (includes CoA to CN), HoT (response to HoTI), CoT (response to CoTI). A Mobile Node uses the Home Test Init (HoTI) message and the Care-of Test Init (CoTI) message to initiate the return routability procedure and request a home keygen token from a correspondent node. The Home Test (HoT) message is a response to the Home Test Init message; The Care-of Test (CoT) message is a response to the Care-of Test Init and is sent from the correspondent node to the Mobile Node. It employs binding management key, Kbm which will be later used for binding. The main advantage of the return routability procedure is that it limits the potential attackers to those having an access to one specific path in the Internet, and avoids forged Binding Updates from anywhere else in the Internet.

The MN must send Fast Neighbor Advertisement (FNA) immediately after attaching to NAR, so that arriving and already buffered packets can be forwarded to the MN right away, once BU to HA&CN as well as return routability tests are done. The Binding Cache Entry (BCE) is kept in both the Home Network and CN. Once the BU to HA and CN is done, these entries are updated in both the BCEs. Once it is done the packets can be sent directly from the CN the NCoA to the MN.

5 Proposed Improved FMIPv6 (IFMIPv6)

The above proposal needs improvement due to lack of continuous media streaming during handover. Our attempt is to transfer prospective address formation capability from MN to AR. The other important attempt is to embed several layer 3 signals including return routability signals into one frame during layer2 handover and redistributes to the access points which are the potential targets. It saves lot of signaling cost and HO latency.

If the capability of forming a new CoA is shifted from MN to PAR or NAR, this would bring down the total handover delays. The Bit P in the signal RtSolPr and PrRtAdv would indicate to use the new proposed scheme. Otherwise it works in a traditional Fast MIPv6 reactive mode.

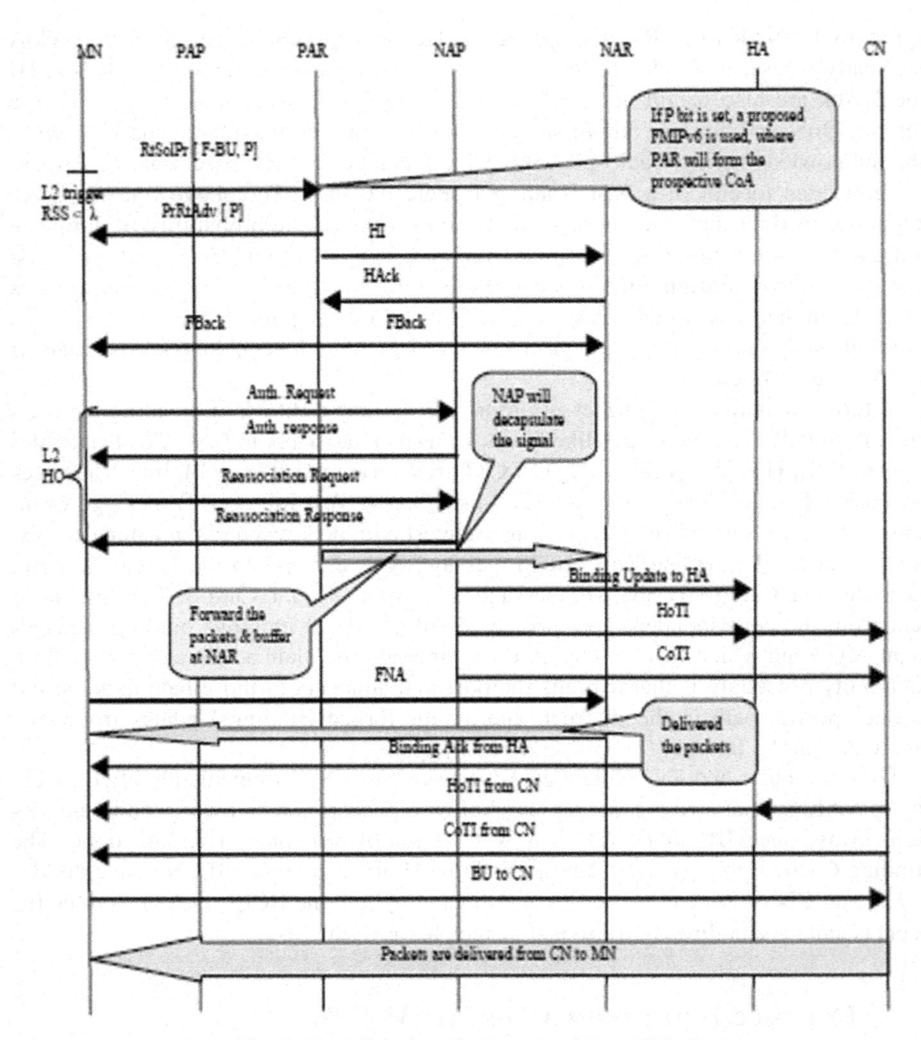

Fig. 3. Signaling flow for Improved Fast MIPv6

Once the F-BU option is received by the PAR, It forms a new CoA on behalf of MN and PAR immediately initiates fast handover in predictive mode. The basic operation of the protocol in predictive mode is as follows,

There will be a handoff decision based on signal strength (RSS), with the threshold value predefined as λ.

- There will be a HO decision made whenever RSS value goes below λ. Even more parameters would be used other than only RSS.
- Upon receiving a layer 2 indication MN initiates a fast HO by sending RtSolPr containing F-BU option and P bit. This will induce PAR to reserve a buffer with IBS size.

- If P bit is set, then proposed fast HO scheme is used. The PAR forms the prospective CoA and immediately sends a HI message to NAR and PrRtAdv [P] to MN.
- Here MN does not need to send F-BU as it has already sent.
- Once HI is received, NAR will verify NCoA uniqueness by the DAD process, which is responded with HAck.
- We can embed IBS as a size parameter to indicate to even NAR buffer also.
- Once the HAck is received, is means that, new CoA is ready to take off. So PAR sends FBack message on both the links i.e. to MN as well as to NAR.
- During layer2 disconnection period, buffer reserve-1 flush data to buffer reserve-2 and trying to continue the session with MN. Here the maximum size of the buffer shall be allowed as the Expected Delay (ED) but never should cross it. It may be nearly equate to Handover Latency (HL).
- At layer 2 also, more optimizations are done proactively. i.e. the MN sends one frame containing several signal messages related to layer 3 handover during the layer 2 handover. i.e.
- MN must be able to encapsulate a BU destined to HA, a HoTI and a CoTI messages to HA/CN in the reassociation request frame.
- A NAP is assumed to decapsulates the reassociation request frame and to forward the BU message to HA, the HoTI and the CoTI messages to CN.
- The proposed scheme can register a new address to HA and CN more quickly than the Fast Handover.
- As soon as MN connects to a new link, MN sends a FNA message so that NAR should get MN's presence. This is the point where buffer reserve-2 flushes data to MN playout buffer. The packets are delivered from these points which are already buffered at NAR.
- Like Fast MIPv6, MN must perform the home registration with Home Agent and correspondent registration with including return routability and binding update.
- The proposed method already has optimized this signaling by encapsulating few return routability signals and binding update to HA.
- Once it is done the packets can be sent directly from the CN the NCoA to the MN.

6 Latency of Handover (Delay) and Signaling Traffic Cost

In MIPv6, handover latency due to more signals and more signaling cost is a major constraint for continuous media streaming during handover. Fast MIPv6, proposed by [4] is an improvement over that in terms of reducing HO delay and signaling cost, still not appropriate to satisfy continuous media streaming. The below analysis shows the signaling cost and HO latency with the help of empirical example.

$$HO_{[Latency\_FMIPv6]} = 4m + 2n + D_{L2} + \mathrm{D}_{IP} + \mathrm{D}_{BU} + \mathrm{D}_{new} \tag{3}$$

$$HO_{[Latency\_IFMIPv6]} = 2\mathrm{m}+2\mathrm{n}+\mathrm{D}_{L2} + \mathrm{D}_{IP} + [\mathrm{D}_{BU} - \Delta] + \mathrm{D}_{new} \tag{4}$$

Where Δ is the reduction in overall delay due to early binding updation with HA.

The Value of

$$\Delta = a+b+0.5c \tag{5}$$

As an empirical example if we assume following data as m= 6ms, n= 2ms, a=5ms, b=8ms, c=10 ms, Δ=18ms, D_{L2} = 20 ms, D_{new} = 10ms, D_{IP} = 525 ms

$$HO_{[Latency\_FMIPv6]} = 659 \text{ ms}$$

$$HO_{[Latency\_IFMIPv6]} = 629 \text{ ms.}$$

The value of Δ = 18ms

FMIPv6: Let D_{FMIPv6} be signaling traffic cost for FMIPv6 Predictive Mode. Then from signaling diagram of FMIPv6 Predictive Mode (Fig.2) it can be calculated as:

$$D_{FMIPv6} = D_{L2} + D_{IP} + D_{BU} \tag{6}$$

$$D_{FMIPv6} = D_{L2} + (5m+2n) + (4a+2b+4c) \tag{7}$$

IFMIPv6: Assuming that during binding update, Cost of packet delivery between NAP and HA = a/2, Cost of packet delivery between NAP and CN = c/2
Let DIFMIPv6 be signaling traffic cost for IFMIPv6. Then from signaling diagram of FMIPv6 (Fig.3) it can be calculated as:

$$D_{IFMIPv6} = D_{L2} + D_{IP} + D_{BU} \tag{8}$$

$$\mathrm{D}_{IFMIPv6} = D_{L2} + (4m+2n) + (3a+2b+2.5c) \tag{9}$$

Considering equation (6) and (7), Where DIP for IFMIPv6 is (4m+2n) as compared to (5m+2n) of FMIPv6. This cost reduces as FBU is sent along with RtSolPr message. Also Binding Update cost reduces as NAP performs half signaling on behalf of MN.

$D_{IFMIPv6} = 104$ ms $\qquad$ $D_{FMIPv6} = 124$ ms

In conclusion, referring fig. 4 we obtain

$D_{FMIPv6} > D_{IFMIPv6}$: Improved by 20 ms.

$HO_{[Latency\_FMIPv6]} > HO_{[Latency\_IFMIPv6]}$: Improved by 30 ms.

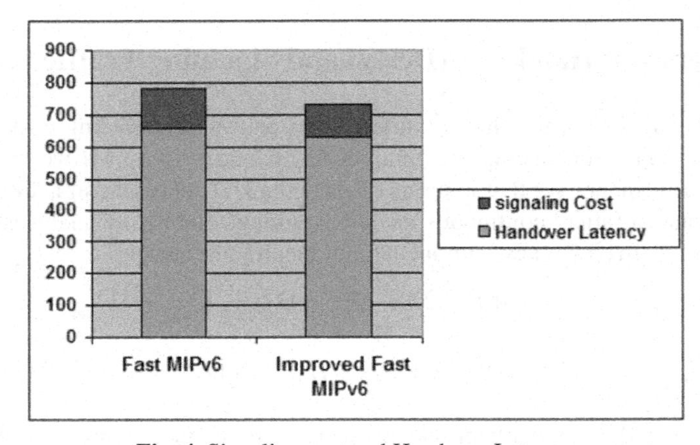

Fig. 4. Signaling cost and Handover Latency

7 Conclusion

The proposed optimizations presented into the existing FMIPv6 model by embedding F-BU signal and by embedding L3 signals into L2 signal reduce the overall Hand Over delay, which is a prime requirement in any real time data delivery. The existing Fast MIPv6 and the proposed method are analyzed analytically and the comparisons are shown. The result shows that Improved FMIPv6 (IFMIPv6) is more efficient than FMIPv6 in reducing overall handover delay as well as packet losses, which will ultimately support streaming continuity.

References

1. McNair, J., Zhu, F.: Vertical handoff in fourth generation multinetwork environment. IEEE Wireless Communications (June 2004)
2. Yan, X., Ahmet, Y., Narayanan, S.: A survey of vertical handover decision algorithms in Fourth Generation heterogeneous wireless networks. Computer Networks 54, 1848–1863 (2010)
3. Johnson, D., Perkins, C., Arkko, J.: Mobility Support in IPv6. Request for Comments: 3775 (June 2004)
4. Koodli, R.: Fast Handovers for Mobile IPv6. Request for Comments: 4068 (July 2005)
5. Nguyen, H.N., Sasase, I.: Downlink queuing model and packet scheduling for providing lossless handoff and QoS in 4G mobile networks. IEEE Trans. on Mobile Computing 5(5), 452–462 (2006)
6. Akyildiz, I.F., Xie, J., Mohanty, S.: A Survey of mobility management in next generation all IP based wireless systems. IEEE Wireless Communications (August 2004)
7. (Tim) Hong, Y.-X.: DAD-Less MIPv6, an improved mechanism for MIPv6, Master of Science in Computer Science at the University of Canterbur (2010)
8. Soliman, H., El-Malki, K., Bellier, L.: Hierarchical MIPv6 mobility management (HMIPv6). RFC (November 2001)
9. Jung, H., Soliman, H., Koh, S., Takamiya, N.: Fast Handover for Hierarchical MIPv6 (F-HMIPv6). RFC (October 2005)
10. Krishnamurthi, G., Chalmers, R.C., Perkins, C.E.: Buffer Management for Smooth Handovers in Mobile IPv6 draft–krishnamurthi –mobileip-buffer6-00.txt, Internet draft (July 13, 2000)
11. Dutta, A., Van den Berg, E., Famolari, D., Fajardo, V., Ohba, Y., Taniuchi, K., Kodama, T., Schulzrinne, H.: Dynamic Buffering Control Scheme for Mobile Handoff. In: The 17th Annual IEEE International Symposium on Personal, Indoor and Mobile Radio Communications, PIMRC 2006 (2006)
12. Zaki, S.M., Razak, S.A.: Mitigating Packet Loss in Mobile IPv6 Using Two-Tier Buffer Scheme. Computer Science Letters 3(2) (June 2011)
13. Bellavista, P., Cinque, M., Cotroneo, D., Foschini, L.: Self-Adaptive Handoff Management for Mobile Streaming Continuity. IEEE Transactions on Network and Service Management 6(2), 80–94 (2009)
14. Yao, W.-M., Chen, Y.-C.: An Enhanced Buffer Management Scheme for Fast Handover Protocol. In: Proceedings of the 24th International Conference on Distributed Computing Systems Workshops, ICDCSW 2004 (2004)

15. Broitman, M., Shilinskii, N., Solovyov, K.: Adaptive Management Algorithms for a Fixed Jitter Buffer. Automatic Control and Computer Sciences 46(1), 12–17 (2012)
16. Berlin Hency, V., Preetha, E., Vadivelan, M., Saranya, K., Sridharan, D.: Enhancement of Handoff Speed by Improving Mobility of Data Packets. IACSIT International Journal of Engineering and Technology 1(2) (June 2009)
17. Bohák, A., Buttyán, L., Dóra, L.: An authentication scheme for fast handover between WiFi access points. In: WICON 2007, Austin, Texas, USA, October 22-24 (2007)
18. Raman, B., Nouri, A., Gangadharan, D., Bozga, M., Basu, A., Maheshwari, M., Milan, J., Legay, A., Bensalem, S., Chakraborty, S.: A General Stochastic Framework for Low-Cost Design of Multimedia SoCs (2012)
19. Li, R., Li, J., Wu, K., Xiao, Y., Xie, J.: An Enhanced Fast Handover with Low Latency for Mobile IPv6. IEEE Transactions on Wireless Communications 7(1) (January 2008)
20. Balan, A., Tickoo, O., Bajic, I., Kalyanaraman, S., Woods, J.: Integrated Buffer Management and Congestion Control for Video Streaming (2003)
21. Liebl, G., Jenkac, H., Stockhammer, T., Buchner, C., Klein, A.: Radio Link Buffer Management and Scheduling for Video Streaming over Wireless Shared Channels (2004)
22. Liebl, G., Jenkac, H., Stockhammer, T., Buchner, C.: Radio Link Buffer Management and Scheduling for Wireless Video Streaming. Telecommunication Systems 30(1/2/3), 255–277 (2005)
23. Scalosub, G., Marbach, P., Liebeherr, J.: Buffer Management for Aggregated Streaming Data with Packet Dependencies. In: 2010 Proceedings IEEE Conference on INFOCOM, March 14-19 (2010)

A Technique to Improve Overall Efficiency in Wireless Cellular Network

Jayashree Shinde and Ruchira Jadhav

EXTC, K.J. Somaiya COE, Vidyavihar, Mumbai
jayhere06@yahoo.com,
adi.the.first@gmail.com

Abstract. In a conventional cellular network, a terminal receives signals not only from the base station of that cell, but also from other cell base stations. This inter-cell interference has more impact on the cell-edge users, in multi cell environment. The inter-cell interference degrades the performance of wireless systems. Using Base Station Cooperation, the ability of a mobile station to receive signals from multiple base stations can be utilized as an opportunity to improve the spectral efficiency and to get higher data rates for cell edge users. In multi cell environment using base station cooperation overall interference can be minimized marginally, whereas the interference within the cooperation region is largely reduced. This leads to a question whether it is worth doing cooperation all the time. However increase in terms of throughput may not always be enough to increase the throughput of each of the user .Hence for such a scenario the distinct cooperation scheme is used. The distinct cooperation is a joint transmission scheme in which the selection criterion is based on user's signal to interference plus noise ratio (SINR) is proposed and the capacity achieved through cooperation is shared equally among the cell-edge users. Here we will analyze the normal operation, full cooperation and distinct cooperation scheme in the downlink environment for cell edge users in cellular network and we will compare and see which method improves the user capacity and data rate.

Keywords: Cooperative transmission, full cooperation, Distinct cooperation, signal to interference plus noise ratio (SINR).

1 Introduction

In order to maintain future competitiveness i.e. the growing demand of higher data rates for broadband services like triple play, online gaming etc, LTE is being standardized. To achieve the higher data rates, the wireless network requires a large capacity. However, due to lack of available radio resources, to achieve a good capacity and Quality of Service (QoS) efficient utilization of channel resources is important. Using a proper frequency reuse, interference is minimized to a tolerable limit. However, this method of using different frequency bands for different cells decreases the spectral efficiency [1]. In full frequency re-use network, the interference degrades the system

S. Unnikrishnan, S. Surve, and D. Bhoir (Eds.): ICAC3 2013, CCIS 361, pp. 393–402, 2013.

performance, and thereby reduces network capacity [2]. In order to overcome this problem, interference mitigation techniques such as interference cancellation, interference coordination and interference randomization are currently investigated within 3GPP . However, the performance improvements offered by these techniques are limited since inter cell interference can not be completely removed [3]. A new scheme to provide high spectral efficiency in downlink direction is cooperative transmission between base stations [4].

The cooperative transmission techniques are used to improve the overall system performance, in particular, to improve the coverage of high data rate and the cell edge user throughput. It allows more than one base stations to transmit data to single user equipment (UE) simultaneously. The analysis in [5] shows, cooperative transmission requires a high-speed backhaul enabling information (data, control, synchronization, and channel state) exchange between the base stations. Further, channel state information is required at the base stations. Both requirements cause significant impact to the existing 3GPP LTE. In this paper, We therefore analyze the preferred cooperation scheme which is a hybrid scheme that adds low complexity, in order to keep the exchange of channel state information between base stations low [7].

The remainder of the paper is organised as follows: section 2 describes overview of normal and full cooperation scheme. Section 3 describes system model and flow chart of distinct cooperation. Section 4 gives results for normal and cooperation schemes with data rate analysis and section 5 gives conclusion.

There can be two modes of operation, No cooperation/Normal operation and Full cooperation. First we will analyze user capacity and data rate under normal operation.

1.1 Normal Operation (NC)

Under normal operation or when there is no cooperative transmission, the signal to interference noise ratio (SINR) in the downlink environment for mobile station MS1 is given by equation (1)

$$\mathrm{SINRnc} = \frac{|h11|^2 * \mathrm{E}\{\mathrm{Y}\}}{\eta^2 + \sum_{k=2}^{12} |h1k|^2 * \mathrm{E}\{\mathrm{Yi}\}} \tag{1}$$

Where h11 represents the channel coefficient between the terminals 1 and base station 1,h1k is the channel coefficient between the terminal 1 and base station k. E{Y} is the average transmit power of Base Station i, E{Yi} is the average power transmitted by interfering base station and η is noise variance. The capacity for terminal MS1 in bits/sec/Hz can be given by the Shannon Capacity as

$$\mathrm{Cnc} = \log_2(1 + \mathrm{a} * \mathrm{SINRnc}) \tag{2}$$

where, a is determined by the SNR gap between the practical coding scheme and the theoretical limit.

1.2 Full Cooperation (FC)

When terminal MS1 is in cooperation with BS1 and BS2, the signals transmitted by both base stations are combined. Hence SINR for the downlink channel will be given by equation (3)

$$\mathrm{SINRcoop} = \frac{|h11|^2 + |h12|^2 * \mathrm{E}\{\mathrm{Y}\}}{\eta^2 + \sum_{k=3}^{12} |h1k|^2 * \mathrm{E}\{\mathrm{Yi}\}} \tag{3}$$

where, $h11$ is the channel coefficient between MS1 and BS1. $h12$ is the channel coefficient between MS1 and BS2.E{Y} is the average power transmitted by serving BS.E{Yi} is the average power transmitted by interfering BS.$h1k$ is the channel coefficient between MS1 and k interfering BSs. η is the noise variance.

The capacity for terminal MS1 under cooperation in bits/sec/Hz will be Given by equation (4)

$$\mathrm{Ccoop} = \beta * \log_2(1 + a * \mathrm{SINRcoop}) \tag{4}$$

where, β defines the proportion of resource sharing among the terminals under cooperation. In our simulation, the value for β is ½.

1.3 System Model

The system model for distinct cooperation scheme is as shown in Figure 1. As shown in the diagram

BS1, BS2	= Base stations
MS1, MS2	= Mobile Stations
h11	= The channel coefficient between terminal 1 and BS 1
h12	= The channel coefficient between terminal 1 and BS 2
h21	= The channel coefficient between terminal 2 and BS 1
h22	= The channel coefficient between terminal 2 and BS 2

Here total 12 base stations in frequency reuse 1 network are considered.

Base stations BS1 and BS2 are under cooperation, to transmit signals to mobile terminals MS1 and MS2.For base station BS1,base station BS2 is one of the interfering base stations among the 12 base stations .Here we have considered 2 base stations under cooperation among 12 base stations.The signals from the serving base station and from the neighbour base station arrives at the mobile terminal at the same time, i.e., signal received by the mobile terminal from the two cooperative base stations are frame synchronized. The frame duration in which the BS1 transmits to MS1 is divided into two sub-frames where the first sub-frame is used for signal transmission to MS1 and the second one to MS2. Similarly, BS2 which is under cooperation with BS1 transmits in the same sequence of BS1.

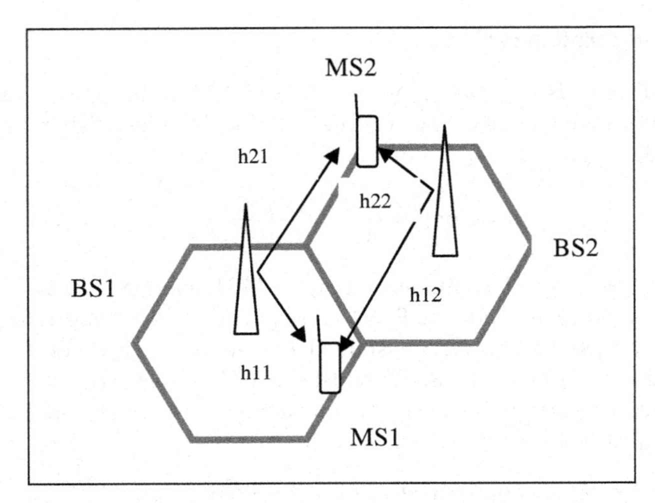

Fig. 1. System model [7]

The received signal at MS1 and MS2 are y1 and y2, and is given by equation 5.

$$\begin{pmatrix} y1 \\ y2 \end{pmatrix} = \begin{pmatrix} h11 & h12 \\ h21 & h22 \end{pmatrix} * \begin{pmatrix} x1 \\ x2 \end{pmatrix} + \begin{pmatrix} n1 \\ n2 \end{pmatrix} + \begin{pmatrix} z1 \\ z2 \end{pmatrix} \quad (5)$$

where hij is the channel coefficient between terminal i and BS j. x1 is transmit signal of BS1 and x2 is that of BS2.zi is the total interference received by MS i due to transmissions from all the base stations other than the one under cooperation (Here BS2) and ni is the additive white Gaussian noise, zi is the interference [7].

1.4 Flow Chart: Cooperation Selection

To improve the SINR, the users in the serving cell and the neighbour cell who decided to cooperate will share the available resource between them equally. Therefore, the individual user throughput is half of the actual capacity of the cooperative transmission as in (5).

The user takes the decision whether to perform cooperation or not, with the measurements of its own channel and the nearest neighbour. The decision is informed to the base station of the serving cell. The serving station intimates the neighbour station whether to do cooperation or not with a single bit information based on the input from the user [8]. The flow for distinct cooperation is shown below.

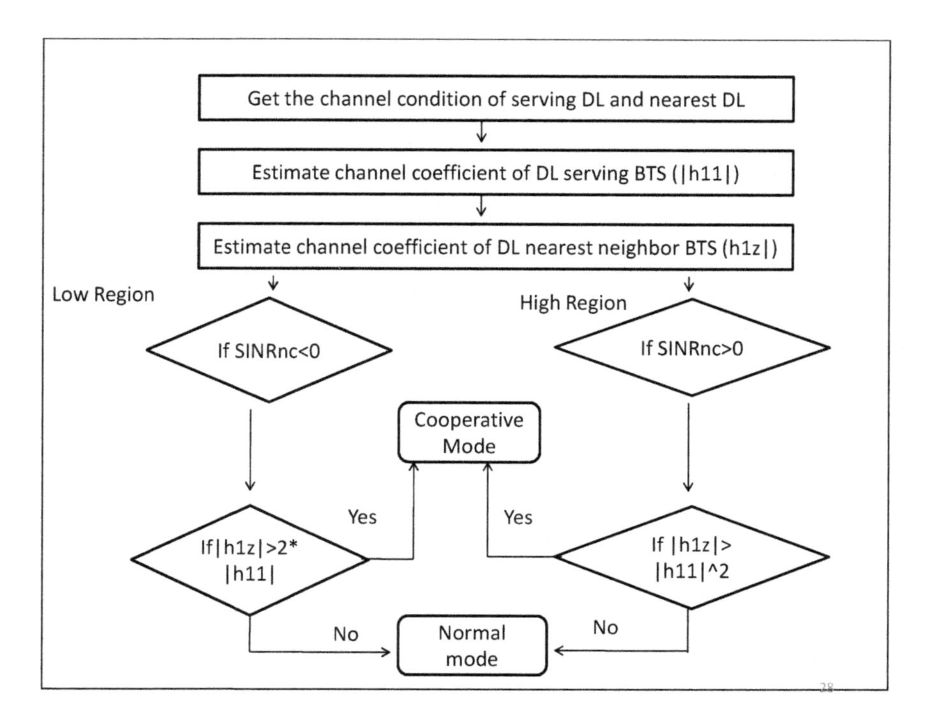

1.5 Results

The distinct Cooperation is a hybrid scheme, where cooperative transmission is performed only if SINRcoop is greater than SINRnc as described in algorithm. As mentioned a small network of 12 base stations with one user per cell edge is considered for simulation. The average power transmitted by serving base station as 10dB, noise variance as 5dB and average interfering power of base stations is 6dB, according to path loss model COST 231is considered. Table 1 show the parameters used in the simulation.

Table 1. Parameters in system-level simulation

Environment	Downlink	Frequency reuse factor	1
Average power Tx by BTS	10 dB	Average power Tx by interfering BTS	6dB
Fast fading model	Metropolitan	Model	COST 231
Noise variance	5 dB	No. of BTS under cooperation	2
Modulation	16QAM	No. of bits/symbol	4
BS Tx antenna	1	MS Rx antenna	1

The observation from the simulation is that almost half the time user capacity with full cooperation is poorer than the capacity with normal operation. But after implementing distinct cooperation, the user capacity is improved, compared to full cooperation. The mean throughput or capacity for cell edge user's under No cooperation, Full cooperation and distinct cooperation scheme is shown in Table 2.

Table 2. Cell edge user SINR and average user throughput values

Type of Scheme	Mean Capacity(bits/sec/Hz)
Normal operation	0.3180
Full Cooperation	0.3408
Distinct cooperation	0.4810

Figure 2 shows the effect of SINR on user capacity under normal operation .

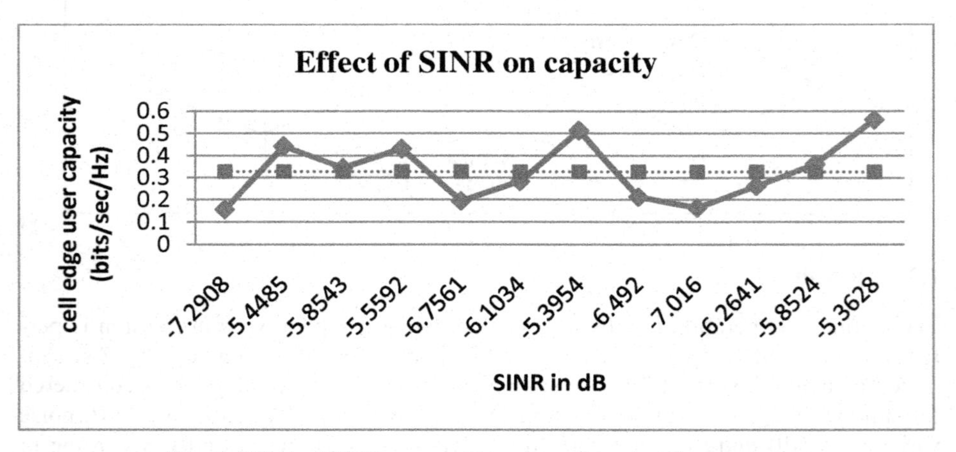

Fig. 2. Effect of SINR on user Capacity under normal operation

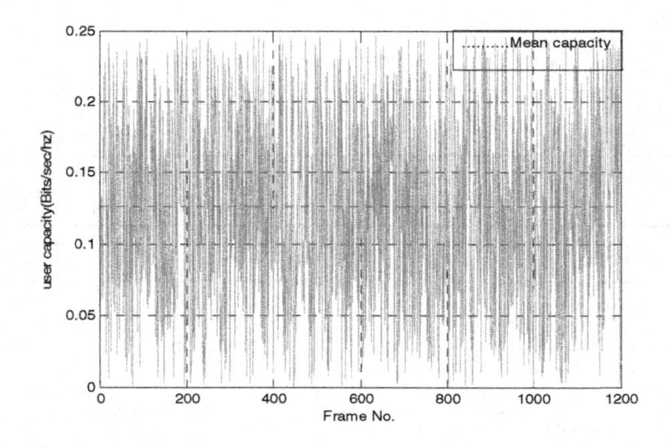

Fig. 3. Average cell edge user capacity under normal operation

The capacity for first 1200 frames is shown and its mean is calculated as shown in figure 3.

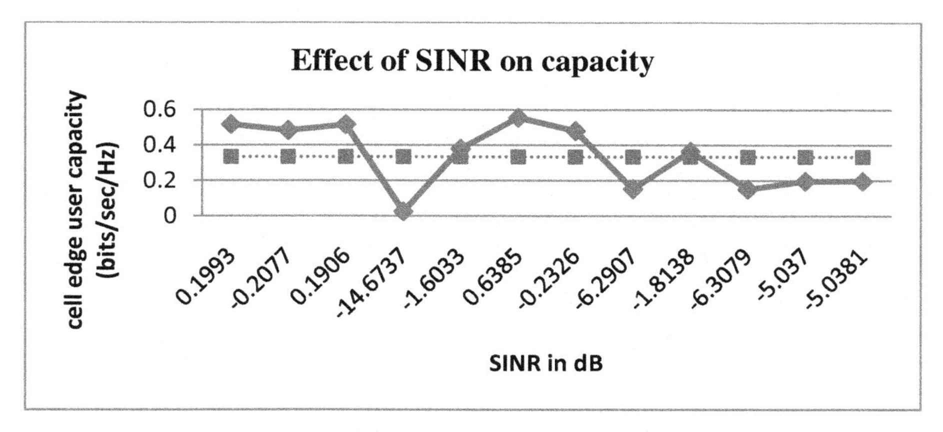

Fig. 4. Effect of SINR on user Capacity under full cooperation

The user throughput variation according to SINR variation under full cooperation scheme is shown in figure 4.

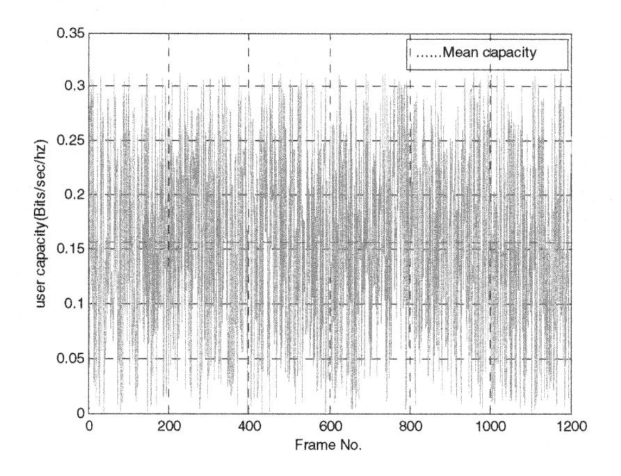

Fig. 5. Average cell edge user capacity under full cooperation

The capacity for first 1200 frames is calculated and its mean is calculated is shown in figure 5.

Figure 6 shows the user capacity variation for first 1200 frames and the average capacity for distinct cooperation is 0.98 bits/sec/Hz.

From Table 2 it is observed that, the user capacity of a cell-edge user under full co-operation is 6.69% better than no operation and the user capacity of a cell-edge user under distinct cooperation is 29.147% better than full cooperation as shown in figure 7.

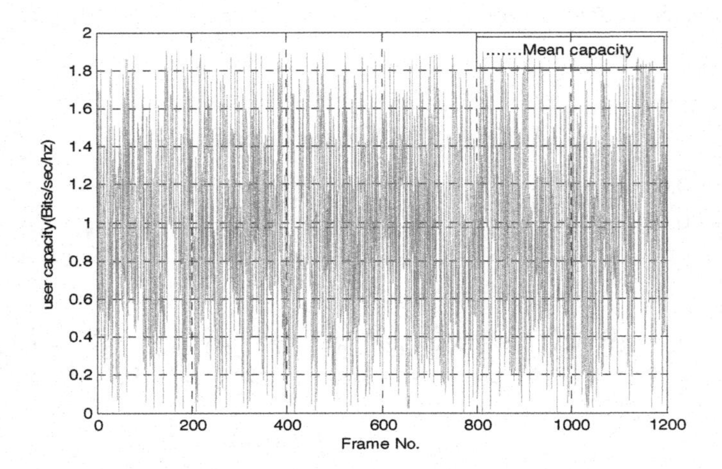

Fig. 6. Average user capacity under distinct cooperation

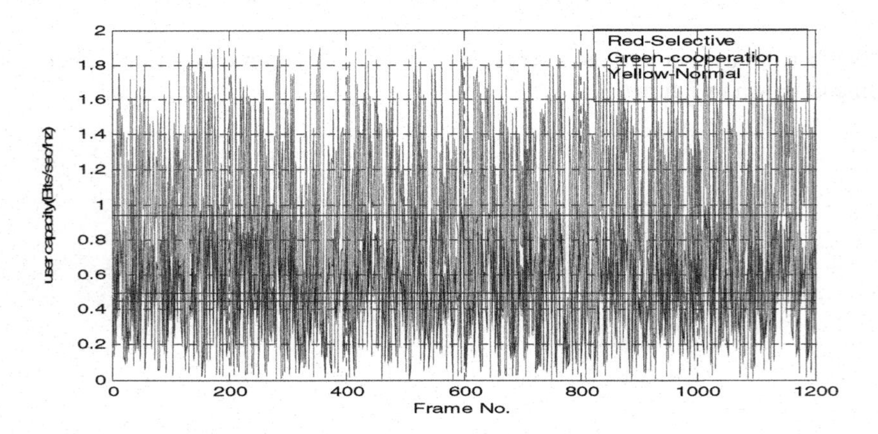

Fig. 7. Analysis of average user capacity

1.6 Data Rate Analysis

To get the data rate under normal operation i.e. when the data is transmitted by single transmitter, 1000 bits are generated with equal probability of 0 and 1.After performing the modulation and generating the AWGN noise with 0dB variance, the noise is added. Considering the Rayleigh channel, the hard decision decoding is performed at the receiver. Finally the time is calculated to transmit the total no. of bits and the data rate is calculated.

Similarly under cooperation of two BTS i.e. when two transmitters are involved to transmit the data, same parameters are used as in the normal operation and data rate is calculated. The data rate comparison of both the schemes is shown in the Table3.

Table 3. Data rate Analysis

Scheme	Data Rate(Bits/Sec/MHz)
Normal Operation	4.177
Cooperation	9.353

From Table 3 it is seen that data rate for normal operation is 4.177MHz .When same data is transmitted by two base stations simultaneously using cooperation, the data rate increases to 9.352 MHz i.e. the data rate gets double.

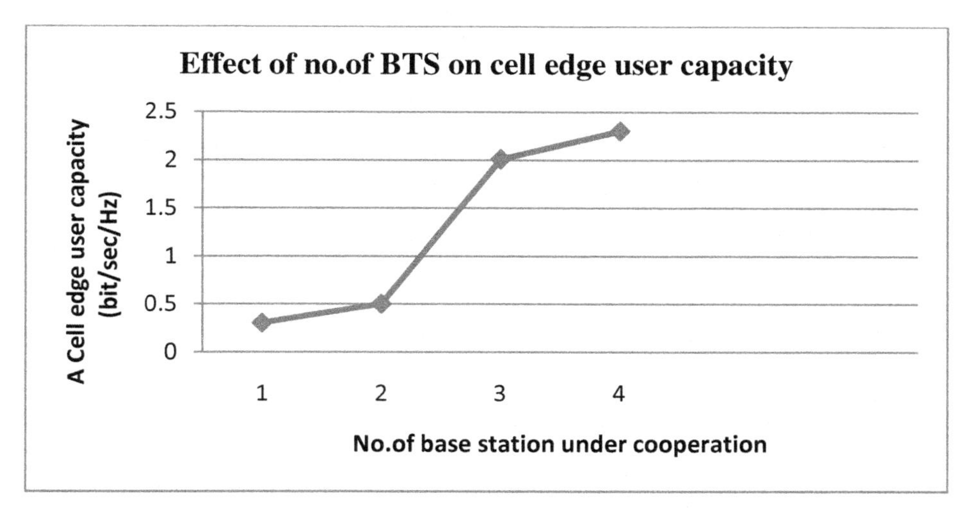

Fig. 8. Effect of no. of BTs in cooperation on a cell edge user capacity

There can be more than one base station involved in cooperation; here we have considered only two base stations. From figure 8 it is seen that if we increase the no. of cooperating base stations a cell edge user capacity starts increasing. But there is limit to this as it adds complexity in the network. So optimum no. of base station for cooperation is taken and generally it is three.

2 Conclusion

In this paper, we simulated a small network with 2-cell under cooperation. The simulation is done for downlink environment. It is seen that by using distinct cooperation, both the capacity and SINR is improved.

The user capacity of a cell-edge user under full cooperation is 6.69% better than no operation. The simulation results show that only few of the time user capacity with full cooperation is better than the capacity with normal operation.

The simulation result shows that distinct cooperation gives the user capacity improvement of 29.14% when compared to full cooperation.

We also analyzed that the data rate gets double under cooperation than normal operation. It is also seen that, optimum no. of base station for cooperation should be taken as it adds complexity in the network.

References

[1] Engstrom, S., Johansson, T., Kronestedt, F., Larsson, M., Lidbrink, S., Olofsson, H.: Multiple reuse patterns for frequency planning in GSM networks. In: 48th IEEE Vehicular Technology Conference, VTC 1998, May 18-21, vol. 3, pp. 2004–2008 (1998)

[2] Sari, H., Sezginer, S.: MIMO techniques and full frequency reuse in mobile WiMAX systems. In: International Conference on Telecommunications, ICT 2009, May 25-27, pp. 8–12 (2009)

[3] Boudreau, G., Panicker, J., Guo, N., Chang, R., Wang, N., Vrzic, S.: Interference coordination and cancellation for 4G networks. IEEE Communications Magazine 47(4), 74–81 (2009)

[4] Nosratinia, A., Hunter, T.E., Hedayat, A.: Cooperative communication in wireless networks. IEEE Communications Magazine 42(10), 74–80 (2004)

[5] Andrews, J.G., Choi, W., Heath, R.W.: Overcoming Interference in Spatial Multiplexing MIMO Cellular Networks. IEEE Wireless Communications Magazine 14(6), 95–104 (2007)

[6] Tamaki, T., Seong, K., Cioffi, J.M.: Downlink MIMO Systems Using Cooperation among Base Stations in a Slow Fading Channel. In: Proceeding of IEEE International Conf. on Communications 2007, pp. 4728–4733 (June 2007)

[7] Ramesh Kumar, M.R., Bhashyam, S., Jalihal, D.: Throughput improvement for cell-edge users using selective cooperation in cellular networks. In: 5th IFIP International Conference on Wireless and Optical Communications Networks, WOCN 2008, May 5-7, pp. 1–5 (2008), doi:10.1109/WOCN.2008.4542516

[8] Tolli, A., Pennanen, H., Komulainen, P.: On the Value of Coherent and Coordinated Multi-Cell Transmission. In: IEEE International Conference on Communications Workshops, ICC Workshops 2009, June 14-18, pp. 1–5 (2009)

[9] Sendonaris, A., Erkip, E., Aazhang, B.: User cooperation diversity, Part II. Implementation aspects and performance analysis. IEEE Transactions on Communications 51(11), 1939–1948 (2003)

[10] Cho, Y.S., Kim, J.: MIMO-OFDM wireless communications with MATLAB, ISBN 978-0-470-82561-7

[11] Andrews, J.G.: Fundamentals of WiMAX Understanding Broadband Wireless Networking, ISBN 0-13-222552-2

Broadband Semi-circular Microstrip Antennas

Amit A. Deshmukh[1], K.P. Ray[2], A.R. Jain[1],
M. Mansi[1], R. Pratiksha[1], and S. Raj[1]

[1] DJSCOE, Vile – Parle (W), Mumbai – 400 056, India
[2] SAMEER, I.I.T. Campus, Powai, Mumbai – 400 076, India
{amitdeshmukh76,kpray}@rediffmail.com

Abstract. The proximity fed broadband semi-circular microstrip antenna is proposed. The analysis to study the effects of slots on the broadband response of slot cut circular patch is presented. The slot does not introduce any additional mode but modifies the resonance frequency of higher order TM_{21} mode of the patch and along with the TM_{11} mode yields broadband response. This configuration yields a BW of nearly 400 MHz at center frequency of around 1000 MHz. Further to increase the gain, a gap-coupled configuration of semi-circular patch with proximity fed slot cut patch is proposed. This configuration gives nearly the same bandwidth but with an increased gain. Further proximity fed gap-coupled rectangular slot cut semi-circular patch configuration is proposed. This configuration yields a bandwidth of more than 500 MHz with broadside radiation pattern and gain of more than 7 dBi over the bandwidth.

Keywords: Semi-circular microstrip antenna, Broadband microstrip antenna, Proximity feeding, Rectangular slot.

1 Introduction

The broadband microstrip antenna (MSA) is more commonly realized by fabricating the slot cut radiating patch on the electrically thicker substrates [1 – 6]. The thicker substrate reduces the quality factor of the cavity below the patch and yields larger bandwidth (BW). The slot cut introduces an additional mode near fundamental mode resonance frequency of the MSA thereby realizing broadband response. These slot cut MSAs are optimized on substrate of thickness 0.06 to $0.08\lambda_0$. For substrate thickness greater than this, the BW is increased by using proximity feeding technique [7, 8]. While designing slot cut MSAs in desired frequency band, the slot length is taken to be either quarter wave or half wave in length. However this simpler approximation of slot length against the wavelength (frequency) does not give closer results for different slot dimensions and its position inside the patch. The analysis to study the effect of U-slot on the broadband and dual band response is reported [9]. The approximate equations in terms of slot and patch dimensions are given. However a clear explanation of the mode introduced by the U-slot and the comparison of calculated frequencies against the simulated and measured results is not given. The analysis to study the effects of slots on the broadband and dual band response of

S. Unnikrishnan, S. Surve, and D. Bhoir (Eds.): ICAC3 2013, CCIS 361, pp. 403–411, 2013.

circular and rectangular MSA is presented [10, 11]. It was observed that the slot does not introduce any additional mode but modifies the resonance frequency of higher order mode of the patch and along with the fundamental mode yields broadband or dual band characteristics. The slot also modifies the directions of surface currents on the patch and aligns them in the same direction as that of the currents at fundamental mode and thereby it gives similar radiation characteristics over the complete BW and at the dual frequencies. The semi-circular MSA (SCMSA) is a compact variation of circular MSA (CMSA). In this paper, broadband proximity fed SCMSA is discussed. The fundamental and higher order modes of the SCMSA are studied. The broadband proximity fed rectangular slot cut SCMSA is proposed. The analysis to study the effects of slots on the broadband response is presented. It was observed that slot reduces the resonance frequency of higher order TM_{21} mode of the patch and due to its coupling with the fundamental TM_{11} mode yields broadband response. The simulated and measured BW of nearly 450 MHz at centre frequency of 1000 MHz is obtained. Further to increase the gain another SCMSA having different patch radius is gap-coupled to proximity fed rectangular slot cut SCMSA. This configuration gives nearly the same BW but with increased overall gain. To further increase the BW, the rectangular slot was cut inside the gap-coupled SCMSA. This configuration gives a BW of more than 500 MHz (> 45%) with broadside radiation pattern with gain of more than 6 dBi over the VSWR BW. To realize maximum efficiency, these gap-coupled SCMSAs were analyzed using air substrate. They were first optimized using IE3D software followed by experimental verifications [12]. In the simulation an infinite ground plane is used. To simulate this effect in measurements, a square ground plane of side length 80 cm is used. The antennas were fabricated using the copper plate and they were suspended in air using the foam spacer supports which were placed towards the antenna corners. The antennas were fed using the N-type connector of 0.32 cm inner wire diameter. The radiation pattern was measured in minimum reflection surroundings with required minimum far field distance between the reference antenna and the antenna under test [13]. The gain measurement was carried out using three antenna method [13].

2 Proximity Fed Scmsa

The proximity fed SCMSA is shown in Fig. 1(a, b). The coupling rectangular strip is placed below the patch. The SCMSA radius is selected such that it resonates in TM_{11} mode at frequency of around 1000 MHz [1]. To realize larger BW and efficiency, the substrate thickness for SCMSA is selected to be 3.0 cm ($0.1\lambda_0$) whereas the coupling strip is placed at substrate thickness of 2.8 cm ($0.093\lambda_0$). The proximity fed SCMSA is simulated using IE3D software and its resonance curve plot is shown in Fig. 1(c). The plot shows two peaks due to TM_{11} and TM_{21} modes. The surface current distribution at these two frequencies is shown in Fig. 1(d, e). At TM_{11} mode the surface currents shows half wavelength variation along the patch diameter and along the patch perimeter. At TM_{21} mode the currents shows half wavelength variation along patch diameter and two half wavelength variation along patch perimeter. Inside this patch, a rectangular slot is cut as shown in Fig. 2(a). The resonance curve plots for varying slot length are shown in Fig. 2(b). The surface current distributions at the modified TM_{11} and TM_{21} modes for slot length of 3.5 cm is shown in Fig. 2(c, d).

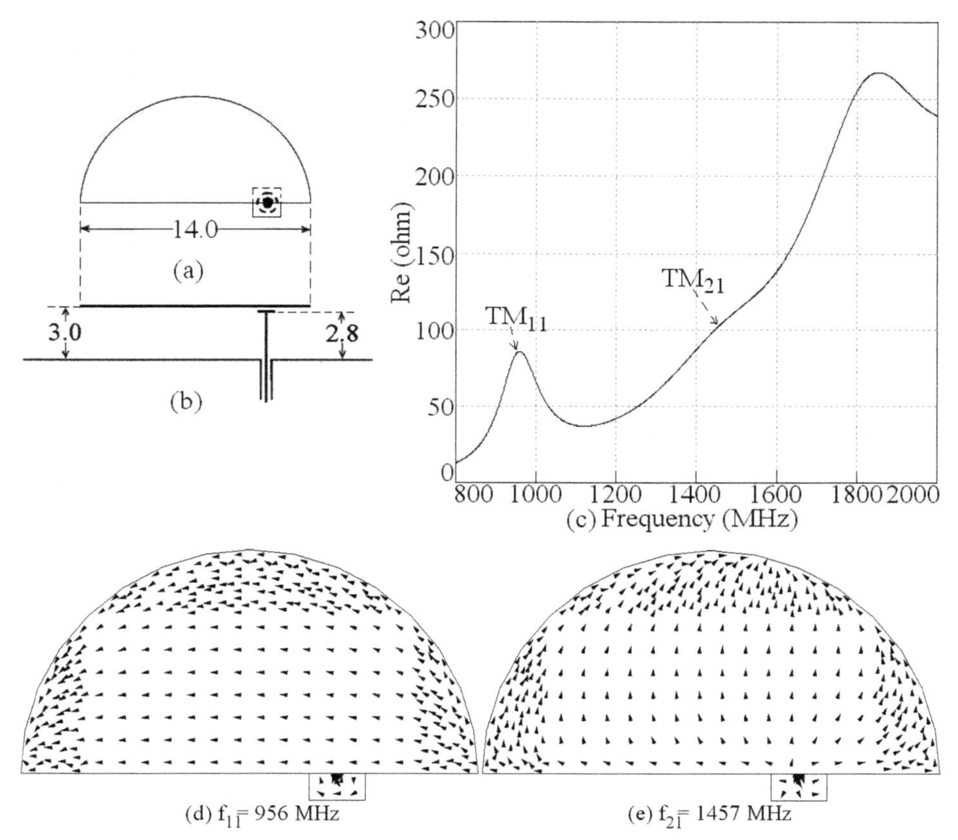

Fig. 1. (a) Top and side views of proximity fed SCMSA, its (c) resonance curve plots and its surface current distribution at (d) TM_{11} and (e) TM_{21} modes

The slot length is parallel to the surface currents at TM_{11} mode whereas it is orthogonal to the currents at TM_{21} mode. Therefore with an increase in slot length the TM_{11} mode frequency remains almost constant and the TM_{21} mode frequency reduces as shown in the resonance curve. With an increasing in slot length the surface currents at TM_{21} mode are re-oriented along the horizontal direction inside the patch. The broadband response is realized when the loop formed due to the coupling between the TM_{11} and TM_{21} modes lies inside the VSWR = 2 circle. This is realized for slot length of 4.4 cm and the input impedance and VSWR plots for the same are shown in Fig. 3(a). The simulated BW is 450 MHz (41.2%) whereas the measured BW is 458 MHz (41.6%). The radiation pattern over the BW is shown in Fig. 3(b – d). The pattern is in the broadside direction with cross-polarization levels less than 12 dB as compared to that of the co-polar levels. The cross polar levels increases towards the higher frequencies of BW due to vertical currents present in the patch due to TM_{21} mode. The antenna gain is more than 7 dBi over the BW as shown in Fig. 3(e). The gain plot shows two peaks due to TM_{11} and TM_{21} modes. To further increase the gain, an additional SCMSA is gap-coupled to the slot cut SCMSA as discussed in the following section.

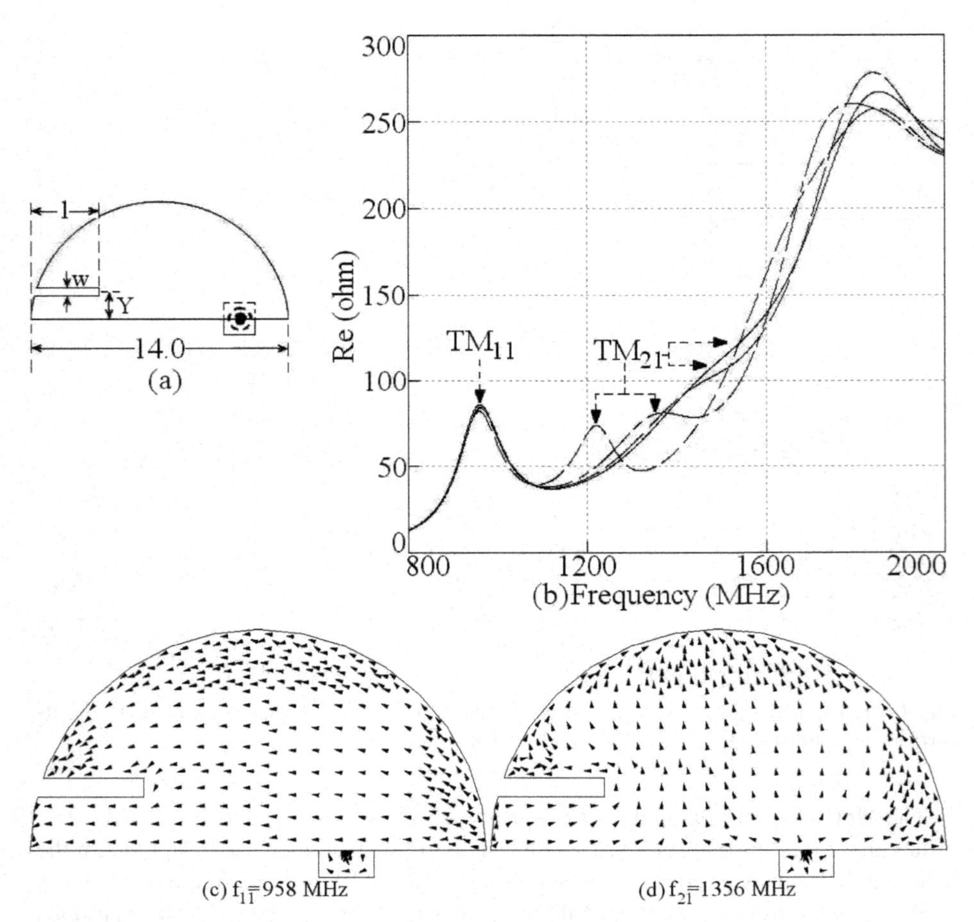

Fig. 2. (a) Slot cut proximity fed SCMSA, its (b) resonance curve plots for, l = (——) 0, (— —) 2.5, (— - —) 3.5, (- - -) 4.5 and its surface current distribution at modified (c) TM_{11} and (d) TM_{21} modes for l = 3.5 cm

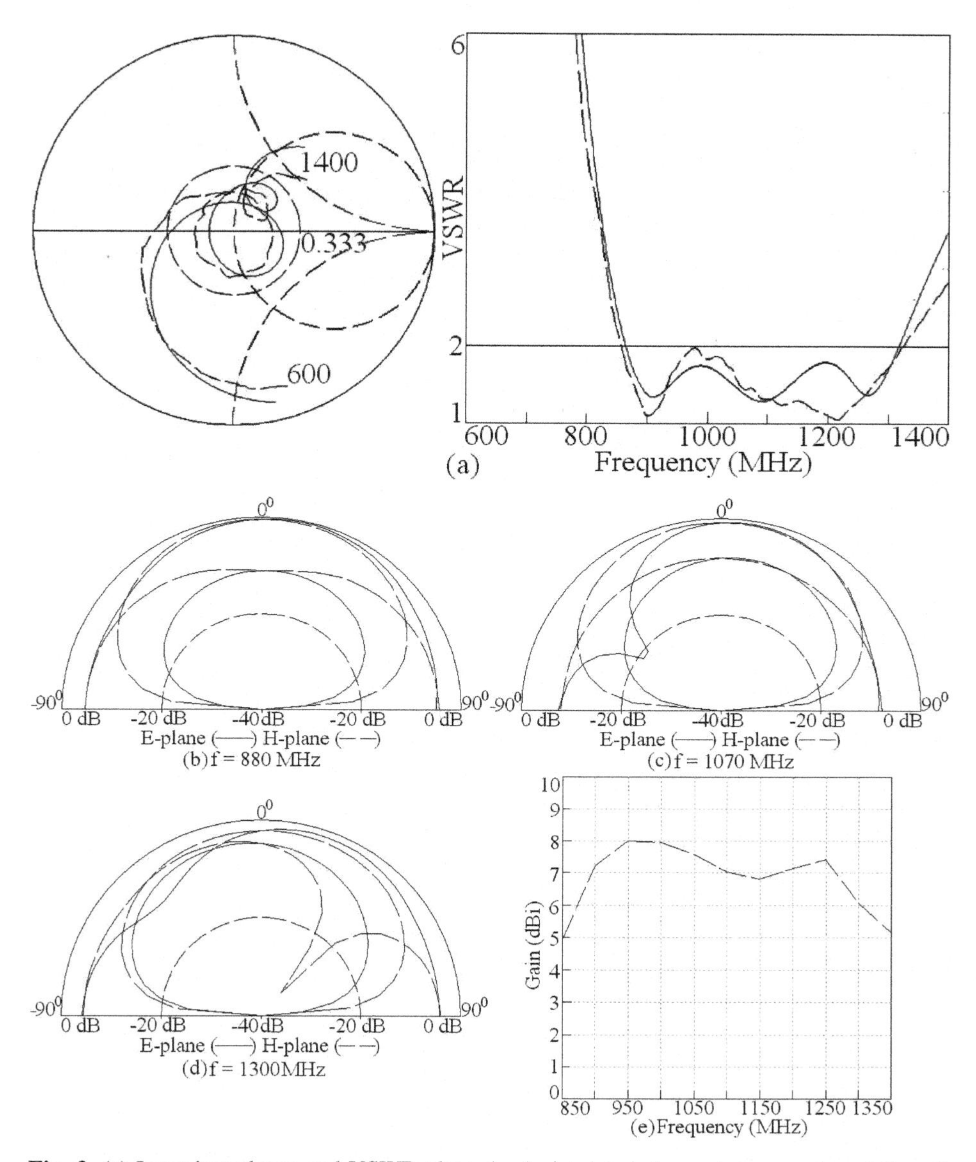

Fig. 3. (a) Input impedance and VSWR plots, (—) simulated, (– – –) measured, and (b – d) radiation pattern and (e) gain over BW for slot cut proximity fed SCMSA

3 Proximity Fed Slot Cut Gap-Coupled Scmsas

The proximity fed slot cut SCMSA gap-coupled to another SCMSA is shown in Fig. 4(a). The dimensions of the gap-coupled SCMSA is taken to be larger than the slot cut SCMSA dimension. Thus in this configuration three resonance frequencies are present, i.e. two due to TM_{11} mode of each of the SCMSAs and one due to modified TM_{21} mode due to the slot on the proximity fed SCMSA. To optimize for broadband response in the same frequency range, the dimensions of gap-coupled SCMSA is optimized such that the three loops will lie inside the VSWR = 2 circle. The optimized input impedance plot is shown in Fig. 4(b). The simulated BW is 431 MHz (43.7%) whereas the measured BW is 449 MHz (44.2%). Due to the closely spaced individual frequencies the increase in BW as compared to the previous configuration is] not present. The radiation pattern over the BW is measured and at the centre frequency of the band, is shown in Fig. 4(c). The gain variation over the BW is shown in Fig. 4(d).

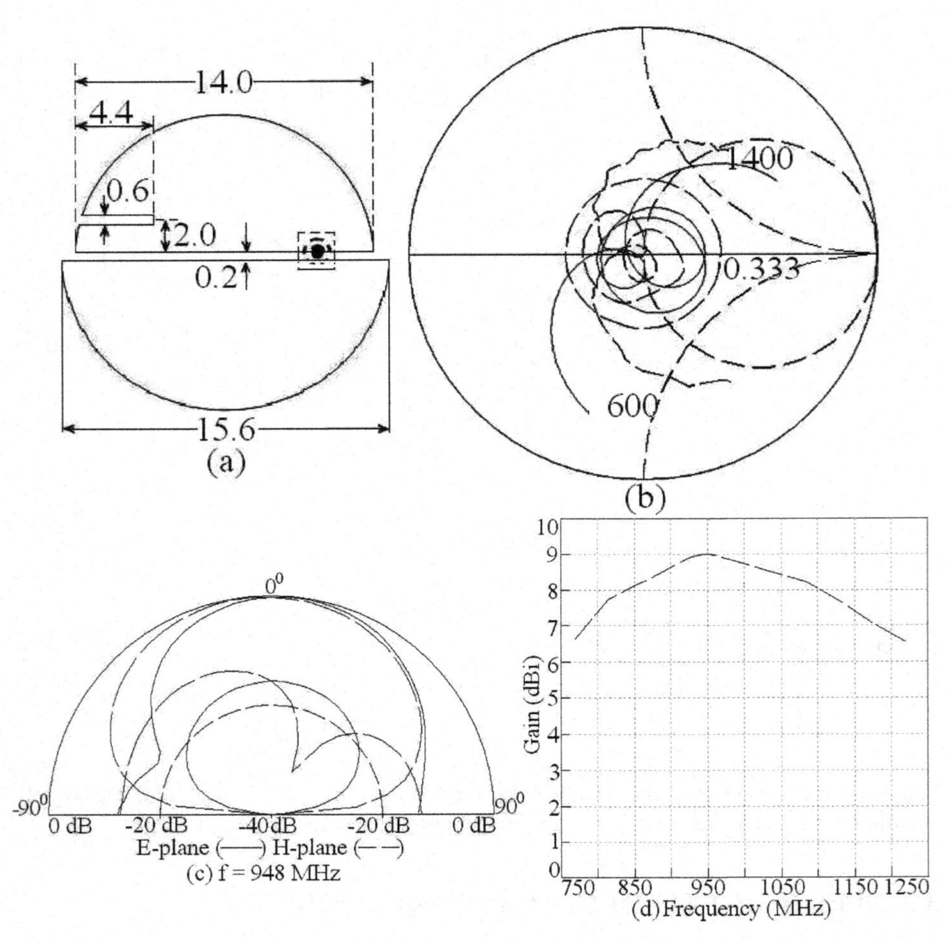

Fig. 4. (a) Proximity fed slot cut SCMSA gap-coupled to SCMSA, its (b) input impedance plot, (—) simulated, (– – –) measured, and its (c) radiation pattern at centre frequency and (d) gain variation over the BW

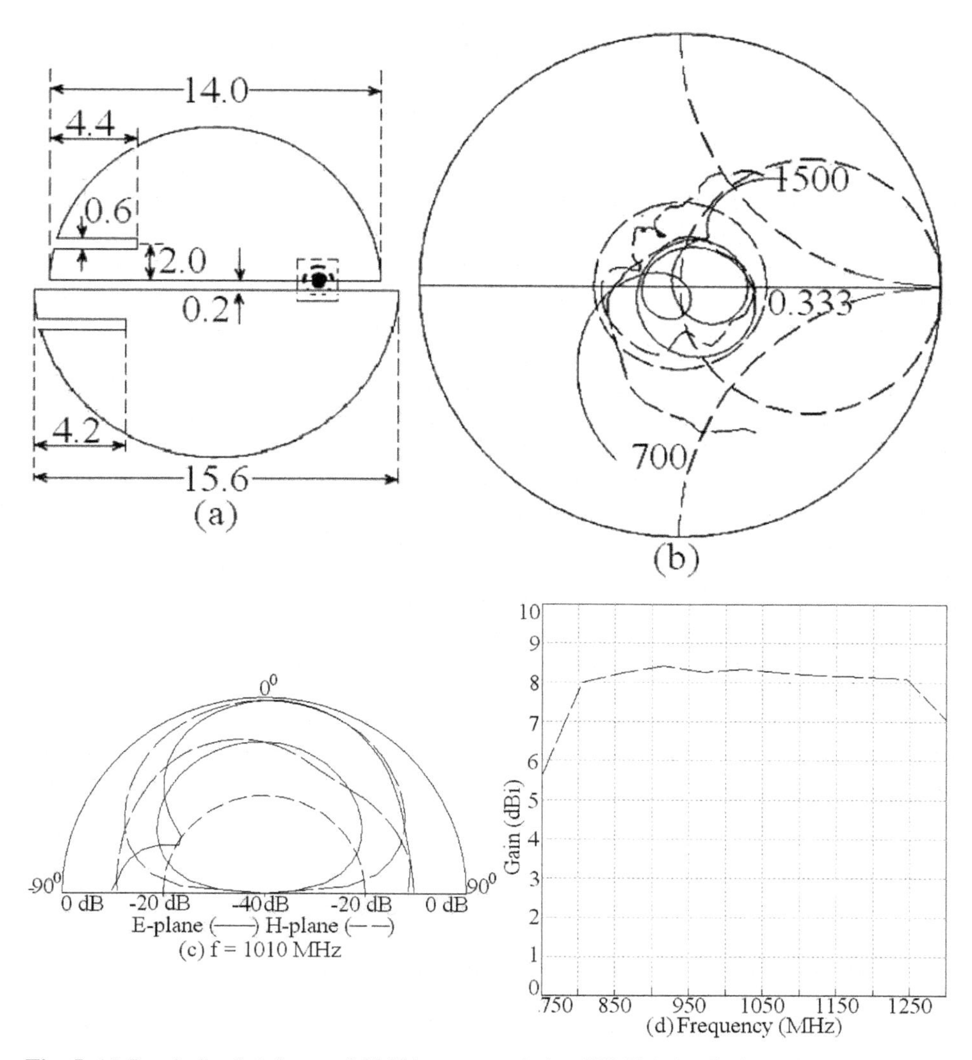

Fig. 5. (a) Proximity fed slot cut SCMSA gap-coupled to SCMSA, its (b) input impedance plot, (—) simulated, (– – –) measured, and its (c) radiation pattern at centre frequency and (d) gain variation over the BW

Table 1. Comparison between various Proximity fed SCMSAs

Configuration shown in	Simulated BW MHz, %	Measured BW MHz, %
Fig. 2(a)	450, 41.2	458, 41.6
Fig. 4(a)	431, 43.7	449, 44.2
Fig. 5(a)	509, 50	514, 50.2

The radiation pattern is in the broadside direction with cross-polar levels less than 12 dB as compared to the co-polar levels. Towards the higher frequencies of the BW, the cross-polarization levels increases as the modified TM_{21} mode is present at those frequencies. The antenna gain is more than 7 dBi over the complete BW with peak gain of 9 dBi. To further increase the BW of gap-coupled antenna the rectangular slot is cut on the edges of the gap-coupled SCMSA as shown in Fig. 5(a). The rectangular slot cut modifies the resonance frequency of TM_{21} mode on the gap-coupled SCMSA. The increase in BW is realized when that frequency comes closer to the frequencies of TM_{11} mode on each of the SCMSAs. This is realized for slot length equal to 4.2 cm. The input impedance plot for the same is shown in Fig. 5(b). Due to four resonant modes present in the configuration, four loops are present in the impedance locus. The simulated BW is 509 MHz (50%) whereas the measured BW is 514 MHz (50.2%). The radiation pattern over the BW is measured and at the centre frequency it is shown in Fig. 5(c). The gain variation over the BW is shown in Fig. 5(d).

Similar to the above configuration the radiation pattern is in the broadside direction with cross polar levels less than 12 dB as compared to the co-polar levels. Since the modified TM_{21} mode is present towards the higher frequencies of BW, the cross-polarization levels increases at those frequencies. The antenna gain is more than 7 dBi over the VSWR BW. As compared to the previous configuration the gain is reduced by nearly 1 dBi. However this configuration shows stable gain characteristics over the complete BW. In all the configuration the radiation pattern is measured using the larger square ground plane, hence the back-lobe radiation is negligible. The antenna characteristics have also been measured using the finite square ground plane of side length 26 cm ($< 1\lambda_0$ with respect to 1000 MHz). This gives similar characteristics as that given by the infinite ground plane. The results for all the configuration are summarized in Table 1.

4 Conclusions

The proximity fed SCMSA is discussed. The higher order modes of proximity fed SCMSA are studied. The rectangular slot cut proximity fed SCMSA is proposed. The analysis to study the effects of slot on the broadband response is presented. The slot modifies the higher order TM_{21} mode resonance frequency of the patch and along with the fundamental TM_{11} mode yields broadband response. The simulated and measured BW of more than 400 MHz is obtained. Further the gain of this configuration is increased by gap-coupling it with another SCMSA having different patch radius. This configuration gives nearly the same BW but with higher gain. The proximity fed gap-coupled configuration of SCMSA wherein the rectangular slot was cut on both the patches is proposed. This configuration yields a BW of more than 500 MHz with broadside radiation pattern and stable gain characteristics. The proposed configuration can find applications in the mobile communication environment in the 800 to 1200 MHz frequency band.

References

1. Garg, R., Bhartia, P., Bahl, I., Ittipiboon, A.: Microstrip Antenna Design Handbook. Artech House, USA (2001)
2. Bhartia, B., Bahl, I.J.: Microstrip Antennas, USA (1980)
3. Kumar, G., Ray, K.P.: Broadband Microstrip Antennas, 1st edn. Artech House, USA (2003)
4. Huynh, T., Lee, K.F.: Single-Layer Single-Patch Wideband Microstrip Antenna. Electronics Letters 31(16), 1310–1312 (1995)
5. Wong, K.L., Hsu, W.H.: A broadband rectangular patch antenna with a pair of wide slits. IEEE Trans. Antennas Propagat. 49, 1345–1347 (2001)
6. Lee, K.F., Yang, S.L.S., Kishk, A.A., Luk, K.M.: The Versatile U-slot Patch. IEEE Antennas & Propagation Magazine 52(1), 71–88 (2010)
7. Cock, R.T., Christodoulou, C.G.: Design of a two layer capacitively coupled, microstrip patch antenna element for broadband applications. In: IEEE Antennas Propag. Soc. Int. Symp. Dig., vol. 2, pp. 936–939 (1987)
8. Deshmukh, A.A., Ray, K.P.: Proximity fed Broadband Rectangular microstrip antenna. Microwave and Optical Technology Letters 53(2), 294–300 (2011)
9. Weigand, S., Huff, G.H., Pan, K.H., Bernhard, J.T.: Analysis and design of Broadband Single Layer Rectangular U-slot Microstrip Patch Antenna. IEEE Transactions on Antennas & Propagation, AP-51(3), 457–468 (2003)
10. Deshmukh, A.A., Ray, K.P., Sonwadkar, T., Varawalla, S., Kataria, R.: Formulation of Resonance Frequency for Multi-Band Circular Microstrip Antennas. International Journal of Microwave and Optical Technology 5(5), 248–256 (2010)
11. Deshmukh, A.A., Ray, K.P.: Analysis of Broadband U-slot cut RMSA. In: Proceedings of AEMC – 2011, Kolkata, India, December 1-4 (2011)
12. IE3D 12.1, Zeland Software. Freemont, USA (2004)
13. Balanis, C.A.: Antenna Theory: analysis and design, 2nd edn. John Wiley & Sons Ltd.

Alternative Mathematical Design of Vector Potential and Radiated Fields for Parabolic Reflector Surface

Praveen Kumar Malik[1], Harish Parthasarthy[2], and M.P. Tripathi[3]

[1] The Electronics and Communication Department
Raj Kumar Goel Institute of Engineering and Technology Ghaziabad U.P, India
malikbareilly@gmail.com
[2] The Electronics and Communication Department,
N.S.I.T, Sector 3, Dwarika, New Delhi, India
harishp@nsit.ac.in
[3] The Electronics and Communication Department, M.A.I.T, Ghaziabad, UP, India
munishprashadtripathi@rediffmail.com

Abstract. This paper is presenting a new and efficient method for computing the charge and current distribution on a circular disk type surface. Integral equation is formulated for the circular loop geometry (taking into the account of Moment Method) to cast the equation for charge and current distribution. The technique involves extension of the conventional Pocklington's integral equation and the uses of antenna's effective parameters. Paper consist of a transformation into a single integral equation, and then into a linear system of algebraic equation. Plot of the current distribution along the antenna are finally given and compared with conventional thin wire antenna using matrix inversion technique.

Index Terms: Charge density, Integral equation, Moment method and plane surface.

1 Introduction

Reflector antennas, in one form or another, have been in use since the discovery of electromagnetic wave propagation in 1888 by Hertz. However the fine art of analyzing and designing reflectors of many various geometrical shapes did not forge ahead until the days of World War II when numerous radar applications evolved[1]. Subsequent demands of reflectors for use in radio astronomy, microwave communication, and satellite tracking resulted in spectacular progress in the development of sophisticated analytical and experimental techniques in shaping the reflector surfaces and optimizing illumination over their apertures so as to maximize the gain. The usual procedure in the analysis of radiation problems is to specify the sources and then require the fields radiated by the sources[2-4].

It has been shown by geometrical optics that if a beam of parallel rays is incident upon a reflector whose geometrical shape is a parabola, the radiation will converge (focus) at a spot which is known as the focal point[5]. A parabola is defined as the locus of a point the ratio of whose distance from a point P and from a line is equal to

S. Unnikrishnan, S. Surve, and D. Bhoir (Eds.): ICAC3 2013, CCIS 361, pp. 412–419, 2013.

unity. The point P is called the focus. Equation of a parabola in the $x-y$ plane can be expressed as $y^2 = 4ax$. Focus of such a parabola is given by $P = (a,0)$. The symmetrical point on the parabolic surface is known as the vertex[6]. Rays that emerge in a parallel. Formation are usually said to be collimated. A paraboloid is a surface obtained by rotating a parabola about the normal to its apex[7]. Equation of parabola can be given as: -

$$x^2 + y^2 = 4az = \rho^2 \tag{1}$$

Where, y= distance at y axis, x= distance at x axis, a= distance of focus point from axis

As, $y^2 = 4ax$ $\quad \rho = \sqrt{4az}$ and $\quad \rho = \sqrt{x^2 + y^2}$

So that
$$x^2 + y^2 - 4az = 0 \tag{2}$$

Parabolic cylinders have widely been used as high-gain apertures fed by line sources. The analysis of a parabolic cylinder (single curved) reflector is similar, but considerably simpler than that of a paraboloidal (double curved) reflector[8,9]. The principal characteristics of aperture amplitude, phase, and polarization for a parabolic cylinder, as contrasted to those of a paraboloid, are as follows:

1. The amplitude taper, due to variations in distance from the feed to the surface of the reflector, is proportional to 1/ρ in a cylinder compared to $\frac{1}{r^2}$ in a paraboloid[10].

2. The focal region, where incident plane waves converge, is a line source for a cylinder and a point source for a paraboloid[11].

3. When the fields of the feed are linearly polarized parallel to the axis of the cylinder, no cross-polarized components are produced by the parabolic cylinder. That is not the case for a paraboloid.

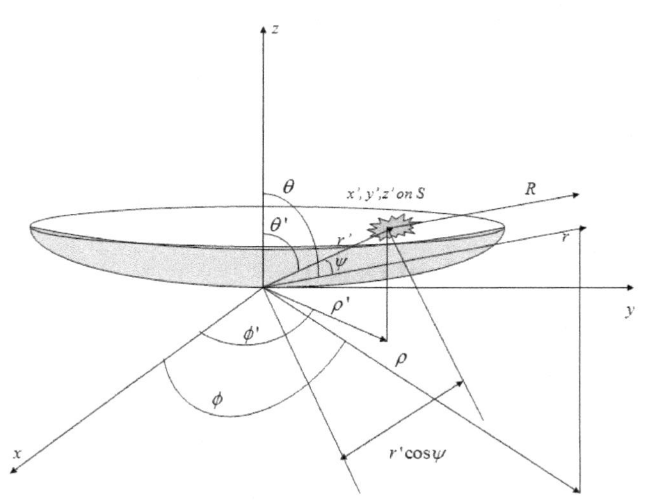

Fig. 1. After rotating a parabola about the normal to its apex

2 Calculation of Infinitesimal Area

From trigonometry, we have

$$x = \rho\cos\phi \qquad y = \rho\sin\phi \qquad \text{and,} \qquad z = \frac{\rho^2}{4a} \tag{3}$$

The equation of the paraboloid can be expressed in parametric form as –

$$r(\rho,\phi) = \rho.\cos(\phi)\hat{x} + \rho.\sin(\phi)\hat{y} + (\frac{\rho^2}{4a})\hat{z} \tag{4}$$

Tangent vector on this surface relative to the parametric coordinate (ρ,ϕ) will be

$$e_\rho = \frac{\delta r}{\delta\rho} = \cos(\phi)\hat{x} + \sin(\phi)\hat{y} + (\frac{\rho}{2a})\hat{z} \text{ and } e_\phi = \frac{1}{\rho}\frac{\delta r}{\delta\phi} = -\sin(\phi)\hat{x} + \cos(\phi)\hat{y} \tag{5}$$

Area element (differential) on the surface is given by –

$$dS(\rho,\phi) = \left|\frac{\delta r}{\delta\rho} x \frac{\delta r}{\delta\phi}\right| d\rho.d\phi \quad \text{Or,} \quad dS(\rho,\phi) = \left|e_\rho \times \rho e_\phi\right| d\rho.d\phi \tag{6}$$

Using cross product, i.e. area under the infinitesimal curve of $\left|e_\rho \times \rho e_\phi\right|$,

$$\text{i.e.} \begin{pmatrix} \hat{x} & \hat{y} & \hat{z} \\ \cos\phi & \sin\phi & \frac{\rho}{2a} \\ -\rho\sin\phi & \rho\cos\phi & 0 \end{pmatrix} e_\rho \times e_\phi = \rho\left|-(\frac{\rho}{2a})\cos(\phi)\hat{x} - (\frac{\rho}{2a})\sin(\phi)\hat{y} + \hat{z}\right| \tag{7}$$

$$\left|e_\rho \times e_\phi\right| = \rho\sqrt{\left(-(\frac{\rho}{2a})\cos(\phi)\right)^2 + \left(-(\frac{\rho}{2a})\sin(\phi)\right)^2 + (1)^2}$$

So that,

$$\left|e_\rho \times e_\phi\right| = \rho\sqrt{1 + \left[\frac{\rho}{2a}\right]^2} \quad \text{Therefore,} \quad dS(\rho,\phi) = \rho\sqrt{1 + \rho^2 / 4a^2}\, d\rho d\phi \tag{8}$$

3 Calculation of Current Density

It is well aware that, electric and magnetic representation in terms of fields generated by an electric current source is J and a magnetic current source is M. The procedure requires that the auxiliary potential functions A and F generated, respectively, by J and M are found first. In turn, the corresponding electric and magnetic fields are then determined (E_A,H_A due to A and E_F ,H_F due to F). The total fields are then obtained by the superposition of the individual fields due to A and F (J and M) [12].

As we know that surface current density is given by: $J_S = \hat{a}_x J_x + \hat{a}_y J_y + \hat{a}_z J_z$ Surface current density for paraboloid can be expressed in parametric form as.

$$J(\rho,\phi) = J_\rho e_\rho + J_\phi e_\phi + J_Z e_Z$$

Where, $J_Z e_Z = 0$

Therefore,

$$J(\rho,\phi) = J_\rho(\cos(\phi)\hat{x} + J_\phi(-\sin(\phi)\hat{x} + \cos(\phi)\hat{y}) + \sin(\phi)\hat{y} + (\frac{\rho}{2a})\hat{z}) \tag{9}$$

Combining them in x, y and z coordinates,

$$J(\rho,\phi) = [J_\rho \cos(\phi) - J_\phi \sin(\phi)]\hat{x} + [J_\rho \sin(\phi) + J_\phi \cos(\phi)]\hat{y} + [\frac{\rho}{2a} J_\rho]\hat{z} \tag{10}$$

4 Calculation of Distance R

Field radiated by a point source varies as, $\frac{e^{-jkr}}{r}$

Where, e^{-jkr} is phase factor and $\frac{1}{r}$ is the amplitude factor

For far field calculation

$$R = r - r'\cos\psi \tag{11}$$

For phase terms (ψ is the angle between the two vectors r and r' as shown)

$R = r$, For amplitude, and

$$r'\cos\psi = \hat{r}.r' \; r' = \rho'\cos(\phi')\hat{x} + \rho'\sin(\phi')\hat{y} + (\frac{\rho'}{2a})\hat{z} \; (r' = \hat{x}x' + \hat{y}y' + \hat{z}z') \tag{12}$$

From
$$\begin{pmatrix} \hat{x} \\ \hat{y} \\ \hat{z} \end{pmatrix} = \begin{pmatrix} \sin\theta\cos\phi & \cos\theta\cos\phi & -\sin\phi \\ \sin\theta\sin\phi & \cos\theta\sin\phi & -\sin\theta \\ \cos\theta & -\sin\theta & 0 \end{pmatrix} \begin{pmatrix} \hat{r} \\ \hat{\theta} \\ \hat{\phi} \end{pmatrix} \tag{13}$$

$$\hat{r} = \hat{x}\sin\theta\cos\phi + \hat{y}\sin\theta\sin\phi + \hat{z}\cos\theta$$

$$\hat{r}.r' = \rho'\sin(\theta)\cos(\phi)\cos(\phi') + \rho'\sin(\theta)\sin(\phi)\sin(\phi') + \frac{\rho'}{2a}\cos\theta$$

$$\hat{r}.r' = r'\cos\psi = \rho'\sin(\theta)\cos(\phi - \phi') + \frac{\rho'}{2a}\cos\theta \tag{14}$$

$$R = r - \hat{r}r'$$

$$R = r - \rho'\sin(\theta)\cos(\phi - \phi') + \frac{\rho'}{2a}\cos\theta \tag{15}$$

This can be written as,

$$R = r - r'\cos\psi$$

Where, $r'\cos\psi = \rho'\sin(\theta)\cos(\phi - \phi') + \frac{\rho'}{2a}\cos\theta$ (16)

5 Calculation of Vector Potential

It is a very common practice in the analysis procedure to introduce auxiliary functions, known as vector potentials, which will aid in the solution of the problems. The most common vector potential functions are the A (magnetic vector potential) and F (electric vector potential). To calculate the E and H field, directly from the electric and magnetic current density (J and M) require higher degree of integration, which introduce the extra computational time. Therefore, calculation of E and H fields are done through vector potential A and F, through J and M. First integration is done to calculate the vector potential A and F, and then differentiation to obtain E and H fields [13].

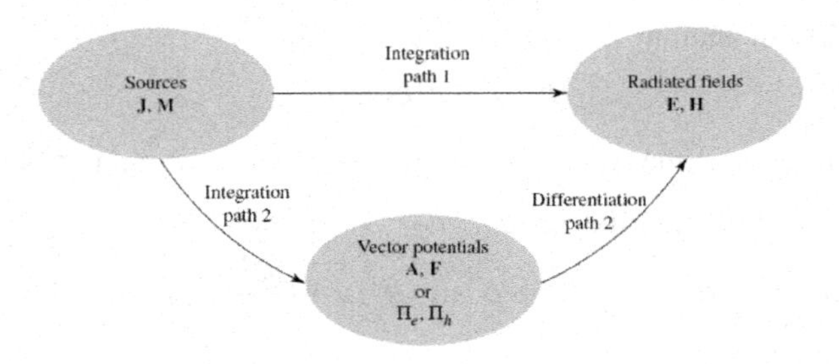

Fig. 2. Calculation of E and H field using J and M field

As per the electromagnetic field theory, we know that magnetic vector potential can be given as:

$$A = \frac{\mu}{4\pi}\iint_S J_s \frac{e^{-jkR}}{R} ds \tag{17}$$

$$A = \frac{\mu e^{-jkr}}{4\pi} \iint_S J_s \frac{e^{+jkr'\cos\psi}}{r} ds$$

$$A = \frac{\mu e^{-jkr}}{4\pi} \iint_S J_s \frac{e^{+jkr'\cos\psi}}{r} \rho\sqrt{1+\rho^2/4a^2}\, d\rho d\phi \tag{18}$$

$$A = \frac{\mu e^{-jkr}}{4\pi} \iint_S \begin{bmatrix} [J_\rho \cos(\phi) - J_\phi \sin(\phi)]\hat{x} + \\ [J_\rho \sin(\phi) + J_\phi \cos(\phi)]\hat{y} + \\ [\frac{\rho}{2a} J_\rho]\hat{z} \end{bmatrix} \frac{e^{+jkr'\cos\psi}}{r} \rho\sqrt{1+\rho^2/4a^2}\, d\rho d\phi \tag{19}$$

$$A = \frac{\mu e^{-jkr}}{4\pi} \iint_S \begin{bmatrix} [J_\rho \cos(\phi) - J_\phi \sin(\phi)]\hat{x} + \\ [J_\rho \sin(\phi) + J_\phi \cos(\phi)]\hat{y} + \\ [\frac{\rho}{2a} J_\rho]\hat{z} \end{bmatrix} \frac{e^{+jk(\rho'\sin(\theta)\cos(\phi-\phi')+\frac{\rho'}{2a}\cos\theta)}}{r} \rho\sqrt{1+\rho^2/4a^2}\, d\rho d\phi \tag{20}$$

With the help of Maxwell equations in homogenous, linear and isotropic medium. We get:

$$E = -j\omega A - \frac{j}{\omega\mu\varepsilon}\nabla(\nabla.A) \quad \text{and} \quad H = -j\omega F - \frac{j}{\omega\mu\varepsilon}\nabla(\nabla.F) \tag{21}$$

6 Numerical Results

Based on above specified steps, a MATLAB program has been written under the following assumption.

1. Input frequency is assumed to be as 3 G Hz.
2. Focus distance in is 1 Meter.
3. Diameter of reflector in meters is 0.5 Meter
4. Offset height of reflector in meters is 0.25
5. Feed location of x axis, y axis and z axis in meters is 0, 0, and 6 respectively.
6. Number of integration points per wavelength is 4.
7. Radial distance to observation point is 1000 meter.

Two different cases, absolute pattern for co polarization and cross polarization has been plot and is being compared with the conventional co polarization and cross polarization. It has been found that radiation pattern is almost similar for traditional radiated fields and the newly designed.

Case 1: Absolute pattern for co and cross polarization for maximum theta angle (degrees) = 45 and phi angle (degrees) = 145 are:

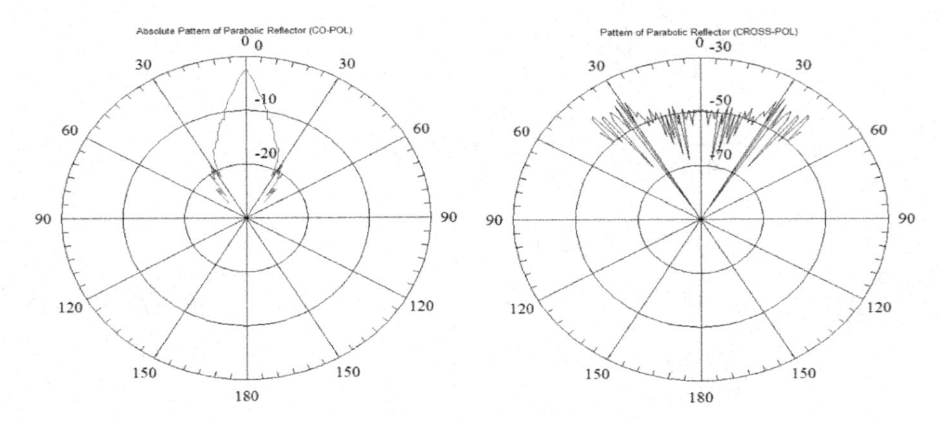

Fig. 3. Co polarization for θ=45 and ϕ=145 **Fig. 4.** Cross polarization for θ=45 and ϕ=145

Case 2: Absolute pattern and cross polarization for maximum theta angle (degrees) = 120 and phi angle (degrees) = 245 are:

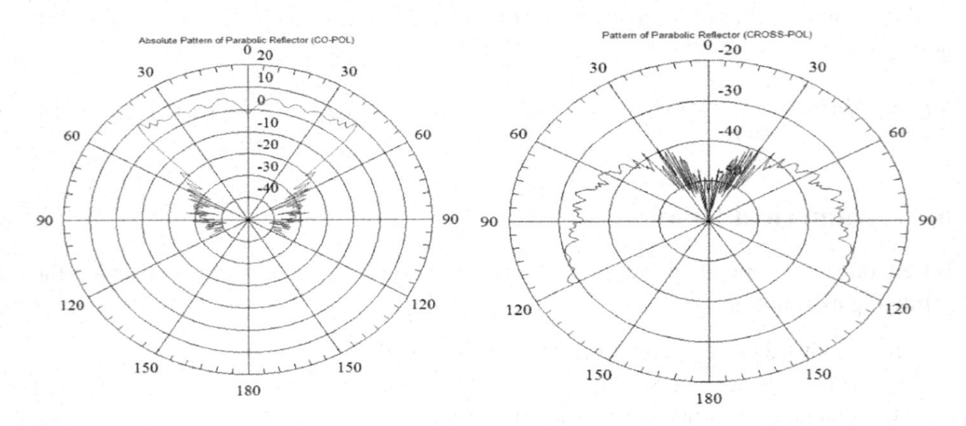

Fig. 5. Co polarization for θ=120 and ϕ=245 **Fig. 6.** Cross polarization for θ=120 and ϕ=245

Acknowledgment. Author would like to thanks ***Prof. Harish Parthsarthy***, department of Electronics and Communication, N.S.I.T, New Delhi for all his support. I am grateful for the encouragement of ***Prof. M P Tripathi***, Retd. Prof. NSIT, New Delhi, that made me write and publish paper. The author will also like to express sincere appreciation and gratitude to department of Electronics and Communication, ***Raj Kumar Goel Institute of Engineering and Technology Ghaziabad*** UP India, which has provided tremendous assistance throughout the work.

Table 1. Description for quantity used

Symbol	Quantity	Numerical Value
ρ	Charge density	(C/m)
θ	Elevation angle	$0 \rightarrow \pi$
ϕ	Azimuthally angle	$0 \rightarrow 2\pi$
Q	Charge C	$1.6\text{x}10^{-19}$ C
E	Electrical field	N C^{-1}
ε_0	Permittivity of Vacuum	8.85 pF m^{-1}
h	Observation point distance	In meter
a	Radius of circular plate	In meter
μ	Permeability	$1 \rightarrow 4\pi \times 10^{-7}$ H/m $= 4\pi \times 10^{-7}$ Wb/(A·m)
μ_r	Relative permeability	$\mu \rightarrow \mu_r$

References

[1] Yee, K.: Numerical solution of initial boundary value problems involving Maxwell's equations in isotropic media. IEEE Trans. Antennas Propag. AP-14(3), 302–307 (1966)

[2] Burke, G.J., Poggio, A.J.: Numerical electromagnetic code (NEC)—Method of moments, Naval Ocean Systems Center, Tech. Doc. 116 (January 1981)

[3] Cortina, R., Porrino, A.: Calculation of impulse current distributions and magnetic fields in lightning protection structures- a computer program and its laboratory validation. IEEE Trans. Magn. 28(2), 1134–1137 (1992)

[4] Forigo, D., Gianola, P., Scotti, R., Vallari, R.: Measurements and numerical evaluation of the electric field in the near-zone of radio base station antennas. In: Proc. IEEE Int. Symp. Antennas and Propagation Society, vol. 3, pp. 338–341 (March 2001)

[5] Blanch, S., Romeu, J., Cardama, A.: Near field in the vicinity of wireless base-station antennas: An exposure compliance approach. IEEE Trans. Antennas Propagat. 50, 685–692 (2002)

[6] Guru, B.S., Hiziroglu, H.R.: Electromagnetic Field Theory Fundamentals, ch. 4. Cambridge University, Cambridge (2004)

[7] Balanis, C.A.: "Antenna Theory": Analysis and design. John Wiley & Sons, New Delhi (2008)

[8] Parthasarthy, H.: Electromagnetism and relativity theory. Application of advance signal analysis, pp. 441–507. I K International Publication (2008)

[9] Parthasarthy, H.: Signal Analysis. Advanced signal analysis and its applications to mathematical physics, pp. 73–147. I K International Publication (2009)

[10] Parthasarthy, H.: Antenna theory, analysis and synthesis. Advanced signal analysis and its applications to mathematical physics, pp. 560–571. I K International Publication (2009)

[11] Parthasarthy, H.: Electromagnetism. Advanced signal analysis and its applications to mathematical physics, pp. 265–321. I K International Publication (2009)

[12] Kraus, J.D.: Antennas and Wave Propagations, 4th edn. McGraw-Hill, New Delhi (2011)

[13] Fuscon, Fundamental of Antenna Theory and Technique. Pearson Education, New Delhi (2011)

Formulation of Resonance Frequency for Shorted Rectangular Microstrip Antennas

Amit A. Deshmukh[1], K.P. Ray[2], A. Joshi[1], K. Aswathi[1], K. Nada[1], and K. Prithvi[1]

[1] DJSCOE, Vile – Parle (W), Mumbai – 400 056, India
[2] SAMEER, I.I.T. Campus, Powai, Mumbai – 400 076, India
{amitdeshmukh76,kpray}@rediffmail.com

Abstract. The compact microstrip antenna is realized by placing the shorting post or shorting plate along the zero field line at the fundamental mode of the conventional microstrip antenna. The shorting post or the plate converts the patch from conventional half wavelength resonator to quarter wavelength resonator. The formulation of resonant length at fundamental and higher order modes of the rectangular microstrip antenna is available. However the formulation of resonant length at various modes for the shorted patch is not reported. Also the equation for fringing field extensions length for using the shorted patch on thicker substrates is not available. In this paper, the fundamental and higher order modes of the rectangular and shorted rectangular patch are studied. By studying the voltage and current distributions of the shorted patch the resonance frequency formulation for the shorted patch is proposed. Further by designing the shorted patch on different substrates thickness and at the different frequencies, the formulation for fringing field extension length is proposed. The frequencies calculated using the proposed formulations agrees closely with the simulated and measured results. Thus the proposed formulations can be used to design shorted patch for given frequency and on given substrate thickness.

Keywords: Rectangular microstrip antenna, Shorted rectangular microstrip antenna, Resonance frequency, Higher order modes.

1 Introduction

The compact microstrip antenna (MSA) is realized by placing the shorting post or the plate along the zero field line at the fundamental mode of the patch [1, 2]. The shorting technique converts the conventional half wavelength resonator to quarter wavelength resonator and reduces the patch size by half. The compact MSA is also realized by cutting the slot at an appropriate position inside the patch [1 – 3]. The slot increases the length of the surface current for the given mode and reduces the frequency for the given patch size. Between these two techniques, the shorting method realizes maximum reduction in the patch frequency. The simplest method to realize broadband MSAs is by fabricating the patch on lower dielectric constant thicker substrates [1 – 3]. The thicker substrate reduces the quality factor of the cavity below the patch and yields broader BW. However for substrate thickness greater than

S. Unnikrishnan, S. Surve, and D. Bhoir (Eds.): ICAC3 2013, CCIS 361, pp. 420–427, 2013.

$0.04\lambda_0$, the BW is limited by the inductance of the feed probe. For these thicker substrates the BW is increased by using proximity feeding technique [4, 5]. The extension in patch length due to the fringing fields present towards the open circuit edges of the patch is reported for thinner substrates ($< 0.03\lambda_0$) [1 – 3]. However while designing the MSAs on thicker substrate the formulations for fringing field extensions are not available. The compact shorted RMSA is realized by placing the shorting post along the zero field line at the fundamental TM_{10} mode of the patch. The resonance frequency formulation for the rectangular MSA (RMSA) at the fundamental and higher order modes of the patch is reported [1 – 3]. When the shorting plate or the post is placed to realize the compact configuration, the field distribution along the patch edges changes as the patch is converted from half wavelength resonator to quarter wave length resonator. The formulation in resonance frequency for the shorted patch at fundamental and higher order modes is not reported.

In this paper, the fundamental and higher order modes of RMSA and shorted RMSA are studied. Further by studying the voltage and current distributions for shorted patch, the formulation in resonance frequency for the shorted plate RMSA is proposed. The parametric study to calculate the extension in patch length due to the fringing fields is presented. The shorted RMSA was designed at different frequencies on different substrate thickness. Using that, the plots of edge extension length against the substrate thickness were developed. Further the resonance frequencies calculated by using the proposed equations and the plots for edge extension length, agrees closely with the simulated and measured results. The proposed analysis is carried out using the IE3D software followed by experimental verification [6]. The proposed formulations will provide the guideline for designing shorted RMSAs in the desired frequency bands and for different substrate thickness.

2 RMSA and Shorted RMSA

The RMSA designed at 1000 MHz for substrate thickness of h = 1.0 cm, is shown in Fig. 1(a, b). The RMSA length (L) and width (W) were calculated such that the patch resonates in TM_{10} mode at 1000 MHz. The fundamental and higher order modes of this RMSA calculated by using resonance frequency equation of RMSA [3], are f_{TM10} = 1027 MHz, f_{TM01} = 909 MHz, f_{TM11} = 1372 MHz, f_{TM02} = 1818 MHz, f_{TM20} = 2055, f_{TM12} = 2088 MHz, and f_{TM21} = 2247 MHz. The patch is fed using the coaxial probe of inner wire diameter 0.32 cm, along the line joining centre of the patch and centre of the width. This RMSA was simulated using IE3D software and input impedance locus for the same is shown in Fig. 1(c). It shows peaks due to TM_{10} (1000 MHz), TM_{02} (1769 MHz) and TM_{12} (2047 MHz) modes. The resonance curve plot for the feed point placed along the diagonal axis inside the patch is also shown in Fig. 1(c). For this feed point location along with the above resonant mode, the TM_{01} (870 MHz) and TM_{11} (1376 MHz) modes are also present. All these frequencies are close to the frequencies obtained using resonance frequency equation of RMSA. This RMSA is shorted along the zero field position of TM_{10} mode, to realize shorted plate RMSA as shown in Fig. 1(d, e). The resonance curve plots for two different feed point locations (C, D) as shown in Fig. 1(e), is shown in Fig. 1(f).

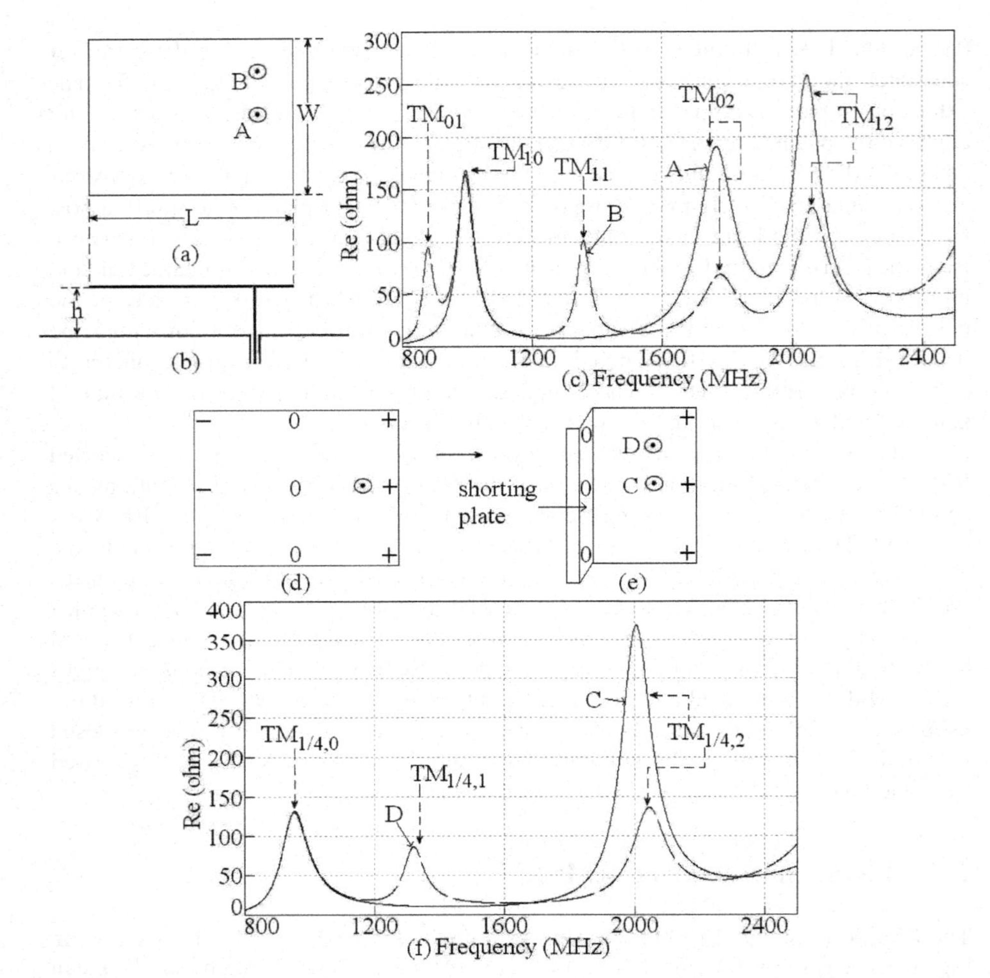

Fig. 1. (a) Top and (b) side views of RMSA, its (c) resonance curve plots for different feed point location, field distributions at fundamental mode for (d) RMSA and (e) shorted plate RMSA and (f) resonance curve plot for shorted plate RMSA

When the feed point is at point 'C', two peaks are observed whereas at point 'D' three peaks are present. The surface current distribution at different peaks for feed point location at 'D' is shown in Fig. 2(a – c). For the shorted plate RMSA, at dominant mode, the shorted length equals quarter wave in length. The surface currents do not show any variations in currents along patch width. Thus the first mode is referred as $TM_{1/4,0}$. At The next peak the surface currents shows quarter wavelength variation along the patch length and half wavelength variation along the patch width. This mode is referred as $TM_{1/4,1}$ mode. At next peak, the surface currents shows quarter wavelength variation along the patch length and two half wavelength variations along patch width. Therefore this mode is referred to as $TM_{1/4,2}$. When the feed point is located at point C, the peak due to $TM_{1/4,1}$ mode is absent as the feed is located at the minimum impedance location of that mode. Thus in the shorted patch there exist a mixed boundary condition along the patch length and therefore the field will show variation in integer multiples of quarter wavelength. However along the patch width similar boundary condition is present. Therefore the field along the width will show integer multiples of half wavelength variation. Thus the resonance frequency formula for the shorted patch can be derived by modifying the resonance frequency equation for conventional half wavelength resonator as given in equation (1).

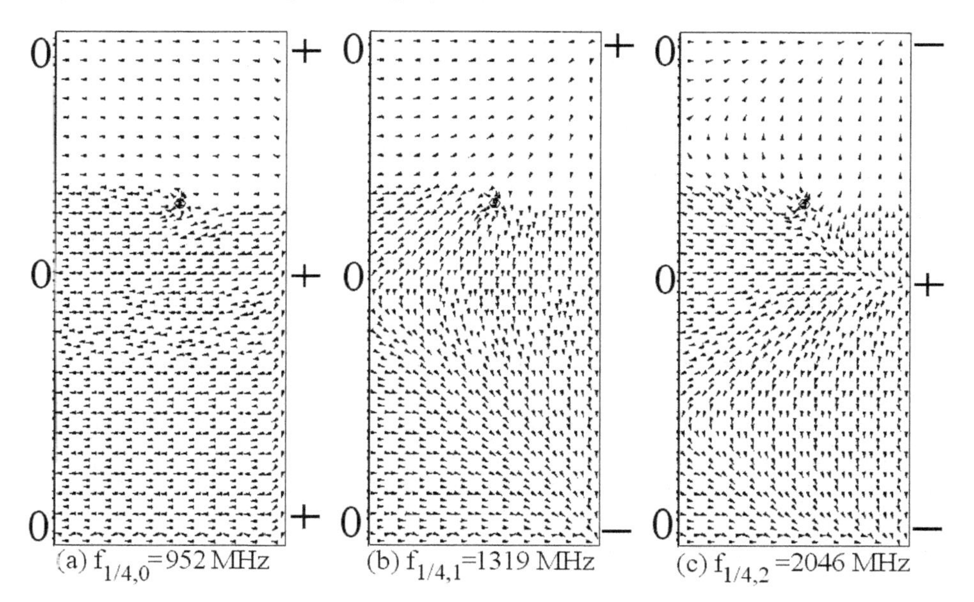

Fig. 2. (a – c) Surface current distributions at three peaks for shorted plate RMSA

$$f_r = \frac{c}{\sqrt{\varepsilon_r}} \sqrt{\left(\frac{m}{L_e}\right)^2 + \left(\frac{n}{2W_e}\right)^2} \tag{1}$$

where,

$c = 3 \times 10^8$ (m/s), velocity of light in free space
L_e = effective patch length, W_e = effective patch width
m (mode index along patch length) = 1/4, 3/4, 5/4, …
n (mode index along patch width) = 0, 1, 2, …

The W_e and L_e are the effective patch dimension calculated by adding the extension in length due to the fringing fields present towards the open circuit edges of the patch. The formulation for the fringing length is available for thinner substrates wherein $h < 0.04\lambda_0$. However for thicker substrates the formulation is not available. To calculate the edge extension in thicker substrates the parametric study is carried out by designing the shorted RMSA for different frequencies and on substrates of different thickness as discussed below.

3 Formulation of Edge Extension for Shorted Plate RMSA

The shorted plate RMSA is designed at different frequencies and for different substrate thickness over 800 to 6000 MHz frequency band. At each of the frequencies, maximum substrate thickness is taken to be less than $0.1\lambda_0$. Thus at 900 MHz, the shorted patch is designed for different substrate thickness in the range of 0.5 cm ($0.02\lambda_0$) to 2.5 cm ($0.075\lambda_0$). In shorted patch the field is zero in magnitude at the shorting point and varies along its length to the maximum value towards the open circuit edge of the patch. Therefore at given frequency, shorted patch length plus the substrate thickness equals quarter wave in length. For 900 MHz frequency with substrate thickness of 0.5 cm, the shorted patch length is taken equal to 8 cm. To realize larger gain the patch width is taken to be 1.2 times the patch length. The patch is simulated using IE3D software and its resonance curve plot was observed. If the peak in the resonance curve is not present at the desired frequency (900 MHz), the patch length is altered and the simulation was again carried out. This procedure is repeated until the peak in the resonance curve is present at around the desired frequency. The shorted patch length is noted for this case. The difference in the value of this total shorted length and the quarter wavelength at the given frequency equals edge extension length (dl). The 'dl' is expressed in terms of substrate thickness. The above procedure of calculating 'dl' is shown using the flowchart in Fig. 3(a). Using this procedure the 'dl/h' is calculated at different frequencies and for different substrate thickness and they are plotted in Figs. 3(b – e) and 4(a, b). It is observed from the plots that with an increase in substrate thickness, 'dl' decreases. For higher frequencies the value of 'dl' is higher. Also the amount of decrease in 'dl' also reduces with the increase in frequency.

To calculate L_e, the 'dl' calculated by using the plots obtained using the above parametric study is used. Thus for above shorted RMSA with h = 1.0 and f = 1000 MHz, the plot for 'dl' at 900 MHz is used. Using the graph (Fig. 3(b)), 'dl' = 0.42h is selected. Along the patch width, the field shows half wavelength variation. The extension in length due to fringing fields is present towards both the open circuit edges of the patch width. To calculate W_e, the formulations reported in [7] for the conventional RMSA on thicker substrates are used. Using these edge extensions and further by using the resonance frequency equation given in (1), the various frequencies for above shorted plate RMSA are calculated.

The various resonance frequencies calculated using proposed equations are, $f_{1/4,0}$ = 1070 MHz, $f_{1/4,1}$ = 1404 MHz, $f_{1/4,2}$ = 2095 MHz. The above shorted plate RMSA was fabricated and measurement was carried out using R & S vector network analyzer using square ground plane of side length 60 cm. The measured frequencies are $f_{1/4,0}$ = 1020 MHz, $f_{1/4,1}$ = 1376 MHz, $f_{1/4,2}$ = 2055 MHz. Thus the calculated frequencies using formulations are closely agreeing with the frequencies obtained using simulations and measurement. The fabricated prototype of above configuration is shown in Fig. 4(c). Thus the proposed formulation of shorted plate RMSA can be used to design the antenna for any given frequency and the substrate thickness.

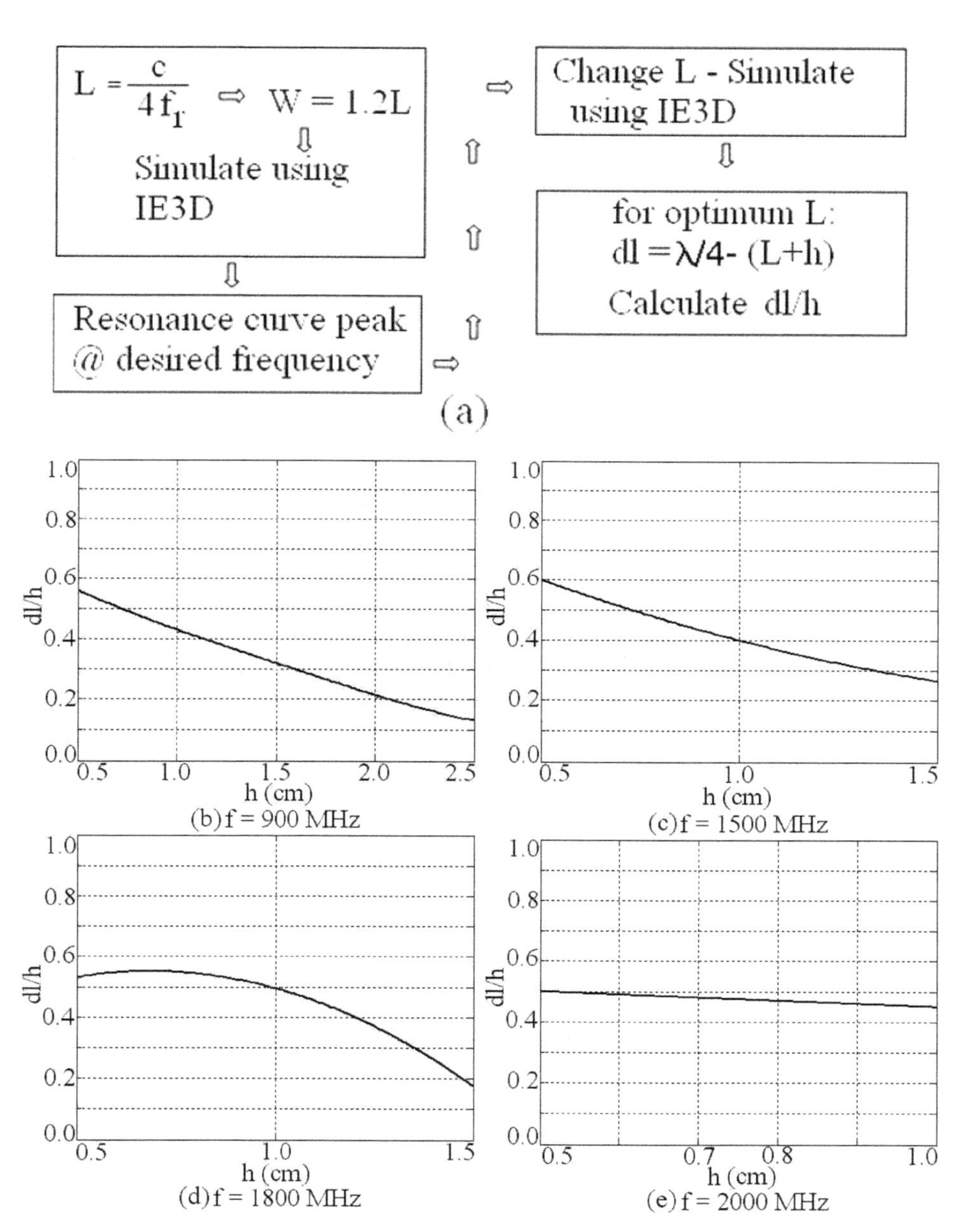

Fig. 3. (a) Flow chart for calculating edge extension for shorted plate RMSA and its (b – e) plots for edge extension against the substrate thickness

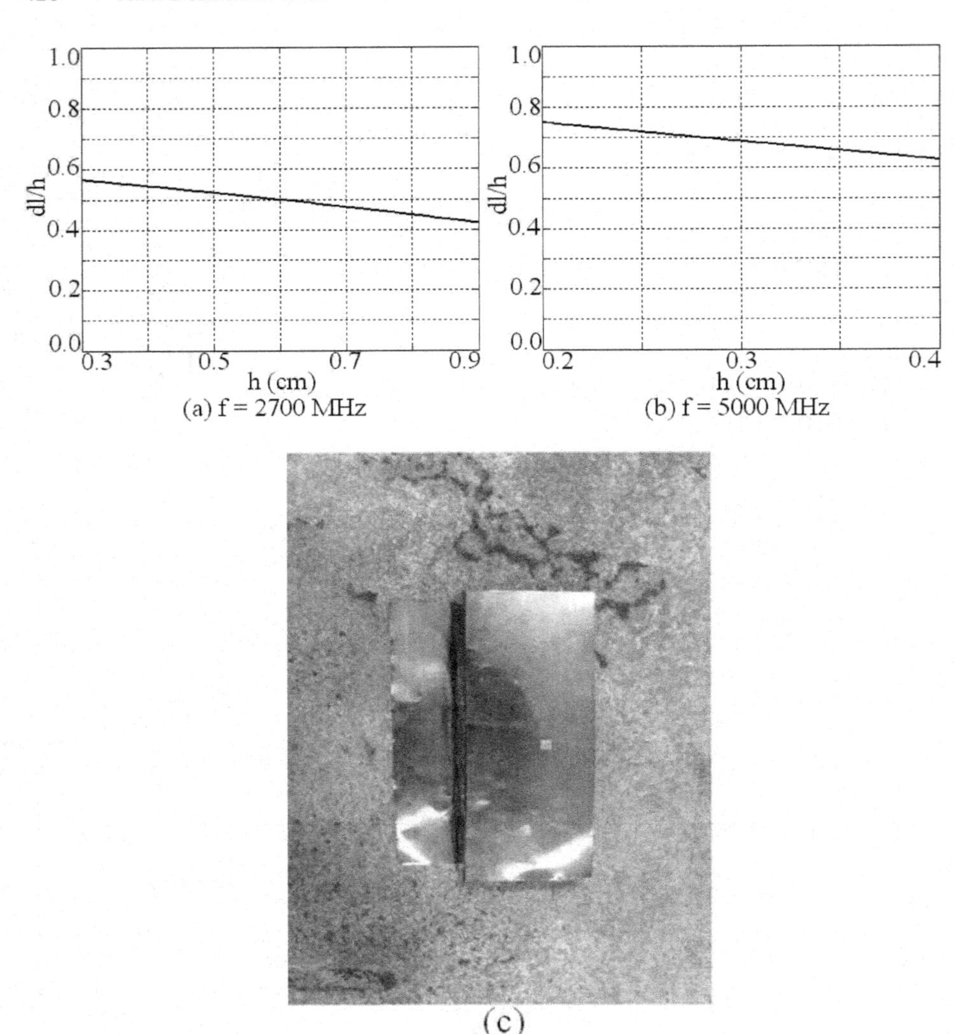

Fig. 4. (a, b) Plots of edge extension against the substrate thickness for shorted plate RMSA and (c) fabricated prototype of shorted plate RMSA

4 Conclusions

The fundamental and higher order modes of RMSA and shorted plate RMSA are discussed. The resonance frequency formulation for RMSA is available. By studying the surface currents and voltage distributions at fundamental and higher order modes of shorted plate RMSA and further by modifying the resonance frequency equation for conventional RMSA, the formulation in resonance frequency for shorted plate RMSA is proposed. By designing the shorted plate RMSA at various frequencies over 800 to 6000 MHz frequency band and for different substrate thickness, the plots for edge extension length were developed. Using the proposed formulation and the plots for edge extension length, the resonance frequencies is calculated for the shorted patch which agrees closely with the simulated and the measured results. Thus the proposed formulations can be used in designing the shorted antennas at any given frequencies and for given substrate thickness.

References

1. Garg, R., Bhartia, P., Bahl, I., Ittipiboon, A.: Microstrip Antenna Design Handbook. Artech House, USA (2001)
2. Bhartia, B., Bahl, I.J.: Microstrip Antennas, USA (1980)
3. Kumar, G., Ray, K.P.: Broadband Microstrip Antennas, 1st edn. Artech House, USA (2003)
4. Cock, R.T., Christodoulou, C.G.: Design of a two layer capacitively coupled, microstrip patch antenna element for broadband applications. In: IEEE Antennas Propag. Soc. Int. Symp. Dig., vol. 2, pp. 936–939 (1987)
5. Deshmukh, A.A., Ray, K.P.: Proximity fed Broadband Rectangular microstrip antenna. Microwave and Optical Technology Letters 53(2), 294–300 (2011)
6. IE3D 12.1, Zeland Software. Freemont, USA (2004)
7. Deshmukh, A.A., Ray, K.P., Desai, V., Shah, D., Kapadia, S.: Analysis of Suspended Rectangular Microstrip Antenna. In: Proceedings of NCET – 2012, Govt. College of Engineering, Goa, India, April 13-14 (2012)

Performance Analysis of PMD/SBS, Nonlinearity and NRZ/RZ Formats in High Speed Optical Fiber Links

B.U. Rindhe[1], Jyothi Digge[2], and S.K. Narayankhedkar[3]

[1] Electronics & Telecommunication Dept.
SIGCOE, Koparkhairane, Navi Mumbai, India
burindhe@yahoo.com
[2] Electronics & Telecommunication Dept.
Terna Engg. College, Nerul, Navi Mumbai, India
jyothijayarajdigge@rediffmail.com
[3] Electronics & Telecommunication Dept.
MGM COET, Kamothe, Navi Mumbai, India
skniitb@yahoo.com

Abstract. The latest innovation and development of optical fibers for 4G technology as the transmission medium of choice for most commercial applications in telecommunication system and computer networks in worldwide. High speed and ultra high capacity of optical communications have emerged as the essential techniques for backbone global information transmission networks. This paper thus describes the various losses for polarization mode dispersion (PMD) and Stimulated Brillouin scattering (SBS) associated with optical fibers using simulation models and its analysis. The fiber nonlinearities and performance of NRZ and RZ formats in high speed optical links. By observing effects of PMD & SBS optical communication and providing a certain technique to minimize the same losses and observe Q factor, BER of 10^{-9} to 10^{-12}, signal amplitude, eye diagram and signal power spectrum with achievement of best results.

Keywords: PMD, SBS, NRZ, RZ, BER, Q-factor, DGD, DCF.

1 Introduction

The transmission rate can exceed 2 Gbps to around 6~8 Gbps and is the highest transmission medium in the world. Recently, in telecommunication field is laying fiber optic cables to provide data super highway to support personal video services. It is expected that the future communications network will consist of one optical fiber as a backbone for WAN in the world with coaxial cable as the backbone within the building [1-3]. The terminator erected around each three stories will be provide a transmission bandwidth to each household at 20 Mbps, at this speed we can use it to watch movie, shopping, a real e-commerce world. Optical fiber can be used as a medium for telecommunication and networking because it is flexible and can be bundled as cables [4]. It is especially advantageous for long-distance communications, because light propagates through the fiber with little attenuation compared to electrical cables. This allows long distances to be spanned with few repeaters. Additionally, the

S. Unnikrishnan, S. Surve, and D. Bhoir (Eds.): ICAC3 2013, CCIS 361, pp. 428–438, 2013.

per-channel light signals propagating in the fiber have been modulated at rates as high as 111 Gigabits per second by NTT, although 10 or 40 Gb/s is typical in deployed systems. Each fiber can carry many independent channels, each using a different wavelength of light i.e. wavelength-division multiplexing (WDM) [14]. Fiber optic sensor: Fibers have many uses in remote sensing. In some applications, the sensor is itself an optical fiber. In other cases, fiber is used to connect a non-fiber optic sensor to a measurement system. Time delay can be determined using a device such as an optical time-domain reflectometer. Ever since the ancient times, one of the principle interest of human beings has been to device a communication systems for sending messages from one distant place to another [5-8]. The fundamental elements of any such communication system are as shown in above fig. 1.

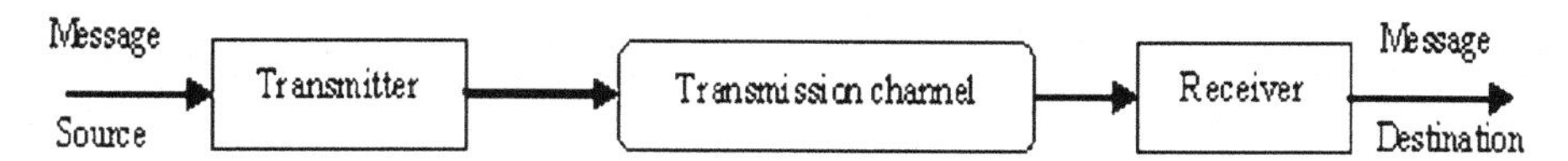

Fig. 1. Basic elements of communication system

This paper thus describes the various losses associated with fiber and its simulation models with its analysis, also the nonlinearity and NRZ/RZ formats in high speed optical fiber links.

2 Modeling of PMD and SBS for Optical Networks

The different wavelength channels will travel at slightly different speeds within the fiber. The even a pulse of light from a laser has some spectral width, each of its wavelength components will travel at slightly different speeds within the fiber. This model accurately accesses the degree of dispersion that affects a wavelength channel-which is comprised of many wavelengths of light centered about its primary wavelength its position with respect to λ_0 must be considered. The transmitter with a center wavelength greater than λ_0 will see positive (or anomalous) dispersion; transmitter with a center wavelength less than λ_0 will experience negative (or normal) dispersion. An optical channel or pulse subjected to positive dispersion will be seen its longer wavelength components travel slower than its shorter counter parts, whereas an optical channel subjected to negative dispersion will seen its longer wavelength components travel faster than its shorter counter parts.

2.1 Polarization Mode Dispersion (PMD)

By using model this example demonstrates the effect that PMD [9-10] has on signal propagation in a fiber and on system performance is shown in fig. 2. For this PMD setup, a transmitter consists of PRBS generator at 40Gbps, CW Laser source at 1550 nm, electrical driver, external modulator, and optical power normalizer. A 40Gbps RZ-modulated signal is launched into a fiber span. The output from the fiber span is inserted into a receiver.

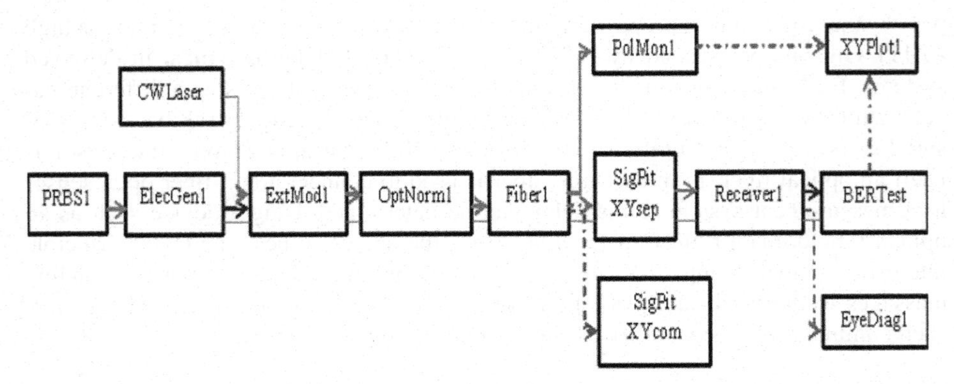

Fig. 2. Model of PMD

2.2 Stimulated Brillouin Scattering (SBS)

The SBS is due to acoustic properties of photon interaction with the medium. When light propagates through the medium, the photons interact with silica molecules during propagation. The photons also interact with themselves and cause scattering effects such as SBS in the reverse direction of propagation along the fiber. In SBS, a low-wavelength wave called Stoke's wave is generated due to the scattering of energy. This wave amplifies the higher wavelengths. The gain obtained by using such a wave forms the basis of Brillouin amplification. The Brillouin gain peaks in a narrow peak near the C-band. SBS is pronounced at high bit rates and high power levels. The margin design requirement to account for SBS is 0.5 dB. The below fig. 3 shows about to study the stimulated Brillouin scattering in the third window region at frequency 1550 nm.This is because SBD can be studied only in limited frequency range and not throughout the fiber operating band. It launch an input signal into two new versions of an SMF-28 fiber. In the top branch of the topology, SBS is included in the model. This set up allows us to determine at what fiber launch power SBS begin to have an effect.

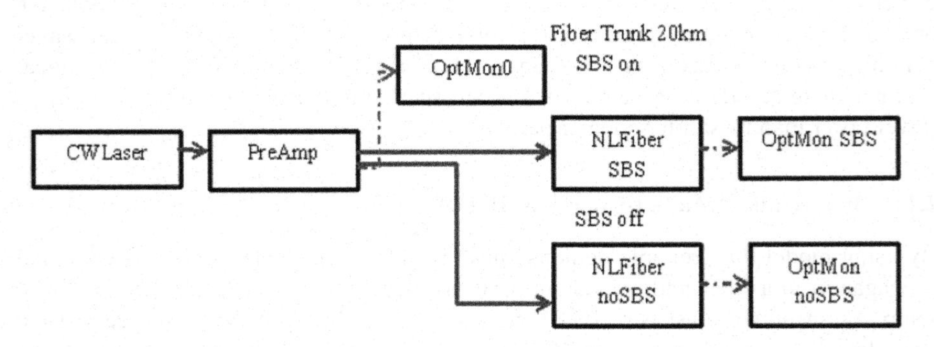

Fig. 3. Model of SBS

3 Modeling of Fiber Nonlinearity and Performance of NRZ and RZ Formats in High Speed Optical Links

The nonlinearity of fiber and its performance of NRZ and RZ formats in high speed optical links studied. This example studies impact of fiber nonlinearity on NRZ (Non return to zero) and RZ (Return to zero) formats [11-13], [15-16] in high-speed optical links. Topology layout for both NRZ and RZ cases with erbium-doped fiber amplifier (EDFA) and dispersion compensating fiber (DCF) is similar to the one shown below in fig. 4 and fig. 5 respectively:

3.1 NRZ

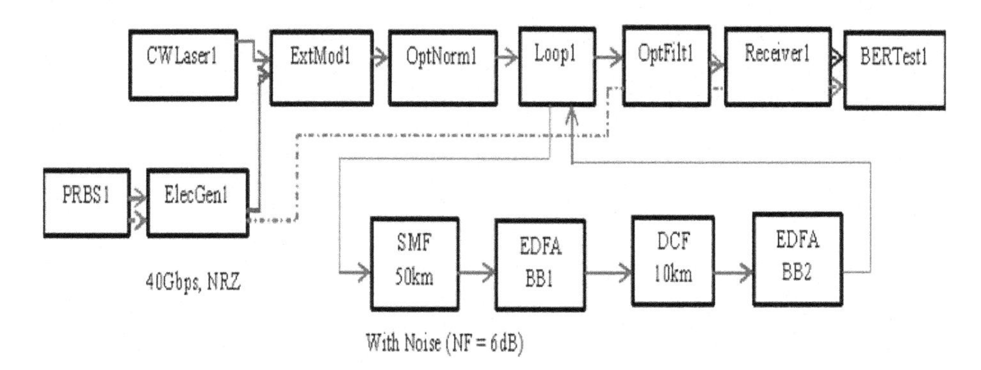

Fig. 4. Model for fiber nonlinearity and performance of NRZ formats in high speed links

3.2 RZ

In both cases as NRZ (non return to zero) and RZ (return to zero), we consider link with noise NF = 6dB and without noise. In case of RZ, the number of spans simulated is double than that in case of the NRZ.

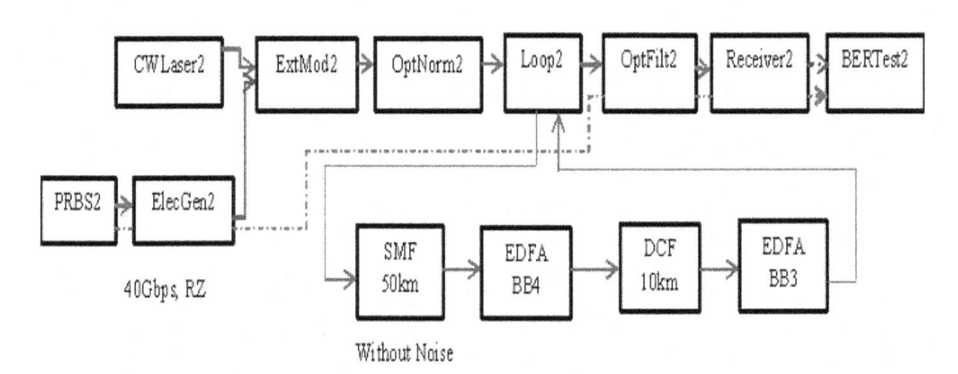

Fig. 5. Model for fiber nonlinearity and performance of RZ formats in high speed links

4 Results and Discussions

4.1 Output of PMD Model

With reference to fig. 2 as PMD model, the performance of polarization mode dispersion (PMD) model shows that as the frequency increases the dispersion increases. It is seen from result that at lower frequency the dispersion across the fiber is less but at higher frequency, the dispersion also increases. Polarization mode dispersion (PMD) is a form of modal dispersion, where the two different polarizations of light in a waveguide, which are normally travel at the same speed, travel at different speeds due to random imperfections and asymmetries, causing random spreading of optical pulses, unless it is compensated which is difficult this ultimately limits the rate at which data can be transmitted over a fiber. It comes to know from fig. 6 that as the frequency increases the dispersion also increases. The fig. 6 shows baseband frequency in Hz vs signal power spectrum in dBm. As it is seen that at lower frequency the dispersion across the fiber is less but at higher frequency the dispersion increases.

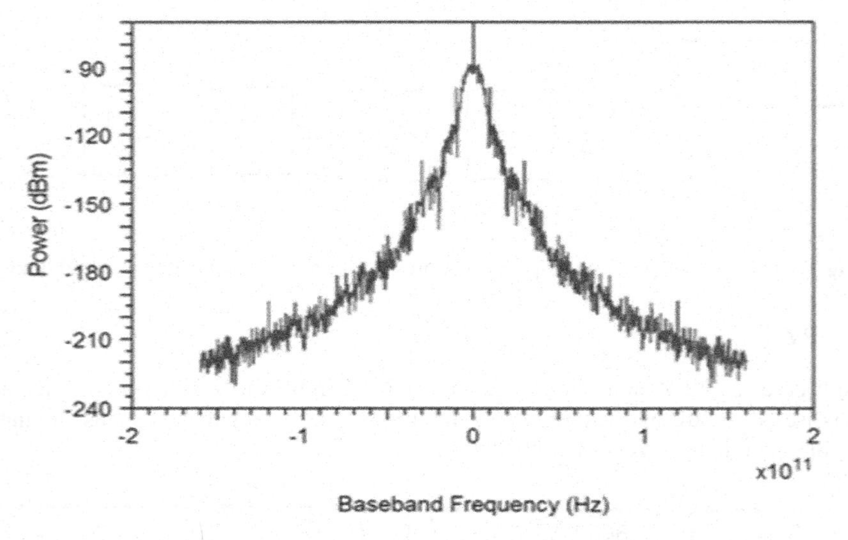

Fig. 6. Baseband frequency v/s signal power spectrum

The PMD is a statistical effect caused by randomly varying fiber birefringence; therefore the simulation results will be different for different settings of the random speed parameter. PMD causes differential group delay (DGD) between x- and y- polarization components during propagation in fiber in fig. 7 and hence eye distortion at the receiver is shown in fig. 8. One can run a parameter scan to obtain DGD and BER/Q for different values of the PMD coefficient and different random seeds. After the simulation run is finished, one can double-click on various analysis blocks to view the signal plot as shown in fig. 8 and eye diagram as shown in fig. 9, DGD, Stokes parameters on Poincare Sphere, and BER/Q values as shown in the following fig. 7.

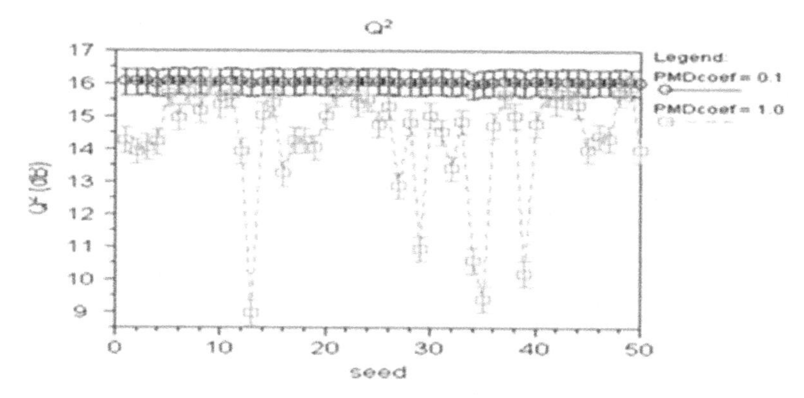

Fig. 7. BER/ Q-factor

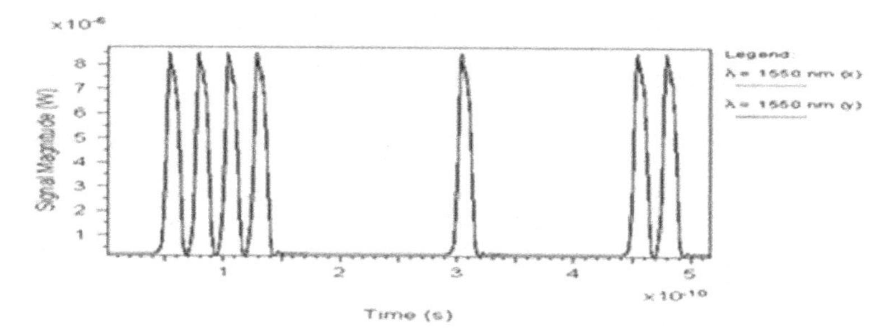

Fig. 8. Signal magnitude v/s time

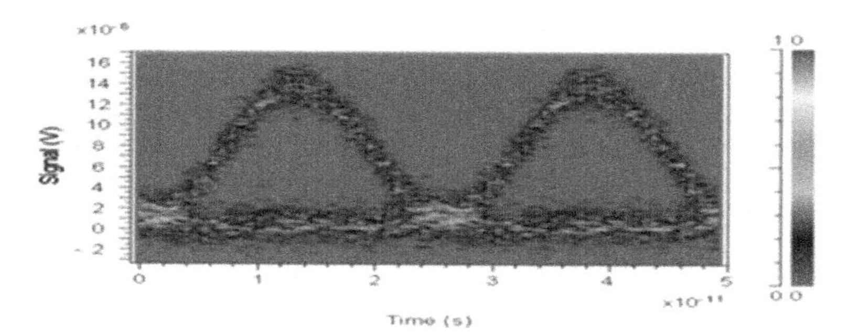

Fig. 9. Eye diagram for PMD at 70km

Best Case. The eye amplitude of the eye diagram shown in fig. 9 is $1.6*10^{-6}$ V at a distance of about 70km.

But however as the distance increases the system performance degrade and at certain distance the eye diagram is totally disturbed which shows there is total loss of information which is shown as follows:

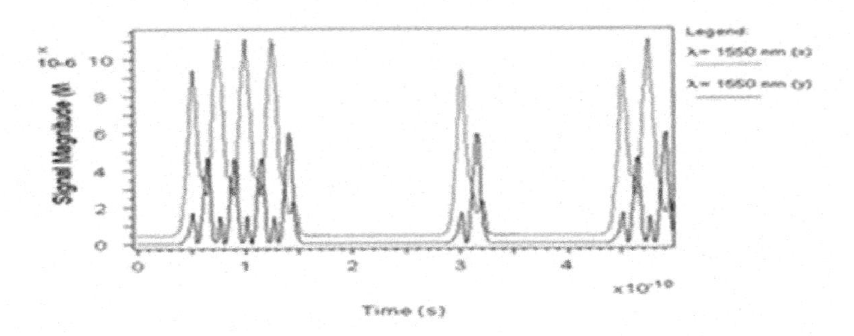

Fig. 10. Signal amplitude v/s time

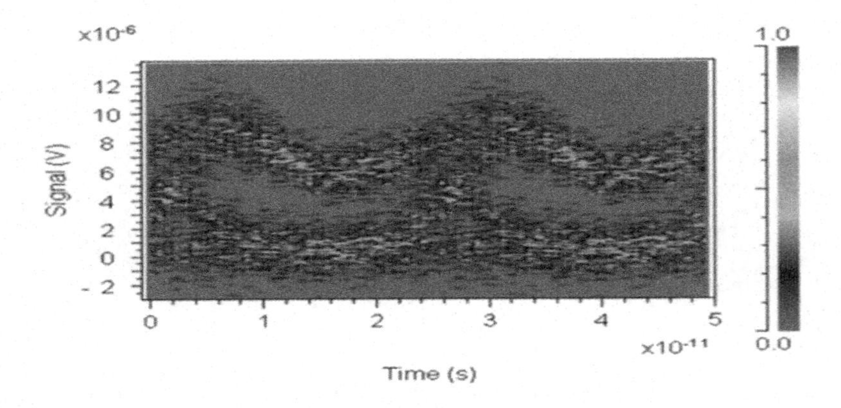

Fig. 11. Eye diagram at more

Worst Case. The PMD-induced differential group delay (DGD) degrades system performance with penalties more severe at higher bit rates and higher PMD coefficients in fibers as shown in fig. 11 & signal magnitude plot as shown in fig. 10.

4.2 Output of SBS Model

With reference to fig. 3 as SBS model setup demonstrates effect of stimulated Brillouin scattering (SBS) on a fiber's transmission performance, namely signal loss. The launch power of a fiber's input signal is increased until SBS becomes important and begins depleting input signal. The below Fig. 12 (a) and (b) shows result with a launch power of approximately 14dBm, the output of the fiber begins to depleted relative to that of the non-SBS fiber. Beyond 14dBm, the output is clearly reduced by the SBS effect.

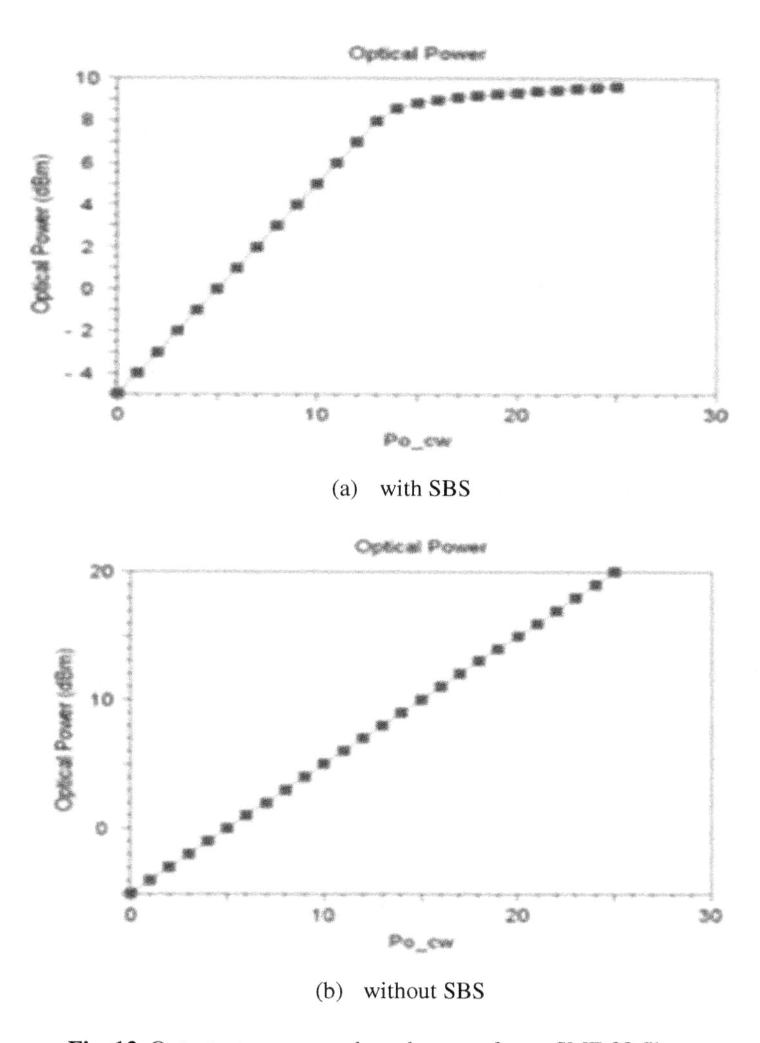

(a) with SBS

(b) without SBS

Fig. 12. Output power versus launch power for an SMF-28 fiber

As shown in Fig. 3 as SBS model depicts the topology used in this example. We launch an input signal into two versions of an SMF-28 fiber. In the top branch of the topology, SBS is included in the model. In the bottom branch, it is not. All other fiber parameters are identical, including the length of 20 km. This setup allows us to determine at what fiber launch power SBS begins to have an effect. Load the example topology and open the parameter scan dialog. The dialog is preset to scan the fiber launch power from 0 to 25dBm in increments of 1dB. When the simulation is completed, display the output power of each fiber as a function of launch power using the monitor blocks. These plots are shown in Fig. 12 (a) and (b).

4.3 Output of Fiber Non-linearity and Performance of NRZ and RZ Formats in High Speed Optical Links

NRZ Model. The below fig. 13 shows the Q^2 vs average fiber input power for NRZ plot at 240km transmission for this topology with linear and nonlinear dispersion compensating fiber (DCF), with and without amplifier noise. As the launched power increases, the performance keeps on improving until the launched power becomes high enough to cause nonlinearities adversely dominate the overall link performance. For this topology, if it can examine the curves corresponding to the cases with and without amplifier noise, it shows that until the launched power is low enough, the link noise plays a dominant role. However, at higher launched powers, the nonlinearities dominate the performance impairments as compared to the influence of link noise. Because of the smaller core diameter of the DCF, optical intensities are enhanced thereby resulting in higher nonlinear effects in the DCF section. The nonlinearities in the DCF, as seen in the following fig. 13 for NRZ can deteriorate the Q-factor by 6 to 9 dB depending upon the launched power.

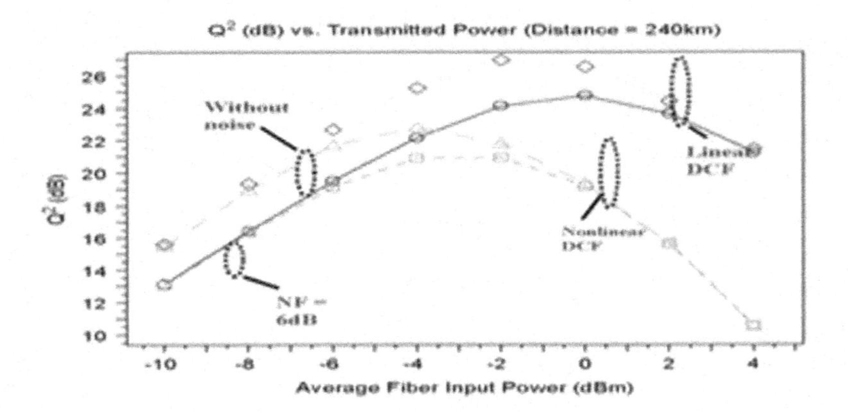

Fig. 13. Q^2 v/s average fiber i/p power for NRZ

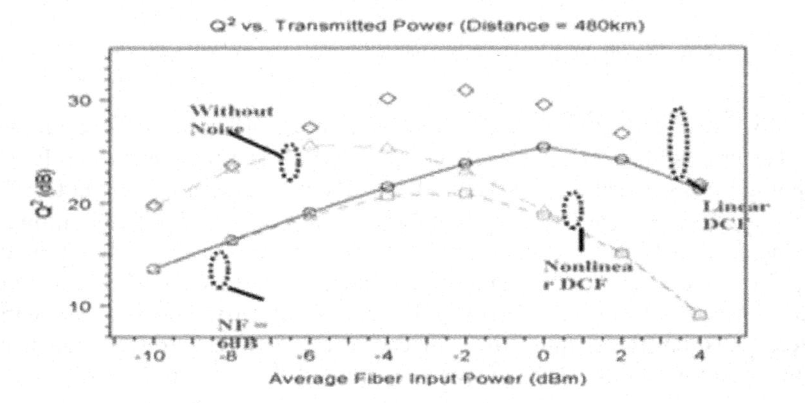

Fig. 14. Q^2 v/s average i/p power for RZ

RZ Model. The above fig. 14 shows Q2 vs average fiber input power for RZ plot at 480-km transmission for this topology with linear and nonlinear DCF, as shown in above fig. 14 for RZ with and without amplifier noise. As the launched power increases, the performance keeps on improving until the launched power becomes high enough to cause nonlinearities adversely dictate the overall link performance. The RZ and NRZ cases qualitatively show resembling behavior except that the transmission distance is now double for RZ.

5 Conclusion

In this paper, we have proposed PMD and SBS model for various losses associated in optical fiber. The impairment induced by a constant DGD scales with the Q^2 (square of bit rate), resulting in drastic PMD degradation for high-speed transmission system effect of stimulated Brillouin scattering (SBS) on a fiber's transmission performance, namely signal loss. The SBS is pronounced at high bit rates and high power levels. The Non linearities & performance of NRZ & RZ formats in high speed optical fiber links are studied in detail. The best & as well as worst case using PMD in optical fiber for high speed of data transmission of 40 Gbps with lower BER & Q factor as 9dB. The various factors broadly include whether the system is power limited or dispersion limited whether it is single channel system or multi-wavelength link, and whether the system operates in liner or non linear region of the Q (or BER) curve. Further work in the continuation to this modeling setup apply to the various types of fiber as the channel rate approaches up to 1Tbps, the achievable capacity becomes a bottle neck.

Acknowledgments. Our thanks to Dr S. V. Dudul, Head of Applied Electronics Dept. of Sant Gadge Baba Amravati University, Amravati, India for providing Lab facility to carry the work and technical support.

References

1. Armstrong, J.: OFDM for Optical Communications. IEEE Journal of Light Wav. Technology 27(3) (2009)
2. Rindhe, B.U., Shah, D., Narayankhedkar, S.K.: OFDM & MC-CDMA for 4G Networks. International Journal for Computer Application IJCA 2011 6(6), 30–37 (2011)
3. Rindhe, B.U., Narayankhedkar, S.K.: BER Analysis for OFDM and MC-CDMA Systems. In: Proceedings of International Conference on ACCT 2012, IETE Mumbai, India (2012)
4. Barbieri, A., Colavolpe, G., Foggi, T., Forestieri, E., Prati, G.: OFDM versus Single Carrier Transmission for 100 Gbps Optical Communication
5. Caplan, D.O.: Laser Communication Transmitter and Receiver Design. Journal on Optical Fiber Communication, rep. 4, 225 (2007)
6. Hu, B.-N., Hua, L., Zhao, R.-M., Pang, H.-W.: Computer Simulation of Optical Fiber Communication System. Inst. of Inf. Sci. & Eng., Hebei Univ. of Sci. & Technol., Shijiazhuang, China

7. Demir, A.: Noise Analysis for Optical Fiber Communication Systems. Dept. of Electr. & Electron. Eng., Koc Univ., Istanbul, Turkey
8. Keiser, G.: Optical Fiber Communication, 4th edn. McGraw-Hill Publications
9. Shieh, W.: PMD-Supported Coherent Optical OFDM systems. IEEE Photonics Technology Letters 19(3) (2007)
10. Corning. Inc.: An Introduction to the Fundamentals of PMD in Fibers. White Paper (2006)
11. Kuschnerov, M., Haushke, F.N., Gourden, E., Piyawanno, K., Lankl, B., Spinnler, B.: Digital Timing Recovery for Coherent Fiber Optic Systems. IEEE Fiber Optics Communication (2008)
12. Agrawal, G.P.: Applications of Nonlinear Fiber Optics. Academic Press
13. Ghassemlooy, Z., Hayes, A.R., Seed, N.L.: Digital Pulse Interval Modulation for Optical Communications. IEEE Comm. Mag. (2009)
14. Zhang, K., Xiang, Q., Wang, Z., Zhu, L., Xu, A., Wu, D., Wang, C., Xie, L.: WDM+EDFA, 12 channels 250 km optical fiber communication system. Dept. of Radioelectron., Peking Univ., Beijing
15. Salz, J.: Coherent light wave communications. AT&T Technical Journal 64, 2153–2209 (1985)
16. Kuschnerov, M., Haushke, F.N., Gourden, E., Piyawanno, K., Lankl, B., Spinnler, B.: Digital Timing Recovery for Coherent Fiber Optic Systems. IEEE Fiber Optics Comm. (2008)

Ultra Long Period Reversible Fiber Gratings as a Pressure Sensor

Sunita Pandit Ugale[1], Vivekanand Mishra[1], and Angela Amphawan[2]

[1] S.V. National Institute of Technology, Surat, Gujarat India
spu.eltx@gmail.com, vive@eced.svnit.ac.in
[2] School of Computing, Universiti Utara Malaysia, 06010 Sintok, Malaysia

Abstract. We report here for the first time the fabrication and characterization of mechanically induced ultralong period fiber gratings (MULPFG) with period size up to several millimeters. In these gratings the coupling of the fundamental guided core mode takes place with cladding modes of high diffraction orders. The transmission characteristic of grating with different external applied pressure has been experimentally verified.

Index Terms: MULPFG, reversible grating, pressure sensing.

1 Introduction

The optical fiber grating is one of the key elements in the established and emerging fields of optical communication systems. Fiber grating is an infiber spectrally selective component and have low losses, high stability, and small size compatible with fiber sizes and low cost. Their applications also spread into the area of optical fiber sensing.

Long period fiber grating (LPFG) is the special case of FBG. It was first suggested by Vengsarkar and coworkers in 1996. Their spectral properties i. e. resonance wavelength, bandwidth etc. can be varied in a wide range. All these features make LPFG as an important component in a variety of light wave applications such as band rejection filter[1], wavelength selective attenuator, dispersion compensator, multichannel filters in WDM applications and gain flatteners for Erbium doped fiber amplifiers[2]. These gratings are also very suitable for various sensing applications. Sharp filtering characteristics, ease of fabrication, direct connectivity to fiber, high sensitivity to external parameters, and easiness in adjusting the resonant wavelength by simply adjusting the grating period are the strong points to push towards the detail study of this device. The period of a typical long period fiber grating (LPFG) ranges from 100μm to 1000μm.

The LPFGs with periods exceeding one millimeter are called ultralong period fiber gratings (ULPFG). The long period size makes fabrication of ULPFG very easy as well as cheap.

ULPFG can be induced optically or mechanically [3] - [5]. Optically induced gratings are permanent, whereas mechanically induced gratings are reversible.

S. Unnikrishnan, S. Surve, and D. Bhoir (Eds.): ICAC3 2013, CCIS 361, pp. 439–443, 2013.

Xuewen Shu et. al. reported fabrication and characterization of ULPFG for the first time in 2002 by using point-by-point writing technique with 244nm UV beam from a frequency doubled Argon ion laser [3]. An ultralong period fiber grating with periodic groove structure(G-ULPFG) fabricated by using an edge-written method with high-frequency CO2 laser pulses is reported by Tao Zhu and co-workers in 2009[4].

We report here, for the first time to our knowledge, the fabrication and characterization of mechanically induced ULPFGs (MULPFG) with periods up to several millimeters. Mechanically induced long period fiber gratings (MLPFG) and MULPFG induced by pressure need neither a special fiber nor an expensive writing device for fabrication. These gratings also offer advantages of being simple, inexpensive, erasable, and reconfigurable and also gives flexible control of transmission spectrum,

2 Theory

ULPFG is a special case of LPFG. In LPFG the core LP_{01} mode is coupled with cladding modes having same symmetry, namely LP_{0m} modes [6]. Whereas in ULPFG the coupling of the fundamental guided core mode to the cladding modes of high diffraction orders takes place [7]. The phase matching condition for a high diffraction order grating is given by (1).

$$\lambda_{res} = \left(n_{eff}^{co} - n_{eff}^{cl,m}\right)\frac{\Lambda}{N} \tag{1}$$

Where λ_{res} is the resonant wavelength, n_{eff}^{co} and $n_{eff}^{cl,m}$ are effective indexes of fundamental core mode and m^{th} cladding mode of Nth diffraction order respectively. Λ is the grating period and N is the diffraction order. N=1 for LPFG.

The resonant wavelength with the variation in the effective indexes of the core and cladding ignoring the dispersion effect is given by (2).

$$\lambda'_{res} = \left(n_{eff}^{co} - n_{eff}^{cl,m}\right)\frac{\Lambda}{N}\times\left[1+\frac{(\delta n_{eff}^{co} - \delta n_{eff}^{cl,m})\times\dfrac{d\lambda_{res}}{d\Lambda}}{(n_{eff}^{co} - n_{eff}^{cl,m})^2}\right] \tag{2}$$

Where λ'_{res} is the resonant wavelength with variation in the effective indexes of core and cladding, δn_{eff}^{co} and $\delta n_{eff}^{cl,m}$ are the effective index changes of the fundamental core mode and m^{th} cladding mode of the Nth diffraction order.

3 Experiment

Reversible MLPFG and MULPFG of different periods ranging from several hundred microns to several millimeters were induced and characterized.

The gratings were induced in single mode fiber SMF28 with core diameter of 9μm and cladding diameter of 125μm. A special probe with unjacketed fiber of 10cm length at the center and APC connectors at both ends was prepared.

The experimental set up as shown in fig. 1 was prepared.

The light from broadband superluminacent LED with output power of -8dBm, center wavelength 1530nm and bandwidth of 69nm was passed through the grating under test and the transmitted signal was analyzed with the help of optical spectrum analyzer covering the wavelength range from 1250nm to 1650nm. MLPFG with a period 1800μm was induced and characterize. The response of this grating to external applied pressure was also studied.

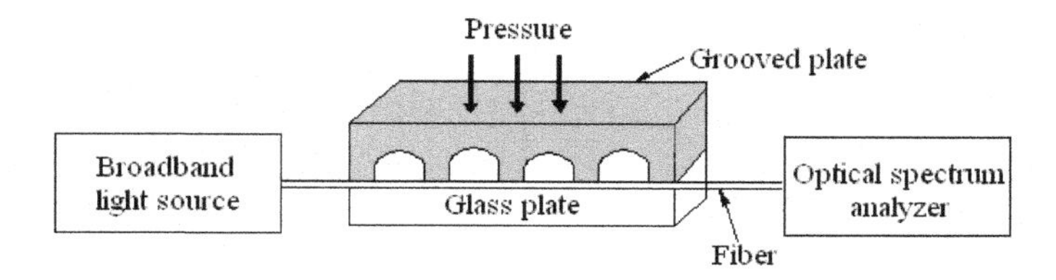

Fig. 1. Experimental set up for inducing MULPFG and its characterization

4 Results

The transmission characteristic for MLPFG with period 1800μm is shown in Fig. 2. and the results are summarized in table I. In MULPFG the transmission loss increases with applied pressure.

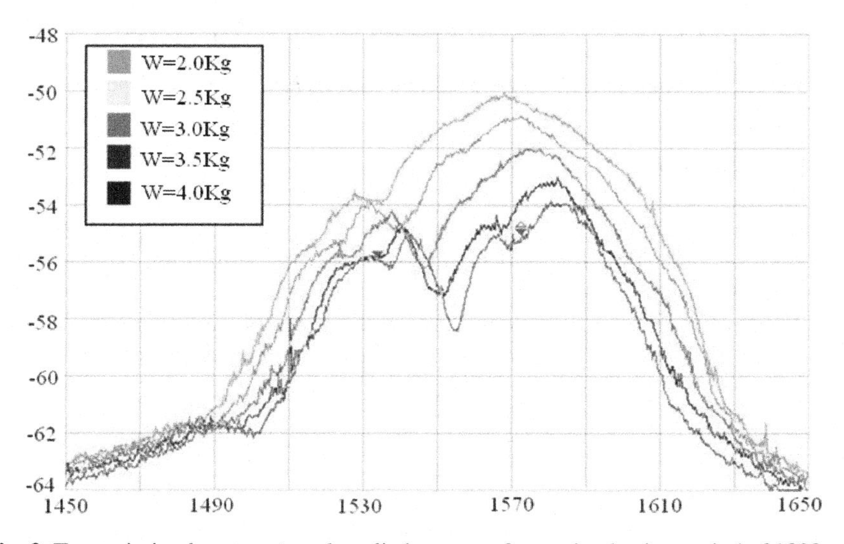

Fig. 2. Transmission loss to external applied pressure for grating having period of 1800μm

Table 1. Transmission loss to applied pressure for grating 3

External applied pressure in kg	Transmission loss in dBm for grating with (period Λ=1800μm)	Resonance wavelength in nm
2.0	53.8	1534
2.5	55.0	1538
3	56.2	1546
3.5	57.2	1550
4.0	58.4	1555

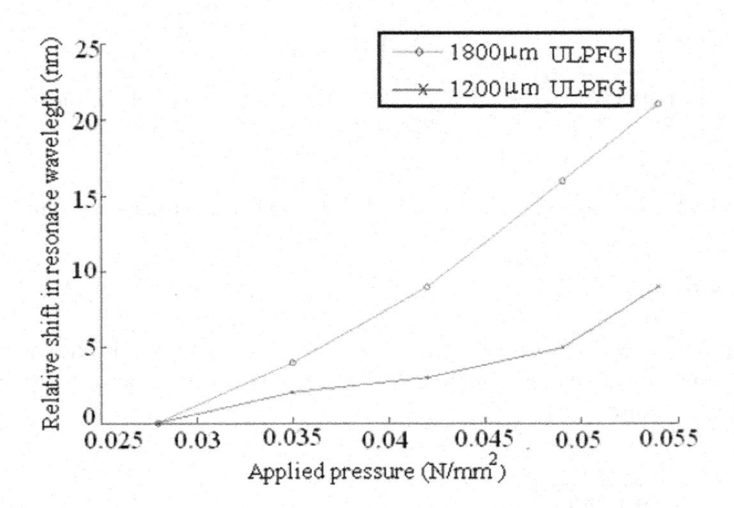

Fig. 3. Relative change in resonance wavelength with change in external applied pressure

5 Conclusion

MULPFG with 1800μm period was induced successfully in SMF28 fiber. we observed significant shift of 20nm in resonance wavelength, thus MULPFG with greater period is more sensitive to external applied pressure. Thus MLPFG can be used as band rejection filter, wavelength selective attenuator, gain flatterers for optical amplifier and MULPFG can be used as a sensor for pressure, temperature or RI.

References

[1] Vengsarkar, A.M., Lemaire, P.J., Judkins, J.B., Bhatia, V., Erdogan, T., Sipe, J.E.: Long-period fiber gratings as band-rejection filters. Journal of Lightwave Technology 14(1), 58–65

[2] Sohn, Song, J.: Gain flattened and improved double pass two stage EDFA using microbending longperiod fiber gratings. Optics Communication 236(1-3), 141–144 (2004)

[3] Shu, X., Zhang, L., Bennion, I.: Fabrication and characterization of ultralong period fiber gratings. Optics Communication 203, 277–281 (2002)
[4] Zhu, T., Song, Y., Rao, Y., Zhu, Y.: Highly sensitive optical refractometer based on edge-written ultralong period fiber grating formed by periodic grooves. IEEE Sensors Journal 9(6) (2009)
[5] Savin, S., Digonnet, M., Kino, G., Shaw, H.: Tunable mechanically induced long-period fiber gratings. Optics Letters 25(10), 710–712 (2000)
[6] Erdogan, T.: Fiber grating spectra. Journal of Lightwave Technology 15(8) (1997)
[7] Zhu, T., Rao, Y.J., Mo, Q.: Simultaneous measurement of refractive index and temperature using a single ultralong period fiber grating. IEEE Photonics Technology Letters 17(12) (2005)

Multicast Traffic Grooming in Sparse Splitting WDM Mesh Networks

Ashok Kumar Pradhan and Tanmay De

Department of Computer Science and Engineering,
National Institute of Technology, Durgapur, India

Abstract. With the growing popularity of multicast applications and the recognition of the potential achievable efficiency gain of the traffic grooming, we face the challenge of optimizing the design of WDM optical networks with sparse splitting multicast traffic grooming. Efficiently grooming low speed connections onto a high capacity wavelength channel can significantly improve the bandwidth utilization in an optical network. In this study, we investigate the problem of sub-wavelength traffic grooming in a WDM optical networks and shows how to take the advantages of multicast capable nodes in grooming these sub-wavelength traffic. The problem of constructing optimal multicast routing trees and grooming their traffics in WDM optical mesh networks is NP-hard. Therefore, we propose an heuristic approach to solve the problem in an efficient manner. The main objective of this paper is to maximize the bandwidth utilization and simultaneously minimize the wavelength usage in a sparse splitting optical network. The problem is mathematically formulated. We have simulated the proposed heuristic approach Multicast Sparse Splitting Traffic Grooming (MSSTG) with different network topologies and compared the performance with Multicast Traffic Grooming with Shortest Path (MTG-SP) algorithm. The simulation results shows that the proposed approach produces better result than existing MTG-SP algorithm.

Keywords: Wavelength Division Multiplexing (WDM), Multicast Routing and Wavelength Assignment (MRWA), Light-tree, Splitting, Traffic Grooming.

1 Introduction

Wavelength Division Multiplexing (WDM) technology provides huge amount of link bandwidth to recent telecommunication networks. In a wavelength routed WDM networks, data is transported in all optical channel from source node to destination node is known as lightpath [1], [2]. This may span multiple fiber links. As WDM technology keeps growing, the transmission rate on a wavelength channel has reached OC-192 (10 Gbps) and is expected to reach OC-768 (40 Gbps) in near future. The traffic granularities may varies a lot, possibly from OC-3 (155 Mbps) to OC-192 (10 Gbps). Hence, there is a mismatch between the capacity of a wavelength channel's transmission speed and bandwidth granularities. The potential inefficiency may be caused by this disparity has motivated the development of techniques that enables the efficient utilization of network resources and maximize the bandwidth utilization by grooming or multiplexing low speed traffic streams onto high capacity optical channel.

S. Unnikrishnan, S. Surve, and D. Bhoir (Eds.): ICAC3 2013, CCIS 361, pp. 444–458, 2013.

Multicasting is a technique to transmit information from a single source to multiple destinations. This is bandwidth efficient because it discards the necessity for the source to send an individual copy of the information to each destination. Multicasting applications such as weather forecasting, on-line video conferences, streamed video broadcast etc are the major applications of the Internet technology. In traffic grooming the topology would be an optimal basic structure to follow their implementations. For such an application, a natural extension of lightpath called light-tree can significantly improve the performance. A light-tree [3], [4], [5], [6] is one to many connections between source to multiple destinations of the network. When wavelength continuity constraint is applied then all the optical links of the light-tree must be assigned with same wavelength. In order to realize a light tree in a WDM optical networks, the node(s) at the branching point(s) of the light tree must be capable of splitting an input light signal into two or more output signals of the same wavelength. A node with such capability is called a multicast capable (MC) node or a splitter. A node without such capability is called a multicast incapable (MI) node. When a signal goes through a splitter, the power of the output signals is degraded by a factor of the number of output signals. To maintain the power level of the signal, a costly active amplification device is required to be installed in the splitter. Therefore, a multicast capable node is much more expensive than a multicast incapable node. A WDM optical networks, in which all nodes are multicast capable is very expensive. Hence, we assume only a few of the nodes have the splitting capacity in the network. A network with few splitting capable nodes is called sparse splitting [7], [2], [6], [8] network. Using optical splitters reduces cost by splitting an incoming signal along many outgoing optical links without using optical-electrical-optical (OEO) conversion. Typical multicast applications need only a fraction of the channel capacity of a light-tree and hence traffic grooming is needed to share the optical bandwidth efficiently.

In this paper, we study the problem of multicast traffic grooming with sparse splitting capable nodes. The source and destination nodes are known in prior, i.e., it is static in nature. The multicast requests are generated randomly between source and all destination nodes. The main objective of our work is to maximize the bandwidth utilization and minimize the number of wavelengths. We have proposed a heuristic Multicast Sparse Splitting Traffic Grooming (MSSTG) algorithm which reduces the number of wavelengths by sharing of the links among various multicast sessions and maximize the bandwidth utilizations of the channels.

The rest of the paper is organized as follows: Section 2 presents the previous work. In Section 3 we have defined the problem in formal notations. The proposed approach of grooming multicast routing and wavelength assignment problem is presented in Section 4. The experimental result and its analysis are described in Section 5. Finally, conclusion is drawn in Section 6.

2 Previous Work

Traffic grooming in WDM optical networks with a mesh topology is relatively new and only a limited effort has recently been directed to this field. Based on whether the connection requests are known a priori, traffic grooming can be broadly classified into two categories: static and dynamic traffic grooming. In static traffic grooming the set of fixed end to end traffic is known a priori. In dynamic traffic grooming, connection requests arrive dynamically and both routing and wavelength assignment are dynamically decided for new connection requests. Yoon et al. in [6] proposed a heuristic algorithm to solve the sparse splitting problem in an efficient manner. It reduces the link cost by constructing a minimal cost tree with multicast capable nodes and increase the traffic grooming effect based on relationship of multicast sessions. BWA algorithm proposed by Billah et al. [7] efficiently constructs multicast routing trees and assign wavelength using First-Fit wavelength conversion capability in the network nodes. By using sparse splitting traffic grooming their algorithm can significantly reduces the maximum number of wavelengths require in a links as well as total number of links needed. B. Mukherjee and K. Zhou in [8] design a sparse splitting WDM mesh networks. Illustrative numerical results from the mathematical formulation as well as heuristics show that, by properly choosing the grooming nodes, a network with sparse-grooming capability can achieve good network performance and the network cost can be significantly reduced. Yan et al. in [2] investigate the problem of Multicast Routing in Sparse Splitting Networks (MR-SSN). The MR-SSN problem is to find a route from the source node of the multicast session to all destinations of the multicast session where, multicast capable nodes are uniformly distributed throughout the network, so that the total number of fibers used in establishing the session is minimized. Lin and Wang in [9] studies the splitter placement problem in a WDM optical networks in which a light forest consisting of a collection of light trees is used to realize a multicast connection. The goal is to place a given number of splitters in the network such that the average per link wavelength resource usage of multicast connections is minimized. R. Mustafa and A. Kamal in [5] proposed an optimal design of WDM optical networks for grooming the multicast sub-wavelength traffic. The objective is to minimize the network cost by minimizing the number of higher layer electronic equipments and, simultaneously, minimizing the total numbers of wavelength used. R. Lin et al. in [3] proposed a light-tree based ILP formulation to minimize the network cost associated with network resources such as higher layer electronic ports and number of wavelengths used. X. Jia et al. [10] proposed two optimization algorithms for minimizing the number of wavelengths. One algorithm minimizes the number of wavelengths through reducing the maximal link load in the system; while the other does it by trying to free out the least used wavelengths. G. Poo and Y. Zhou [4] addressed the multicast wavelength assignment (MC-WA) problem in wavelength-routed WDM networks with full light splitting and wavelength conversion capabilities. It makes the multicast wavelength assignment more flexible, covering different switching schemes and different assignment strategies. The authors in [11], propose a Multi-Objective Evolutionary algorithm to solve a grooming problem. Maximize the traffic throughput, minimize the number of transceivers, and minimize the average propagation delay or average hop counts are three objectives considered in this paper. The authors in [12] propose two multicast grooming methods, namely, single hop (SH) and multiple hops (MH). In SH, an existing light-tree with sufficient available bandwidth and the same source and destination nodes as the new multicast request is chosen. This approach is simple to implement but it is not efficient. MH provides bandwidth to the new request in two logical hops and would be more likely to succeed than SH. A multicast tree decomposition (MTD) scheme [13] was proposed to improve the blocking performance by grooming a request partially over several existing light-trees. The authors in [14] propose maximize lightpath sharing multi-hop (MLS-MH) grooming algorithm to support dynamic traffic grooming in sparse grooming networks. The objective of this paper is to maximize the utilization of lightpaths.

3 Problem Description

In this section, we consider the sparse splitting traffic grooming problem, where few of the nodes are splitting capable. Suppose that, $G(V_{MC}, V_{MI}, E)$ is a bi-directional graph, where V_{MC} is the set of multicast capable nodes, V_{MI} is the set of multicast incapable nodes and E is the set of edges between the nodes. A multicast session includes one source and many destinations. The root of the tree T, denoted by v_0, is the source of the multicast session and $v_0 \in (V_{MC}, V_{MI})$. All leaf nodes are the destinations of the multicast session. However, a destination in the multicast session is not necessarily a leaf node in the tree T. It may also be an intermediate node. In tree T, each node v_i, except the root, has an incoming link denoted by e_i and $e_i \in E$. Each link can carry traffic up to wavelength channel W and each wavelength channel have capacity C. Each MC node $v_{mc} \in V_{MC}$ has capability of splitting the light signal and each MI node $v_{mi} \in V_{MI}$ has no splitting capacity. Wavelength continuity constraint is imposed in all multicast sessions. That means there is no wavelength converter is used. In each multicast session source and destination nodes are randomly generated and sub-wavelength bandwidth requirements are assigned in each multicast session.

We introduce the following notations to define our problem:

λ: set of all possible wavelengths in the network, $\lambda = \{\lambda_1, \lambda_2, \lambda_3, ..., \lambda_\Lambda\}$. Where, $|\Lambda| = W$ is the number of all possible wavelength channels.

Λ_{req}: set of wavelengths required to establish the connections.

C: capacity of the wavelength channel.

M: set of n number of multicast session requests, $M = \{m_1, m_2, m_3, ..., m_n\}$.

ω_m: wavelength assigned to multicast request m.

$m_i(s_i, D_i, c_i)$: a tuple of the ith multicast session request. Where s_i is the source node, D_i is the set of destination nodes, and c_i is the bandwidth requirement of ith multicast session.

c_m:bandwidth requested by multicast session request m.

Objective:

Our objective is to maximize the bandwidth utilization in each wavelength channels. We defined bandwidth utilization as follows:

$$Maximize \quad \left[\frac{\sum_{m=1}^{M} c_i}{\Lambda_{req} \times C}\right] \times 100(\%) \tag{1}$$

$$Where, \quad \Lambda_{req} = \sum_{m=1}^{M} \delta_m \tag{2}$$

$$\delta_m = 1, \quad if\ \omega_m \neq \omega_n,\ 1 \leq n < m \leq M$$
$$= 0, \quad otherwise$$

Constraints:
We assume the following constraints in this work.

- In each multicast session only one multicast tree is generated.
- In each multicast tree a single wavelength is assigned that is supported by multiple sub-wavelength channels.
- Cumulative bandwidth of all sub-wavelength must be less than or equal to the total wavelength capacity.
- Each multicast sessions may be supported by multiple sub-light trees.

Assumptions:
We use the following assumptions in this work.

- Traffic demands is static in nature.
- Network nodes do not have wavelength conversion capacity. So, a light-tree must use the same wavelength from source to all destination nodes.
- Each network node can act as both source node as well as destination nodes.
- Granularity of low speed traffic request is $x \in \{1,3,12,48\}$, which means that bandwidth capacity can be one of the OC-1, OC-3, OC-12 and OC-48 between various source and destination nodes.

4 Proposed Approach

As we know that the splitters are very costly devices, hence few of the nodes should be splitting capable such that the total network cost can be reduced. The main objective of our work is to groom the traffic in such a fashion that we can maximize the bandwidth utilization.

In this section, we propose a heuristic algorithm called Multicast Sparse Splitting Traffic Grooming (MSSTG) as shown in algorithm 1. Here, our proposed approach is divided into two phases similar to existing algorithm in [6]. In first phase of the proposed approach, multicast-tree is generated. In second phase grooming and wavelength assignment is done. In multicast-tree generation phase, a multicast tree is constructed from the routing information maintained by each node. The route from the source to all destination nodes is determined by all pair shortest path algorithm. Since, we assume that the network have sparse splitting capability, an MI node can not have more than one child node in the generated multicast tree. So, we limit the connectivity degree of MI nodes to two. Once the route is determined for each node, the multicast

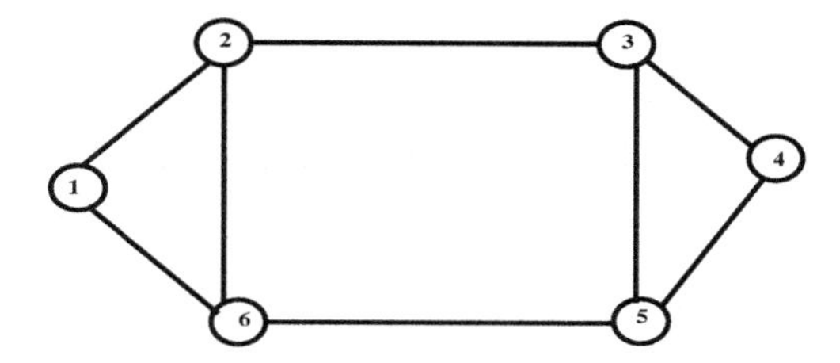

Fig. 1. 6 Node Network

trees for multicast sessions is generated by combining individual route from source to all destination nodes. In the following subsection, we explain the construction of multicast tree by using the proposed approach MSSTG.

Limit the degree of connectivity of each MI node is two. In our proposed model, MI node must have at least one MC node at its neighbor in the physical network. So, each MI node can reach an MC node by one hop. If an MI node has more than one MC node as neighbors, select the two MC nodes with the highest node degree as the neighbor. The nodes with the same degree are selected arbitrarily. Computes the route by running all pair shortest path on the modified network to determine the path between the MC nodes and all other nodes of the network. The union of all explored paths leading each of the destinations form the multicast tree.

Once the multicast tree is generated, the trees are groomed such that cumulative bandwidth requirements for the groomed sessions are less than or equal to the available capacity of each wavelength in the network. To groom the sessions efficiently the traffic is sorted according its bandwidth requirement, number of destinations in the session and number of common destinations in all sessions with same source. Construct multicast tree for each request whose required bandwidth is of a highest wavelength capacity. If there are multiple such multicast trees then it can be done in any order. Calculate the number of source and destination nodes common in two consecutive multicast sessions. If the multicast trees have same source and destination nodes and their cumulative bandwidth is less than total wavelength capacity then groom these two multicast trees. Those sessions which have not all the common source and destinations and one is a sub multicast tree of other and cumulative bandwidth is less than the total wavelength capacity, then larger sub-multicast tree will groom first than smaller sub-multicast tree. Finally, First-Fit wavelength assignment technique is used for wavelength assignment to all multicast tree and sub-multicast trees.

Example:

We have considered a six node standard network as shown in Fig.1 to illustrate the working principle of MSSTG algorithm. Here, we have consider 10 randomly generated multicast sessions, as shown in Table 1. Let us assume that the capacity C of a wavelength is OC-48 and bandwidth requirement of a multicast requests can be OC-1, OC-3, OC-12 or OC-48.

In this heuristic approach MSSTG, we must sort the multicast sessions in such a fashion that at first we shall assign the bandwidth to the multicast trees of highest wavelength capacity. Then remaining sessions having common source and destination

Algorithm 1. Multicast Sparse Splitting Traffic Grooming (MSSTG)

```
Input : G(V_MI, V_MC, E), A set of multicast session requests R = {m_1, m_2, ..., m_n}, where
        n = |R| with a wavelength capacity C
Output: A set of multicast sessions
1  Construct a multicast tree for each of the multicast request whose required bandwidth is
   of highest wavelength capacity
2  for i = 1 to n do
3      S = φ   /*Let S to be an Empty set/*
4      S ← t_i; D ← D_i;   /* Add t_i into S
5      for j = i+1 to n do
6          for all m_i, m_j ∈ R | (i ≠ j) ∧ (s_i = s_j) do
7              P = D ∩ D_j   /*Number of destinations common in both
               sessions/*
8  for j = 1 to W do
9      for k = 1 to |E| do
10         ω_k^j = C
11 for i = 1 to n do
12         /*Grooming and wavelength Assignment*/
13     t = getMulticastTree(m_i)   /*Generate tree by combining shortest paths
       between all source-destination pairs*/
14     for j = 1 to W do
15         for k = 1 to |t| do
16             if ω_k^j ≥ c_i then
17                 Continue
18         if k > |t| then
19             Assign wavelength λ_j to m_i
20             for k = 1 to |t| do
21                 ω_k^j = ω_k^j − c_i
22             Break
```

Function getMulticastTree(m_i)

```
Input : G(V_MI, V_MC, E) and a multicast session request m_i
Output: A multicast tree t
1  t_i = φ
2  limit the degree of v_mi ≤ 2 from G
3  Arrange the v_mc nodes in an increasing order of their out-degree
4  Build the modified network G_m
5  for k = 1 to |m_i| do
6      p_k = shortestPath(s(m_i), d_k)
7      t = t ∪ p_k
8  return t
```

Table 1. Ten multicast requests for the 6 nodes network

Index	Bandwidth (OC)	Source	Destination
1	48	5	2, 1, 3, 4
2	3	6	5, 4
3	12	6	5, 4, 3
4	48	4	1, 6, 2
5	12	6	5, 4, 3
6	1	2	3, 4
7	1	6	2, 5
8	12	1	4, 6
9	1	1	2, 4
10	3	6	2, 5

Table 2. Ten multicast requests for the 6 nodes network using MSSTG approach

Index	Bandwidth (OC)	Source	Destination
1	48	5	1, 2, 3, 4
2	48	4	1, 2, 6
3	12	6	3, 4, 5
4	12	6	3, 4, 5
5	3	6	4, 5
6	3	6	2, 5
7	1	6	2, 5
8	12	1	4, 6
9	1	1	2, 4
10	1	2	3, 4

nodes shall groom together and sessions with higher bandwidth will groom prior to the session request seeking lower bandwidth. To sort the sessions in this fashion, we have used heuristic approach which is based on number of destinations and bandwidth requirement in each session. By using the heuristic approach MSSTG, we have generated 10 sorted multicast sessions as shown in Table 2. Let us consider the 1st multicast tree $(\{5\},\{1,2,3,4\},48)$, here the source node is 5 and destination nodes are $\{1,2,3,4\}$ and bandwidth requirement is OC-48. The shorted multicast tree is as shown in Fig.2. Here, the first MC node is 5 as its out-degree is 3 and second MC node is 6 as its out-degree is 2. The rest of the nodes are MI nodes. In the same fashion, in other sorted multicast trees we can select the MC and MI nodes. In the second multicast tree $(\{4\},\{1,2,6\},48)$, the source nodes is 4 and destination nodes are $\{1,2,6\}$, and the bandwidth requirement is OC-48. Here, the 1st two multicast trees having maximum bandwidth capacity i.e., equal to the wavelength capacity of the network. Hence, in each multicast tree a single wavelength channel say (W_1, W_2) shall assign. In 3rd and 4th multicast sessions, the multicast trees are generated as $(\{6\},\{3,4,5\},12)$ and $(\{6\},\{3,4,5\},12)$. Here, both the multicast trees have common source and destination nodes and sum of their bandwidth capacity is OC-24 which is

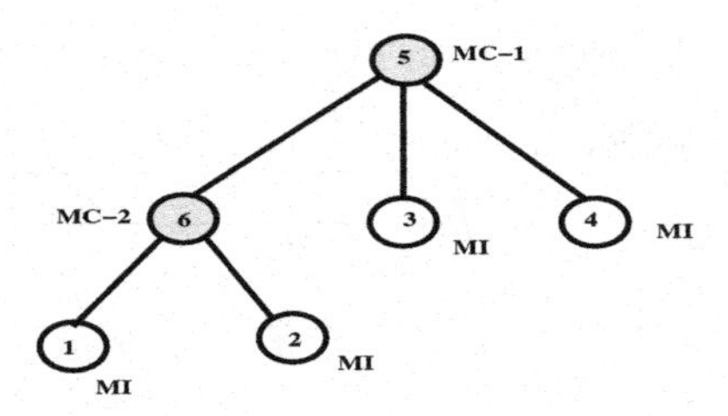

Fig. 2. Sparse splitting multicast tree

less than OC-48, hence a single wavelength say (W_3) shall assign. In the 5th multicast session, the multicast tree is $(\{6\},\{4,5\},3)$, which is a sub multicast tree of the 3rd and 4th multicast trees and its cumulative bandwidth OC- $(24+3) <$ OC-48. Hence, it shall groom with these two multicast tree to form a single multicast tree. In the same fashion multicast tree generation, grooming and wavelength assignment will take place for remaining multicast sessions. That means those multicast sessions having common sessions will groom first and those session which are sub-set of the multicast session requests will groom later. This will reduces the total wavelength requirement and increases the bandwidth utilization.

5 Analysis of Results

In this section, we present and discuss the result obtained using proposed heuristic Multicast Sparse Splitting Traffic Grooming (MSSTG) algorithm. Our main objective is to maximize the bandwidth utilization by maximize the common link sharing among the multiple routes by using sparse traffic grooming and simultaneously reducing the number of wavelength channel requirements. We use the 14 node NSF network and 17 node German network for our simulation studies as shown in Fig.3 and Fig.4. We assume capacity of wavelength is OC-48, and the required bandwidth of multicast request randomly chosen to be one of OC-1, OC-3, OC-12 and OC-48. In Fig. 3, we assume that there are 5 nodes act as a splitting node and in Fig. 4, there are 7 nodes act as a splitting node for our simulation purpose. Here, we have compared the performance of our proposed algorithm Multicast Sparse Splitting Traffic Grooming (MSSTG) with existing Multicast Traffic Grooming with Shortest Path (MTG-SP) algorithm [6]. The performance metrics we have considered here are: bandwidth utilization, wavelength requirement and splitting capacity. The average value of 100 iterations of simulation on randomly generated multicast requests is as shown in this section.

Fig.5 and Fig.6 shows the results of both the algorithms MSSTG and MTG-SP and it demonstrate the relationships between the number of wavelengths requirement to the bandwidth request. Here, we have considered the maximum session size is 60% of the network size. Our proposed approach produces better result than existing algorithm. Here, in our proposed approach MSSTG lesser number of wavelengths are used with the increase of number of multicast traffic bandwidth than the existing algorithm MTG-SP.

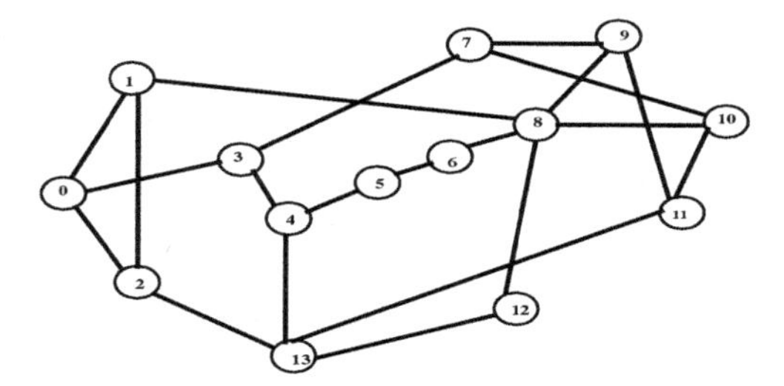

Fig. 3. 14-node NSF network

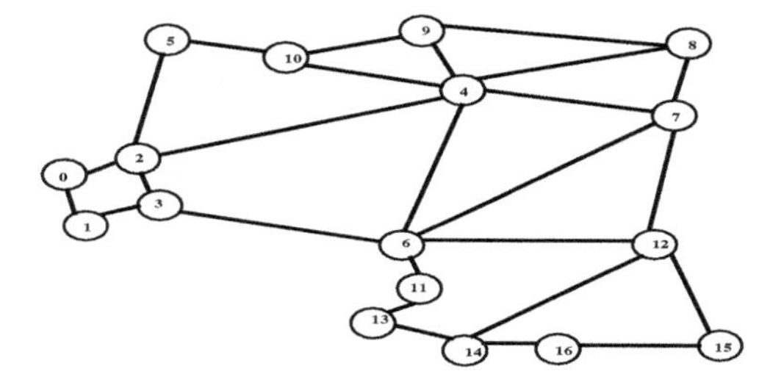

Fig. 4. 17-node German network

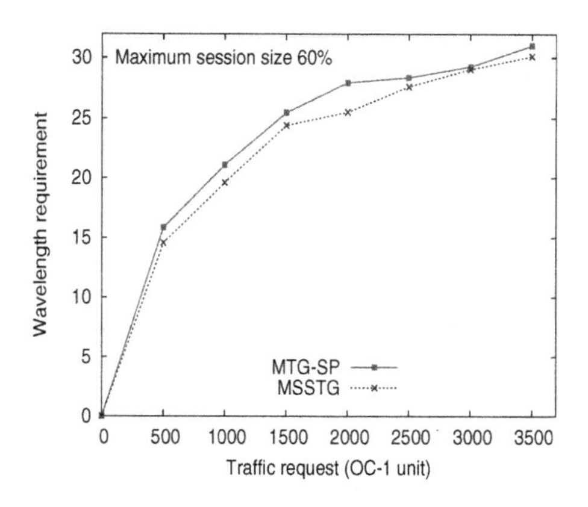

Fig. 5. Relationship of Requested bandwidth with Wavelength requirement for 14-node NSF network

In Fig.7 and Fig.8, we have compared the performance of bandwidth utilization with maximum session size (20% to 100% of the network size) keeping number of sessions fixed at 100. When the maximum session size is 2, the probability of grooming is very less. Due to this, bandwidth utilization is negligible. When the session size increases more number of multicast trees are generated between source to various destination nodes. So that grooming probability increases rapidly with the increase of multicast sessions size.

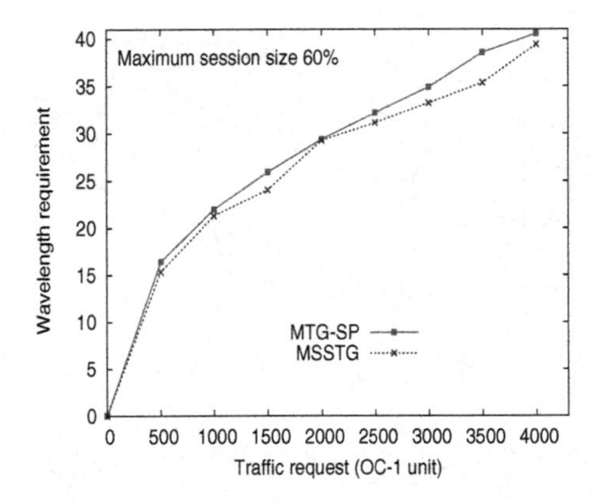

Fig. 6. Relationship of Requested bandwidth with Wavelength requirement for 17-node German network

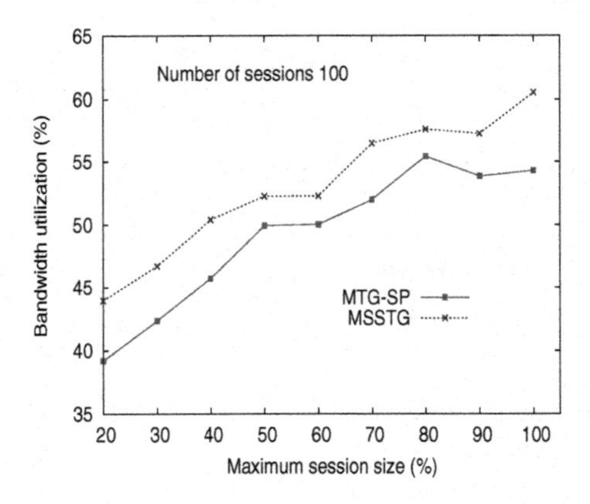

Fig. 7. Relationship of Bandwidth utilization with Maximum session size (%) for 14-node NSF network

Hence, in our proposed approached bandwidth utilization increases with the increase of session size compared to MTG-SP. In Fig.9 and Fig.10, we have compared the results between bandwidth utilization with number of sessions varied [0,100] keeping maximum session size 60% of the total network size. Here, our proposed algorithm MSSTG produces better bandwidth utilization than MTG-SP. It is because, bandwidth utilization depends on bandwidth request by multicast sessions and number of wavelengths required to establish all the connections. Here, in our approach bandwidth request by multicast sessions increases with the increase of number of sessions. As a result, the proposed algorithm improves the bandwidth utilization efficiently both in NSF network and German network compared with MTG-SP algorithm.

Fig.11 and Fig.12 compare the wavelengths required by the two algorithms. Here, we have considered the number of sessions varies from [0,100] keeping the maximum

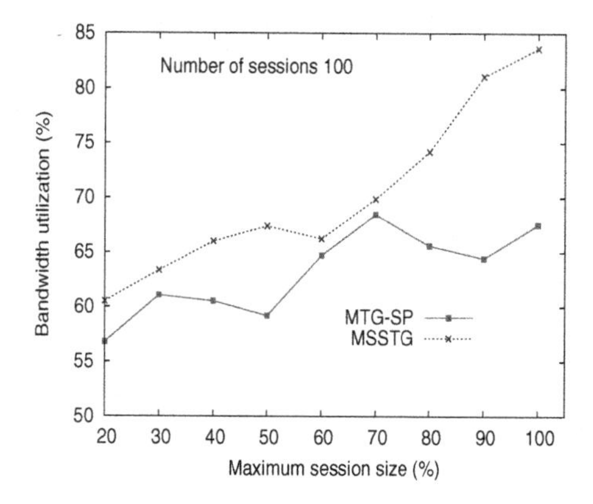

Fig. 8. Relationship of Bandwidth utilization with Maximum session size (%) for 17-node German network

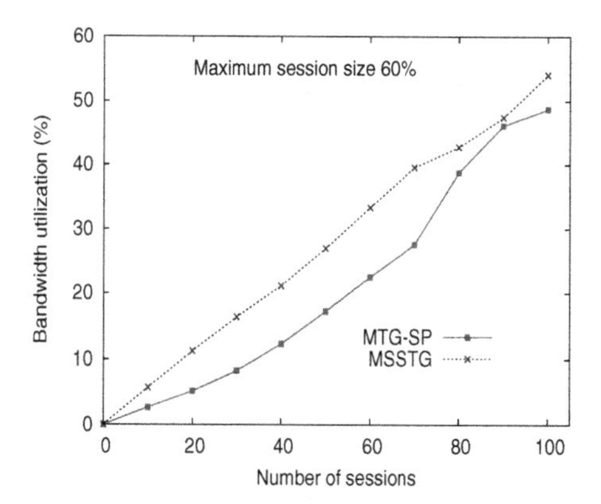

Fig. 9. Relationship of Bandwidth utilization with Number of sessions for 14-node NSF network

session size 60% of the network size. MSSTG algorithm groom the multicast trees having common source and destination nodes as well it grooms the sub-multicast trees that contains sub set of the another multicast tree. Here, our proposed approach produces slightly better result in wavelength utilization than the existing algorithm (MTG-SP) with the increase of number of sessions.

Here, one of our motto is to minimize the splitting capacity, i.e., number of light signal emitted from all the splitting node, for all multicast requests generated. In Fig.13 and Fig.14, we have compared the performance of splitting capacity with maximum session size (20% to 100% of the network size) keeping the number of sessions 100. In this case MSSTG produces slightly better result than MTG-SP when the session size increases. This is due to fact of common sharing of the links increases with the increase of session size.

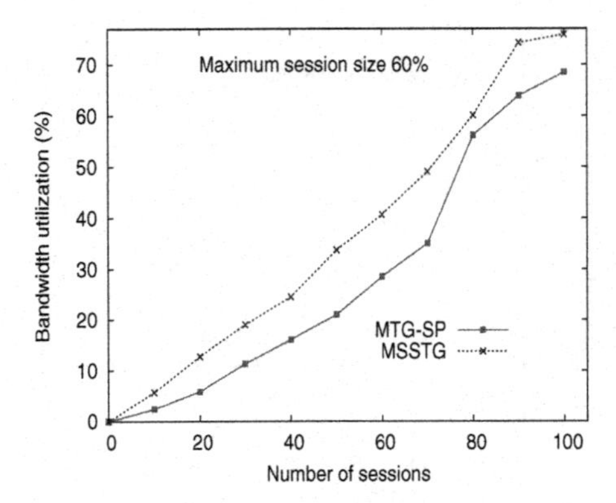

Fig. 10. Relationship of Bandwidth utilization with Number of sessions for 17-node German network

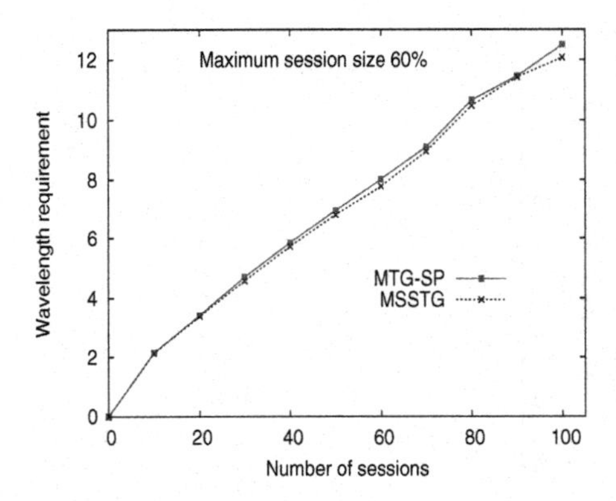

Fig. 11. Relationship of Wavelength requirement with Number of sessions for 14-node NSF network

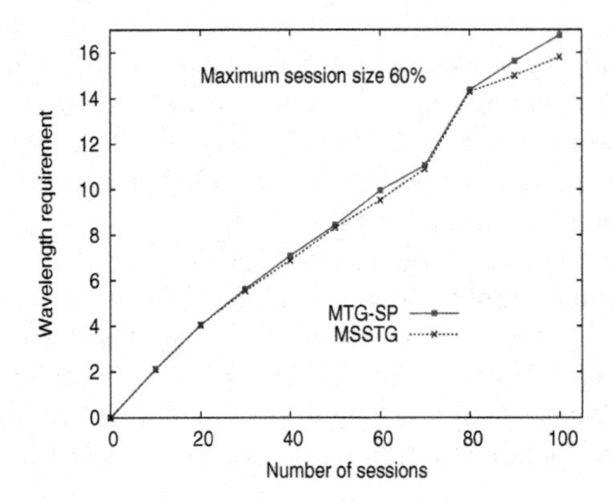

Fig. 12. Relationship of Wavelength requirement with Number of sessions for 17-node NSF network

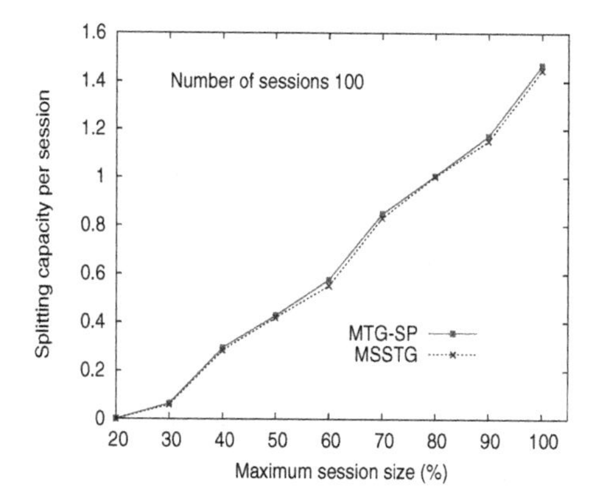

Fig. 13. Relationship of Splitting capacity per session with Maximum session size (%) for 14-node NSF network

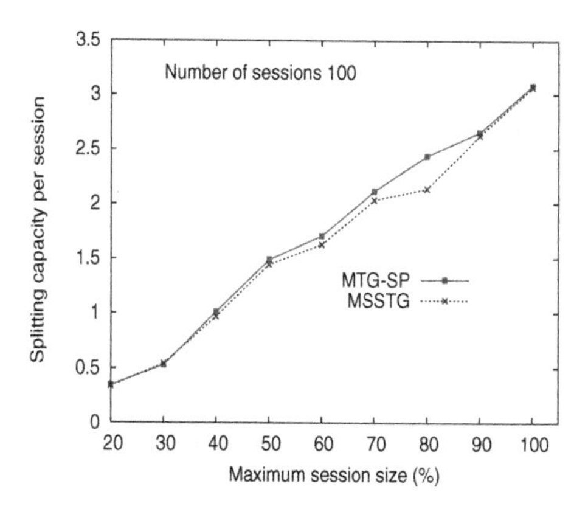

Fig. 14. Relationship of Splitting capacity per session with Maximum session size (%) for 17-node German network

6 Conclusion

In this work, we address the problem of grooming for multicast traffic in WDM optical mesh networks with sparse splitting nodes of the network. Our heuristic algorithm MSSTG remarkably increases the bandwidth utilization as well as minimize the wavelength usage. The proposed heuristic algorithm constructing the modified tree using multicast capable nodes and multicast incapable nodes, improve the relationship of multicast sessions by considering the multicast trees to all the sessions and thus increase the grooming effect. While we have minimized the wavelength usage at the mean time we have also reduced the splitting capacity. This is due the sharing of the links among the various multicast sessions. When the duplication among the nodes increase in various randomly generated multicast sessions in a network it will produce more better result in terms of bandwidth utilization, wavelength usage and splitting capacity.

References

1. Chiu, A.L., Modiano, E.H.: Traffic grooming algorithms for reducing electronic multiplexing costs in wdm ring networks. Journal of Lightwave Technology 18(1), 2–12 (2000)
2. Yan, S., Deogun, J.S., Ali, M.: Routing in sparse splitting optical networks with multicast traffic (2003)
3. Lin, R., Zhong, W.D., Bose, S., Zukerman, M.: Design of wdm networks with multicast traffic grooming. Journal of Lightwave Technology 29(16), 2337–2349 (2011)
4. Poo, G.S., Zhou, Y.: A new multicast wavelength assignment algorithm in wavelength-routed wdm networks. IEEE Journal on Selected Areas in Communications 24(4), 2–12 (2006)
5. Ul-Mustafa, R., Kamal, A.: Design and provisioning of wdm networks with multicast traffic grooming. IEEE Journal on Selected Areas in Communications 24(4), 37–53 (2006)
6. Yoon, Y.-R., Lee, T.-J., Chung, M.Y., Choo, H.: Traffic Groomed Multicasting in Sparse-Splitting WDM Backbone Networks. In: Gavrilova, M.L., Gervasi, O., Kumar, V., Tan, C.J.K., Taniar, D., Laganá, A., Mun, Y., Choo, H. (eds.) ICCSA 2006, Part II. LNCS, vol. 3981, pp. 534–544. Springer, Heidelberg (2006)
7. Billah, A., Wang, B., Awwal, A.: Multicast traffic grooming in wdm optical mesh networks. In: Proceedings of Global Telecommunications Conference, GLOBECOM 2003, vol. 5, pp. 2755–2760. IEEE (2003)
8. Zhu, K., Zang, H., Mukherjee, B.: Design of wdm mesh networks with sparse grooming capability. In: Proceedings of Global Telecommunications Conference, GLOBECOM 2002, vol. 3, pp. 2696–2700. IEEE (2002)
9. Lin, H.C., Wang, S.W.: Splitter placement in all-optical wdm networks. In: Proceedings of Global Telecommunications Conference, GLOBECOM 2005, vol. 1, pp. 306–310. IEEE (2005)
10. Jia, X.H., Du, D.Z., Hu, X.D., Lee, M.K., Gu, J.: Optimization of wavelength assignment for qos multicast in wdm networks. IEEE Transactions on Communications 49(2), 341–350 (2001)
11. Prathombutr, P., Stach, J., Park, E.: An algorithm for traffic grooming in wdm optical mesh networks with multiple objectives. In: Proceedings of the 12th International Conference on Computer Communications and Networks, ICCCN 2003, pp. 405–411 (2003)
12. Khalil, A., Hadjiantonis, A., Assi, C., Shami, A., Ellinas, G., Ali, M.: Dynamic provisioning of low-speed unicast/multicast traffic demands in mesh-based wdm optical networks. Journal of Lightwave Technology 24(2), 681–693 (2006)
13. Lu, C., Nie, X., Wang, S., Li, L.: Efficient dynamic multicast traffic grooming algorithm on wdm networks 6022, 809–818 (2005)
14. Yao, W., Li, M., Ramamurthy, B.: Performance analysis of sparse traffic grooming in wdm networks. In: Proceedings of IEEE International Conference on Communications (2005)

Performance Analysis of LMS Filter in Linearization of Different Memoryless Non Linear Power Amplifier Models

J.N. Swaminathan[1], P. Kumar[2], and M. Vinoth[1]

[1] Chettinad College of Engineering & Technology, Puliyur-CF, Karur, India
swaminathan@chettinadtech.ac.in,
m.vinoth87@yahoo.com
[2] K.S.Rangasamy College of Technology,Thiruchengode, India
Kumar_ksrct@yahoo.co.in

Abstract. The Memoryless non linearity nature of the Power amplifier varies according to its output power gain function. The nature of the amplifier can be classified into diffrent models by using the gain function. The Non linearity of the PA can be reduced by Linearization Techniques. Pre-distortion is one of the method used to linearize the PA. Here we are going to use LMS algorithm in Predistorter and evaluates its error estimating capability in diffrent power amplifier models using 16-QAM modulation technique. We measured ACPR for different models using Matlab software.

Keywords: LMS- Least Mean Square, PA- Power Amplfier, QAM- Quadrature Ampliture Modulation.

1 Introduction

The Non Linearity of the power amplifier can be expressed using voltera series .

$$H_0 = a_1H_i+a_2H_i^2+a_3H_i^3+a_4H_i^4+ a_5H_i^5+\ldots\ldots \quad (1)$$

The Memor less Non Linearity of Power amplifier models are cubic polynomial, hyperbolic tangent, saleh, Ghorbani and Rapp model. Each model has been designed according to the output power function F(n) of the PA. The details of the each power amplifier model and their amplitude distortion equation(AM-AM) and thier phase distortion equation(AM-PM) has been explained below.

$$\text{For a 16-QAM signal } H_i = H_1\cos\omega_1 t + H_2\cos\omega_2 t \quad (2)$$

The memorless non linear equation for diffrent model are discussed one by one. Here cubic polynomial [7] and hyperbolic tangent are used in low power and medium power amplifiers

$$H_{AM/AM} = H_i - \frac{{H_i}^3}{3} \qquad \text{(Cubic Polynomial)} \quad (3)$$

S. Unnikrishnan, S. Surve, and D. Bhoir (Eds.): ICAC3 2013, CCIS 361, pp. 459–464, 2013.

$$H_{AM/AM} = \tanh(H_i) \quad \text{(Hyperbolic Tangent)} \tag{4}$$

The Phase distortion measurment is common for both cubic & hyperbolic tangent

$$H_{AM/PM} = \text{upper input power limit} - \text{lower input power limit} \tag{5}$$

The third one is saleh model [7] which is suitable for microwave TWT power amplifier models.

$$H_{AM/AM} = \frac{\alpha * H_i}{1 + \beta * H_i^2} \tag{6}$$

$$H_{AM/PM} = \frac{\alpha * H_i^2}{1 + \beta * H_i^2} \tag{7}$$

The Ghorbani model [1] and Rapp model is suitable for the Solid state power amplifiers.

$$H_{AM/AM} = \frac{x_1 h^{x_2}}{1 + x_3 h^{x_2}} + x_4 h \tag{8}$$

$$H_{AM/PM} = \frac{y_1 h^{y_2}}{1 + y_3 h^{y_2}} + y_4 h \tag{9}$$

& for rapp model is

$$H_{AM/AM} = \frac{H_i}{1 + \left(\left(\frac{H_i}{O_{sat}}\right)^{2s}\right)^\wedge\left(\frac{1}{2s}\right)} \tag{10}$$

2 Predistortion and Error Estimation

There are three diffrent types of linearization techniques 1. Feed back method 2. Feed forward method 3. Predistortion method. Among these three methods we used predistortion method to linearize the power amplifier. The principle of predistortion and how it is linearizing the PA is explained below

For a input signal H_i (t)

The output with memory less non linearity [15] is $H_{out}(t)$ is given by

$$H_{out}(t) = b_1 H(t + Г_1) + b_3 H^3(t + Г_3) + b_5 H^5(t + Г_5) + \ldots \tag{11}$$

Our desired signal is

$$b_1 H(t + Г_1) \tag{12}$$

but it is coming along with the intermodulation distortion. To remove IM distortion LMS filter estimates the error

$$E(n) = - b_3 H^3(t + Г_3) + b_5 H^5(t + Г_5) + \ldots \tag{13}$$

The error E(n) [3] will be predistorted with the input signal H(n) which results in the predistortion function Z(n)

$$Z(n) = b_1 H_p(\omega t) + b_3 H_p^3 (\omega t + \Phi_3) + b_5 H_p^5 (\omega t + \Phi_5) \tag{14}$$

Here H_p will be given by

$$H_p(t) = b_1H(\Gamma)\cos\omega t + \frac{3}{4}\, b_3H^3(\Gamma)\cos(\omega t + \Phi_3) + \frac{5}{8}\, b_5H^5(\Gamma)\cos(\omega t + \Phi_5)$$
$$+ \frac{35}{64}\, b_7H^7(\Gamma)\cos(\omega t + \Phi_7) \quad (15)$$

The predistortion function Z(n) will be given as the input to the PA , the predistorted error in input signal and Non linear products will cancel each other which results in a linear fuction as output i.e our desired signal. Here F(n) is the memory less nonlinear function of PA.

$$H_{out}(n) = F(n)\,.Z(n) \quad (16)$$

Finally

$$H_{out}(n) = K\, H(t + \Gamma_1) \quad (17)$$

Here K is the Complex linear gain at the output of the power amplifier

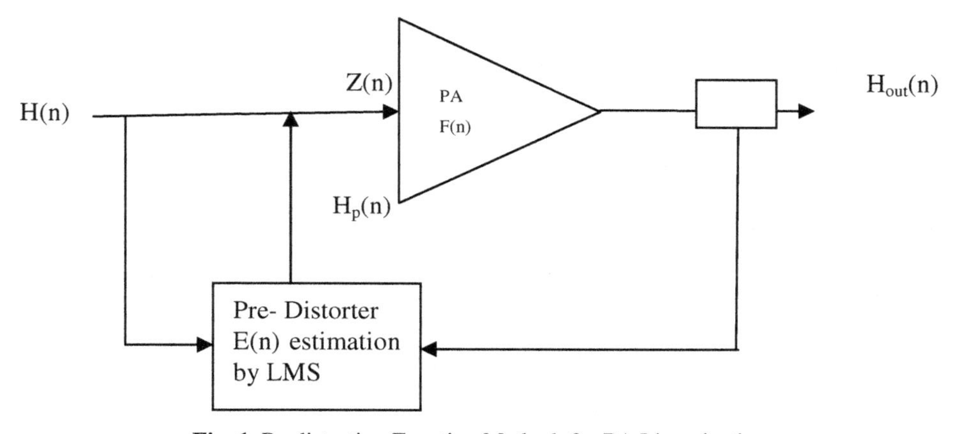

Fig. 1. Predistortion Function Method for PA Linearization

The above figure perfectly explains the working principle of pre-distortion the same principle has been applied for all the Power amplifier models listed in the first topic. We applied the Normalised LMS filter [5], so the fast error estimation capability of the filter is mainly depends upon the step size of the filter.

3 Analysis of Error Estimation Capability of LMS

The results are taken using matlab software. Here we took ACPR as the measurement for analysing the performance of the LMS filter for various modeling.

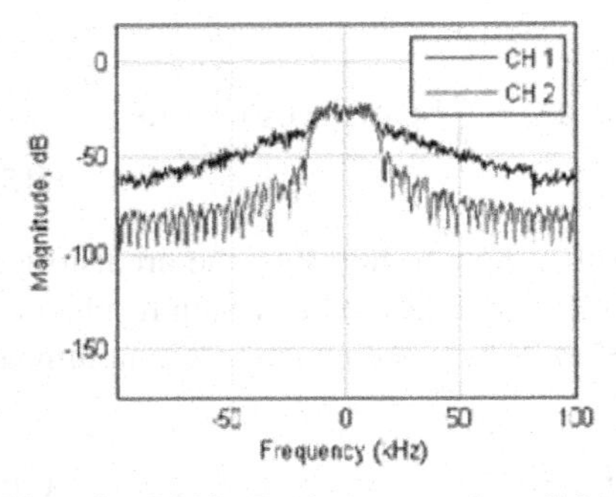

Fig. 2. LMS Performance for Cubic Polynomial

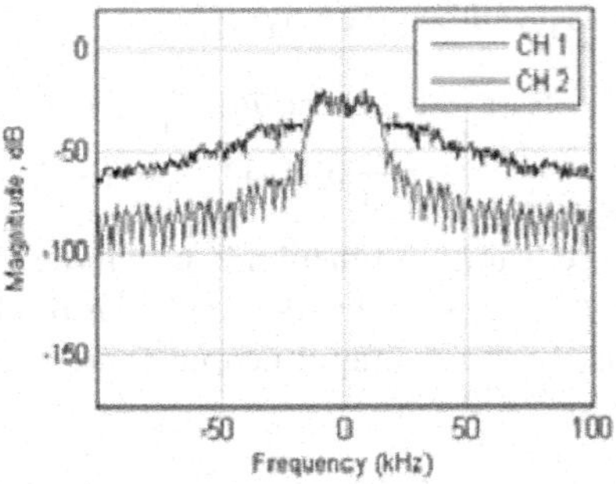

Fig. 3. LMS Performance for Hperboic tangent

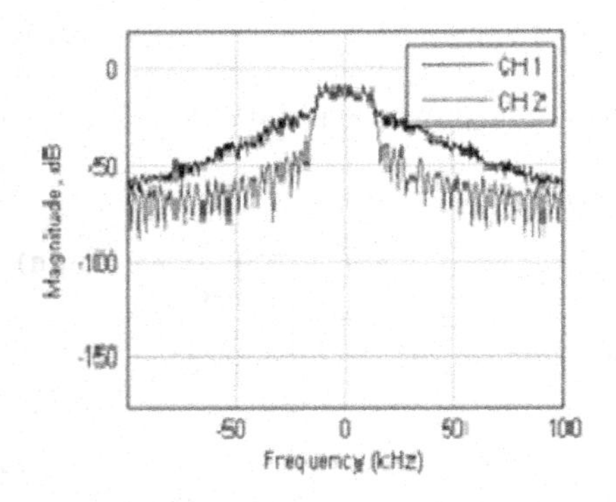

Fig. 4. LMS Performance for Saleh

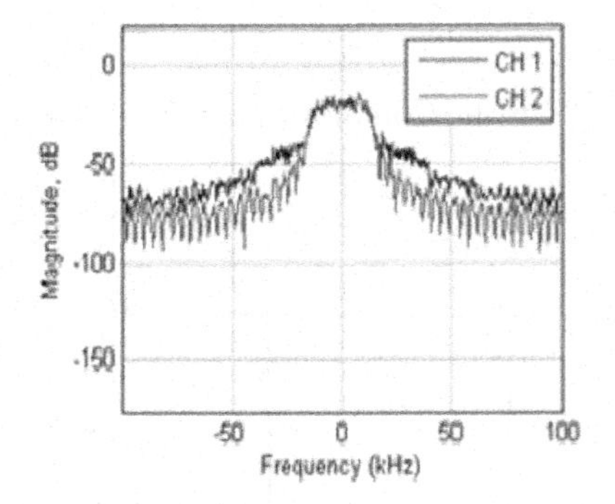

Fig. 5. LMS Performance for Ghorbani

In the above Fig. 2 & Fig. 3 shows the performance of LMS for cubic and hperbolic tangent model both the output looking almost same but in Fig.3 a slight dip in the peak power shows the granular noise has been newly introduced for hyperbolic tangent model. The filter performance in saleh model(TWT PA Fig.4) is quiet good all the intermodulation products has been sucess fully compressed. In Ghorbani model (Fig. 5) which is for SSPA type the filter performed well only for IM3 and IM7 only.

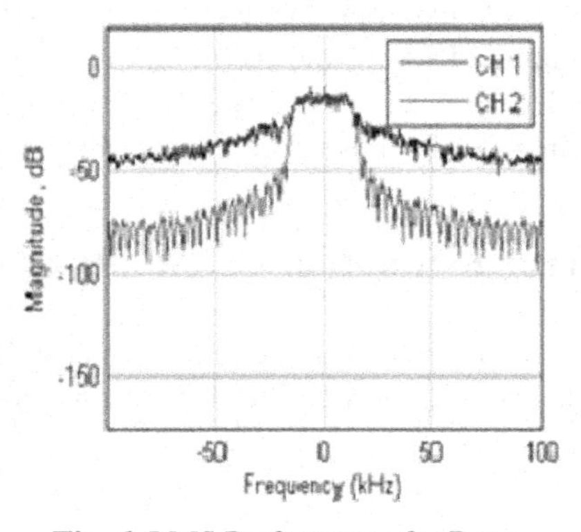

Fig. 6. LMS Performance for Rapp

In Fig.6 the performance of the filter is quiet un natural almost a 35db compression of ACPR has been found but it is not possible in practical conditions.

Table 1. ACPR Level for all models before and after Linearization

PA Model	Before Pre-distortion	After Pre-distortion
Cubic Polynomial	-30db	-47db
Hperbolic Tangent	-33db	-52db
Saleh	-29db	-50db
Ghorbani	-35db	-51db
Rapp	-25db	-57db

4 Conclusion

We applied the technique to all power amplifier models and found it is suitable for saleh (Fig.4) and ghorbani model (Fig.5) i.e. the error estimation process using LMS filter is highly suitable for TWT type Power amplifiers and Solid state power amplifiers.

References

1. Aghasi, A., Ghorbani, A., Amindavar, H., Karkhaneh, H.: Solid state Power Amplifier Linearization Using the Concept of Best Approximation in Hilbert Spaces. IEEE Trans. (2006)
2. Hetrakul, P., Taylor, D.P.: The effects of transponder nonlinearity on binary CPSK signal transmission. IEEE Trans. Commun. COM-24, 546–553 (1976)
3. D'Andrea, N.A., Lottici, V., Reggiannini, R.: RF Power Amplifier Linearization Through Amplitude and Phase Predistortion. IEEE Trans. Commun. 44(11) (1996)
4. Tsimbinos, J., Lever, K.V.: Nonlinear System Compensation Based on Orthogonal Polynomial Inverses. IEEE Trans. Circuit Syst. 48(4), 406–417 (2001)
5. Ghorbani, A., Sheikhan, M.: The Effect of Solid State Power Amplifiers (SSPAs) Nonlinearities on MPSK and M-QAM Signal Transmission. In: Sixth Int'l Conference on Digital Processing of Signals in Comm., pp. 193–197 (1991)
6. Broadband Wireless Access Working Group (Rev.0), IEEE 802.16.lpp 00/15
7. `http://www.mathworks.com/access/helpdesk/help/toolbox/commblks/ref/memorylessnonlinearity.html`
8. Gerald, C.F., Wheatley, P.O.: Applied Numerical Analysis. Addison-Wesley, Reading (1999)
9. Blachman, N.M.: The signalxsignal, noisexnoise, and signal noise output of a nonlinearity. IEEE Trans. Inform. Theory IT-14, 21–27 (1968)
10. The Uncorrelated Output Components of a Nonlinearity. IEEE Trans. Inform. Thoery IT-14, 250–255 (1968)
11. Dudley, D.G.: Mathematical Foundations for Electromagnetic Theory. IEEE, Reading (1994)
12. Aghvami, A.H.: Performance Analysis of 16-ary QAM Signalling Through Two-Link Nonlinear Channels in Additive Gaussian Noise. IEEE Proc. 131(4 Pt. F) (1984)

13. Badr, M.I., El-Zanfally, R.H.: Performance Analysis of 49-QPRS Through Nonlinear Satellite Channel in the Presence of Additive White Gaussian Noise. In: Fifteenth National Radio Science Conference, Helwan, Cairo, Egypt, February 24-26 (1998)
14. Lagarias, J.C., Reeds, J.A., Wright, M.H., Wright, P.E.: Convergence Properties of the Nelder-Mead Simplex Method in Low Dimensions. SIAM Journal of Optimization 9(1), 112–147 (1998)
15. Cripps, S.C.: Advanced Technologies in RF Power Amplifier Design. Artech Publications

FPGA Implementation of ZigBee Baseband Transceiver System for IEEE 802.15.4

Bijaya Kumar Muni and Sarat Kumar Patra

Dept. of Electronics & Comm. Engg.
NIT Rourkela, India
{muni.bijaya,patra.sarat}@gmail.com
http://www.nitrkl.ac.in

Abstract. The ZigBee standard is based on the application for IEEE 802.15.4 LR WPAN for control and monitoring of wireless sensor network, short range low data rate wireless communication system. the ZigBee standard along with IEEE 802.15.4 standard combined together to form complete ZigBee networking protocol.the physical layer and MAC layer are are defined by IEEE 802.15.4, and other upper layers like network layer, application layer etc are defined by ZigBee standard. ZigBee is particularly designed for low power short range wireless communication system. The work presented in this paper is the design and implementation of ZigBee baseband transceiver for IEEE 802.15.4. The design includes the spread spectrum of the input data bit '0' and '1' according to the quasi orthogonal symbol to chip mapping employed in standard IEEE 802.15.4. The unipolar input data bits '0' and '1' are transmitted with out converting to bipolar signal. The spreading method Direct sequence spread spectrum are implemented in the design of the transceiver, and later the design includes Offset Quadrature phase shift keying modulation with half sine pulse shaping. Half sine pulse shaping is done to reduce Inter symbol Interference.The design is carried out by the hardware description language Verilog HDL and implemented in the VIRTEX 5 FPGA board. The analog continuous sample values used in the design are stored in the memory of the board and used as per the clock signal. The simulation of the design is done by the help of Mentor Graphics Modelsim simulator and the synthesis is done by XILINX ISE SYNTHESIZER.

Keywords: ZigBee transceiver, IEEE 802.5.4., LR WPAN.

1 Introduction

In the past few years the field of data communication has been evolved starting from analog communication to digital communication system for transmission and reception of message signal. Different wireless standards have been developed by the institute of electrical and electronics engineers(IEEE) such as IEEE 802.3, IEE 802.11, IEEE 802.16 and IEEE 802.15 etc. These standards defines various layers of Open systems interconnection (OSI) model [1]. IEEE 802.15.4 standard

S. Unnikrishnan, S. Surve, and D. Bhoir (Eds.): ICAC3 2013, CCIS 361, pp. 465–474, 2013.

[1] is developed for low data rate short range wireless communication system such as Wireless Personal Area Network(WPAN). IEEE 802.15.4 defines the Physical layer and MAC layer of ZigBee Networking Protocol and upper layers such as networking, application layer are defined by ZigBee Standard.Now a days this standard has a huge impact in the area of wireless sensor network applications. The ZigBee transceiver for IEEE 802.15.4 standard has tremendous application in the area of short range wireless communication system such as military applications.

ZigBee protocol combined with IEEE 802.15.4 is the final product of the combinational effort of the bodies IEEE and ZigBee alliance group [1][3]. It is developed for new networking technology for small range of frequencies in Industrial, Scientific and Medical(ISM) band. This is applicable for the simplest industry and home applications. The applications are:

- Indoor patient monitoring system in hospital by connecting a ZigBee device in the body of the patient
- Testing the physical strength or architectural strength of a building by fitting different wireless sensors in the different part of a building.
- ZigBee offers wireless application for low data rate that is maximum of 250 Kbps, low power consumption and low cost short range wireless application.

Bluetooth and IEEE 802.11b standard is developed for short range wireless communication with different data rates mention in table. IEEE 802.11b is the wireless local area networks(WLAN) standard, is the extension of wired local area network standard IEEE 802.3. IEEE 802.11b developed to widen the ethernet network by integrating a WLAN device to the wired network.

Along with zigbee the IEEE 802.11 standard also operates with 2.4 GHz band[1]. The IEEE 802.11b has a high data rate of 11 Mbps for which it is mainly used for wireless internet connection. The indoor range of the standard is 30 to 100 meters [4]. Bluetooth has a indoor range of 2 to 10 meters and the data rate of less than 3 Mbps. As compared to all these standards, ZigBee has lower data rate with indoor coverage of 100 meters and low power consumption. Hence the battery life time of ZigBee devices is more with low data rate less complexity. The Zigbee standard is suitable for the transmission and reception of simple commands which is again a suitable application in defence. Keeping in view this ZigBee standard provides the most cost efficient and power efficient solution compared to Bluetooth and IEEE 802.11.[7]

The three frequency bands for ZigBee standard are 868 MHz, 915 MHz and 2.4 GHz[1,2]. The ZigBee specifications for 2.4 GHz band are given in table 1.

The transceiver design includes symbol to chip conversion, serial to parallel conversion along with pulse shaping of even bit stream and odd bit stream. In reception the design includes even bit and odd bit separation along with the decoding of input data bit. In decoding the chip to symbol process obtained by matching bit to bit of chip sequence at the receiver with the chip sequence transmitted by the transmitter.

Table 1. ZigBee specification for 2.4GHz

Parameter	Specification
Operating Frequency	2.4 GHz
Channel spacing	5 MHz
Data rate	250 Kbps
Chip rate	2 Mbps
Spread spectrum	Direct sequence spread spectrum
Modulation	OQPSK with Half-sine pulse shaping

2 Spreading and Despreading

IEEE 802.15.4 [1] standard uses the spreading methods to increase the effect of receiver sensitivity, jamming resistance, and also to reduce the effect of multipath. The spreading method used here is the Direct sequence spread spectrum(DSSS) [5] as shown in Table 1. In this standard every four incoming data bit is combined together to form a four bit symbol and this four bit symbol is mapped into an unique 32 chip sequence in a look up table. Hence there are sixteen different 32 bit chip sequences as shown in Table 2. The chip sequences are also known as Pseudo noise random sequence (PN sequence). The chip sequence is the collection of zeros and ones randomly(random noise) which leads up to some extent that this chip sequences may become similar to each other. To avoid the similarity between any two chip sequences, the standard follows a procedure or algorithm of doing the cross-correlation between the chip sequences. The cross-correlation is calculated by multiplying the sequences together and then calculating the summation of the result. The 32 bit chip sequence contains a sequence of zeros and ones in unipolar form. Before calculation of the cross correlation the unipolar sequence is converted to bipolar sequence ie a '0' is represented by '-1' and '1' is as it is. If $X(n)$ and $Y(n)$ are two sequences then the cross correlation of $X(n)$ and $Y(n)$ calculated by using the following equation 1.

$$r_{xy}(0) = \sum_{n=-\infty}^{n=\infty} x(n)y(n) \tag{1}$$

The above equation calculates the cross correlation of $X(n)$ and $Y(n)$ where neither of these sequences shifted. Hence it defines that the two sequences are dissimilar to each other. In this case the sequences are known as orthogonal to each other and termed as orthogonal sequences. The sixteen pseudo noise sequences used in the standard IEEE 802.15.4 are not completely orthogonal to each other and called as quasi orthogonal or nearly orthogonal. The cross correlation of other Pseudo Noise(PN) sequences are calculated by shifting one sequence with a margin of time period k hence the equation is written as:

$$r_{xy}(k) = \sum_{n=-\infty}^{n=\infty} x(n)y(n-k) \tag{2}$$

By using the above equation the standard IEEE 802.15.4 has different sets of symbol to chip mapping for three bands of frequency such as 868 MHz, 915 MHz and 2.4 GHz. All the sixteen combination of symbol to chip mapping for 2.4 GHz frequency band are mentioned in Table 2.

Table 2. Symbol to Chip Mapping for 2.4 GHz Band

Data Symbol(Binary)	**Chip values** ($C_0\ C_1\ \ldots\ C_{31}$)
0000	11011001110000110101001000101110
1000	11101101100111000011010100100010
0100	00101110110110011100001101010010
1100	00100010111011011001110000110101
0010	01010010001011101101100111000011
1010	00110101001000101110110110011100
0110	11000011010100100010111011011001
1110	10011100001101010010001011101101
0001	10001100100101100000011101111011
1001	10111000110010010110000001110111
0101	01111011100011001001011000000111
1101	01110111101110001100100101100000
0011	00000111011110111000110010010110
1011	01100000011101111011100011001001
0111	10010110000001110111101110001100
1111	11001001011000000111011110111000

3 Proposed Zigbee Transmitter Design

The baseband transmitter implementation is shown in Figure 1,with the various blocks of the transmitter section such as bit to symbol mapping, symbol to chip conversion, serial to parallel conversion, half-sine pulse shaping etc are shown.[6]

The input bit stream in digital form like digital '1' and digital '0'. The four input bits are combined to form a four bit symbol. The input bit is at a data rate of 250 Kbps which means in every 4 micro second one digital input is coming ie either '0' or '1'. So combining four input bits one symbol forms and as per the standard IEEE 802.15.4 the symbol to chip sequence is taken. The symbol to chip conversion performs DSSS spread spectrum baseband modulation to avoid loss at the baseband. The four bit symbol is mapped to respective 32 bit chip sequence in a lookup table. Hence there are sixteen different combination of 32 chip sequences or pseudo noise random sequence (PN sequence) as shown in Table 2. The 32 bit chip sequence are in unipolar form, that is digital '1' or digital '0' but during the DSSS spreading there is a loss during binary bit '0'.So in the verilog HDL design spreading occurs as per the bipolar chip sequences. The serial chip sequence data stream obtained becomes a 2 Mbps.

The 32 bit chip sequences represents the different four bit symbol are modulated using OQPSK modulation scheme for 2.4 GHz unlicensed band. The even

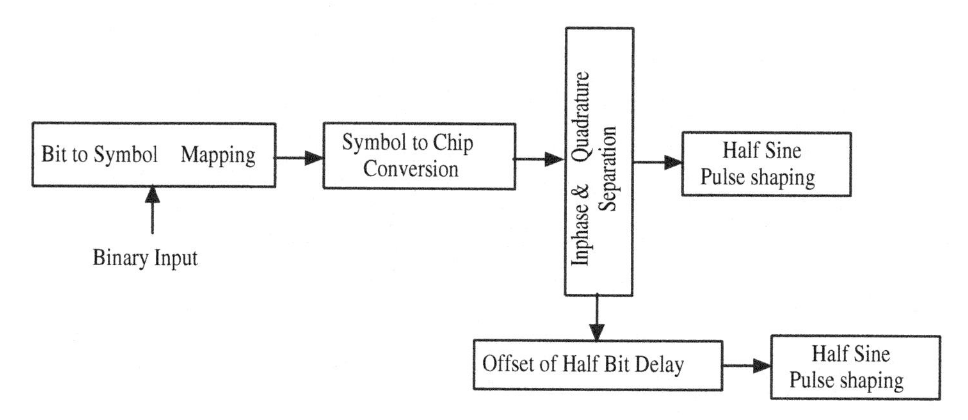

Fig. 1. ZigBee baseband Transmitter

and odd indexed chips are separated into two streams having two bit period extended at the same time the even bitstream is having half a bit delay to the odd bitstream. The even and odd bitstreams are transmitted along with the pulse shaping stage to reduce the inter-symbol interference (ISI). Finally the resultant signal having even and odd stream are transmitted by the RF transmitter.

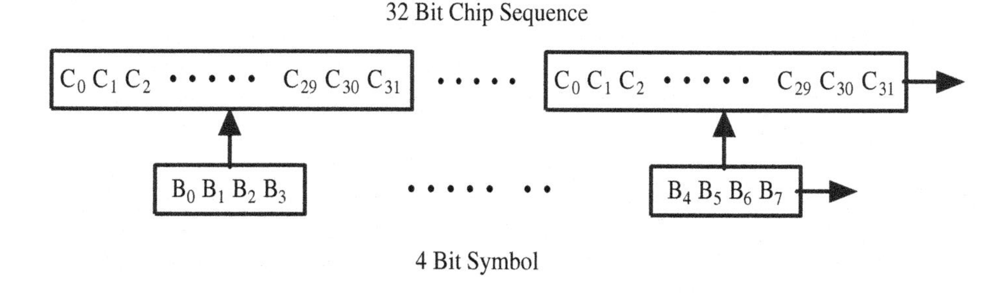

Fig. 2. 4 Bit Symbol

4 Proposed Zigbee Receiver Design

The figure 3 shows the design of ZigBee baseband receiver, which includes the sample detection, extraction of even bit stream and odd bit stream, addition of odd bit stream data and decoding of input digital signal. [6]

Assuming the reception to be synchronous, the pulse shaping output at the transmitter is fed to the input of the receiver. The pulse shaping of even bit and odd bit are sampled with the ten samples as given in table 3. For ten samples the sampling time for each sample is 100 ns because the chip duration is 1000 ns so to accommodate ten samples in one bit, the sampling time is 100 ns taken.

The samples are considered as the decimal values. It can also be the integer values to multiply with every bit of even stream and odd stream to get the ZigBee

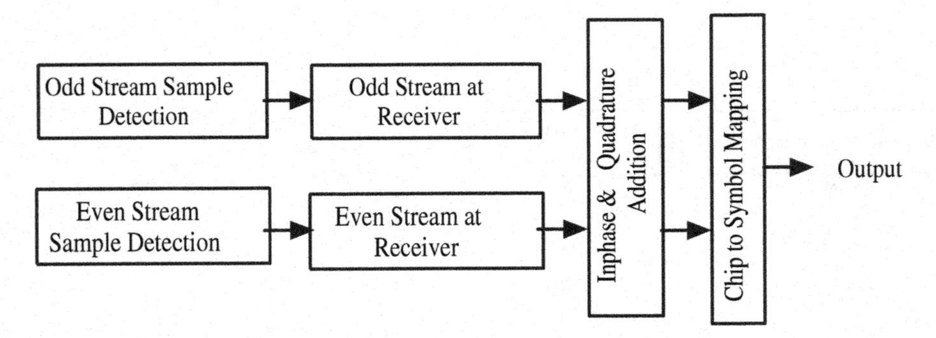

Fig. 3. ZigBee Baseband Receiver

Table 3. Samples for Pulse shaping

Sampling time in ns	Sample value for bit 1	Sample value for bit 0
100	0	0
200	16	-16
300	32	-32
400	48	-48
500	56	-56
600	64	-64
700	56	-56
800	48	-48
900	32	-32
1000	16	-16

pulse shaping. In the Verilog FPGA design the samples are stored in a memory and then the samples are read from the memory according to the parallel stream chip bit '1' and '0' simultaneously. Since we need to multiply the samples with the bipolar chip having the amplitude varying from '+1' and '-1' for the pulse shaping filter. Notably the pulse shaping filter is used to reduce the Inter Symbol Interference (ISI) by converting the pulses into the half sine shapes. The chip bit here are unipolar so for the chip '0' the negative samples stored in the memory are read instead of multiplying the samples with the '-1' because chip '-1' is represented as '0'. Similarly for chip '1' the positive samples stored in the memory are read instead of multiplying the samples with the chip '1'.

5 Field Programmable Gate Array (Fpga) Implementation

Field programmable Gate Arrays are pre-fabricated silicon devices which can be programmed to make any kind of digital circuits or systems. The FPGA is advantageous over Application specific Integrated circuits (ASIC)[8]. The ASICs are designed for particular application which cant be changed further. The ASICs are designed by the Cadence tool for VLSI design and then fabricated at a FAB

Lab. The development of ASIC needs a longer time and a a fully furnished FAB Lab to manufacture, which is again too expensive. In comparison to ASIC, the FPGA are configured after the fabrication. It can also be reconfigured by different designing tools. This reconfiguration is done through hardware description language like verilog-HDL [2],[9] and VHDL. After that synthesis of the design, the bit stream file is downloaded to the FPGA. In ASIC the error correction is not possible after fabrication. For error correction in ASIC design, the design fabrication process has to start from the beginning again which is a lengthy process and not cost effective. But in FPGA the error correction is easy in contrast to ASIC because the designer will have to do some changes in the HDL[8] coding of the design. This is the reason, for which we have gone for FPGA design of the baseband transceiver. And later the RF part of the transceiver can be integrated with the baseband design of the Transceiver.

6 Decoding or Despreading

The serial data after addition of I-phase and Q-phase data reforms into a 2 Mbps data transmitted by the transmitter. This serial data includes the input data along with noise added by various stages of the transceiver mainly by the communication channel. The decoding process extracts the input data from noise.

The serial data is stored in a 32 bit register as per the clock signal used in the verilog-HDL design of the transceiver. The 32 bit PN sequence used at the transmitter for spreading is reused at the receiver for despreading.

The sixteen PN sequences used at the transmitter and the serial data stored in 32 bit register are XNORed to each other bit by bit to check the mismatch of the data bit. Sixteen different counters or flag registers of six bits are used to count minimum up to 32 in decimal, because the 32 bits of the serial data are checked with all sixteen quasi orthogonal PN sequences at a time. In XNOR operation, when both the bits of serial data and PN sequences are same the counters are increased by one and if both the bits are different the counters stores the same value. So after finishing the XNOR operation of all the 32 bits of the serial data the counter values are checked for the maximum value. The counter which stores

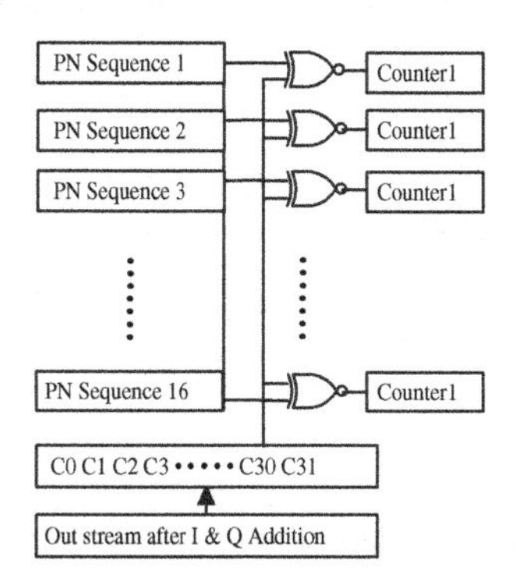

Fig. 4. Decoding process

the maximum value as comparing to all other counters indicates the highest level of matching. Hence the respective PN sequence with maximum counter value is referred as the chip sequence which was transmitted by the transmitter. Later on the different symbols are extracted from a look up table which contains the chip to symbol mapping. In receiver the MSB of decoded symbol is fed to the output.

7 Results

The paper presents the analysis of ZigBee baseband transceiver for IEEE 802.15.4 standard. The modelsim Simulation result shows that the input data bit recovered successfully by the decoding process. The simulated code synthesized by ISE Synthesizer and dumped into VIRTEX 5 FPGA board for implementation.

The internal pulse shaping signals, inphase and Qphase data stream of the design is presented in the Figure 5 to 7 and the board result is shown in Figure 8.

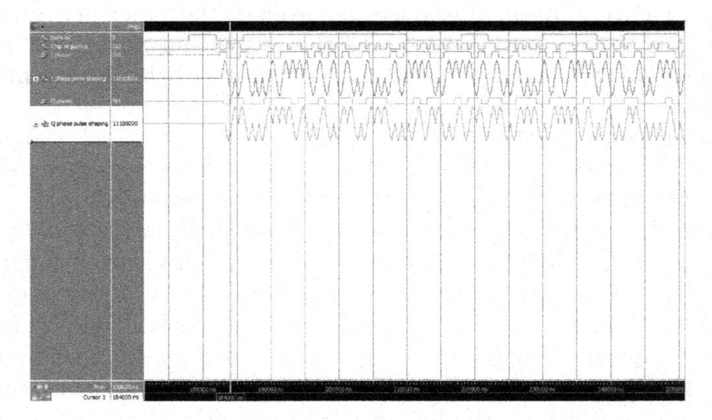

Fig. 5. Transmitter simulation in Modelsim

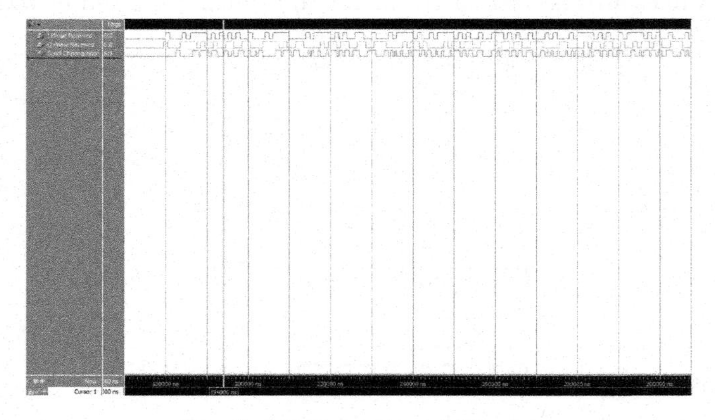

Fig. 6. I Q Detection at Receiver

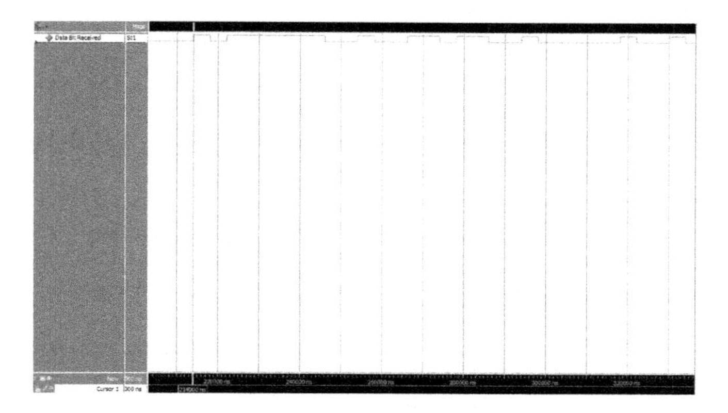

Fig. 7. Data bit received

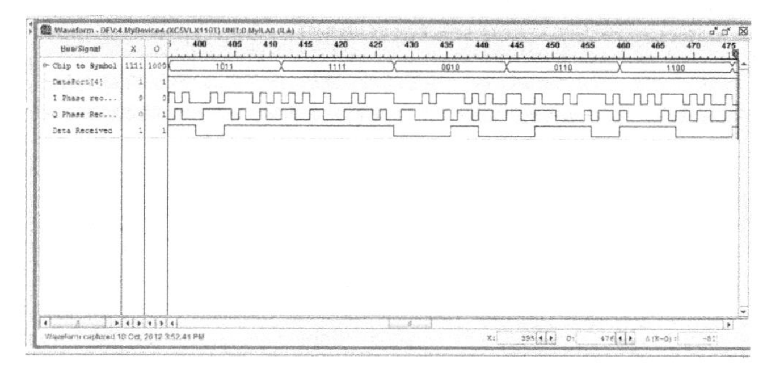

Fig. 8. VIRTEX5 Board Results

8 Conclusion

The paper presents the verilog HDL design and circuit implementation of mixed signal ZigBee baseband transceiver for IEEE 802.15.4 standard. The paper describes the behaviour of bit to symbol, symbol to chip, pulse shaping, chip to symbol in verilog HDL. The different blocks are integrated in the design to complete the base band module of the ZigBee baseband transceiver. The verilog design code was initially simulated by the mentor graphics modelsim simulator and then synthesized by the XILINX ISE synthesizer. The bit stream downloaded successfully on to VIRTEX 5 FPGA board for the implementation of the transceiver. The functionality of the transceiver design matches with its theoretical expectations. As long as the receiver is synchronized with the transmitter the extraction of the input bits from the decoding module of the design does not face any problem.

References

1. Ieee standard for local and metropolitan area networks– part 15.4: low-rate wireless personal area networks (lr-wpans)
2. Coffman, K.: Real world FPGA design with Verilog. Prentice Hall Modern Semiconductor Design Series. Prentice Hall PTR (2000)
3. Farahani, S.: ZigBee wireless Network and Transceivers. Newnes, USA (2008)
4. Lee, J.S., Su, Y.W., Shen, C.C.: A comparative study of wireless protocols: bluetooth, uwb, zigbee, and wi-fi, Piscataway, NJ, USA, pp. 46–51 (2007)
5. Madhow, U.: Fundamentals of Digital Communication. Cambridge University Press (2008)
6. Oh, N.J., Lee, S.G.: Building a 2.4-ghz radio transceiver using ieee 802.15.4. IEEE Circuits and Devices Magazine 21(6), 43–51 (2005)
7. Park, C., Rappaport, T.: Short-range wireless communications for next-generation networks: Uwb, 60 ghz millimeter-wave wpan, and zigbee. IEEE Wireless Communications 14(4), 70–78 (2007)
8. Ramachandran, S.: Digital VLSI Systems Design: A Design Manual for Implementation of Projects on FPGAs and ASICs Using Verilog. Springer (2007)
9. Zeidman, B.: Introduction to cpld and fpga design. The Chalkboard Network (2001)

Area and Speed Efficient Arithmetic Logic Unit Design Using Ancient Vedic Mathematics on FPGA

Sushma R. Huddar[1], Sudhir Rao Rupanagudi[1], Venkatesh Janardhan[2], Surabhi Mohan[3], and S. Sandya[4]

[1] WorldServe Education, Bangalore, India
worldserve.tchw2@gmail.com, sudhir@worldserve.in
[2] TCS, Bangalore, India
venkatesh1.j@tcs.com
[3] Atria Institute of Technology, Bangalore, India
surabhimohan06@gmail.com
[4] Dept. of ECE, RNSIT, Bangalore, India
sandya9prasad@gmail.com

Abstract. High speed and area efficient multiplier architecture plays a vital role in Arithmetic Logic Unit (ALU) design, especially when it comes to low power implementation of Central Processing Units, Microprocessors and Microcontrollers. In this paper, we propose a novel, area efficient and high speed architecture to implement a Vedic mathematics multiplier based on Urdhva Tiryakbhyam methodology. The same was utilized within an ALU and its performance was compared with the existing ALU designs. It was found that the proposed methodology is around 2 times faster than the existing ALU architectures. Also a significant 3% decrease in area, in comparison with existing designs was obtained. The proposed algorithm was designed and implemented using Verilog RTL on a Xilinx Spartan 3e FPGA and compared with other algorithms. The results prove that the proposed architecture could be a great boon for high speed and low area based processor design.

Keywords: Vedic Mathematics, Arithmetic Logic Unit, Urdhva Tiryakbhyam Sutra, Multiplier design, FPGA, Low area multiplier, High Speed Multiplier.

1 Introduction

The main component of a Central Processing Unit (CPU) , microprocessor or microcontroller is the Arithmetic Logic Unit also known as the ALU [1-2]. The ALU performs several arithmetic and logical operations. One of the primary sources of concern in ALU design is achieving low area, high speed and less power consumption for a wide variety of applications. A few of the arithmetic operations performed by the ALU include addition, subtraction and multiplication. Among these operations, multiplication is one of the most fundamental and arduous function which is

S. Unnikrishnan, S. Surve, and D. Bhoir (Eds.): ICAC3 2013, CCIS 361, pp. 475–483, 2013.

employed in several mathematical and signal processing computations [3]. The multiplier is an integral component of a wide variety of expeditious circuitry like FIR filters, image processors, Digital Signal Processors etc. [4].

Conventionally, algorithms such as the Booth's algorithm [5] and modified Booth's [6] algorithm are utilized for multiplier implementation in a standard ALU. But, it is a well known fact that these algorithms involve several iterative and time consuming operations like shifting of the multiplier and multiplicand bits, arithmetic operations like addition and also involve comparisons. These iterations increase linearly with the number of bits to be multiplied. This leads to a significant decrease in speed and also increase in power consumed [7]. Apart from this, another significant flaw in the ALU design is that the conventional ALU contains three separate architectures in order to perform addition, subtraction and multiplication respectively [8-9] which lead to increased on-chip area. This is a major concern in processor architectures especially where the three operations would not occur simultaneously and low area is a requirement.

In this paper, we present a high speed and low area multiplier for the ALU based on ancient Vedic mathematic formulae. Also, we further modify the same architecture in order to overcome the various flaws observed in the conventional ALU. We achieve this by aggregating the adder unit, subtraction unit & the Vedic maths multiplier unit into a single unit called the Multiplier Adder and Subtractor(MAS) unit. The novel MAS unit is implemented by modifying the architecture in one of the most popular Vedic mathematics sutra (formula) for multiplication called the Urdhva Tiryagbhyam Sutra which will be explained in detail in the further sections.

Section 2 describes the existing architecture of the Urdhva method of multiplication for an 8 bit ALU. Section 3 further continues to define and describe the MAS method of multiplication and its implementation in an 8 bit ALU. Section 4 and 5 highlight the results obtained by this novel architecture and also scopes what future lies ahead for research in this field.

The complete design and implementation of the conventional ALU and MAS based ALU was performed on Xilinx Spartan 3e FPGA – XC3S100E – 5CP132. All results in terms of area and speed were computed and compared utilizing the same board.

2 Vedic Mathematics

Vedic mathematics is one of the ancient forms of mathematics which originated in India [10]. An interesting feature of Vedic mathematics is the simplicity in its approach to solve a mathematical problem, thereby leading to faster computation of results. It was discovered around 1500 B.C. and was later rediscovered between 1911 to 1918 by a renowned Sanskrit scholar and mathematician - Sri Bharti Krishna Tirthaji [11]. He reorganized and formulated the entire Vedic mathematics into sixteen sutras, of which Urdhva Tiryakbhyam sutra is one of the most popular technique used for multiplication [12].

2.1 Urdhva Tiryakbhyam Multiplier

Urdhva Tiryakbhyam means vertically crosswise [10]. As the name suggests, this method computes each digit of the final product of multiplication in a vertical and crosswise fashion. This involves only two operations – multiplication of individual digits of the multiplier and multiplicand and addition of these partial products obtained. Also, these processes are carried out in parallel, for each digit of the final product. This in turn yields results much faster than the conventional Booth multiplier which involves several iterative computations. Since the Urdhva Tiryakbhyam method utilizes only single digit multiplication operation and addition operation, it further leads to reduction in computational complexity. It was further found that the Urdhva Tiryakbhyam method can also be applied to binary numbers and that this methodology used, is the same as that used for integers. The only difference is that, for binary numbers, the operations mentioned above i.e. the multiplication operation would now reduce to bitwise "AND" operation and for the addition operation utilization of "full adders" or "half adders" would suffice. This further reduces the area occupied by the architecture required for the multiplier. The Urdhva Tiryakbhyam method can be best explained with the help of the illustration below:

Let A and B be two eight bit numbers which are supposed to be multiplied. The eight bits of A and B are represented by A_7 to A_0 and B_7 to B_0 respectively, where A_7 and B_7 are the Most Significant Bits (MSB) and A_0 and B_0 represent the Least Significant Bits (LSB). P_{15} to P_0 are obtained upon application of the Urdhva Tiryakbhyam method of multiplication and are represented by equations 1 to 15. These products inturn concatenated provide us the final result as shown in equation 16.

$$P_0 = A_0 * B_0 \tag{1}$$

$$C_1P_1 = (A_1 * B_0) + (A_0 * B_1) \tag{2}$$

$$C_3C_2P_2 = (A_2 * B_0) + (A_0 * B_2) + (A_1 * B_1) + C_1 \tag{3}$$

$$C_5C_4P_3 = (A_3 * B_0) + (A_2 * B_1) + (A_1 * B_2) + (A_0 * B_3) + C_2 \tag{4}$$

$$C_7C_6P_4 = (A_4 * B_0) + (A_3 * B_1) + (A_2 * B_2) + (A_1 * B_3) + (A_0 * B_4) + C_3 + C_4 \tag{5}$$

$$C_{10}C_9C_8P_5 = (A_5 * B_0) + (A_4 * B_1) + (A_3 * B_2) + (A_2 * B_3) + (A_1 * B_4) + (A_0 * B_5) + C_5 + C_6 \tag{6}$$

$$C_{13}C_{12}C_{11}P_6 = (A_6 * B_0) + (A_5 * B_1) + (A_4 * B_2) + (A_3 * B_3) + (A_2 * B_4) + (A_1 * B_5) + (A_0 * B_6) + C_7 + C_8 \quad (7)$$

$$C_{16}C_{15}C_{14}P_7 = (A_7 * B_0) + (A_6 * B_1) + (A_5 * B_2) + (A_4 * B_3) + (A_2 * B_5) + (A_1 * B_6) + (A_0 * B_7) + C_9 + C_{11} \quad (8)$$

$$C_{19}C_{18}C_{17}P_8 = (A_7 * B_1) + (A_6 * B_2) + (A_5 * B_3) + (A_4 * B_4) + (A_3 * B_5) + (A_2 * B_6) + (A_1 * B_7) + C_{10} + C_{12} + C_{14} \quad (9)$$

$$C_{22}C_{21}C_{20}P_9 = (A_7 * B_2) + (A_6 * B_3) + (A_5 * B_4) + (A_4 * B_5) + (A_3 * B_6) + (A_2 * B_7) + C_{13} + C_{15} + C_{17} \quad (10)$$

$$C_{25}C_{24}C_{23}P_{10} = (A_7 * B_3) + (A_6 * B_4) + (A_5 * B_5) + (A_4 * B_6) + (A_3 * B_7) + C_{16} + C_{18} + C_{20} \quad (11)$$

$$C_{27}C_{26}P_{11} = (A_7 * B_4) + (A_6 * B_5) + (A_5 * B_6) + (A_4 * B_7) + C_{19} + C_{21} + C_{23} \quad (12)$$

$$C_{29}C_{28}P_{12} = (A_7 * B_5) + (A_5 * B_6) + (A_5 * B_7) + C_{22} + C_{24} + C_{26} \quad (13)$$

$$C_{30}P_{13} = (A_7 * B_6) + (A_6 * B_7) + C_{25} + C_{27} + C_{28} \quad (14)$$

$$P_{14} = (A_7 * B_7) + C_{29} + C_{30} \quad (15)$$

$$P_{15} = (A_7 * B_7) \quad (16)$$

It can also be seen from equation 1 to 15, that each stage of the Urdhva multiplication process involves the generation of a carry which would be propagated to the remaining stages. The intermediate carry bits obtained have been represented by C_{30} to C_1. The carry obtained in equations 14 and 15 have been neglected since they are redundant. The equations mentioned above can also be graphically represented with the help of a stick diagram as shown in Fig. 1., where each dot represents the bits of the multiplier and the multiplicand, and each step represents the above equations.

A hardware implementation of the same using AND gates, full adders and half adders has been shown in Fig. 2. Modification of the same to incorporate a combined multiplier, adder and subtraction (MAS) unit has been discussed in the following section. Utilizing the same in an ALU has also been discussed.

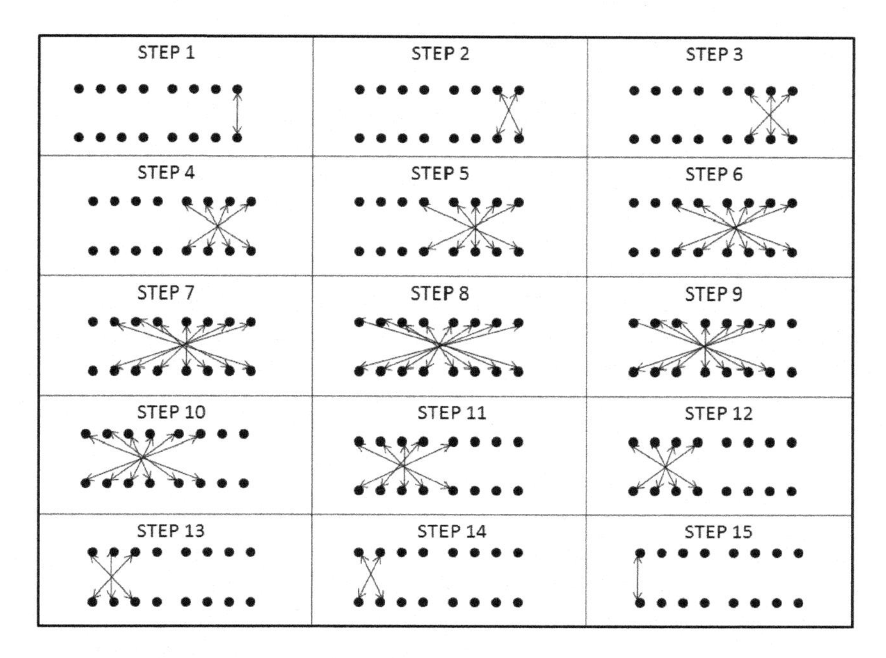

Fig. 1. Illustration of Urdhva Tiryakbhyam Sutra for multiplication using stick diagram

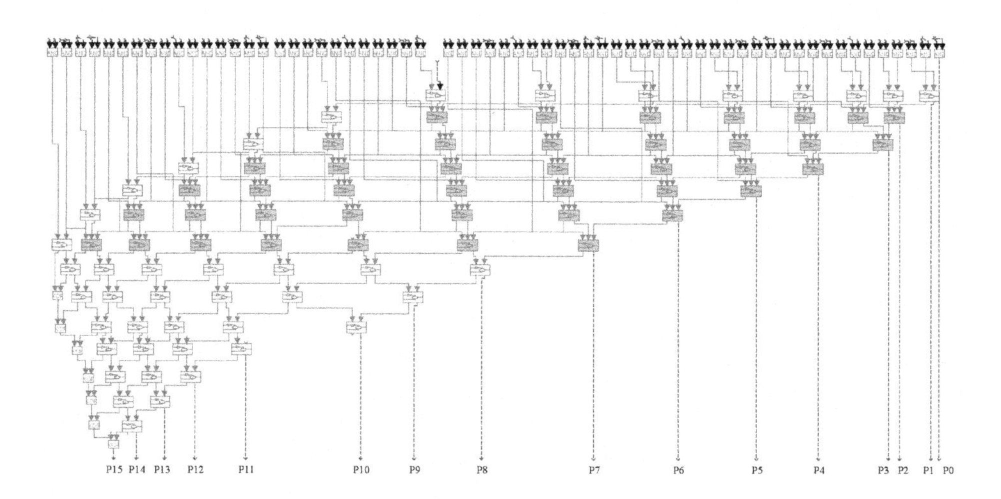

Fig. 2. Hardware architecture of Urdhva Tiryakbhyam multiplier where the blocks in gray represent the full adders and the blocks in white represent the half adders

3 ALU Design

A conventional ALU which consists of the Vedic multiplier, an adder unit, a subtraction unit, a logical block and a shifter unit has been shown in Fig. 3. Even though the speed of the ALU mentioned here is comparatively faster than its predecessors due to the inclusion of a Vedic maths based multiplier, the area is still of concern. The arithmetic unit consists of three separate blocks for multiplication, addition and subtraction which leads to an overall increase in the area of the ALU. Also, it must be noted, as shown in Fig.2, that the Vedic multiplier itself is a combination of a series of full adders and half adders arranged in a "carry save" fashion. Reusing them in order to perform addition and subtraction whenever needed would be a practical way to reduce the "on-chip" area of the design. This forms the basis for the MAS based ALU design.

3.1 MAS Based ALU Design

As seen in Fig. 4. the MAS based ALU consists of the multiplier, adder and a subtraction unit merged into a single unit. A control word decides which operation is to be performed. A set of multiplexers embedded within the multiplier as shown in Fig. 5 decide which operation is to be performed. If the MAS unit is to be used in adder/subtraction mode, the initial bitwise "AND" gates are switched off and bypassed with the help of the first set of multiplexers. Since the latter few full adders and half adders are not required to compute the sum or the difference, they are switched off with another set of multiplexers much earlier itself. Therefore, it can be clearly seen, that by introducing a set of multiplexers, the existing Urdhva Architecture can not only be used as a multiplier but it can be also be reused as an adder or a subtraction unit, thereby reducing the area of the chip. Introducing multiplexers does not slow down the working of the circuit by a large extent. A comparison of the conventional architecture of multipliers and ALU's, along with the MAS based multipliers and ALU's in terms of area and speed have been computed and presented in the next section.

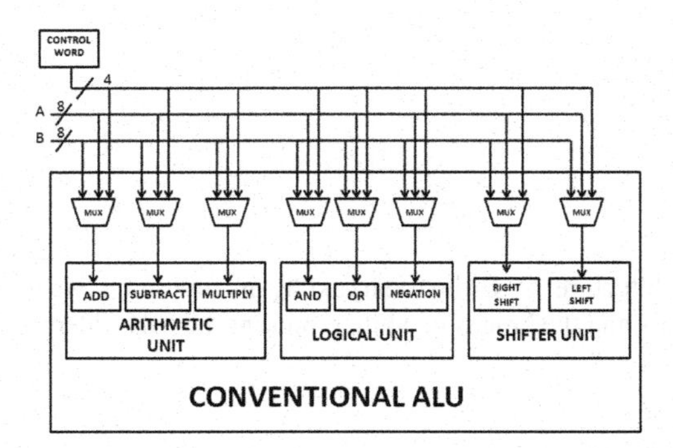

Fig. 3. Architecture of the conventional ALU clearly depicts separate units for multiplication, addition and subtraction

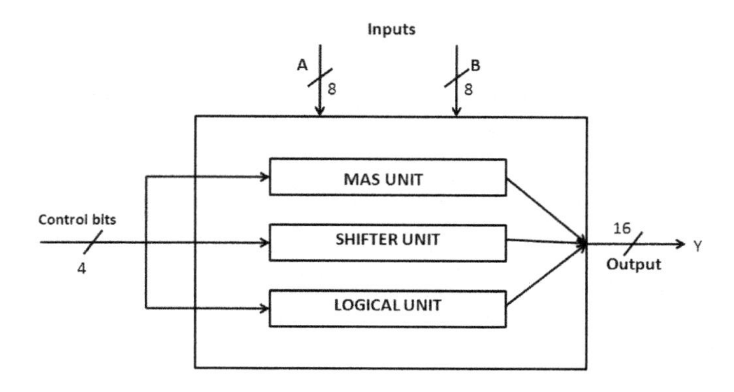

Fig. 4. ALU incorporating the MAS unit

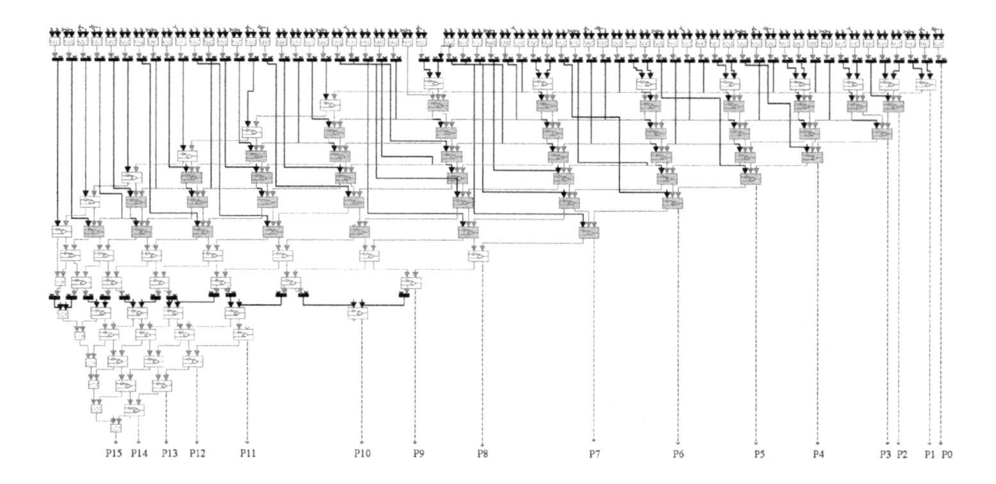

Fig. 5. Hardware architecture of the MAS unit – the black boxes represent the multiplexers

4 Results

In order to perform a comparative study, various multipliers – Booth's, modified Booth's, Urdhva multiplier and also the Urdhva based MAS multiplier were implemented using Verilog language as the RTL design language and synthesized using Xilinx Project Navigator 12.1 for FPGA. Unoptimized speed and area parameters were compared. A Spartan 3e FPGA was used for the experiments, having a speed grade of -5 and package CP132. The results have been tabulated in Table 1.

Table 1. Comparison of area occupied,frequency of operation and time consumed by different types of multipliers on Spartan 3e FPGA

Algorithm used	**LUTs Used**	**Total LUTs present**	**% of area occupied**	**Frequency (MHz)**	**Time (ns)**
Booth algorithm	55	1920	2.86	32.84	30.448
Modified Booth algorithm	213	1920	11.093	45.982	21.747
Urdhva-Tiryakbhyam	185	1920	9.63	61.869	16.163
MAS	207	1920	10.78	57.213	17.479

It can be clearly seen from Table 1. that in terms of speed, the Urdhva method performs exceptionally well but the MAS based multiplier falls short by just 1 ns which is highly negligible, in turn proving to be an efficient design. It can also be seen that the Urdhva and the MAS based multiplier are almost 2 times faster than the conventional Booths algorithm. In terms of area, though the MAS based multiplier occupies a slightly larger area than the conventional Urdhva method of multiplication, the resultant area occupied by the ALU utilizing these multipliers is the matter of concern for this paper. This has been highlighted next.

Table 2. gives us a comparison of the area and speed parameters of various ALU's.

Table 2. Comparison of LUT usage and percentage of area occupied on Spartan 3e FPGA for various ALU architectures

Algorithm used	No. of LUT's used	Total LUT's present	Total % area occupied
Booth algorithm	251	1920	13.07
Modified Booth Algorithm	406	1920	21.14
Urdhva Tiryakbhyam	430	1920	22.39
MAS	371	1920	19.32

A comparison of ALU architectures utilizing the Booth's, modified Booth's, Urdhva and MAS methods in terms of area was performed. Since the multiplier, adder and subtraction unit are all now incorporated as a single unit with the help of few multiplexers, it is distinctly seen that there is a significant 3% decrease in the area occupied by the MAS based ALU in comparison with the Urdhva based ALU. Area comparison with Booth and modified Booth based ALU architectures are not necessary, keeping in mind high speed requirement of the multiplier in the ALU, which is possible only utilizing the Urdhva based multiplier.

5 Conclusion

In this paper we have presented a novel methodology to reduce the area of an ALU considerably using an MAS Urdhva based Multiplier. Also, from the results obtained it can be clearly seen that the speed of this multiplier is relatively fast compared to the other popular designs of multipliers. It therefore, can be concluded that the MAS based ALU design would be a good choice for CPU's, Microcontrollers and Microprocessors of the future. Further research can be performed with the other algorithms of Vedic mathematics and a much more area or speed efficient design could be explored.

References

1. Arrathoon, R., Kozaitis, S.: Architectural and performance considerations for a 10^7-instruction/sec optoelectronic central processing unit. Optics Letters 12(11), 956–958 (1987)
2. Zhang, H., Wang, Z.-Q., Leu, W., Tan, Z.: The Design of Arithmetic Logic Unit based on ALM
3. Patterson, H.: Computer Organization and Design, 4th Revised edn., pp. 220–230. Morgan Kaufmann (2011)
4. Sriraman, L., Prabakar, T.N.: Design and Implementation of Two Variable Multiplier Using KCM and Vedic Mathematics. In: 1st International Conference on Recent Advances in Information Technology. IEEE (2012)
5. Booth, A.D.: A Signed Binary Multiplication Technique. J. Mech. Appl. Math. 4, 236–240 (1951)
6. Baugh, C.R., Wooley, B.A.: A Two's Complement Parallel Array Multiplication Algorithm. IEEE Trans. Computers 22(12), 1045–1047 (1973)
7. Israel, K.: Computer Arithmetic Algorithms, 2nd edn., pp. 141–149. A.K. Peters Limited (2001)
8. Tsoozol, P., Börcsök, J., Schwarz, M.: Instruction based Built-in-Testing of Instruction Select Multiplexer in Arithmetic Logic Unit. In: 25th International Congress on Condition Monitoring and Diagnostic Engineering. Journal of Physics: Conference Series, vol. 364. IOP Publishing (2012)
9. Ramalatha, M., Deena Dayalan, K., Deborah Priya, S.: High Speed Energy Efficient ALU Design using Vedic Multiplication Techniques. In: Advances in Computational Tools for Engineering Applications. IEEE (2009)
10. Jagadguru Swami Sri Bharati Krisna Tirthaji Maharaja: Vedic Mathematics: Sixteen Simple Mathematical Formulae from the Veda, pp. 5–45. Motilal Banarasidas Publishers, Delhi (2009)
11. Sandesh, S.S., Banakar, R.M., Siddamal, S.: High Speed Signed Multiplier for Digital Signal Processing Applications. Signal Processing, Computing and Control (2012)
12. Thapliyal, H., Srinivas, M.B.: An efficient method of elliptic curve encryption using Ancient Indian Vedic Mathematics. In: 48th IEEE International Midwest Symposium on Circuits and Systems, vol. 1, pp. 826–828 (2005)

Implementation and Evaluation of Performance Parameters of Zigbee Transreceiver

Jagruti Nagaonkar[1] and Srija Unnikrishnan[2]

[1] Department of Electronics,
Fr. Conceicao Rodrigues College of Engineering,
University of Mumbai, India
jagruti@frcrce.ac.in
[2] Conceicao Rodrigues College of Engineering,
Bandra, Mumbai-50, India
srija@frcrce.ac.in

Abstract. ZigBee is an IEEE 802.15.4 standard for data communications with business and consumer devices The ZigBee standard provides network, security, and application support services operating on top of the IEEE 802.15.4 Medium Access Control (MAC) and Physical Layer (PHY) wireless standard.. ZigBee is a low-cost, low-power, wireless mesh networking standard. ZigBee has been developed to meet the growing demand for capable wireless networking between numerous low power devices. In industry ZigBee is being used for next generation automated manufacturing, with small transmitters in every device on the floor, allowing for communication between devices to a central computer. This new level of communication permits finely-tuned remote monitoring and manipulation. We have developed Matlab/simulink model for Zigbee transreceiver using spread spectrum modulation and studied its performance in terms of Bit Error Rate (BER) under different performance conditions.

Keywords: Zigbee, SNR, AWGN, Network topologies, Spread spectrum Modulation.

1 Introduction

Wireless personal area network (WPAN) and wireless local area network (WLAN) technologies are growing fast with the new emerging standards being developed. Bluetooth was most widely used for short range communications. Now, ZigBee is becoming as an alternative to Bluetooth for devices with low power consumption and for low data rate applications.

The Bluetooth standard is a specification for WPAN. Although products based on the Bluetooth standards are often capable of operating at greater distances, the targeted operating area is the one around the individual i.e. within a 10m diameter. Bluetooth utilizes a short range radio link that operates in the 2.4GHz industrial scientific and medical(ISM) band similar to WLAN. However, the radio link in

S. Unnikrishnan, S. Surve, and D. Bhoir (Eds.): ICAC3 2013, CCIS 361, pp. 484–491, 2013.

Bluetooth is based on the frequency hop spread spectrum. We know that Bluetooth occupies only 1MHz, the signal changes the centre frequency or hops at the rate of 1600Hz.

The ZigBee standard has adopted IEEE 802.15.4 as its Physical Layer (PHY) and Medium Access Control (MAC) protocols. Therefore, a ZigBee compliant device is compliant with the IEEE 802.15.4 standard as well.

2 Relationship between ZigBee and IEEE 802.15.4 Standard

ZigBee wireless networking protocols are shown in Figure 1. ZigBee protocol layers are based on the Open System Interconnect (OSI) basic reference model. As shown in Figure 1 the bottom two networking layers are defined by IEEE 802.15.4 which also defines the specifications for PHY and MAC layers of wireless networking, but it does not specify any requirements for higher networking layers. The ZigBee standard defines only the networking, applications and security layers of the protocol and adopts IEEE 802.15.4 PHY and MAC layers as a part of the ZigBee networking protocol. Therefore, ZigBee-compliant device conforms to IEEE 802.15.4 as well.

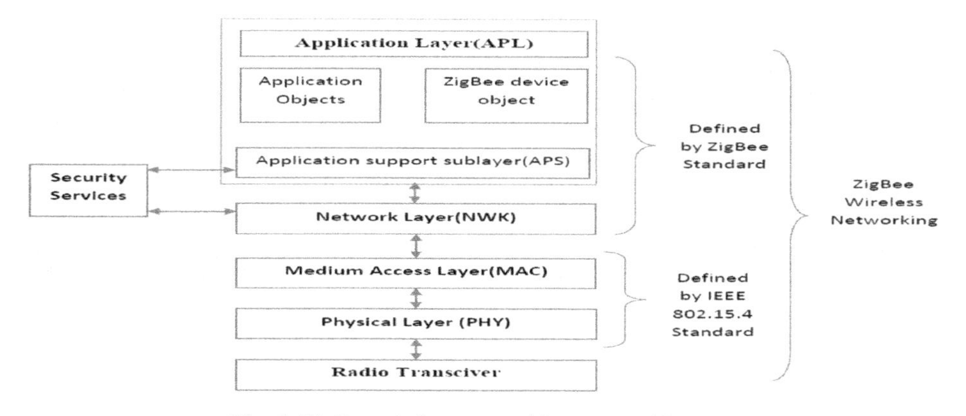

Fig. 1. ZigBee wireless networking protocol Layers

3 Proposed Scheme of Implementing Zigbee Transreceiver Using Spread Spectrum Modulation

This model includes a spreader, a de-spreader, a modulator, a demodulator, and an Additive White Gaussian Noise (AWGN) channel. A random integer generator block generates a number randomly .Then, this integer is taken as input to the spreader block, which spreads it into bit stream. Following that, the bit-stream is taken as an input to the OQPSK modulation block modulated stream using the AWGN block. The latter is then passed through the OQPSK demodulation block before being de-spread.

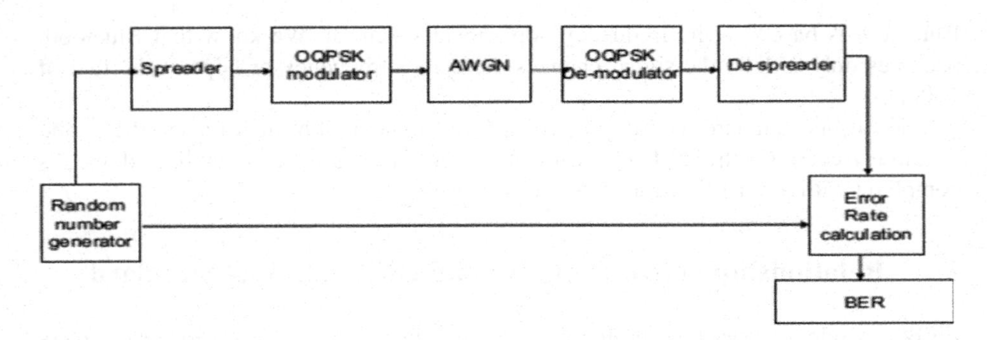

The BER of the received data is calculated as follow: The received bits are sent to the de-spreader which converts them back to an integer. Then, the integer-to-bit converter converts the received integer to a bit-stream.

Finally, the recovered bit stream is compared with the original one and the BER is calculated.

4 Result and Analysis

Simulations were performed to study the Bit error rate (BER) versus Eb/No (SNR) of the designed models. The results show the effects of various communication parameters on the BER for Zigbee model. From graph it can be seen that BER decreases as SNR increases and as data rate is increased BER is increased for particular value of SNR.As number of bits per symbol (M) are increased BER is increased for particular SNR. For the BPSK modulation, varying the data rate has little impact on the performance with respect to the BER. From graph of received power Vs range it can be observed that Zigbee can receive power considerably in the range of 10-15meters.

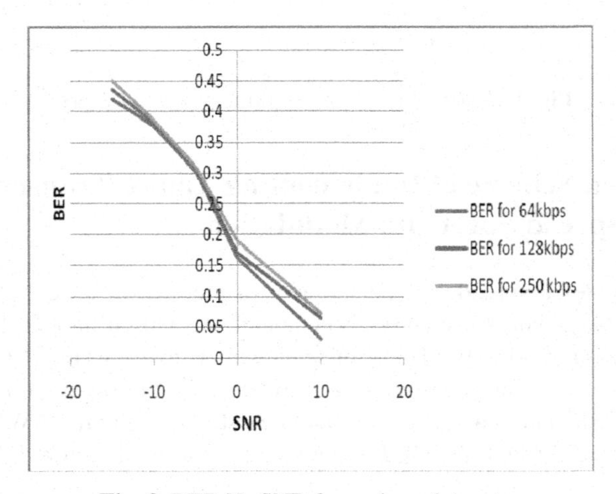

Fig. 2. BER Vs SNR for various data rates

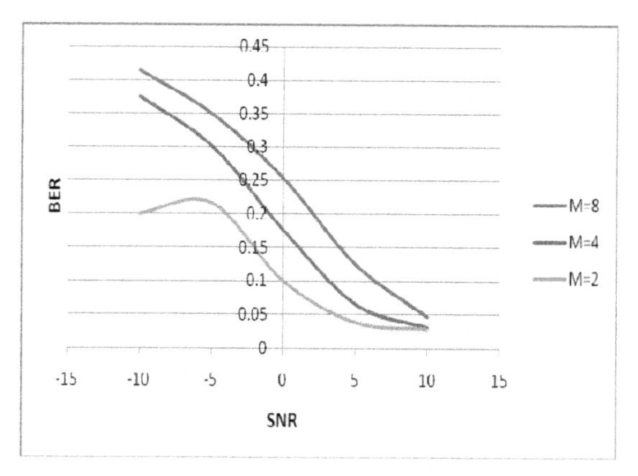

Fig. 3. BER versus the SNR for different number of bits per symbol (M)

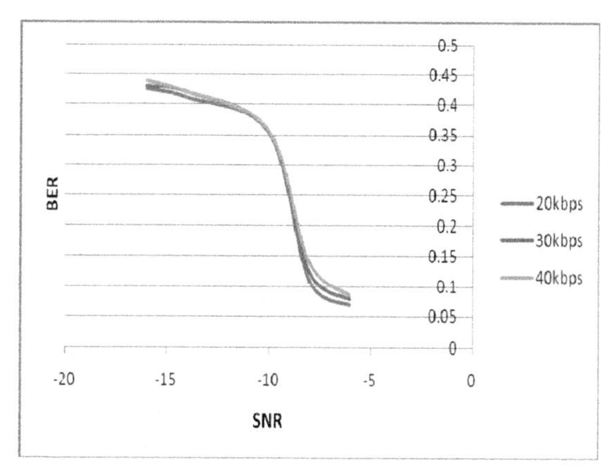

Fig. 4. BER versus SNR for the BPSK model

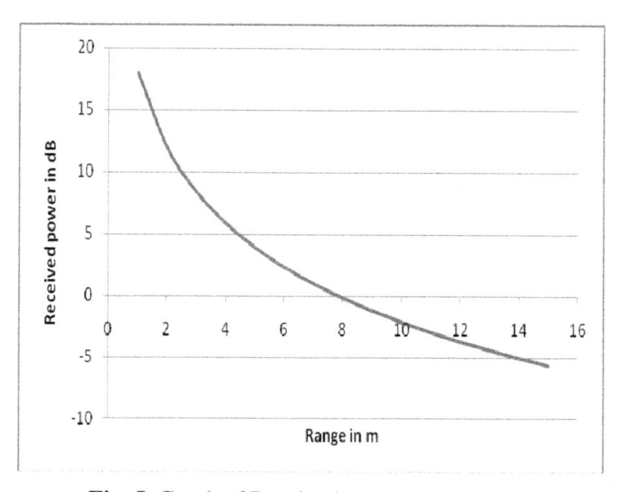

Fig. 5. Graph of Received power versus range

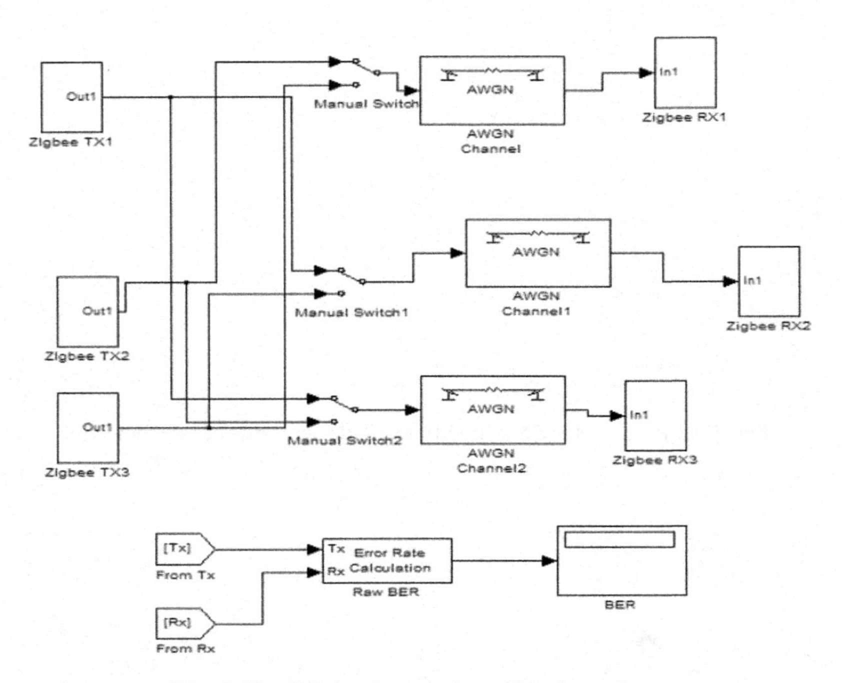

Fig. 6. Simulink implementation of Mesh topology

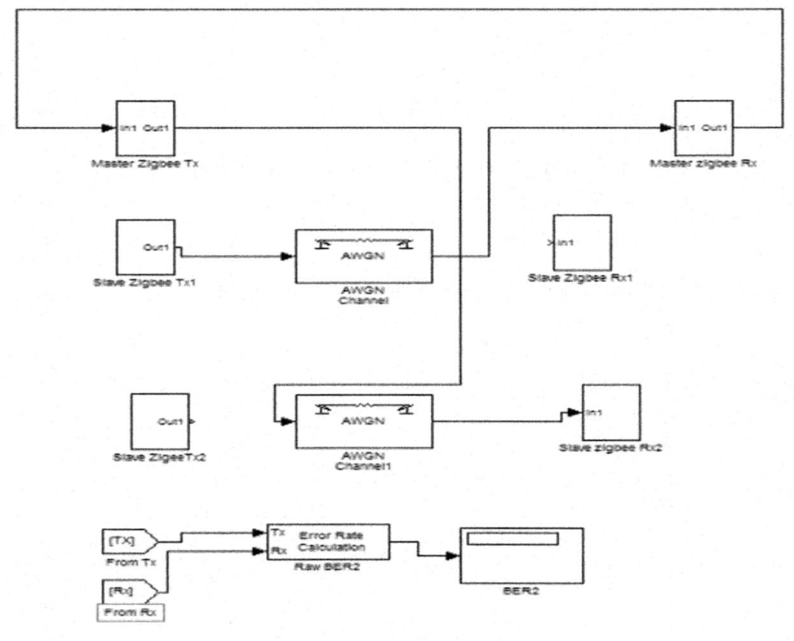

Fig. 7. Simulink implementation of Star topology

5 Network Topologies

ZigBee supports 3 types of topologies: star topology, peer-to-peer topology and cluster tree.

6 Simulation Results

Bit Error Rate Vs SNR

BER versus SNR graph is plotted using Bit Error Rate Analysis Tool in MATLAB/SIMULINK. We can see from the graph that BER decreases as SNR increases. If we compare between BER versus SNR graph for Mesh and Star topology we can see BER is more for star topology than that of mesh for a particular value of SNR.

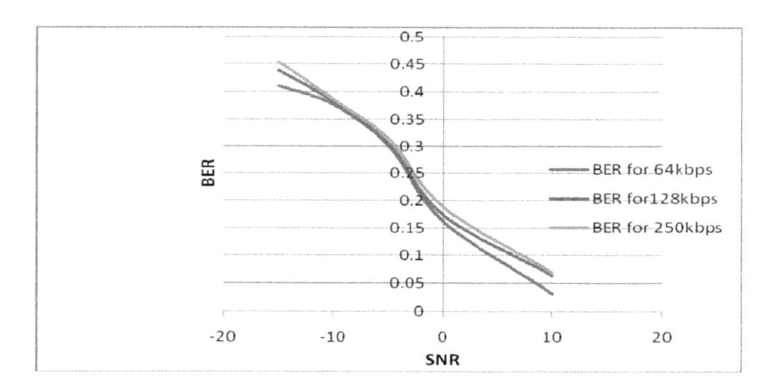

Fig. 8. BER Versus SNR for different data rates for Mesh topology

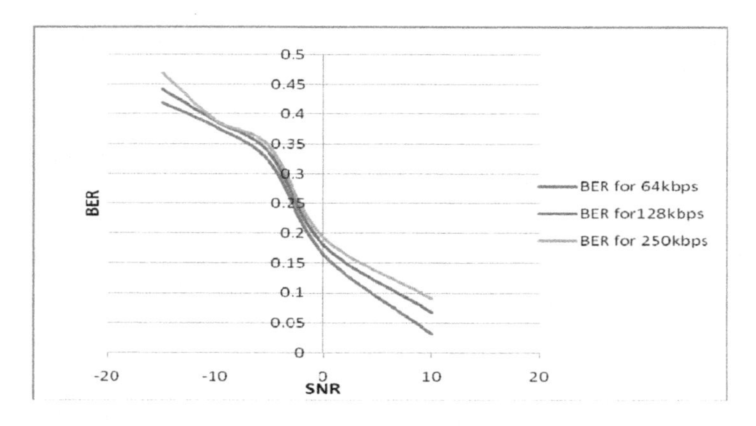

Fig. 9. BER Versus SNR for different data rates for Star topology

7 Conclusion

The work presented here gives implementation of Zigbee transreceiver in Simulink with view to find out optimum operating parameters and operating conditions. The simulation results show how the BER versus the SNR values were affected when varying communication parameters such the input data rate and number of bits per symbol.

For the BPSK modulation, varying the data rate has little impact on the performance with respect to the BER .From graph of received power Vs range it can be concluded that Zigbee can operate satisfactorily in the range of 10 to 15 meters.

If we compare between BER versus SNR graph for Mesh and Star topology we can see BER is more for star topology than that of mesh for a particular value of SNR.

References

[1] Shuaib, K., Alnuaimi, M., Boulmalf, M., Shuaib, K., Alnuaimi, M.: Performance Evaluation of IEEE 802.15.4:Experimental and Simulation results. Journal of Communications 2(4) (June 2007)

[2] Haykin, S.: Digital Communication. Wiley India (P.) Ltd., New Delhi (2007)

[3] IEEE standard 802.11 Wireless Medium Access Control and Physical Layer Specifications (1999)

[4] Kanna, R.: Design of Zigbee Transmitter Receiver IEEE 802.15.4 using Matlab/Simulink

[5] Bluetooth SIG, Specifications of the Bluetooth System – Version 1.2 (November 2003)

[6] Shoemake, M.B.: Broadcom white paper, IEEE 802.11g: The New Mainstream Wireless LAN Standard. Wi-Fi (IEEE 802.11b) and Bluetooth Coexistence Issues and Solution 2.4 GHz ISM Band, Texas Instrument white paper (2003)

[7] http://www.maxstream.net

[8] http://en.wikipedia.org/wiki/zigbee

[9] Anwar, A.-K., Lavagno, L.: Simulink modelling of the 802.15.4 physical layer for model based design of wireless sensors. In: Third International Conference on Sensor Networks 2009 (2009)

[10] Golmie, N., Cypher, D.: Performance Evaluation of low rate WPAN for sensor and medical Application. In: Military Communication Conference (2004)

[11] Petrova, M., Mahonen, P.: Performance Study of IEEE 802.15.4 Using Measurment and Simulation. In: WCNC Proccedings (2006)

[12] IEEE, IEEE Std 802.15.4-2006, Part 15.4: Wireless Medium Access Control (MAC) and Physical layer (PHY) Specifications for Low-Rate Wireless Personal Area Networks (WPANs) (2007)

[13] Bougard, B., Catthoor, F., Daly, C., Chandrakasan, A., Dehaene, W.: Energy efficiency of IEEE 802.15.4 Standard in dense wireless microsensor network Modelling and improvement perspectives. In: Design, Automation and Test in Europe, DATE (2005)

[14] Cristi, R.: Wireless Communications with Matlab and Simulink: IEEE 802.16 (WiMax) physical Layer. Naval Postgraduate School, Monterey, CA 93943 (August 2009)

[15] Han, S., Lee, S., Lee, S., Kim, Y.: Co existence Performance Evaluation of IEEE 802.15.4 Under IEEE 802.11BI Interference in Fading Channels. In: 18th Annual IEEE International Symposium on Personal, Indoor and Mobile Radio Communications, PIMRC 2007 (2007)

[16] Agrawal, J.P., Farook, O., Sekhar, C.R.: SIMULINK Laboratory Exercises In Communication Technology. In: Proceedings of the 2005 American Society for Engineering Education Annual Conference & Exposition. American Society for Engineering Education (2005)

Improved Technique to Remove ECG Baseline Wander – Application to Pan & Tompkins QRS Detection Algorithm

Rameshwari Mane[1], A.N. Cheeran[1], and Vaibhav D. Awandekar[2]

[1] Veermata Jijabai Technological Institute, Mumbai, Maharashtra
[2] A3 Remote Monitoring Technologies Private Limited SINE, IIT Bombay, Mumbai, Maharashtra

Abstract. Electrocardiographic analysis plays an important role in clinical diagnosis. The pre-processing of Electrocardiogram (ECG) signal consists of low-frequency baseline wander (BW) correction and high-frequency artifact noise reduction from the raw ECG signal. Baseline wander in ECG signal is the biggest hurdle in visualization of correct waveform and computerized detection of wave complexes based on threshold decision. This paper compares three methods of baseline wander removal, first using high pass filter, second FFT and third Wavelet transform. The ECG data is taken from standard MIT-BIH Arrhythmia database for simulation in Python 2.6. Simulation result shows that Wavelet transform removes BW completely.

This paper also applies correction to BW of Pan & Tompkins QRS detection algorithm using Wavelet transform. This method helps in reducing the BW to a near zero value and inspecting the morphology of the wave components in ECG.

Keywords: Electrocardiogram (ECG), QRS, Baseline Wander (BW), Fast Fourier Transform (FFT), Wavelet transform.

1 Introduction

The electrocardiogram (ECG) is a time-varying signal reflecting the ionic current flow which causes the cardiac fibers to contract and subsequently relax. Baseline wander is the extragenoeous low-frequency high bandwidth component, can be caused by Perspiration (effects electrode impedance), Respiration or Body movements. Baseline wander can cause problems to analysis, especially when examining the low-frequency ST-T segment.

Baseline removal has been addressed in many different ways in literature. Baseline removal using high pass IIR filter is proposed in [1]. Digital filtering directly through the Fast Fourier Transform is used for baseline removal by simply substitution of zero-magnitude components at frequencies outside the band limits [2].In [3] Short-Time Fourier transform (STFT) is used to remove the presence of baseline wander in

S. Unnikrishnan, S. Surve, and D. Bhoir (Eds.): ICAC3 2013, CCIS 361, pp. 492–499, 2013.

ECG signals, which can then be removed using a time varying filter. The baseline drift is efficiently removed by zeroing the scaling coefficients of the discrete wavelet transform of Daubechies-4 wavelet [4], [5]. Wavelet transform can remove the low frequency components without introducing distortions in the ECG waveform.

The Pan & Tompkins QRS detection algorithm [1] is the most widely used and the often cited algorithm for the extraction of QRS complexes from the ECG signal. The algorithm is based on the application of several filters.

Due to the variation in the baseline, the detection of QRS complexes will be more difficult. In [6] ECG baseline is corrected using a median filter to make it horizontal in order to apply a fixed thresholding.

This paper suggests correcting the baseline using wavelet transform to improve the performance of the Pan & Tompkins QRS detection algorithm.

The rest of this paper is organized as follows: Section 2 presents the comparison of above three methods. Section 3 describes Pan & Tompkins QRS detection algorithm. The correction of the baseline wander is given in section 4. Finally, the section 5 & 6 summarizes the result & conclusion of this work.

2 Comparison of Three Methods

2.1 IIR High Pass Filter

In this method second order IIR high pass filter is used for removal of baseline wander. The cut off frequency of this filter is about 5Hz.The signal is then passed through low pass filter to remove high frequency noise. These filters are implemented in Python and simulation result is obtained.

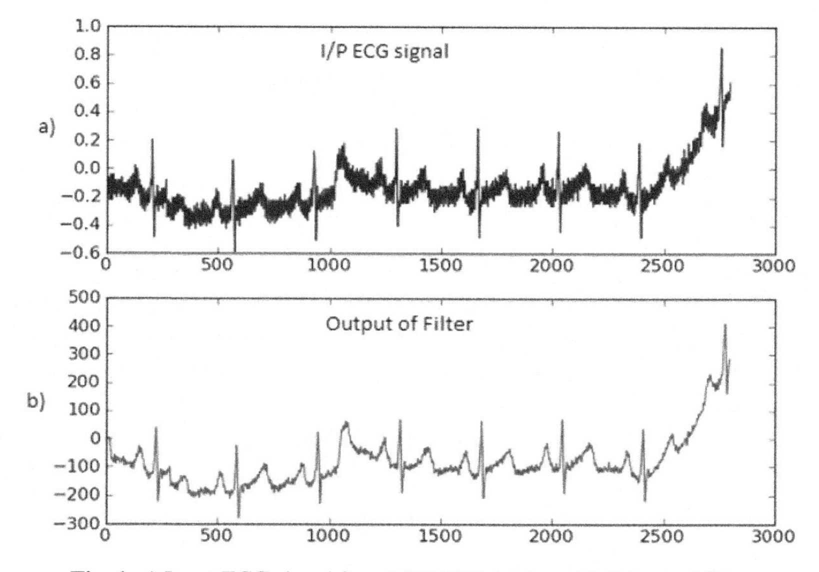

Fig. 1. a) Input ECG signal from MIT/BIH database b) Output of filter

Fig.1. shows that BW is not removed completely. Only IIR filter is simple to implement and requires less number of coefficients.

2.2 FFT

This method involves taking an FFT of the data, then zeroing the coefficients of frequencies below 1Hz and above 35Hz and then taking an inverse fast Fourier transform (IFFT). Here band pass filter is implemented in Python by substitution of zero-magnitude components at frequencies outside the band limits. The result in Fig.2 shows some improvement compared with high pass filter, but BW is not removed properly.

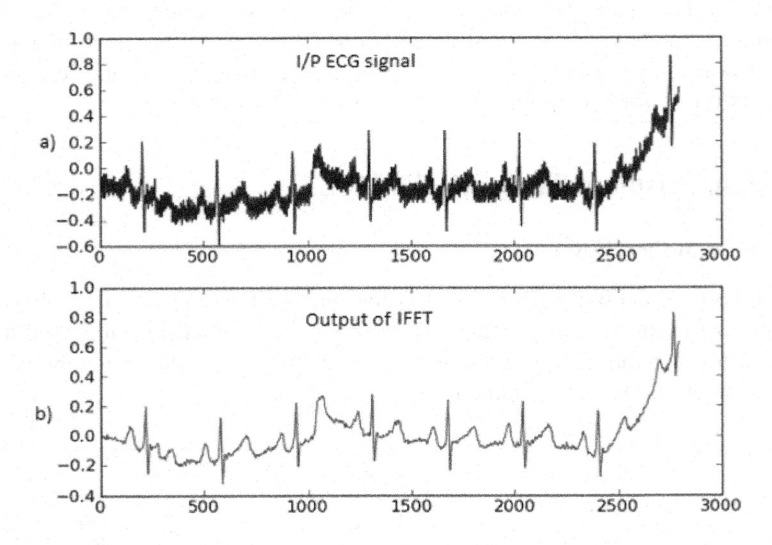

Fig. 2. a) Input ECG signal from MIT/BIH database b) Output of IFFT

2.3 Wavelet Transform

Wavelet transform uses only the zeroing of the scaling coefficients to remove the baseline wander. By setting the scaling coefficients of the low frequency components to zero, we are able to remove the BW. An additional benefit of using the wavelet transform is that it can be applied to remove noise and power line interference from the ECG signal.

Wavelet transform decomposition and its reconstruction is implemented in Python using Db8 wavelet and level 8 decomposition. Result in Fig.3. shows complete removal of BW.

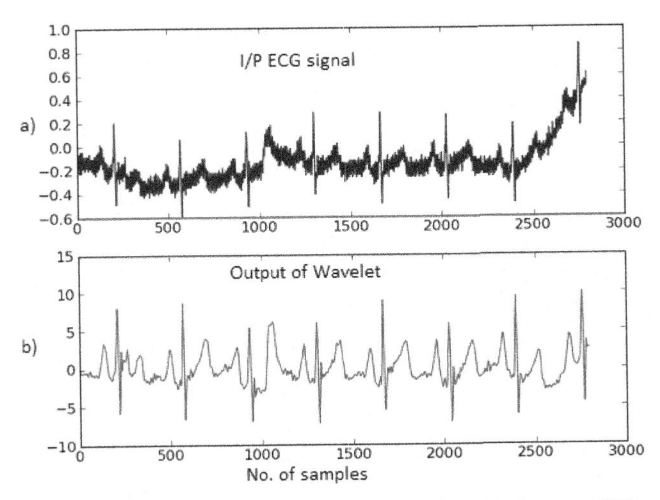

Fig. 3. a) Input ECG signal from MIT/BIH database b) Output of Wavelet

3 Pan & Tompkins QRS Detection Algorithm

The Pan & Tompkins QRS detection algorithm is based upon digital analyses of slopes, amplitude, and width. The methodology followed is that the ECG is passed through a special digital bandpass filter, to reduce the noise. Then the filtered signal is passed through derivative, squaring and window integration phases. Finally, an adaptative threshold is applied in order to increase the detection sensitivity. The different steps of the algorithm are shown in the Fig.4.

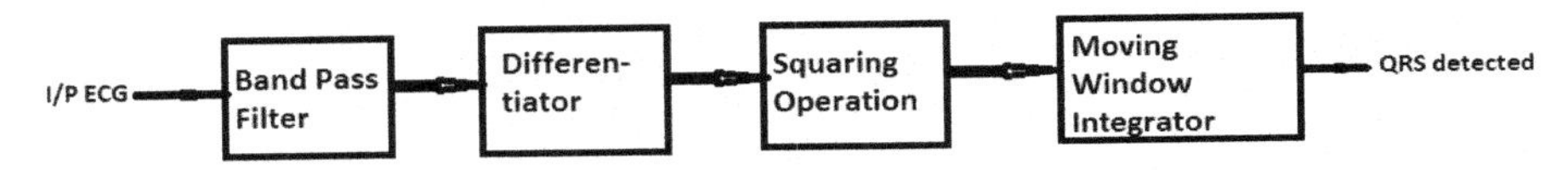

Fig. 4. A graphical representation of the algorithm

In the first step, the signal passes through a bandpass filter composed by a low pass and a high pass filter in cascade, in order to reduce the influence of muscle noise, 60 Hz interference, base wander, and T-wave interference.

The desirable pass band to maximize the QRS energy is approximately 5-15 Hz. The low pass filter is described by the following transfer function:

$$H(Z) = \frac{Y(Z)}{X(Z)} = \frac{1 - 2Z^{-6} + Z^{-12}}{1 - 2Z^{-1} + Z^{-2}} \tag{1}$$

and the high pass one is given by:

$$H(Z) = \frac{Y(Z)}{X(Z)} = \frac{-1 + 32Z^{-16} + Z^{-32}}{1 + Z^{-1}} \tag{2}$$

Where X (Z) and Y (Z) are respectively the Z-transform of the input x (n) and the output y (n) signals. After filtering, the signal is differentiated to provide the QRS complex slope information using the following transfer function:

$$H(Z) = \frac{1}{8}[2 + Z^{-1} - Z^{-3} - 2Z^{-4}] \tag{3}$$

Then the signal is squared point by point making all data point positive and emphasizing the higher frequencies, given by this relation:

$$X = Y^2 \tag{4}$$

After squaring, the signal passes through a moving-window integrator in order to obtain wave-form feature information. It is calculated from:

$$H(Z) = \frac{1}{N}[Z^{N-1} + Z^{N-2} + ... + Z^{N-(N-1)} + 1] \tag{5}$$

Where N is the number of samples in the width of the integration window. The choice of N is very important, generally, it should be the approximately the same as the widest possible QRS complex. If the window is too wide, the integration waveform will merge the QRS and T complexes together, and if it is too narrow, some QRS complexes will produce several peaks in the integration waveform.

This algorithm is implemented in Python and simulation results are obtained.

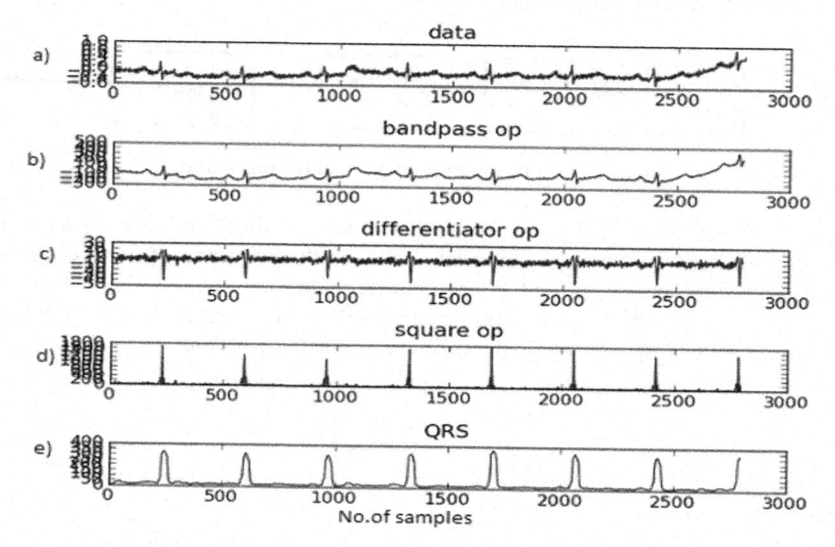

Fig. 5. (a) original signal (from MIT/BIH database), (b) band pass filter, (c) differentiation, (d) squaring, (e) Integration i.e. QRS detection

Fig.5 shows the output at each stage of the algorithm. The output of bandpass filter shows that BW is not removed completely, thus QRS complex detected shows higher amplitude due to the presence of BW.

4 Correction of the ECG Baseline Wander

The proposed method shown in Fig.6 uses wavelet decomposition and reconstruction to remove BW from ECG and then the signal is processed by the steps of Pan – Tompkins algorithm. Simulation results are obtained in Python using MIT-BIH Arrhythmia database.

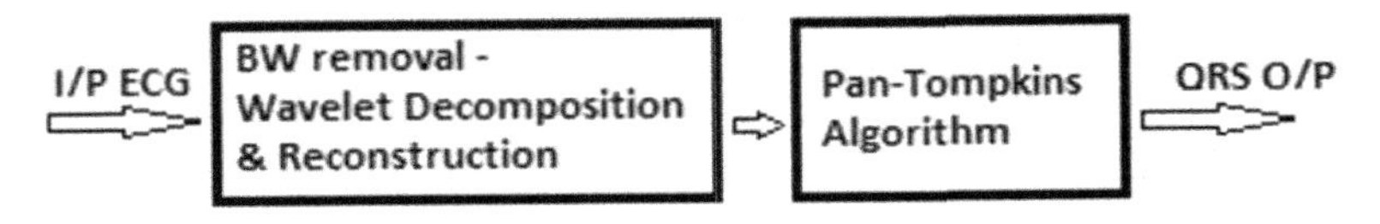

Fig. 6. Block Dia. Of proposed method

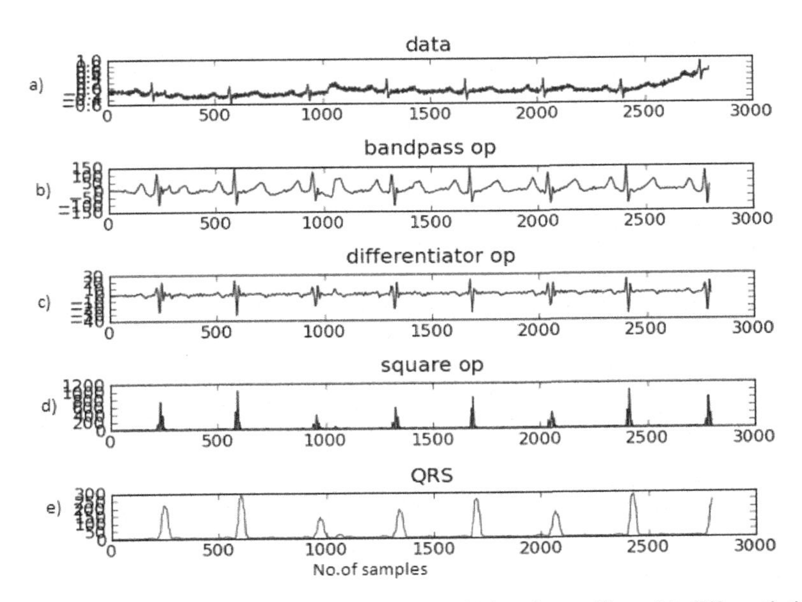

Fig. 7. (a) original signal (from MIT/BIH database), (b) band pass filter, (c) differentiation, (d) squaring, (e) Integration i.e. QRS detection

Fig.7 shows that bandpass output contains almost zero BW. Exact QRS complex can be extracted using proposed method.

5 Result

For comparison QRS obtained from two methods are superimposed in Fig.8. Red indicates QRS with wavelet after correcting the baseline and blue indicates QRS without wavelet. The final result shows significant difference in amplitude of two.

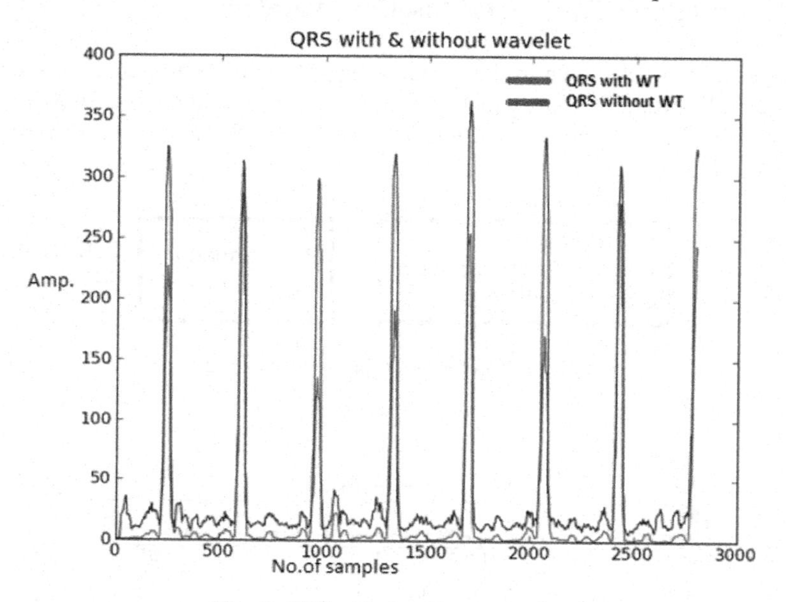

Fig. 8. QRS with & without wavelet

6 Conclusion

In this paper, first three methods for BW removal are compared, among which wavelet analysis gives the best result. Then the method of Pan & Tompkins is used for the extraction of the QRS complexes.

The ECG baseline of Pan & Tompkins algorithm is corrected using Wavelet transform. Eventually this correction is used for detection of wave complexes based on threshold decision. Thus the morphology of the wave components can be clearly extracted with a drift free ECG signal.

Complete removal of Baseline Wander is important in feature extraction of ECG signal especially when examining the low-frequency ST-T segment which helps in the detection of different cardiac arrhythmias.

References

1. Pan, J., Tompkins, W.J.: A real-time QRS Detection Algorithm. IEEE Trans. Biomed. Eng. BME-32, 230–236 (1985)
2. Pan, C.: Gibbs Phenomenon Removal and Digital Filtering Directly through the Fast Fourier Transform. IEEE Transactions on Signal Processing 49(2), 1053–1058 (2001)

3. Pandit, S.V.: ECG baseline drift removal through STFT. In: Proceedings of 18th Annual International Conference of the Engineering in Medicine and Biology Society, vol. 4, pp. 1405–1406 (1996)
4. von Borries, R.F., Pierluissi, J.H., Nazeran, H.: Wavelet Transform-Based ECG Baseline Drift Removal for Body Surface Potential Mapping. In: Proceedings of the 2005 IEEE, Engineering in Medicine and Biology 27th Annual Conference, Shanghai, China, September 1-4 (2005)
5. Vega-Martínez, G., Alvarado-Serrano, C., Leija-Salas, L.: ECG Baseline Drift Removal Using Discrete Wavelet Transform. In: 8th International Conference on Electrical Engineering Computing Science and Automatic Control (CCE), pp. 1–5 (2011)
6. Debbabi, N., El Asmi, S., Arfa, H.: Correction of ECG baseline wander Application to the Pan & Tompkins QRS detection Algorithm. In: 5th International Symposium on I/V Communications and Mobile Network (ISVC), pp. 1–4 (2010)

Image Classification Using Advanced Block Truncation Coding with Ternary Image Maps

Sudeep D. Thepade[1], Rik Kamal Kumar Das[2,3], and Saurav Ghosh[2]

[1] Dept. of Computer Engg., Pimpri Chinchwad College of Engineering, Pune, India
Sudeepthepade@gmail.com
[2] A.K. Choudhury School of Information Technology, University of Calcutta, Kolkata, India
rikdas78@gmail.com
[3] Dept. of Information Technology, Xavier Institute of Social Service, Ranchi, Jharkhand, India
sauravghoshcu@gmail.com

Abstract. Incredible escalation of Information Technology leads to generation, storage and transfer of enormous information. Easy and round the clock access of data has been made possible by virtue of world wide web. The high capacity storage devices and communication links facilitates the archiving of information in the form of multimedia. This type of information comprises of images in majority and is growing in number by leaps and bounds. But the usefulness of this information will be at stake if maximum information is not retrieved in minimum time. The huge database of information comprising of multiple number of image data is diversified mix in nature. Proper Classification of Image data based on their content is highly applicable in these databases to form limited number of major categories. The novel ternary block truncation coding (Ternary BTC) is proposed in the paper, also the comparison of Binary block truncation coding (Binary BTC) and Ternary Block Truncation Coding is done for image classification. Here two image databases are considered for experimentation. The proposed ternary BTC is found to be better than Binary BTC for image classification as indicated by higher average success rate.

Keywords: CBIC, BTC, Ternary Image Maps, Color Space, RGB.

1 Introduction

Storage and manipulation of image data is evolving as a new challenge with every passing day. Easy access of image data from an image database is passing through a phase of optimization and upgraded efficiency as the number of images is growing astronomical. Image compression is an effective method used to store and transfer images from one place to another. These image data may be used in various fields like military services, criminology, entertainment, education, etc.[1] Efficient use of image data requires ready access and prompt retrieval. The method of Image Classification catalyzes this efficiency and act as a precursor for rapid image retrieval.

S. Unnikrishnan, S. Surve, and D. Bhoir (Eds.): ICAC3 2013, CCIS 361, pp. 500–509, 2013.

A supervised and systematized database with limited number of major classes is the most useful tool for increased competence in searching of images from an image database.

2 Content Based Image Classification

Content based Image Classification uses various content feature vector extraction methods to precisely symbolize the content of each image.[1,2] The process of feature extraction (FE) is followed by similarity measurement (SM) where feature vector is used to measure a distance between the query image and each image in the database used to classify the images based on maximum resemblance.[7,8] Feature extraction can be performed in two different ways, viz., feature extraction in spatial domain and feature extraction in transform domain. The methods used for feature extraction in spatial domain include histogram[6], BTC [7] and VQ [9,11,12]. Extensive uses of transform domain methods are found in image retrieval techniques as high energy compaction is obtained in transformed images[13,17,18]. But using transforms for feature extraction in content based image classification is time consuming. In spatial domain, pure image pixel data is used to reduce the size of the feature vector. This improves the performance of image classification. The direct Euclidian distance between the query image Q and the image in the database P may be represented by the equation1.

$$ED = \sqrt{\sum_{i=1}^{n} (Vpi - Vqi)^2} \quad (1)$$

Previous approaches of image classification have considered K-means clustering with Block Truncation Coding using color moments. Six assorted color spaces are considered to perform block truncation coding based image classification in a more recent approach[1]. Another approach has proposed the method of Bit Plane Slicing to boost the performance of block truncation coding based image classification[2]. All the above mentioned approaches are based on binary block truncation coding BTC based feature vector extraction. This paper proposes a novel ternary block truncation coding based feature vector extraction method for image classification and gives the results comparison of image classification using binary btc and ternary btc.

2.1 Block Truncation Coding (BTC)

BTC was developed way back in the year 1979 at the early stages of image processing. It was an algorithm developed for gray scale images and later it was extended for color images[1]. BTC follows nxn segmentation (typically 4x4) of non overlapping blocks for an image. Then the blocks are coded one at a time. Mean and standard deviation of each block is calculated to form the new values of the reconstructed blocks. The value of mean and standard deviation of the reconstructed blocks remains same as the original one[2,10].

2.2 RGB Color Space

The description of color is given by a computer by emitting Red, Green and Blue Phosphor [4]. Three co ordinates are required to specify a color space. The position of the color is described by the parameter inside the used color space. The color space comprising of base colors viz., Red, Green and Blue is considered in this work [2,3]. If these base colors are mixed in full intensities, white color is produced and if none of the colors is used it results to black color. Required proportions of these colors may be mixed to create any desired color. RGB color space has been considered in the proposed methodology[5].

3 Image Feature Generations Using Block Truncation Coding for Classification

3.1 Binary Block Truncation Coding

R,G and B are considered as three independent color components for the considered image categories. A threshold value is calculated for each of the color components. The three resultant threshold values are named as TR, TG and TB. The equations for computation of TR, TG and TB are given in equation 2,3 and 4.

$$TR = (1/m*n)*\sum_{i=1}^{m}\sum_{j=1}^{n} R(i,j) \tag{2}$$

$$TG = (1/m*n)*\sum_{i=1}^{m}\sum_{j=1}^{n} G(i,j) \tag{3}$$

$$TB = (1/m*n)*\sum_{i=1}^{m}\sum_{j=1}^{n} B(i,j) \tag{4}$$

The image is partitioned into two intensity values of higher intensity and lower intensity by comparing each pixel value with the derived mean value for each color component R, G and B. The values of the pixels which are higher than or equal to the threshold value of the corresponding color space are assigned with value 1 and the pixels possessing values lower than the consequent threshold values are assigned with value 0. Three bitmaps thus calculated are named as BMr, BMg and BMb respectively. The equations are given in equation 5, 6 and 7.

$$BMr(i,j) = \begin{cases} 1, if....R(i,j) >= TavR \\ 0,....if...R(i,j) < TavR \end{cases} \tag{5}$$

$$BMg(i,j) = \begin{cases} 1, if \ldots R(i,j) >= TavG \\ 0, \ldots if \ldots R(i,j) < TavG \end{cases} \tag{6}$$

$$BMb(i,j) = \begin{cases} 1, if \ldots R(i,j) >= TavB \\ 0, \ldots if \ldots R(i,j) < TavB \end{cases} \tag{7}$$

The image features are generated for binary BTC used in image classification are given in the equations from 8 to 13. Finally the feature vector would be represernted ads a vector {bUR,bLR,bUG,bLG and bUB,bLB}.

$$bUR = \frac{1}{\sum_{i=1}^{n}\sum_{j=1}^{m}(BMr(i,j))} * \sum_{i=1}^{n}\sum_{j=1}^{m}(BMr(i,j) * R(i,j)) \tag{8}$$

$$bLR = \frac{1}{m*n - \sum_{i=1}^{n}\sum_{j=1}^{m}(BMr(i,j)} * \sum_{i=1}^{n}\sum_{j=1}^{m}(1 - BMr(i,j) * R(i,j) \tag{9}$$

$$bUG = \frac{1}{\sum_{i=1}^{n}\sum_{j=1}^{m}(BMg(i,j))} * \sum_{i=1}^{n}\sum_{j=1}^{m}(BMg(i,j) * G(i,j)) \tag{10}$$

$$bLG = \frac{1}{m*n - \sum_{i=1}^{n}\sum_{j=1}^{m}(BMg(i,j)} * \sum_{i=1}^{n}\sum_{j=1}^{m}(1 - BMg(i,j) * G(i,j) \tag{11}$$

$$bUB = \frac{1}{\sum_{i=1}^{n}\sum_{j=1}^{m}(BMb(i,j))} * \sum_{i=1}^{n}\sum_{j=1}^{m}(BMb(i,j) * B(i,j)) \tag{12}$$

$$bLG = \frac{1}{m*n - \sum_{i=1}^{n}\sum_{j=1}^{m}(BMb(i,j)} * \sum_{i=1}^{n}\sum_{j=1}^{m}(1 - BMb(i,j) * B(i,j) \tag{13}$$

3.2 Ternary Block Truncation Coding

One overall luminance threshold is calculated from the threshold values TR, TG and TB computed for generation of binary image maps in equation 2, 3 and 4. The luminance threshold is used to calculate the ternary image bitmaps. The equation for calculating the luminance threshold is given in equation 14.

$$T = \frac{TR+TG+TB}{3} \tag{14}$$

Individual color threshold intervals (lower-Tshl and higher-Tshh) are calculated for each component (R, G and B). The equations are shown in 15, 16 and 17.

$$Tshrl = TR - |TR - T| \text{ , } Tshrh = TR + |TR - T| \tag{15}$$

$$Tshgl = TG - |TG - T| \text{ , } Tshgh = TG + |TG - T| \tag{16}$$

$$Tshbl = TB - |TB - T| \text{ , } Tshbh = TB + |TB - T| \tag{17}$$

Computation of the individual color plane global ternary image maps TMr, TMg and TMb are shown in equation 18, 19 and 20. A value 'one' is allotted to the corresponding pixel position if a pixel value of respective color component is greater than the respective higher threshold interval (Tshh). A lesser pixel value than the respective lower threshold interval (Tshl), corresponds to a value of 'minus one' for the consequent pixel position of the image map; else it gets a value 'zero'.[14]

$$TMr(i,j) = \begin{cases} 1 & \text{if} \quad R(i,j) > Tshrh \\ 0 & \text{if } Tshrl <= R(i,j) <= Tshrh \\ -1 & \text{if} \quad R(i,j) < Tshrl \end{cases} \tag{18}$$

$$TMg(i,j) = \begin{cases} 1 & \text{if} \quad G(i,j) > Tshgh \\ 0 & \text{if } Tshgl <= G(i,j) <= Tshgh \\ -1 & \text{if} \quad G(i,j) < Tshgl \end{cases} \tag{19}$$

$$TMb(i,j) = \begin{cases} 1 & \text{if} \quad B(i,j) > Tshbh \\ 0 & \text{if } Tshbl <= B(i,j) <= Tshbh \\ -1 & \text{if} \quad B(i,j) < Tshbl \end{cases} \tag{20}$$

The image features are generated for binary BTC used in image classification are given in the equations from 21 to 29. Finally the feature vector would be represented ads a vector {tUR,tMR,tLR,tUG,tMG,tLG and tUB,tMB,tLB}.

$$tUR = \frac{1}{\sum_{i=1}^{m}\sum_{j=1}^{n} TMr(i,j), iffTMr(i,j)=1} * \sum_{i=1}^{m}\sum_{j=1}^{n} R(i,j), iffTMr(i,j)=1 \tag{21}$$

$$tMR = \frac{1}{\sum_{i=1}^{m}\sum_{j=1}^{n} TMr(i,j), iffTMr(i,j)=0} * \sum_{i=1}^{m}\sum_{j=1}^{n} R(i,j), iffTMr(i,j)=0 \tag{22}$$

$$tLR = \frac{1}{\sum_{i=1}^{m}\sum_{j=1}^{n} TMr(i,j), iff TMr(i,j) = -1} * \sum_{i=1}^{m}\sum_{j=1}^{n} R(i,j), iff TMr(i,j) = -1 \tag{23}$$

$$tUG = \frac{1}{\sum_{i=1}^{m}\sum_{j=1}^{n} TMg(i,j), iff TMg(i,j) = 1} * \sum_{i=1}^{m}\sum_{j=1}^{n} G(i,j), iff TMg(i,j) = 1 \tag{24}$$

$$tMG = \frac{1}{\sum_{i=1}^{m}\sum_{j=1}^{n} TMg(i,j), iff TMg(i,j) = 0} * \sum_{i=1}^{m}\sum_{j=1}^{n} G(i,j), iff TMg(i,j) = 0 \tag{25}$$

$$tLG = \frac{1}{\sum_{i=1}^{m}\sum_{j=1}^{n} TMg(i,j), iff TMg(i,j) = -1} * \sum_{i=1}^{m}\sum_{j=1}^{n} G(i,j), iff TMg(i,j) = -1 \tag{26}$$

$$tUB = \frac{1}{\sum_{i=1}^{m}\sum_{j=1}^{n} TMb(i,j), iff TMb(i,j) = 1} * \sum_{i=1}^{m}\sum_{j=1}^{n} B(i,j), iff TMb(i,j) = 1 \tag{27}$$

$$tMB = \frac{1}{\sum_{i=1}^{m}\sum_{j=1}^{n} TMb(i,j), iff TMb(i,j) = 0} * \sum_{i=1}^{m}\sum_{j=1}^{n} B(i,j), iff TMb(i,j) = 0 \tag{28}$$

$$tLB = \frac{1}{\sum_{i=1}^{m}\sum_{j=1}^{n} TMb(i,j), iff TMb(i,j) = -1} * \sum_{i=1}^{m}\sum_{j=1}^{n} B(i,j), iff TMb(i,j) = -1 \tag{29}$$

4 Implementation

Intel core 2 duo processor with 1 GB RAM has been used to execute image classification using binary and ternary image maps with BTC. The experiment is carried out in Matlab 7.11.0(R2010b). One thousand queries are fired on two different databases to find out the classification success rate. Performance comparison of the proposed image classification techniques is done by comparing the average classification success rate of all these queries.

4.1 Databases

Two databases are considered to carry out the performance analysis of the proposed methodology. The datasets are the Ponce Group 3D photographic database and a generic database. The Ponce Group dataset comprises of several public databases. The considered database is the 3D photography dataset consisting of 100 images and divided into five different categories. Fig.1 shows the sample of various categories considered for the 3D dataset from Ponce group image database [15].

Fig. 1. Sample Images from 3D Photography Dataset

The generic dataset consists of nine different categories with each group comprising of hundred images. The nine different categories are Ganeshji, Sea Beaches, Sunflower, Candles, Dinosaurs, Elephants, Roses, Horses and Mountains. Some categories of the generic database are directly taken from the Wang database.[16]. A glimpse of the database used is represented in Fig.2.

Fig. 2. Sample Image Database from generic database

5 Performance Comparison

The performance of both the binary and the ternary method are tested by firing 100 queries on Ponce group 3D photography dataset and 900 queries on generic database respectively. Comparison of feature vector of the query image and the database images is done with Mean Square Error (MSE) method. Calculation of MSE is represented in equation 30.

$$MSE = \frac{1}{MN}\sum_{y=1}^{M}\sum_{x=1}^{N}\left[I(x,y) - I'(x,y)\right]^{2} \tag{30}$$

Where I and I' are the two images used for comparison using the MSE method.

Success Rate of the queries classified denotes the performance measure of the proposed technique. The equation is given in equation 31.

$$SuRate = \frac{No.\ of\ \ queries\ \ classified}{Total\ \ no.\ of\ \ queries\ \ considered\ \ for\ \ classification} \tag{31}$$

The comparison of average percentage success rate of the two classification methods on various categories of images considered in the two different databases is given in Fig.3 and Fig.4.

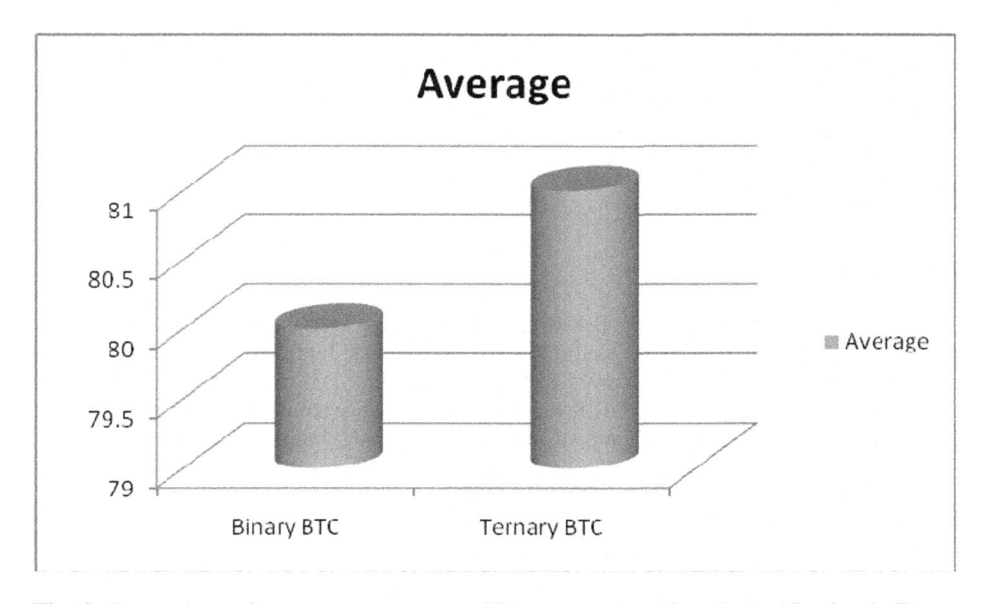

Fig. 3. Comparison of average success rate of binary to ternary based classification in Ponce Group 3D database

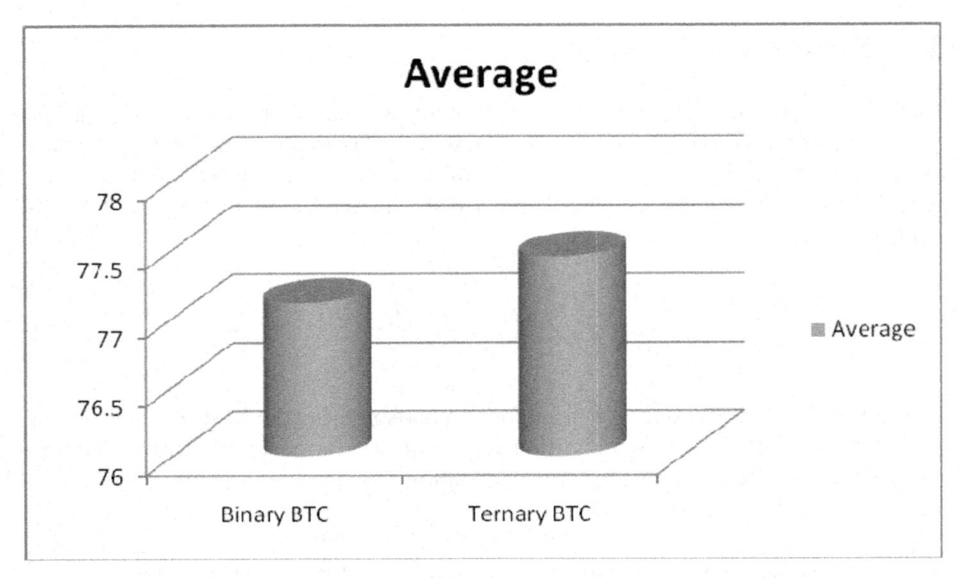

Fig. 4. Comparison of average success rate of binary to ternary based classification in generic database

6 Conclusion

A comparison of binary btc and ternary btc is done for image classification. The results establish the fact that ternary btc based classification yields better classification results compared to the binary one. The considerable improvement of results with ternary image map based image classification may extend considerable amount of support towards extensive application of image classification in diversified fields of knowledge and technology.

References

1. Kekre, H.B., Thepade, S., Das, R.K.K., Ghosh, S.: Image Classification using Block Truncation Coding with Assorted Color Spaces. International Journal of Computer Applications 44(6), 9–14 (2012) ISSN:0975-8887
2. Kekre, H.B., Thepade, S., Das, R.K.K., Ghosh, S.: Performance Boost of Block Truncation Coding based Image Classification using Bit Plane Slicing. International Journal of Computer Applications 47(15), 45–48 (2012) ISSN:0975-8887
3. Zakariya, S.M., Ali, R., Ahmad, N.: Combining Visual Features of an Image at Different Precision Value of Unsupervised Content Based Image Retrieval. In: Computational Intelligence and Computing Research, ICCIC (2011)
4. Maheshwari, M., Silakari, S., Motwani, M.: Image Clustering Using Color and Texture. In: 2009 First International Conference on Computational Intelligence Communication Systems and Networks, pp. 403–408 (2009)
5. Gonzalez, R.C., Woods, R.E.: Digital Image Processing, 3rd edn., pp. 407–413. Prentice Hall, Upper Saddle River (2008) ISBN 0-13-168728-X

6. Kekre, H.B., Thepade, S.D.: Rendering Futuristic Image Retrieval System. In: National Conference on Enhancements in Computer, Communication and Information Technology, EC2IT 2009, March 20-21, K.J. Somaiya College of Engineering, Vidyavihar, Mumbai-77 (2009)
7. Kekre, H.B., Thepade, S.D.: Using YUV Color Space to Hoist the Performance of Block Truncation Coding for Image Retrieval. In: IEEE International Advanced Computing Conference 2009 (IACC 2009), Thapar University, Patiala, India, March 6-7 (2009)
8. Kekre, H.B., Thepade, S.D., Athawale, A., Shah, A., Verlekar, P., Shirke, S.: Energy Compaction and Image Splitting for Image Retrieval using Kekre Transform over Row and Column Feature Vectors. International Journal of Computer Science and Network Security (IJCSNS) 10(1) (January 2010) ISSN: 1738-7906, http://www.IJCSNS.org
9. Kekre, H.B., Sarode, T.K., Thepade, S.D.: Image Retrieval by Kekre's Transform Applied on Each Row of Walsh Transformed VQ Codebook. In: ACM-International Conference and Workshop on Emerging Trends in Technology (ICWET 2010), Thakur College of Engg. and Tech., Mumbai, February 26-27 (2010) (invited); The paper is invited at ICWET 2010. Is uploaded on online ACM Portal
10. Kekre, H.B., Thepade, S.D.: Image Retrieval using Augmented Block Truncation Coding Techniques. In: ACM International Conference on Advances in Computing, Communication and Control (ICAC3 2009), Fr. Conceicao Rodrigous College of Engg., Mumbai, January 23-24, pp. 384–390 (2009); Is uploaded on online ACM Portal
11. Kekre, H.B., Sarode, T., Thepade, S.D.: Color-Texture Feature based Image Retrieval using DCT applied on Kekre's Median Codebook. International Journal on Imaging (IJI) 2(A09), 55–65 (2009) ISSN:0974-0627, http://www.ceser.res.in/iji.html
12. Kekre, H.B., Sarode, T., Thepade, S.D., Suryavanshi, V.: Improved Texture Feature Based Image Retrieval using Kekre's Fast Codebook Generation Algorithm. In: International Conference on Contours of Computing Technology (Thinkquest 2010), Babasaheb Gawde Institute of Technology, Mumbai, March 13-14. Springer (2011); The paper is available online on Springerlink
13. Kekre, H.B., Thepade, S.D., Athawale, A., Shah, A., Verlekar, P., Shirke, S.: Performance Evaluation of Image Retrieval using Energy Compaction and Image Tiling over DCT Row Mean and DCT Column Mean. In: International Conference on Contours of Computing Technology (Thinkquest 2010), Babasaheb Gawde Institute of Technology, Mumbai, March 13-14. Springer (2011); The paper is available online on Springerlink
14. Kekre, H.B., Thepade, S.D., Banura, V.K.: Performance Comparison of Texture Pattern based Image Retrieval using Haar Transform with Binary and Ternary Image Maps. In: IJCA Proceedings on International Conference and Workshop on Emerging Trends in Technology (ICWET), vol. (3), pp. 23–28. Foundation of Computer Science (2011)
15. http://www-cvr.ai.uiuc.edu/ponce_grp/data/ (last referred on September 25, 2012)
16. Image Database, http://wang.ist.psu.edu/docs/related/Image.orig (last referred on September 25, 2012)
17. Kekre, H.B., Thepade, S.D., Banura, V.K.: Performance Comparison of Image Retrieval Using Fractional Coefficients of Transformed Image Using DCT, Walsh, Haar and Kekre's Transform
18. Kekre, H.B., Thepade, S.D., Athawale, A., Shah, A., Verlekar, P., Shirke, S.: Kekre Transform over Row Mean, Column Mean and Both Using Image Tiling for Image Retrieval. IJCEE 2(6), 964–971 (2010) ISSN: 1793-8163

Low Cost Braille Embosser Using Speech Hypothesis

Neerja Sonawane, Amey Laddad, and Sanjay Gandhe

Department of E&TC, SITRC, Nasik, University of Pune
{ns24x7,amey.laddad,stgandhe}@gmail.com

Abstract. This paper, present an innovative method to create a Braille printer (Braille Embosser), which would be controlled by voice. By giving a voice command, the printer would provide the required embossed character. The technique would be most helpful for visually disabled people to perform their typing task. The main objective is to create a speech controlled printer using various electrical and mechanical domains such as digital signal processing, analog circuit design, and interfacing the electrical bobbins.

Keywords: Braille Embosser, Electrical bobbins, Signal processing, Interfacing Card.

1 Introduction

Since its invention in the early nineteenth century, Braille has remained vital to the literacy of people who are blind, and it continues to thrive despite the predictions of some to the contrary. Braille embossers, now more widely available, can produce reams of paper Braille. Because the existing technology makes it possible to produce Braille more easily, it is often used in cash-strapped education settings by people who are not necessarily knowledgeable about Braille itself. On the other hand, the work of knowledgeable transcribers, still extremely important, can be far more efficient with the use of this technology. Translation software and Braille embossers, combined with the ability to scan documents and the availability of electronic source files from publishers, has created the potential to greatly speed the transcription of Braille books [2]. Transcribers are now able to invest less time in entering text and more time in preparing the proper structure and format books that will be translated. Greater ease of Braille production correlates positively with a greater availability of Braille textbooks, even in higher education. Thus, the stage is set for quicker, cheaper Braille. The basic concept of the proposed system, "Low Cost Braille Embosser Using Speech Hypothesis", is that, we would be making mechanical electronic embosser which would put punch on paper kept below it with our voice commands. The voice inputs would be given through a mic. The inputs taken from the mic is given to the amplifier. The amplifier converts the weak signal to a stronger signal. A binary sequence of this command is sent to the PCs port. The DSP processor understands the meaning of the given inputs using sound hypothesis and then takes the appropriate decision regarding the recognition of the voice input. Subsequently, the recognized letter is given as input to an interfacing card, which is connected to the microprocessor with the help of

S. Unnikrishnan, S. Surve, and D. Bhoir (Eds.): ICAC3 2013, CCIS 361, pp. 510–524, 2013.

LPT/COM Port. Based on the word that matches the best the program sends a signal to the transistor arrangement on the interfacing card, which also comprises of a relay network [9] .This relay network excites the bobbin arrangement, which in turn perform the operation of punching the paper with a specific pattern, corresponding to the Braille equivalent of the input english alphabet.

It would be most helpful to visually disabled people in performing their typing task. Also it is reliable, easy to implement and user-friendly [6].

Though there exist embossers for blind at a very high cost, but none of them use speech to control a printer, that too at a minimal expense.

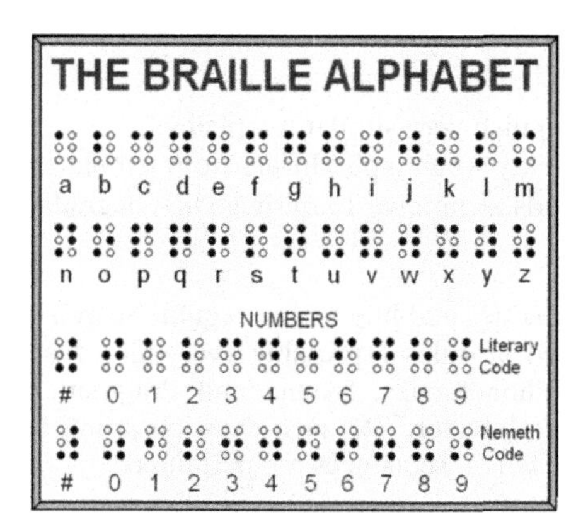

Fig. 1. Braille Character Chart

If the input is given, then it determines from the microprocessor and interface card what is to be done, which electrical bobbins are excited and the punch regarding it is embossed as per the Braille Character Chart is shown in Figure 1.

Braille embossers, transfer computer generated text into embossed Braille output. Braille translation programs convert text scanned in or generated via standard word processing programs into Braille, which can be printed on the embosser [7]. Talking and large-print word processors are software programs that use speech synthesizers to provide auditory feedback of what is typed. Large-print word processors allow the user to view everything in large text without added screen enlargement. Individuals with learning disabilities often use these special-featured word processors to assist them with their spelling and grammar and/or to provide the auditory feedback they require to be able to write. Scanners with optical character recognition can read printed material and store it electronically on computers, where it can be read using speech synthesis or printed using Braille translation software and Braille printers[5]. Such systems provide independent access to journals, syllabi, and homework assignments for students who are blind. Some hardware and software vendors also provide Braille or ASCII versions of their documentation to support computer users who are blind.

Various speech recognition techniques can be used for recognition of the spoken alphabet [1]. These are discussed below:

Following Technique are used for Automatic Speech Recognition Techniques:

1. Template matching
2. Knowledge-based (or rule-based) approach
3. Statistical approach: Noisy channel model + machine learning

1. Template-Based Approach
Stores examples of units (words, phonemes), then finds the example that most closely fits the input. Extract features from speech signal, then it's a complex similarity matching problem, using solutions developed for all sorts of applications ok for discrete utterances, and a single user

Disadvantages:

- Hard to distinguish very similar templates
- Degrades quickly when input differs from templates
- Therefore needs techniques to mitigate this degradation.

2. Rule-Based Approach
It uses knowledge of phonetics and linguistics to guide search process.

Templates are replaced by rules expressing everything that might help to decode: Phonetics, Phonology, Phonotactics, Syntax and Pragmatics. Typical approach is based on "blackboard" architecture. At each decision point, lay out the possibilities. Apply rules to determine which sequences are permitted.

Poor performance due to:

- Difficulty to express rules
- Difficulty to make rules interact
- Difficulty to know how to improve the system

3. Statistics-Based Approach
It can be seen as extension of template- based approach, using more powerful mathematical and statistical tools Collect a large corpus of transcribed speech recordings Train the computer to learn the correspondences ("machine learning")

At run time, apply statistical processes to search through the space of all possible solutions, and pick the statistically most likely one [8].

Machine learning is done using Acoustic and Lexical Models; it analyzes training data in terms of relevant features. It learns from large amount of data. Different possibilities different phonic sequences for a given word different combinations of elements of the speech signal for a given phone/phoneme. Combine these into a Hidden Markov Model expressing the probabilities [4].

The Noisy Channel Model:
It uses the acoustic model to give a set of likely phonic sequences. It uses the lexical and language models to judge which of these are likely to result in probable word sequences. The trick is having sophisticated algorithms to juggle the statistics. A bit like the rule-based approach except that it is all learned automatically from data.

The following factors are considered while designing automatic speech recognition system:

Robustness – Graceful degradation, not catastrophic failure
Portability – Independence of computing platform
Adaptability – To changing conditions (different mic, background noise, new speaker, new task domain, even new language also)
Out-of-Vocabulary (OOV) Words – Systems must have some method of detecting OOV words, and dealing with them in a sensible way.
Spontaneous Speech – disfluencies (filled pauses, false starts, hesitations, ungrammatical constructions etc) remain a problem.
Prosody –Stress, intonation, and rhythm convey important information for word recognition and the user's intentions (e.g., sarcasm, anger).

A. Algorithm used in the proposed system:

The algorithm used for speech recognition has a number of approaches out of which one has to choose wisely considering the advantages and disadvantages of every approach [3].

The flow chart for implementing the embosser which consists of pattern storage and pattern recognition is as shown in Figure.3.

2 Proposed System Architecture

Figure 2 shows the basic component of the proposed system architecture. The hardware consists of power supply unit the interfacing card and mechanical assembly of the Braille Embosser. A mic is used to speak out the alphabets. Processor includes DSP processor within it, which used to convert the word spoken by user into text form (English). The interfacing card provides an easy way of interfacing Braille printer to microprocessor. It provides terminals for the connections and conditions the signals for use in printer. Braille Printer is used to print the Braille words on sheet. It is operated by relays which are controlled by signals given by processor

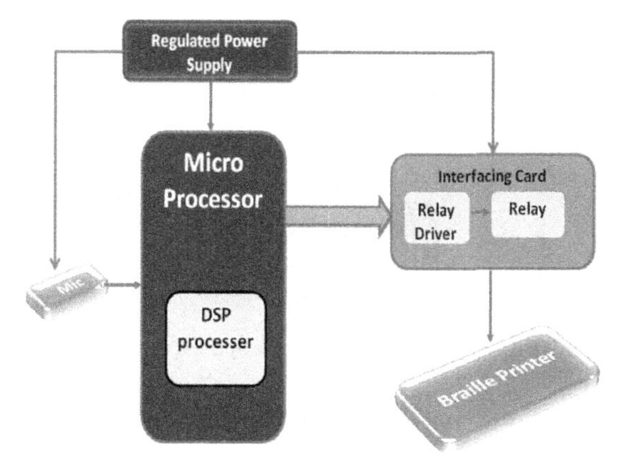

Fig. 2. Block Diagram of Braille Embosser Using Speech Hypothesis

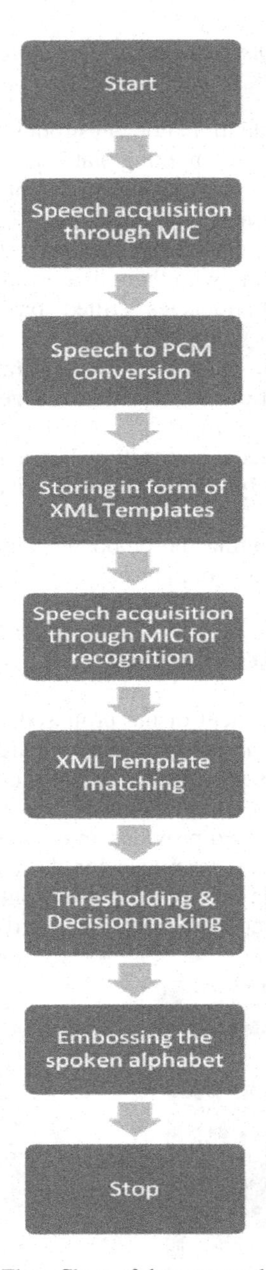

Fig. 3. Flow Chart of the proposed system

The Braille Embosser mainly consists of series of Bobbins
The printer would provide the required embossed character using an electro-mechanical arrangement comprising of bobbins. The bobbins would punch a hole through the paper kept below it as per the voice commands at the input.

- *Ding Dong Bell Mechanism*

Figure 4 shows bobbin mechanism in which, the solenoid piston consists of an iron core mounted to a non-magnetic metal bar. When there is no power to the electromagnet, a spring pushes the piston to the left, and the iron core extends outside of the wire coil. When you turn the electromagnet on (by pressing the doorbell button), the iron core is drawn to the magnetic field, so it slides into the center of the coiled wire.

A typical electric solenoid actuator is shown above. It consists of a coil, armature, spring, and stem. The coil is connected to an external current supply. The spring rests on the armature to force it downward. The armature moves vertically inside the coil.

When current flows through the coil, a magnetic field forms around the coil. The magnetic field attracts the armature toward the center of the coil. As the armature moves upward, the spring collapses and a hole is punched in the paper. When the circuit is opened and current stops flowing to the coil, the magnetic field collapses. This allows the spring to expand. A major advantage of solenoid actuators is their quick operation and solenoid actuators have only two positions i.e.fully open and fully closed.

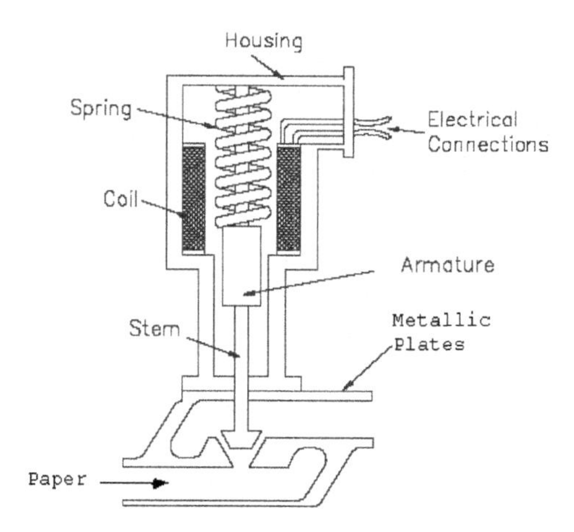

Fig. 4. Bobbin mechanism

3 Proposed System

A. Flow of the Proposed System

1. Login of the form is password protected. User has to speak out the password in order to get logged in. When the user speaks the password "NOTEPAD", the "text detection" form is loaded as shown in figure 5.

Fig. 5. Login Form (password 'Notepad')

2. The next part text detection. "RESET" button is used for the initialization. When we press "ENABLE SPEECH RECOGNITION" button, the recognition STARTS. The SPEAKER speaks the desired letter in the mic[10]. The program takes the voice input, converts it to text as shown in Figure 6.

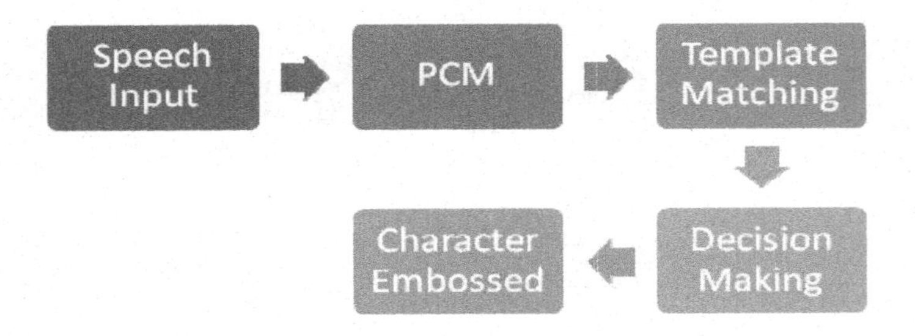

Fig. 6. Process flow of Speech-to-text and Text-to-Braille conversion

For example: If the user speaks the alphabet 'x', the program detects it and displays the detected letter onto the screen, as shown in Figure 7.

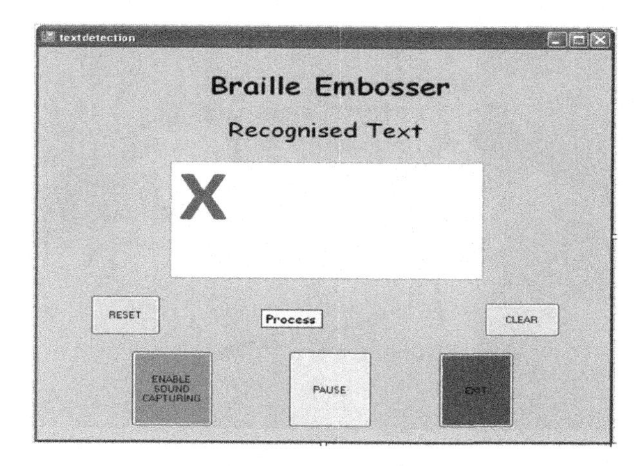

Fig. 7. Text detection form with detected text 'x'

3. The letter recognised is then coded to its Braille equivalent and the output is given to the LPT port.The LPT port controls the relay driver circuits as shown in Figure 8 and either energizes or de-energizes the relays according to the detected character.The energizes relays provide supply to the bobbins and gets activated.

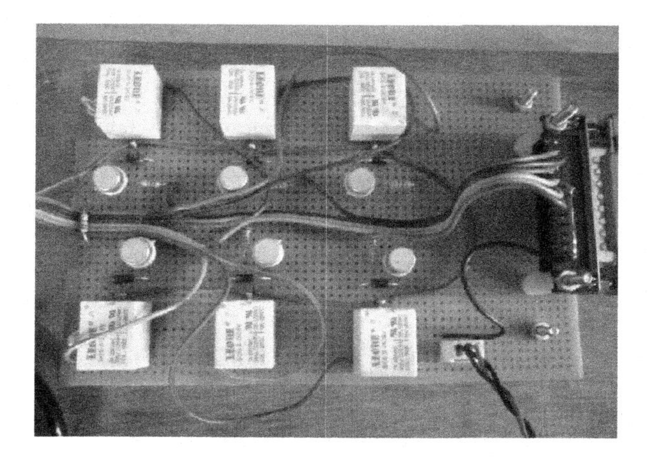

Fig. 8. Interfacing Card

4. The bobbins go down as per the detected character as shown in Figure 9.

Fig. 9. Bobbin arrangement on printer

The holes are punched in the paper as shown in Figure 10:

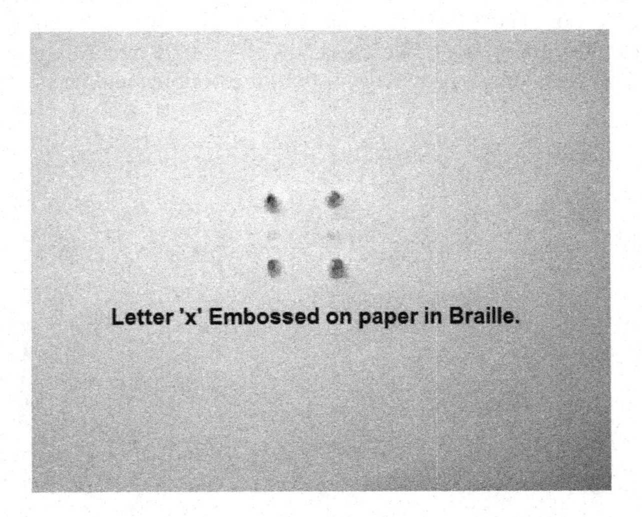

Fig. 10. Embossed letter 'x' on paper

4 Result and Analysis

The software used for the development of the proposed system is Visual Basics .Net along with Microsoft Speech SDK 5.1. The proposed system is implemented in Noiseless Environment and Noisy Environment and performance analysis is done in both the environment and results are compared.

A. Noiseless Environment:
This is the ideal environment for any speech processing based application. As the input signal is free from any interfering noise, there is no question of incorrect detection of the input signal by the speech-to-text conversion block or any mistake in the embossing of the characters.

Hence, considering the above case where the speaker inputs a letter 'b', the speech -to-text conversion system output is obtained as shown in Figure 11.

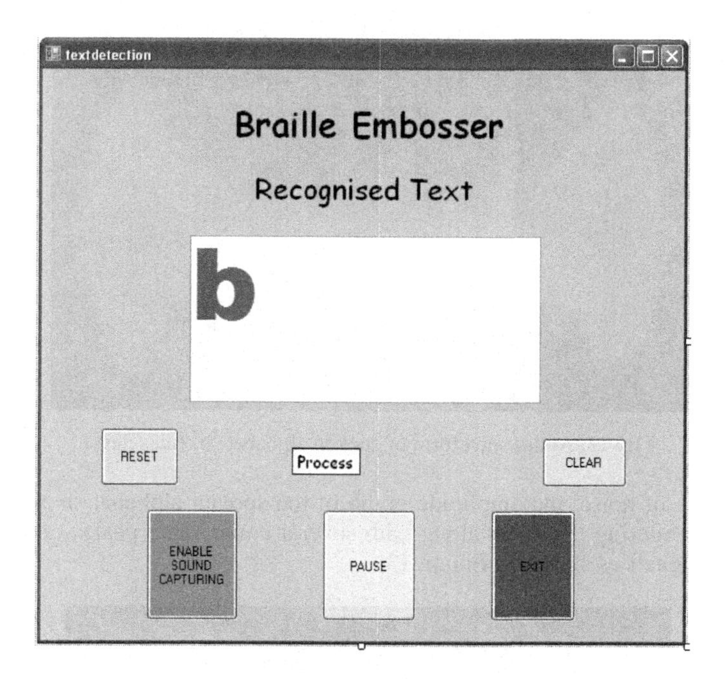

Fig. 11. Text detection form for detection of noiseless 'b'

The character 'b' thus detected then energizes a set of relays, which in turn trigger the appropriate bobbins to emboss the Braille equivalent of 'b' on the paper.

The performance improvement measures, which were actually meant for the noisy environment, can also be implemented in the noiseless environment to further enhance the performance. These are as follows:

- Training the speaker to ensure a clear and high pitch input signal.
- Use of a High Fidelity mic.

B. Noisy Environment:
In this Environment along with the speaker's voice signal, noise signal is also present at the input. The sources of these noises could be many people talking simultaneously in the same room, vehicular noises emerging in heavy traffic zone due to honking of horns and other mundane activities. These noises impact the proposed system adversely in the following ways.

As the punching operation is very sensitive to the input signal, even a small amount of noise at the input may cause random punching to output a embossed character altogether different than the spoken character. Thus the probability of faulty recognition increases, decreasing the efficiency of the system.

Consider a case where the speaker inputs a letter 'b' to the speech-to-text conversion system as shown below in Figure 12.

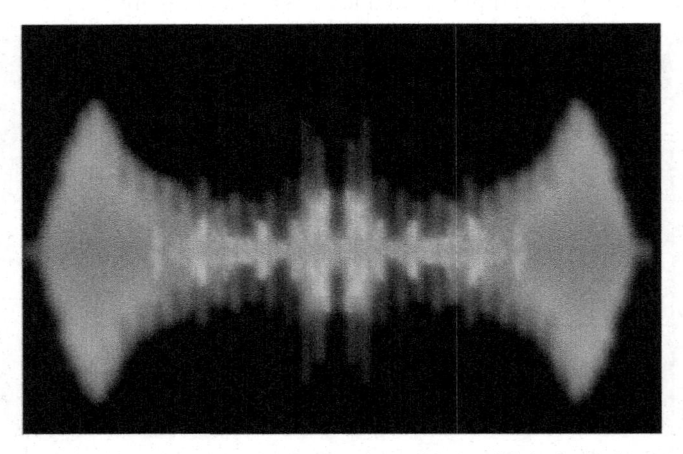

Fig. 12. Audio waveform of spoken alphabet 'b' (noiseless)

In presence of noise, the amplitude graph of the spoken alphabet shows a peak at the instant of uttering the letter along with several comparable peaks, corresponding to the noise signals as shown in Figure 13.

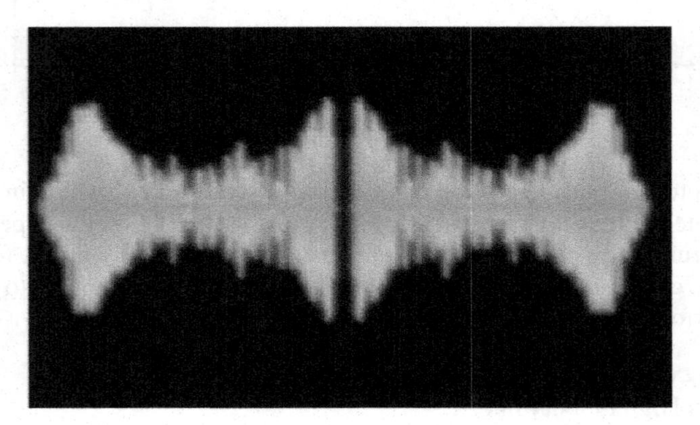

Fig. 13. Audio waveform of spoken alphabet 'b' (noisy)

Due to this, an altogether different alphabet will be displayed on the output screen as the output of the speech- to-text conversion block.

The faulty character detected then energizes a wrong set of relays, which in turn trigger the wrong bobbins to produce the incorrect character in Braille.

In order to improve the performance of the proposed system in presence of noise we have considered the issue while writing the program. In case of a random output the most closely matching character is displayed on the screen post speech-to-text conversion. However, this character does not energize the relay network, thus avoiding the punching of paper in case of a doubtful input signal.

Suppose the speaker utters the letter ‘b’ in the mic, amidst a noisy environment. The speech-to-text conversion software recognizes the noisy ‘b’ signal as a potential match to the ‘noiseless b’ signal. However, as the match could not be established clearly, the speech-to-text converted output will display the phrase, “b please” as shown in Figure 14.

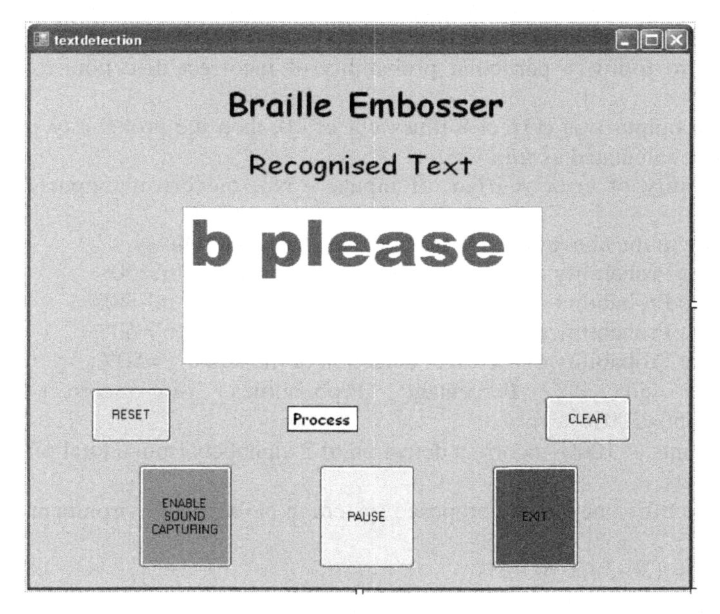

Fig. 14. Text detection form for detection of noisy ‘b’

Thus speaking in the non-technical sense, the speech processor requests the speaker to provide the ‘b’ signal in a less noisy environment, to ascertain the exact match for the signal, rather than the closest match. The uncertainty with regard to the ‘noisy b’ signal, avoids the embossing of the input letter on the paper. This mechanism ensures that the paper is punched only if the identity of the character is completely established, thus avoiding the embossing of incorrect character.

In order to enhance the performance in presence of noise, the following measures can also be taken:

- Training the speaker to ensure a clear and high pitch input signal.
- Use of a High Fidelity mic.

C. Performance Analysis:

Other than the external noise present in the environment, the performance of the proposed system also depends on the following errors:

- Pronunciation errors
- Dialect, Accent related errors
- Homophone related errors
- Errors due to Lack of clarity

D. Noiseless Environment:
The inputs were given several times, and the output was recorded. Though the probability of error is very small in the noiseless environment, still the above factors affect our proposed system adversely leading to the incorrect recognition of the alphabets: H, M, P and T.

There were 10 inputs given and the output was recorded. However it has been observed that the recognition of the above letters does not suffer a 100% error, rather they appear to follow a particular probability of incorrect detection, as mentioned below:

When the output was correct 8 times out of 10, then the probability of error was taken as 20% calculated as follows:

% probability of error = [(No. of inputs – No. of correct outputs)/No. of inputs]*100

According to the above formula the calculated error is follows:

Percentage Probability of incorrect detection of the letter 'h'=50%.

Percentage Probability of incorrect detection of the letter 'm'=40%.

Percentage Probability of incorrect detection of the letter 'p'=60%.

Percentage Probability of incorrect detection of the letter 't'=50%.

Sum of all the Percentage Probabilities of incorrect detection =50+40+60+50=200%

This amounts to 100% incorrect detection of 2 alphabets from a total of 26 possible input alphabets.

Hence, the efficiency of the proposed system in Noiseless Environment can be calculated as:

Efficiency= (26-2)/26=0.9231

Percentage Efficiency= Efficiency*100=92.31%

Thus the proposed system offers a high efficiency service in Noiseless Environment.

Noisy Environment:
It is observed that the impact of the factors mentioned at the start of the article, degrades the performance of the proposed system to a greater extent in the noisy environment as compared to the noiseless environment.

This can be deduced from the fact that the no. of alphabets detected incorrectly in the Noisy Environment rises to seven alphabets as compared to only two alphabets in the Noiseless Environment. These seven alphabets are:

B, H, I, M, P, T and Y.

In this case the Percentage Probability of incorrect detection for each of the seven letters is considered to be 100%, unlike the assumption of partial probabilities made in the Noiseless Environment.

This is attributed to the fact that the degradation in the performance is more in the Noisy Environment.

Hence, the efficiency of the proposed system in a Noisy Environment can be calculated as:

Efficiency= (26-7)/26=0.7308
Percentage Efficiency= Efficiency*100=73.08%
Thus the efficiency of the proposed system in the Noisy Environment degrades by 19.23% with regard to that of the Noiseless Environment.

5 Conclusion and Future Scope

The conclusion based on result and analysis can be summed up as below.

• The software works well in a quiet environment. When there is noise in the environment the necessary changes can be considered and incorporated in the software. The analysis shows that the efficiency of recognition in noiseless environment is 92.31% (ideal conditions), 84.62% (normal conditions) and in noisy environment 73.08%.
• We can remove the inaccuracies to some extent using the programming. These may include the changes in template stored and templates recognized. This will improve the efficiency by 19.23%.
• Language barrier is always there for the use of this technology because of the mismatch between the local accent and method used for speech recognition. It is complex and expensive to generate software for the local languages or dialects. We can implement a printer which accommodates efficient speech processing and increase versatility. Thus designing a multi-lingual, intelligent printer can be done in future.
• Tactile teaching models can be made as teaching aids for better understanding of diagrams, visual science and mathematical concepts. These are easily available now a day. We can raise the Braille printer by adding the tactile techniques and graphical tools. We can use the following technique and ease the printing task of books in Braille:

Scanning printed material to e-text. Proof reading the e-text. Then translation of e-text to Braille using the software can be done. Printing the translated text on the paper can also be done.

References

[1] Hifny, Y., Renals, S.: Speech Recognition Using Augmented Conditional Random Fields. IEEE Trans. on Audio, Speech, and Language Processing ASLP-17(2), 365 (2009)
[2] A system for converting print into Braille. IEEE Trans. Rehab. Eng. 5, 121–129 (1997)
[3] Huang, X., Acero, A., Hon, H.-W.: Spoken Language Processing: A Guide to Theory, Algorithm and System Development. Prentice-Hall, Upper Saddle River (2001)
[4] Bilmes, J.: What HMMs can do. IEICE Trans. Inf. Syst. E89-D(3), 869–891 (2006)
[5] American Association of Workers for the Blind and Association for the Education of the Visually Handicapped, English Braille, American Edition. American Printing House for the Blind, Louisville (1970)

[6] Sullivan, J.E.: Conversion of print format for Braille. In: Proc. 6th International Workshop on Computer Applications for the Visually Handicapped, pp. 1–14 (1990)
[7] Blenkhorn, P.: A system for converting Braille into print. IEEE Trans. Rehab. Eng. 3, 215–221 (1995)
[8] Jelinek, F.: Statistical Methods for Speech Recognition. MIT Press, Cambridge (1997)
[9] Gildea, R.A.J., Hubner, G., Werner, H. (eds.): Computerised Braille Production. In: Proceedings of the International Workshop in Munster (Germany). Rechenzentrum der Universität, Münster (March 1973/1974)
[10] Microsoft Corporation, Microsoft office 97/visual basic programmer's guide, Microsoft Corporation (1996)

ISSR: Intensity Slicing and Spatial Resolution Approaches for Moving Object Detection and Tracking under Litter Background

M.T. Gopala Krishna[1], M. Ravishankar[1], and D.R. Rameshbabu[2]

[1] Department of Information Science & Engineering,
Dayananda Sagar College of Engineering, Bangalore, India, 560078
[2] Department of Computer Science & Engineering,
Dayananda Sagar College of Engineering, Bangalore, India, 560078
{gopalmtm,ravishankarmcn,bobramysore}@gmail.com

Abstract. Moving object Detection in video sequences is one among the foremost indispensable challenges in Image and video processing. Its conjoint research areas are activity monitoring and video surveillance application. However, still beneath the biological process stage needs robust approaches once applied in an unconstrained environment. Several detection algorithms have higher performance under the static background, however decline results under background with fake motions. Detecting and Tracking of multiple moving objects in presence of Litter background like leaves movement of trees, water waves, fountain, window curtain movement and change of illumination in video sequences is a challenging problem. Because of these little movements within the background, it affects the performance of the automated tracking system. To overcome the above said problem, an approach consisting of Intensity Slicing and Spatial Resolution is considered to attenuate the results caused by the Litter Background. A modified 3-frame difference technique is employed to detect a moving object. Then, Adaptive Thresholding is used to segment the object from the background and to track the object. Results are compared with the existing well known traditional techniques. The proposed technique is tested on standard PETS datasets and our own collected video datasets. The experimental results prove the feasibility and usefulness of the proposed technique.

Keywords: Video surveillance, Spatial Resolution, Intensity Slicing, Object Detection, Object Tracking, Litter Background, Adaptive threshold.

1 Introduction

In recent years, intelligent object detection and tracking is indispensable challenges in intelligent video surveillance systems, which are widely used in many applications such as in security, robotics, authentication systems, traffic management, video communication/expression and user interfaces by gestures. Many imaging techniques [1, 2] are proposed for detection, tracking and identification of the moving objects.

S. Unnikrishnan, S. Surve, and D. Bhoir (Eds.): ICAC3 2013, CCIS 361, pp. 525–536, 2013.

Lipton et al [3] proposed a real time tracking system for extracting moving targets from a real-time video, classifying them into different categories and then robustly track them. Moving targets are detected using pixel wise frame difference between consecutive image frames. Sugandi et al [4] suggested that the tracking of moving objects by reducing spatial resolution of an image. In this, the low resolution image is being used to overcome the problem of illumination changes and small movement in the background. Stauffer et al [5] proposed a visual monitoring system that observes tracking objects and learn patterns of activity from those observations. In regard to the present work, many researchers have reported methods to solve the problem of object detection, object tracking and object identification. However, detection and tracking of the moving object is essential and to be further improved. As there are many obstacles such as a sudden or gradual change of illumination, temporal cluttered motion and small movement in the background may cause inaccurate detection. With all these, tracking has been a tough task in the presence of litter background and varying illumination condition. Generally, there are two approaches for motion detection: region based approach and boundary based approach. In region based approach, the general techniques applied are background subtraction, frame difference and optical flow. These methods are sensitive to illumination changes and small movement in the background such as leaves movement of the trees, window curtain movement, fountain, and sea tides. In case of boundary-based approach, many of the researchers have used edge based optical flow, level sets, and active contour. The result of these methods is inaccurate when the image is not smoothed sufficiently. An intelligent automated tracking system should able to detect changes caused by a new object, whereas non-stationary background regions such as leaves movement of the trees or a waving flag in the wind should be identified as a part of the background. To perform all these, the present system proposes a new method for detecting and tracking of the moving objects under Litter Background based on Intensity Slicing and Spatial Resolution. This new method can be applied to remove the scattering noise and the small movement in the background. In this, proposed system is modified the work carried by [4, 6], difference image frames are obtained from the Modified Three frames differencing techniques and then Adaptive Thresholding is performed instead of empirical Thresholding. The obtained thresholded image is ANDed followed by Morphological operations. The proposed method is tested on standard PETS datasets and our own collected video datasets. The results of the proposed method are better than the results obtained in [4, 6]. The present system is robust because low resolution image is insensitive to change of illumination and can remove small movements in the background.

2 Proposed Method

The proposed Moving object detection technique comprises two main steps. The primary step is preprocessing consisting of gray scaling, median filtering, Spatial Resolution & Intensity Slicing. In this stage smoothened input images are obtained as result. The second step is filtering, it is applied to remove the noise contained within the image and is performed by using morphological filter. To the obtained filtered image, connected component labeling is performed. The entire process of moving object detection is illustrated in Fig. 1.

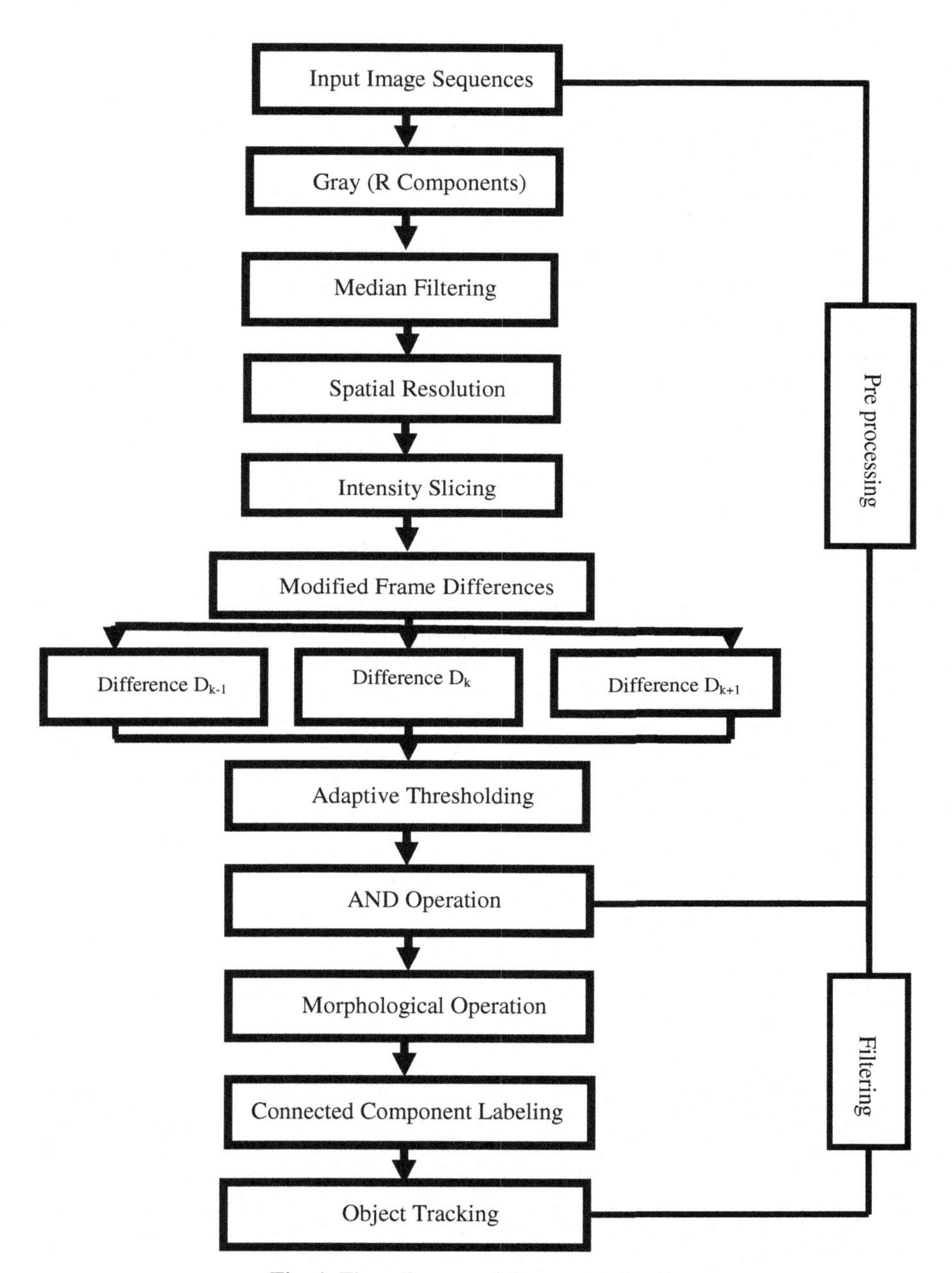

Fig. 1. Flow diagram of the proposed system

2.1 Preprocessing

In the first step of moving object detection process, a video image is captured using a video camera. In order to reduce the processing time, an image is converted into a grayscale image for further processing. Though many algorithms are based on gray images but they are not as good as the process which only uses the single component of the image (such as R, G or B component). In this article, the proposed system is selected R component of each frame to detect moving objects in order to shorten the runtime. Then, smoothing method is performed by employing a median filter with m×m pixels. The present work considers an input video image as a un-stationary background consisting of leaves movement of the trees, window curtain movement etc,. The un-stationary background usually considers a faux motion aside from the motion of the object of interest and may cause the failure in detecting a moving object. To overcome this drawback, the resolution of the image as in [4, 7] is reduced. In the present work, a low resolution image is obtained by reducing the Spatial Resolution of the image by averaging pixel values of its neighbors, including itself and by keeping the image size as 320x240. The low resolution images are obtained as 160x120, 80x60 & 40x30 pixels respectively and are shown in Fig 2.

The proposed work also used the concept of Intensity Slicing to detect the sensibly moving objects. Instead of considering all the 8 bit's information of an image, in this only 4 bit's (slices 4,5,6 & 7) most significant information is considered by ignoring the remaining 4 bit's (slices 0, 1, 2 & 3) information containing noise. While reducing the Intensity Resolution, the number of bits used to represent the image is reduced from k=7 to k=1 by using Eqn. (1) and Intensity Slicing images are shown in Fig. 3. So, in order to detect moving objects, the Spatial and Intensity reduced resolution image is employed to reduce the scattering noise and faux motion within the background due to non-stationary background. The noise owing to the non-stationary backgrounds have little motion region and it disappears in spatial and intensity reduced image.

$$IS(k) = \mathrm{mod}(floor(img / 2^{k-1}, 2)) \tag{1}$$

Where IS: Intensity Slicing, img: original image, k: plane number, mod: modulus operation (reminder).

(a) (b) (c) (d)

Fig. 2. (a) Original image (b), (c) & (d) Resulted low resolution images with the number of pixels 160× 120, 80× 60 & 40× 30 respectively.

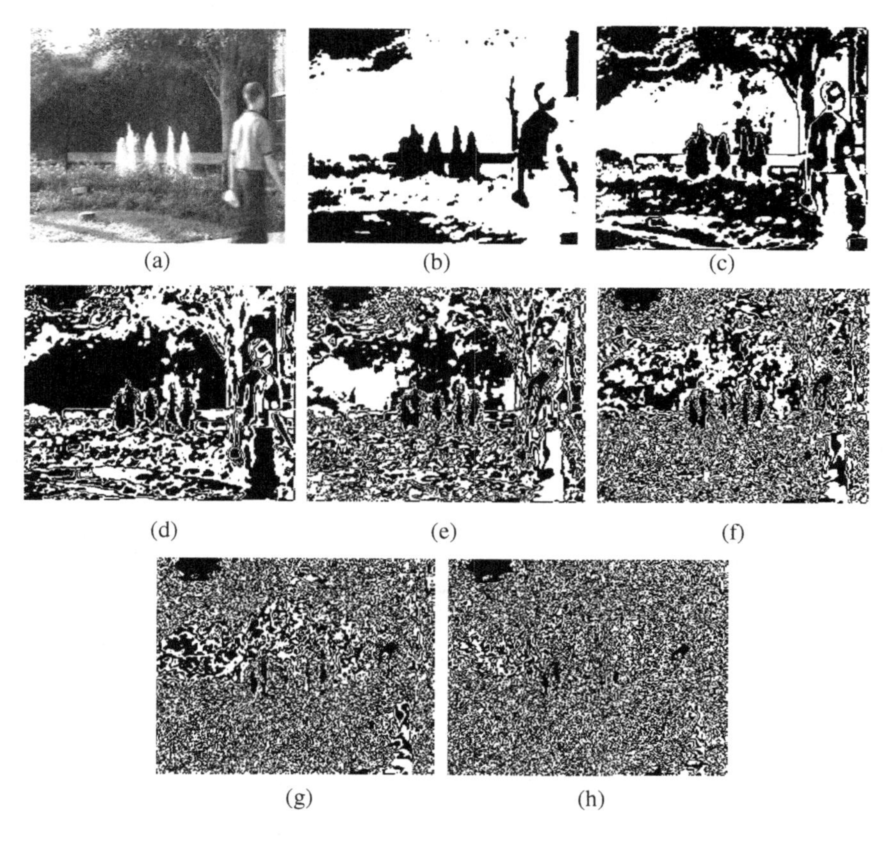

Fig. 3. (a) Original 8-bit image (b), (c), (d), (e), (f), (g) & (h) Resulted Intensity Resolution images respectively

2.1.1 Modified Frame Differencing for Moving Object Detection

As mentioned in [8, 9, 10, 11], Frame Difference computes the difference image between successive frame images. In this paper, the Modified Frame Difference method is proposed to detect the moving objects. The Modified frame difference method is performed on three successive frames [f_k, f_{k+1}, & f_{k-1} (k=1, 2,..)] between frame f_k and f_{k-1}, between frame f_k and f_{k+1} and between $f_{k-1}$and f_{k+1}. The resultant Modified frame difference is three different images (Difference images D_{K-1}, D_{K+1} & D_K [k=1,2..]) obtained as D_{K-1}, D_{K+1} & D_K of Eqns. (2), (3) and (4) respectively and is shown in Fig. 4 (a). Then an adaptive threshold technique is performed to segment foreground object from the background of a video image. By considering threshold value T on the difference images D_{K-1} , D_{K+1} & D_K as defined in Eqn. (8), which is shown in Fig. 4 (b). The process is further followed by applying AND operator on to the adaptive thresholded images as in Eqn. (9). The reason behind performing AND operation is to find out the similarity between the three difference images D_{K-1}, D_{K+1} and D_K respectively. As a result of this operation, the uncorrelated noise present in the three frames is removed and its result is shown in Fig 5. Finally, the resulted Modified Frame Difference image is further processed by applying Morphological filter to track the moving objects of a video sequences in the filtering stage.

$$D_{K-1} = \left| f_k - f_{k-1} \right| \tag{2}$$

$$D_{K+1} = \left| f_k - f_{k+1} \right| \tag{3}$$

$$D_K = \left| f_{k-1} - f_{k+1} \right| \tag{4}$$

$$D_{k'}(x,y) = \begin{cases} 1, if & D_{k'}(x,y) > T \\ 0, & otherwise \end{cases} \tag{5}$$

$$V = \frac{1}{r * c} \sum_{x=1}^{r} \sum_{y=1}^{c} D_{k'}(x,y) \tag{6}$$

$$\sigma = \sqrt{\frac{1}{r * c} \sum_{x=1}^{r} \sum_{y=1}^{c} (D_{k'}(x,y) - V)^2} \tag{7}$$

$$\mathrm{T} = 1.5 * \sigma \tag{8}$$

$$Mp = D_{k-1} \cap \mathrm{D}_{\mathrm{k}+1} \cap \mathrm{D}_{\mathrm{k}} \tag{9}$$

Where $D_{k'}(x,y)$ is the difference of pixel (x, y) between three successive frames, [r, c] is the size of frames, K'= k-1, k +1 and k, T=Adaptive Threshold and Mp=Motion Map.

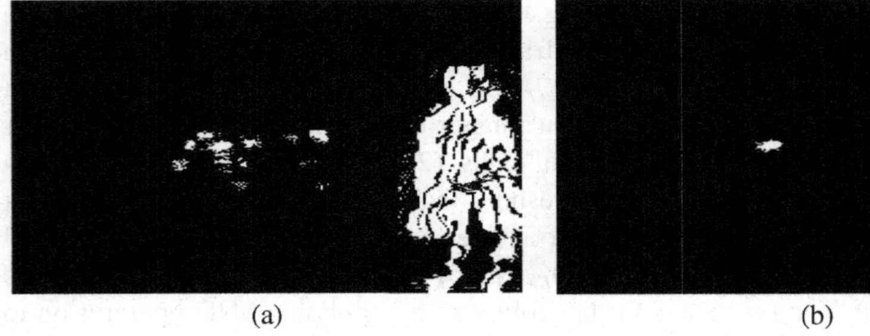

(a)

Fig. 4. (a) Resulted Modified Frame Difference image

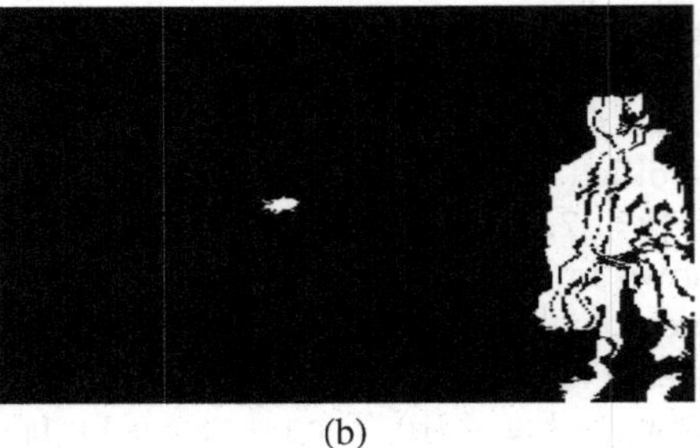

(b)

Fig. 4. (b) Resulted Adaptive Threshold image

Fig. 5. Resulted image of AND Operations

2.2 Filtering

In this stage, the resulted Modified Frame Difference image is further processed by employing Morphological filter to track the moving objects of a video sequences.

2.2.1 Moving Object Tracking

The detected foreground object is not clearly segmented from the background of an image, hence, Dilation and erosion are performed to clearly segment the foreground object from the background and is shown in Fig 6. Finally, Connected Component labeling is employed to track the moving objects. In this algorithm Connected Component Labeling scans the image pixel by pixel from top left to bottom right in order to identify connected pixel regions by comparing with the four neighbors.

Fig. 6. Resulted image of Morphological Erosion & Dilation operations

Fig. 7. Tracking of Moving Object

After obtaining the labeled moving objects, then the boundary of the moving objects are extracted using rectangular area. A rectangular area is created by searching each position's left & the right boundary (x_{min} and x_{max}) and also the top & the bottom boundary (y_{min} and y_{max}) of each moving object. The process of tracking of a moving object with a rectangular area is shown in Fig 7.

3 Experimental Results and Comparative Study

The proposed system is implemented in Pentium IV 1 GHz processor with MATLAB 10. The system is experimented on standard PETS Datasets and also on our own collection of video sequences with various low resolutions and different numbers of moving objects as shown in Fig 8, 9, & 10. These video sequences show different moving objects in an environment of outdoor and indoor in the presence of non stationary background such as leaves of tree, Fountain, water waves, window curtain movement, and change of illumination in the background. The proposed system is able to detect and track moving objects of indoor and outdoor environments of PETS datasets and also of our own collected video sequences efficiently by reducing the Spatial and Intensity resolution of an image. A comparative study is carried out between the proposed method and well known existing traditional approach as shown in Fig 8. The Figure shows the successive moving object's detection and tracking of a scene of PETS dataset, in that Column 2 shows the results obtained for traditional Frame Difference method, Column 3 shows the results obtained for the proposed method. The detected moving objects of the proposed method are comparatively better than existing traditional approach in the Litter Background. We also noticed that

Table 1. Percentage of Detection and False Alarm Rate

Input Sequences	Approaches	Detection Rate	False Alarm Rate
Fountain-1	Traditional Frame Difference Method	88.89%	36.00%
	ISSR (**Proposed)**	**90.91%**	**28.57%**
Water Surface	Traditional Frame Difference Method	75.00%	45.45%
	ISSR (**Proposed)**	**97.50%**	**15.22%**
Curtains	Traditional Frame Difference Method	81.08%	55.22%
	ISSR (**Proposed)**	**93.18%**	**46.75%**
Moving Trees	Traditional Frame Difference Method	86.96%	42.86%
	ISSR**(Proposed)**	**85.71%**	**23.08%**
Light watch	Traditional Frame Difference Method	66.67%	66.67%
	ISSR**(Proposed)**	**84.38%**	**27.03%**
Fountain-2	Traditional Frame Difference Method	66.67%	50.00%
	ISSR**(Proposed)**	**90.91%**	**20.00%**
Glass Shadows	Traditional Frame Difference Method	78.95%	54.55%
	ISSR**(Proposed)**	**92.31%**	**07.69%**

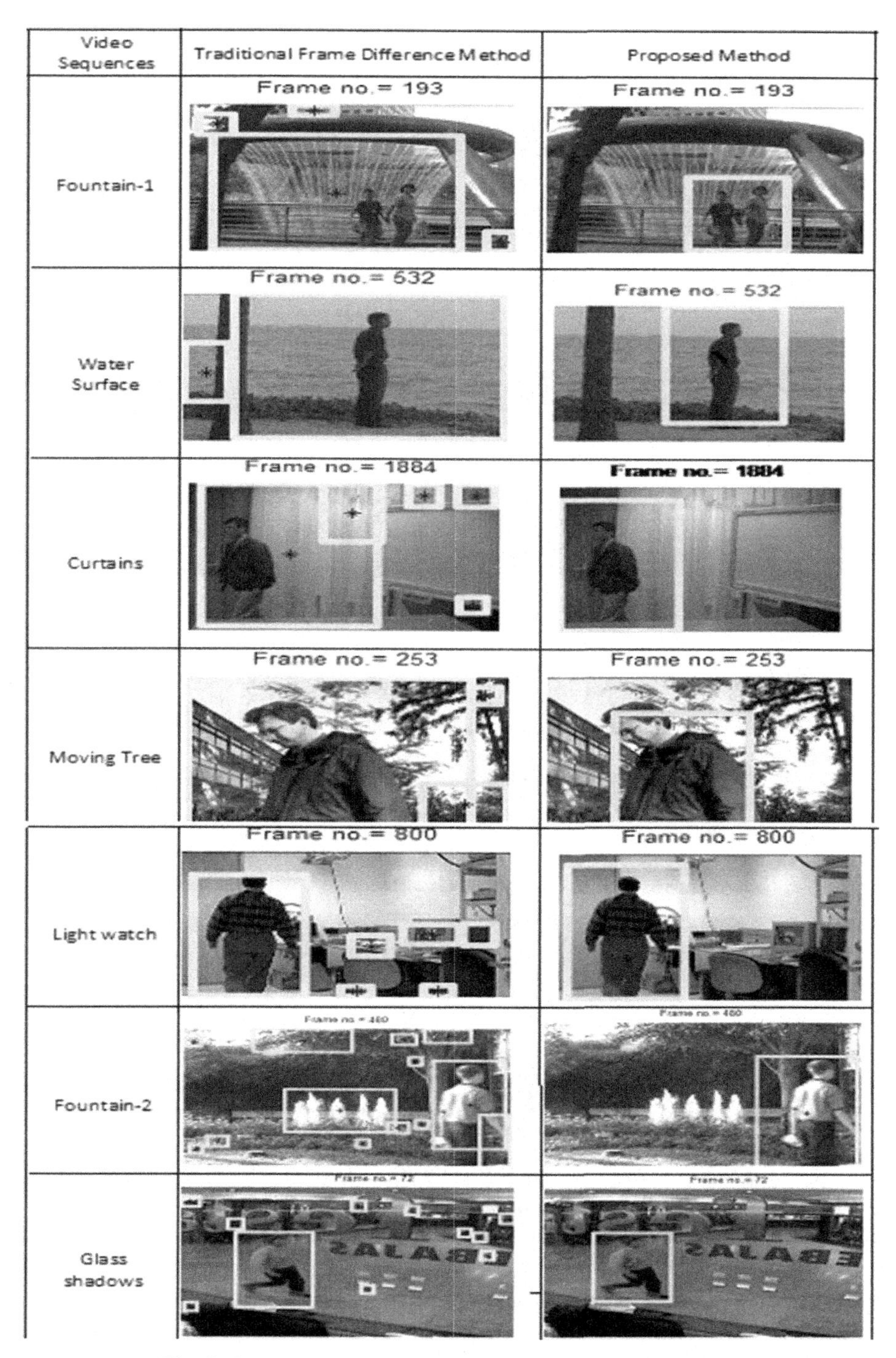

Fig. 8. Comparative Experimental Results of PETS datasets

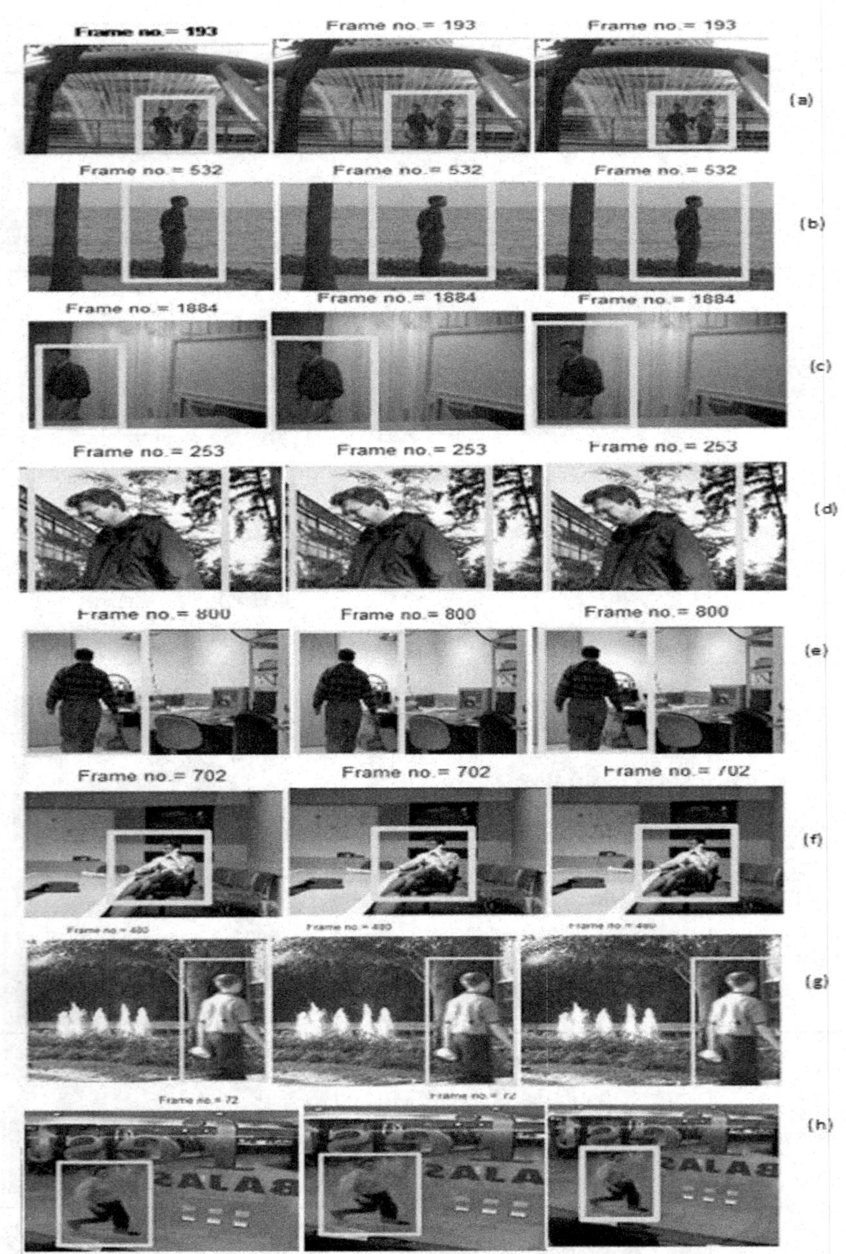

(a), (b), (c), (d), (e), (f), (g) & (h) of Reduced Spatial Resolution to 1/2,1/4/&1/8 pixels and each pixel is represented with 4 bits.

Fig. 9. Tracking of object in an input video of PETS dataset in the presence of un-stationary background such as Fountain, water waves, ,Window Curtain , leaves of Tree, Light Variations, and Glass shodow respectively

(a), (b),(c) &(d) of Reduced spatial resolution to 1/2,1/4/&1/8 pixels and each pixel is represented with 4 bit.

Fig. 10. Tracking of object in an input video of our own collected dataset in the presence of unstationary background

the false alarm regions obtained from existing approach are high compared to the proposed method. Table1 shows the Detection rate and false alarm rate of the existing traditional algorithms along with the proposed algorithm. From the table, it is clear that the percentage of false alarms obtained for the proposed algorithm is considerably decreased compared to existing traditional approaches. The proposed system successfully detects and tracks moving object by removing the noise when the resolution factor decreases by 1/2, 1/4 and 1/8 as shown in Fig. 9 (rows (a) to (h)) and Fig.10 (rows (a) to (d)).

4 Conclusions

As an alternative to the existing traditional approach, the Modified Frame Differencing is proposed to detect and track the moving object with an elimination of scattered noise and faux motions in the background by employing Spatial Resolution and Intensity Slicing Approaches. The Proposed Method is compared with well known existing traditional approach and the comparison result shows that the proposed method performs better than conventional method. The output of tracking the moving object has been improved with a complete removal of the unrelated noise such as a small movement in the background and only extracting the moving objects. Furthermore, the performance of the moving object detection needs more improvement in specific considerations such as occlusion between objects, which will be considered in future work.

References

1. Collins, R., Lipton, A., Kanade, T., Fujiyoshi, H., Duggins, D., Tsin, Y., Tolliver, D., Enomoto, N., Hasegawa: System for video surveillance and monitoring, Technical Report CMU-RI-TR-00-12, Robotics Institute, Carnegie Mellon University (2000)
2. Yilmaz, A., Javed, O., Shah, M.: Object tracking: a survey. ACM Computing Survey 38(13), 1–45 (2006)
3. Lipton, A.J., Fujiyoshi, H., Patil, R.S.: Moving Target Classification and Tracking from Real-Time Video. In: Proc. Fourth IEEE Workshop on Applications of Computer Vision, pp. 8–14 (1998)
4. Sugandi, B., Kim, H.S., Tan, J.K., Ishikawa, S.: Tracking of Moving Object by using Low Resolution Image. In: Proc. International Conference on Innovative Computive, Information and Control, p. 408 (2007)
5. Stauffer, C., Grimson, W.E.L.: Learning Patterns of Activity Using Real-Time Tracking. IEEE Trans. Pattern Analysis and Machine Intelligence, 747–757 (2000)
6. Panda, D.K., Meher, S.: Robust Real-Time Object Tracking Under Background Clutter. In: 2011 International Conference on Image Information Processing, ICIIP 2011 (2011)
7. Gonzalez, R.C., Woods, R.E.: Digital Image Processing. Addison-Wesley Longman Publishing Co., Inc., Boston (2001)
8. Shahbe, M.D., Salih, Q.A.: Image Subtraction for Real Time Moving Object Extraction. In: Proc. International Conference on Computer Graphics, Imaging and Visualization (CGIV 2004), pp. 41–45 (2004)
9. Liu, Y., Al, H., Xu, G.: Moving object detection and tracking Based on Background Subtraction. In: Proc. Society of Photo-Optical Instrumentation Engineers (SPIE), vol. 4554, pp. 62–66 (2001)
10. Prabakhar, B., Kadaba, D.V.: Automatic Detection and Matching of Moving Object. CRL Technical Journal 3(3), 32–37 (2001)
11. Koay, S.Y., Ramli, A.R., Lew, Y.P., Prakash, V., Ali, R.: A Motion Region Estimation Technique for Web Camera Application. In: Student Conference on Research and Development Proceedings, pp. 352–355 (2002)

Novel Algorithm for Image Processing Based Hand Gesture Recognition and Its Application in Security

N. Dhruva[1], Sudhir Rao Rupanagudi[1], and H.N. Neelkant Kashyap[2]

[1] WorldServe Education, Bangalore, India
sudhir@worldserve.in
[2] National Institute of Technology (NIT), Trichy, India
neel.kash91@gmail.com

Abstract. The concept of hand gesture recognition has been widely used in communication, artificial intelligence and robotics. It is a staple method of interaction especially for the deaf and the blind. Techniques for recognizing hand gestures are in great demand. Many algorithms have been discovered for this purpose, each of them having their own advantages and disadvantages. In this paper we present a novel algorithm for hand recognition using image processing and explore its application in security based systems. The algorithm was tested for different gestures on over 50 samples and an accuracy of 95.2% was achieved. In terms of speed, the algorithm is 1.5 times faster than its contemporaries. An alternative to finding the centroid based on a less computationally complex algorithm has also been explored. The algorithm was implemented in MATLAB programming language on MATLAB 7.13 build R2011b and SIMULINK.

Keywords: American Sign Language, bounding box, centroid calculation, finger detection, hand gesture recognition, image processing, security system, segmentation.

1 Introduction

Nowadays, most of the security systems rely on biometric techniques. Biometrics is one of the most inviolable tools ever used, since it involves parts of the human body for authentication [1], [2]. It can neither be forged nor stolen. Each biometric technique is related to a distinct part of the human body and each individual is uniquely identified by measuring their physical characteristics or behavioral traits. The biometrics generally used are fingerprint analysis, iris and retina scan, voice analysis and signature authorization [3]. Furthermore, the above mentioned approaches are advantageous and compliant in terms of cost when put to use in situations which demand complex high level security [4]. However, there are possibilities that the mentioned techniques fail sometimes [5]. For example, one cannot use finger recognition if one's fingerprints are disfigured. Voice and signature can also be duplicated and forged. Also the algorithms used for these methods are not simple [6], [7], [8]. With more computations required, these methodologies would therefore require more

S. Unnikrishnan, S. Surve, and D. Bhoir (Eds.): ICAC3 2013, CCIS 361, pp. 537–547, 2013.

computational capacity, memory and power. Thus using these methods for simple security purposes is not yielding and is disadvantageous.

In this paper we introduce a novel algorithm for implementing a security system using American Sign Language. For advanced security, any other type of hand gestures can also be used, including those invented by the user himself. One of the chief advantages of this method is its high accuracy, low cost and simplicity in implementation. The system is also highly secure and less vulnerable to misuse not only due to its dependency on the individual's hand characteristics but also due to the unique gestures signaled by the person. Also, because hand gestures are being used, it is highly useful for the visually impaired since no visualization of letters or numbers is involved as in the case of keypad based security systems.

A few existing algorithms involved in hand gesture recognition have been elaborated in Section 2. The experimental setup for the novel sign language based security system has been presented in Section 3. Section 4 describes the proposed algorithm and its implementation. The results for the same have been tabulated in Section 5.

2 Existing Methodologies

The general technique for hand gesture recognition involves a sequence of image processing steps [9]. The first major step is image segmentation where the hand region is identified from the image. The Otsu's Algorithm is the most famous algorithm used for segmentation [10]. An optimum value of threshold is determined to segment different regions in the image. Consider a gray scale image with gray levels from 0 to 255. If the image has to be segmented into N levels, N categories are defined, each spanning across a few gray values between predefined thresholds. These thresholds distinguish different regions in the image. The probability and the mean value of each category are found. The optimum threshold value is determined by using the method of 'maximal between class' variance methodology [11].

The second step is to find the centroid of the hand for finger identification. The existing methodology [12] for centroid calculation involves calculating the image moment M_{ij} as represented in (1).

$$M_{ij} = \sum_i \sum_j x^i y^j I(x, y) \quad (1)$$

where I(x,y) is the intensity of the binary image. The coordinates of the centroid (X,Y) is determined using (2) and (3).

$$X = \frac{M_{10}}{M_{00}} \quad (2)$$

$$Y = \frac{M_{10}}{M_0} \quad (3)$$

It can be seen that both the Otsu's method and the centroid calculating algorithm involve several computationally complex steps which impede the fast execution of the hand gesture recognition algorithm.

In Section 4 a simple and efficient method to segment and find the centroid for a particular hand is presented.

3 Experimental Setup

The experimental setup for the proposed hand recognition algorithm consists of a thermocol box with a single face open as shown in Fig. 1. Image acquisition is made possible with the help of a simple web camera, specifications of which are given below:

- Image sensor: 1/7" CMOS sensor.
- Image resolution: 16.0 Mega pixels interpolated.
- Image size: 640x480 pixels.
- Frame rate: Up to 30 frames per second.
- Power consumption: 160 mW.

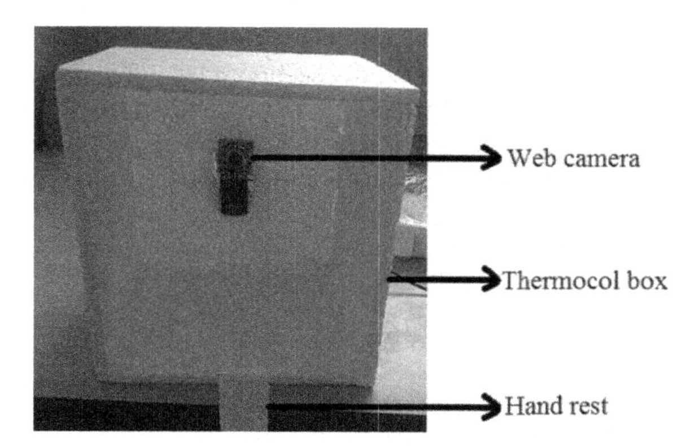

Fig. 1. Thermocol box used to conduct the experiment

Provision is made in the thermocol box to fit the web camera which captures the open face of the thermocol box. Thus the camera is directly in front of the hand when the gesture is shown. A modification to the existing setup can also be performed, by encasing the hand in turn keeping the hand gestures confidential.

The working principle of the setup is simple. The user has to signal the password using hand gestures. These gestures are recognized as numbers using an algorithm and cross checked with a predefined password. The camera is interfaced with SIMULINK on which the algorithm runs. The input to the algorithm is a sequence of images (video) from the camera. The authentication and verification status of the password is also displayed on SIMULINK.

There are three primary advantages of using this setup. Firstly, since the box is closed on four sides, it minimizes the effect of noise in the image acquired due to environmental changes and changes like external light intensity as well. Secondly, it protects the web camera from damage due to external factors. The third most important factor is that it is a highly cost-effective solution.

As seen in the diagram there is an optional hand rest once again made of thermocol, on which the wrist could be placed and the gestures could be shown. This helps to keep the hand stationary.

4 Novel Algorithm for Gesture Recognition

The complete algorithm for hand gesture recognition and number interpretation can be divided into the steps mentioned below. The hand gestures were based on American Sign Language which consists of 150 symbols represented using hand gestures. It is a highly popular form of sign language around the world and hence chosen for the experiments [13].

4.1 Color Space Conversion

The video input from the web camera is treated as a sequence of images at 30 frames per second. Each image is then subjected to color conversion in order to assist in segmentation, which is explained later on in this section. Generally, in color conversion the RGB values of each pixel is converted to a YCbCr color space where Y is the intensity value, Cr is the Chrominance red, Cb is the Chrominance blue component [14].

On experimenting with 50 images it was found that the Cr value of each pixel was sufficient to uniquely represent the pixels which belong to the hand as opposed to those which belong to the background. The equation for color conversion to Cr values has been indicated by (4).

$$Cr = 0.5 * R - 0.418688 * G - 0.081312 * B \tag{4}$$

One of the most important advantages of color conversion to Cr is that, instead of utilizing 3 bytes of memory for storing the RGB values of a color pixel, only 1 byte of memory is sufficient after color conversion. This assists in reducing memory usage for a hardware implementation and improvises speed for algorithm execution.

4.2 Image Segmentation

In order to recognize the hand gesture, the hand must be extracted from the background. This process is referred to as image segmentation. In this implementation we use threshold based segmentation [15]. In this method a threshold value θ is selected based on experimentation. Pixel values less than θ are marked as background, which is represented by '0' and values greater than the threshold are marked as foreground (hand), which is represented as '1'. This is shown in (5).

$$img(i,j) = \begin{cases} 0; \text{ if } Cr(i,j) < \theta \\ 1; \text{ if } Cr(i,j) > \theta \end{cases} \tag{5}$$

where 'img' is the binary image obtained from the 'Cr' image, through (5). 'θ' is the selected threshold value. 'i' and 'j' are the indices representing the coordinates of the row and the column of the present pixel respectively, where $i \in (1,640)$ and $j \in (1,480)$.

Several experiments were conducted to obtain the threshold value. It was found that a θ value of 137 could segregate the hand satisfactorily from the background. Those pixel values which are greater than 137 belong to the foreground (hand). The others belong to the background. As represented in (4), pixel values are now reduced to a single bit – 1 for white and 0 for black. This in turn reduces the memory occupancy further and improvises the efficiency for the further stages.

It must be noted that the threshold value of Cr may change depending on the lighting conditions of the surroundings. But since the setup is used in a fixed onetime installation, this condition does not impede the working of the algorithm.

4.3 Obtaining the Reference Point

The first step in gesture recognition is determining a reference point of the hand. Let this reference point be denoted as ρ. It is necessary to find ρ of the hand because this would serve as a frame of reference to detect the position and status of all the fingers of the hand. In other words, even if the hand moves horizontally, ρ also shifts accordingly. Thus ρ remains at a fixed position with respect to the hand irrespective of the position of the hand. This makes it easy to locate the position of the fingers with respect to this reference point.

The first step in calculating ρ is to find the exact location of the hand. In order to perform this, the binary image obtained above containing the hand is scanned from right to left across its width, starting from the bottom right hand corner. This technique of scanning is followed based on the assumption that the base of the hand always lies at the bottom of the image. As mentioned previously, since the background is black and the hand is white, the very first black to white transition represents the starting point of the hand.

The second step in the reference point calculation is based on the assumption that ρ always lies δ row above the starting point obtained. Through experiments conducted on different hand images, δ was found to lie within the range of 50 to 60 rows above. Scanning is now performed from right to left on this δ row.

The extremum points of the hand are determined based on transition in color during the scanning of this row. An initial transition from black to white represents the right extremum point (A) of the hand. Similarly the left extremum point (B) is obtained on a pixel transition from white to black. This is shown in Fig. 2. Since abnormalities in the hand image exist due to noise and other defects, the intermediate white to black transitions are ignored.

The exact location of ρ is considered to be the arithmetic mean of A and B as represented by (6). Let the coordinates of point A and B be (x, y) and (x', y') respectively. Then the coordinates of the reference point ρ is (X, Y), where

$$X = \frac{(x + x')}{2} \tag{6}$$

$$Y = \frac{(y + y')}{2} \tag{7}$$

4.4 Finger Identification

Once ρ is found out, it becomes easy to identify the tentative positions of the fingers .The fingers are spotted using a bounding box technique. Relative positions of the bounding boxes with respect to the reference point are calculated and are plotted - each box enclosing approximately the fingertips of each finger as shown in Fig. 2. The dimension of each bounding box was 65x100 pixels. But this can be varied relative to the size of the hand. The position of the bounding boxes for each finger with respect to ρ was determined by experimental observations on 50 images. The position of the bounding boxes does not change because the hand size and the distance of the hand from the camera remains the same.

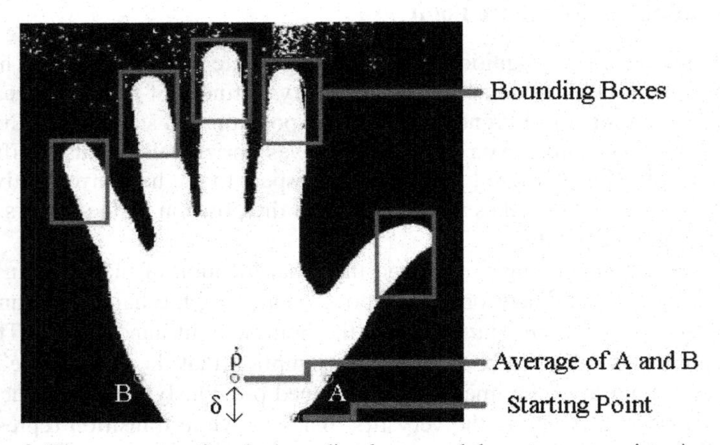

Fig. 2. Figure representing the bounding boxes and the extremum points A and B

In order to find if a finger is open, the approximate number of white pixels in each bounding box is counted. If the total number of white pixels (N_{white}) in each box is greater than a threshold θ_N, then the fingertip is said to be present. In other words the finger is open. In case the number of pixels falls less than θ_N, the finger is pronounced closed. This is expressed in (8).

$$Status\ of\ the\ Finger = \begin{cases} Open;\ if\ N_{white} > \theta_N \\ Closed\ ;\ if\ N_{white} < \theta_N \end{cases} \tag{8}$$

where N_{white} is the total number of white pixels in a bounding box. Thus by finding the number of fingers open, the number associated with the hand gesture is determined.

The ASL chart shown in Fig. 3 is used for this purpose. The number obtained is passed on to the password checking block.

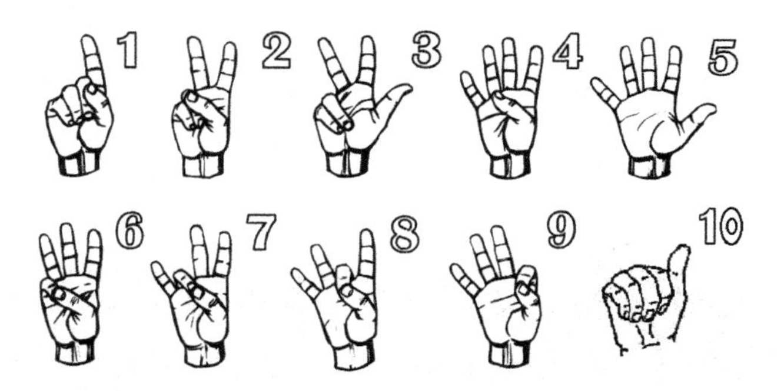

Fig. 3. American Sign Language Chart - **Image Source:** www.lifeprint.com

4.5 Password Checking Block

At the time of deployment, a password is initially stored before making use of the setup. Each time the system is accessed, the user must first signal a delimiter to show that he/she is ready to input the password using hand gestures. The delimiter used can also be made unique for each user thus increasing the security level of the system. Each number in the signaled password is crosschecked with the predefined password on the fly.

If the signaled password is authenticated correctly, the user is intimated and granted access. A maximum number of attempts and the intimation methods can be customized depending on the security level required.

5 Results

The following are the results of the experiments that were conducted to test the validity of the algorithm. The algorithm was designed, developed and executed in SIMULINK. Each block in the SIMULINK model performs a specific task mentioned in the algorithm.

Fig. 4a shows the image as acquired by the camera, consisting of all the ten hand gestures that are to be recognized using this algorithm. This image is color converted to its corresponding Cr values as represented in Fig. 4b. The next step in the algorithm is image segmentation as was explained in detail in Section 3. The binary image obtained after segmentation of the Cr image is shown in Fig. 4c.

From Fig. 4c it can be seen that the hand has been extracted perfectly from the background and is seen distinctly. From this step onwards only this binary image is subjected to further image processing.

Each of these gestures was subjected to the reference point identification and finger identification algorithms. The resultant image consists of bounding boxes confining the finger regions, as depicted in Fig. 5. Based on the number of white pixels in these boxes, it is adjudged whether the finger is open or closed.

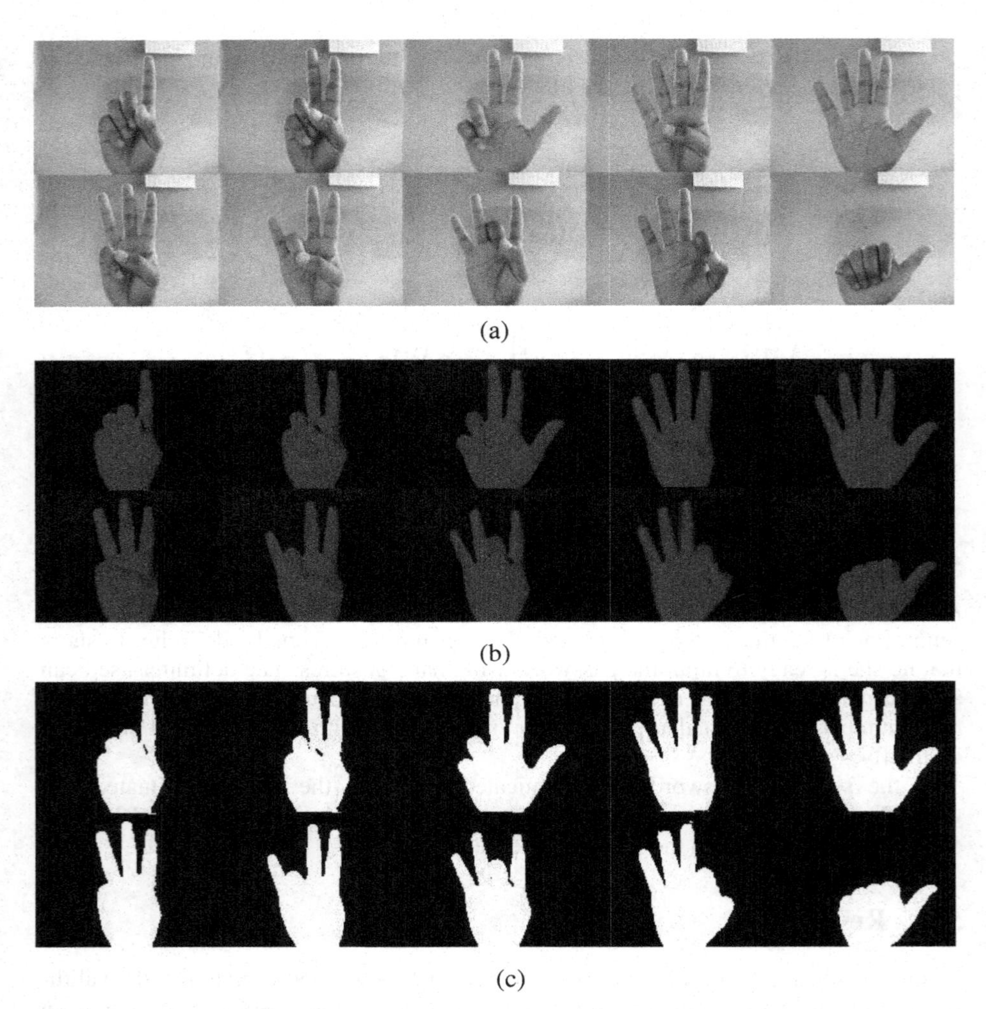

Fig. 4. Image of the hand after various stages of the algorithm (a) Color image as captured by web camera (b) Color converted image (c) Binary image after segmentation

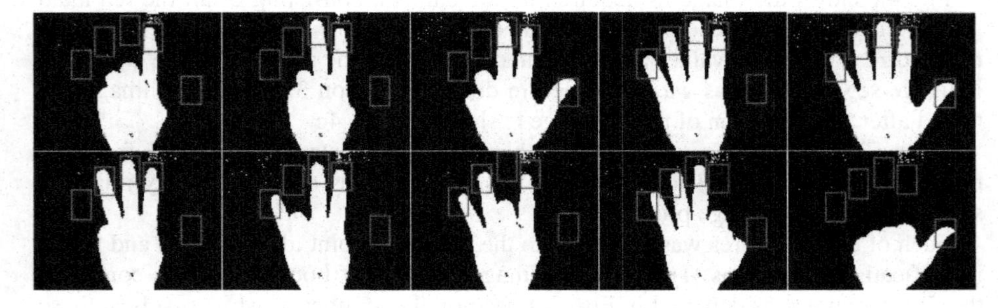

Fig. 5. Final binary image with the bounding boxes

5.1 Efficiency Comparison

Efficiency comparison has been performed by determining the time taken to execute the aforementioned algorithm and comparing it with the time taken by popular general algorithms for hand gesture recognition.

The time taken to execute the image segmentation section of the algorithm in MATLAB was compared with the Otsu's thresholding algorithm which is the most popular method of image segmentation [11]. Furthermore the execution time for finding the centroid was also compared with the technique - Method of Moments as used in [12]. In order to obtain accurate efficiency comparisons, the same image of the hand was fed as input to both the algorithms. The results have been tabulated in Table.

Table 1. A Comparison of the Execution Time For The Proposed Algorithm And General Existing Algorithms

	Our Algorithm	General Algorithm
Color Conversion and Segmentation	377ms	539ms [11]
Finding the reference point ρ	2.5ms	14ms [12]
Total time of execution	379.5ms	553ms

It can be clearly seen from Table 1 that the novel algorithm described takes only 68.62% of the total time taken by the general algorithm. So it is appreciably faster than the general techniques being used for gesture recognition.

5.2 Accuracy Chart

Each gesture was tested with 50 trials. The accuracy rates have been tabulated in Table 2.

Table 2. Success Rates Of Identification Of Hand Gestures Using The Proposed Algorithm

Gesture	Hits	Success Percentage
1	46	92%
2	49	98%
3	48	96%
4	48	96%
5	49	98%
6	47	94%
7	43	86%
8	44	88%
9	43	86%
10	49	98%

Thus, it can be seen from Table 2 that the gross success percentage is 95.2%. In other words we can conclude that the accuracy achieved is significantly high. It can be observed that the successful hits achieved for smaller numbers is slightly higher than the larger ones. This is because the orientation and the position of the fingers in the gestures corresponding to the larger numbers slightly shift. But this can be taken care of by using alternative hand gestures or by the correct placement of the hand.

6 Conclusion and Future Work

The main advantages of the novel algorithm for hand gesture recognition presented in this paper are its accuracy and speed. From the results presented in Section 4, it can be observed that its execution time is around 1.45 times faster than the general existing algorithm. The accuracy is also appreciable with an average success rate of 95.2%. The application of this concept in security systems has been the topic of focus in this paper. Since this algorithm has been designed to be used in fixed installations, it can be used almost anywhere.

Apart from security systems, detection of hand gestures can be used in controlling machines and processes. For example, a handicapped person on a wheelchair can control its movement and direction by simply showing corresponding gestures. It can also be of great help to bridge the communication gap between common man and hearing impaired individuals.

Numbers are not the only symbols that can be represented using hand gestures. Its scope can definitely be extended to English alphabets and perhaps phrases as well. Since the number of alphabets and phrases is quite large, many gestures have to be developed that are different and distinct to some extent, so that they can be recognized using 2D imaging itself. Otherwise, 3D imaging will have to be used, which would make the algorithms more challenging.

References

1. Ratha, N.K.: Enhancing security and privacy in biometrics-based authentication systems. IBM Systems Journal 40, 614–634 (2001); IEEE Journals & Magazines
2. Ko, J.G., Moon, K.Y.: Biometric security scheme for privacy protection. In: Advanced Software Engineering and Its Applications (ASEA 2008), pp. 230–232 (2008)
3. Vielhauer, C.: Biometric User Authentication for IT Security: From Fundamentals to Handwriting, 1st edn. Springer, New York (2006)
4. Phillips, P.J., Martin, A., Wilson, C.L., Przybocki, M.: An Introduction Evaluating Biometric Systems. Computer 33, 56–63 (2000); IEEE Journals & Magazines
5. Prabhakar, S.: Biometric recognition: security and privacy concerns. Security & Privacy 1, 33–42 (2003); IEEE Journals & Magazines
6. Bhanu, B., Tan, X.: Fingerprint indexing based on novel features of minutiae triplets. IEEE Transactions on Pattern Analysis and Machine Intelligence 25, 616–622 (2000); IEEE Journals & Magazines

7. Cano, P., Batle, E., Kalker, T., Haitsma, J.: A review of algorithms for audio fingerprinting. In: IEEE Workshop on Multimedia Signal Processing, pp. 169–173. IEEE Conference Publications (2002)
8. Vijaya Kumar, B.V.K., Savvides, M., Venkataramani, K., Chunyan, X.: Spatial Frequency Domain Image Processing for Biometric Recognition. In: International Conference on Image Processing, vol. 1, pp. 53–56 (2002)
9. Ghotkar, A.S., Kharate, G.K.: Hand Segmentation Techniques to Hand Gesture Recognition for Natural Human Computer Interaction. International Journal of Human Computer Interaction (IJHCI) 3, 15–25 (2012)
10. Otsu, N.: A threshold selection method from gray-level histograms. IEEE Systems, Man, and Cybernetics Society 9(1), 62–66 (1979)
11. Qiuyu, Z., Fan, C., Liu, X.: Hand Gesture Detection and Segmentation Based on Difference Background Image with Complex Background. In: International Conference on Embedded Software and Systems (ICESS), pp. 338–343 (2008)
12. J-Apiraksa, A., Pongstiensak, W., Kondo, T.: A Simple Shape-Based Approach to Hand Gesture Recognition. In: International Conference on ECTI-CON, pp. 851–855 (2010)
13. Starner, T., Pentland, A.: Real Time American Sign Language Recognition From Video Using Hidden Markov Models. In: International Symposium on Computer Vision, pp. 265–270 (1995)
14. Saravanan, C.: Color Image to Grayscale Image Conversion. In: Second International Conference Computer Engineering and Applications (ICCEA), pp. 196–199 (2010)
15. Gonzalez, R.C., Woods, R.E.: Digital Image Processing. Prentice Hall (2002)

ROI Based Iris Segmentation and Block Reduction Based Pixel Match for Improved Biometric Applications

Shrinivas Rao B. Kulkarni[1], Ravindra S. Hegadi[2], and Umakant P. Kulkarni[1]

[1] Department of Computer Science and Engineering, SDMCET, Dharwad, Karnataka, India
[2] Department of Computer Science, SolapurUniversity, Solapur, Maharashtra, India
{sbkulkarni_in,upkulkarni}@yahoo.com, rshegadi@gmail.com

Abstract. Iris recognition is one of the important authentication mechanism used in many of the applications. Most of the applications capture the eye image, extracts the iris features and stores in the database in digitized form. The existing methods use normalization process in the iris recognition this is comparatively takes more time to overcome this drawback we proposed method which extracts the iris data by separating the iris region from the pupil and sclera, using roipoly and inverse roipoly. Extracted features are matched with the stored repository by using reduced pixel block algorithm. The experiment is carried out on CASIA-IrisV3-Interval and result shows an improvement in feature extraction by 45% and matching by 91% compared to the existing method.

Keywords: ROI, roipoly, inverse-roipoly, Edge detection, Dimensionality Reduction, Reduced Block.

1 Introduction

The idea of using the iris as biometric is 100 years old but the technology advancement is in automating the iris recognition [2]. Biometric comes from the Greek words bios means life and metrics means measure [1]. Biometric involves recognizing individuals based on the feature extracted from their physiological (face, palm prints, iris, fingerprint etc) and behavioral (voice) characteristics. The randomness of iris pattern is one of the most reliable biometric characteristic. Image is constructed by the pixel. Lyon Richard F describes that "The number of pixels in an image is the number of image samples and gives useful information that bound to an image resolution" [10].

To extract iris from eye image, edge detection must be employed to find the edges in the eye image. Since edge separates two different regions, an edge point is considered. Edge point is a point where the local intensity of the pixel in the image varies rapidly than in the neighbor points, which are close from the edge; such a point could be characterized as a local maximum of the gradient of the image intensity at that pixel [13].

Iris part is the region of interest. Region of interest (ROI) is the part of the image for which the observer of the image shows interest in the particular region of the image.

S. Unnikrishnan, S. Surve, and D. Bhoir (Eds.): ICAC3 2013, CCIS 361, pp. 548–557, 2013.

The region of the interest shown by the observer in viewing the image is determined by the image itself or also by the observer own sensitivity [3].

The proposed method uses canny edge detection along with Hough Transform algorithm to identify the edges that separates the iris from pupil and sclera. The two rings are as shown in Figure 1 are the references to extract the iris by applying the ROI.

In this paper we proposed roi and inverse roi along with dimensionality reduction and block reduction methods to reduce the computation time there by achieving improvement in feature extraction and matching time.

The remaining sections of this paper are organized as follows: Section 2 gives Related work. Section 3 deals with the proposed methodologies. Section 4 Experimental Data and Result Analysis, Finally paper ends with acknowledgement, conclusions and references.

2 Related Work

Feature extraction is the core part of image processing; images are classified by shape, color and texture etc. Iris features are invariant to image scaling and rotation and partially invariant to change in illumination and 3D view point [9]. Extracted iris features should be significant, compact and fast [13].

Significant: The iris image signature is an effective representation of the original iris image.
Compact: Quick computation in similarity measurement.
Fast: Algorithm must be quick responsive to the large dataset.

Many researches proposed different approaches to extract iris features. Following are feature extraction techniques these are related to proposed techniques;

Local binary patterns (LBP) are adopted to represent texture patterns of iris image. Originally LBP is introduced by Ojala. He considered 3X3 neighborhood pixel with the centre pixel value and considering binary bit string [8].

Pixel path can be found by tracking the intensity variations in horizontal stripes preprocessed iris texture. The pre processed iris texture of a person I, I_i is divided into n different horizontal stripes. $I_i \longrightarrow \{I_{i1}, I_{i2} \dots\dots I_{in}\}$, of height h pixels, each texture is of dimension lXh, where l is the length of preprocessed iris texture. Next two pixel paths represent light and dark intensity variations. These paths are created for each texture strip I_{ij}.

$$PL_{ij} = \{PL_{ij0}, PL_{ij1} \dots\dots PL_{ijl}\} \tag{1}$$

$$\mathrm{PD_{ij}} = \{\mathrm{PD_{ij0}}, \mathrm{PD_{ij1}} \dots\dots \mathrm{PD_{ijl}}\} \tag{2}$$

The Light and Dark paths are given in equation (1) and (2) respectively.The element of light and dark pixel of each path is defined as;

$PL_{ij0} \longleftarrow \frac{h}{2}$ And $PD_{ij0} \longleftarrow \frac{h}{2}$

The elements PL_{ij1} and PD_{ij1} are calculated by examining the three directly neighboring pixel values of PL_{ijo} and PD_{ijo} $(PL_{ijo} = PD_{ijo})$ in the next pixel column. Then PL_{ij1} is set to y value of the maximum and PD_{ij1} is set to the y value of the minimum of these three values. Maximum and minimum value corresponds to brightest and darkest grey scale values of pixels [14].

Textural feature extraction algorithm uses 1-D log polar Gabor transform, like the Gabor transform the log Gabor transform is also based on polar co-ordinates. Another feature is Topological feature extraction using Euler Number. Euler number is defined as the difference between the number of connected components and the number of holes. Euler numbers are invariant to rotational translation, scaling and polar transformation of the image. Further these two features are fused and fed to the Support Vector Machine (SVM) match score fusion to test accept / reject [15].

Circular Dark and brightness and contrast method is used to detect the boundary between the pupil and iris, Adam Czajka et. they proposed method which is sensitive to circular dark shapes and unresponsive to other dark areas as well as light circles, such as specular reflections [6],. They achieved this by modifying Hough Transform that uses the directional image to consider the image gradient, rather than considering the edge image which neglects the gradient direction. Here the boundary between the iris and the sclera is detected using Daugman's integro differential operator. Dal lto cho et. al. they used brightness and contrast [7]. Corneal Specular reflection technique is used to detect pupil region.

3 Proposed Method

The method proposed includes the following steps.

1. Segmentation
2. Generation of Tyre Model
3. Noise Removal
4. Dimensionality reduction
5. Matching

Finally the proposed algorithms for the above methods are discussed.

3.1 Segmentation

Canny edge detection and boundary between sclera/iris, pupil/iris are detected for iris localization [5],[12], along with Circular Hough transform is used to detect the edges in the iris which differentiates the inner and outer part of the iris. The centre coordinates x_c and y_c, and the radius r, which define circle as given in the equation (3). Further roipoly and inverse roipoly are applied to separate the iris region from rest of eye image.

$$x_c^2 + y_c^2 = r^2 \tag{3}$$

The Hough transform is a standard computer vision algorithm that can be used to determine the parameters of simple geometric objects, such as lines and circles, present in an image. The circular Hough transform can be employed to deduce the radius and centre coordinates of the iris and pupil regions. Initially, an edge map is generated by calculating the first derivatives of intensity values in an eye image and applied threshold to get the result [11].

A maximum point in the Hough space will correspond to the radius and centre coordinates of the circle defined by the edge points. Wildes et al. and Kong and Zhang also make use of the parabolic Hough transform to detect the eyelids, approximating the upper and lower eyelids with parabolic arcs [16], which is given in equation (4).

$$\left(-(x-h_j)sin\theta+(y-k_j)cos\theta\right)^2 = a_j(x-h_i)sin\theta+(y-k_j)cos\theta \qquad (4)$$

where a_j controls the curvature, (h_j, k_j) is the peak of the parabola and θ is the angle of rotation relative to the x-axis.

3.2 Generation of Tyre Model

The need for generation of tyre model is to reduce the time of iris feature extraction. The generated pattern is called tyre model because it looks like a vehicle tyre with inner (pupil) segmented circle forming the rims of tyre and iris forming the tyre.

Masking the Exterior Part of the Outer Segmented Circle: To mask the exterior part of outer circle, x and y co-ordinates of the exterior circle and radius r are considered from equation (3) in section 3.1. This becomes an input to ROI (invert roipoly) to mask an image (blackening) as shown in figure 1.

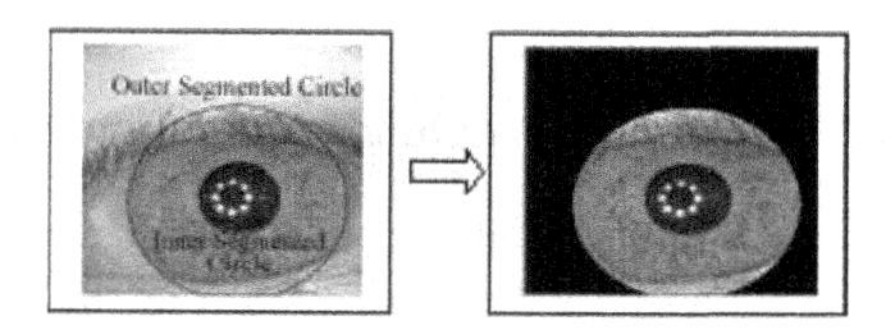

Fig. 1. Masking the Exterior Part of the Outer Segmented Region

Masking the Interior Part of Inner Segmented Circle: To mask the interior circle, the x and y co-ordinates of the circle are taken along with the radius r of the circle from equation (3) in section 3.1. Then roipoly is applied to mask inner circle, as shown in figure 2.

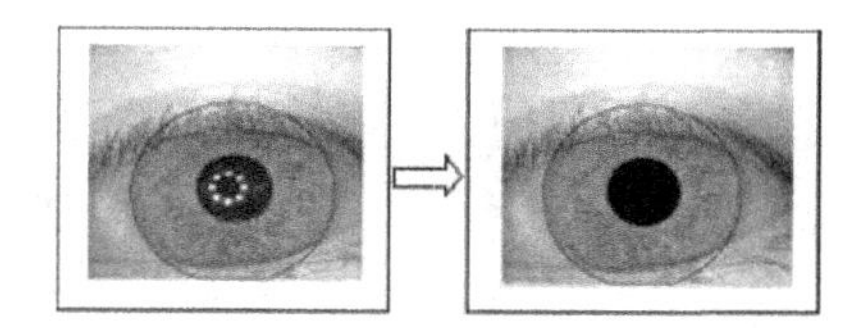

Fig. 2. Masking the Interior Part of Inner Segmented Region

Cropping the iris region : Tyre model is generated by combining figure 1 and 2. In the tyre model the Human Iris part looks like a tyre by masking the rest of the iris part as shown in figure 3.

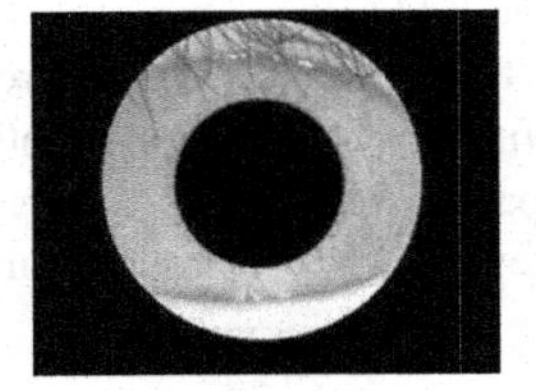

Fig. 3. Tyre model generation

3.3 Noise Removal

The presence of eyelash and eyelids introduce noise in the iris data. The noise is removed by identifying noise pixels and is replaced by zero. Noise removal is shown in figure 4.

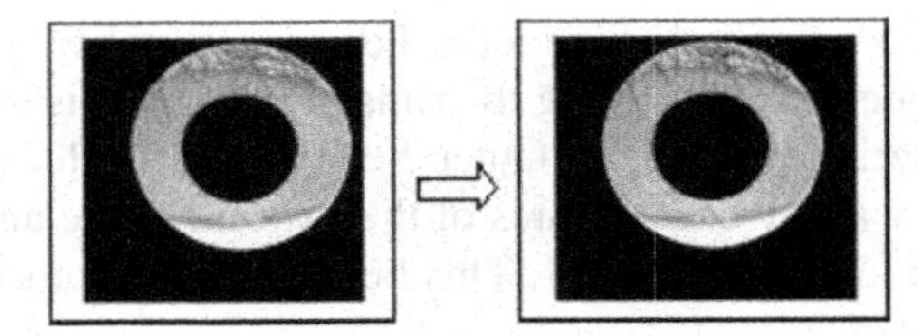

Fig. 4. Masking the Non-iris part of the tyre model

3.4 Dividing Tyre Model into Four Quadrants (Dimensionality Reduction)

Sqncher-Reillo and Sanchez-Avila, "detected iris boundaries using and integro differential type operator and then divided the iris into four portions (Top, Bottom, Left, Right) and top and bottom portions are discarded due to occlusion" [2]. Ma et. al. chooses a different part of the iris. They use the three-quarters of the iris region closest to the pupil. They used circular symmetric feature (CSF) based on Gabor filter, to extract features [2]. Hence only one quadrant comparison of the elements are enough to determine whether the iris is of the same eye or different. Quadrant approach (Dimensionality reduction) is shown in figure 5. This improves the computation time in match score.

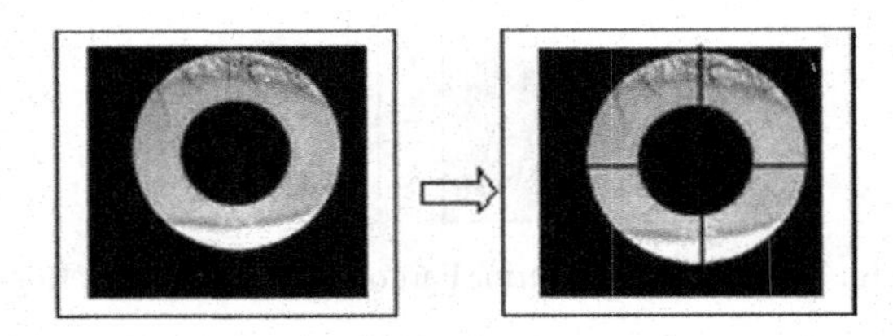

Fig. 5. Dividing Tyre Model into Four Quadrants

3.5 Iris Matching Technique (Reduced Block Matching)

Separation of second quadrant from the input segmented iris is shown in Figure 6. The figure shows quadrant part of iris. Similarly the reduced block (smallest block among sample block and repository) is checked with the matching quadrant. Finally direct block of pixel matching is adopted.

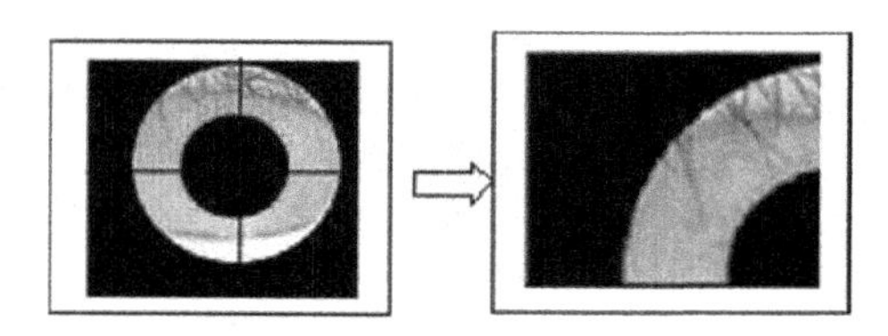

Fig. 6. Comparison of the Second Quadrant iris features

3.6 Algorithms

Algorithm: Tyre model generation by ROI technique.

```
Input: Iris Image
Output: Segmented and Tyre model iris Image.
1.  Read image.
2.  Find Inner and Outer Circle of Iris.
3.  Apply inverse roipoly to mask outer region of iris.
4.  Apply roipoly to mask inner region of pupil for the
    image generated in step 3.
5.  Write image
6.  Take reference image.
7.  Subtract reference image with segmented image.
8.  Write image (Tyre model)
```

To mask the circle, set 1 to mask upper half of iris, set 2 to lower half, set 3 to full circle on inverse roipoly.

Algorithm: Noise removal from tyre model.

```
Input: Tyre Model Image
Output: Tyre Model Image with noise free.
1.  Read Tyre Model Image.
2.  Compute Size of an Image.
3.  Store size in 'a' and 'b'.
4.  For Each pixel value of  iris image,
    a.  Replace 0 if pixel value is less than 100.
5.  Write noise free image.
```

Algorithm: Block Reduction Pixel Match

```
Input: Registered Image and Sample Image
Output: Match or Mismatch score.
1.  Read Sample Image S.
2.  Apply tyre generation and noise removal algorithm.
3.  Divide the image S into four quadrants (block).
4.  Repeat step 1 to 3 for Registered image R.
5.  Compare 2nd quadrant of block of R and Block S for the
    smallest block size(base array).
6.  Consider base array block by taking centre of pupil.
    a.Traversing to compare R and S images from center of
      pupil (x, y) co-ordinates to (1, 1).
    b.For each pixel in base block compare with the block
      of S or R image that represents the smallest block.
    c.  Check for threshold (T) for match score.
    d.If match score is > T then iris is of same person,
      else both the iris are different.
8. Print match score result.
```

In this algorithm steps 1-3 are used to reduce the iris feature by extracting quadrant of iris data. This reduces the size of iris data.

Two cropped iris images need to be compared are taken as input. Dimensionality reduction is applied to get the quadrant information of iris, along with center of iris as the center or intersection point of four quadrants. Only second quadrant value of both images are traversed and stored in separate arrays (block1, block2). Traversing is from center i.e., (x, y) co-ordinates of iris-center up to (1, 1) i.e., first row and first column of the arrays. Only these values are stored into separate arrays and are sufficient to be considered for match score.

The centre of angular deformed iris will not lie in the centre as that of the centre of the frontal iris. Therefore direct block matching is not possible here. However Pixel to pixel mapping can be done by keeping threshold. We calculate the size of both arrays block1 and block2. The smallest array is considered as base array. Co-ordinates of all corners of only base array are considered in matching these two blocks of arrays. The remaining pixel values of larger block are not considered in match score. This saves the computation time in matching. There by reducing the matching time. Now the pixel values can be matched using direct block matching as both the blocks that are going to be matched will have same size.

Block1 and Block2 are traversed from their centers up to their opposite corners. If both of the pixel values are same and they are not equal to zero then match score is incremented. If both the pixel values are not same then mismatch (no match) score is incremented.

4 Experimental Data and Result Analysis

The proposed work is carried out on CASIA-IrisV3-Interval dataset which has total 249 subjects, 395 classes and number of images 2639 [4]. Most of the images were captured in two sessions, with at least one month interval. Here a threshold of 30 is set by empirical method conducting an experiment.

If the number of pixel matching is greater than threshold (T) 30 then we conclude that both the iris images are same. Hence match is found and vice versa.

Table 1. Comparison of Feature Extraction Time

Image No	Template Creation Time (in Sec.)	Proposed method (Tyre Model) Creation Time (in Sec.)
1	0.205238	0.156609
2	0.343570	0.208841
3	0.342297	0.119250
4	0.273172	0.123280
5	0.261796	0.127477
6	0.261796	0.154128
7	0.258083	0.143749
8	0.254656	0.262652
9	0.269566	0.128226
10	0.301718	0.134126
11	0.286580	0.137143
12	0.256067	0.129896
13	0.280027	0.141200
14	0.258010	0.133168
15	0.249300	0.207320
16	0.275385	0.136315
17	0.269311	0.127943
18	0.261355	0.132078
19	0.184208	0.121404
20	0.290805	0.137921
AVERAGE TIME	**0.269147**	**0.148136**

Table 2. Comparison of matching results

Test Image	Matching Image	Libor Masek Matching Time (in sec)	Block Reduction Pixel Match Time (in sec)
S1002L03	S1002L04	0.016128	0.00150009
S1007L05	S1007L04	0.016263	0.0016139
S1008L05	S1008L04	0.01766	0.00189597
S1011L01	S1011L02	0.017666	0.00147465
S1013L02	S1013L03	0.016544	0.00124836
S1013L02	S1022L03	0.016133	0.00138851
S1002L02	S1013L02	0.016345	0.00132468
S1001L09	S1002L02	0.01609	0.00106225
S1002L07	S1001L05	0.016016308	0.0013827
S1002L01	S1007L05	0.01638	0.00171075

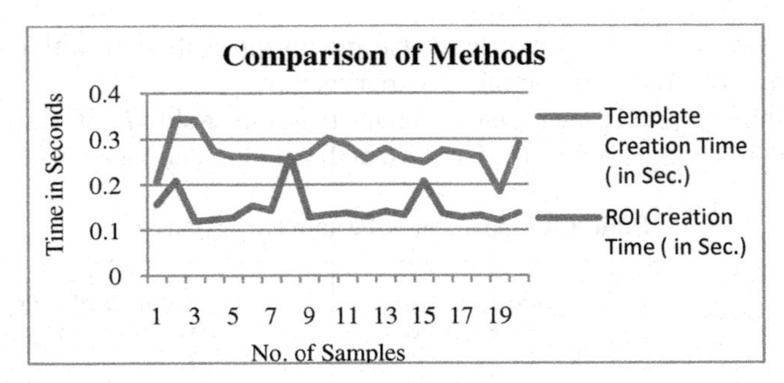

Fig. 7. Comparison of methods

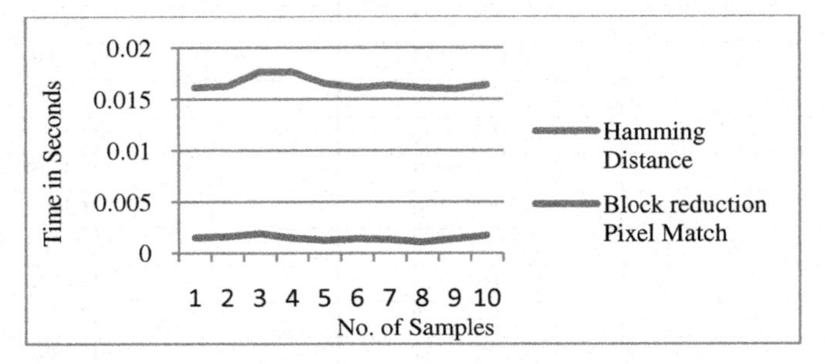

Fig. 8. Comparison of Matching

Experiment is carried out on with HP Compaq nx6320 with 2 GB Ram, Intel Core 2 Duo processor with 1667 MHz processor. Table 1 and Table 2 are the sample data taken from experiment. Figure 7 and figure 8 show the improvement in the proposed method.

5 Conclusion

Proposed method extracts the iris data and generates the tyre model using region of interest i.e., roipoly and inverse roipoly. Result shows that, an improvement of feature extraction time is by 45% and matching time by 91.16% as compared to Libor Masek method.

Acknowledgments. Author acknowledges CASIA for providing iris image database collected by Institute of Automation, Chinese Academy of Sciences [4]. Author thanks to Ms.Kavitha Acharya, Ms.Muktha.M, Ms.Nireeksha Shetty for partial implementation support. The work is partially supported by the Research Grant from AICTE, Govt. of India, Reference no: 8023/RID/RPS-114(PVT)/2011-12 Dated December,24-2011.

References

1. Srinivasan, B., Narendra Kumar, V.K.: Implementation of Efficient Secured Biometric Identification System Using Iris Technology. International Journal of Advanced Engineering and Application (January 2011)

2. Bowyer Kevin, W., Karen, H., Flyann Patrick, J.: Image understanding for iris biometrics: A survey. Computer Vision and Image Understanding 110, 281–307 (2008)
3. Caron, Y.P., Vincent Makris, N.: Use of power law modules in detecting Region of Interest. Pattern Recognition 40, 2521–2529 (2007)
4. Chinese Academy of Sciences — Institute of Automation (CASIA) Iris Database, `http://www.cbsr.ia.ac.cn`
5. Bhavana, C., Shailja, S.: Iris recognition system using canny edge detection for Biometric Identification. International Journal of Engineering Science and Technology (IJEST) 3(1) (January 2011)
6. Adam, C., Przemek, S., Marcin, C., Pacut, R.: Iris Recognition with Match on Card. In: EUSIPCO, Pozman (2007)
7. Cho, D.H., et al.: Real-time Iris Localization for iris Recognition in Cellular Phone. In: ACIS(SNDP/SAWN 2005). IEEE (2005)
8. Savithiri, G., Murugan, A.: Perfromance analysis of Half Iris Feature Extraction using GW, LBP and HOG. International Journal of Computer Applications (0975-8887) 22(2) (May 2011)
9. Mansi, J., Kumar, K.V.: Iris based human recognition system. International Journal of Biometrics and Bioinformatics (IJBB) 5(1) (2011)
10. Lyon Richard, F.: Reprint- Paper EI 6069-1 Digital Photography II – Invited Paper. In: IS&T/SPIE Symposium on Electronic Imaging, California, USA, pp. 15–19 (2006)
11. Libor, M.: Recognition of Human Iris Patterns for Biometric Identification, The University of Western Australia (2003)
12. Miyazawa, K., Ito, K., Aoki, T., Kobayashi, K., Nakajima, H.: A Phase-Based Iris Recognition Algorithm. In: Zhang, D., Jain, A.K. (eds.) ICB 2005. LNCS, vol. 3832, pp. 356–365. Springer, Heidelberg (2005)
13. Makram, N., Lahouari, G., Ahmed, B.: An effective and fast iris recognition system based on a combined multiscale feature extraction technique. Pattern Recognition 41, 868–879 (2008)
14. Christian, R., Andres, U.: Secure Iris Recognition Based on Local Intensity Variations. In: Campilho, A., Kamel, M. (eds.) ICIAR 2010, Part II. LNCS, vol. 6112, pp. 266–275. Springer, Heidelberg (2010)
15. Mayank, V., Richa, S., Afzel, N.: Improving Iris Recognition Performance Using Segmentation, Quality Enhancement, Match Score Fusion and Indexing. IEEE Transactions on Syatems, Man and Cybernetics – Part B: Cybernetics (2008)
16. Wildes, R., Asmuth, J., Green, G., Hsu, S., Kolczynski, R., Matey, J., McBride, S.: A system for automated iris recognition. In: Proceedings IEEE Workshop on Applications of Computer Vision, Sarasota, FL, pp. 121–128 (1994)

Simulation and Implementation of Electronic Cochlea

Nibha Desai[1], Deepak V. Bhoir[2], and Sapna Prabhu[2]

[1] Shah and Anchor Kutchhi Engineering College, University of Mumbai, India
nibha_29@yahoo.com
[2] Fr Conceicao Rodrigues College of Engg.
University of Mumbai, India
{Bhoir,sapna}@fragnel.edu.in

Abstract. Cochlear implants are devices designed to provide a measure of hearing to the deaf. It can help a person to hear effectively.Cochlear Implants have classified based on the no. of channels used in the cochlea. Cochlear Implant consists of signal processor, external transmitter, internal receiver and electrode array. The signal processing strategy has been used for transmitting the original speech signal to electrical stimuli. In this paper, we mainly focus on the Continuous Interleaved Sampling (CIS) technique having high stimulation rate and least channel interaction for each channel used in cochlea. This technique is modeled using the Xilinx's System Generator. Translation of the Simulink model into a hardware realization has been done using System generator. The Verilog code for each blocks of CIS processor is developed and simulation results are verified. Further, the design can be downloaded into the Spartan3 FPGA device for hardware requirements.

Keywords: Cochlea, Cochlea Implant, CIS, FIR, FPGA.

1 Introduction

Cochlear Implant is a surgically implanted electronic device that provides a sense of sound to a person who is profoundly deaf or severely hard of hearing. This prosthetic device has invented using various signal-processing techniques over last 25 years [1]. Cochlea is a part of inner ear. It is often described as looking like a snail [2]. Electronic cochlea has two parts: an internal part called cochlear implant and an external part called a speech processor. There are two types of CIs based on no. of electrodes used in cochlea. i) Single-channel Implants (SCIs) ii) Multichannel Implants (MCIs). Multichannel cochlea implant produces superior speech recognition performance compared with single-channel stimulation. An array of electrode is inserted into cochlea in MCI, so that different auditory nerve fibers can be stimulated at different places in cochlea.

In this paper, we focus on CIS technique as effective signal processing strategy. The goal is to get high pulse rate and least channel interaction for each channel. This strategy is modeled and implemented using Xilinx's System generator. Xilinx System Generator allows the design of hardware system starting from a graphical high-level

S. Unnikrishnan, S. Surve, and D. Bhoir (Eds.): ICAC3 2013, CCIS 361, pp. 558–569, 2013.

Simulink environment. System Generator extends the traditional HDL design with the use of available System Generator blocks and subsystems. This reduces the time necessary between the control design derivations and hardware implementation. The ability to realize a system design quickly and directly as a real-time embedded system greatly facilitates the design process.

Section 2 gives scheme of implementing CIS model for four channels. Section 3 gives simulation and implementation results followed by conclusion and references.

2 Implementation Scheme of CIS Strategy

2.1 Design Methodology Using Xilinx's System Generator

The used tools are MATLAB with Simulink from Mathworks, System generator for DSP and ISE from Xilinx present such capabilities (Fig.1).The system generator environment allows for the Xilinx line of FPGAs to be interphase directly with Simulink.

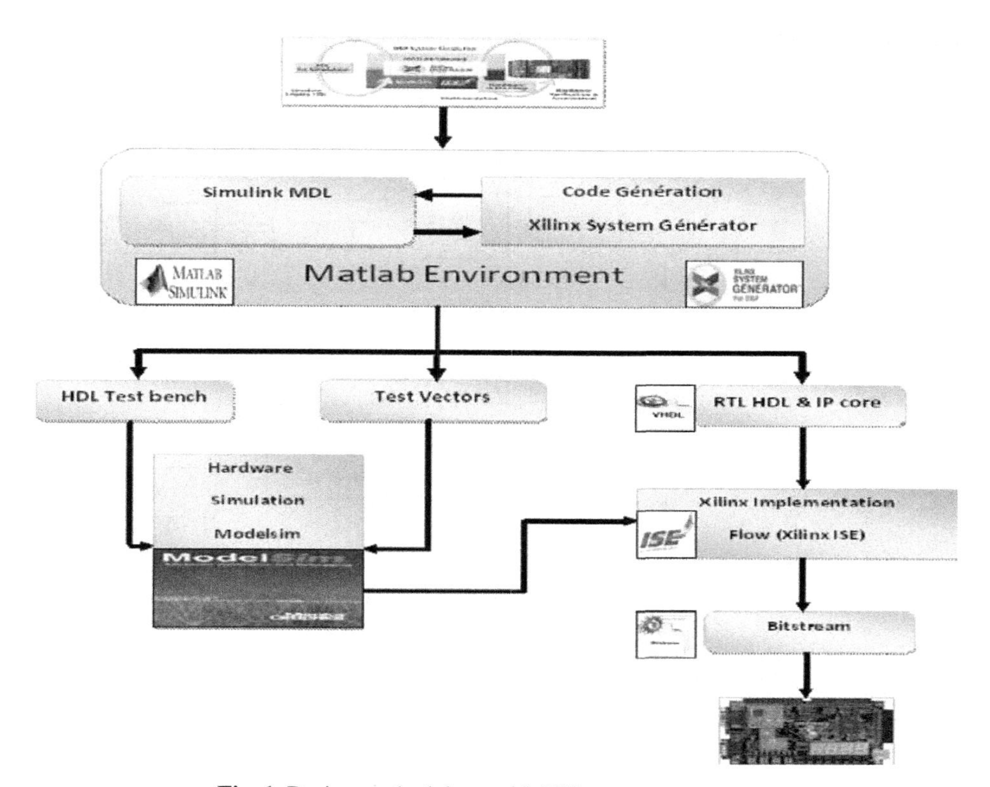

Fig. 1. Design methodology with Xilinx system generator

Xilinx system generator is MATLAB-Simulink based design tool. MATLAB is interactive software for numerical computations and powerful operations. Simulink is an additional MATLAB toolbox that provides for modeling, simulating and analysing systems. Complex digital circuits have been developed using multiple HDL Modules [3].HDL Modules simulate using Modelsim simulator. The Xilinx's ISE is powerful design environment that is working in the background when implementing system generator blocks. The ISE environment consists of a set of programme modules those are utilized to simulate and implement designs in a FPGA or CPLD target device.

2.2 Proposed Scheme for CIS Strategy

Fig.2 shows the architecture of CIS Strategy. The speech signal is first pre-emphasized and then passed through a bank of band pass filters. The envelopes of filtered waveforms are extracted by full wave rectification and low pass filtering (200 or 400Hz cutoff frequency). The envelope outputs are finally compressed and then used to modulated biphasic pulses. A logarithmic compression function is used to ensure that the envelope outputs fit the patient's dynamic range of electrically evoked hearing. Then, Compressed output waveforms are used to modulated biphasic pulses. Output of Modulator is electrical pulses, which stimulates no. of electrodes present in auditory neurons.

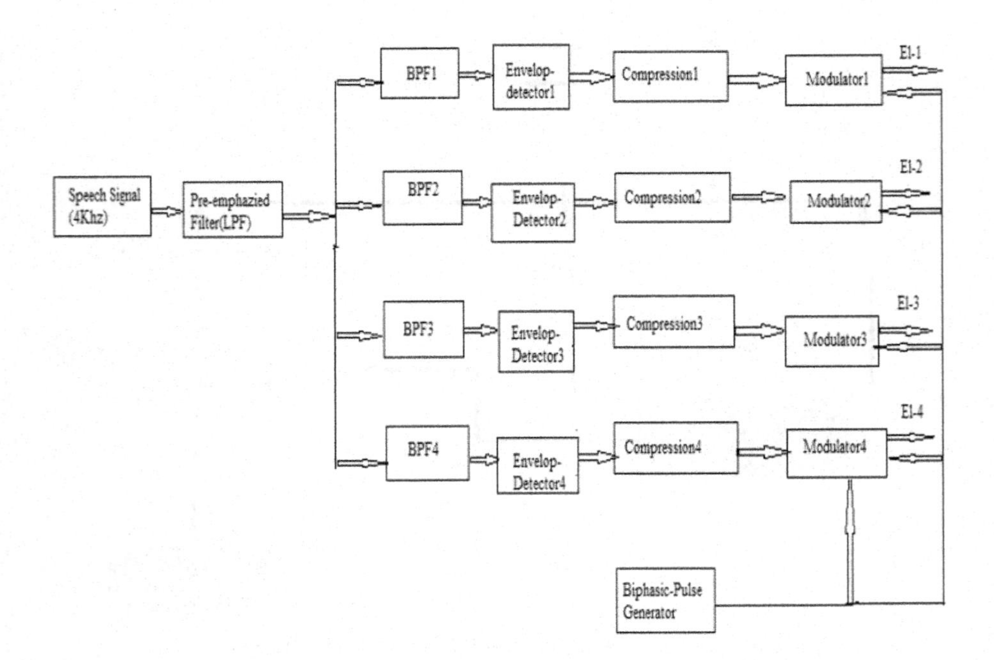

Fig. 2. Block diagram of CIS

CIS approach addressed the channel interaction issue by using non-simultaneous interleaved pulses. Trains of biphasic pulses are delivered to the electrodes in non-overlapping (non-simultaneous) fashion: i.e. such that only one electrode is stimulated at a time (Fig.3). The amplitudes of the pulses are delivered by extracting the envelopes of band passed waveforms. CIS parameter like i) pulse rate and pulse duration, ii) stimulation order iii) Compression function can be varied to optimize speech recognition performance.

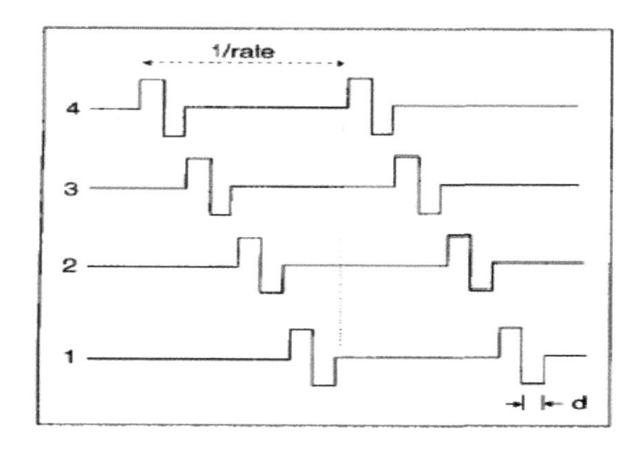

Fig. 3. Interleaved pulses used in the CIS strategy. The period between pulses on each channel (1/rate) and the pulse duration (d) per phase are indicated.

2.3 Modeling of CIS Strategy in System Generator

Xilinx's system generator provides a set of Simulink blocks for several hardware operations that could be implemented on various Xilinx FPGAs. Here, CIS Strategy for Four channels design is utilized suitable hardware for Spartan3 FPGA. System generator provides some blocks to transform data provided from the software side of the simulation environment (Simulink) and the hardware side (System generator block) [4]. Hardware implementation of floating-point arithmetic increases the complexity. For this reason Xilinx's system generator uses fixed-point format to represent all numerical values in the system.

"Gateway In" Module, which converts data types in Simulink, such as integer, double and fixed-point into fixed-point type accepted by system generator, is the only interface connecting Simulink and System Generator. The "Gateway Out" Module used to transform the data generated from the System Generator in fixed-point format into floating point format required by Simulink. Output gateways automatically detect the fixed-point format from the system output and do not require any modification.

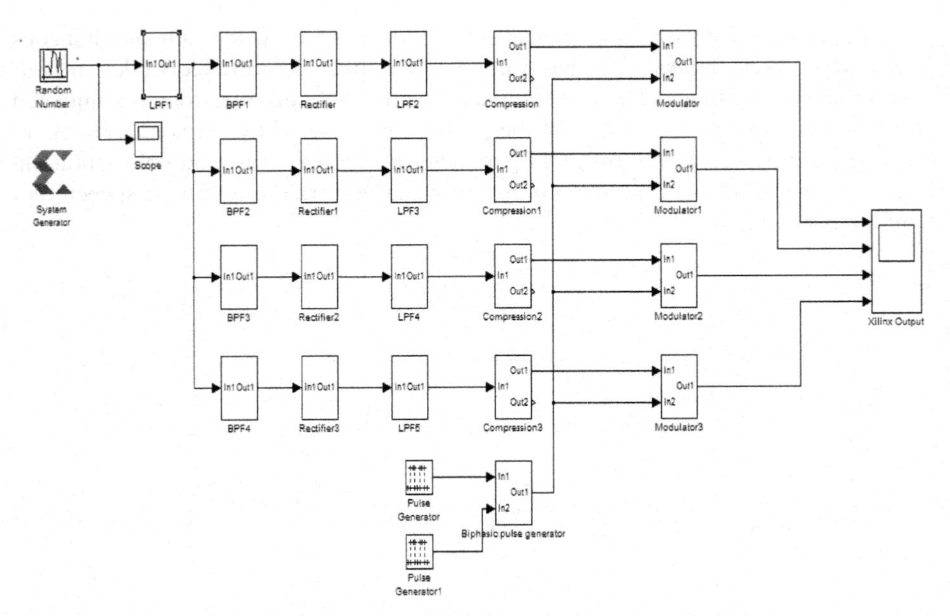

Fig. 4. Four-channel CIS Scheme in system generator

Fig.4. shows implementation of CIS Scheme for four channels. Modeling of each block has shown following.

2.4 Pre-emphasized Filter

Pre emphasized filter is used to reduce wide noise generated in Microphones. Here, Low-pass filter with cut-off frequency (400Hz) is used. In Simulink, FDA tool is used to design filer, which generates co-efficient of filter. With this FDA tool, FIR compiler is designed in system generator. The design diagram with Xilinx's Block sets has shown in Fig.5.

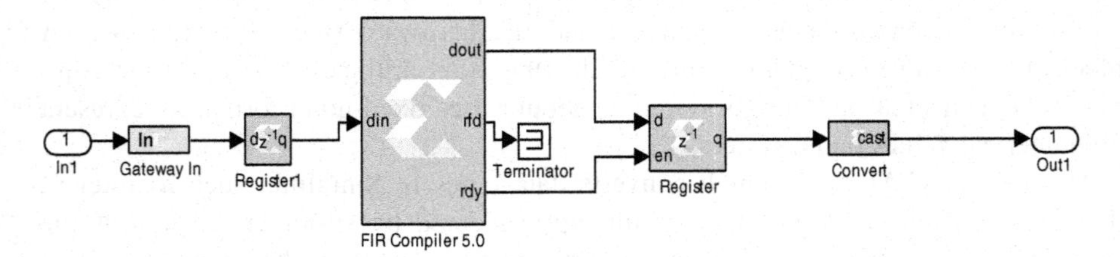

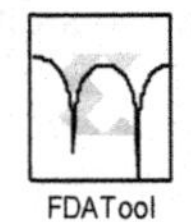

Fig. 5. LPF in System Generator

2.5 Band Pass Filter

The cochlea acts as a spectrum analyzer, decomposing complex sound into its frequency components. The speech processor part is the heart of the device that models electrically, cochlea of the ear.

In this project, we have select 4 KHz speech signal as input to the filter banks. Four logarithmically spaced band pass filters are used to divide 4KHz frequency into specified range: 0.1-0.7Khz, 0.7-1.4Khz, 1.4-2.3Khz, 2.3-5Khz. FIR (Finite Impulse Response) filter is chosen. Even though the characteristic of cochlea filter is nonlinear, the linear phase FIR filter is used here to overcome the disadvantage of IIR filter due to interaction of speech signal from different channels. FIR has several advantages over IIR filters. FIR filters do not have poles and are unconditionally stable. FIR does not accumulate errors since they depend on only a finite number of past input samples. The order of filter is chosen as 4'th which is sufficient for the speech recognition. Simulink's FDA tool is used to design filter and filter response with input frequency is created using system generator. Block diagram with Xilinx's block sets is same as LPF accept the filter co-efficients.

2.6 Full Wave Rectifier and Low Pass Filter (Envelope Detector)

Full wave rectifier and LPF extract the envelopes of filtered waveform. LPF used for smoothing the envelop waveforms. The rectifier is used to complete a modulus operation. Using system generator's model in Fig.6, 'a' is the original data and 'b' is opposite to 'a'. When 'a'>'b', 'a' is selected by the multiplexer. When 'a' ≤ 'b', 'b' is selected. In this way, the output is always kept non-negative [5].

The LPF is then used to smooth the rectified output waveforms. The FIR's architecture of LPF is the same as pre-emphasized filter.

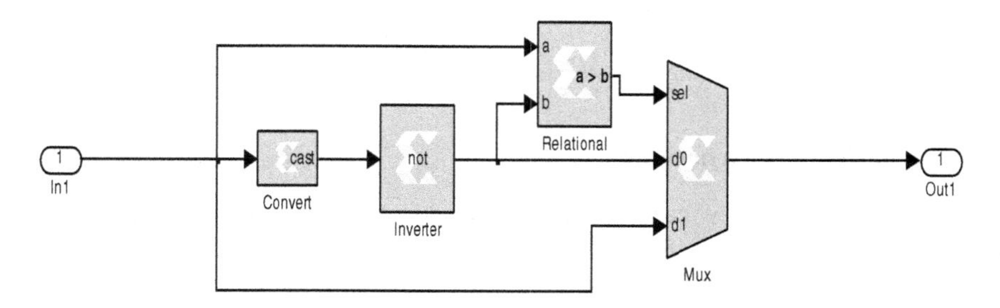

Fig. 6. Full wave rectifier using Xilinx's blocksets

2.7 Compression

The compression of envelope outputs is an essential component of the CIS processor because it transforms acoustical amplitudes into electrical amplitudes. This transformation is necessary because the range of acoustic amplitudes in conversational speech is considerably larger than the implant patient's dynamic range. Dynamic range is defined here as the range in electrical amplitudes between threshold (barely audible level) and loudness uncomfortable level (extremely loud). In conversational speech, the acoustic amplitudes may vary over range of 30dB. Implant listeners, however, may have a dynamic range as small as 5dB. For that reason, the CIS processor compress, the acoustic amplitudes to fit the patient's electrical dynamic range. The logarithmic function is commonly used for compression because it matches the loudness between acoustic and electrical amplitudes. Logarithmic compression function of the form,

$$Y = A\ log(x) + B \tag{1}$$

is used, where x is the acoustic amplitude (output of envelope detector). A and B are constants and Y is the (Compressed) electrical amplitude [6]. This function is designed using system generator and values A and B are selected constants value from Simulink block, which can vary for four channels and define the compression ratio. The compression block has designed with CORDIC LOG processor set which finds natural log for input x. The block designed as per following Fig.7.

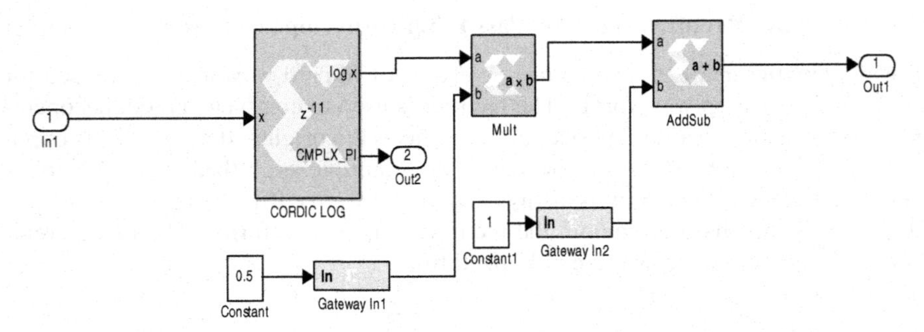

Fig. 7. Logarithmic compression function in System generator

2.8 Biphasic Pulse Generator

The CIS processor takes care for the channel interaction issue by using non-simultaneous, interleaved pulses. Trains of biphasic pulses are delivered to the electrodes in a non-overlapping (non-simultaneous) fashion .i.e. such that only one electrode is stimulated at a time extracting the envelopes of band-passed.

Simulink's two pulse generator blocks are used. One for generating with +Ve amplitude pulse and second for generating with –Ve amplitude pulse. After generation of these pulses, it is required to add them to describe as Biphasic pulses. Once pulses are delivered for one channel, same pulses are delayed with previous channel pulse such that they are not interacting with each other. Same way for four-channel biphasic generator with delayed circuit is designed in System Generator. Fig.8. has shown biphasic pulse generator in system generator.

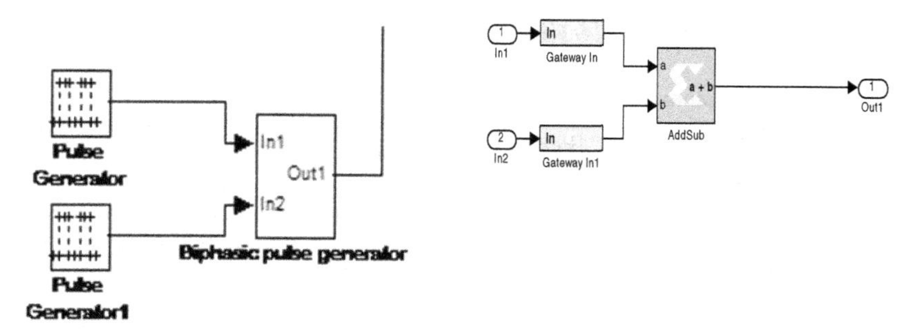

Fig. 8. Biphasic pulse generator in System generator

2.9 Modulator

Modulator modulates the output of compression waveforms and Biphasic pulse generated from generator for each channel. Therefore, output of each modulator is four biphasic non-simultaneous pulses, which stimulate the nerve fibers and sound is interpreted by the brain.

3 Simulation and Synthesis Results

3.1 Simulation Results in Simulink

CIS parameters for different freuqncies range and period are observed. This comparision are shown in following two tables for different biphasic pulse rate.

1) Biphasic Pulse Rate = 10 Khz

Table 1. Summary of CIS Parameter for Biphasic pulse frequency 10Khz

Frequencies (100Hz-10Khz)	Period=3		Period=4		Period=6		Period=8	
	Pulse rate (PPS)	Channel interaction	Pulse rate (PPS)	Channel interaction	Pulse rate (PPS)	Channel interaction	Pulse rate (PPS)	Channel interaction
100Hz	3333	Ch1-Ch4	2500	No	1666	No	1250	No
300Hz	3333	Ch1-Ch4	2500	No	1666	No	1250	No
600Hz	3333	Ch1-Ch4	2500	No	1666	No	1250	No
900Hz	3333	Ch1-Ch4	2500	No	1666	No	1250	No
1Khz	3333	Ch1-Ch4	2500	No	1666	No	1250	No
2Khz	3333	Ch1-Ch4	2500	No	1666	No	1250	No
4Khz	3333	Ch1-Ch4	2500	No	1666	No	1250	No
5Khz	3333	Ch1-Ch4	2500	No	1666	No	1250	No
6Khz	3333	Ch1-Ch4	2500	No	1666	No	1250	No
8Khz	3333	Ch1-Ch3	2500	No	1666	No	1250	No
10Khz	3333	Ch2-Ch3	2500	No	1666	No	1250	No

2) Biphasic Pulse Rate = 16 Khz

Table 2. Summary of CIS Parameter for Biphasic pulse frequency 16 Khz

Frequencies (100Hz-10Khz)	Period=3		Period=4		Period=6		Period=8	
	Pulse rate (PPS)	Channel interaction	Pulse rate (PPS)	Channel interaction	Pulse rate (PPS)	Channel interaction	Pulse rate (PPS)	Channel interaction
100Hz	5000	All	3333	Ch3-Ch4	2500	Ch1-Ch3	2000	No
300Hz	5000	All	3333	Ch3-Ch4	2500	Ch1-Ch3	2000	No
600Hz	5000	All	3333	Ch3-Ch4	2500	Ch1-Ch3	2000	No
900Hz	5000	All	3333	Ch3-Ch4	2500	Ch1-Ch3	2000	No
1Khz	5000	All	3333	Ch3-Ch4	2500	Ch1-Ch3	2000	No
2Khz	5000	All	3333	Ch3-Ch4	2500	Ch1-Ch3	2000	No
4Khz	5000	All	3333	Ch3-Ch4	2500	Ch1-Ch3	2000	No
5Khz	5000	All	3333	Ch3-Ch4	2500	Ch1-Ch3	2000	No
6Khz	5000	All	3333	Ch3-Ch4	2500	Ch1-Ch3	2000	No
8Khz	5000	All	3333	Ch3-Ch4	2500	Ch1-Ch3	2000	No
10Khz	5000	All	3333	Ch3-Ch4	2500	Ch1-Ch3	2000	No

1. Stimulation rate:
The stimulation rate is the pulse rate. It defines the number of pulses per sec (pps) delivered to each electrode. It is observed from above results that as we increse the period of biphasic pulses, stimulation rate is reduced but at the same time least channel interaction is achieved and vice-versa. Also, it is observed that if biphasic pulse rate is changed from 10 Khz to 16 Khz, stimulation rate for electrode is incresed and vice-versa. Biphasic pulse frequency 16 Khz (Periode=8) at 4Khz speech frequency constant pulse rate is achieved as 2000 pps, with no channel interaction. This stimulation rate is sufficient for speech reorganization performance.

For ex. for ch.1

$(2 - 1.5) * 10^{-3} = x$. Pulse rate = $1/x$ = 2000 pps.

2. Stimulation order:
The stimulation order refers to the order with which the electrodes are stimulated. The stimulation order can be varied to minimize possible interaction between channels. Here the electrodes are stimulated in an apex-to-base order. i.e., first stimulate electrode 1, then 2, etc., and lastly, 4.

3. Channel interaction:
It is verified from Fig 9 that electrical pulses for each channel are not interacting with each other and channel interaction problem is resolved by introducing biphasic pulses in CIS strategy.

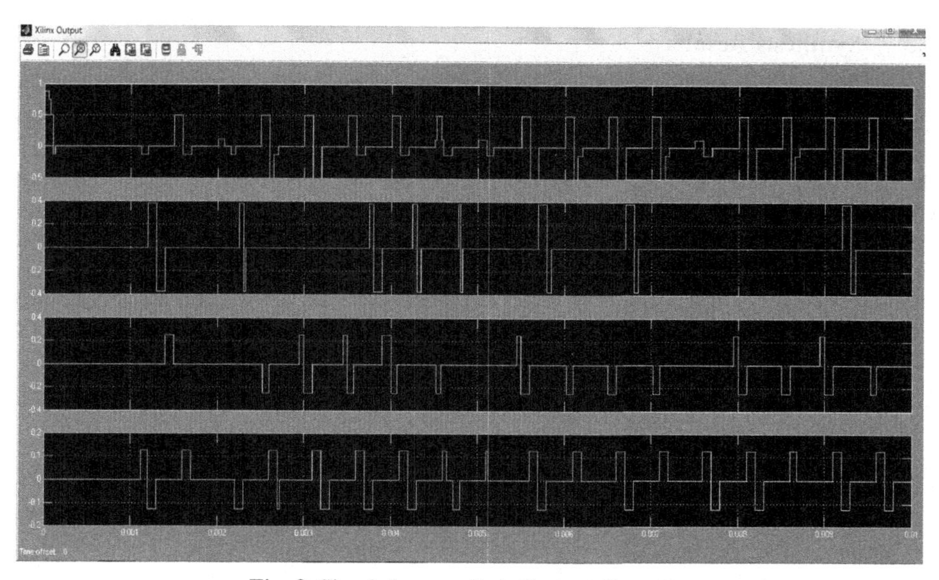

Fig. 9. Simulation results in System Generator

3.2 Simulation Results in Simulator

For generated Verilog codes for each channel, it is required to check the functionality. In the Modelsim simulator, simulation results are checked as shown in Fig.10. It has shown that, each channel has non-overlapping biphasic pulses.

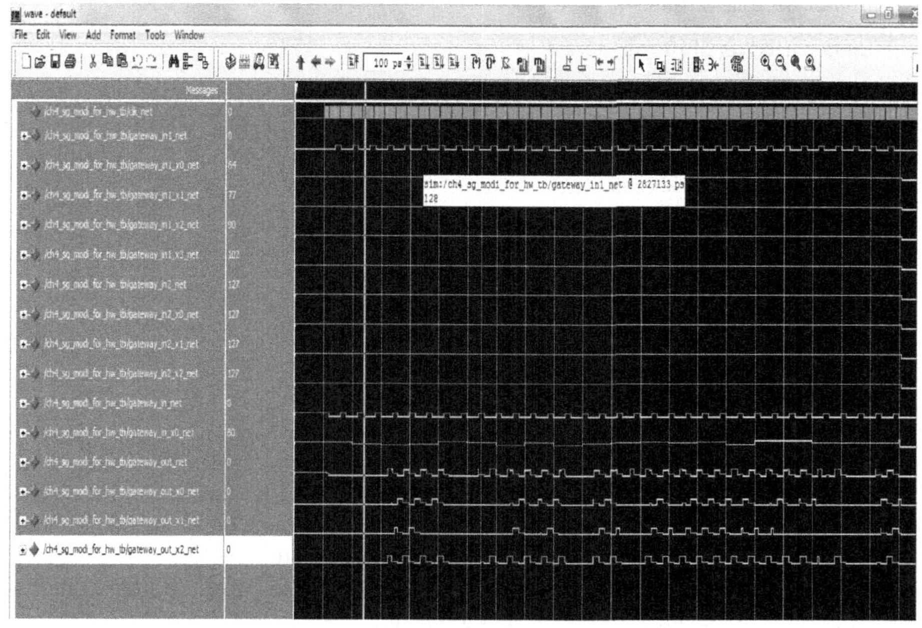

Fig. 10. Simulation result in Modelsim

3.3 Synthesis Results

Synthesis is the process to convert high-level abstraction of the design into low-level abstraction. The synthesis result gives the information about amount of minimum hardware is required for selected design. Synthesis is done after simulation process. In the synthesis result, total device utilization details are generated. The four channel CIS schemes fit in the Xilinx's Spartan3 XC3S2000-5fq900 FPGA Device. Once, the synthesis is done successfully, it is required to add User Constraints file (.ucf) with project. This file gives the information about I/O pins. To generate UCF file, Xilinx's Plan-ahead 12.3 Software is used. After synthesis process, design can transform into FPGA. In the Table 4 Detailed summary of synthesis results have shown for CIS scheme.

Table 3. Detailed summary of synthesis result

Logic Utilization	Used	Available	Utiliztion
Number of Slice Flip Flops	2,771	40,960	6%
Number of 4 input LUTs	2,226	40,960	5%
Number of occupied Slices	2,039	20,480	9%
Number of Slices containing only related logic	2,039	2,039	100%
Total Number of 4 input LUTs	2,548	40,960	6%
Number used as logic	1,869		
Number used as a route-thru	322		
Number used as Shift registers	357		
Number of bonded IOBs	105	565	18%
IOB Flip Flops	8		
Number of MULT18X18s	35	40	87%
Number of BUFGMUXs	1	8	12%
Average Fanout of Non-Clock Nets	1.78		

4 Conclusion

The procedure of Modeling and implementation for CIS Strategy is provided here in detail. CIS approach addressed the channel interaction issue by using non-simultaneous interleaved pulses. Trains of biphasic pulses are delivered to the electrodes in non-overlapping fashion. Better performance would be obtained for high pulse rate (2000 pps), since high rate-stimulation can better represent fine speech signal. The system modeled in Xilinx's System Generator with desired Xilinx's Blocksets for four channels and in Simulation result, it is verified that at each channel, electrical pulses are arrived in non-overlapping fashion. HDL code is generated, the functionality of each block in CIS scheme is again verified using Simulator. Synthesis result is generated successfully for Spartan3 XC3S2000-5fq900. This experimental set-up can be extended for more no. of channels. In the implementation stage, bit-stream file can be downloaded in to FPGA for suitable hardware requirement.

References

1. Wilson, B.: Signal processing. In: Tyler, R. (ed.) Cochlear Implants: Audiological Foundations, pp. 35–86. Singular Publishing Group, Inc. (1993)
2. Silva, C.: Cochlear Implants. In: Proceedings of Electrical Insulation Conference and Electrical Manufacturing Expo, pp. 442–446 (2005)
3. Ownby, M., Mahmoud, W.H.: A Design Methodology for Implementing DSP with Xilinx System Generator for Matlab. In: IEEE International Symposium on System Theory, pp. 404–408 (2003)
4. Using Xilinx System Generator 10.1 for Co-Simulation on Sparatan 3E Starter Board, Ahmed Elhossini (January 24, 2010)
5. Wu, S., Mai, S., Zhang, C.: FPGA Implementation of CIS Speech Processing Strategy for Cochlear Implants, Graduate School at Shenzhen, Institute of Microelectronics, Tsinghua University Shenzhen, Guangdong 518055, P. R. China, Beijing, 100084, P. R. China
6. Loizou, P.C.: Signal Processing Techniques for Cochlear Implants. IEEE Transaction on Engineering in Medicine and Biology, 34–46 (May/June 1999)

Face Detection in Color Images Based on Explicitly-Defined Skin Color Model

Ravi Subban and Richa Mishra

Department of Computer Science, School of Engineering and Technology,
Pondicherry University, Puducherry, India
{sravicite,richamishra.0312}@gmail.com

Abstract. Detecting human faces in color images plays an important role in real life application such as face recognition, human computer interface, video surveillance and face image database management. This paper proposes four new methods for human skin segmentation and one facial feature detection using normalized RGB color space. The linear piecewise decision boundary strategy is used for human skin detection techniques. Both adaptive and non-adaptive skin color models are used for skin detection. The rule based technique is used to detect facial features like lips and eye. These techniques are very effective in detecting facial features. The experimental results show that both the non-adaptive and adaptive skin color models produce better results as compared to other skin detection techniques proposed.

Keywords: Face Detection, Normalized RGB, Facial Features, Adaptive Approach, Non-adaptive Approach.

1 Introduction

Human face detection in color images is an important research area in the fields of pattern recognition and computer vision. It is a preprocessing step in the fields of automatic face recognition, video conference, intelligent video surveillance, advance human-computer interaction, medical diagnosis, criminal and terrorist identification [6]. It remains elusive because of variations in illumination, pose, expression, and visibility, which complicates the face detection process, especially under real-time constraints [5]. Face detection techniques are used to determine the presence and location of a face in an image, by distinguishing it from the background of the image, or from all other patterns present in the image.

Face localization can either be a preprocessing step or an integral part of face detection activity. A number of approaches are in use for detecting facial regions in color images [17]. One of the important approaches used to detect face regions in color images is through skin detection, because skin-tone color is one of the most important features of human faces. The skin regions in a color image can be segmented by processing one pixel at a time sequentially and independently. Skin segmentation means differentiating skin regions from non-skin regions in a color image [28].

S. Unnikrishnan, S. Surve, and D. Bhoir (Eds.): ICAC3 2013, CCIS 361, pp. 570–582, 2013.

Although there are many color spaces that can be used for skin detection, the normalized RGB has been selected in this paper because this color space is more suitable for skin detection and facial feature detection. It is also used in this paper due to the evidences that human skin color is more compactly represented in chromaticity space than in other color spaces [3].This color space is useful in particular to detect lip regions and hence other facial features. Face detection techniques can be classified into two main categories: feature-based and image-based. The complete information of face is one of the necessary requirements of face detection technique. Feature-based techniques depend on feature derivation and analysis to gain the required knowledge about faces [5]. Face detection is inadequate without detecting its facial features. It generally includes salient points, which can be traced easily, like corner of eyes, nostrils, lip corners, ear corners, etc. The facial geometry is generally estimated based on the position of eyes. On the other hand, image-based techniques treat face detection as a general pattern recognition problem. It uses training algorithms to classify regions into face or non-face classes [5]. Jones [7] construct a generic color model as well as separate skin and non-skin models using RGB color space over large image datasets that are available on the World Wide Web. Jayaram [4] also has used large dataset for the making the comparison for combinations of nine color spaces, the presence or the absence of the illuminance component, and the two coloring models.

The methods described in this paper use feature-based technique to identify the human face regions in the color images. This paper proposed all the methods used for the face detection based on feature-based technique in terms of their effect on the color images of the database that is commonly used.

The remainder of the paper is organized as follows: Section 2 describes the previous work done on human skin color detection and face detection using normalized RGB. Section 3 focuses on skin detection techniques. Section 4 proposes four novel approaches that can be used for the human skin detection. Section 5 proposes facial feature detection technique. The section 6 focuses on experimental results and describes about the effect of methods on the color images. Finally, section 7 concludes the paper by suggesting some future enhancements.

2 Related Work

Brand et al. [8] assessed the merits of three different approaches to pixel-level human skin detection over RGB color space. The concept of one of the approaches used for skin detection is referred in this paper using normalized RGB instead of RGB. Pankaj et al. [11] has presented a fundamental unbiased study of five different color spaces in addition to normalized RGB, for detecting foreground objects and their shadows in image sequence. Caetano et al. [2] proposed a novel method to model the skin color in the chromatic subspace, which is by default normalized with respect to illumination. They used two image face databases, one for white people and another for black people, with two Gaussians, rather than with a single Gaussian. Boussaid et al. [14] has achieved pixel classification on readout in the normalized color space based on a statistical skin color model. Vezhnevets et al. [20] provided the description, comparison and evaluation results of popular methods for skin modeling and detection.

Bayoumi et al. [16] presented an online adapted skin-color model to handle the variations of skin-color for different images under different lighting conditions. Zarit et al. [19] presented a comparative evaluation of pixel classification performance of two skin detection methods in five color spaces. Hsieh et al. [18] proposed an adaptive skin color model which is based on face detection. Skin colors were sampled from extracted face region where non-skin color pixels like eyebrow or eyeglasses could be excluded. Gaussian distributions of normalized RGB were then used to define the skin color model for the detected people. Andreeva [29] has used nonparametric skin distribution modeling and parametric skin distribution modeling in addition to explicitly defined skin region method.

3 Skin Detection Techniques

Face detection in color images through skin detection is the preliminary step in the biometric technology like face recognition. It uses skin color to identify the human face. The methods belonging to the feature-based technique are based on texture and skin-color analysis. The human face texture can be used to separate human face from other objects. This technique is used as an auxiliary method because there may be other objects similar to human face texture. Human skin color has been used as an effective feature in face detection tasks. Several approaches have been used for skin color pixel classifier. They include piecewise linear classifier [26], Gaussian classifiers [3], [26], [27], Bayesian classifiers [13], [25], [26], neural networks [12], probabilistic approach [9], statistical approach [15]. Piecewise linear classifiers are used in this paper.

The technique determines whether the color of each pixel describes human skin or non-skin regions helping in finding the areas that can have facial regions. In order to discriminate the human skin from non-skin pixels, various skin color methods have been used from the past few decades [24]. Explicitly-defined skin region is one of the efficient methods used for detecting skin regions from the color images using different color spaces. In this method, the boundaries of skin cluster are explicitly defined in some color spaces. One of the main advantages of using this method is its simplicity that attracts many researchers. The simplicity of skin detection rules leads to construction of a very rapid classifier without much computational work [20].

Most of the color spaces can be used for the identification of skin or non-skin pixel. In this paper, normalized RGB color space is used to achieve high skin detection rates for the color images. The normalization process is very important for "skin-color" and "face" detection because the skin-colors of different people belonging to different ethnicity will be different. Still this color space can detect skin color of all people with different skin color with ease.

The normalized RGB can be represented as follows:

$$\mathrm{Base} = \mathrm{R} + \mathrm{G} + \mathrm{B} \tag{1}$$

$$r = R/Base \tag{2}$$

$$g = G/Base \tag{3}$$

$$b = B/Base \tag{4}$$

Normalized RGB is easily obtained from RGB values by the above normalization procedure. As the sum of the three normalized components is known (R + G + B = 1), the third component does not hold any significant information and can be omitted, reducing the space dimensionality. The remaining components are treated as "pure colors", for the dependence of r and g on the brightness of the source RGB color is diminished by the normalization.

There are two approaches that can be used for segmenting skin regions in the color images when the normalized RGB color space is used. The two approaches are called adaptive and non-adaptive skin color models. Chai [23] has used adaptive method to discriminate between skin and non skin color pixels.

The segmentation performance of the above methods can be measured in terms of the true negative and false positive. In true negative, non skin pixel is classified as skin pixel and in case of false positive, skin pixel is classified as non skin pixel. After segmenting the skin regions, any one of the several approaches commonly applied for face detection can be used. They include neural networks [21], geometrical and information analysis [22].

4 Proposed Methods

4.1 Adaptive Skin Color Model

The adaptive skin color model will use the threshold values using trial and error method for lower and upper bound on the red and green chrominance components. By changing these threshold values the skin regions can be detected. Many researchers have proposed different adaptive skin color models based on the explicit skin cluster method for detecting skin regions in color images using normalized RGB color space. For the evaluation purpose, the commonly used face databases, consisting of more than thousand face images are used for the human skin detection and face detection. Some of the sample images used from extended M2VTS face data base [30] is shown in figure 3(a). The methods has also tested on some other databases, namely Caltech Face Database [31], Indian Face Database [32], and Postech faces'01 [33].

There are two adaptive skin color models which are discussed in the following subsection. The outputs obtained by using these newly proposed novel methods are compared in the experimental results section using the standard face database. There are two adaptive skin color models which are discussed in the following subsection.

Method I

On the basis of mean and standard deviation of normalized red and green values, human skin pixels are determined by using the rules specified as inequalities. In this method, samples of human skin from different images of the database were collected and used. For every pixel in the skin samples the values of r, g and R are calculated, and then the mean and standard deviation of r, g and R in all skin samples are calculated. The mean values of the r, g and R components are denoted by μ_r, μ_g and μ_R respectively and the standard deviation of the r, g and R components are denoted by σ_r, σ_g and σ_R respectively.

Now, the above information is used with the normalized RGB color space to detect the human skin pixels. The first step of the skin detection is to compute r, g and R for every pixel of the input image. If the pixel values of r, g and R satisfy the following inequalities (5), then the pixel is considered as a skin pixel, otherwise, the pixel is classified as non-skin pixel. The value of α determines how accurate the skin detector will be and its value is to be determined by using the trial and error method.

$$\begin{aligned} \sigma_r - \alpha\mu_r < r < \sigma_r + \alpha\,\mu_r \\ \sigma_g - \alpha\mu_g < g < \sigma_g + \alpha\mu_g \\ \sigma_R - \alpha\mu_R < R < \sigma_R + \alpha\mu_R \end{aligned} \tag{5}$$

The output is produced in the form of binary mask i.e. ones represent the skin region and zeros represent the non-skin region.

Method II

In the method, the mean and standard deviation of Base are calculated in lieu of R that is used in the above method. The following inequalities (6) are used of the identification of skin and non-skin pixels in the method. The pixels that satisfy the inequalities are considered as a skin.

$$\begin{aligned} \mu_r - \alpha\sigma_r + \mu_r/\sigma_r < r < \mu_r + \alpha\sigma_r + \mu_r/\sigma_r \\ \mu_g - \alpha\sigma_g + \mu_g/\sigma_g < g < \mu_g + \alpha\sigma_g + \mu_g/\sigma_g \\ \mu_{Base} - \alpha\sigma_{Base} + \mu_{Base}/\sigma_{Base} < Base < \mu_{Base} + \alpha\sigma_{Base} + \mu_{Base}/\sigma_{Base} \end{aligned} \tag{6}$$

4.2 Nonadaptive Skin Color Model

The paper has proposed two nonadaptive skin color model that uses decision rules for the extraction of skin region of human face. The nonadaptive skin color model use a pre-defined inequalities which are given as rules as shown below.

Method I

It is the simple threshold as well as the simple skin color method. In this method, the skin pixels are explicitly identified by defining the boundaries of skin cluster in the specified color space. For every normalized green value g and the RGB red value R, there is a lower bound for discriminating the pixels whereas for every normalized red value, both upper and lower bound are applied for discriminating the pixels. If the pixel values of r, g and R satisfy the following inequalities (7), the pixel is considered as a skin.

$$(t1 > r > t2) \cap (g > t3) \cap (R > t4) \tag{7}$$

Method II

The method uses the concept of r/g ratio. The first step of the method is started with the calculation of the ratio of normalized red values r and normalized green values g. The ratio is denoted by a. The next step is to add all the components of normalized RGB, make a square of it and denote it in the name of bb. After that the values of c and d are calculated by using the following equations:

$$a = r/g \tag{8}$$
$$bb = (r + g + b)^2 \tag{9}$$
$$c = (r * g)/bb \tag{10}$$
$$d = (r * b)/bb \tag{11}$$

If the pixel values of a, c, d and g satisfy the following inequalities (12), the pixel is considered as a skin.

$$(a > t9) \cap (c > t10) \cap (d > t11) \cap (g > t12) \tag{12}$$

5 Facial Feature Detection

As mentioned previously, the purpose of extracting skin-color regions is to reduce the search time for possible face regions on the input image. The main goal of the paper is to find out the facial features accurately. The method proposed by Soriano [1] helps in detecting both skin region and facial features. The author had suggested quadratic equations in order to reduce the sensitivity to changing light. The skin color distribution would gather in r-g plane. The quadratic equations are given as follows:

$$Q+ = -1.3767r^2 + 1.0743r + 0.1452 \tag{13}$$
$$Q- = -0.776r^2 + 0.5601r + 0.1766 \tag{14}$$
$$W = (r - 0.33)^2 + (g - 0.33)^2 \tag{15}$$

The final equation for skin color detection is formulated as specified in the equation (16).

$$S = \begin{cases} Q+ > g > Q- \\ W > 0.0004 \\ 0.6 > r > 0.2 \end{cases} \tag{16}$$

It is the possibility that some skin detecting equations produce better result for some images as compared to others. There is also another equation [10] that doesn't produce better result for the database used. In this paper, the above equation has been slightly modified because it is not producing better result for the database used by us. So by making a change to the inequalities, the better results are obtained by us for the images. The modified equation (17) is given as follows:

$$S = \begin{cases} Q+ > g > Q- \\ W > 0.0999 \\ r > 0.387 \cap g < 0.24 \cap b > 1.963 \cup R > 145 \end{cases} \tag{17}$$

There may be possibilities that the above methods don't produce good result for some images. This is done because of the effect of variations of lights in color images that leads to missed skin pixels.

Based on the observation, the above quadratic polynomials help in detecting the facial features like eyes, and lips. In this paper, another quadratic polynomial is defined as the discriminant function for facial feature pixels.

The quadratic polynomial of the lower boundary of the Q- defined in eq. (14) can be reused with a slight modification. The modified polynomial proposed by Soriano [1] is defined as follows:

$$f = -0.776r^2 + 0.5601r + 0.165 \tag{18}$$

The following rule is modified by us and is now defined for detecting facial feature pixels:

$$F = \begin{cases} g \leq f \\ r \geq 0.48 \cap g \geq 0.275 \cap B \geq 35 \end{cases} \tag{19}$$

where F = 1 means the pixel shows the facial feature.

6 Experimental Results and Discussions

All the methods used in this paper are implemented using IDL (Interactive Data Language) language which is ideal software for image processing. The practical implementation of the above methods produces better results. The steps for skin detection are as follows: transforming the color of pixel to another color space i.e. from RGB color space to normalized RGB, using only two color components in the classification, and classifying by modeling the distribution of skin color [4]. Two skin color models are used for skin detection in this paper: adaptive skin color model and non-adaptive skin color model. Some of the two adaptive and two nonadaptive skin color models are proposed for the evaluation of human skin detection. After detecting human skin regions in the color images, facial features are detected using the same normalized RGB color space. The results obtained for human skin and face detections are much better. For the evaluation purpose, the commonly used face databases, consisting of more than thousand face images are used for the human skin detection and face detection. Some of the sample images used from extended M2VTS face data base [30] is shown in figure 3(a). The methods has also tested on some other databases, namely Caltech Face Database [31], Indian Face Database [32], and Postech faces'01 [33].

The first method in the adaptive skin color model produces better results in terms of facial feature for most of the images. Background is classified correctly. There is no problem with the clothes and hair for most of the images except the images having some colors close to the skin color. The accuracy of skin detector depends on the value of α as shown in fig 3.b. The second method also produces better result for all of the images but if it is compared with the first method it has some false positives in terms of clothes as shown in fig 3.c. The clothes which are light in color or in other words close to the skin-tone are considered as a skin by this method.

Table 1. Skin Detection Rates of Adaptive Skin Color Models

S.No.	*Methods*	*No. of images producing better results*	*Percentage of detection rates*	*Percentage of false detection rates*
1.	Method I	43	98	0.02
2.	Method II	42	95	0.04

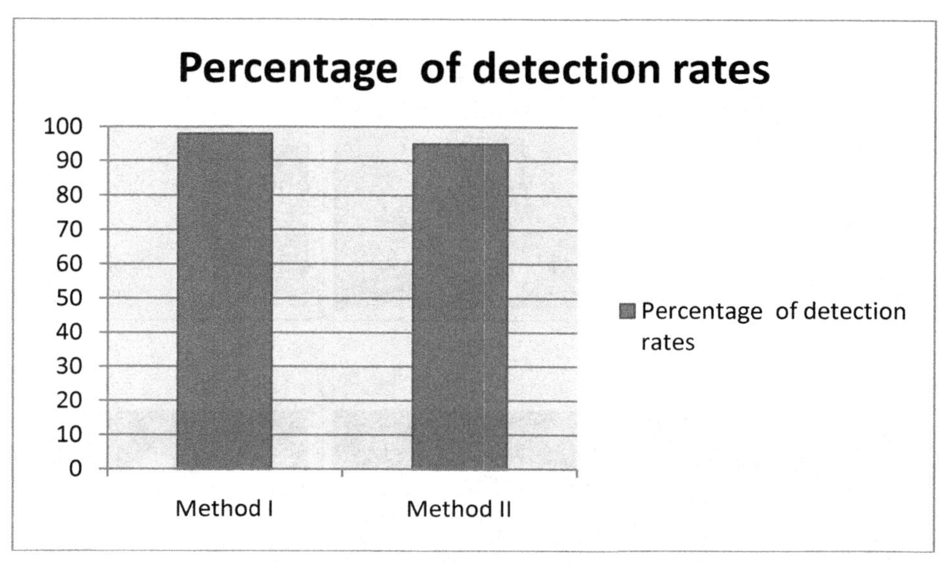

Fig. 1. Graph showing skin detection rate of adaptive methods

Table 2. Skin Detection Rates of Non-Adaptive Skin Color Models

S.No.	*Methods*	*No. of images producing better results*	*Percentage of detection rates*	*Percentage of false detection rates*
1.	Method I	42	95	0.04
2.	Method II	43	98	0.02

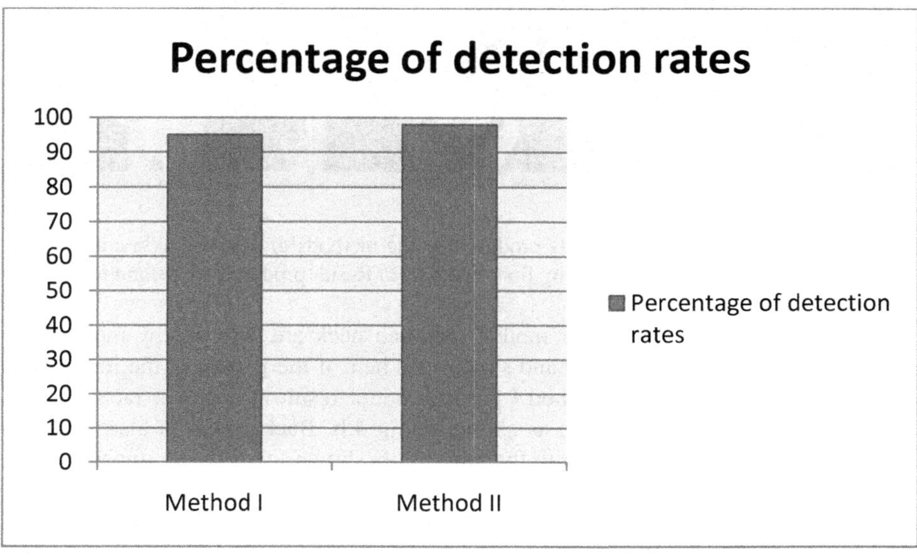

Fig. 2. Graph showing skin detection rate of nonadaptive methods

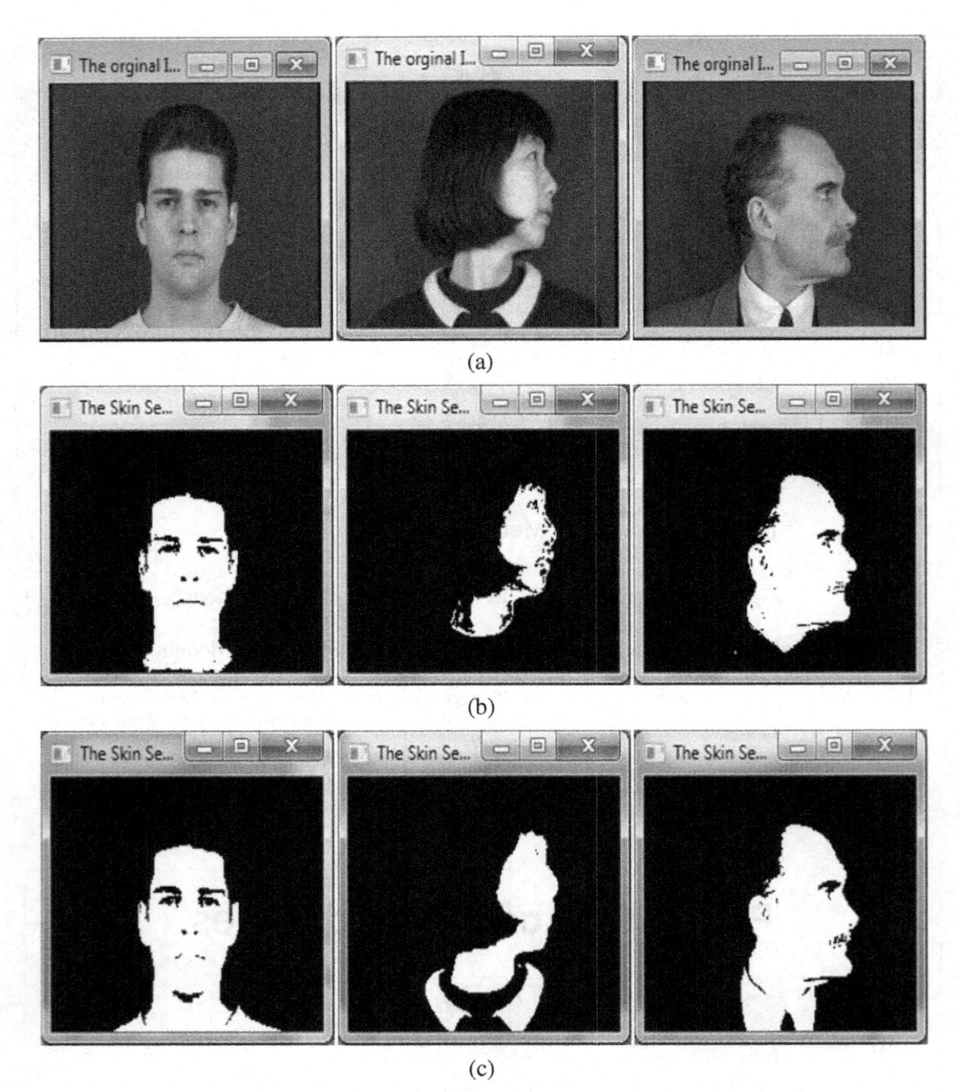

Fig. 3. Sample images and the results produces by the methods of adaptive skin-color model: (a) Input image; (b) Result produced by first method; (c) Result produced by second method

In the non-adaptive skin color model, face and neck are detected by most of the methods but clothes, background and sometimes hair of the person in the image may not be easily classified. The method I produce better result in terms of facial feature detection for most of the images as shown in fig 4.b. Background is also correctly classified. There is no problem with the hair colors shown in all of the images except the golden hairs. If the color of the hair and clothes are the same as skin color, the hair or clothes are detected as skin regions resulting in false positives. The second method produces much better result in terms of facial feature detection for all the images.

Background is correctly classified. The major problem with this method is in detecting the golden hair as skin if the skin color and hair color are the same. The result of this method is shown in fig 4.c.

After extracting the skin-color region of human face, the next step is to identify the facial features like eyes and lips by using face detection technique. The detection of facial feature pixels can be done in parallel with the detection of skin-color pixels. This technique detects nostrils, eyes, mouth, etc as shown in fig. 5. The technique produces the results with 82% accuracy.

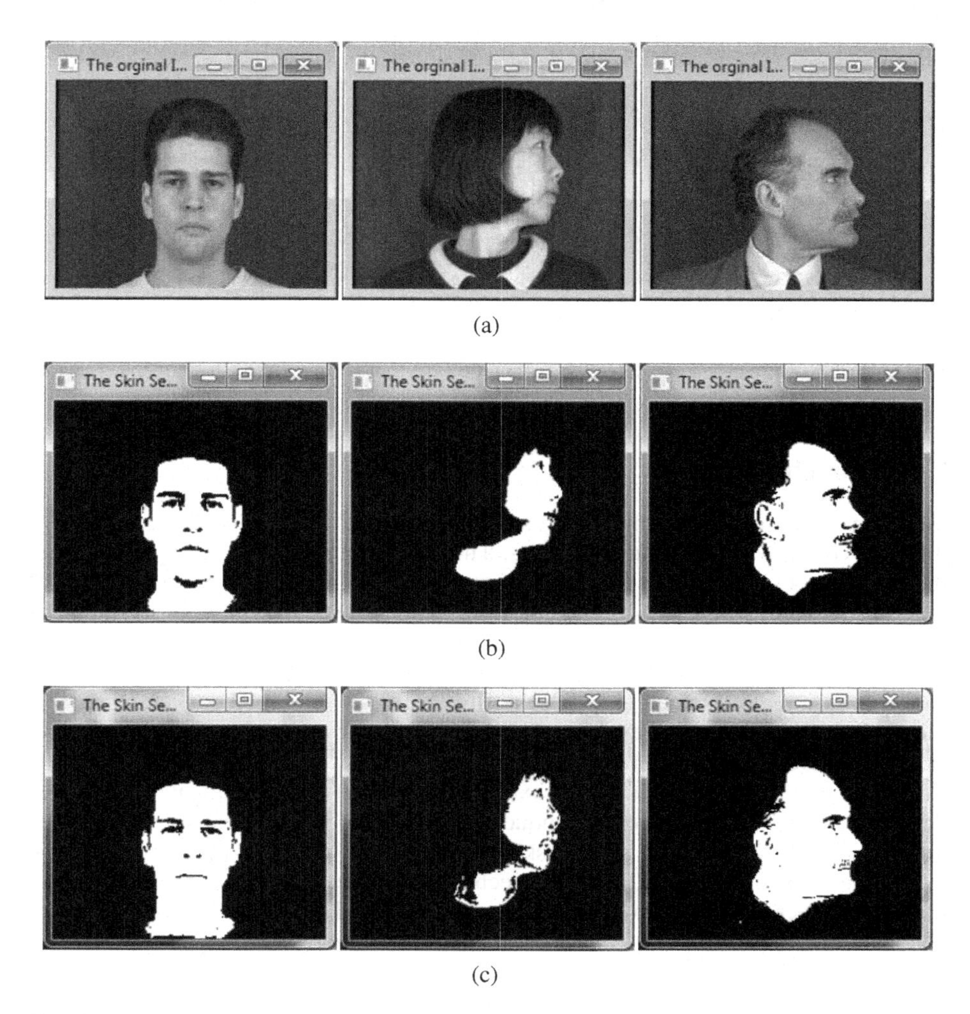

Fig. 4. Sample images and the results produces by the methods of nonadaptive skin-color model: (a) Input image; (b) Result produced by first method; (c) Result produced by second method

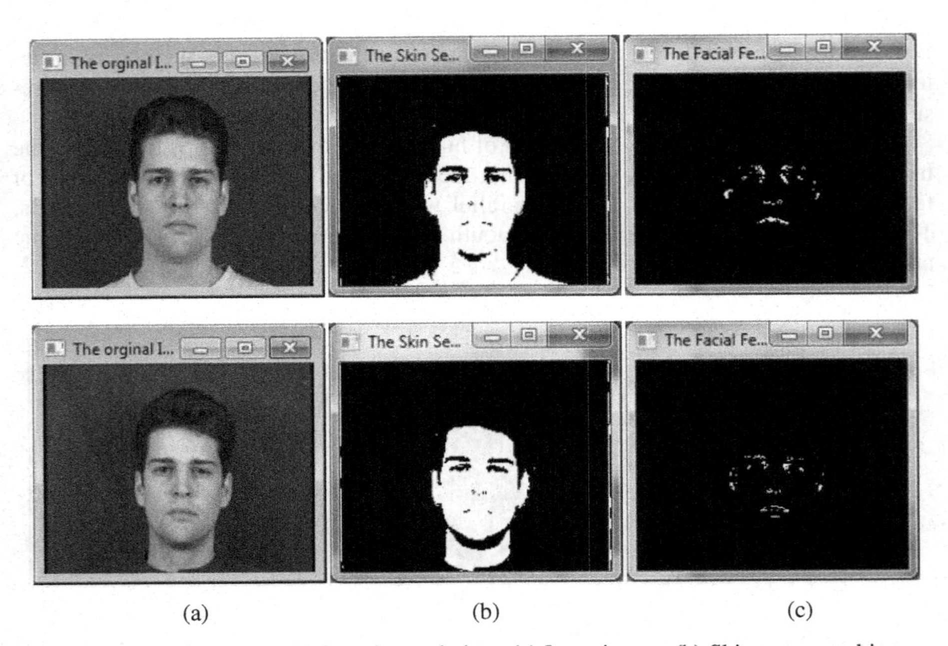

(a) (b) (c)

Fig. 5. Result of facial feature detection technique (a) Input image; (b) Skin segmented image; (c) Facial feature

7 Conclusion

This paper focused on a comparative study of adaptive and non-adaptive skin color model used for skin detection techniques. The experimental results show that the methods belong to adaptive and non-adaptive skin color model respectively produces much better results in terms of skin and facial feature detection. Most of the methods proposed as adaptive and non-adaptive skin color models produce better results. If the color of the hair or clothes is the same as that of skin color, the hair or clothes are detected as skin regions resulting in false positives. If these two methods are compared, the non-adaptive skin color model produces better result as compared to the adaptive skin color model which is shown by comparing the table 1 and 2. After extracting the skin-color region of human face, the next step is to identify the facial features like eyes and lips by using face detection technique. The detection of facial feature pixels can be done in parallel with the detection of skin-color pixels. This technique detects nostrils, eyes, mouth, etc as shown in fig. 5. The technique produces the results with 82% accuracy. This paper can be improved by incorporating special methods that may be used to detect face regions in the color image containing one or more faces.

References

1. Soriano, M., Huovinen, S., Martinkauppi, B., Laaksonen, M.: Using the Skin Locus to Cope with Changing Illumination Conditions in Color-Based Face Tracking. In: Proceedings by IEEE Nordiac Signal Processing Symposium (NORSIG 2000), Kolmarden, Sweden, pp. 383–386 (2000)

2. Caetano, T.S., Barone, D.A.C.: A probabilistic Model for the Human Skin Color. In: Proceedings of the 11th International Conference on Image Analysis and Processing, ICIAP 2001 (2001)
3. Caetano, T.S., Olabarriaga, S.D., Barone, D.A.C.: Performance Evaluation of Single and Multiple-Gaussian Models for Skin Color Modeling. In: Proceedings of the XV Brazilian Symposium on Computer Graphics and Image Processing (SIBGRAPI 2002). IEEE (2002)
4. Jayaram, S., Schmugge, S., Shin, M.C., Tsap, L.V.: Effect of Color Space Transformation, the Illuminance Component, and Color Modeling on Skin Detection. In: Proceedings of the IEEE Computer Society Conference on Computer Vision and Pattern Recognition, CVPR 2004 (2004)
5. Hota, R.N., Venkoparao, V., Bedros, S.: Face Detection by using Skin Color Model based on One Class Classifier. In: 9th International Conference on Information Technology, ICIT 2006 (2006)
6. Liu, R., Zhang, M., Ma, S.: Design of Face Detection and Tracking System. In: 3rd International Congress on Image and Signal Processing (CISP 2010), pp. 1840–1844 (2010)
7. Jones, M.J., Rehg, J.M.: Statistical Color Models with Application to Skin Detection. Proceeding of IEEE, 1–7 (1999)
8. Brand, J., Mason, J.S.: A Comparative Assessment of Three Approaches to Pixel-level Human Skin-Detection. Proceedings by IEEE, 1056–1059 (2000)
9. Pujol, A., Lumbreras, F., Varona, X., Villanueva, J.J.: Locating People in Indoor Scenes for Real Applications. Proceedings of IEEE, 632–635 (2000)
10. Soriano, M., Martinkauppi, B., Huovinen, S., Laaksonen, M.: Skin Detection in Video under Changing Illumination Conditions. Proceedings of IEEE, 839–842 (2000)
11. Kumar, P., Sengupta, K., Lee, A.: A Comparative Studies of Different Color Spaces for Foreground and Shadow Detection for Traffic Monitoring System. Proceedings by IEEE, 100–105 (2002)
12. Chen, L., Zhou, J., Liu, Z., Chen, W., Xiong, G.: A Skin Detector Based on Neural Network. Proceedings of IEEE, 615–619 (2002)
13. Gomez, G.: On Selecting Colour Components for Skin Detection. Proceedings of IEEE, 961–964 (2002)
14. Boussaid, F., Chai, D., Bouzerdoum, A.: A Current-Mode VLSI Architecture for Skin Detection. Proceeding by IEEE, 629–632 (2003)
15. Jedynak, B., Zheng, H., Daoudi, M.: Statistical Models for Skin Detection. In: Proceedings of the 2003 Conference on Computer Vision and Pattern Recognition Workshop (2003)
16. Bayoumi, F., Fouad, M., Shaheen, S.: Based Skin Human Detection in Natural and Complex Scenes. Proceedings by IEEE, 568–571 (2004)
17. Ravindra Babu, T., Mahalakshmi, K., Subrahmanya, S.V.: Optimal Skin Detection for Face Localization using Genetic Algorithms. Procedings of IEEE (2011)
18. Hsieh, C.-C., Liou, D.-H., Lai, W.-R.: Enhanced face-Based Adaptive Skin Color Model. Journal of Applied Science and Engineering 15(2), 167–176 (2012)
19. Zarit, B.D., Super, B.J., Quek, F.K.H.: Comparison of Five Color Models in Skin Pixel Classification
20. Vezhnevets, V., Sazonov, V., Andreeva, A.: A Survey on Pixel-Based Skin Color Detection Techniques
21. Mostafa, L., Abdelazeem, S.: Face Detection Based on Skin Color Using Neural Networks. In: GVIP 2005 Conference, December 19-21. CICC, Cairo (2005)
22. Bayoumi, F., Fouad, M., Shaheen, S.: Feature-Based Human Face Detection. In: 21st National Radio Science Conference (NRSC 2004) (NTI), March 16-18 (2004)

23. Boussaid, F., Chai, D., Bouzerdoum, A.: On-Chip Skin Detection for Color CMOS Imagers. In: Proceedings of the International Conference on MEMS, NANO and Smart Systems (2003)
24. Terrillon, J.-C., Akamatsu, S.: Comparative Performance of Different Chrominance Spaces for Color Segmentation and Detection of Human Faces in Complex Scene Images. In: Vision Interface 1999, Trois-Rivières, Canada May 19-21 (1999)
25. Sigal, L., Sclaroff, S., Athitsos, V.: Skin Color-Based Video Segmentation under Time-Varying Illumination. IEEE Transactions on Pattern Analysis and Machine Intelligence 26(7) (July 2004)
26. Phung, S.L., Bouzerdoum, A., Chai, D.: Skin Segmentation Using Color Pixel Classification: Analysis and Comparison. IEEE Transactions on Pattern Analysis and Machine Intelligence 27(1) (January 2005)
27. Zhao, X., Boussaid, F., Bermak, A.: Characterization of a 0.18 μm CMOS Color Processing Scheme for Skin Detection. IEEE Sensors Journal 7(11), 1471–1474 (2007)
28. Yogarajah, P., Condell, J., Curran, K., Kevitt, P.M., Cheddad, A.: A Dynamic threshold approach for Skin Segmentation in Color Images. Int. J. of Biometrics (2010)
29. Vezhnevets, V., Andreeva, A.: A Comparative Assessment of Pixel-based Skin Detection Methods. GML Computer Vision Group (2005)
30. Extended M2VTS database, http://www.ee.surrey.ac.uk/CVSSP/xm2vtsdb/
31. Caltech Face Database, http://www.vision.caltech.edu/html-files/archive
32. Indian Face Database, http://vis-www.cs.unmass.edu/~vidit/IndianFaceDatabase/
33. Postech Faces 2001, http://dblp.uni-trier.de/db/conf/fgr/fg2008

Detection of Drowsiness Using Fusion of Yawning and Eyelid Movements

Vidyagouri B. Hemadri and Umakant P. Kulkarni

Department of Computer Science and Engineering, SDMCET, Dharwad- 580 002, India
{vidya_gouri,upkulkarni}@yahoo.com

Abstract. Use of technology in building human comforts and automation is growing fast, particularly in automobile industry. Safety of human being is the major concern in vehicle automation. Statistics shows that 20% of all the traffic accidents are due to diminished vigilance level of driver and hence use of technology in detecting drowsiness and alerting driver is of prime importance. In this paper, method for detection of drowsiness based on multidimensional facial features like eyelid movements and yawning is proposed. The geometrical features of mouth and eyelid movement are processed, in parallel to detect drowsiness. Harr classifiers are used to detect eyes and mouth region. Only the position of lower lip is selected to check for drowsiness as during yawn only lower lip is moved due to downward movement of lower jaw and position of the upper lip is fixed. Processing is done only on one of the eye to analyze attributes of eyelid movement in drowsiness, thus increasing the speed and reducing the false detection. Experimental results show that the algorithm can achieve a 80% performance for drowsiness detection under varying lighting conditions.

Keywords: Face detection, Eyelid movement, Yawn detection, Drowsiness detection.

1 Introduction

The increasing number of traffic accidents due to a diminished vigilance level of driver has become a serious problem for society. Accidents related to driver hypo-vigilance are more serious than other types of accidents, since sleepy drivers often do not take evasive action prior to a collision [1, 2]. Monitoring the driver's level of vigilance and alerting the driver when he is not paying adequate attention to the driving has become matter of concern in order to prevent accidents. The prevention of such accidents is a major focus of effort in the field of active safety research.

Many researchers are working on the development of monitoring systems using specific techniques. The best detection techniques are based on physiological phenomena like brain waves, heart rate, pulse rate and respiration. These techniques are intrusive, causing annoyance. A driver's state of vigilance can also be characterized by indirect behaviors of the vehicle like lateral position, steering wheel movements. Although these techniques are not intrusive, they are subjected to several limitations as the vehicle type, driver experience, geometric characteristics and state of the road. People in fatigue show some visual behavior easily observable from changes in their facial features like eyes, head and face.

S. Unnikrishnan, S. Surve, and D. Bhoir (Eds.): ICAC3 2013, CCIS 361, pp. 583–594, 2013.

In this paper a novel algorithm to detect drowsiness based on eyelid movements and yawning is proposed. The Viola Jones algorithm is used to detect face and Harr classifiers are used to detect eyes and mouth region. The mouth geometrical features and eyelid movement is processed, in parallel, to detect drowsiness. The fusion of these data allows the system to detect both drowsiness and inattention, thus improving the reliability of existing unidimensional features based algorithms. Experimental results show that this algorithm can achieve an 80% performance for drowsiness detection under good lighting conditions and 50% detection under poor lighting conditions.

This paper is organized as follows: in section 2 the related work about the detection of driver fatigue is presented. Section 3 describes the proposed method. Experimental results are shown on section 4 and finally section 5 presents the conclusion and future studies.

2 Related Work

Boon-Giin Lee and Wan-Young Chung describes a method to monitor driver safety by analyzing information related to fatigue based on eye movement monitoring and bio-signal processing [3]. Video sensors are used to capture the driver image and a bio-signal sensor to gather the driver photoplethysmograph signal. A monitoring system is designed in Android-based smartphone where it receives sensory data via wireless sensor network and further processes the data to indicate the current driving aptitude of the driver. A dynamic Bayesian network framework is used for the driver fatigue evaluation.

The drowsiness prediction by employing support vector machine with eyelid related parameters extracted from EOG data is proposed by Hu Shuyan, Zheng Gangtie [4]. The dataset is firstly divided into three incremental drowsiness levels, and then a paired t-test is done to identify how the parameters are associated with drivers' sleepy condition. With all the features, a SVM drowsiness detection model is constructed. The validation results show that the drowsiness detection accuracy is quite high especially when the subjects are very sleepy.

Shabnam Abtahi, Behnoosh Hariri, Shervin Shimiohammadi presents a method for detecting drivers' drowsiness and subsequently alerting them based on the changes in the mouth geometric features [5]. The driver's face is continuously recorded using a camera that is installed under the front mirror. The face is detected and tracked using the series of frame shots taken by the camera followed with detection of the eyes and the mouth. The mouth geometrical features are then used to detect the yawn. The system will alert the driver of his fatigue and the improper driving situation in case of yawning detection.

Tianyi Hong, Huabiao Qin proposes a method based on eye state identification of driver's drowsiness detection in embedded system using image processing techniques [6]. This method utilizes face detection and eye detection to initialize the location of driver's eyes, after that an object tracking method is used to keep track of the eyes, finally, identify drowsiness state of driver can be identified with PERCLOS by identified eye state. Experiment results show that it makes good agreement with analysis.

Danghui Liu, Peng Sun, YanQing Xiao, Yunxia Yin proposed drowsiness detection algorithms based on eyelid movement [7]. Driver's face is detected using cascaded classifiers algorithm and the diamond searching algorithm used to trace the face. From a temporal difference image, a simple feature is extracted and used to analyze rules of eyelid movement in drowsiness. Three criterions such as the duration of eyelid closure, the number of groups of continuous blinks and the frequency of eye blink are used to judge whether a driver is drowsy or not. Experimental results show that this algorithm achieves a satisfied performance for drowsiness detection.

Advanced driver assistance system for automatic driver's drowsiness detection based on visual information and artificial intelligent is presented by Marco Javier Flores, Jose Maria Armingol and Arturo de la Escalera [8]. Face and the eyes are located and tracked to compute a drowsiness index. This system uses advanced technologies for analyzing and monitoring driver's eye state at real-time and real-driving conditions.

Antoine Picot, Sylvie Charbonnier and Alice Caplier presents an algorithm for driver's drowsiness detection based on visual signs that can be extracted from the analysis of a high frame rate video [9]. A data mining approach is used to evaluate the relevance of different visual features on a consistent database, to detect drowsiness. Fuzzy logic that merges the most relevant blinking features (duration, percentage of eye closure, frequency of the blinks and amplitude-velocity ratio) is proposed.

Smith, Shah, and Lobo present a system for analyzing human driver alertness [10,11]. The system relies on estimation of global motion and color statistics to robustly track a person's head and facial features. The system classifies rotation in all viewing directions, detects eye/mouth occlusion, detects eye blinking and eye closure, and recovers the 3D gaze of the eyes. The system tracks both through occlusions due to eye blinking, and eye closure, large mouth movement, and also through occlusion due to rotation.

Azim Eskandarian and Ali Mortazavi describes the performance of a neural network based algorithm for monitoring the driver's steering input, and correlations are found between the change in steering and state of drowsiness in case of truck drivers [12]. The algorithm develops the design of a drowsiness detection method for the commercial truck using artificial neural network classifier, trained and tested in a simulated environment.

Taner Danisman, Ian Marius Bilasco, Chabane Djeraba, Nacim Ihaddadene presents an automatic drowsy driver monitoring and accident prevention system based on monitoring the changes in the eye blink duration [13]. This method detects visual changes in eye locations using the horizontal symmetry feature of the eyes and detects eye blinks via a standard webcam in real-time. The eye blink duration is the time spent while upper and lower eye-lids are connected. The pattern indicates a potential drowsiness prior to the driver falling asleep and then alerts the driver by voice messages. Results showed that the proposed system detects eye blinks with 94% accuracy with a 1% false positive rate.

Effective driver fatigue detection method based on spontaneous pupillary fluctuation behavior is described by Xingliang Xiong, Lifang Deng, Yan Zhang, Longcong Chen [14]. Using infrared-sensitive CCD camera 90s long infrared video of pupillogram is captured. Edge detection algorithm based on curvature characteristic of pupil boundary is employed to extract a set of points of visible pupil boundary, particularly the eyelids or eyelashes are occluding the pupil. These points are adapted to fit a circle by using least squares error criterion, the diameter of this fitted circle is the approximated diameter of the pupil in current frame of video. Finally, the values of pupil diameter variability (PDV) in 90s long video was calculated.

Regan *et al* [15] defined driver distraction and driver inattention and taxonomy is presented in which driver distraction is distinguished from other forms of driver inattention. The taxonomy and the definitions are intended to provide a common framework for coding different forms of driver inattention as contributing factors in crashes and incidents, so that comparable estimates of their role as contributing factors can be made across different studies, and to make it possible to more accurately interpret and compare, across studies, the research findings for a given form of driver inattention. Authors suggested that the driver inattention as insufficient or no attention to activities critical for safe driving, and that driver diverted attention (which is synonymous with "driver distraction) is just one form of driver inattention. The other forms of driver inattention they have labeled tentatively as driver restricted attention, driver misprioritised attention, driver neglected attention and driver cursory attention.

Fan et al [16] described a novel Gabor-based dynamic representation for dynamics in facial image sequences to monitor human fatigue. Considering the multi-scale character of different facial behaviors, Gabor wavelets were employed to extract multi-scale and multi-orientation features for each image. Then features of the same scale were fused into a single feature according to two fusion rules to extract the local orientation information. AdaBoost algorithm was exploited to select the most discriminative features and construct a strong classifier to monitor fatigue. Authors tested on a wide range of human subjects of different genders, poses and illuminations under real-life fatigue conditions. Findings indicated the validity of the proposed method, and an encouraging average correct rate was achieved.

Most of these algorithms [4-9, 12-14] are unidimensional in which case detection of drowsiness is based on single parameter. The reliability of the system can be improved by the fusion of these data and allowing the system to detect both drowsiness and inattention. A novel drowsiness detection algorithm based on fusion of eyelid movements and yawning is proposed. Processing of the mouth geometrical features and eyelid movement is done, in parallel to detect drowsiness. To increase the speed of detection only the position of lower lip is selected and only one of the eyes is analyzed for criterions, such as opening and closing of eye, duration of opening and closing and frequency of blink. Experimental results show that algorithm can achieve a satisfied performance for drowsiness detection.

3 Proposed Approach

The proposed algorithm consists of different phases, to properly analyze changes in the facial attributes like, changes in the mouth geometrical features of the driver and changes in the status of the eyes. These phases are categorized as follows.

1. Face detection
2. Face tracking
3. Eye detection and Mouth detection
4. Fusion of the data
5. Drowsiness detection

The sample videos are collected with different intensities or captured using webcam for fixed interval of time. The face region is detected from the videos after being framed using Viola and Jones algorithm in opencv [16, 17]. After detecting the face, mouth and eye regions of the face are detected and processed in parallel as shown in figure 1.

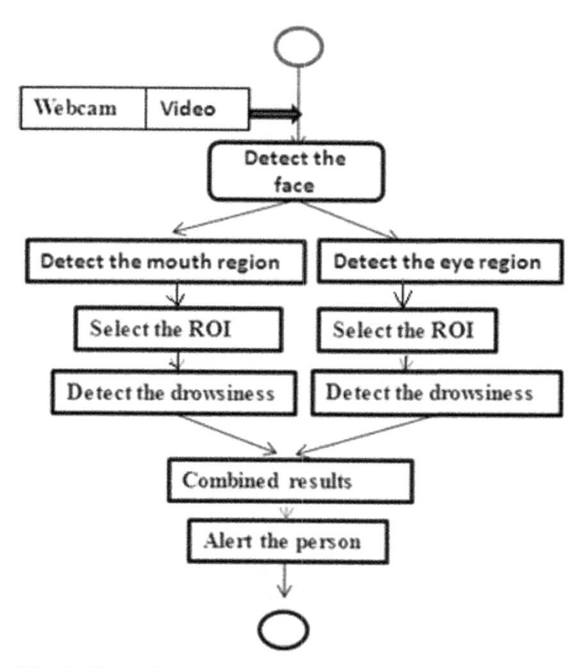

Fig. 1. Flow diagram of detection of drowsiness system

3.1 Yawn Detection

Mouth part is extracted, binarized and subjected for edge detection using canny edge detection. Mouth region is selected based on the face region, converted to gray scale and canny edge detection is applied to find lips edge. This edge detection depends on the different threshold value for different light intensity and resolution of the camera. The method uses multiple thresholds to find edges. Upper threshold is used to find start point of an edge. The path of the edge is then traced through the image pixel by pixel and marking the edge if it is above the lower threshold. The pixel point is not marked if the value falls below the lower threshold. The algorithm for yawn detection is described below.

```
Procedure drowsy detect
  1. Capture the video
  2. Form the frames from the input video
  3. Detect the face contour. Viola and Jones algorithm
     is used to detect face as shown in  figure 2.
  4. Call Yawn detection and eye detection algorithms.
  5. Collect the returned value from the above two
     algorithms.
  6. If Drowsy then
        Person is alarmed by a voice alert.
     End if
End of algorithm
```

```
Procedure Yawn Detection
  1. Detect mouth region as shown in figure 3.
  2. Convert the above image to gray scale.
  3. Using Canny edge detectors edges of lips are
     detected.
  4. The template matching is used to match the resultant
     image
  5. If the template match with any of four white Objects
     then Mouth opened
       else Mouth closed
       End if
  6.  If mouth found open in four continuous frames in
      two seconds then  yawn Detected
      End if
  7.  If Yawning then return drowsy End if
End of algorithm
```

Fig. 2. Detection of face

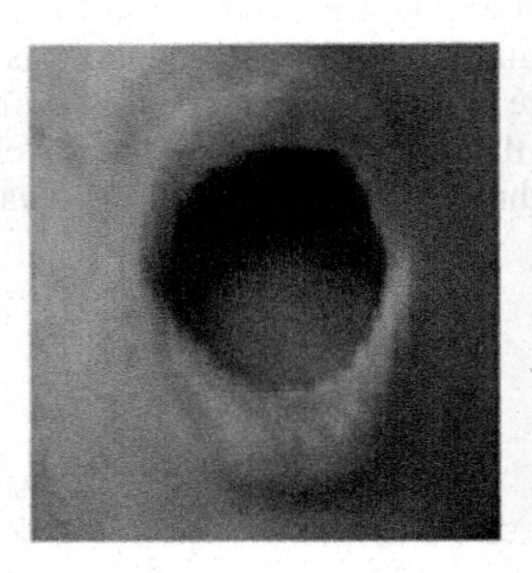

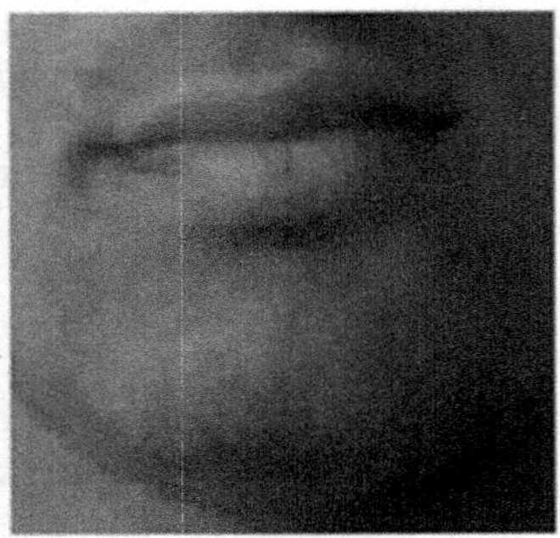

Fig. 3. Detection of mouth

The result of applying canny edge detection is as shown in figure 4.

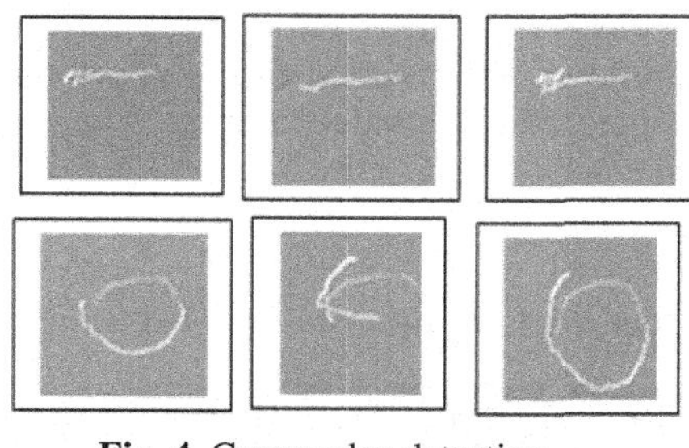

Fig. 4. Canny edge detection

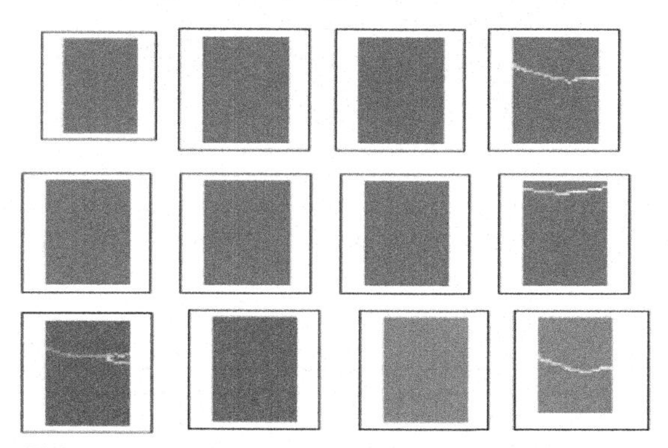

Fig. 5. Detection of downward moment of mouth

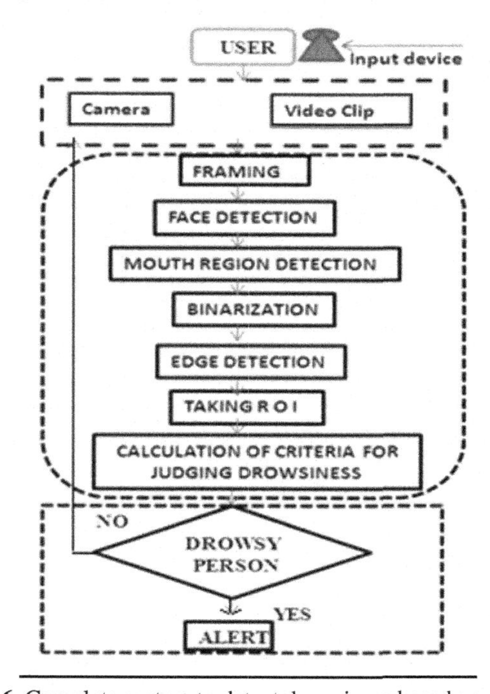

Fig. 6. Complete system to detect drowsiness based yawn

Only the position of lower lip is selected to check for drowsiness as during yawn only lower lip is moved down and position of the upper lip is fixed. In figure 5 are shown few regions obtained after the values are set. The complete system is divided into two modules which are shown in figure 6.

3.2 Eye Detection

Once the face region is detected, the eyes are detected using Harr classifiers.

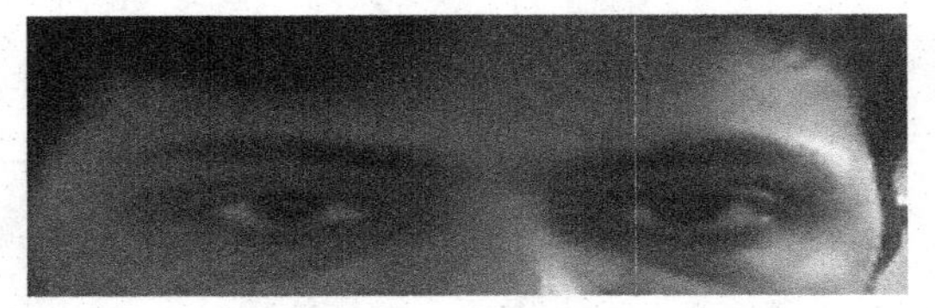

Fig. 7. Detection of eye

The object detector will put the list of detected eyes and, rectangle can be drawn for each eye. Processing is done only on one of the eyes. This approach can be used if

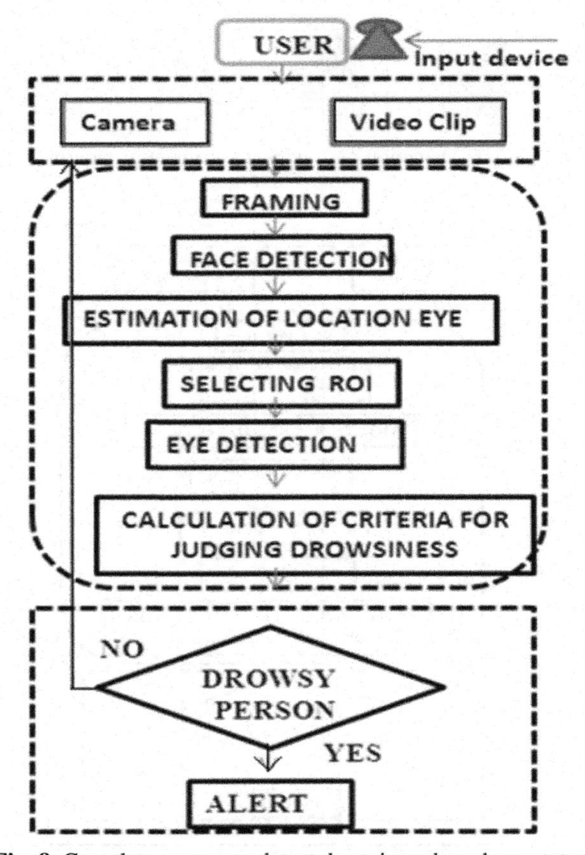

Fig. 8. Complete system to detect drowsiness based eye status

one of the eyes is partially covered with hair. Eye blinking is monitored continuously by observing the status of the eye such as open or closed. In closed state, eye is not detected and is detected only in the open state. This feature is used to find three criterions such as duration of eye closure, duration of opening and frequency of blink. Normally, the eye blinks 10 times per minute, the interval is about two to six seconds, and the duration of each blink is about 0.15 to 0.25 seconds. If eye is closed for more than 8 frames then the person is in fatigue. When a person is drowsy the frequency of his eye blink becomes slower. Contrarily, when he just becomes drowsy from normal spirit state, the frequency of his eye blink is faster during the conversion stage. If the person feels drowsy, to avoid this, he tries to keep his eye open for longer duration. If eye blink of a person is continuous with more frequency, that is if it is larger than 10 times per minute then he is in a fatigue state and he is alarmed. The algorithm for eye detection is described below. The complete system to detect drowsiness based eye parameters is shown in figure 8.

```
Procedure Eye Detection
   1. Eyes are detected using Harr classifiers
   2. Monitor the blinking of the eye by observing the
      status of the eye such as open or closed
   3. If eye is detected then it is in OPEN state
      else it is in CLOSED state.
      End if
   4. If (eye is closed for more than 8 frames) OR (eye
      is open for more than 30 frames) OR (if frequency
      of blink  is larger than 10 times per minute )
     then              Return drowsy
     End if
End of algorithm
```

4 Experimental Results

Nearly fifty videos under different lighting conditions were collected, such as during day time and night time. Proposed algorithm detects the face first. If face is not

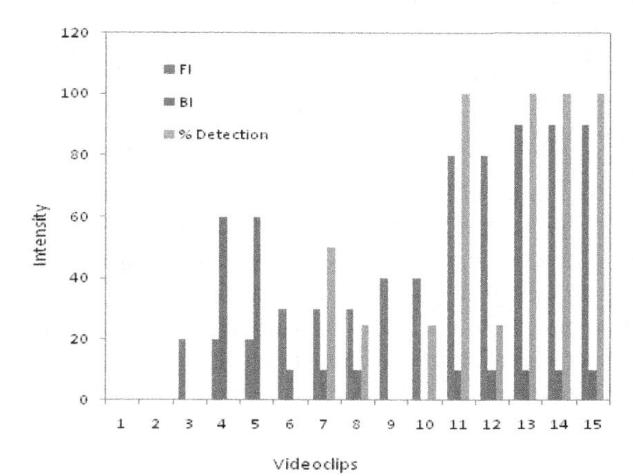

Fig. 9. % Drowsiness Detection under different light conditions (partial data)

Table 1. Detection of drowsiness under different lighting conditions

Input Video clip	FI (In LUX)	BI (In LUX)	Duration of frame yawn found	Duration of eye Closure	Duration of Opening	% of Drowsy
1	0	0	0	2	8	0
2	0	0	0	3	7	0
3	20	0	0	3	9	0
4	20	60	0	3	8	0
5	20	60	0	5	8	0
6	30	10	0	3	7	0
7	30	10	3	3	7	50
8	30	10	2	3	8	25
9	40	0	0	4	7	0
10	40	0	2	3	20	25
11	80	10	6	4	8	100
12	80	10	2	6	8	25
13	90	10	7	3	7	100
14	90	10	7	8	4	100
15	90	10	7	9	3	100
16	130	30	6	9	3	100
17	130	30	6	10	2	100
18	140	40	6	9	3	100
19	140	40	7	9	3	100
20	140	40	2	8	4	25
21	140	100	7	3	7	100
22	140	40	2	5	25	25
23	160	10	8	2	35	100
24	160	10	7	8	3	100
25	160	10	2	4	5	25
26	160	40	7	3	5	100
27	160	40	2	4	6	25
28	190	20	6	2	35	100
29	190	20	2	6	4	25
30	200	70	6	10	2	100

detected, the algorithm fails. Following the detection of the face, the algorithm next finds the mouth and eye region in the face. Detection of the mouth part is followed with yawn detection and detection of eyes followed with attributes of the eyes. The two algorithms are processed in parallel and reporting is done based on the two results. The following shows the result of applying algorithm on different images captured during the experiment under different intensity conditions, FI, front intensity, the intensity on the face during video recording and BI, back intensity, the intensity at the background of the face.

There will be 25% detection if the yawning or eye closure is found in 2 continuous frames in a second,, 50% detection if the yawning or eye closure is found in 3 or 4 continuous frames in a second, 75% detection if mouth or eye closure is found open in 5 to 6 continuous frames and 100% detection if mouth or eye closure is found open in 6 or more continuous frames in 2 second. The Table 1 shows the part of the result of drowsiness detection under different luminance condition. The result shows that the detection rate is 80% under lighting conditions with light intensity greater than 50 lux and less than 8000 lux. The algorithm fails under a very dark (less than 50) or very bright (more than 8000) lighting conditions.

5 Conclusion

In this paper, the algorithm to detect drowsiness based on multidimensional biometric features like mouth geometrical features and eyelid movement is proposed. The position of the lower lip is used to check for yawning. The three parameters of the eye such as duration of closure, duration of opening and frequency of blink is used to check for drowsiness based on eyelid movement. The processing is done only on one of the eye, thus to increase speed of detection and to detect under different hair style conditions. The algorithm works satisfactory under reasonable lighting conditions (50 lux to 8000 lux) and fails to detect under very dark or very bright lighting conditions((less than 50 lux, more than 8000 lux). The algorithm fails to detect drowsiness under few adverse conditions such as different skin colors, wearing spectacles and any lighting conditions.

Acknowledgements. This project was carried under Research Promotion Scheme grant from All India Council for Technical Education (AICTE), project Ref. No: 8023/RID/RPS-114(Pvt)/ 2011-12. Authors wish to thank AICTE, New Delhi.

References

1. Bergasa, L.M., Nuevo, J., Sotelo, M.A., Barea, R., Lopez, M.E.: Real-time system for monitoring driver vigilance. IEEE Transactions on Intelligent Transportation Systems 7(1), 63–77 (2006) ISSN: 1524-9050
2. Ueno, H., Kaneda, M., Tsukino, M.: Development of drowsiness detection system. In: Proc. Vehicle Navigation and Information Systems Conf., pp. 15–20 (1994)
3. Lee, B.-G., Chung, W.-Y.: Driver Alertness Monitoring Using Fusion of Facial Features and Bio-Signals. IEEE Sensors Journal 12, doi:10.1109/JSEN.2012.2190505
4. Hu, S., Zheng, G.: Driver drowsiness detection with eyelid related parameters by Support Vector Machine. Expert Systems with Applications 36, 7651–7658 (2009), doi:10.1016/j.eswa.2008.09.030
5. Abtahi, S., Hariri, B., Shimiohammadi, S.: Driver Drowsiness Monitoring Based on Yawning Detection, 978-1-4244-7935-1/11/IEEE
6. Hong, T., Qin, H.: Drivers Drowsiness Detection in Embedded System, 1-4244-1266-8/07/IEEE

7. Liu, D., Sun, P., Xiao, Y., Yin, Y.: Drowsiness Detection Based on Eyelid Movement. In: 2010 Second International Workshop on Education Technology and Computer Science. IEEE (2010) 978-0-7695-3987-4/10, doi:10.1109/ETCS.2010.292
8. Flores, M.J., Armingol, J.M., de la Escalera, A.: Real-Time Drowsiness Detection System for an Intelligent Vehicle. In: 2008 IEEE Intelligent Vehicles Symposium (2008), 978-1-4244-2569-3/08/IEEE
9. Picot, A., Charbonnier, S., Caplier, A.: Drowsiness detection based on visual signs: blinking analysis based on high frame rate video, 978-1-4244-2833-5/10/IEEE
10. Smith, P., Shah, M., da Vitoria Lobo, N.: Monitoring head/eye motionfor driver alertness with one camera. In: Proc. 15th Int. Conf. Pattern Recognition, Barcelona, Spain, vol. 4, pp. 636–642 (2000)
11. Smith, P., Shah, M., da Vitoria Lobo, N.: Determine driver visual attention with one camera. Intelligent Transportation Systems 4, 205–218 (2003)
12. Eskandarian, A., Mortazavi, A.: Evaluation of a Smart Algorithm for Commercial Vehicle Driver Drowsiness Detection. In: Proceedings of the 2007 IEEE Intelligent Vehicles Symposium (2007), 1-4244-1068-1/07/IEEE
13. Danisman, T., Bilasco, I.M., Djeraba, C., Ihaddadene, N.: Drowsy Driver Detection System Using Eye Blink Patterns, 978-1-4244-8611-3/10/IEEE
14. Xiong, X., Deng, L., Zhang, Y., Chen, L.: Objective Evaluation of Driver Fatigue by Using Spontaneous Pupillary fluctuation, 978-1-4244-5089-3/11/IEEE
15. Regan, M.A., Hallett, C., Gordon, C.P.: Driver distraction and driver inattention: Definition, relationship and taxonomy. Accident Analysis and Prevention 43, 1771–1781 (2011)
16. Fan, X., Sun, Y., Yin, B., Guo, X.: Gabor-based dynamic representation for human fatigue monitoring in facial image sequences. Pattern Recognition Letters 31, 234–243 (2010)
17. Viola, P., Jones, M.: Rapid object detection using a boosted cascade of simple features. In: Proceedings of the IEEE Computer Society Conference on Computer Vision and Pattern Recognition, vol. 12, pp. I-511–I-518 (2001)
18. Intel Open Source Computer Vision Library, http://www.intel.com/technology/computing/opencv/index.htm

Design of Novel Algorithm and Architecture for Gaussian Based Color Image Enhancement System for Real Time Applications

M.C. Hanumantharaju[1], M. Ravishankar[1], and D.R. Rameshbabu[2]

[1] Department of Information Science & Engineering,
Dayananda Sagar College of Engineering, Bangalore, India
[2] Department of Computer Science & Engineering,
Dayananda Sagar College of Engineering, Bangalore, India
{mchanumantharaju,ravishankarmcn,bobrammysore}@gmail.com

Abstract. This paper presents the development of a new algorithm for Gaussian based color image enhancement system. The algorithm has been designed into architecture suitable for FPGA/ASIC implementation. The color image enhancement is achieved by first convolving an original image with a Gaussian kernel since Gaussian distribution is a point spread function which smoothes the image. Further, logarithm-domain processing and gain/offset corrections are employed in order to enhance and translate pixels into the display range of 0 to 255. The proposed algorithm not only provides better dynamic range compression and color rendition effect but also achieves color constancy in an image. The design exploits high degrees of pipelining and parallel processing to achieve real time performance. The design has been realized by RTL compliant Verilog coding and fits into a single FPGA with a gate count utilization of 321,804. The proposed method is implemented using Xilinx Virtex-II Pro XC2VP40-7FF1148 FPGA device and is capable of processing high resolution color motion pictures of sizes of up to 1600×1200 pixels at the real time video rate of 116 frames per second. This shows that the proposed design would work for not only still images but also for high resolution video sequences.

Keywords: Gaussian color image enhancement, Serpentine memory, 2D Gaussian convolution, Logarithm, Field Programmable Gate Array.

1 Introduction

Digital image enhancement [1] refers to accentuation, sharpening of image features such as edges, boundaries or contrast to make a graphic display more useful for display and analysis. The enhanced images with better contrast and details are required in many areas such as computer vision, remote sensing, dynamic scene analysis, autonomous navigation and medical image analysis. In the recent years, color image enhancement has been becoming an increasingly important research area with the widespread use of color images. Numerous methods [2]

S. Unnikrishnan, S. Surve, and D. Bhoir (Eds.): ICAC3 2013, CCIS 361, pp. 595–608, 2013.

are available in the literature for color image enhancement. The color image enhancement can be classified into two categories according to the color space: color image enhancement in RGB color space and color image enhancement based on transformed space.

The paper is organized as follows: Section 2 gives a brief review of previous work done. Section 3 describes the proposed Gaussian based color image enhancement algorithm. Section 4 provides detailed system architecture for hardware realization of color image enhancement algorithm. The results and discussions follow these. Conclusion arrived at are presented in the final section.

2 Related Work

Digital Signal Processors (DSPs) [3][4] have been employed for enhancement of images which provides some improvement compared to general purpose computers. Only marginal improvement has been achieved since parallelism and pipelining incorporated in the design are inadequate. This scheme uses optimized DSP libraries for complex operations and does not take full advantage of inherent parallelism of image enhancement algorithm. The neural network based learning algorithm [5] provides an excellent solution for the color image enhancement with color restoration. The hardware implementation of these algorithms parallelizes the computation and delivers real time throughput for color image enhancement. However, its window related operations such as convolution, summation and matrix dot products in an image enhancement architecture demands enormous amount of hardware resources.

Hiroshi Tsutsui et al. [6] proposed an FPGA implementation of adaptive real-time video image enhancement based on variational model of the Retinex theory. The authors have claimed that the architectures developed in this scheme are efficient and can handle color picture of size 1900×1200 pixels at the real time video rate of 60 frames per sec. The authors have not justified how high throughput has been achieved in spite of time consuming iterations to the tune of 30. Abdullah M. Alsuwailem et al. [7] proposed a new approach for histogram equalization using FPGAs. Although efficient architectures were developed for histogram equalization, the restored images using this scheme are generally not satisfactory.

An efficient architecture for enhancement of video stream captured in non-uniform lighting conditions was proposed by Ming Z. Zhang et al. [8]. The new architecture processes images and streaming video in the HSV domain with the homomorphic filter and converts the result back to HSV. This leads to an additional computational cost and, the error rate is high for the RGB to HSV conversion process. Digital architecture for real time video enhancement based on illumination reflection model was proposed by Hau T. Ngo et al. [9]. This scheme improves visual quality of digital images and video captured under insufficient and non-uniform lighting conditions. Bidarte et al. [10] proposed spatial-based adaptive and reusable hardware architecture for image enhancement. However, the histogram modification used in this scheme treats all regions of the image equally and often results in poor local performance, which in turn affects the image details. The modified luminance based multiscale retinex algorithm proposed by Tao et al. [11] achieves optimal enhancement result with minimal complexity of hardware implementation. However, the algorithm works fine so long as the background is dark and the object is bright.

The limitations mentioned earlier are overcome in the proposed method in an efficient way. To start with, the input image is convolved with 5×5 Gaussian kernel in order to smooth the image. Further, the dynamic range of an image is compressed by replacing each pixel with its logarithm. In the proposed method, the image enhancement operations are arranged in an efficient way adding true color constancy at every step. It has less number of parameters to specify and provides true color fidelity. In addition, the proposed algorithm is computationally inexpensive. In the proposed scheme, an additional step is necessary to solve the gray world violation problem as is the case with the implementation reported in Ref. [11].

In the present work, in order to test the developed algorithm, standard test images have been used and results are favorably compared with that of other researchers. In order to evaluate the performance of the proposed algorithm, the metric Peak Signal to Noise Ratio (PSNR) has been used.

3 Proposed Gaussian Based Image Enhancement Algorithm

In order to account for the smoothness, lightness, color constancy and dynamic range properties of Human Visual System (HVS), Gaussian based image enhancement algorithm has been proposed. The basic operational sequence of the proposed Gaussian based color image enhancement algorithm is shown in Fig. 1. To start with, the original image (which is of poor quality and needing enhancement) is read in RGB color space. The color components are separated followed by the selection of window size as 5×5 for each of the R, G and B components. In each color component, the selected window is convolved with 5×5 Gaussian kernel in order to smooth the image. Next, Logarithmic operation is accomplished in order to compress the dynamic range of the image. Finally, Gain/Offset adjustment is done in order to translate the pixels into the display range of 0 to 255. The separated R, G and B components are combined into composite RGB in order to obtain the enhanced image.

3.1 Convolution Operation

Convolution is a simple mathematical operation which is essential in many image processing algorithms. The color image enhancement algorithm proposed in this paper uses 5×5 pixel window, where a block of twenty five pixels of original image is convolved with Gaussian kernel of size 5×5. The two dimension (2D) Gaussian function is defined by Eqn. (1):

$$g(x,y) = \frac{1}{2\pi\sigma^2} e^{-\frac{x^2+y^2}{2\sigma^2}} \tag{1}$$

where σ is the standard deviation of the distribution; x and y are spatial coordinates.

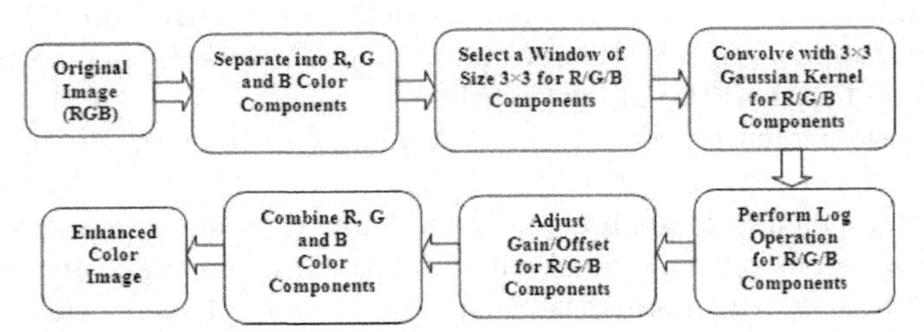

Fig. 1. Flow Sequence of the Proposed Gaussian based Color Image Enhancement Algorithm

The basic principle of Gaussian filters is to use 2D distribution as a point spread function. This property is adopted in image enhancement application and is achieved by convolving the 2D Gaussian distribution function with the original image. A Gaussian kernel can be a 3 × 3 or a 5 × 5 matrix as shown in Fig. 2.

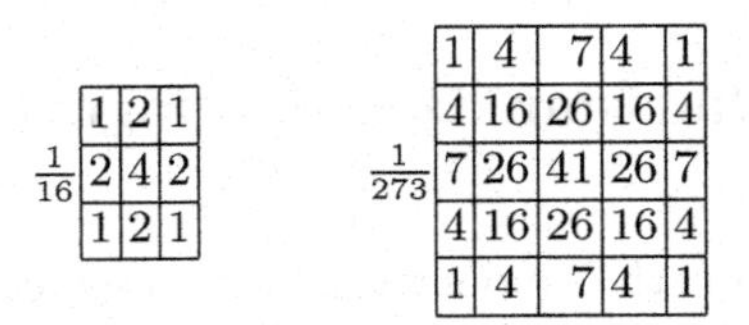

Fig. 2. Gaussian Kernels: 3 × 3 kernel and 5 × 5 kernel

The Gaussian convolution matrix is given by Eqn. (2).

$$G(x, y) = I(x, y) \otimes g(x, y) \tag{2}$$

where ⊗ denotes convolution, g(x, y) is a Gaussian kernel, I(x, y) is the original image and G(x, y) is the convolved output.

Mathematically, 2D convolution can be represented by the following Eqn. (3):

$$G(x, y) = \sum_{i=1}^{M} \sum_{j=1}^{N} I(i, j) \times g(x - i, y - j) \tag{3}$$

where m = 5 and n = 5 for 5 × 5 Gaussian kernel. In this work, the convolution mask is chosen as 5 × 5 in order to enhance the implementation speed, at the same time minimizing blocking artifacts. The convolution operation for a mask of 5 × 5 is given by Eqn. (4)

$$P(x, y) = \frac{\sum_{i=0}^{4} W_i \times P_i}{\sum_{i=0}^{4} W_i} \tag{4}$$

where W_i indicates the 5 × 5 Gaussian mask, P_i is the 5 × 5 sliding window in the input image and P is the Gaussian convolved pixel. As an example, Fig. 3

illustrated the hardware implementation of convolution with mask as 3×3. The convolution mask of 5×5 implemented in this work is similar to that shown in Fig. 3.

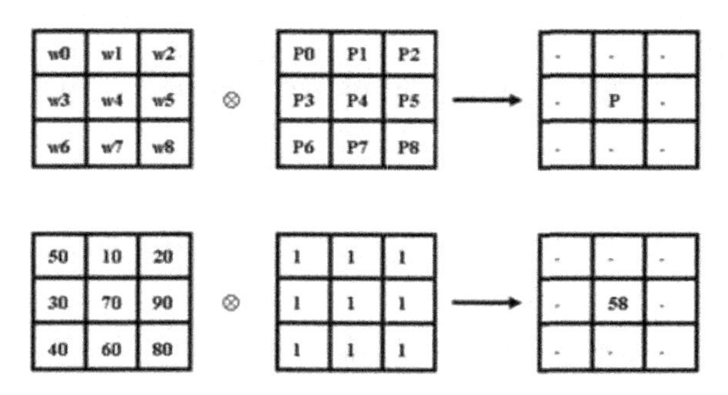

Fig. 3. Two Dimensional Convolution Scheme

Next, the log transformation operation applied on the image compresses the dynamic range of gray level input values to manageable levels for further processing. The logarithmic processing on a 2D image is carried out by using Eqn. (5):

$$G_L(x, y) = K \times \log_2 [1 + G(x, y)] \tag{5}$$

where K is a constant arrived as 1.5 after conducting a number of experiments with various test images. Further, the logarithm computed image is scaled by a scaling factor of 32 for getting pixel values in the range of 0 to 255. The final step in proposed approach is gain/offset correction. This gain/offset correction is accomplished by using Eqn. (6).

$$I'(x, y) = \frac{d_{max}}{G_{Lmax} - G_{Lmin}} [G_L(x, y) - G_{Lmin}] \tag{6}$$

where d_{max} is the maximum intensity which is chosen as 255 for an image with 8-bit representation, $G_L(x, y)$ is the log-transformed image, G_{Lmin} is the minimum value of log transformed image, G_{Lmax} is the maximum value of log transformed image, $I'(x, y)$ is the enhanced image and, x and y represent spatial coordinates.

4 Architecture of Proposed Color Image Enhancement System

This section presents the architectures of the proposed color image enhancement system. It is composed of several components such as serpentine memory, 2D convolution, logarithm and gain/offset corrections.

4.1 System Overview

The block diagram of a top level Gaussian based color image enhancement system shown in Fig. 4, which performs a color image enhancement operation as implied in the name. The detailed signal descriptions of this block are provided in Table 1. The inputs "rin", "gin", and "bin" are the red, green, and blue pixels, each of size 8-bits. The input pixels are valid at the positive edge of the clock. The outputs "ro", "go" and "bo" represent the enhanced pixels of proposed image enhancement system. The "pixel_valid" signal is asserted, When the enhanced pixel data is valid. The enhancement system can be reset at any point of time by asserting the asynchronous, active low signal, "reset_n".

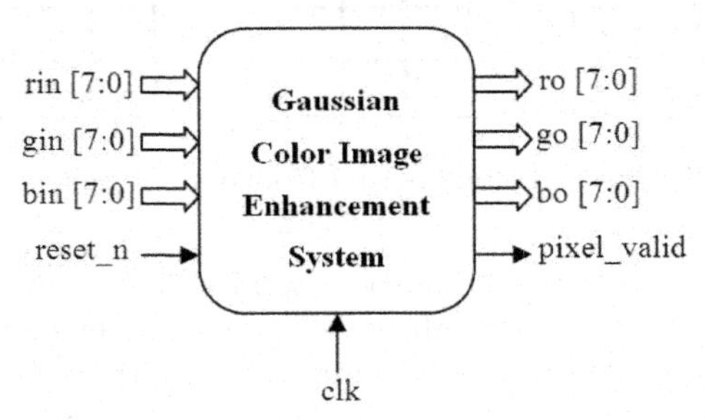

Fig. 4. Block Diagram of Top Level Gaussian Based Color Image Enhancement System

Table 1. Signal Description for the Top Module of Gaussian Based Image Enhancement System

Signals	Description
clk	This is the global clock signal
reset_n	Active low system reset
rin [7:0]	Red color component
gin [7:0]	Green color component
bin [7:0]	Blue color component
ro [7:0]	Enhanced red color component
go [7:0]	Enhanced Green color component
bo [7:0]	Enhanced Blue color component
pixel_valid	Valid signal for enhanced RGB pixel

The architecture proposed for color image enhancement system consists of twelve modules, each color component (R/G/B) comprising four basic modules: Sliding window, 2D Gaussian Convolution, Logarithm Base-2, and Gain/Offset Correction as shown in Fig. 5. Pipelining and parallel processing techniques have been adopted in the design in order to increase the processing speed of the proposed system.

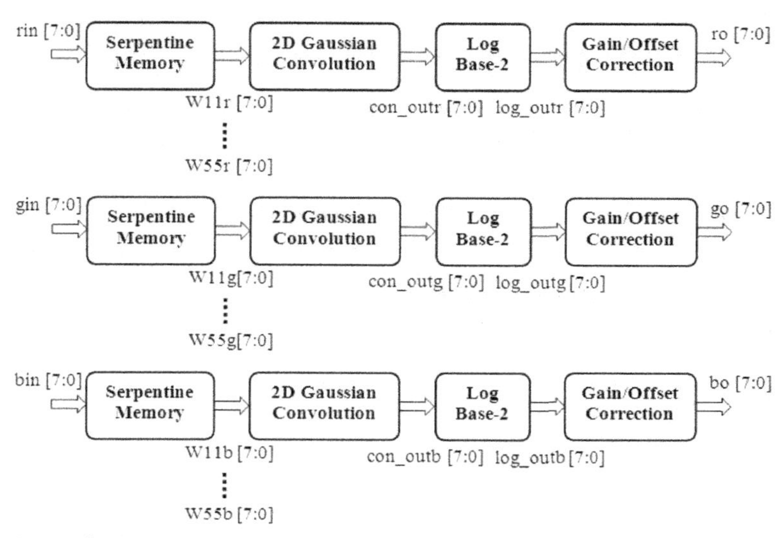

Fig. 5. Architecture of Gaussian Based Color Image Enhancement System

4.2 Serpentine Memory Architecture

The signal diagram of a serpentine memory for each of the color components R, G and B is presented in Fig. 6(a). The serpentine memory is commonly referred to as a sliding window. The proposed method uses 5×5 sliding window due to its ease of implementation. The pixel values of the input image "pixel_in" of width 8-bits are imported serially into the sliding window module on the positive edge of clock, "clk". Detailed hardware architecture for 5×5 sliding window function or serpentine memory that uses row buffers is shown in Fig. 6(b).

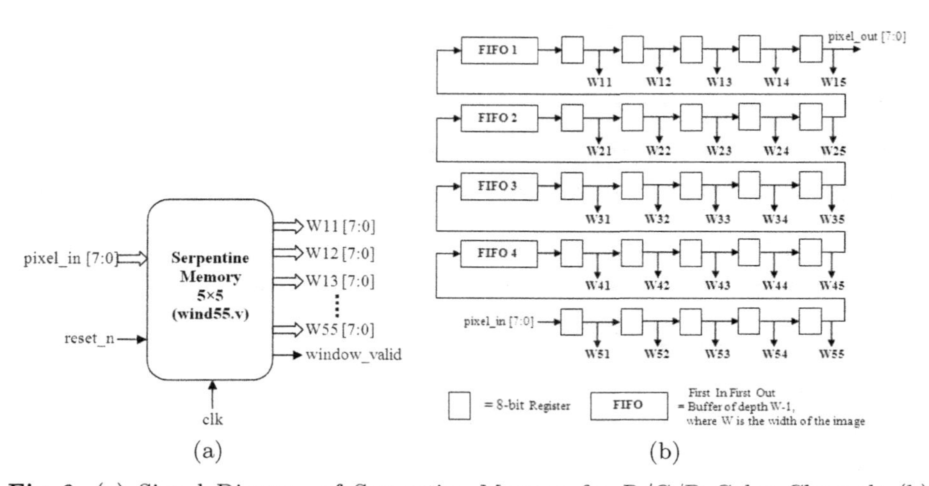

Fig. 6. (a) Signal Diagram of Serpentine Memory for R/G/B Color Channels (b) Schematic Diagram of a 5×5 Serpentine Memory for R/G/B Color Channels

In this work, one pixel is read from memory in one clock cycle with no latency. The pixels are read row by row in a raster scan order. For a 5×5 sliding window, four First-In-First-Out (FIFO) buffers are used. The FIFO buffers are used to reduce the memory access to one pixel per clock cycle. The depth of the FIFO buffer is chosen as (W-3), where W is the width of the image.

4.3 Architecture of Gaussian Convolution Processor

The design of the 2D Gaussian convolution processor in hardware is more difficult than that of the sliding window. The convolution algorithm uses adders, multipliers and dividers in order to calculate its output. The signal diagram of the 2D Gaussian convolution processor is shown in Fig. 7.

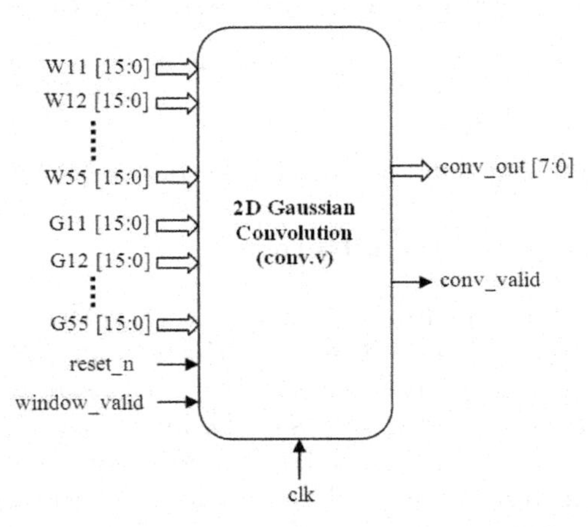

Fig. 7. Signal Diagram of 2D Gaussian Convolution Processor for R/G/B Color Channels

The input pixels W11 to W55 are scaled to 16 bits in order to match the Gaussian coefficients. The input pixels are valid on the positive edge of the "clk" with "window_valid" asserted. The coefficient values of the convolution mask employed in most of the image processing applications remain constant for the entire processing. Therefore, constant coefficient multiplier is employed in order to implement convolution efficiently. The twenty five Gaussian coefficients G11 [15:0] to G55 [15:0] represent the convolution mask shown earlier in Fig. 2. The output signal "conv_out" is 8-bits in width and is valid only on the positive edge of signal, "conv_valid".

4.4 Logarithm of Base-2 Architecture

The signal diagram of logarithm base-2 module for R/G/B color channels is shown in Fig. 8(a). The convolution output comprises the input for the logarithm block. The architecture for logarithmic module is shown in Fig. 8(b).

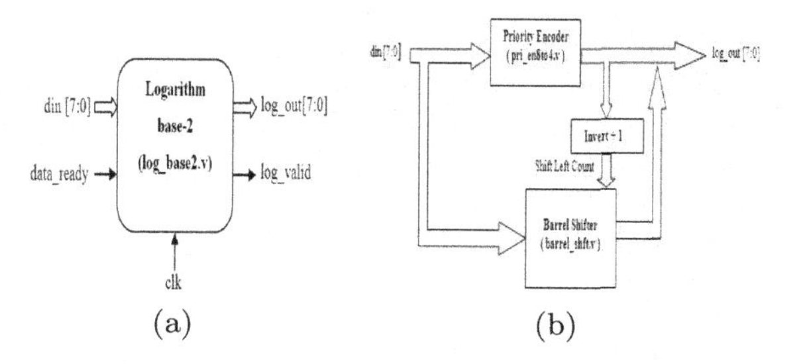

Fig. 8. (a) Signal Diagram of Logarithm Base-2 Module for R/G/B Color Channels (b) Architecture of Logarithm Base-2 Processor for R/G/B Color Channels

The log of base-2 is computed in two steps: Integer portion computation and fractional portion computation. Accordingly, the architecture consists of two modules.

4.5 Gain/Offset Correction

The gain/offset correction is accomplished using Eqn. (6). The hardware implementation of gain and offset correction uses a multiplier and subtractor module. This step steers pixels into the correct display range: 0 to 255. The multiplier design presented in this work incorporates a high degree of parallel circuits and pipelining of five levels. The multiplier performs the multiplication of two 8-bits unsigned numbers n1 and n2 as shown in Fig. 9(a). The multiplier result is of width 16-bits. The detailed architecture for the multiplier is shown in Fig. 9(b).

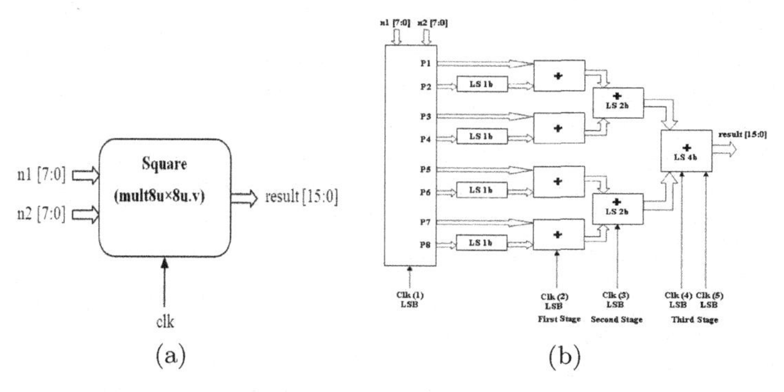

Fig. 9. (a) Signal Diagram of a Multiplier Block for R/G/B Color Channels(b) Detailed Architecture of Pipelined Multiplier Design for R/G/B Color Channels

5 Results and Discussions

Over the years, many image enhancement techniques have been reported, major ones being Multiscale Retinex with Color Restoration. The Multiscale Retinex

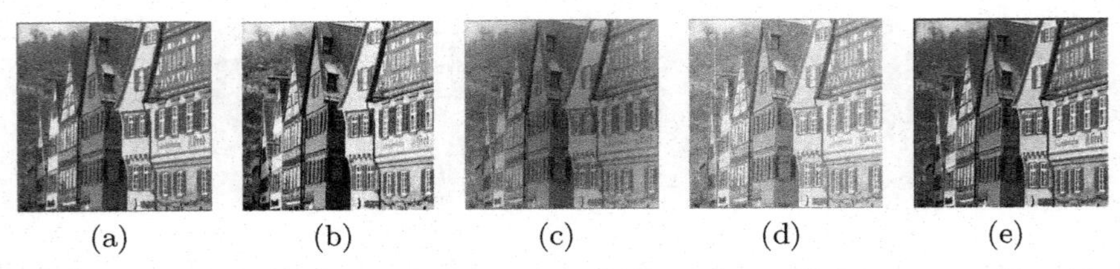

(a) (b) (c) (d) (e)

Fig. 10. Comparison of Reconstructed Pictures Using Image Enhancement Algorithms: (a) Original Image of resolution 256 × 256 pixels (b) Image enhanced based on Histogram Equalization of Ref. [12], PSNR: 31.2 dB (c) Multiscale Retinex with Color Restoration of Ref. [13], PSNR: 30 dB (d) Improved Multiscale Retinex in HSV Color Space of Ref. [14], PSNR: 29.5 dB (e) Proposed Gaussian Color Image Enhancement, PSNR: 31.8 dB.

is the most popular technique that works well under most lighting conditions. In this paper, the most popular image enhancement techniques namely, histogram equalization[12], MSRCR [13] and improved MSRCR [14] are chosen in order to validate the proposed scheme.

The image enhancement algorithms mentioned earlier have been coded in Matlab by the present authors and the reconstructed images are shown in Fig. 10. The original image of resolution 256 × 256 pixels is shown in Fig. 10(a). The image enhancement using conventional histogram equalization of R, G and B channels is shown in Fig. 10(b). The histogram equalization method improves the global contrast of the original image. However, the overexposed regions of the original image are highly enhanced in this approach. Image enhancement based on Multiscale Retinex with Color Restoration is shown in Fig. 10(c). The enhanced image obtained by MSRCR is better compared with histogram equalization. MSRCR method fails to produce good color rendition for a class of images that contain violation of gray world assumption.

The image enhanced based on improved multiscale retinex using hue-saturation-value (HSV) color space shown in Fig. 10(d) improves the visual quality better than MSRCR method. However, this approach is complex from computation point of view. The limitations expressed in the above algorithms are conspicuous by its absence in the proposed Gaussian based color image enhancement method as can be seen from the result displayed in Fig. 10(e). The proposed algorithm outperforms the other image enhancement techniques in terms of quality of the reconstructed picture. The images enhanced by our method are clearer, more vivid, and more brilliant than that achieved by other researchers.

The proposed FPGA implementation of Gaussian based Color Image Enhancement System has been coded and tested in Matlab (Version 8.1) first in order to ensure the correct working of the algorithm. Subsequently, the complete system has been coded in Verilog HDL so that it may be implemented on an FPGA or ASIC. The proposed scheme has been coded in RTL compliant Verilog and the hardware simulation results are compared with Matlab results described earlier in order to validate the hardware design. The system simulation has been done using ModelSim (Version SE 6.4) and Synthesized using

Xilinx ISE 9.2i. The algorithm has been implemented on Xilinx Virtex-II Pro XC2VP40-7FF1148 FPGA device. In the proposed work, window size is chosen as 5×5 since the enhanced image looks more appealing than that for other sizes.

Elaborate experiments were conducted on over a dozen varieties of images and consistently good results have been obtained for the proposed Verilog implementation. As examples, three poor quality images have been enhanced using the Gaussian based color image enhancement hardware system and is presented in Fig. 11. Test images (a), (d) and (g) of Fig. 11 show the original image of resolution 256×256 pixels. The images (b), (e) and (h) of Fig. 13 show the enhanced images using the proposed method based on Matlab approach. The images in Fig. 11 (c), (f) and (i) show the enhanced images using the proposed Gaussian method based on hardware (Verilog) approach.

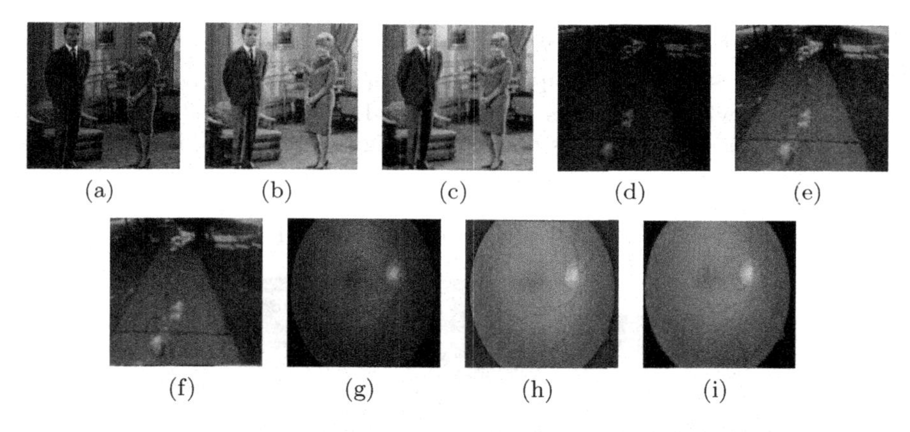

Fig. 11. Experimental Results for Gaussian based Color Image Enhancement System: (a) Original couple image (256×256 pixels) (b) Reconstructed couple image using Matlab, PSNR: 36.9 dB (c) Reconstructed couple image using Verilog, PSNR: 36.2 dB. (d) Original dark road image (256×256 pixels) (e) Reconstructed dark road image using Matlab, PSNR: 40.8 dB (f) Reconstructed dark road image using Verilog, PSNR: 40.3 dB. (g) Original eyeball image (256×256 pixels) (h) Reconstructed eyeball image using Matlab, PSNR: 36.8 dB (i) Reconstructed eyeball image using Verilog, PSNR: 36.1 dB

5.1 Verilog Simulation Results Using Modelsim

The ModelSim simulation waveforms for inputting the image pixel data for all the three color components is shown in Fig. 12(a). The enhancement process starts when the "din_valid" signal is asserted at the positive edge of the clock. The RGB image is separated into R, G and B data and the algorithm is applied concurrently to all the color components. The reconstructed pixels are issued at the output pins "dout" with "dvm" signal asserted as presented in Fig. 12(b). The outputs are issued out continuously one pixel every clock cycle after a latency of 535 clock cycles.

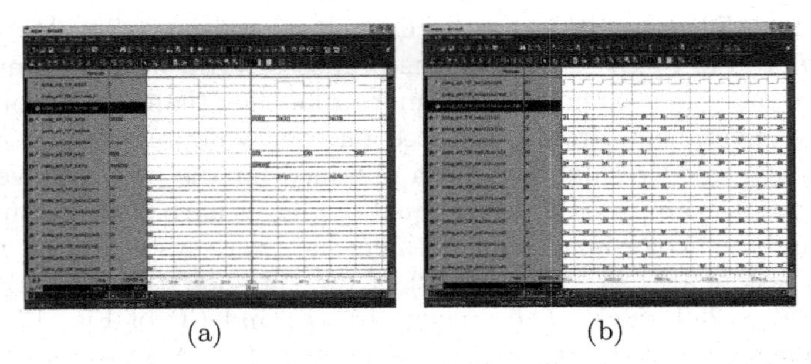

(a) (b)

Fig. 12. (a) Waveforms for Gaussian Based Color Image Enhancement System: Inputting Image Data (b) Waveforms for Gaussian based Color Image Enhancement System: Enhanced Pixel Data

5.2 Place and Route Results

The design was Synthesized, Placed and Routed using Xilinx ISE 9.2i. The FPGA target device chosen was Xilinx Virtex-II Pro XC2VP40-7FF1148. The synthesis and place and route results for the design are presented in the following. The ISE generated RTL view of the top level module "Gaussian_IE" is shown in Fig. 13(a) and the zoomed top view modules are presented in Fig. 13(b). The detailed zoomed view of R/G/B processor is shown in Fig. 14.

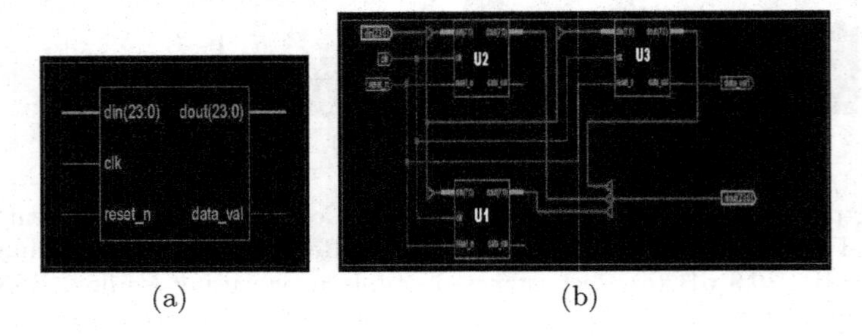

(a) (b)

Fig. 13. Experimental Results for Gaussian based Color Image Enhancement System: (a) RTL View of the Top Module "Gaussian_IE" (b) Zoomed View of "Gaussian_IE" Module where **U1:** Red Color, **U2:** Green Color and **U3:** Blue Color Component Processors

The device utilization and the performance summary for Gaussian based color image enhancement system are presented in Fig. 15. The system utilizes about 321,804 gates as reported by the ISE tool. The maximum frequency of operation is 224.840 MHz. This works out to a frame rate of 117 per second for a picture size of 1600 $\times$ 1200 pixels since the design has the capability of processing one pixel every clock cycle ignoring initial latency of 535 clock cycles. As a result, it can work on any FPGA or ASIC without needing to change any code. As ASIC, it is likely to work for higher resolutions beyond 4K format at real time rates. This shows that the design would work for not only still images but also for high resolution video sequences.

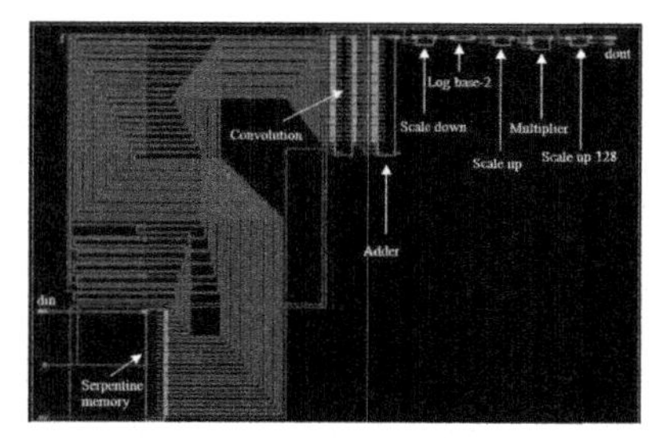

Fig. 14. Zoomed View of U1 or U2 or U3 Module

GAUSSIAN_IE Project Status			
Project File:	Gaussian_IE.ise	Current State:	Programming File Generated
Module Name:	coleng_enh_TOP	• Errors:	No Errors
Target Device:	xc2vp40-7ff1148	• Warnings:	[illegible] Warnings
Product Version:	ISE 9.2i	• Updated:	Sun Jan 22 09:55:22 2012

Device Utilization Summary				
Logic Utilization	**Used**	**Available**	**Utilization**	**Note(s)**
Number of Slice Flip Flops	29,337	38,784	75%	
Number of 4 input LUTs	1,422	38,784	3%	
Logic Distribution				
Number of occupied Slices	15,423	19,392	79%	
Number of Slices containing only related logic	15,423	15,423	100%	
Number of Slices containing unrelated logic	0	15,423	0%	
Total Number of 4 input LUTs	2,847	38,784	7%	
Number used as logic	1,422			
Number used as a route-thru	348			
Number used as Shift registers	1,077			
Number of bonded IOBs	51	804	6%	
IOB Flip Flops	48			
Number of PPC405s	0	2	0%	
Number of GCLKs	1	16	6%	
Number of GTs	0	12	0%	
Number of GT10s	0	0	0%	
Total equivalent gate count for design	321,804			
Additional JTAG gate count for IOBs	2,448			

Fig. 15. FPGA Resource Utilization for Gaussian Based Color Image Enhancement Design

The timing summary for the design as reported by Xilinx ISE tool is as follows:
Speed Grade : -7
Minimum period: 4.448ns (Maximum Frequency: 224.840 MHz)
Minimum input arrival time before clock: 1.418ns
Maximum output required time after clock: 3.293ns
Clock period: 4.448ns (frequency: 224.840 MHz)
Total number of paths /destination ports: 68988 / 30564
Delay: 4.448ns (Levels of Logic = 12)

6 Conclusion

A new algorithm and architecture for Gaussian based color image enhancement system for real time applications has been presented in this paper. The Gaussian convolution used in this scheme not only smoothes the image but also removes the noise present in the image. Further, the pixels are processed in log-domain in order compress the dynamic range. Finally, linear operation such as gain/offset correction is used in order to obtain image in display range. The Verilog design of the color image enhancement system is implemented on Xilinx Virtex-II Pro XC2VP40-7FF1148 FPGA device and is capable of processing high resolution videos up to 1600×1200 pixels at 117 frames per second. The implementation was tested with various images and found to produce high quality enhanced images. The design realized using RTL compliant Verilog fits into a single FPGA chip with a gate count utilization of about 321,804.

References

1. Jain, A.K.: Fundamentals of Digital Image Processing. Prentice-Hall, Inc., Upper Saddle River (1989)
2. An, C., Yu, M.: Fast Color Image Enhancement based on Fuzzy Multiple-scale Retinex. In: 6th International Forum Strategic Technology (IFOST 2011), vol. 2, pp. 1065–1069 (2011)
3. Hines, G., Rahman, Z., Jobson, D., Woodell, G.: Single-scale Retinex using Digital Signal Processors. In: Proceedings of Global Signal Processing Conference, vol. 27. Citeseer (2004)
4. Hines, G., Rahman, Z., Jobson, D., Woodell, G.: DSP Implementation of the Retinex Image Enhancement Algorithm. In: Proceedings of SPIE Visual Information Processing XIII, vol. 5438 (2004)
5. Zhang, M., Seow, M., Asari, V.K.: A High Performance Architecture for Color Image Enhancement using a Machine Learning Approach. International Journal of Computational Intelligence Research-Special Issue on Advances in Neural Networks 2(1), 40–47 (2006)
6. Tsutsui, H., Nakamura, H., Hashimoto, R., Okuhata, H., Onoye, T.: An FPGA Implementation of Real-time Retinex. In: World Automation Congress (WAC 2010) Video Image Enhancement, pp. 1–6 (2010)
7. Alsuwailem, A., Alshebeili, S.: A New Approach for Real-time Histogram Equalization using FPGA. In: IEEE Proceedings of 2005 International Symposium on Intelligent Signal Processing and Communication Systems (ISPACS 2005), pp. 397–400 (2005)
8. Zhang, M., Seow, M., Tao, L., Asari, V.K.: Design of an Efficient Architecture for Enhancement of Stream Video Captured in Non-uniform Lighting Conditions. In: International Symposium on Signals, Circuits and Systems (ISSCS 2007), vol. 2, pp. 1–4 (2007)
9. Ngo, H., Zhang, M., Tao, L., Asari, V.K.: Design of a Digital Architecture for Real-time video, Enhancement based on Illuminance-Reflectance Model. In: 49th IEEE International Midwest Symposium on Circuits and Systems (MWSCAS 2006), vol. 1, pp. 286–290 (2006)
10. Bidarte, U., Ezquerra, J., Zuloaga, A., Martin, J.: VHDL Modeling of an Adaptive Architecture for Real-time Image Enhancement. In: Fall VIUF Workshop, pp. 94–100 (1999)
11. Tao, L., Asari, V.K.: Modified Luminance based MSR for Fast and Efficient Image Enhancement. In: 32nd Proceedings Applied Imagery Pattern Recognition Workshop, pp. 174–179 (2003)
12. Cheng, H., Shi, X.: A Simple and Effective Histogram Equalization Approach to Image Enhancement. Digital Signal Processing 14(2), 158–170 (2004)
13. Jobson, D., Rahman, Z., Woodell, G.A.: A Multiscale Retinex for Bridging the gap between Color Images and the Human Observation of Scenes. IEEE Transactions on Image Processing 6(7), 965–976 (1997)
14. Shen, C., Hwang, W.: Color Image Enhancement using Retinex with Robust Envelope. In: 16th IEEE International Conference on Image Processing (ICIP 2009), pp. 3141–3144 (2009)
15. Ramachandran, S.: Digital VLSI Systems design: A Design Manual for Implementation of Projects on FPGAs and ASICs using Verilog, 1st edn. Springer (2007)

Content Based Video Retrieval Using Color Feature: An Integration Approach

Brijmohan Daga

Fr. Conceicao Rodrigues College of Engineering, Mumbai, India
bsdaga@yahoo.com

Abstract. As large amount of visual Information is available on web in form of images, graphics, animations and videos, so it is important in internet era to have an effective video search system. As there are number of video search engine (blinkx, Videosurf, Google, YouTube, etc.) which search for relevant videos based on user "keyword" or "term", But very less commercial video search engine are available which search videos based on visual image/clip/video. In this paper we are recommending a system that will search for relevant video using color feature of video in response of user Query.

Keywords: CBVR, Video Segmentation, Key Feature Extraction Color Feature Extraction, Classification.

1 Introduction

As in internet era most difficult task is to retrieve the relevant information in response to a query. To help a user in this context various search systems/engines are there in market with different features. In web search era 1.0 the main focus was on text retrieval using link analysis. It was totally read only era. There was no interaction in between the user and the search engine i.e. after obtaining search result user have no option to provide feedback regarding whether the result is relevant or not. In web search era 2.0 the focus was on retrieval of data based on relevance ranking as well as on social networking to read, write, edit and publish the result. Due to Proliferation of technology the current search era based on contextual search. Where rather than ranking of a page focus is on content based similarity to provide accurate result to user.

The CBVR (Content Based Video Retrieval) have received intensive attention in the literature of video information retrieval since this area was started couple of years ago, and consequently a broad range of techniques has been proposed. The algorithms used in these systems are commonly divided into four tasks:

- Segmentation
- Extraction
- Selection
- Classification

The segmentation task splits the video into number of chunks or shots. The extraction task transforms the content of video into various content features. Feature extraction is the process of generating features to be used in the selection and classification tasks. A feature is a characteristic that can capture a certain visual property of an

S. Unnikrishnan, S. Surve, and D. Bhoir (Eds.): ICAC3 2013, CCIS 361, pp. 609–625, 2013.

image either globally for the whole image, or locally for objects or regions. Feature selection reduces the number of features provided to the classification task. Those features which are assisting in discrimination are selected and which are not selected is discarded. The selected features are used in the classification task [2]. The figure 1 shows the content based video search systems with four primitive tasks.

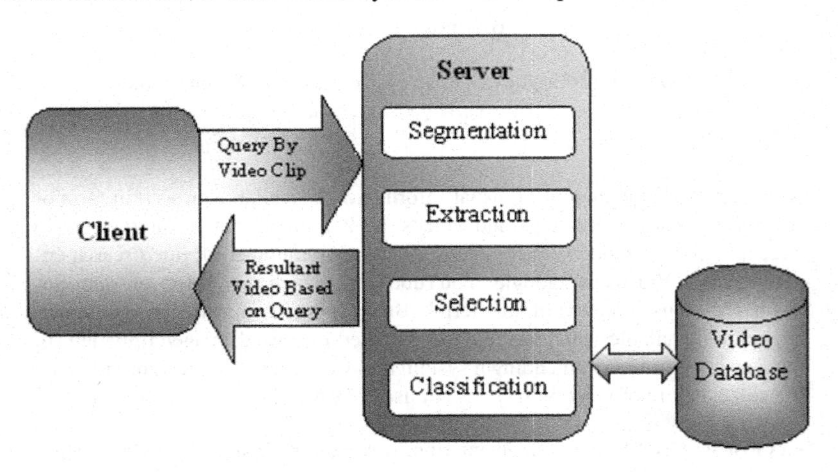

Fig. 1. Content Based Video Search System and its tasks

Among these four activities feature extraction is critical because it directly influence the classification task. The set of features are the result of feature extraction. In past few years , the number of CBVR systems using different segmentation and extraction techniques, which proves reliable professional applications in Industry automation, social security, crime prevention, biometric security, CCTV surveillance [1], etc.

2 Proposed System

Due to rapidity of digital information (Audio, Video) it become essential to develop a tool for efficient search of these media. With help of this paper we are proposing a Video Search system which will provide accurate and efficient result to a user query. The proposed system is a web based application as shown in fig.1 which consists of following processing:

1. Client Side Processing:
From client machine user can access the Graphical User Interface of the system. User can access and able to perform three tasks:

- Register the video
- Search the video
- Accept the efficient result from server.

2. Server Side Processing:
The core processing will be carried out at server side to minimize the overhead on client. Client will make a request for similar type of videos by providing query by video clip. On reception of this query by video clip, server will perform some

processing on query video as well as on videos in its database and extract the video which are similar to query video. After retrieving the similar videos from the database, server will provide the list to the client in prioritized order. To accomplish this following operations are carried out at the server.

2.1 Video Segmentation and Key Frame Extraction

The idea of segmenting an image into layers was introduced by Darrell and Pentland[3], and Wang and Adelson [5]. Darrel and Pentland [3] used a robust estimation method to iteratively estimate the number of layers and the pixel assignments to each layer. They show examples with range images and with optical flow. Wang and Adelson [5] is the seminal paper on segmenting video into layers. Affine model is fitted to blocks of optical flow, followed by a K-means clustering of these affine parameters. This step involves splitting and merging of layers, and therefore is not fixed. After the first iteration, the shape of regions is not confined to aggregate of blocks but is taken to a pixel level within the blocks. The results presented are convincing, though the edges of segments are not very accurate, most likely due to the errors in the computation of optical flow at occlusion boundaries Bergen et. al. [7] presents a method for motion segmentation by computing first the global parametric motion of the entire frame, and then finding the segments that do not fit the global motion model well. Irani and Peleg[6] incorporate temporal integration in this loop. A weighted aggregate of a number of frames is used to register the current image with the previous one. The object that is currently being compensated for thus becomes sharply into focus and everything else blurs out, improving the stability of the solution.

In order to extract valid information from video, process video data efficiently, and reduce the transfer stress of network, more and more attention is being paid to the video processing technology.

The amount of data in video processing is significantly reduced by using video segmentation and key-frame extraction. Fig. 2 shows the basic framework of key frame extraction from a video .The explanation of each term is as given below:

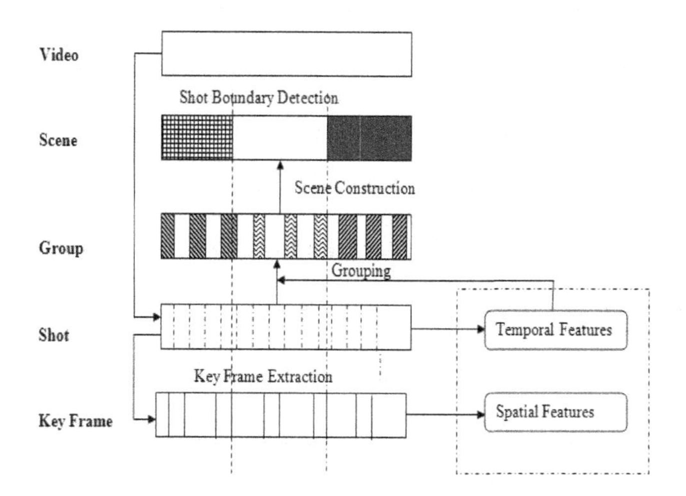

Fig. 2. The basic framework of the key frame extraction

A shot is defined as the consecutive frames taken by a single camera without any significant change in the color content[8]. It shows a continuous action in an image sequence. Key frame is the frame which represents the salient visual contents of a shot [9]. The key frames extracted must summarize the characteristics of the video, and the image characteristics of a video can be tracked by all the key frames in time sequence. Depending on the complexity of the content of the shot, one or more frame can be extracted.

Video Scene is defines as collection of semantically related and temporally adjacent shots, depicting and conveying a high level concept or story. While shots are marked by physical boundaries, scenes are marked by semantic boundaries. Video group is an intermediate entity between the physical shots and semantic scenes and serves as the bridge between the two. For implementation of this phase we will use the Java Media Framework Plug in API's.

2.2 Color Feature Extraction

A key function in the content based video search system is feature extraction. A feature is a characteristic that can capture a certain visual property of an image either globally for the whole image, or locally for objects or regions. Color is an important feature for image representation which is widely used in image retrieval. This is due to the fact that color is invariance with respect to image scaling, translation and rotation [10].

Color is an important feature for image representation which is widely used in image retrieval. This is due to the fact that color is invariance with respect to image scaling, translation, and rotation [8]. The human eye is sensitive to colors, and color features are one of the most important elements enabling humans to recognize images [9]. Color features are, therefore, fundamental characteristics of the content of images. The algorithms which we are used to extract the color features of key frame are Average RGB,GCH, LCH, Color Moments.

i) Average RGB

Colors are commonly defined in three-dimensional color spaces. The color space models [13] can be differentiated as hardware-oriented and user-oriented. The hardware-oriented color spaces, including RGB and CMY are based on the three-color stimuli theory. The user-oriented color spaces, including HLS, HCV, HSV and HSB are based on the three human percepts of colors, i.e., hue, saturation, and brightness [11]. The RGB color space (see Figure 3) is defined as a unit cube with red, green, and blue axes; hence, a color in an RGB color space is represented by a vector with three coordinates. When all three values are set to 0, the corresponding color is black. When all three values are set to 1, the corresponding color is white [12].

The color histograms are defined as a set of bins where each bin denotes the probability of pixels in the image being of a particular color. A color histogram H for a given image is defined as a vector:

$$H=\{H[0],H[1],......H[i]........H[N]\}$$

Where i represents a color in the color histogram and corresponds to a sub-cube in the RGB color space, H[i] is the number of pixels in color i in that image, and N is the number of bins in the color histogram, i.e., the number of colors in the adopted color model.

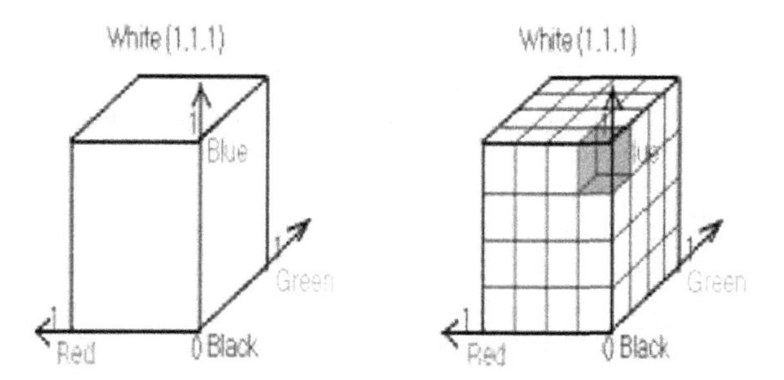

Fig. 3. RGB Color Space

ii) Global Color Histogram
As we have discussed, the color histogram depicts color distribution using a set of bins. Using the Global Color Histogram (GCH)[14], an image will be encoded with its color histogram, and the distance between two images will be determined by the distance between their color histograms. The following example (see Figure 4) shows how a GCH works.

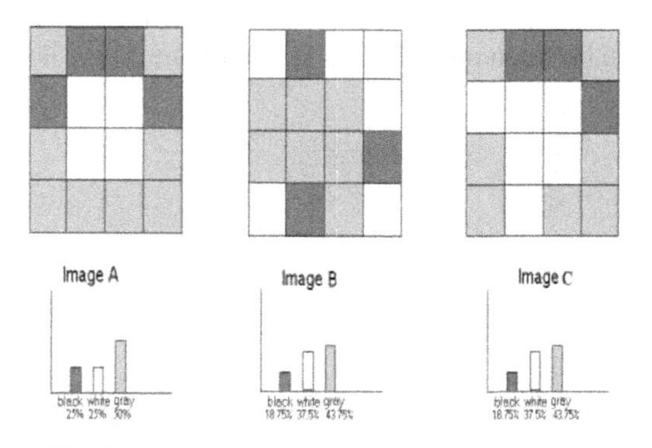

Fig. 4. Three images and their color histograms

In the sample color histograms there are three bins: black, white, and gray. We note the color histogram of image A:{25% ,25%,50%},the color histogram of image B:{18.75%, 37.5%, 43.75%} and image C has the same color histogram as image B. If we use the Euclidean distance metric to calculate the histogram distance, the distance between images A and B for GCH is:

$$d_{GCH}(A,B) = \sqrt{(a_1 - b_1)^2 + (a_2 - b_2)^2 + (a_3 - b_3)^2} \quad (1)$$

The distance between images A and C equals the distance between images A and B and the distance between images B and C is zero. The GCH is the traditional method for color-based image retrieval. However, it does not include information concerning the color distribution of the regions, so the distance between images sometimes cannot show the real difference between images.

iii) Local Color Histogram

This approach (referred to as LCH)[13][14] includes information concerning the color distribution of regions. The first step is to segment the image into blocks and then to obtain a color histogram for each block. An image will then be represented by these histograms. When comparing two images, we calculate the distance, using their histograms, between a region in one image and a region in same location in the other image. The distance between the two images will be determined by the sum of all these distances. If we use the square root of Euclidean distance as the distance between color histograms, the distance metric between two images Q and I used in the LCH will be defined as:

$$d_{LCH}(Q,I) = \sum_{k=1}^{M} \sqrt{\sum_{i=1}^{N} \left(H_Q^k[i] - H_I^k[i]\right)^2} \tag{2}$$

where M is the number of segmented regions in the images, N is the number of bins in the color histograms, and $H^k_Q[i]$ $H^K_I[i]$ is the value of bin i in color histogram H^k_Q H^K_I which represents the region k in the image Q(I).

The following examples use the same images A, B and C in Figure 5 to show how a LCH works and illustrate how we segment each image into 4 equally sized blocks. For the LCH, the distance between image A and B (see Figure 5) is calculated as follows:

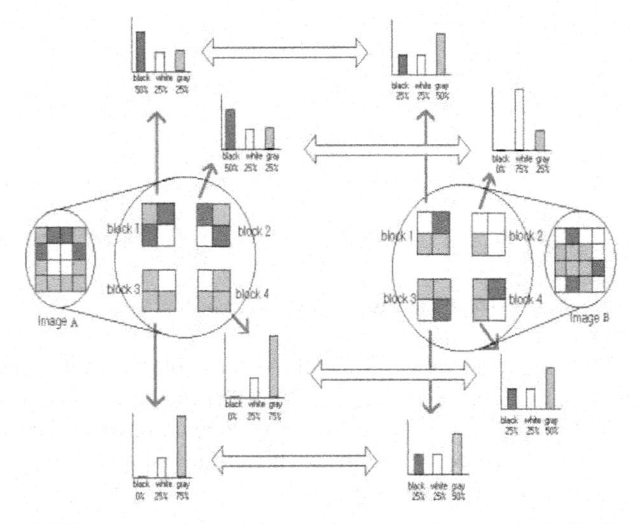

Fig. 5. Using LCH to compute the distance between images A & B $d_{LCH}(A,B)$ =1.768 & $d_{GCH}(A,B)$ =0.153

In some scenarios, using LCHs can obtain better retrieval effectiveness than using GCHs. The above examples show that the LCH overcomes the main disadvantage of the GCH, and the new distances between images may be more reasonable than those obtained using the GCH. However, since the LCH only compares regions in the same location, when the image is translated or rotated, it does not work well.

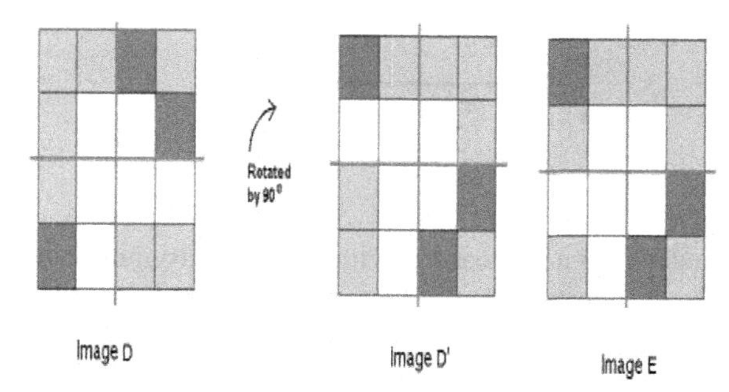

Fig. 6. An example showing that the LCH fails to compare images D and E

For example, in Figure 6 if image (is rotated by 90o) , we get image D'. We can then see that image is very similar to image E with only two blocks being different. The distance between images D and E should be equal to the distance between images D' and E. However, using LCH, the distance between images D and E will be greater than the distance between images D' and E. The reason for this discrepancy is that the LCH compares blocks only in the same location, but not necessarily in the proper locations. For example, using LCH, the north-west block in image D is compared with the north-west block in image E but, in fact, the north-west block in image D should be compared with the north-east block in image E. LCHs incorporate spatial information by dividing the image with fixed block segmentation. Some other approaches combine spatial information with color, using different partition schemes [15][16][17].

iv) Color Moment

Color moments[18] are measures that can be used to differentiate images based on their features of color. Once calculated, these moments provide a measurement for color similarity between images. These values of similarity can then be compared to the values of images indexed in a database for tasks like image retrieval. The basis of color moments lays in the assumption that the distribution of color in an image can be interpreted as a probability distribution. Probability distributions are characterized by a number of unique moments. These are mean, standard deviation and skewness. Moments are calculated for each of these channels in an image. An image therefore is characterized by 9 moments where 3 moments for 3 color channels, which are as follows.

Moment 1:Mean

$$E_i = \sum_{N}^{j=1} \frac{1}{N} P_{ij} \tag{3}$$

Mean can be understood as the average color value in the I age.

Moment 2: Standard Deviation

$$\sigma_i = \sqrt{\left(\frac{1}{N}\sum_{N}^{j=1}(P_{ij} - E_i)^2\right)} \tag{4}$$

The standard deviation is the square root of the variance of the distribution.

Moment 3: Skewness

$$S_i = \sqrt[3]{\left(\frac{1}{N}\sum_{N}^{j=1}(P_{ij} - E_i)^3\right)} \quad (5)$$

Skewness can be understood as a measure of the degree of asymmetry in the distribution.

A function of the similarity between two image distributions is defined as the sum of the weightedifferences between the moments of the two distributions. Formally this is:

$$D_{mom}(I_1, I_2) = \sum_{i=1}^{r} w_{i1}\left|E_i^1 - E_i^2\right| + w_{i2}\left|\sigma_i^1 - \sigma_i^2\right| + w_{i3}\left|S_i^1 - S_i^2\right| \quad (6)$$

Where,(H,I): are the two image distributions being compared

v) Color Coherence Vector

In Color Coherence Vector approach [19], each histogram bin is partitioned into two types, coherent and incoherent. If the pixel value belongs to a large uniformly-colored region then is referred to coherent otherwise it is called incoherent. In other words, coherent pixels are a part of a contiguous region in an image, while incoherent pixels are not. A color coherence vector represents this classification for each color in the image [20].The coherence measure classifies pixels as either coherent or incoherent.

Coherent pixels are a part of some sizable contiguous region, while incoherent pixels are not. A color coherence vector represents this classification for each color in the image. Color coherence vector prevent coherent pixels in one image from matching incoherent pixels in another. This allows fine distinctions that cannot be made with color histograms.

2.3 Integration Approach

As Global color histogram method does not include information concerning the color distribution of the regions, Local Color Histogram fails when the image is translated or rotated, Color moment method do not encapsulate information about spatial correlation of colors. So to obtain the exact and accurate result from a video database we are using an integrated approach of color feature extraction methods where we are providing options to user to select any of combination of above described techniques. With this approach the feature vectors in different feature classes are combined into one overall feature vector. The system compares this overall feature vector of the query image to those of database images, using a predetermined similarity measurement. We are recommending here that to obtain a more relevant result user should select all the options provided in GUI.

The system is extracting the frames from a given video using Java Media Framework (JMF) Library. Input video supported by system are MPEG, AVI.

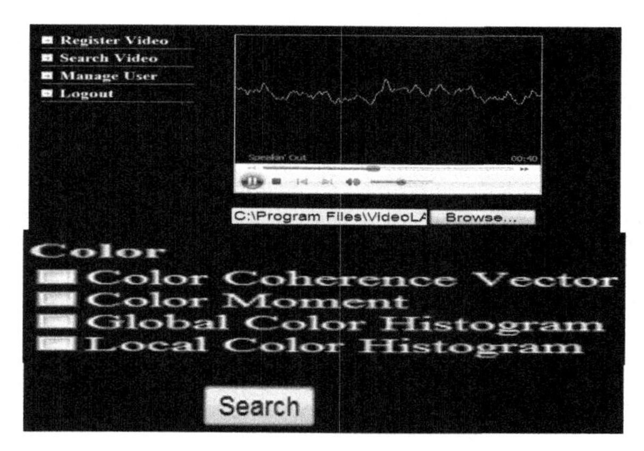

Fig. 7. GUI at client side for search

Kindly refer the Appendix A for results 1. Video to Frame Extraction and 2. Video to Key Frame Extraction.

After extraction of each frame from inputted video, System is extracting the key frames by using the clustering based on Average RGB value.

Appendix A: Table 1 shows the every 5th frame of inputted video i.e car1.avi. Table 2 shows the every even frame of inputted video i.e car2.avi. Table 3 shows every even frame of inputted video i.e car3.avi Table 4 shows the result of key frame extraction of Car1.avi, Car2.avi and Car3.avi. Table 5 shows Query video with Key frames.

2.4 Matching and Retrieval

In retrieval stage of a Video search system, features of the given query video is also extracted. After that the similarity between the features of the query video and the stored feature vectors is determined. That means that computing the similarity between two videos can be transformed into the problem of computing the similarity between two feature vectors [21]. This similarity measure is used to give a distance between the query video and a candidate match from the feature data database as shown in Figure. 8.

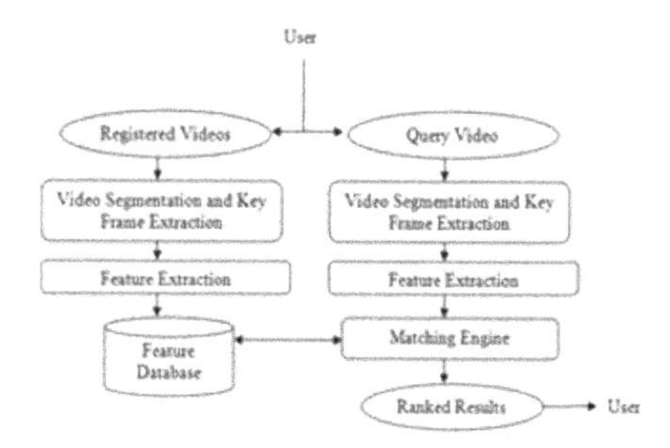

Fig. 8. Matching and retrieval of CBVR

Kindly refer the Appendix B for color feature extraction Table 7 shows the behavior of color algorithms (CC: Color coherence, CM: color Moment, GC: Global color histogram, LC: Local color Histogram, AR: Average RGB) in response of query video (car4.avi).

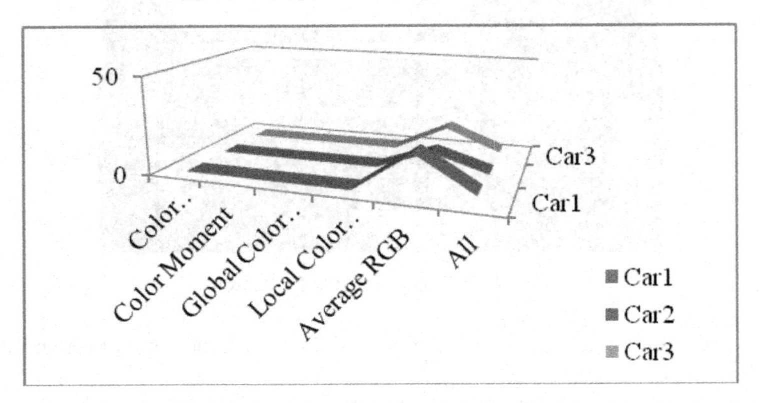

Fig. 9. Comparison of Color Feature Extraction Methods for query(car4.avi)

3 Conclusion

The proposed systems facilitate the segmentation of the elementary shots in the long video sequence proficiently. Subsequently, the extraction of the color features using color histogram, color moment and color coherence vector is performed and the feature library is employed for storage purposes. One of advantage of this systems is user not only search for a relevant video but one can register his/her videos on the web server.

References

1. Choras, R.S.: Image Feature Extraction Techniques and Their Applications for CBIR and Biometrics Systems. International Journal of Biology and Biomedical Engineering 1(1), 7–16 (2007)
2. Yang, Y., Lovel, B.C., et al.: Content-Based Video Retrieval (CBVR) System for CCTV Surveillance Videos. In: Proceeding DICTA 2009 Proceedings of the 2009 Digital Image Computing: Techniques and Applications, pp. 183–187. IEEE Computer Society, Washington, DC (2009)
3. Darrel, A.P.A.: Cooperative robust estimation using layers of support. T.R. 163, MIT Media Lab, Vision and Modeling Group (February 1991)
4. Bergen, J.R., Anandan, P., Hanna, K., Hingorani, R.: Hierarchical Model-Based Motion Estimation. In: Sandini, G. (ed.) ECCV 1992. LNCS, vol. 588, pp. 237–252. Springer, Heidelberg (1992)
5. Adelson, E.H., Wang, J.Y.A.: Representing moving images with layers. IEEE Transactions on Image Processing (September 1994)

6. Peleg, S., Irani, M.: Motion analysis for image enhancement: Resolution, occlusion and transparency. Journal of Visual Communication and Image Representation 4(4), 324–335 (1993)
7. Black, M.: Combining intensity and motion for incremental segmentation and tracking over long image sequences. In: ECCV (1992)
8. Bhute, A.N., Meshram, B.B.: IVSS: Integration of Color Feature Extraction Techniques for Intelligent Video Search Systems. In: Proceeding of Int'l Conf. ICECT, Kanyakumari, India (April 2012)
9. Zhuang, Y., Rui, Y., Huang, T.S., Mehrotra, S.: Adaptive key frame extraction using unsupervised clustering. In: Proc. IEEE Int. Conf. on Image Proc. (1998)
10. Wu, J., Wei, Z., Chang, Y.: Color and Texture Feature For Content Based Image Retrieval. International Journal of Digital Content Technology and its Applications 4(3) (June 2010)
11. Oraintara, S., Nguyen, T.T.: Using Phase and Magnitude Information of the Complex directional Filter Bank for Texture Image Retrieval. In: Proc. IEEE Int. Conf. on Image Processing, vol. 4, pp. 61–64 (October 2007)
12. De Valois, R.L., De Valois, K.K.: A multi-stage color model. Vision Research 33(8), 1053–1065 (1993)
13. Gonzalez, R.C., Woods, R.E.: Digital Image Processing, 3rd edn. Addison-Wesley, Reading (1992)
14. Wang, S.: A Robust CBIR Approach Using Local Color Histograms. Technical report, Department of computer science, University of Alberta, Canada (2001)
15. Smith, J., Chang, S.: Tools and techniques for color image retrieval. In: Proc. of the SPIE Conference on the Storage and Retrieval for Image and Video Databases IV, San Jose, CA, USA, pp. 426–437 (1996)
16. Stricker, M.A., Dimai, A.: Color indexing with weak spatial constraints. In: Proc. of the SPIE Conference on the Storage and Retrieval for Image and Video Databases IV, San Diego, CA, USA, pp. 29–40 (February 1996)
17. Hsu, W., Chua, T.S., Pung, H.K.: An integrated color-spatial approach to content-based imageretrieval. In: Proc. of the ACM Multimedia 1995, pp. 305–313 (1995)
18. Stricker, M.A., Orengo, M.: Similarity of Color Images. In: Storage and Retrieval for Image and Video Databases III, pp. 381–392 (1995)
19. Pass, G., Zabih, R.: Histogram Refinement for Content-Based Image Retrieval. In: IEEE Workshop on Applications of Computer Vision, pp. 96–102 (1996)
20. Han, J., Ma, K.: Fuzzy Color Histogram and Its Use in Color Image Retrieval. IEEE Trans. on Image Processing 11, 944–952 (2002)
21. Shanmugam, T.N., Rajendran, P.: An Enhanced Content-Based Video Retrieval System based on query clip. International Journal of Research and Reviews in Applied Sciences (December 2009)

Appendix A

1. Video to Frame Extraction

The system is extracting the frames from a given video using Java Media Framework (JMF) Library. Input video supported by system are n MPG, AVI. Table 1 shows the every 5th frame of inputted video i.e car1.avi.

Table 1. Car1.avi Frame Extraction

Input video	Every 5th Frame of Input video
Car1.avi Length : 9 Sec. Size **in bytes:** **1402998** Total Frame extracted: 225	5.jpg 10.jpg 15.jpg 20.jpg 25.jpg 30.jpg 35.jpg 40.jpg 45.jpg 50.jpg 55.jpg 60.jpg 65.jpg 70.jpg 75.jpg 80.jpg 85.jpg 90.jpg 95.jpg 100.jpg 105.jpg 110.jpg 115.jpg 120.jpg 125.jpg 130.jpg 135.jpg 140.jpg 145.jpg 150.jpg 155.jpg 160.jpg 165.jpg 170.jpg 175.jpg 180.jpg 185.jpg 190.jpg 195.jpg 200.jpg 205.jpg 210.jpg 215.jpg 220.jpg 225.jpg

In this way, three sample videos are registered on the server titled as car1.avi, car2.avi and car3.avi.

Table 2 shows the every even frame of inputted video i.e car2.avi.

Table 2. Car2 Frame extraction

Input video	**Every Even Frame of Input video**
Car4.avi Length: Sec. Size **in bytes: 461648** Total frame extracted: 83	2.jpg 4.jpg 6.jpg 8.jpg 10.jpg 12.jpg 14.jpg 16.jpg 18.jpg 20.jpg 22.jpg 24.jpg 26.jpg 28.jpg 30.jpg 32.jpg 34.jpg 36.jpg 38.jpg 40.jpg 42.jpg 44.jpg 46.jpg 48.jpg 50.jpg 52.jpg 54.jpg 56.jpg 58.jpg 60.jpg 62.jpg 64.jpg 66.jpg 68.jpg 70.jpg 72.jpg 74.jpg 76.jpg 78.jpg 80.jpg 82.jpg 83.jpg

Table 3 shows the every even frame of inputted video i.e car3.avi

Table 3. Car3 Frame extraction

Input video	Every Even Frame of Input video
Car5.avi **Length** 6 Sec **Size in Bytes** **991150** **Total Frame Extracted** 166	4.jpg 8.jpg 12.jpg 16.jpg 20.jpg 24.jpg 28.jpg 32.jpg 36.jpg 40.jpg 44.jpg 48.jpg 52.jpg 56.jpg 60.jpg 64.jpg 68.jpg 72.jpg 76.jpg 80.jpg 84.jpg 88.jpg 92.jpg 96.jpg 100.jpg 104.jpg 108.jpg 112.jpg 116.jpg 120.jpg 124.jpg 128.jpg 132.jpg 136.jpg 140.jpg 144.jpg 148.jpg 152.jpg 156.jpg 160.jpg 164.jpg 166.jpg

2. Video To Key Frame Extraction

   After extraction of each frame from inputted video, System is extracting the key frames by using the clustering based on Average RGB value. Table 4 shows the result of key frame extraction of Car1.avi, Car2.avi and Car3.av1.

Table 4. Registered Video with Key Frames

Registered video	Key Frame Extracted
Car1.avi	1.jpg 2.jpg 3.jpg 4.jpg 5.jpg 6.jpg 7.jpg
Car2.avi	1.jpg
Car3.avi	1.jpg 2.jpg 3.jpg 4.jpg 5.jpg 6.jpg 7.jpg 8.jpg 9.jpg 10.jpg

The system has extracted the 7 frames out of 225 frames for car1.avi, 1 frame out of 83 for Car2.avi and 10 frames out of 166 for Car3.avi. To search similar type of videos user is providing a query by video clip and to provide the result system is extracting the frames of query video followed by key frame extraction.

Table 5. Query video with Key frames

Query video	Key Frame Extracted
Car4.avi	1.jpg 2.jpg

Appendix B

Table 6 shows the behavior of color algorithms (CC: Color coherence, CM: color Moment, GC: Global color histogram, LC: Local color Histogram, AR: Average RGB) in response of query video (car4.avi).

Table 6. Search result when Color Feature Extraction algorithms are selected

Color Algorithms					Priority		
CC	**CM**	**GC**	**LC**	**AR**	**I**	**II**	**III**
√					Car2 0.01733622	Car3 0.02727861	Car1 0.06116247
	√				Car2 0.02548367	Car1 0.07100401	Car3 0.08799920
		√			Car2 0.01984108	Car3 0.03524798	Car1 0.08130584
			√		Car2 0.04764891	Car3 0.06438684	Car1 0.11297473
				√	Car2 10.60982918	Car3 13.88653719	Car1 20.93979997
√	√				Car2 0.02140994	Car3 0.05763890	Car1 0.06608324
√		√			Car2 0.01858865	Car3 0.03126329	Car1 0.07123415
√			√		Car2 0.03249257	Car3 0.04583272	Car1 0.08706860
√				√	Car2 5.31358270	Car3 6.95690790	Car1 10.50048122
	√	√			Car2 0.02266237	Car3 0.06162359	Car1 0.07615493
	√		√		Car2 0.03656629	Car3 0.07619302	Car1 0.09198937
	√			√	Car2 5.31765642	Car3 6.98726820	Car1 10.50540199
		√	√		Car2 0.03374500	Car3 0.04981741	Car1 0.09714029
		√		√	Car2 5.31483513	Car3 6.96089258	Car1 10.51055291
			√	√	Car2 5.32873904	Car3 6.97546202	Car1 10.52638735
√	√	√			Car2 0.02088699	Car3 0.05017526	Car1 0.07115744

Table 6. (*Continued*)

√	√		√		Car2 0.03015627	Car3 0.05988822	Car1 0.08171374
√	√			√	Car2 3.55088302	Car3 4.66727167	Car1 7.02398882
√		√	√		Car2 0.02827541	Car3 0.04230448	Car1 0.08514768
√		√		√	Car2 3.54900216	Car3 4.64968792	Car1 7.02742276
√			√	√	Car2 3.55827144	Car3 4.65940088	Car1 7.03797906
	√	√	√		Car2 0.03099122	Car3 0.06254467	Car1 0.08842819
	√	√		√	Car2 3.54900216	Car3 4.64968792	Car1 7.02742276
	√		√	√	Car2 3.54900216	Car3 4.64968792	Car1 7.02742276
		√	√	√	Car2 3.55910639	4.66205734	Car1 7.04469351
√	√	√	√		Car2 0.02757747	Car3 0.05372816	Car1 0.08161176
√	√	√		√	Car2 2.66812254	Car3 3.50926574	Car1 5.28831807
√	√		√	√	Car2 2.67507449	Car3 3.51655046	Car1 5.29623529
	√	√	√	√	Car2 2.67570071	Car3 3.51854280	Car1 5.30127114
√		√	√	√	Car2 2.67366385	Car3 3.50336265	Car1 5.29881075
√	√	√	√	√	Car2 2.14402781	Car3 2.82028996	Car1 4.25324940

From Theory to Practice: Collaborative Coverage Using Swarm Networked Robot in Localisation

Paramita Mandal[1,*], Ranjit Kumar Barai[1], Madhubanti Maitra[1], and Subhasish Roy[2]

[1] Electrical Engineering Department
Jadavpur University, Kolkata-700032, India
[2] Electrical Engineering Department
Bengal Engineering & Science University, Shibpur
Howrah-711103, West Bengal, India
{eie.paramita,ranjit.k.barai,subhasish.ece06}@gmail.com,
madhubanti.maitra66@hotmail.com

Abstract. In this paper, we try to implement collaborative coverage using swarm networked robot. The collaborative robots will reduce time to cover area using enhanced reusability in technique. By reusing parameters in localization operation we proposed coverage. This type of coverage are useful in farms, open grounds etc. A feasible solution has been found and simulation results for a swarm of mobile robots are also presented that uses coverage and positioning algorithms.

Keywords: swarm intelligence, swarm robotics, localization, collaborative area coverage.

1 Introduction

Indeed, ants are most fascinating social insects. These tiny creatures live in societies as complex as ours, and are the most abundant and resilient creatures on the earth. The communities of these myopic creatures capable of short-range interactions achieve amazing feats of global problem solving and pattern formation. It should come as no surprise that ant-behavior became a source of algorithmic ideas for distributed systems where a computer, or a robot, is the "individual" and the network, or a swarm of robots, plays the role of the "colony". The flow of ideas from insects to machines is becoming increasingly popular, see, e.g., [2]. Basic lessons learned from the ants are achieving reliability through redundancy, relying on decentralization via some "individual utility" optimization and transforming the environment into a grand shared database by active marking. A specific advantage of ants is their robustness against changes in the environment. Covering is an important problem in robotics, having applications from floor cleaning to lawn mowing and field demining. Covering has many variations, which are related to traversal, patrolling, mapping and exploration. Hazon et al. in [2] have analytically showed that the coverage will be completed as long as a single robot is able to move. The results presented therein show empirical coverage-time by running the algorithm in two different environments

S. Unnikrishnan, S. Surve, and D. Bhoir (Eds.): ICAC3 2013, CCIS 361, pp. 626–635, 2013

and several group sizes. Rekleitis et al. [3] presented an algorithmic solution for distributed complete coverage, and path planning problem. There the coverage is based on a single robot using *Boustrophedon decomposition*. The robots are initially distributed through space and each robot allocated a virtually bounded area to cover. Communication between robots is available without any restrictions. Canadair et al. [4] presented a new algorithm based on the *Boustrophedon cellular decomposition* of particular interest of providing a solution that guarantees the complete coverage of the free space by traversing an optimal path in terms of the distance travelled. They presented an algorithm that encodes the areas (cells) to be covered as edges of the Reef graph. The optimal solution to the *Chinese Postman Problem (CPP)* is used to calculate an Euler tour, which guarantees complete coverage of the available free space while minimizing the path of the robot. Proof of correctness is provided together with experimental results in different environments. A coverage-based path planning algorithm for multiple robotic agents with the application on the automated inspection of an unknown 2D environment is studied by Wang et al [5]. The proposed path planning algorithm determines a motion path that a robotic agent will follow to sweep and survey all areas of the unknown environment, which is enclosed by the known boundary. The proposed path planning algorithm has been tested and evaluated on the problem of planning path for two types of robotic agents– flying agents and crawling agents in a two-tier hierarchical mission planner to cover various unknown 2D environments. Winward et al. [6] presented a team of automated vehicles safely and effectively they must be coordinated to avoid collisions and deadlock situations. Unexpected events may occur during the operation which may affect vehicles' velocities, so the coordination method must be robust with respect to these events. The method's computation speed and solution quality are evaluated through simulation, and compared with two other methods based on common path coordination techniques. Distributed coverage of environments with unknown extension using a team of networked miniature robots is verified analytically and experimentally has been proposed by Rutishauser et al [7]. The proposed algorithm is robust to positional noise and communication loss, and its performance gracefully degrades for communication and localization failures to a lower bound, which is given by the performance of a non-coordinated, randomized solution. Amstutz et al. [8] proposed the system performance is systematically analyzed at two different microscopic modeling levels, using *agent-based, discrete-event and module based, realistic* simulators. Finally, results obtained in simulation are validated using a team of Alice miniature robots involved in a distributed inspection case study. An inspired control strategy by the hunting tactics of ladybugs to simultaneously achieve sensor coverage and exploration of an area with a group of networked robots is presented by Schwager et al [9]. The controller is distributed in that it requires only information local to each robot and adaptive in that it modifies its behavior based on information in the environment. Stability is proven for both cases *adaptive and non adaptive with a Lyapunov-type proof.* Slotine et al. [10] presented a decentralized controller is presented that causes a network of robots to converge to an optimal sensing configuration, while simultaneously learning the distribution of sensory information in the environment. Convergence and consensus is proven using a *Lyapunov-type proof.* The controller with parameter consensus is shown to perform better than the basic controller in numerical simulations. The control law is *adaptive* in that it uses sensor measurements to learn the distribution of sensory information in the environment. It is *decentralized* in that it requires only information local to each robot. These techniques are suggestive of broader applications of adaptive control methodologies to decentralized control problems in unknown dynamic environments proposed by Rus et al. [11].

In this investigation a novel coverage control strategy has been proposed where the controller can be used by groups of robot swarm to carry out tasks such as environmental monitoring and clean-up, automatic surveillance of rooms, buildings or towns, or search and rescue. This methodology may be useful for scientific studies on geological and ecological scales, and provide tools for a host of security and surveillance applications. Here, we introduce a novel multi-robot on-line coverage algorithm, based on *approximate cell decomposition*. We introduce the algorithm in a form that is feasible for practical implementation on multiple robot platforms with potentially limited computational resources. The algorithm is shown to operate in realistic situations in the presence of noise on sensor measurements and actuator outputs and with asynchronous operation among the robots in the group using enhanced reusability in localization. Simulation results are analyzed with performance metrics to quantify the performance.

The rest of the paper is organized as follows. The technical approach for the problem under consideration has been elaborated in section 2, In Section 3; we describe our algorithm and discuss challenges related to practical implementation. In Section 4 we give results of two simulation studies and show experimental snapshots. Conclusions and discussion on the overall results are given in Section 5.

2 Technical Approach

The main task of the first robot is to position itself with respect to poles placed at each corner of the rectangle. In figure1 with the help of camera it takes images of the colored poles and identifies which colored poles are at what places. It then with the help of sharp IR sensors gets the knowledge of its distance from the other poles.

Positioning

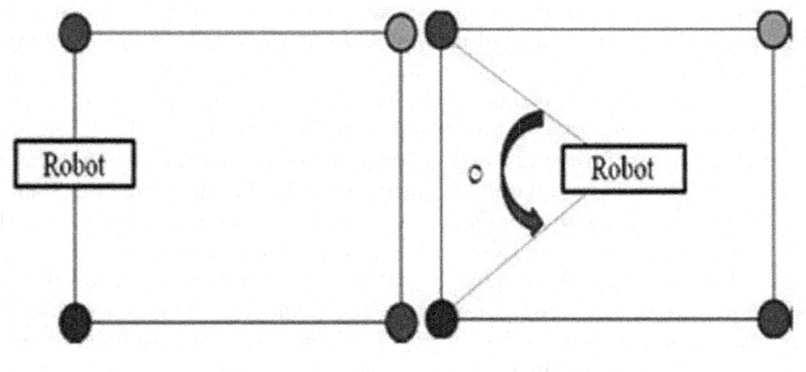

Fig. 1. **Fig. 2.**

In figure 2 the robot with the help of triangle formed (two colored poles and robot at the vertices of triangle) finds the distance between two poles and similarly finds the distance between other poles also and thus gets the knowledge of the whole rectangle. For a triangle with length of sides a, b, c and angles of α, β, γ respectively, given two known lengths of a triangle a and b, and the angle between the two known sides γ (or the angle opposite to the unknown side c), to calculate the third side c, the following formula can be used: The law of Sines or sine rule states that the ratio of the length of a side to the sine of its corresponding opposite angle is constant, that is Thus the robot finds its position in the rectangle.

Coverage

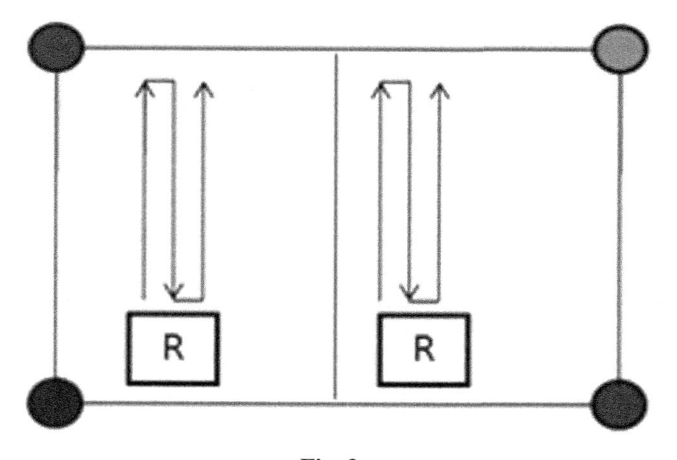

Fig. 3.

Robot after finding its position and having the knowledge of the rectangle divides the area into parts depending upon the total robots employed for covering of the area which in our case is two. In figure3 after dividing the area into quadrants, it assigns the each quadrant to other robots for covering. New robots who enters into rectangular area also position itself with the help of algorithm described above and goes into its prescribed part of the rectangle which the first robot instruct it through data sent by Zigbee communication.

Since robot has got the knowledge of the area which it has to cover, it just sweeps through the area step by step and at the same time checking whether it does not goes into other robot's territory. It does this by checking it position by algorithm described earlier.

Assumptions and limitations of this proposed work is, cleaning area of the robots is assumed to be a rectangular area. At each corner of the cleaning area poles of different colors are installed for the positioning of robots. Zigbee communication is broadcast communication protocol. Due to broadcast in nature particular robot cannot be addressed. After some interval it senses location using camera for better positioning [17].

3 Proposed Technique

In our proposed technique, Collaborative coverage robot is meant for covering a rectangular space using multiple robots thus reducing the time required. There will be a master robot which control slave robots and directs them to cover in specific non-overlapping areas. It can be used in various environments like rooms, grounds, farms etc.

3.1 Enhanced Reusability in Localization

Localization means, where the robot is with respect to a fixed reference frame, set at the initial position of the robot. Now, the entire technique is built with the modules. Modules can be reused in localization to localize area like farms, open grounds etc. The modules that can be reused are: i) Calibrate ii) Detect poles iii) MATLAB-Bot communication iv)Rectangular motion area coverage. According to Bug report, this reusable phenomenon can be described as follows:

3.1.1 Bug Report

Initial position should be such that red/blue/green (in that order) poles should be in detectable 180 degrees of servo motor rotation. To remove, we can first detect red pole, align the bot so that red is roughly close to 0-15 degrees and move back till green pole is in range. Functions to do it are already made. It should be roughly 25 lines of code to implement it. [Algo: turn left till red pole is detected (function: Detect Red Pole (angle)) then move back (function: move with argument 'b') till green pole is detected (function Detect Green Pole)].

It sometimes sends the wrong value of angle of green pole. We have not able to detect the reason. Same code is running for red/blue poles without error.

Aligning takes some time. Algo can be more complex to reduce it.

3.2 Challenges

3.2.1 Image Processing

The pole color should be detected only if it is large enough and in center of image. This is because sharp sensor is mounted on rotating servo motor and it should be aligned to pole if it has to measure the distance correctly.

Hardware is not accurate so we have to align the bot on red/blue line or red/green line (so that it should be on the line and also perpendicular to it) using only two poles was difficult. Function has been made for it, which is pretty accurate even though it takes time.

Remembering position of bot at each point is difficult and without it we cannot specify the starting point and area to cover. To overcome this, we divide the area in parts (each part of width=diameter of bot, and length=side of rectangle). After this we align the bot to starting point and specify how many parts should be cleaned to its left. We align it whenever it reaches the other end of rectangle. It's not very accurate but it works if area is large.

3.2.2 Communication

Synchronizing the bot and Matlab such that, Matlab reads from Zigbee exactly when bot sends. To overcome this, bot sends in a loop - 10 times at regular intervals. Also flush the read buffer in matlab every time we read.

Sending the command to bot which involves numbers (for example move 50cm) was difficult. We had to first send character 'n' to specify that whatever comes next is a number. Then after that send a command which instructs the bot as to (what to do with the number, like '8' means move forward etc).Given bellow the flowchart of the algorithm.

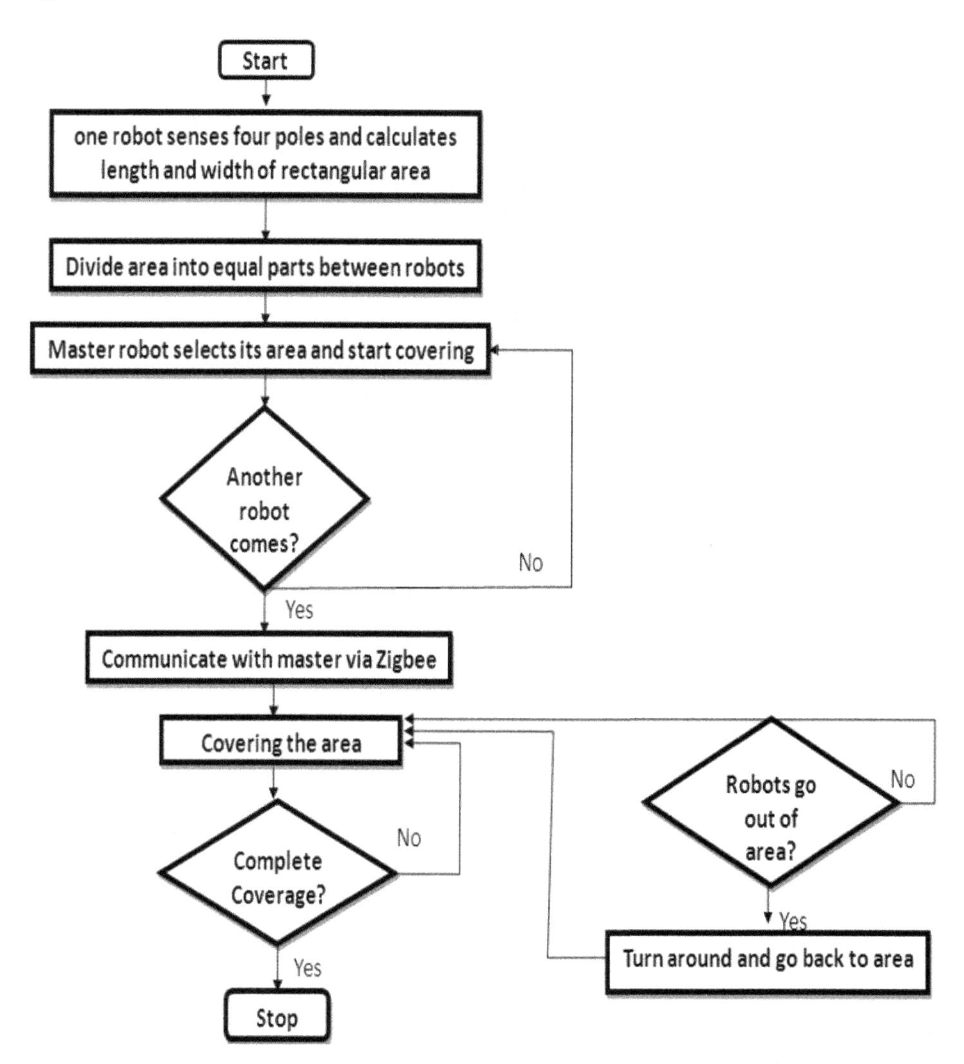

4 Simulation Results

Setup and any extensions implemented on the robot are i) AVR Studio 4 ii) Matlab 7.7.0.471(R2008b). Additional hardware used: Camera - To take image and further to process it for the detection of poles. Servo Motor - To detect poles when a robot enters into the area and to position itself. Sharp sensor –To detect the distance of a robot from the poles. Zigbee – To communicate with other slave robots. Camera should be mounted on servo motor, and make sure "90 degree" of servo motor should be in front of robot.

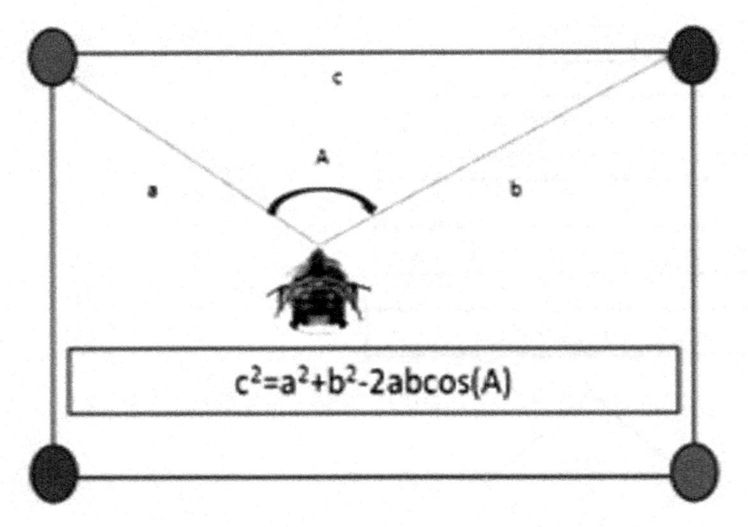

Fig. 4. Robot Positioning

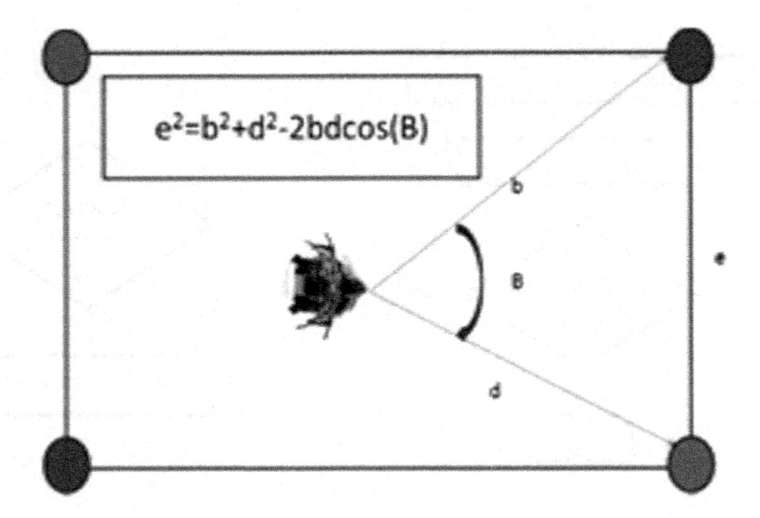

Fig. 5. Robot Positioning

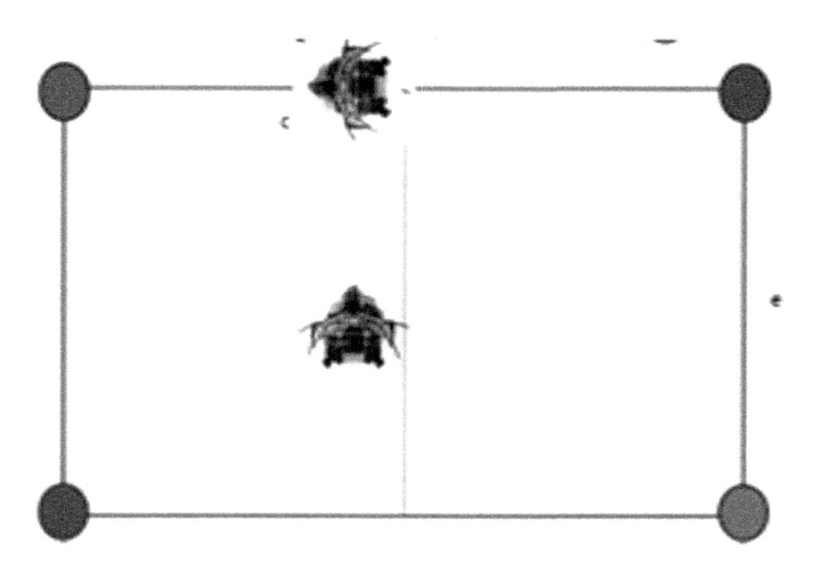

Fig. 6. Robot Positioning

Fig. 7. Robot Positioning

Covering: It also covers the area which it has divided and assigned to it. Sweeps the area as shown in the figure bellow:

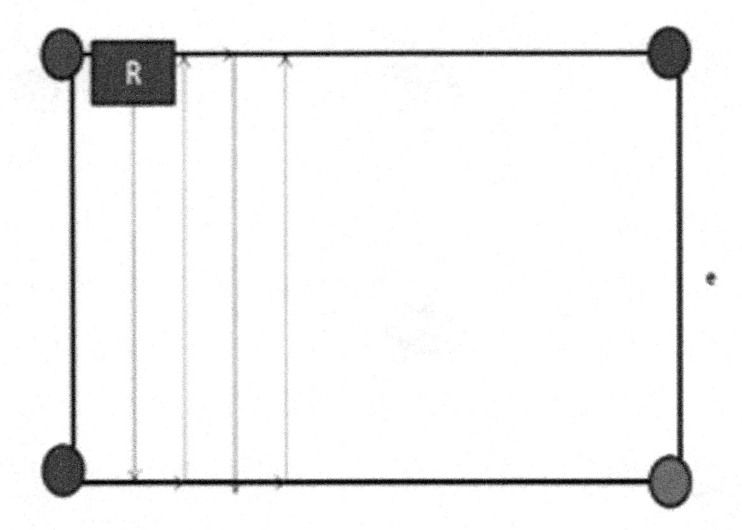

Fig. 8. Robot covering the area

Figure8 Robot covering the area. Here, the main results of robot positioning and coverage are depicted in the figure 4, 5,6,7,8 respectively. For completing the task Detect different color poles, Sends angles and distances to Matlab code, Calculate parameters of rectangular area, Divides the area and calculates starting point of each area, Moves to the red-blue line and calibrates itself, Goes to starting point on red-blue line and calibrates itself.

5 Conclusions and Future Work

This paper shows how multi robots can be included to increase parallelism. If area can be of any shape, instead of Zigbee communication, Bluetooth can be used. We also used to learn image processing and problems with the Zigbee communication and solutions. We expect that the proposed technique will find broader application beyond the problem chosen here. The foundation of this algorithm is that, sometimes hardware may not work as desired in that case the robot able to position itself. Covers the area which it has divided and assigned to it and can be extended to include more robots.

References

1. Wagner, I.A., Bruckstein, A.M.: From Ants to A(ge)nts: A Special Issue on Ant-Robotics. Annals of Mathematics and Artificial Intelligence 31, 1–5 (2001)
2. Hazon, N., Mieli, F., Kaminka, G.A.: Towards Robust On-line Multi-Robot Coverage. In: Proc. of the IEEE Int. Conf. on Robotics and Automation, pp. 1710–1715 (2006)
3. Rekleitis, I., New, A.P., Choset, H.: Distributed Coverage of Unknown/Unstructured Environment by Mobile Sensor Network. In: Multi-Robot Systems, From Swarms to Intelligent Automata, vol. III(IV), pp. 145–155 (2005)

4. Mannadiar, R., Rekleitis, I.: Optimal Coverage of a Known Arbitrary Environment. In: IEEE Int. Conf. on Robotics and es, pp. 5525–5530 (2010)
5. Wang, X., Syrmos, V.L.: Coverage Path Planning for Multiple Robotic Agent-Based Inspection of an Unknown 2D Environment. In: 17th Mediterranean Conf. on Control & Automation, pp. 1295–1300 (2009)
6. Winward, G., Flann, N.S.: Coordination of Multiple Vehicles for Area Coverage Tasks. In: Proc. of the IEEE/RSJ Int. Conf. on Intelligent Robots and Systems, pp. 1351–1356 (2007)
7. Rutishauser, S., Correll, N., Martinoli, A.: Collaborative coverage using a swarm of networked miniature robots. Robotics and Autonomous Systems 57, 517–525 (2009)
8. Amstutz, P., Correll, N., Martinoli, A.: Distributed Boundary Coverage with a Team of Networked miniature Robots using a Robust market-based Algorithm. Springer Science + Business Media B.V. 2009, Ann. Math. Artif. Intell. 52, 307–333 (2008)
9. Schwager, M., Bullo, F., Skelly, D., Rus, D.: A Ladybug Exploration Strategy for Distributed Adaptive Coverage Control. In: IEEE Int. Conf. on Robotics and Automation, pp. 2346–2353 (2008)
10. Schwager, M., Slotine, J.-J., Rus, D.: Consensus Learning for Distributed Coverage Control. In: 2008 IEEE Int. Conf. on Robotics and Automation, pp. 1042–1048 (2008)
11. Schwager, M., Rus, D., Slotine, J.-J.: Decentralized, Adaptive Coverage Control for Networked Robots. In: IEEE Int. Conf. on Robotics and Automation, pp. 3289–3294 (2007)
12. Wang, B.: Coverage Control in Sensor Networks. Springer-Verlag London Limited (2010) ISBN 978-1-84996-058-8
13. Bai, X., Li, S., Jiang, C., Gao, Z.: Coverage Optimization in Wireless Mobile Sensor Networks. In: The 5th Int. Conf. on Wireless Communications, Networking and Mobile Computing, Wicom 2009, pp. 1–4 (2009)
14. Hazon, N., Kaminka, G.A.: Redundancy, Efficiency and Robustness in Multi-Robot Coverage. In: Proc. of the IEEE Int. Conf. on Robotics and Automation, pp. 735–741 (2005)
15. Correll, N., Martinoli, A.: Robust Distributed Coverage using a Swarm of Miniature Robots. In: IEEE Int. Conf. on Robotics and Automation, pp. 379–384 (2007)
16. Simon, G., Molnar, M., Gonczy, L., Cousin, B.: Robust k-Coverage Algorithms for Sensor Networks. IEEE Trans. on Instrumentation and Measurement 57, 1741–1748 (2008)
17. Mnadal, P., Barai, R.K., Maitra, M.: Collaborative Coverage Using Swarm Networked Robot. In: The Proc. of the IEEE Int. Conf. on Advances in Engineering, Science and Management, pp. 301–304 (2012)

A Frequency Response Matching Method for PID Controller Design for Industrial Processes with Time Delay

Somnath Pan and Md. Nishat Anwar

Department of Electrical Engineering, Indian School of Mines, Dhanbad – 826004, India
somnath_pan@hotmail.com, nishatnith@gmail.com

Abstract. This paper presents a frequency domain PID controller design technique that start with specifying the desired closed-loop specifications of an industrial process with time delay. Frequency response of the desired system is matched with that of the control system to be designed at the specified frequency points. Resultant linear algebraic equations are solved to find the PID controller. Examples, taken from literature, are illustrated to compare this design method with other design methods found in literature.

Keywords: PID controller, Frequency response matching, Process control, Time delay system.

1 Introduction

Low order simple controllers are always preferred for the industrial control systems. The PID controllers are the most widely accepted in the industrial environment, especially for the industrial process control systems due to its familiarity and ease of use. After the introduction of the seminal paper by Ziegler and Nichols [1], a large number of papers have been published either to modify the Ziegler-Nichols tuning method or to introduce new ones. A paper of special credit goes to Cohen and Coon [2] showing the achievement of quarter-amplitude decay ratio through the adjustment of dominant poles of the control system. Both the frequency domain and the time domain methods that consider various intricate properties of the control systems through sophisticated and complex mathematics have been evolved to give excellent controller property.

In the papers of Persson [3] and Astrom [4] on the method of dominant pole placement for the PID controller, dominancy of poles placed is not assured and eventually, the overall response may be sluggish or even unstable. Such drawback has been removed by Wang, et al [5]. Internal model control (IMC) has been used for PID controller design by Rivera et al [6] and a simplified IMC methods have been introduced by Fruehauf et al [7] and Skogestad [8] the for PID tuning. Techniques based on optimization of error directly or in some way may be found in [9]-[11]. In frequency domain design method, various properties of frequency response have been used in [12], [13].

S. Unnikrishnan, S. Surve, and D. Bhoir (Eds.): ICAC3 2013, CCIS 361, pp. 636–646, 2013.

However, though the models of the industrial processes are not normally so accurate, the dynamics of the processes are of low order. For such industrial processes, rather, simple to understand and mathematically less-involved methods which can give acceptable response in terms of practical requirement would be attractive and beneficial. Such a method is proposed in this paper which is based on frequency response matching of the controlled system with that of a model system. The model system shows the behaviour or response which is desired by the designer and the model may be called as the reference model. A low order model up to second order can be chosen from the industrial specifications given in time domain (such as settling time, peak overshoot, etc.) or in frequency domain (such as gain margin and phase margin, etc.). Then, the frequency response of the reference model is matched with that of the control system with the PID controller at two frequency points to evaluate the parameters of the PID controller.

The choice of such two frequency points is important. For the purpose of good frequency matching over the effective region of frequency response, we follow the concept of low frequency matching around the frequency $\omega=0$. In sequel, a good matching is achieved in the overall range of the frequency response and an acceptable PID controller is resulted. Its set-point response and the load disturbance rejection are highly comparable if not better in various aspects than that of the controllers designed by the prevailed design methods taken from the literature.

The proposed design method has been described in Section 2. In Section 3 the method has been illustrated through examples taken from the literature and a comparison with the existing methods has been given there. Section 4 concludes the paper.

2 The Design Method

Consider a plant which is described by the transfer function,

$$G(s) = \frac{N(s)}{D(s)} e^{-sL} = G_1(s) e^{-sL} \tag{1}$$

where, $N(s)/D(s)$ is a rational transfer function and L is the time delay of the system. The poles and zeros of G1(s) are considered to be in the left hand side of s-plane. The system considered may have monotonic (Overdamped) or oscillatory (underdamped) time response to step input.

We are considering the conventional unity negative output feedback control system with a PID controller as given by

$$C(s) = K_P + \frac{K_I}{s} + K_D s \tag{2}$$

where, K_P, K_I and K_D are the proportional, integral and derivative constants of the controller that are to be determined by the proposed design method.

The first step of the design method is to choose a reference model M(s) which will encompass the desired specification of the overall control system. The desired industrial specification for control system can be stated in the time domain such as, settling time, peak overshoot etc. It may be described in the frequency domain such as gain margin, phase margin, band width etc. For a feasible and practical operation of the designed control system, the reference model should be such that, the plant would be able to response accordingly under the excitation generated by the controller.

The reference model should obviously have the same time delay term of the plant i.e., e^{-sL} as shown below

$$M(s) = \frac{P(s)}{Q(s)} e^{-sL} \tag{3}$$

where, $P(s)/Q(s)$ is in the form of rational polynomial.

For a first order D(s) of the plant, Q(s) of the reference model can be chosen as an appropriate first order term, whereas for D(s) of second order or of higher order, Q(s) is to be chosen in second order form. The choice of $P(s)/Q(s)$ is important as because it manifests the dominant characteristics of the plant. For simplicity, P(s) is taken to be 1.

Now, for the purpose of design, we are relying on frequency response matching of the reference model with the control system. This may be written as

$$\left.\frac{C(s)G(s)}{1+C(s)G(s)}\right|_{s=j\omega} \equiv M(s)|_{s=j\omega} \tag{4}$$

where, the LHS expression is equivalent with the RHS expression in terms of frequency response and the unknown parameters of the controller C(s) are to be evaluated.

Equation (4) can be written as

$$C(s)G(s)|_{s=j\omega} \equiv \left.\frac{M(s)}{1-M(s)}\right|_{s=j\omega} \equiv M_o(s)|_{s=j\omega} \tag{5}$$

where, $M_o(s)$ may be called as equivalent open loop reference model.

Equation (5) gives

$$C(s)|_{s=j\omega} \equiv \left.\frac{M_o(s)}{G(s)}\right|_{s=j\omega}$$

Or, explicitly

$$\left.K_P + \frac{K_I}{s} + K_D s\right|_{s=j\omega} \equiv \left.\frac{M_O(s)}{G(s)}\right|_{s=j\omega} \tag{6}$$

where, the expression in RHS is fully known.

For the purpose of equating the LHS to the RHS of Equation (6) in terms of frequency response, minimum two frequency values are to be selected for evaluation of three unknown K_P, K_I and K_D. For low frequency matching which is suitable for most of the industrial systems, specially the process control systems, the frequency points for matching will be selected around $\omega=0$. However, for a theoretical frequency response, the range of ω is from 0 to ∞ and for this infinite range, choosing the frequency of 'low value' should be consistent with the effective range of frequency response. Here, the 'low frequency' values are selected with the following concept.

If τ is the dominant time constant of the plant, the corresponding frequency is $2\pi/\tau$, and $20\pi/\tau$ can be assumed as the effective range of dominant frequency response of the plant. Hence, the low frequency values, for the purpose of matching, can be selected at around 0.01 times of the effective range. Such frequency points for matching give good result for the most of the plants.

By putting such a low frequency value ω_1 in Equation (6) and separating the real and imaginary parts, we get the following two linear algebraic equations.

$$K_P = Re\left[\frac{M_O(j\omega_1)}{G(j\omega_1)}\right] \tag{7}$$

$$-\frac{K_I}{\omega_1} + K_D\omega_1 = Im\left[\frac{M_O(j\omega_1)}{G(j\omega_1)}\right] \tag{8}$$

Equation (7) gives the value of K_P directly. Putting another low frequency point ω_2 in Equation (6) and considering the imaginary part, we get

$$-\frac{K_I}{\omega_2} + K_D\omega_2 = Im\left[\frac{M_O(j\omega_2)}{G(j\omega_2)}\right] \tag{9}$$

Now, solving Equations (8) and (9), K_I and K_D are evaluated. Hence, the PID controller is now completely determined.

3 Examples

Example 1:
We consider a FOPDT (first order plus dead time) industrial process, taken from the literature [12], with the following transfer function.

$$G(s) = \frac{e^{-5s}}{10s + 1}$$

This plant has the time constant of 10 sec and the time delay of 5 sec. The plant has the settling time of 44.1 sec and the rise time of 22 sec with the gain margin as 11.6 dB and the phase margin as –180 deg.

For the purpose of designing a PID controller, we first choose a reference model that is expected to be followed by the close loop plant. The reference model of is chosen as

$$M(s) = \frac{e^{-5s}}{5s + 1}$$

This reference model has the settling time of 24.6 sec with the rise time 11 sec and the gain margin and the phase margin are 7.09 dB and –180 deg, respectively.

Following the concept of proposed design method, the low frequency points selected are ω_1=0.03 rad/sec and ω_2=0.06 rad/sec for the purpose of frequency matching between the reference model and control system. Following Equations (7)-(9), the values of the controller parameters are obtained as given below.

$$K_P = 1.12$$
$$K_I = 0.1$$
$$K_D = 1.2$$

Performance of the controller obtained by the proposed method has been compared with that of the controller obtained by the methods of Ziegler-Nichols [1], Zhuang-Atherton [9], internal model control (IMC) [6], and Wang et al [12] in terms of process output in Figure 1, controller output in Figure 2 and bode magnitude plot in Figure 3. A summary of performances of these controllers is shown in Table 1. In case of time response, both the set-point response (for unit step input) and load disturbance response (for 0.5 unit step load disturbance applied at the plant input) have been shown.

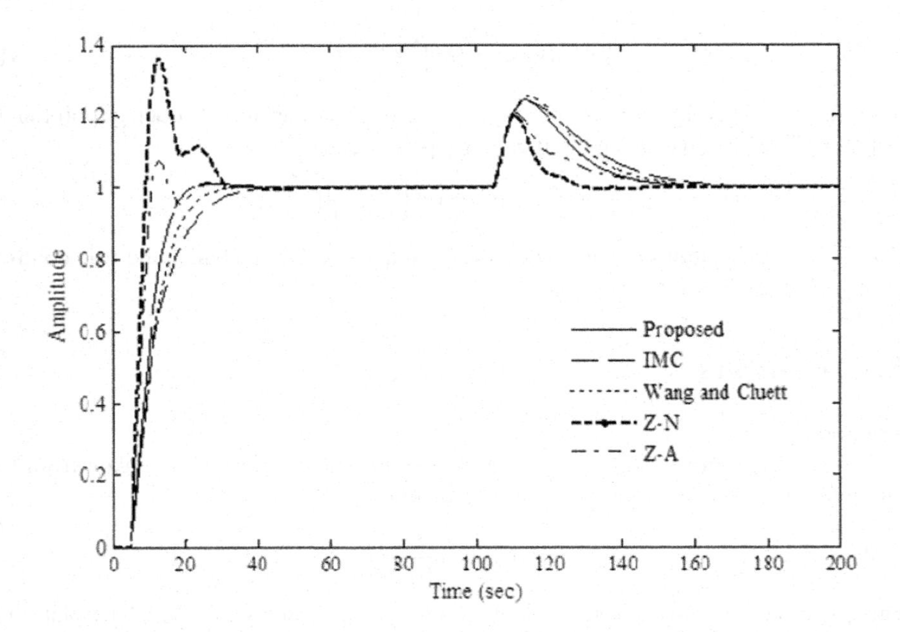

Fig. 1. Comparison of process output for Example 1

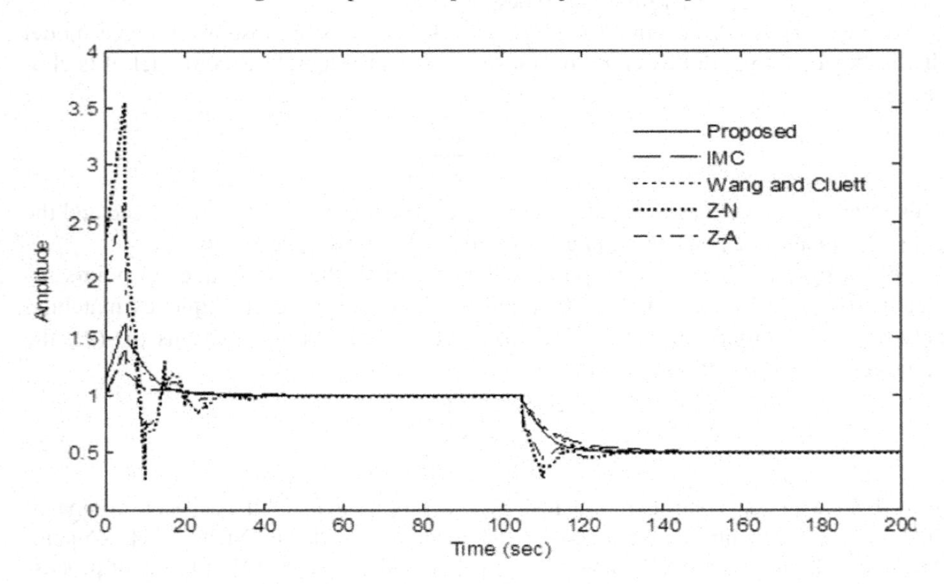

Fig. 2. Comparison of controller output for Example 1

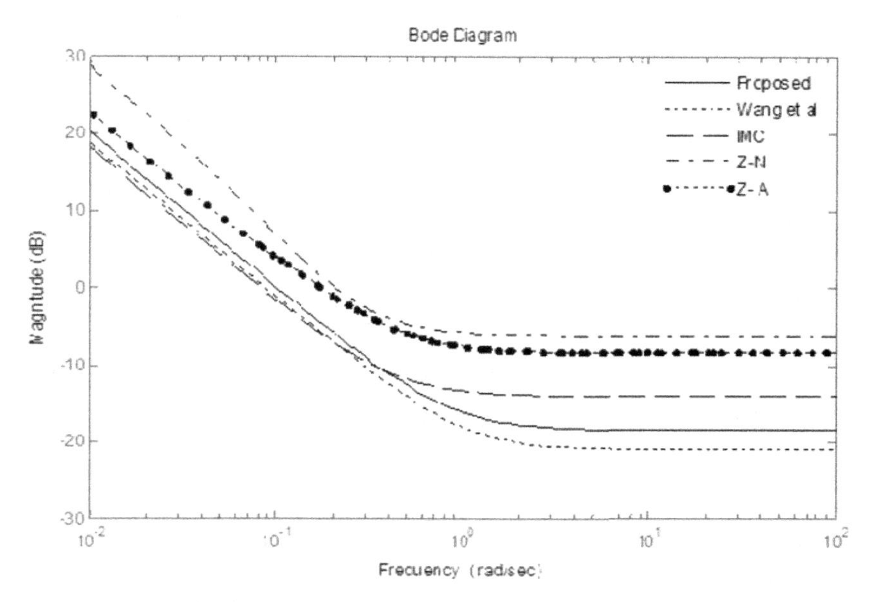

Fig. 3. Comparison of bode magnitude plot for Example 1

Table 1. Performance comparison among different methods for Example 1

Method	K_P	K_I	K_D	Set-point response				Load disturbance response		Frequency response	
				Peak overshoot (%)	Time to peak overshoot (sec)	Rise time (sec)	Settling time (sec)	Maximum process output	Settling time (sec)	Gain margin (dB)	Phase margin (deg)
Proposed	1.12	0.1	1.2	1.2	29	11.4	19.6	0.24	45.6	11.2	68.0
IMC	0.966	0.08	0.88	Overdamped		14.9	31	0.25	64.7	12.3	70.6
Wang et al.	1	0.08	2	Overdamped		20.2	41.5	0.23	73.2	11.8	78.1
Z-N	2.275	0.26	5.91	37.4	10	2.19	27.2	0.18	35.8	4.64	49.9
Z-A	1.993	0.13	3.82	16	10	3.29	34.1	0.20	68.1	5.83	67.2

It is observed from Figure 1 and Table 1 that the proposed method gives the best set-point response while the load disturbance response is better than the other methods except the Ziegler-Nichols method whereas the set-point response by the Ziegler-Nichols method is the worst. Figure 2 shows the proposed method gives a comparable controller output with other methods with the fact that, the peak value as well as the variation of the controller output are maximum in case of Ziegler-Nichols method.

Example 2:

We consider a FOPDT industrial process from the literature [12], as given by following transfer function.

$$G(s) = \frac{e^{-50s}}{10s + 1}$$

The nominal plant has the settling time of 89.1 sec and rise time of 22 sec with gain margin as 1.08 dB and phase margin as -180 deg.

For the purpose of designing a PID controller, we first choose a reference model that would be followed by the close loop plant. Here, it is to be noted that the time delay of the plant is very large (5 times) compared to the time constant of the plant. In such case, designing a controller to achieve a very strict response is more difficult [14], [15]. To illustrate this, we choose four reference models as shown in Table 2 along with their performance specifications. Here, it is to be observed that the reference models show more strict performance (becoming faster) as the time constants of them are decreasing.

Table 2. Performance of various reference models

Reference model	Settling time (sec)	Rise time (sec)	Gain margin (dB)	Phase margin (deg)
$M_1(s) = e^{-50s}/(30s + 1)$	167	65.9	4.42	-180
$M_2(s) = e^{-50s}/(20s + 1)$	128	43.9	2.8	-180
$M_3(s) = e^{-50s}/(15s + 1)$	109	33	1.93	-180
$M_4(s) = e^{-50s}/(10s + 1)$	89.1	22	1.08	-180

Following the concept of the proposed design method, the low frequency points selected are ω_1=0.03 rad/sec and ω_2=0.06 rad/sec, for the purpose of frequency matching between the reference model and the control system. We obtained the PID controllers as shown in Table 3.

Performance of the controllers designed by the proposed method have been compared with controllers obtained by the methods of Cohen-Coon (C-C) [2], Zhuang-Atherton [9], internal model control (IMC) [6], and Wang et al [12] and a summary of performances is shown in Table 4. In case of time response, both the set-point response (for unit step input) and the load disturbance response (for 0.5 unit step load disturbance applied at the plant input) have been considered.

Table 3. Various PID controllers designed for Example 2

Reference model	PID controller designed		
	K_P	K_I	K_D
$M_1(s)$	0.33	0.012	1.87
$M_2(s)$	0.41	0.014	3.72
$M_3(s)$	0.51	0.016	4.99
$M_4(s)$	0.53	0.018	6.67

Table 4. Comparison of the controllers obtained by different methods for Example 2

Method	Controller parameters			Set-point response				Load disturbance response		Frequency response	
	K_P	K_I	K_D	Peak overshoot (%)	Time to peak overshoot (sec)	Rise time (sec)	Settling time (sec)	Maximum process output	Settling time (sec)	Gain margin (dB)	Phase margin (deg)
Proposed 1	0.335	0.012	1.878	Overdamped		67.4	166	0.49	204	8.87	65.4
Proposed 2	0.412	0.014	3.72	2.4	133	44.8	150	0.49	214	8.35	64.8
Proposed 3	0.51	0.016	4.99	15	100	34.1	202	0.49	238	6.04	62.9
Proposed 4	0.53	0.018	6.67	23.2	100	30.8	355	0.49	254	3.51	58.3
Wang et al.	0.275	0.011	1.023	Overdamped		88.6	220	0.48	281	9.65	67.0
IMC	0.538	0.015	3.84	15.3	100	33.5	206	0.49	266	6.1	64.9
C-C	0.517	0.008	4.92	Overdamped		215	459	0.49	520	6.16	87.6
Z-A	0.508	0.015	6.74	12.9	150	35.3	401	0.49	251	3.42	67.1

It is observed from the Table 4 that, as the time constant of the reference model is decreasing (i.e., the desired specification is becoming stricter) the improvement of the performance of the designed control system is not improving after the 'Proposed 2' controller in the list. Hence, it may be noted that, an arbitrarily strict reference model would not be followed by the plant.

Finally, the 'Proposed 2' controller is selected for further comparison with the above mentioned methods. Figure 4 and 5 shows the process output comparison, both for the set-point response (unit step input) and the load disturbance response (for 0.5 unit step load disturbance applied at the plant input). Figure 6 shows the controller output comparison, whereas, Figure 7 shows comparison of the Bode magnitude plot. It is observed that performance of the 'Proposed 2' controller is better than the controllers obtained by the other methods in terms of set-point response, load disturbance response and controller output. Performance in terms of frequency response of the 'Proposed 2' controller is better than the other methods except the method of Wang et al which is marginally better.

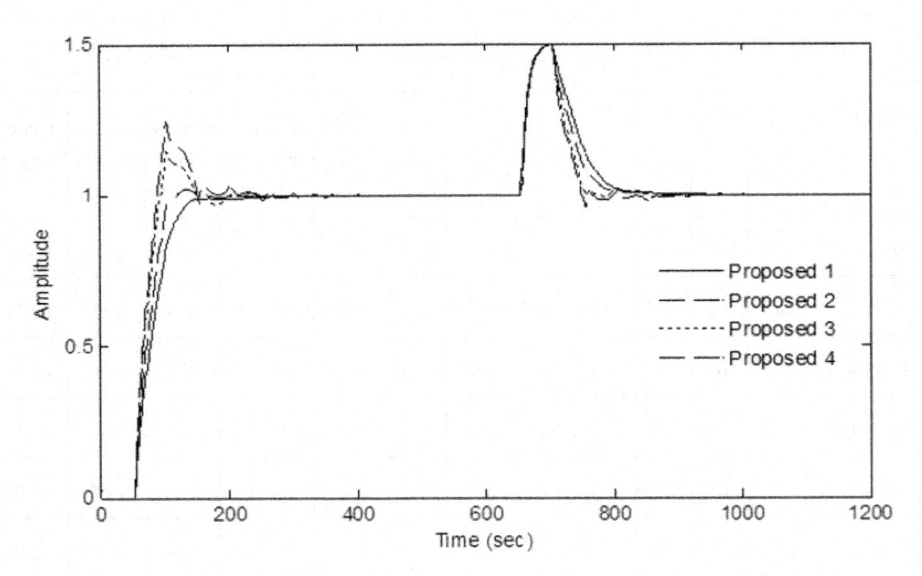

Fig. 4. Comparison of process output for the proposed controllers for Example 2

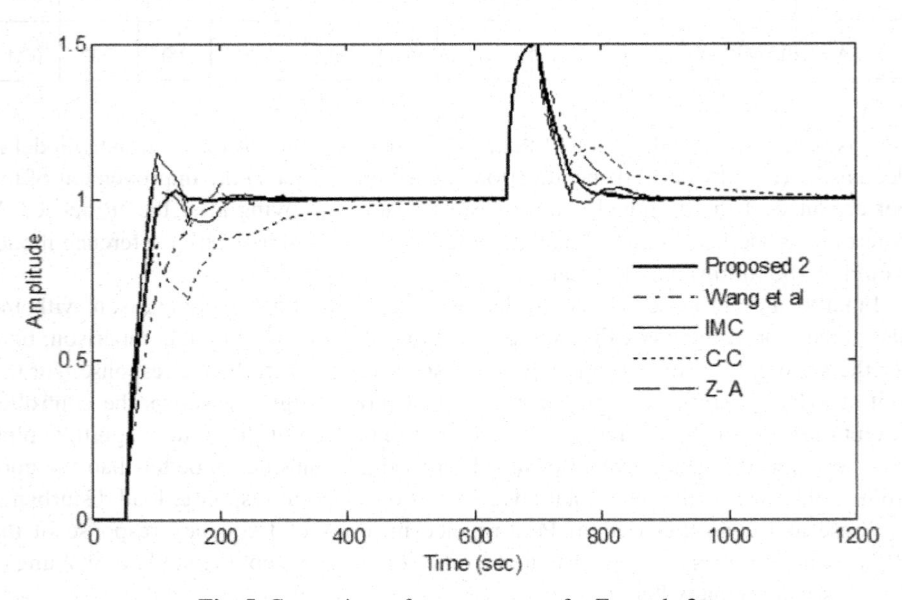

Fig. 5. Comparison of process output for Example 2

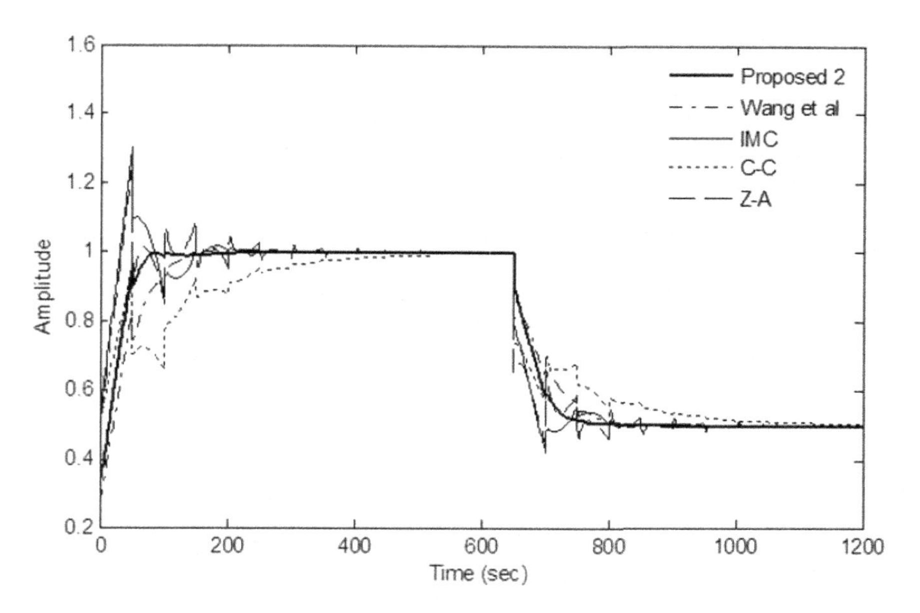

Fig. 6. Comparison of controller output for Example 2

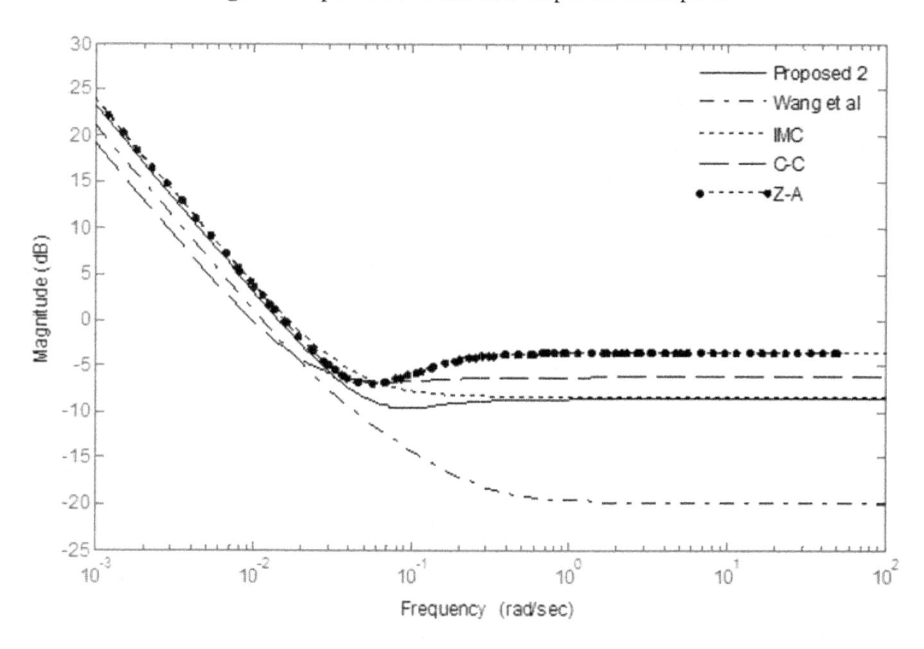

Fig. 7. Comparison of bode magnitude plot for Example 2

4 Conclusion

A simple frequency domain matching method has been described for design of PID controller for industrial processes with time delay. Considering the plant dynamics a suitable low order reference model is chosen. The frequency response matching in between the reference model and the control system to be designed is done at some low frequency points to arrive at linear algebraic equations, solution of which gives the controller parameters. For the purpose of matching, there is no requirement of elaborate frequency response analysis or any optimization technique. A simple but meaningful criterion has been provided for selecting the frequency points. Though matching is performed only at two frequency points, a good overall matching is obtained resulting an acceptable time and frequency domain responses. The proposed method has been favorably compared with the controller design methods prevailed in the literature through examples taken from the literature.

References

1. Ziegler, J.G., Nichols, N.B.: Optimum Settings for Automatic Controllers. Trans. ASME 64, 759–768 (1942)
2. Cohen, G.H., Coon, G.A.: Theoretical Considerations of Retarded Control. Trans. ASME 75, 827–834 (1953)
3. Persson, P., Astrom, K.J.: Dominant Pole Design – A Unified View of PID Controller Tuning. In: Dugard, L., M'Saad, M., Landau, I.D. (eds.) Adaptive System in Control and Signal Processing 1992: Selected Papers from the Fourth IFAC Symposium, Grenoble, France, July 1-3, pp. 377–382. Pergamon Press, Oxford (1993)
4. Astrom, K.J., Hagglund, T.: PID Controller Theory, Design and Tuning, 2nd edn. Instrument Society of America. Research Triangle Park, North Carolina (1995)
5. Wang, Q.G., Zhang, Z., Astrom, K.J., Chek, L.S.: Guaranteed Dominant Pole Placement With PID Controllers. Journal of Process Control 19, 349–352 (2009)
6. Rivera, D.E., Morari, M., Skogestad, S.: Internal Model Control 4. PID Controller Design. Ind. Eng. Chem. Process Des. Dev. 25, 252–265 (1986)
7. Fruehauf, P.S., Chien, I.L., Lauritsen, M.D.: Simplified IMC-PID Tuning Rules. ISA Transactions 33, 43–59 (1994)
8. Skogestad, S.: Simple Analytic Rules for Model Reduction and PID Controller Tuning. Journal of Process Control 13, 291–309 (2003)
9. Zhuang, M., Atherton, D.P.: Automatic Tuning of Optimum PID Controllers. IEE Proceeding-D 140, 216–224 (1993)
10. Panagopoulos, H., Astrom, K.J., Hagglund, T.: Design of PID Controllers Based on Constrained Optimisation. IEE Proc. Control Theory Application 149, 32–40 (2002)
11. Sekara, T.B., Matausek, M.R.: Optimization Of PID Controller Based on Maximization of The Proportional Gain Under Constraints on Robustness and Sensitivity to Measurement Noise. IEEE Transaction on Automatic Control 54, 184–189 (2009)
12. Wang, L., Barnes, T.J.D., Cluett, W.R.: New Frequency-Domain Design Method for PID Controllers. IEE Proc. Control Theory Application 142, 265–271 (1995)
13. Karimi, A., Garcia, D., Longchamp, R.: PID Controller Tuning Using Bode's Integrals. IEEE Transaction, Control System Technology 11, 812–821 (2003)
14. Astrom, K.J., Hang, C.C., Persson, P., Ho, W.K.: Towards Intelligent Control. Automatica 28, 1–9 (1992)
15. Astrom, K.J., Hagglund, T.: Advanced PID control. ISA-The Instrumentation, Systems and Automation Society (2006)

Uncertainty Quantification and Sensitivity Analysis for a Nonlinear Bioreactor

T. Babu[1] and N. Pappa[2]

[1] Dept. of ICE, St. Joseph's College of Engg., Chennai-119
[2] Dept. of Instrumentation Engg, MIT Campus, Anna University, Chennai-44
tbabume@gmail.com

Abstract. Every process is having uncertainty, affects the performance of the system, even breakdown. Hence the uncertainty has to be analyzed seriously. In this paper, the parametric uncertainty is determined using multiple experiment method for a highly non linear plant. An unstable bioreactor is considered as a nonlinear process for the analysis. It is having various uncertainties in the process parameter. Substrate feed and surrounding temperature are the sources of external parametric uncertainty; also the specific growth rate and dilution rate are the internal parametric uncertainty. Overall uncertainty is determined using Monte Carlo method and the results were compared with the standard methods. Simulation results addresses that the effect of various uncertainties in the biomass concentration. A robust PID controller was designed for both stable and unstable conditions using Particle swarm optimization (PSO) algorithm, which control the system effectively when the uncertain parameter varies within the specified range. Controller performances were analyzed for various uncertainties range using MATLAB Simulink and estimated the sensitivity of each parameter individually and combined.

Keywords: Monte Carlo uncertainty analysis, bioreactor, Particle swarm optimization.

1 Introduction

Study of uncertainties in the system component and their effect on the performance of the system has attracted the researchers in the model based controller design field. Uncertainty is a quantification of the doubt about the measurement result. The following stages are involved in the uncertainty analysis [1], [2] of a model (a) estimation of uncertainties in model inputs and parameter (b) estimation of the uncertainty in model outputs resulting from the uncertainty in model inputs and model parameters (c) characterization of uncertainties associated with different model structures and model formulations and (d) characterization of the uncertainties in model predictions resulting from uncertainties in the evaluation data.

Even though many parameters could be measurable up to any desired precision, but, at least in principle, there are significant uncertainties [3] associated with their estimates. Some uncertainties arise from measurement; it involves (a) random errors

S. Unnikrishnan, S. Surve, and D. Bhoir (Eds.): ICAC3 2013, CCIS 361, pp. 647–658, 2013.

in analytic devices (b) systematic biases that occur due to imprecise calibration and (c) inaccuracies in the assumptions used to infer the actual quantity of interest from the observed readings of an alternate variable. Further, uncertainty in model application arises from uncertainties associated with measurement data used for the model evaluation.

In order to estimate the uncertainties associated with model formulation, several different models, each corresponding to a different formulation of the mathematical problem corresponding to the original physical system, have to be developed. Development of several alternative computational models with different model coefficients can require substantial time and effort.

Monte Carlo (MC) method is the most widely used technique for uncertainty analysis. This method involves random sampling from the distribution of inputs and successive model runs until a statistically significant distribution of output is obtained. They can be used to solve problems with physical probabilistic structures, such as uncertainty propagation in models or solution of stochastic equations. Monte Carlo method is also used in the solution of problems that can be modeled by the sequence of a set of random steps that eventually converge to a desired solution. Problems such as optimization and the simulation of movement of fluid molecules are often addressed through Monte Carlo simulations.

Interval mathematics is used to address data uncertainty that arises (a) due to imprecise measurements (b) due to the existence of several alternative methods, techniques, or theories to estimate model parameters. In many cases, it may not be possible to obtain the probabilities of different values of imprecision in data; in some cases only error bounds can be obtained. This is especially true in case of conflicting theories for the estimation of model parameters, in the sense that probabilities cannot be assigned to the validity of one theory over another. In such cases, interval mathematics is used for uncertainty estimation.

The objective of interval analysis is to estimate the bounds on various model outputs based on the bounds of the model inputs and parameters. In the interval mathematics approach, uncertain parameters are assumed to be "unknown but bounded", and each of them has upper and lower limits without a probability structure, every uncertain parameter is described by an interval. If a parameter of model is known to be between, then the representation of the process is called interval system [7]. Special arithmetic procedures for calculation of functions of intervals used in this method are described in the literature [9].

In this paper bioreactor is considered for the uncertainty analysis due to its nonlinear nature. The outcome of the reactor is depending on many factors like Biomass concentration, substrate feed, substrate concentration, Feed flow rate, Dilution rate, Volume of the reactor etc. Uncertainties in process parameters listed above affect the rate of growth and quality of the product. A detailed study has been made for the sensitivity of each parameter.

Frequently the bioreactor behaves as unstable process; suitable controller has to be identified for the smooth operation. In this paper Particle swarm Optimization (PSO) [10] based technique is attempted to obtain the controller parameter.

Uncertainty in each process parameter is estimated using multiple experiment method, by tabulating the output of the reactor for various inputs. From the different input and its corresponding output, sensitivity of each parameter is calculated.

2 Process Description

Bioreactors[13] are used in many applications including industries concerned with food, beverages, pharmaceuticals, waste treatment and alcohol fermentation. They are inherently nonlinear. A typical diagram of the bioreactor is shown in figure 1. New cells are produced by the reactor by consuming substrate.

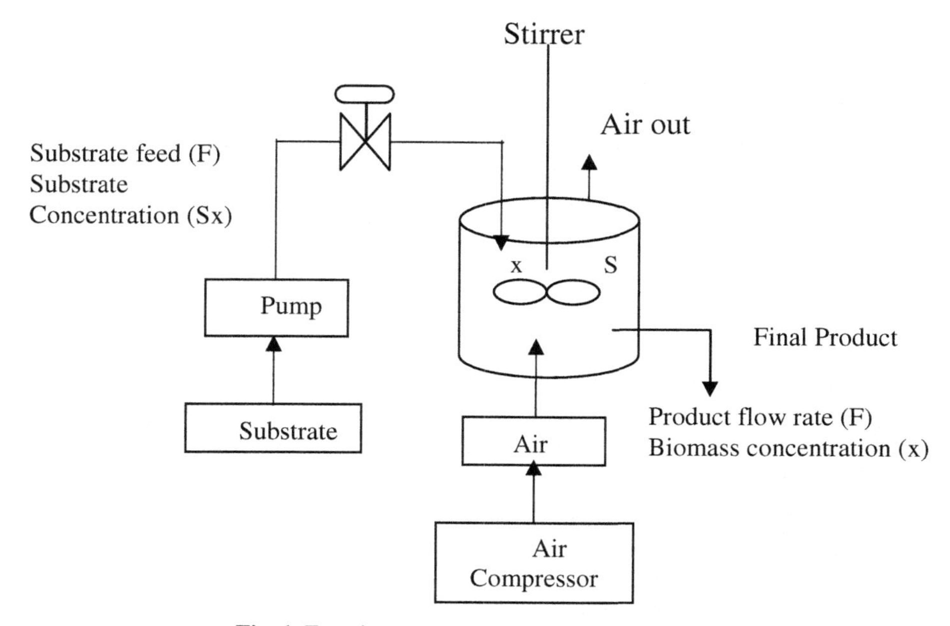

Fig. 1. Functional Block diagram of a bioreactor

After adding living cells to a reactor containing substrate, micro organisms will start to grow up. The growth of micro organism depends on many factors like substrate feed, dilution rate and biomass concentration etc. Cell growth may take place with number of phases [12] like, lag phase (no cell growth), exponential growth phase (cell growth is too high, but at some stage, amount of substrate will be limiting for the growth), stationary phase (cell count neither increase nor decrease) and death phase (The number of cells decreases due to food shortage.). Fig.1 shows different phases in cell growth.

The influent flow rate is equal to the effluent (output) flow rate Q. Hence, the volume V is constant. The influent has a substrate concentration Sin [mass/volume] and substrate concentration Xin [mass/volume]. The rate of accumulation of biomass is obtained from a mass balance equation. Model is derived with various assumptions, like biomass has a specific growth rate, the total amount of produced biomass per unit time in a reactor with volume V is VX, compare with (1). Since the reactor is completely mixed, the outflow concentration of biomass is equal to the concentration in the tank.

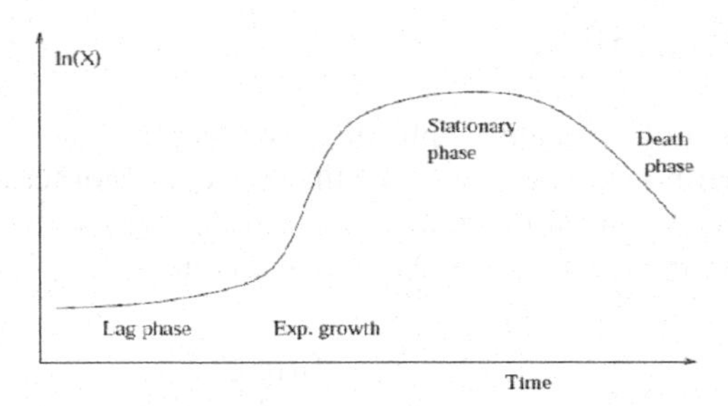

Fig. 2. Typical bacterial growth curve in a reactor after adding substrate

Uncertainty in the measurement of above manipulated variable creates the deviation between the expected output and actual output. A main drawback in biotechnological process control is the problem to measure physical and biochemical parameters with higher rate of accuracy.

3 Process Modeling

A model can be used to explain some fundamentals properties of bioreactors and also give suitable background for understanding of process [11]. The modeling (Mass balance and energy balance) equations of a bioreactor are obtained with the following assumption [12]

- Reactor contents are perfectly mixed
- Reactor operating at a constant temperature
- The feed is sterile
- The feed stream and reactor contents have equal and constant density
- The feed and product stream have the same flow rate
- The microbial culture involves a single biomass growing on single substrate

When the real process parameter goes beyond the assumptions, the output will vary in a significant manner. Hence a tolerance value of each parameter was calculated using the uncertainty quantification method.

The dynamic model of a process is

$$\frac{dx_1}{dt} = (\mu - D)x_1$$

$$\frac{dx_2}{dt} = D(x_{2f} - x_2) - \frac{\mu x_1}{y} \tag{1}$$

Where x1 = biomass (cell) concentration (mass of cells/volume)

x2 = substrate concentration (mass of substrate/volume)

D= dilution rate (Volumetric flow rate/reactor volume) – manipulated input.

μ= specific growth rate and y = yield (mass of cells formed / mass of substrate consumed)

$$\mu_s = \frac{\partial \mu}{\partial x_{2s}} = \frac{\mu_{\max} k_m}{(k_m + x_{2s})^2} \tag{2}$$

Where μ_{max} is maximum specific growth rate.

Km is limiting substrate concentration.

Linearised model of a bioreactor is

$$A = \begin{bmatrix} \mu_s - D_s & x_{1s}\mu_s \\ -\frac{\mu_s}{Y} & -D_s - \frac{\mu_s x_{1s}}{Y} \end{bmatrix}$$

$$B = \begin{bmatrix} -x_{1s} \\ x_{2fs} - x_{2s} \end{bmatrix}$$

$$C=[1\ 0],\ D=[0] \tag{3}$$

The bioreactor can be operated in three different modes like Equilibrium mode 1, 2 and 3. In Equilibrium mode 1, all the cells are washed out from the reactor. Hence usually reactors will not be operated in this phase[13]. Equilibrium mode 2, produce cells but it is a unstable phase, but reactor can be operated with some constraints with a suitable controller or compensator and in Equilibrium mode 3, produce cells with stable phase, hence reactor can be operated and produce the cell according to the requirement. Table 1 show that the operating conditions of a bioreactor for stable and unstable modes.

Table 1. Bioreactor operating mode

Sl. No	Steady state	Biomass Concentration	Substrate Concentration	Stability
1	Equilibrium mode 1 (wash out)	X1s=0	X2s=4	stable
2	Equilibrium mode 2 (nontrivial)	X1s=0.995103	X2s=1.512243	unstable
3	Equilibrium mode 3 (trivial)	X1s=1.530163	X2s=0.174593	stable

For a stable operating point [13], the state equations with state variables are dilution rate and biomass concentration with the nominal values of μmax= 0.53hr-1 , km= 0.12g/liter, k1=0.4545, Y=0.4, Dilution rate Ds=0.3hr-1, feed substrate concentration x2fs=4.0 g/liter.

The linearised model at the equilibrium state 2 is

$$A = \begin{bmatrix} 0 & -0.0679 \\ -0.75 & -0.1302 \end{bmatrix} B = \begin{bmatrix} -0.9951 \\ 2.4878 \end{bmatrix} C = \begin{bmatrix} 1 & 0 \end{bmatrix} D = \begin{bmatrix} 0 & 0 \end{bmatrix} \tag{4}$$

The linearised model at the equilibrium state 3 is

$$A=\begin{bmatrix}0 & 0.9056\\ -0.75 & -2.5640\end{bmatrix} B=\begin{bmatrix}-1.5301\\ 3.8255\end{bmatrix} C=\begin{bmatrix}1 & 0\end{bmatrix} D=\begin{bmatrix}0 & 0\end{bmatrix} \tag{5}$$

Hence before conducting multiple experiment method of uncertainty analysis, process has to be operated either equilibrium mode 2 or 3. To operate the reactor, controller parameter has to be identified. In this paper Particle swarm optimization (PSO) based controller design is attempted to determine the controller parameter for equilibrium mode 2 or 3.

4 Particle Swarm Optimization Algorithm

The PSO method is a swarm intelligence technique for solving the optimization type problem. It is a population based search algorithm where each individual is referred to as particle. Each particle in PSO flies through the search space with an adaptable velocity that is dynamically modified according to its own flying experience and also to the flying experience of the other particles. In PSO each particles strive to improve themselves by imitating traits from their successful peers. Further, each particle has a memory and hence it is capable of remembering the best position in the search space ever visited by it. The position corresponding to the best fitness is known as pbest and the overall best out of all the particles in the population is called gbest [10]. The modified velocity and position of each particle can be calculated using the current velocity and the distances from the pbest to gbest as shown in the following formula.

$$v_i^{(t+1)} = w^t\, v_i^{(t)} + C_1 * r_1 * (p_{best} - S_i^{(t)}) + C_2 . r_2 . (g_{best\,g} - S_i^{(t)}) \tag{6}$$

$$W^t = (W_{max} - Iter)x\left[\frac{(W_{max} - W_{min}}{Iter_{max}}\right] \tag{7}$$

Steps in PSO based PID Controller Optimization

Step 1 Assign values for the PSO parameters
Initialize: swarm (N) and step size; learning rate (C1, C2) dimension for search space (D); inertia weight (W); Initialize random values and current fitness.
R1= rand (D, N); R2= rand (D, N); Current fitness = 0*ones (N,1)

Step 2 Initialize Swarm Velocity and Position.
Current position = 10*(rand (D,N)-0.2), Current velocity = 0.5*rand (D,N)

Step 3 Evaluate the objective function of every particle and record each particle's P_i^t and G_i^t. Evaluate the desired optimization fitness function in D – dimension variables

Step 4 Compare the fitness of particle with its Pit and replace the local best value as given below. for i = 1: N

If current fitness (i) < local best fitness (i);
Then local best fitness = current fitness;
local best position = current position (i);
end

Same operation to be performed for G_i^t

Step 5 Change the current velocity and position of the particle group according to Eqn.1and 2.

Step 6 Step 2 to step 5 are repeated until the predefined value of the function or the number of iterations have been reached. Record the optimized Kp, Ki, Kd values.

Step 7 Perform closed loop test with the optimized values of controller parameters and calculate the Time Domain Specification for the system.

If the values are within the allowable limit, consider the current Kp, Ki, Kd values.

Otherwise perform the retuning operation for Ki ,by replacing the optimized numerical values or Kp and Kd.

5 Uncertainty Quantification

In this bioreactor most of the parameters like substrate concentration and substrate feed rate are uncertain. Due to the perturbation in this process parameter, biomass growth rate and concentration will vary. In this work, uncertainty in cell growth due to each component of the reactor parameter is assessed by introducing a manual variation like 5% to 10% acted together and individually. Also the sensitivity of process output for the variation in each component is studied by analyzing the bio mass concentration and production.

The methodology of conventional estimation of uncertainty can be illustrated using a simple measurement equation with y as a continuous function of x1 and x2. y is approximated using a polynomial approximation or a 2nd order Taylor's series expansion about the means:

$$y = f(\bar{x}_1, \bar{x}_2) + \frac{\partial f}{\partial x_1}(x_1 - \bar{x}_1) + \frac{\partial f}{\partial x_2}(x_2 - \bar{x}_2) \tag{8}$$

The standard deviations s(x1) and s(x2) are referred to, by the Guide to the Expression of Uncertainty in Measurement (GUM), as the standard uncertainties associated with the input estimates x1 and x2. The standard uncertainty in y and can be obtained by Taylor [2].

$$u(y) = \sigma(y) = \sqrt{\frac{1}{n}\sum_{1}^{n}(y_i - y)^2}$$
$$= \sqrt{\left(\frac{\partial f}{\partial x_1}\right)^2 \sigma(x_1)^2 + \left(\frac{\partial f}{\partial x_2}\right)^2 \sigma(x_2)^2 + 2\frac{\partial f}{\partial x_1}\frac{\partial f}{\partial x_2}\sigma(x_1)\sigma(x_2)} \tag{9}$$

This equation gives the uncertainty as a standard deviation irrespective of whether or not the measurements of x1 and x2 are independent and of the nature of the probability distribution. Equation 3 can be written in terms of the correlation coefficient,ρx1x2

$$u(y)=\sqrt{\left(\frac{\partial f}{\partial x_1}\right)^2 \sigma(x_1)^2 + \left(\frac{\partial f}{\partial x_2}\right)^2 \sigma(x_2)^2 + 2\frac{\partial f}{\partial x_1}\frac{\partial f}{\partial x_2}\rho_{x_1x_2}\sigma(x_1)\sigma(x_2)} \tag{10}$$

The partial derivatives are called sensitivity coefficients, which give the effects of each input quantity on the final results (or the sensitivity of the output quantity to each input quantity). The term, expanded uncertainty is used in GUM to express the % confidence interval about the measurement result within which the true value of the measured is believed to lie and is given by:

$$\Upsilon(\psi) = \tau\,\upsilon(\psi) \tag{11}$$

Where 't' is the coverage factor on the basis of the confidence required for the interval y ± U(y). For a level of confidence of approximately 95% the value of t is 2 for normal distributed measurement. In other words, y is between y ± 2σ(y) with 95% confidence.

Fig. 3 shows, bioreactor output when the substrate varies between 3.9 units to 4.1 units and k1 (Substrate concentration) varies between 0.25units and 0.5 units, results biomass concentration varied between 0.985 and 1.015 units. Similar experiment was conducted by introducing possible variation in k1 and substrate individually. From the set of experimental data the Type A [5] uncertainty was calculated.

Similar to the above method, Type B uncertainty was calculated by accounting the perturbation in each component, like variation in K1, substrate concentration separately. Finally overall uncertainty was determined using (8)

$$U_{ov} = U_a + U_b \tag{12}$$

The Level of confidence of the uncertainty analysis has been made with student 't' test. Table 2 show that the data obtained from the simulation when the process parameters were varied with the given range.

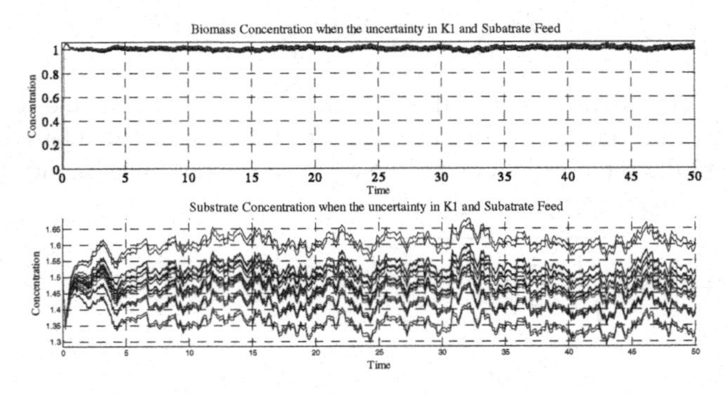

Fig. 3. Substrate uncertainty [3.9, 4.1] and k_1 (Substrate Concentration) uncertainty [0.25 .5]

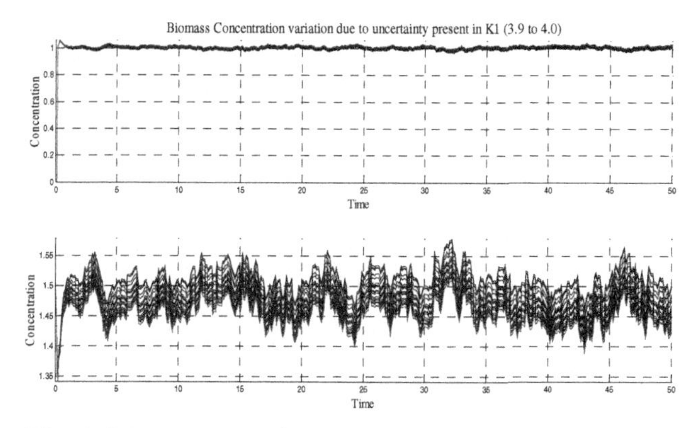

Fig. 4. Substrate uncertainty [3.9, 4.1] with nominal value of 4

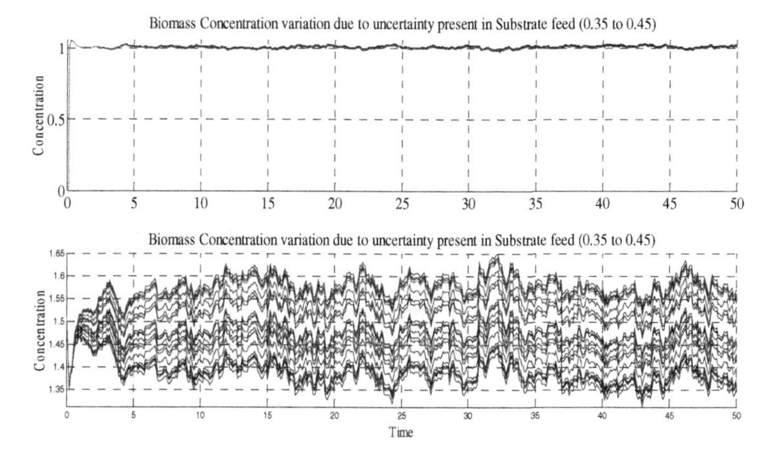

Fig. 5. K1 (substrate Concentration) Variation [0.35, 0.45] alone

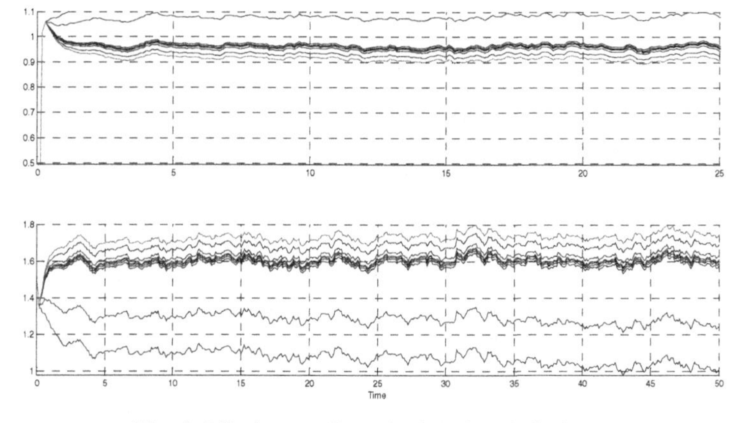

Fig. 6. Dilution rate Perturbation [0.1, 0.4] alone

Table 2. Dilution rate uncertainty [.35,.45]

Sl. No	Substrate feed	Concentration
1.	0.4008	0.1945
2.	0.3762	0.2849
3.	0.3529	0.3490
4.	0.4230	0.2298
5.	0.4079	0.3375
6.	0.3959	0.1638
7.	0.4047	0.3938
8.	0.3732	0.3215
9.	0.4124	0.3574
10.	0.3896	0.3016
Mean	0.3937	0.2934
σ	0.021066	0.075103

Table 3. Dilution rate uncertainty [0.1, 0.4] with nominal value of 0.25

Sl. No	Dilution rate uncertainty	Biomass Concentration
1.	0.3694	0.97
2.	0.1676	0.925
3.	0.3324	0.968
4.	0.2307	0.952
5.	0.3935	0.965
6.	0.1817	0.96
7.	0.3934	1.093
8.	0.3667	1.19
9.	0.2780	0.96
10.	0.3652	0.91
Mean	0.3079	0.989
σ	0.087067	0.085623

Table 4. Substrate uncertainty [3.9 4.0] and K_1 Uncertainty [0.1, 0.4] with nominal value of 4

Sl. No	Substrate concentration	Dilution rate uncertainty	Biomass Concentration
1.	0.2657	0.4266	1.115
2.	0.3658	0.3687	0.958
3.	0.2973	0.3946	0.9
4.	0.3355	0.4209	1.18
5.	0.3581	0.3776	1.105
6.	0.3428	0.4155	1.11
7.	0.1839	0.3619	1.108
8.	0.2996	0.4460	1.08
9.	0.2524	0.4085	1.11
10.	0.2097	0.4251	1.02
Mean	0.2911	0.4045	1.069
σ	0.062436	0.027807	0.08434

When the K1 varies from 3.9 to 4 and substrate feed varies from 0.35 to 0.45 then the biomass concentration is varied with 0.995 and 1.029, substrate concentration is varied from 1.3 to 1.65. Similarly If K1 alone varies from 3.9 to 4.0 then biomass concentration is varied from 0.9872 and 1.024 and, substrate concentration is varied from 1.3988 to 1.579. If Substrate Feed Alone varies, then biomass concentration is varied from 1.005 and 1.0172.

6 Conclusion

In this work overall uncertainty has been estimated with the help of multiple experiment method. Also level of confidence of the uncertainty is quantified. The uncertainty obtained here is verified with the help of student 't' test. Uncertainty is a quantification of the possibility of deviation between the expected value and the obtained value. From this method the uncertainty in substrate concentration was calculated by estimating the biomass. Also sensitivity of each process parameter can be calculated by analyzing the output of the system when its input varies within bound.

References

1. Lew, J.S., Horta, L.G.: Uncertainty quantification using interval modeling with performance sensitivity. Elsevier Science 338(1-2), 330–336 (2007)
2. ISO, Guide to the Expression of Uncertainty in Measurement, 1st edn. (1993) ISBN 92-67-10188-9
3. Stern, F., Muste, M., Beninati, M.-L., Eichinger, W.E.: Summary of experimental uncertainty assessment methodology with example. IIHR Technical Report No. 406
4. Coleman, H.W., Glenn Steele, W.: Experimentation, Validation, and Uncertainty Analysis for Engineers
5. Adams, T.M.: A Guide for Estimation of Measurement Uncertainty in Testing (July 2002)
6. Coleman, H.W., Steele, W.G.: Experimentation and Uncertainty Analysis for Engineers, 2nd edn. John Wiley & Sons (1999)
7. Hargreaves, G.I.: Interval Analysis in MATLAB. Numerical Analysis Report (416) (December 2002)
8. Hsu, C.-C., Chang, S.-C., Yu, C.-Y.: Tolerance design of robust controllers for uncertain interval systems based on evolutionary algorithms. IET Control Theory Appl. 1(1), 244–252 (2007)
9. Tan, N., Atherton, D.P.: Stability and performance analysis in an uncertain world. IEE Computing and Control Engg. Journal, 91–101 (April 2000)
10. Rajinikanth, V., Latha, K.: Optimization of PID Controller Parameters for Unstable Chemical Systems Using Soft Computing Technique. International Review of Chemical Engineering 3, 350–358 (2011)
11. Jana, A.K.: Chemical process modeling computer Simulation. Printice hall of India (2008)
12. Carlsson, B.: An introduction to modeling of bioreactors (March 24, 2009)
13. Wayne Bedquette, B.: Process Control Modeling, Design and Simulation. PHI Learning Pvt. Ltd. (2010)

Design and Implementation of a Direct Torque Control Space Vector Modulated Three Phase Induction Motor Drive

Shrishell Muchande, Arvind Kadam, Keerthi Unni, and Sushil Thale

Fr. Conceicao Rodrigues Institute of Technology,
Vashi, Navi Mumbai, MH, India – 400703
shirish_muchande@rediffmail.com, arvind_hk@ymail.com,
keerthy101@yahoo.com, ssthale@gmail.com

Abstract. Electric motor drives find a wide range of applications ranging from very precise, position controlled drives used in the field of robotic applications to variable speed drives for industrial process control. By use of fast switching power electronic converters, the input power is processed and applied to the motor. With the recent developments in the area of power electronics and due to availability of low cost, high performance digital signal processors, the motor drives becomes an affordable entity for most of the industrial applications. In variable speed drive applications the dc machines have been replaced by ac drives. Many control techniques have been developed and used for induction motor drives namely the scalar control, vector or field- oriented control, direct torque and flux control, and adaptive control. Amongst them, the direct torque and flux control is found to be independent of the machine rotor parameters and offers the added advantage of eliminating the need for any speed or position sensors. The direct torque and flux control is also called direct torque control (DTC) and is an advanced scalar control technique. The basic DTC scheme eliminates the need for PI regulators, co-ordinate transformations, current regulators and PWM signal generators. Further, DTC offers a good torque control in steady state and transient operating conditions. This paper presents the details of the design and implementation of a direct torque controlled space vector modulated drive for an induction motor using Texas Instrument's Piccolo series TMS320F28069 digital signal controller.

Keywords: Induction motor, Scalar Control, Vector Control, Direct Torque Control, Flux estimators, PWM, Duty Cycle, transient response, Space Vectors, Voltage Source Inverter.

1 Introduction

Because of low cost, simple construction, high reliability and low maintenance, induction machine is widely used for industrial commercial and domestic applications The induction motor could be considered as a higher order non-linear system with a considerable complexity [1-2]. The speed of induction motor can be controlled by

S. Unnikrishnan, S. Surve, and D. Bhoir (Eds.): ICAC3 2013, CCIS 361, pp. 659–672, 2013.

changing the pole, frequency or the stator voltage. With the use of power electronics, it is possible to vary the input frequency of a motor. This makes induction motor suitable for variable speed drives. The development of high power fast switching components like Insulated Gate Bipolar Transistor (IGBT) and MOSFETs and efficient control methods (using Digital Signal Processors DSPs) makes the control of induction motor easier and precise [3]. The control for high performance induction motor drives includes the scalar control and the vector control method.

In scalar control, instead of alignment in space, only the magnitude of the controlled variables like flux, current and voltage etc. are controlled. During the steady state conditions, scalar control improves the performance and efficiency of the machine whereas during transient condition, because of inherent coupling between the torque and flux, the response of machine is sluggish. The disadvantage of scalar control is that it gives a slow transient response. Vector control method controls magnitude as well as alignment of controlled variables. Generally the current is split up into torque producing component and the flux producing component. These two components are decoupled i.e. they are kept orthogonal to each other. Because of this, the change in torque command does not change the flux value. This improves the transient response of the motor. Vector control gives the high performance and more efficient control features. In vector control the system variables need to be transformed into d-q reference frame which rotates synchronously with the stator field. In synchronously rotating reference frame the stator currents appear as a dc which makes the analysis of control variables simpler [2].

The direct torque control of induction machines can be considered as an alternative to field oriented control of induction motors. Implementation of DTC needs a controller, torque & flux calculator and voltage sourced inverter (VSI) [2]. In DTC control, it is possible to achieve fast torque response, selection of optimal switching vector and direct and independent control of torque and flux. The harmonic losses can be reduced by reducing inverter switching frequency [4]. Recently, based on the traditional DTC method, a new control scheme called Direct Torque Control – Space Vector Modulated (DTC – SVM) has been developed. This eliminates the drawbacks of traditional DTC. This technique was claimed to have nearly comparable performance with vector controlled drives. Basically, the DTC-SVM strategies are the methods, which operate with constant switching frequency. Compared to the FOC scheme, DTC scheme is simple, eliminates the need for feedback current control and frame transformation, and gives a fast and good dynamic torque response [3]. Feedback signal processing in DTC is somewhat similar to stator flux-oriented vector control. Hysteresis-band control generates flux and torque ripple and switching frequency does not remain a constant. Stator flux vector and torque estimation is required, but the values are not sensitive to rotor parameters.

2 Literature Survey

R. Arunadevi el al. [5] has proposed the direct torque control of induction motor using Space Vector Modulation (SVM). The stator flux vector, and the torque produced by the motor, T_{em}, can be estimated using the following equations [5].

$$\bar{\lambda} = \int (\bar{V_S} - R_S \cdot \bar{I_S})\, dt \tag{1}$$

$$T_{em} = \frac{3}{2} \cdot \frac{P}{2} (\bar{\lambda_S} \times \bar{I_S}) \tag{2}$$

Once the current stator flux magnitude and output torque are known, the change required in order to reach the demanded values by the end of the current switching period can be determined. Over a short time period, the change in torque is related to the change in current, which can be obtained from the equation (3). The voltage $\bar{E}$ can also be determined by using the stator flux and current vectors.

$$\Delta I_S = \frac{\bar{V} - \bar{E}}{L'_S} \Delta t \tag{3}$$

The change in torque can be obtained as

$$\Delta T_{em} = \frac{3}{2} \cdot \frac{P}{2} \cdot \frac{\Delta t}{L'_S} [\bar{\lambda_S} \times (\bar{V} - \bar{E})] \tag{4}$$

The change in the stator flux,

$$\Delta \lambda_S = (\bar{V} - R_S \cdot \bar{I_S}) \cdot \Delta t = \bar{V} \cdot \Delta t \tag{5}$$

The equations can be solved to find the smallest voltage vector, required to drive both the torque and flux to their reference values. The required stator voltage can be calculated by adding on the voltage drop across the stator resistance calculated using the current measured from the last cycle. The voltage required, to drive the error in the torque and flux to zero, is calculated directly. This calculated voltage is then produced using Space Vector Modulation. If the inverter is not capable of generating the required voltage then the appropriate voltage vector which will drive the torque and flux towards the reference value is chosen and held for the complete cycle.

To improve the performance of VSI fed DTC drive many modifications are suggested by various researchers [6-11]. Vinay Kumar and Srinivasa Rao [12] have proposed the new algorithm for direct torque & flux control of three phase induction motor based on the control of slip speed and decoupled between amplitude and angle of reference stator flux for determining required stator voltage vectors. The hysteresis comparators and voltage-switching table of conventional DTC gets eliminated here. But, the reference value of stator flux linkage needs to be known in advance. This method also requires calculation of the reference for voltage space vector and SVPWM technique to be done. In this technique, the torque control is achieved by maintaining the reference stator flux to be constant. The instantaneous electromagnetic torque is given by [12].

$$T_e(t) = \left[\frac{3}{2} \cdot P \cdot \frac{L_m^2}{R_r \cdot L_S^2} \cdot |\Psi_S^2|\right] \cdot \left[1 - e^{-e/\tau}\right] \cdot \left[\frac{\Delta}{\Delta t} \cdot \theta_{slip}\right] \tag{6}$$

The instantaneous electromagnetic torque control is performed by changing the value of slip angle which is controlled by using direct stator flux control method.

Dinkar Prasad et. al. [13] has developed a low cost hardware prototype which is capable of giving higher PWM switching frequency. The DTC scheme is implemented without using sector calculator and switching table formulation, which is the unique feature of this scheme. Gdaim et al. [14] have proposed two intelligent approaches to improve the direct torque control of induction machine; fuzzy logic control and artificial neural networks control.

In conventional direct torque control (CDTC), the switching frequency is variable which depends on the mechanical speed, stator flux, stator voltage and the hysteresis band of the comparator. The distortion in torque and current is due to sector changes. Several solutions have been proposed to keep constant switching frequency. Ahmed. A. Mahfouz et. al. [1] has proposed a new modified DTC with a space vector modulator (SVM), and fuzzy logic controller (FLC). The SVM and FLC are used to obtain constant switching frequency and decoupling between torque and flux respectively. The two hysteresis controllers are replaced with FLC and the look-up table replaced by space vector modulator (SVM). The FLC block based on are torque error e_T, flux error e_F, and the stator flux position information gives space voltage vector V_S^* and its position angle θ_v at the output. Based on the output of FLC, SVM block generates constant frequency pulses to trigger the inverter. Space vector modulated direct torque control using field programmable gate array (FPGA) is presented in [15]. To get the desired load angle, the stator flux Ψs is changed by selecting proper stator voltage vector (V_S).

3 Problem Definition

The operation of the conventional DTC is very simple but it produces high ripple in torque due to considered non-linear hysteresis controllers. The sampling frequency of conventional DTC is not constant and also only one voltage space vector is applied for the entire sampling period. Hence the motor torque may exceed the upper/lower torque limit even if the error is small. The problems usually associated with DTC drives which are based on hysteresis comparators are [16-17]:

1. Because of the use of hysteresis operations the switching frequency is variable.
2. Due to inherent inaccuracy in stator flux estimations, the drive performance can get severely degraded.
3. During start up and at low speed values, there is the difficulty in controlling the startup current and there exists inaccuracies in the calculation of currents due to motor parameter variation.

The objective of this research work is to attempt the implementation of Direct Torque Control (DTC) of Induction Motor (IM) with sinusoidal PWM (SPWM). The implementation consists of designing the hardware setup for DTC based induction motor drive using DSP. The motor of rating 0.37 KW has been selected. The algorithm and programming required for DTC has to be implemented through the DSP. The main aim of the research work includes running a motor with an inverter using PWM module of the DSC, sensing the current and speed values from the motor on real time basis, calculation of torque and flux values through the DSC and establishing closed loop control of the motor speed by comparison with the reference values. The sensing of the current and voltage must happen on real time basis so that the torque and the flux calculations remain accurate. Also, the reference frame transformations followed by the integrations and the comparisons should be performed at each and every sample which requires a controller which can do computations extensively. Hence the digital signal controller from Texas Instruments TMS320F28069 is used instead of a microcontroller, as it can do mathematical calculations promptly.

4 Analysis of DTC – SVM

The flux linkage equations in terms of the currents are [16]:

$$\Psi_{qs}^{e} = L_{ls} \cdot I_{qs}^{e} + L_{m} \cdot \left(I_{qs}^{e} + I_{qr}^{e}\right) \tag{7}$$

$$\Psi_{ds}^{e} = L_{ls} \cdot I_{ds}^{e} + L_{m} \cdot \left(I_{ds}^{e} + I_{dr}^{e}\right) \tag{8}$$

$$\Psi_{qr}^{e} = L_{lr} \cdot I_{qr}^{e} + L_{m} \cdot \left(I_{qs}^{e} + I_{qr}^{e}\right) \tag{9}$$

$$\Psi_{dr}^{e} = L_{lr} \cdot I_{dr}^{e} + L_{m} \cdot \left(I_{ds}^{e} + I_{dr}^{e}\right) \tag{10}$$

The torque equations of induction motor are given by [16]:

$$T_e - T_l = J \cdot \frac{d}{dt} \cdot \omega_r \tag{11}$$

$$T_e = \left(\frac{3}{2}\right) \cdot \frac{P}{2} \cdot \left(\Psi_{sd}^{e} \cdot I_{sq}^{e} - \Psi_{sq}^{e} \cdot I_{sd}^{e}\right) \tag{12}$$

where, T_l is load torque, ω_r is electrical speed. The torque equation with the corresponding variables in stationary frame:

$$T_e = \left(\frac{3}{2}\right) \cdot \frac{P}{2} \cdot \left(\Psi_{ds}^{s} \cdot I_{qs}^{s} - \Psi_{qs}^{s} \cdot I_{ds}^{s}\right) \tag{13}$$

Consider a three phase two level voltage source inverter where E is the dc link voltage, and $S_a^+, S_b^+ and\ S_c^+$ are the states of the upper switches. The load is a balanced induction motor; the phase voltages generated by the voltage source inverter and applied to the stator are constrained by [16]:

$$V_{as} + V_{bs} + V_{cs} = 0 \tag{14}$$

Where, V_{as}, V_{bs}, V_{cs} are the phase voltages of inverter. The phase voltages in terms of S_a, S_b, S_c can be expressed as [2]:

$$V_{as} = \frac{2S_a - S_b - S_c}{3} E \tag{15}$$

$$V_{bs} = \frac{S_a + 2S_b - S_c}{3} E \tag{16}$$

$$V_{cs} = \frac{-S_a - S_b + 2S_c}{3} E \tag{17}$$

$$\overline{V_S^S} = \frac{2}{3} \cdot E \cdot \left[S_a + S_b \cdot e^{j2\pi/3} + S_c \cdot e^{j4\pi/3}\right] \tag{18}$$

Equation (18) represents the space vector of phase voltages generated by the PWM inverter. Taking the 8 values $\overline{V_k}$ $(k = 0\ to\ 7)$ the space vectors with, $(k = 1\ to\ 6)$, have the same amplitude $(2E/3)$ and phase angles equal to $[(k-1) \cdot \pi/3]$. Remaining two vectors $(k = 0\ \&\ 7)$ coincides with the zero space vectors.

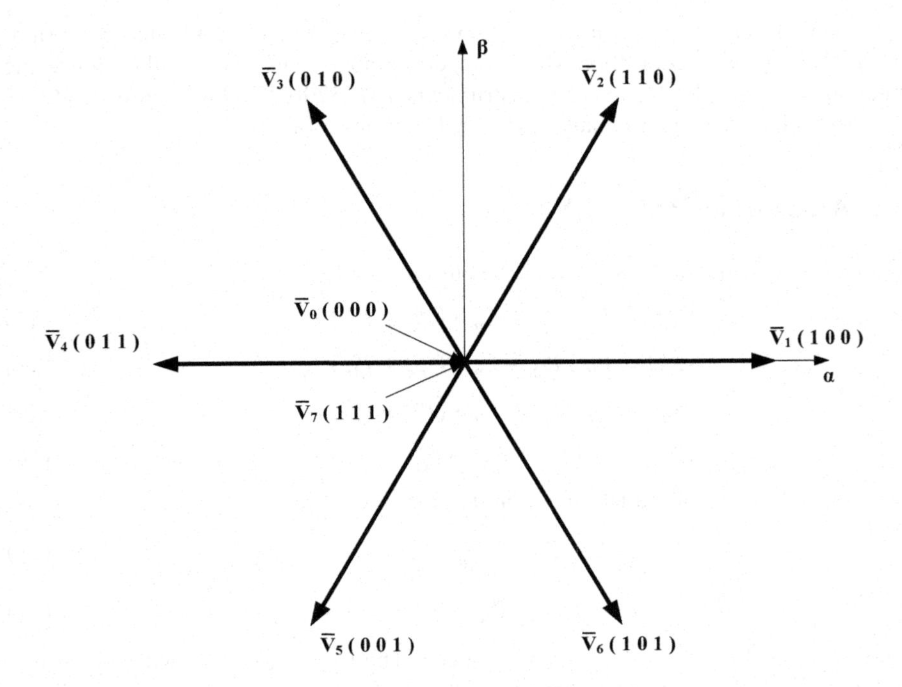

Fig. 1. Inverter output voltage phase vectors [2]

The inverter output voltage phase vectors are shown in Fig 1. The space vector modulations technique differs from the conventional sine PWM methods. There are no separate modulators used for each of the three phases. Instead of them, the reference voltages are given by space voltage vector and the output voltages of the inverter are considered as space vectors. There are eight possible output voltage vectors, six active vectors $\overline{V_1}$ to $\overline{V_6}$, and two zero vectors $\overline{V_0}$& $\overline{V_7}$ (Fig.2). The reference voltage vector is realized by the sequential switching of active and zero vectors.

$$V_S^* \cdot T_S = V_0 \cdot T_0 + V_1 \cdot T_1 + V_2 \cdot T_2 + V_7 \cdot T_7 \tag{19}$$

$$T_S = T_0 + T_1 + T_2 + T_7 \tag{20}$$

Where, V_S^* is the desired space vector, T_S is the switching time, $T_0, T_1, T_2$and T_7 are the time intervals, and V_1, V_2 are the vectors adjacent to V_S^*.

In the Fig. 2, the reference voltage vector V_s^* and eight voltage vectors, which correspond to the possible states of inverter, are shown. The six active vectors divide a plane for the six sectors I - VI. In the each sector the reference voltage vector V_s^* is obtained by switching on, two adjacent vectors for appropriate time. As shown in Fig.2, the reference vector V_s^* can be obtained by the switching vectors of V_1, V_2 and zero vectorsV_0, V_7.

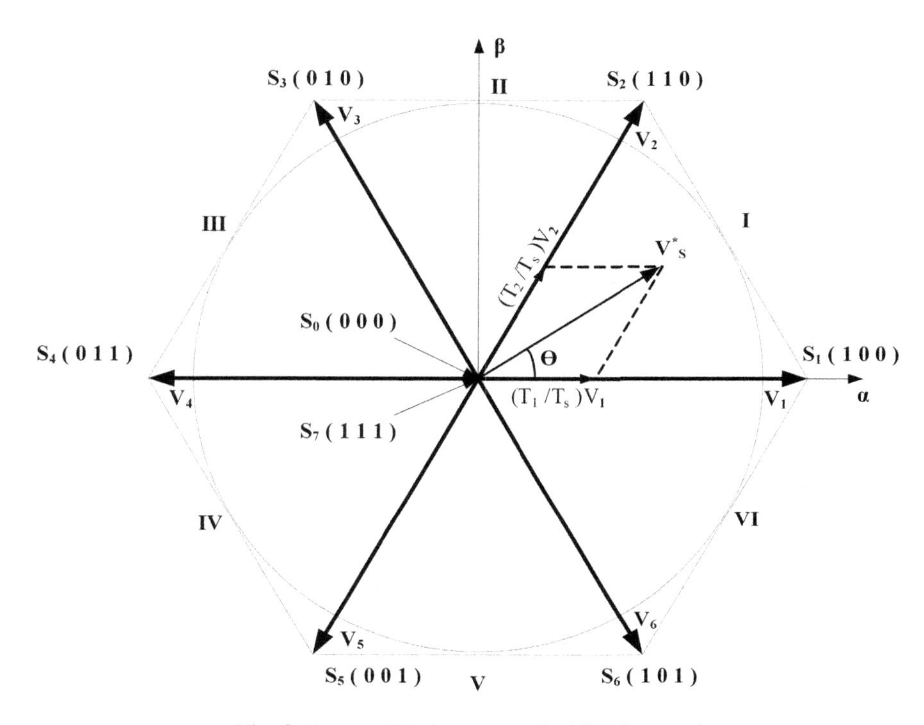

Fig. 2. States of the inverters using SVM control

5 Hardware Implementation

The IM drive based on DTC-SVM control structure can be operated in scalar control mode. The system configuration is shown in Fig. 3.

Initially, the torque and the flux values are estimated from the voltage and current values that are sensed. The torque and the flux values are then compared with the reference values. There are two hysteresis controllers implemented in software; one for the torque and another for the flux. The controller outputs are used to select the appropriate voltage vectors using Space Vector Modulation. The PWM signals are generated according to the voltage vectors and this controls the inverter output. The inverter then makes the motor to remain in the designed speed.

The hardware implementation of the system includes the design of three phase inverter with the gate driver circuit, design of voltage and current sensing circuit, the design of PCB artwork of the same. The conventional topology of a three leg inverter is selected for implementation. The nameplate details of the induction motor are given in Table 1.

The operational switching frequency for induction motor drive is typically selected between 2 kHz to 20 kHz depending upon the power rating of the motor. Here the switching frequency selected is 10 kHz.

$$\text{DC Link voltage, } V_{dc} = \sqrt{2} \times V_{LL} = 311\,V$$

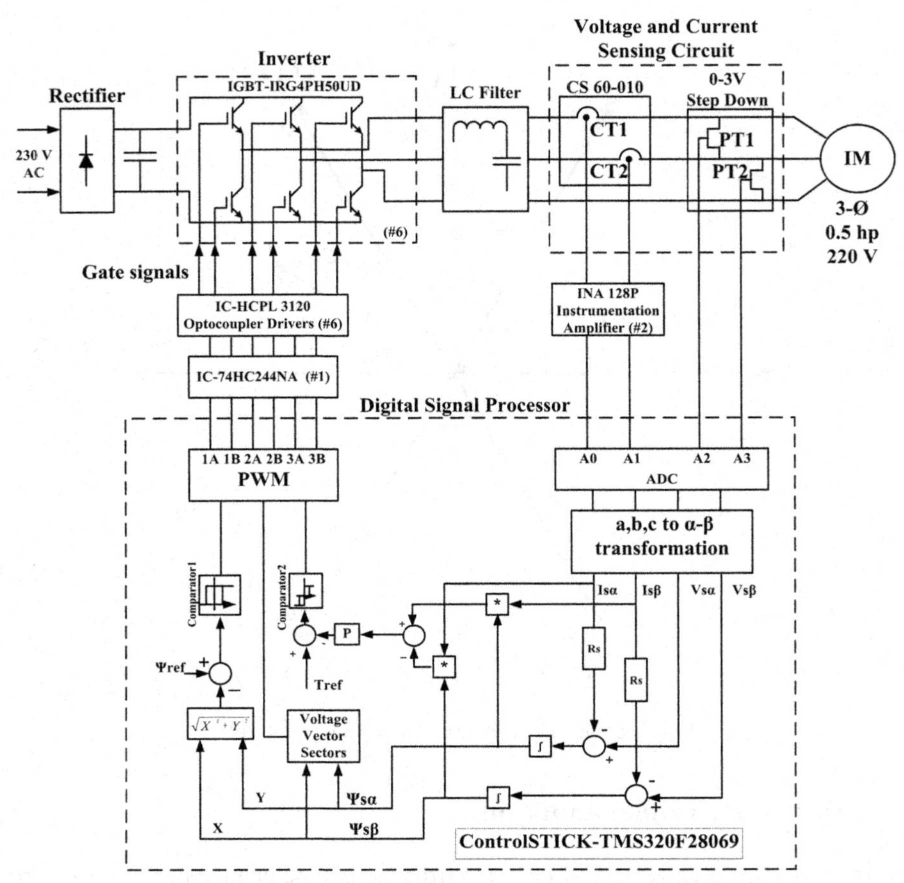

Fig. 3. Block diagram of implemented control algorithm [4]

Table 1. Specifications of Induction Motor

Type of Motor	3 Φ 50Hz Squirrel Cage Induction Motor
Rating in hp/kW	0.5 hp/0.37 kW
Voltage Rating	220 Volts Line to Line
Full Load Line Current / pf	1.05 Amps / 0.74
Full Load Efficiency	74%
Starting Current (Full voltage)	3.6 Amps
Full Load Torque	0.26 Nm
Stator Resistance (R_s)	10.9 Ω
Stator Self Inductance L_s	0.125 H
Nominal Speed	1440 rpm

The current rating of the device selected to be at least 10 times of the nominal current requirement of the motor at operating temperature of 70°C and 10 kHz switching frequency i.e. 10A. The net power capacity of the inverter is calculated as below,

$$Power\ rating\ of\ inverter = \frac{Power\ rating\ of\ motor}{(Effi.of\ motor\ \times Effi.of\ the\ inverter)}$$

Assuming efficiency of inverter as 85 %, the power rating of the inverter is calculated as,

$$= \frac{370\ W}{(0.74\ \times 0.85)}$$

$$= 588\ W$$

The gate driver selected is IC HCPL-3120 suitable to drive the IGBT IRG4PH50UD. The required isolated power supplies for the gate driver ICs is derived from 230/15V transformer with eight windings, out of which six are rated for 200mA continuous current rating and two for 1 Amps. The six power supplies are used for inverter and two for auxiliary circuits like level shifter, amplifiers in sensing board. The buffer IC 74HC244NA is used as the interface between the DSP controller and gate driver. The detailed list of components selected for 3Ø inverter and the specifications of the designed systems are listed in Table 2 and Table 3 respectively.

Table 2. Component Selection for 3Φ Inverter

Sr. No	Component	Type
1.	Switches	IGBT-IRG4PH50UD
2.	Gate Driver	HCPL3120
3.	Buffer	74HC244NA
4.	Regulator	LM 7815
5.	Fuse	2A
6.	Rectifier	DB107

Table 3. System Specifications

Sr. No.	Parameters	Specifications
1.	Power capacity	600W
2.	Operating Voltage and frequency	3Φ, 220V,50Hz
3.	DC link Capacitor	700μF/900V
4.	Inverter switching frequency	10kHz
5.	Filter Inductance	6mH
6.	Filter Capacitors (3 in delta and 3 in star connection)	50μf /440V ac each

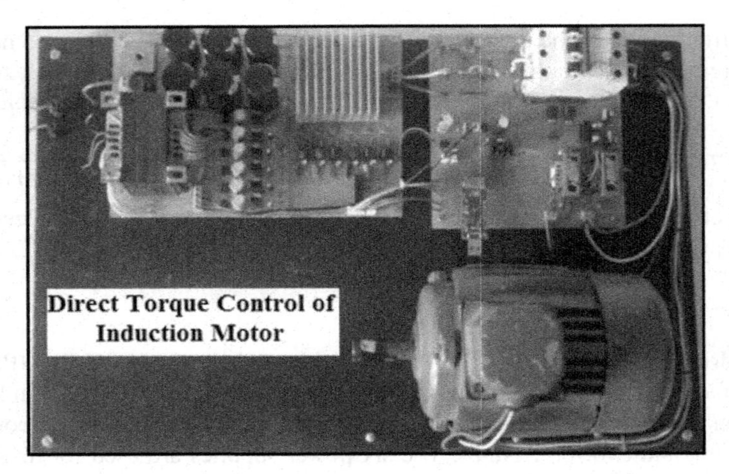

Fig. 4. Laboratory setup (a) without step up transformer

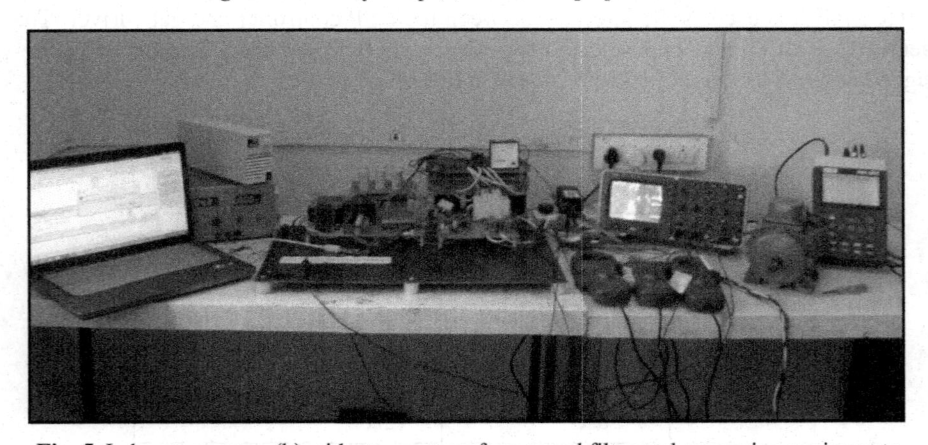

Fig. 5. Laboratory setup (b) with step up transformer and filter and measuring equipments

The hardware set up shown in Fig. 4 and Fig. 5 consists of a Voltage Source inverter (VSI) and 0.37kW induction motor. The induction motor is fed by the VSI through an MCB mounted on the sensing board. The two current sensors, CS60-010 (Coil craft), and two PTs (step-down transformer) are used to sense the currents and voltages. DTC control algorithm is implemented in the drive based on DSP controller TMS320F28069.

6 Hardware Results

The ePWM module of the DSC generates complimentary pulses for inverter with appropriate dead band. The dead band of 0.5 µSec each at both falling and rising edge of PWM is selected. This is done by assigning the dead band registers, a value calculated as per the clock setting of TMS320F28069. Fig. 6 shows the complementary PWM pulses for upper and lower switch of first leg of inverter.

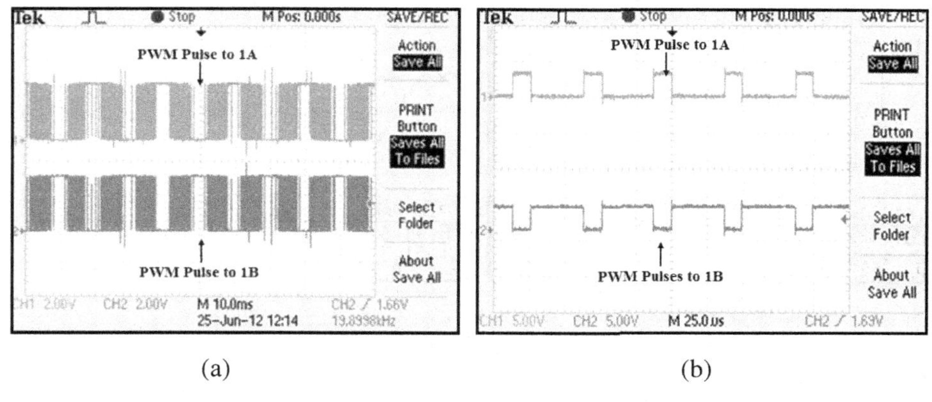

(a) (b)

Fig. 6. (a) PWM pulse of 1A& 1B at a frequency of 10 KHz and (b) Expanded view

The current sensors produce an output voltage, between 10mV to 100mV, proportional to the actual currents flowing in the system. The current sensor output is directed to an instrumentation amplifier INA129 to amplify it to the suitable level. The DTC-SVM algorithm is implemented in Code Compose Studio (CCS-4) platform. The PWM pulses generated using SVM and the motor terminal voltages are as shown in the Fig 7.

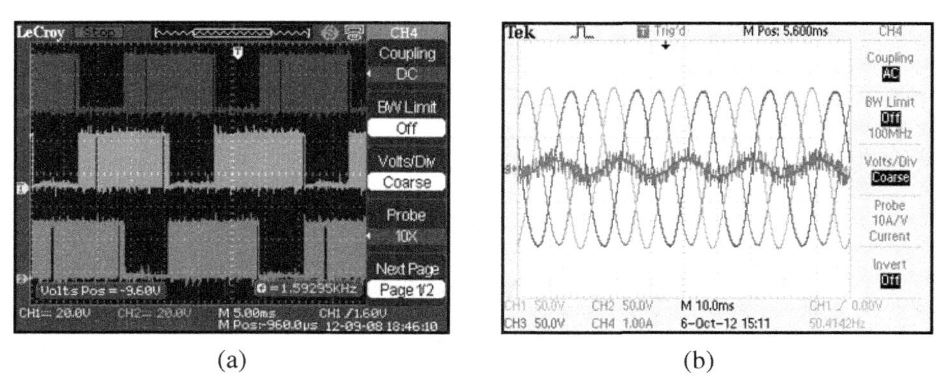

(a) (b)

Fig. 7. (a) Space Vector Modulated PWM gate pulses for inverter (b)Motor's 3 phase output terminal voltages and current after filtering

Fig. 8 shows the motor current and terminal voltages measured using Power analyzer. The test was carried out with low DC link voltage. The test was conducted to verify the THD contents in the output. The current harmonics show the significant amplitude at 5th and 7th harmonics compared to the rest. The harmonic spectrum of voltage shows the dominance of 3rd and 5th harmonics produced by the energizing component of transformer used for boosting the voltage.

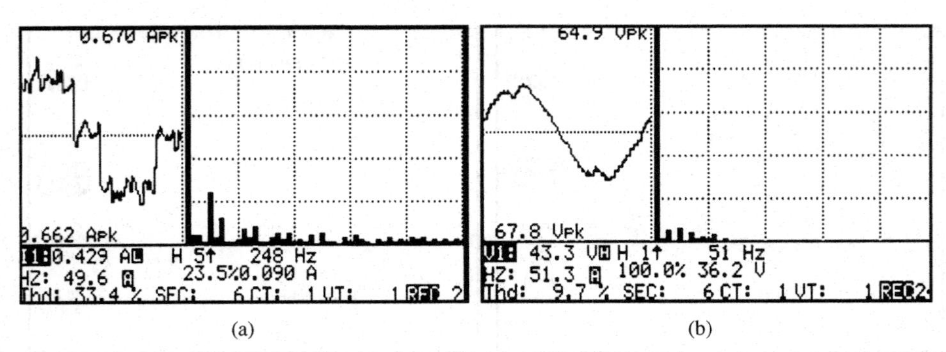

Fig. 8. Induction motor phase A terminal voltage and line current waveforms recorded by Meco's Power and Harmonic Analyzer PHA5850

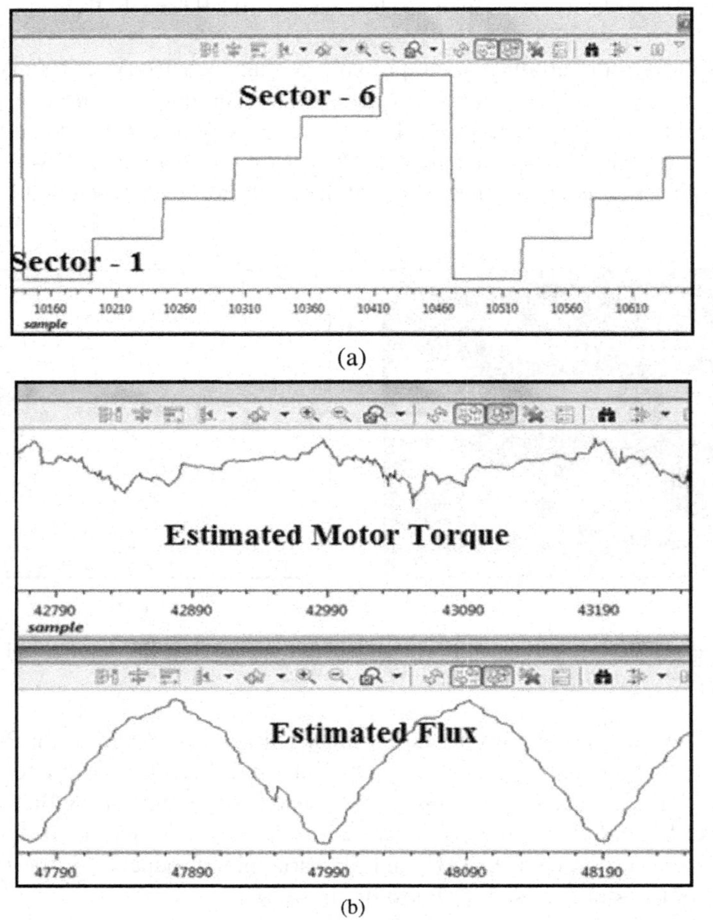

Fig. 9. Screen shot of CCS-4 platform indicating (a) sector selection of space vector modulation (b) torque and flux estimation based on measurement of actual motor terminal voltages and currents

Fig. 9 shows the screen shot of CCS-4 indicating the SVM implementation, where the reference voltage vector position is identified in terms of sector which repetitively moves from 1 to 6 and repeats. The position of the reference vector is generated using torque error and flux error generated in DTC algorithm. The reference torque is given as input to the system in the form of predefined variable set to a particular value. The estimation of flux and torque is done at the sampling rate of 60μsec through an timer interrupt and interrupt service routine (ISR), allows accommodation of fast dynamic response to torque changes in DTC.

7 Conclusion and Future Scope

Amongst all the control methods for induction motor drives, direct torque and flux control method provides a simplified, dynamically fast and machine rotor parameters independent control. It also eliminates the use of any speed or position sensors, reducing the cost of the system. The basic DTC scheme which is characterized by the absence of PI regulators, co-ordinate transformations, current regulators and PWM signals generators, makes it possible to use lower end processors to be used for control further reducing the cost of the drive. The DTC allows a good torque control in steady state and transient operating conditions to be obtained. The modification of basic DTC to accommodate features like accuracy of control, reduction in torque jitter, speed control loop and lesser dependency on accurate measurement of motor voltages and currents, various researchers proposed the modified form of DTC. This caused an increase in complexity of implementation of DTC, but retained most of the inherent features of DTC intact. As the complete hardware is ready for an induction motor drive, the work can be extended to implement following features;

- Reduction in torque ripple by means of reducing the torque error band which will require accurate and noise free measurement of current and voltages.
- Reduction of torque ripple through modification of control strategy with introduction of suitable torque and flux controller

References

1. Mahfouz, A.A., Sarhan, G.M.: Microcontroller implementation of direct torque control. International Journal of Numerical Modeling: Electronic Networks, Devices and Fields 18(1), 85–94 (2005)
2. Arya, M.K., Wadhwani, S.: Development of direct torque control model with using SVI for three phase induction motor. International Journal of Engineering Science and Technology (IJEST) 3(8), 6650–6655 (2011) ISSN: 0975-5462
3. Buja, G.S., Kazmierkowski, M.P.: Direct Torque Control of PWM Inverter-Fed AC Motors-A Survey. IEEE Transactions on Industrial Electronics 51(4) (August 2004)
4. Silva, N.M., Martins, A.P., Carvalho, A.S.: Torque and speed modes simulation of DTC-Controlled induction motor. In: Proceedings of the 10th Mediterranean Conference on Control and Automation, MED 2002, Lisbon, Portugal, July 9-12 (2002)

5. Arunadevi, Ramesh, N., Joseline Metilda, A., Sharmeela, C.: Analysis of Direct Torque Control using Space Vector Modulation for Three Phase Induction Motor. Recent Research in Science and Technology 3(7), 37–40 (2011)
6. TMS320x2806x Technical Reference Guide, Texas Instruments, Literature Number: SPRUH18C (December 2011), http://www.ti.com/lit/ug/spruh18c/spruh18c.pdf
7. Casadei, D., Serra, G., Tani, A.: Implementation of a Direct Torque Control Algorithm for Induction Motors based on Discrete Space Vector Modulation. IEEE Transactions on Power Electronics 15(4) (July 2000)
8. Lee, K.B., Blaabjerg, F.: An Improved DTC-SVM Method for Sensorless Matrix Converter Drives Using an Over-modulation Strategy and a Simple Nonlinearity Compensation. IEEE Transactions on Industrial Electronics 54(6), 3155–3166 (2007)
9. Swierczynski, D., Kazmierkowski, M.P.: Direct torque control of permanent magnet synchronous motor (PMSM) using space vector modulation (DTC-SVM)-simulation and experimental results. In: IEEE 2002 28th Annual Conference of Industrial Electronics Society, IECON 2002, Spain, vol. 1, pp. 751–755 (November 2012)
10. Casadei, D., Profumo, F., Serra, G., Tani, A.: FOC and DTC: two viable schemes for induction motors torque control. IEEE Transactions on Power Electronics 17(5), 779–787 (2002)
11. Tang, L., Rahman, M.F.: A new direct torque control strategy for flux and torque ripple reduction for induction motors drive by using space vector modulation. In: IEEE 32nd Annual Power Electronics Specialists Conference, PESC 2001, Vancouver, Canada, vol. 3, pp. 1440–1445 (June 2001)
12. Kumar, V., Rao, S.: Modified Direct Torque Control of Three-Phase Induction Motor Drives With Low Ripple in Flux and Torque. Leonardo Journal of Sciences 10(18), 27–44 (2011)
13. Prasad, D., Panigrahi, B.P., Gupta, S.S.: Digital simulation and hardware implementation of a simple scheme for direct torque control of induction motor. Energy Conversion and Management 49(4), 687–697 (2008)
14. Gdaim, S., Mtibaa, A., Mimouni, M.F.: Direct Torque Control of Induction Machine based on Intelligent Techniques. International Journal of Computer Applications (0975 – 8887) 10(8) (November 2010)
15. Rajendran, R.: FPGA Based Implementation of Space Vector Modulated Direct Torque Control for Induction Motor Drive. International Journal of Computer and Electrical Engineering 2(3), 589–594 (2010)
16. Bose, B.K.: Modern power electronics and ac drives, 5th edn. Pearson Education, Singapore (2005)
17. Idris, N.R.N., Yatim, A.H.M., Azli, N.A.: Direct Torque Control of Induction Machines with Constant Switching Frequency and Improved Stator Flux Estimation. In: 27th Annual Conference of IEEE Industrial Electronics, Colorado, USA, vol. 2, pp. 1285–1291 (2001)

Passivity Based Control for Lateral Slosh

G. Gogte, Ch. Venkatesh, D. Tiwari, and N.M. Singh

Veermata Jijabai Technological Institute(V.J.T.I),
Electrical Engineering Department, Mumbai
chanti.venky47@gmail.com

Abstract. Liquid sloshing is a problem that affects several industries. This paper presents a passivity based control strategy for slosh control. The lateral sloshing is represented by a simple pendulum and the tank carrying the liquid is approximated to an integrator system. Both the systems are independently passive. We find a feedback such that the closed loop formed on interconnection of the two systems is passive and hence stable. The controller designed is based on the energy balance of the system and is hence independent of the fluid properties.

Keywords: Slosh, Underactuated, Passivity.

1 Introduction

Slosh refers to the movement of liquid inside another object. The liquid must have a free surface to constitute a slosh dynamics problem, where the dynamics of liquid can interact with the container to alter the system dynamics significantly. Important examples of slosh include propellant slosh in spacecraft tanks and rockets (especially the upper stages) and cargo slosh in ships and trucks transporting liquids (like petrol or oil). It is common to refer to liquid motion in a completely filled tank (i.e. liquid without free surface) as 'fuel slosh'.

Sloshing of liquids in partially filled tanks is a problem faced in various fields like auto mobile industry, shipping industry, launching of space vehicles, packaging industry and so on. It can be thought of as a transfer of liquid from one side of the container to the other in form of a wave which is excited by angular and/or translational motion of the tank.

In launch vehicles or spacecraft, sloshing can be induced by tank motions resulting from guidance and control system commands or from changes in vehicle acceleration like those occurring when thrust is reduced by engine cutoff or when vehicle encounters wind shears or gusts. If the liquid is allowed to slosh freely, it can produce forces that can cause additional vehicle accelerations. These accelerations are sensed and responded to by the guidance and control system, forming a closed loop that can lead to instability. This instability may result in premature engine shutdown. The failure of several booster vehicles has been attributed to inadequate slosh suppression. The effects of slosh on spacecraft can be seen in details in [1].

Liquid slosh poses a problem in other industries too. Sloshing of liquid during packaging can lead to improper sealing affecting the quality and the shelf life of the product. Vehicles carrying large liquid cargo face the risk of dangerous

S. Unnikrishnan, S. Surve, and D. Bhoir (Eds.): ICAC3 2013, CCIS 361, pp. 673–681, 2013.

overturn due to the wave of liquid. In automobiles fuel slosh induced forces and movements during turning and braking can affect the vehicle's performance.

Slosh has proved to be a serious problem in many fields resulting in several accidents. Thus slosh problem has been studied and various slosh control strategies have been proposed. The traditional method of slosh control was inclusion of physical barriers in form of baffles or dividing the liquid container into small compartments. This limited the movement of liquid to small amplitudes of negligible frequencies. Bladders were also added to limit the liquid motion. These methods of slosh control were effective to some extent but could not suppress the effects of slosh effectively. These techniques also added to the weight and the cost of the system. Thus various active control strategies have been forwarded in the literature for slosh control. In [2] slosh is investigated from an active feedback control perspective and feedback controllers using LQG synthesis are designed for attenuating the response of fluid to an external disturbance acceleration acting on the tank. An infinite impulse response (IIR) filter is used in [3] to alter acceleration profile of a liquid in an open container carried by a robot arm so that liquid remains level except for a single wave at the beginning and the end of the motion. An H^∞ feedback control has been designed for control of liquid container transfer with high speed is seen in [4]. In [5], a trajectory has been designed to suppress residual vibration without direct measurements along a 3 D transfer path. A minimum time optimal controller for moving rectangular container in packaging industry is seen in [6]. A control scheme combining partial inverse dynamics controller with a PID controller tuned using metaheuristic search algorithm for slosh control in tilting phases of pouring process is seen in [7]. Sliding mode control using partial feedback linearisation is seen in [8]. Robust control for slosh free motion using sliding modes is seen in [9].

Passivity is a useful tool for the analysis of nonlinear systems [10], [11]. It relates well to the Lyapunov stability method. Passivity theory is used in various forms like interconnection and damping assignment passivity based control(IDA-PBC) [13], [14], backstepping, forwarding and feedback passivation [11] for controlling underactuated mechanical systems. While using passivity based approach, dynamic systems are viewed as energy transformation devices. This allows the decomposition of nonlinear systems into simpler subsystems, which upon interconnection add up their energies to determine the full systems behaviour. The controller may be viewed as another dynamical system interconnected with the process to modify it's behaviour. The major advantage of this strategy is that it does not make use of any particular structural property of the mechanical system but depends on the fundamental property of energy balance.

This paper presents a feedback passivity controller for lateral sloshing in a moving container. The fundamental mode is considered for purpose of design. The controller design is inspired from [12] where passivity based control is used for control of a 2-D SpiderCrane system. Section 2 describes the dynamics of the lateral slosh. Section 3 gives an idea about passivity of systems and its relation with system stability. Section 4 presents the controller design. Section 5 gives the simulation results for the proposed controller.

2 Slosh Dynamics

The accurate analysis of liquid slosh requires application of the Navier-Stokes equations which are difficult to formulate and solve. Thus sloshing is represented by equivalent mechanical model to ease control development. The most commonly used mechanical models in the literature are the simple pendulum and the mass-spring-damper model. This paper will consider the fundamental mode of lateral sloshing phenomenon in a moving container represented by a simple pendulum. The slosh mass is m and length l. Slosh angle ϕ represents the pendulum angle. Rest of the liquid mass (i.e. the mass of liquid that does not slosh) and mass of tank form the rigid mass. The system is thus like a moving rigid mass coupled with a simple pendulum as shown in Fig 1.

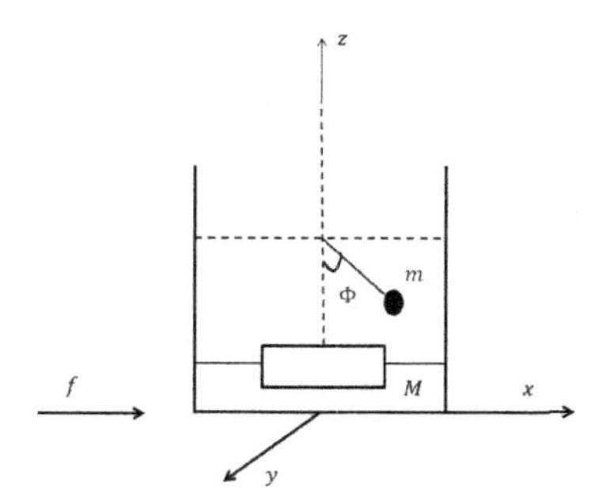

Fig. 1. Slosh mass modelled by pendulum

Following assumptions are made regarding the system:

1. The liquid is incompressible, irrotational and inviscid.
2. There is no source, no sink or damping elements like baffles.
3. The tank is assumed to be rigid.

The system parameters are:

m: mass of pendulum
M: total mass
l: length of the pendulum
f: force applied for translational motion
x: displacement of rigid mass
X: displacement of m in horizontal direction
Z: displacement of m in vertical direction
ϕ: pendulum angle or slosh angle
c: damping co-efficient

Referring to Fig 1,

$$X = l \sin \phi + x \tag{1}$$

$$Z = l - l \cos \phi \tag{2}$$

The potential energy of the system is:

$$V = -mgl\cos\phi \tag{3}$$

The kinetic energy of the system is:

$$T = \frac{1}{2}M(\dot{x})^2 + \frac{1}{2}m(\dot{x})^2 + \frac{1}{2}m(\dot{Z})^2 \tag{4}$$

Thus Lagrangian of the system is: $\mathcal{L} = T - V$.

Lagrangian equations in x and ϕ give the dynamical equations of the system:

$$M\ddot{x} + ml\cos\phi\ddot{\phi} - ml\dot{\phi}^2\sin\phi = f \tag{5}$$

$$ml\cos\phi\ddot{x} + ml^2\ddot{\phi} + c\dot{\phi} + mgl\sin\phi = 0 \tag{6}$$

The objective is to suppress the slosh angle ϕ as the container moves from a given initial position to a desired position using a single control input u. Since there are two degrees of freedom x and ϕ and one input, the system is underactuated by one degree.

3 Passivity Theory

Consider a dynamical system given by $\dot{x} = f(x,u)$ and $y = h(x,u)$ where $f : \mathbb{R}^n \times \mathbb{R}^p \to \mathbb{R}^n$ is locally Lipschitz, $h : \mathbb{R}^n \times \mathbb{R}^p \to \mathbb{R}$ is continuous, $f(0,0) = 0$ and $h(0,0) = 0$. The system is said to be passive if there exists a positive semidefinite function $V(x)$ such that

$$u^T y \geq \dot{V} = \frac{\partial V}{\partial x} f(x,u) \tag{7}$$

Passivity is the property that the increase in storage V is not higher than the supply rate. Thus any increase in storage due to external sources alone.

A passive system has a stable origin. Any system may be stabilised using a feedback that forces the system to be passive.Consider the following system: $\dot{x} = f(x,u)$, $y = h(x)$. Assume the system is

1. Passive with a radially unbounded positive semidefinite storage function.
2. Zero state detectable

Then $x = 0$ can be stabilised with some feedback $u = -\varphi(y)$ where φ is any locally Lipschitz function such that $\varphi(0) = 0$ and $y^T\varphi(y) > 0$, $\forall y \neq 0$.

Some important properties of passive systems that will be used in controller design are:

1. The feedback connection of two passive systems is passive.
2. The parallel connection of two passive systems is passive.

Consider the system represented by $\dot{x} = f(x) + G(x)u$. Suppose there exists a radially unbounded positive semidefinite C^1 storage function $V(x)$ such that $\frac{\partial V}{\partial x} f(x) \leq 0$, $\forall x$. Assume an output $y = h(x) = [\frac{\partial V}{\partial x} G(x)]^T$. Therefore $\dot{V} = \frac{\partial V}{\partial x} f + [\frac{\partial V}{\partial x} G(x)]^T u \leq y^T u$. Thus we get a system with input u and output y which is passive and hence stabilisable. Thus any system having positive semidefinite energy function may be stabilised using feedback.

4 Controller Design

The tank system with the forces applied can be approximated to a double integrator system connected with the pendulum as shown in Fig 2:

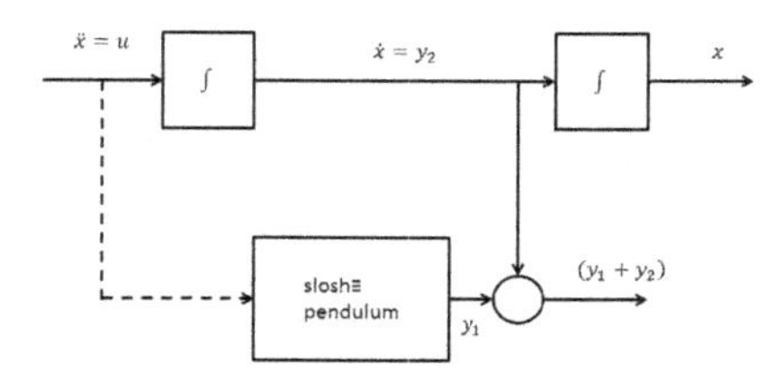

Fig. 2. Slosh approximated as double integrator systems

The input to the first system is the acceleration of the rigid mass i.e. $\ddot{x}$ while its output $\dot{x}$ is the input to the second integrator system. A single integrator system being passive, we consider velocity as the output and not the displacement. The system visualisation is on the basis of collocated partial linearisation as seen in [15].

The total force in the system f be expressed as a sum of the force exerted by the actuator $\tilde{f}$ and the force generated due to slosh dynamics $\alpha(\phi, \dot{\phi})$.

$$f = \tilde{f} + \alpha(\phi, \dot{\phi}) \tag{8}$$

Comparing (5) with (8)

$$\tilde{f} = M\ddot{x} \tag{9}$$

$$\alpha(\phi.\dot{\phi}) = ml \cos\phi\ddot{\phi} - ml\dot{\phi}^2 \sin\phi \tag{10}$$

Therefore

$$\ddot{x} = \frac{\tilde{f}}{M} \tag{11}$$

Let $\frac{\tilde{f}}{M} = u$. Thus

$$\ddot{x} = u \tag{12}$$

The total energy E of the system may be expressed as sum of the kinetic and potential energy of the pendulum along with the energy of the integrator.

$$E = \frac{1}{2}ml^2\dot{\phi}^2 - mgl\cos\phi + \frac{1}{2}\dot{x}^2 \tag{13}$$

where g is the acceleration due to gravity. Thus

$$\dot{E} = \dot{x}\ddot{x} + ml^2\dot{\phi}\ddot{\phi} + mgl\sin\phi\dot{\phi} \tag{14}$$

From (6) we get:

$$\ddot{\phi} = \frac{-1}{ml^2}(ml\cos\phi u + c\dot{\phi} + mgl\sin\phi) \tag{15}$$

Therefore

$$\dot{E} = (\dot{x} - ml\dot{\phi}\cos\phi)u - c\dot{\phi}^2 \tag{16}$$

Let $-ml\dot{\phi}\cos\phi = y_1$ and $\dot{x} = y_2$. Therefore

$$\dot{E} = (y_1 + y_2)u - c\dot{\phi}^2 \tag{17}$$

Select an energy function V

$$V = E + \frac{1}{2}K_p[\int -ml\dot{\phi}\cos\phi dt + (x - x_d)]^2 \tag{18}$$

$$\dot{V} = \dot{E} + K_p[\int y_1 dt + (x - x_d)](\dot{x} - ml\dot{\phi}\cos\phi) \tag{19}$$

Therefore

$$\dot{V} = (y_1 + y_2)u - c\dot{\phi}^2 + K_p[\int y_1 dt + (x - x_d)](y_1 + y_2) \tag{20}$$

i.e.

$$\dot{V} = -c\dot{\phi}^2 + [K_p \int y_1 dt + K_p(x - x_d) + u](y_1 + y_2) \tag{21}$$

Let

$$u = -K_p \int y_1 dt - K_p(x - x_d) - K_d(y_1 + y_2) \tag{22}$$

Substituting (22) in (21)

$$\dot{V} = -c\dot{\phi}^2 - K_d(y_1 + y_2)^2 \tag{23}$$

From (23) $\dot{V}$ is negative semidefinte if $c \geq 0$ and $K_d \geq 0$ so that V is a positive semidefinite function. Thus for the liquid in the container there exists a function V which satisfies (7). The system becomes passive under the action of control u.

5 Simulation Results

A passivity based control for a lateral slosh problem has been simulated using following parameters:

Mass of rigid part $M = 6\ kg$.
Slosh mass $m = 1.32\ kg$.
Pendulum length $l = 0.052126\ m$.
Gravity $g = 9.8\ m/s^2$.
Coefficient of damping $c = 3.0490e - 004\ kgm^2/s$.

The desired lateral position of the tank is 10 m.. The initial container position is 0.1 m. while the initial slosh angle position is 1.8 *degrees*.

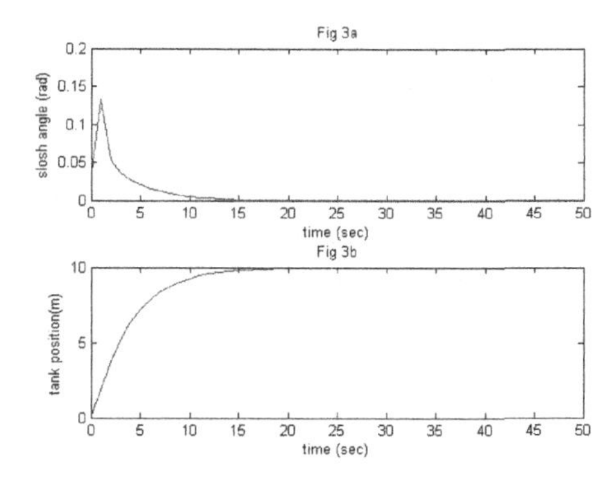

Fig. 3. Slosh angle and lateral position of container

Fig 3a shows the slosh angle position w.r.t. time. The slosh angle reduces to zero as desired in around 15 seconds. Also the desired lateral position is achieved within the same time as shown in Fig 3b. This result is obtained for $K_p = 1$ and $K_d = 4$.

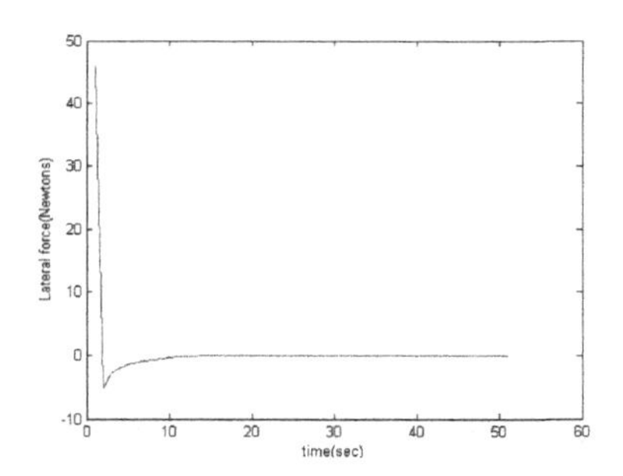

Fig. 4. Lateral force

Fig 4 shows the variation in the lateral force exerted on the container. The direction of the force is initially opposite to that of the displacement and then in the direction of the displacement. The force reduces to zero as the desired position is reached.

Fig 5 shows the simulation results for $K_p = 0.9$ and $K_d = 2$. The reduction in damping results in the slosh angle reduing to zero faster i.e. in around 10 seconds. The reduction in K_d also results in an increase in the overshoot as seen in Fig 5a. The desired position is also reached in around 10 seconds as seen in Fig 5b.

Fig 6 shows the lateral force in the system for $K_p = 0.9$ and $K_d = 2$. The force reduces to zero as desired position is reached as in the case of simulation.

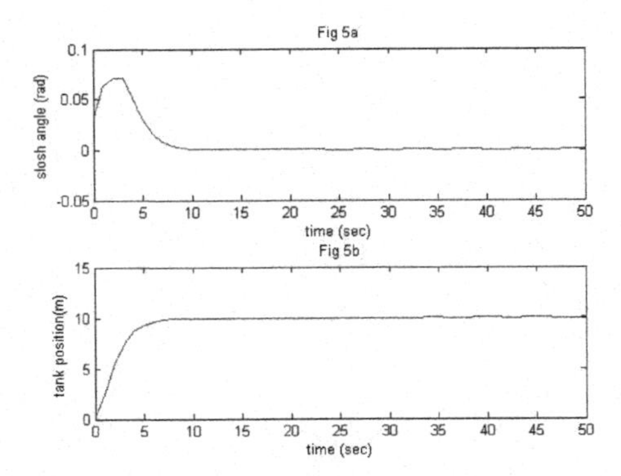

Fig. 5. Slosh angle and lateral position of container

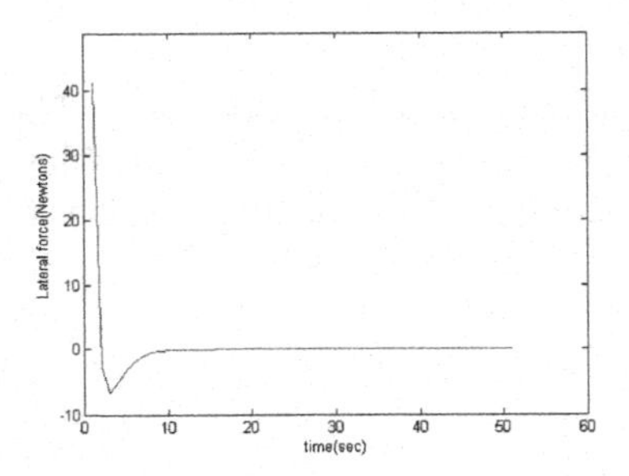

Fig. 6. Lateral force

6 Conclusion

The passivity based control technique is an effective method for one dimensional slosh problem. The same control algorithm can be applied to several fluids. The controller has to be tuned to find suitable values of gain K_p and damping K_d for different values of damping co-efficient 'c'. Thus control is independent of not only the physical properties of fluid like viscosity but also the properties of tank like dimensions and material of construction. Future work aims at developing a passivity based control strategy for systems involving damping components like baffles as well as fluids with mixed phases.

References

1. Slosh Suppression, NASA Space Vehicle Design Criteria (Structures) NASA SP-8031
2. Venugopal, R., Bernstein, D.S.: State space Modelling and Active Control of Slosh. In: Proc. of 1996 IEEE International Conference on Control Applications, pp. 1072–1077 (September 1996)

3. Feddema, J., Dohrmann, C., Parker, G., Robinett, R., Romero, V., Schmitt, D.: Robotically Controlled Slosh-free motion of an Open Container of Liquid. In: Proc. of 1996 IEEE International Conference on Robotics and Automation, pp. 596–602 (April 1996)
4. Yano, K., Terashima, K.: Robust Liquid Transfer Control for Complete Sloshing Suppression. IEEE Trans. on Control Systems Technology 9(3), 483–493 (2001)
5. Yano, K., Terashima, K.: Sloshing Suppression Control of Liquid Transfer Systems Considering a 3-D Transfer Path. IEEE/ASME Trans. on Mechatronics 10(1), 8–16 (2005)
6. Grundelius, M., Bernhardsson, B.: Control of Liquid Slosh in an Industrial Packaging Machine. In: Proc. of 1999 IEEE International Conference on Control Applications, pp. 1654–1659 (August 1999)
7. Tzamtzi, M., Koumboulis, F.: Sloshing Control in Tilting Phases of the Pouring Process. World Academy of Science, Engineering and Technology 34, 307–314 (2007)
8. Kurode, S., Bandyopadhyay, B., Gandhi, P.: Sliding Mode Control for Slosh-free Motion of a Container using Partial Feedback Linearization. In: IEEE International Workshop on Variable Structure Systems, Antalya, Turkey, pp. 101–107 (June 2008)
9. Kurode, S., Bandyopadhyay, B., Gandhi, P.: Robust Control for Slosh-free Motions using Sliding Modes. In: 36th Annual Conference on IEEE Industrial Electronics Society, pp. 2408–2413 (November 2010)
10. Khalil, H.: Nonlinear Systems, 3rd edn. Prentice Hall
11. Sepulchre, R., Jancović, M., Kokotović, P.: Constructive Nonlinear Control. Springer (1997)
12. Gogte, G., Venkatesh, C., Kazi, F., Singh, N.M., Pasumarthy, R.: Passivity Based Control of Underactuated 2-D SpiderCrane Manipulator. In: MTNS 2012, Melbourne, Australia (2012)
13. Ortega, R., Spong, M., Estern, F., Blankenstein, G.: Stabilization of a Class of Underactuated Mechanical Systems Via Interconnection and Damping Assignment. IEEE Transactions on Automatic Control 47(8), 1218–1233 (2002)
14. Acosta, J., Ortega, R., Astolfi, A., Mahindrakar, A.: Interconnection and Damping Assignment Passivity-Based Control of Mechanical Systems With Underactuation Degree One. IEEE Transactions on Automatic Control 50(12), 1936–(1955)
15. Spong, M.: Partial Feedback Linearization of Underactuated Mechanical Systems. In: Proceedings of the IEEE/RSJ/GI International conference on Intelligent Robots and Systems, pp. 314–321 (1994)
16. Spong, M.: Swing up Control of the Acrobot. In: IEEE International Conference on Robotics and Automation, pp. 2356–2361 (1994)
17. Ortega, R., Schaft, A., Mareels, I., Maschke, B.: Putting Energy Back in Control. IEEE Control Systems Magazine, 1–16 (February 2001)
18. Ortega, R., Loriá, A., Nicklasson, P., Sira-Ramirez, H.: Passivity-based Control of Euler-Lagrange Systems: Mechanical. Electrical and Electromechanical Applications

Performance Improvement of Hysteresis Current Controller Based Three-Phase Shunt Active Power Filter for Harmonics Elimination in a Distribution System

Gayadhar Panda[1], Santanu Kumar Dash[1], Pravat Kumar Ray[2], and Pratap Sekhar Puhan[3]

[1] Dept. of Electrical Engineering, IGIT Sarang, Orissa
[2] Dept. of Electrical Engineering, NIT Rourkela, Orissa
[3] Dept. of Electrical Engineering, BIET Bhadrak, Orissa
p_gayadhar@yahoo.com, santanu4129@gmail.com,
rayp@nitrkl.ac.in, psp12_puhan@rediffmail.com

Abstract. Performance investigation of Shunt Active Power Filter for harmonic elimination is an interdisciplinary area of interest for many researchers. This paper presents performance improvement of 3-phase Shunt Active Power Filter with Hysteresis Current Control technique for elimination of harmonic in a 3-phase distribution system. The insulated gate bipolar transistor based Voltage Source Inverter with a dc bus capacitor is used as active power filter. The shunt active filter employs a simple method called synchronous detection technique for reference current gene-ration. Two level hysteresis current controller is proposed for voltage source inverter switching signal generation which helps in tuning of Active Power Filter. Shunt Active power filter generates compensating signal which reduces the harmonic content of the system and improve the system performance. Shunt active power filter is modeled and investigated under non-linear load condition using MATLAB/Simulink.Various simulation results are presented to verify the improvement of performance by using shunt active power filter with hysteresis current controller for the distribution system.

Keywords: Shunt active power filter, hysteresis current controller, harmonic elimination, voltage source inverter, distribution system.

1 Introduction

The advanced technology of power semiconductor leads to frequent use of power electronic converter circuits in many different types of industrial equipment such static var compensators (SVCs), adjustable speed drives (ASDs) and uninterruptible power supplies (UPSs).These non-linear loads inject harmonics in the distribution system. The presence of harmonics in the system results in several effects like increased heating losses in transformers, motors, and lines, low power factor, torque pulsation in motors, and poor utilization of distribution wiring and plant. These serious problems make the harmonic elimination necessary. A passive filter has been

S. Unnikrishnan, S. Surve, and D. Bhoir (Eds.): ICAC3 2013, CCIS 361, pp. 682–692, 2013.

a viable approach. Conventionally, to reduce harmonics passive LC filters and to improve power factor of the AC loads capacitor banks were used [1-6]. However, the performance of the passive scheme has a limitation since the addition of the passive filter interfaces with the system impedance and causes resonance with other networks. Since the conventional passive filter is not able to provide a complete solution, recently some Active Power Filters (APFs) have been widely investigated for the compensation of harmonics in electric power system [7-9]. The active power filter topology can be connected in series or shunt and combination of both. Shunt active power filter (SAPF) is more popular than series filter because most of the industrial applications require current harmonics compensation. Shunt active power filters are developed to suppress the harmonic currents and compensate reactive power simultaneously. The shunt active power filters are operated as a current source parallel with the nonlinear load. A compensation current equal but opposite to the harmonic and reactive current drawn by the non linear loads is generated by suitably controlling the converters present in the active power filter so as to make the supply current sinusoidal and in phase with the supply voltage [10].

The shunt active power filter can be executed with current source inverter (CSI) or voltage source inverter (VSI). The voltage source inverter has three coupling inductors in series in the ac-side and an energy storage capacitor on the dc side. Similarly, the current source inverter has three coupling capacitors in parallel in the ac-side and an energy storage inductor in series with the dc side [10, 11]. In general the voltage source inverter type is preferred for the shunt active power circuit due to the lower losses in the dc side and it is possible to create multilevel inverter topologies for higher power level applications. The current control techniques of the voltage source inverters can be broadly classified into linear and nonlinear control scheme [12-15].

The linear current controllers are characterized by constant switching frequency. On the other hand, the non-linear current controllers are characterized by varying switching frequencies. It provides fast response to transient conditions and hence the non-linear hysteresis current control is the most sought after method of current control [14-17]. It provides advantages like automatic peak current limitation, simple hardware implementation, load parameter independence and unconditional stability.

This paper describes three-phase shunt active power filter for current harmonic compensation under non-linear load conditions. The three phase active power filter is implemented with current controlled voltage source inverter (VSI). The VSI switching signals are derived from 2-level hysteresis current controller (HCC). The SAPF system is investigated under non-linear load condition.

The remaining sections of this paper are organized as follows. The studied system is described in section two. Control scheme is described in section three. The synchronous detection is described in section four. Then the results with discussion of simulation study are provided in section five. Finally some conclusions are drawn and future works are suggested in section six.

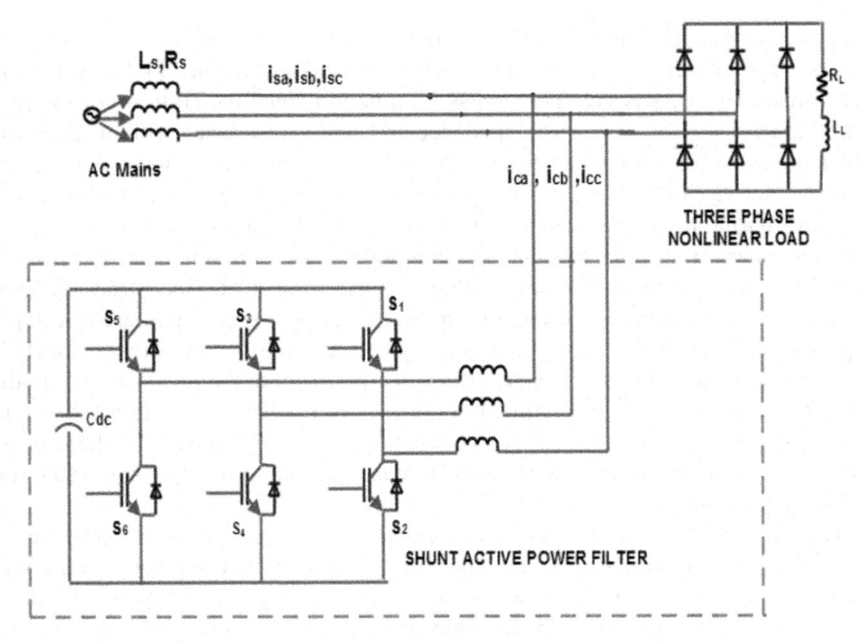

Fig. 1. Connection diagram of a voltage source active power filter

2 Three-Phase Shunt Active Power Filter Model

The shunt active power filter shown in Fig. 1 consists of a three-phase full-bridge voltage source PWM inverter, a DC bus capacitor Cdc and three coupling inductors in series in the ac-side. The shunt active power filter is controlled to draw and inject compensating current, ic to the distribution system and cancel the harmonic currents on the AC side of a non-linear load. Besides that, it has the capability of damping harmonic resonance between an existing passive filter and the supply impedance.

3 Control Scheme

Hysteresis current control scheme is more advantageous for the control of active power filter can be used to generate the switching signals of the inverter. It creates an environment for fastest control with minimum hardware and has excellent dynamics. Conventional hysteresis current controllers produce bad harmonic performance, by varying the hysteresis band its performance can be improved to get a fixed switching frequency.

There are different types of Hysteresis current controllers available like two-level hysteresis current controller and three-level hysteresis current controller. This paper covers two-level hysteresis current controller for the proposed active power line conditioner.

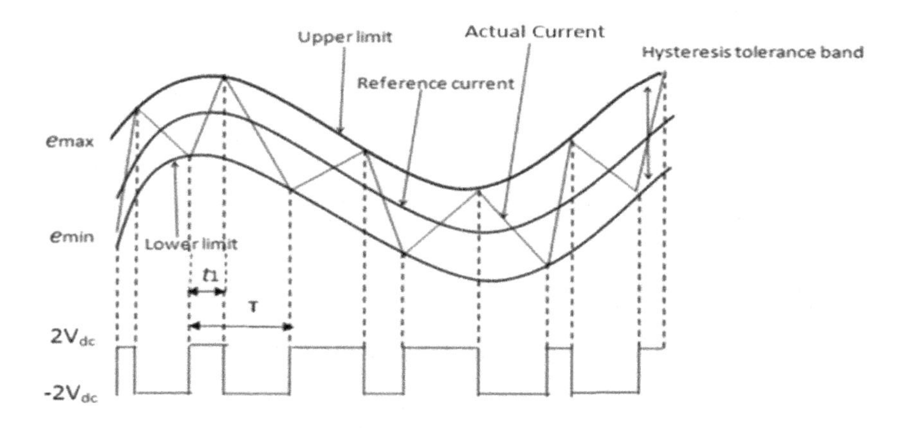

Fig. 2. Hysteresis Band

Fig. 2 Shows the two levels hysteresis current control indicating the upper hysteresis band, lower hysteresis band, actual and reference current. Conventional HCC operates the PWM voltage source inverter by comparing the current error against fixed hysteresis bands. The difference between the reference current and the current being injected by inverter means the current error. When this current error exceeds the upper limit of the hysteresis band the lower switch of the Inverter arm is turn on and the upper switch is turned OFF and vice-versa.

At any point of time if we resolve the rate of change of phase current, this can be written as

$$\frac{dI}{dt} = \frac{\Delta I}{\Delta t} = \pm\frac{2V_{dc}}{L} \Rightarrow \Delta t = \frac{\Delta IL}{\pm 2V_{dc}} \tag{1}$$

In the equation (1) $\pm 2V_{dc}$ depends on switching state of inverter , rate of change of inverter current is represented as ΔI ,rate of change in current in time period is Δt .In the hysteresis band complete switching cycle is from $0 \rightarrow t_1 \rightarrow T$

So the equation can be written as for $0 \rightarrow t_1$ (i.e. Δt = t$_1$)

$$t_1 = \frac{+\Delta IL}{2V_{dc}} \tag{2}$$

Again for the period $t_1 \rightarrow T$ (i.e. Δt = T - t$_1$)

$$T - t_1 = \frac{-\Delta IL}{-2V_{dc}} \tag{3}$$

By combining both the period equation total switching time can be determined, it is given as:

$$f_s = \frac{1}{T} = \frac{V_{dc}}{\Delta ILV_{dc}}^2 \rightarrow f_{\max} = \frac{V_{dc}}{\Delta IL} \tag{4}$$

Where the inverter maximum switching frequency can be represented as $f_{\max}$.

4 Estimation of Reference Source Current

Different control algorithms are proposed for APF but a Synchronous detection method is used for harmonic detection to calculate reference current for shunt active power filter due to its simplicity. The balanced three phase source currents can be obtained after compensation. The equal current distribution method of this control scheme is implemented in this research work. The following steps are used for generation of reference signal. Assumption for this method is

$$I_{sa} = I_{sb} = I_{sc}$$

Where I_{sa}, I_{sb}. and I_{sc} represent the peak values of source current in phase a, b and c respectively.

Considering 3-phase balance supply and nonlinear load connected in the proposed system as shown in Fig. 1, voltage and current expression can be written as follows:

3-phase source voltages are given by

$$V_{sa}(t) = V_{ma} \sin wt \tag{5}$$

$$V_{sb}(t) = V_{mb} \sin(wt - 120^0) \tag{6}$$

$$V_{sc}(t) = V_{mc} \sin(wt - 240^0) \tag{7}$$

3-phase current drawn by load given by

$$I_{La}(t) = \sum_{n=1}^{\infty} I_{an} \sin(wt - \phi_{an}) \tag{8}$$

$$I_{Lb}(t) = \sum_{n=1}^{\infty} I_{bn} \sin(wt - \phi_{bn} - 120^0) \tag{9}$$

$$I_{Lc}(t) = \sum_{n=1}^{\infty} I_{cn} \sin(wt - \phi_{cn} - 240^0) \tag{10}$$

Step 1:

The 3-phase instantaneous power ($p_{3\phi}$) in the proposed system can be written as

$$\begin{aligned} p_{3\phi} &= v_{sa} i_{La} + v_{sb} i_{Lb} + v_{sc} i_{Lc} \\ &= p_a + p_b + p_c \\ &= V_{ma} \sin wt \sum_{n=1}^{\infty} I_{an} \sin(wt - \phi_{an}) + \\ & V_{mb} \sin(wt - 120^0) \sum_{n=1}^{\infty} I_{bn} \sin(wt - \phi_{bn} - 120^0) + \\ & V_{mc} \sin(wt - 240^0) \sum_{n=1}^{\infty} I_{cn} \sin(wt - \phi_{cn} - 240^0) \end{aligned} \quad (11)$$

Step 2:

The instantaneous power is passed through low pass filter (LPF), which blocks higher order frequency component and only fundamental component is obtained from the output of LPF.

$$\begin{aligned} P_{fund} &= V_{ma} \sin wt I_{a1} \sin(wt - \phi_{a1}) + \\ & V_{mb} \sin(wt - 120^0) I_{b1} \sin(wt - \phi_{b1} - 120^0) + \\ & V_{mc} \sin(wt - 240^0) I_{c1} \sin(wt - \phi_{c1} - 240^0) \\ &= \frac{V_{ma} I_{a1}}{2} [\cos\phi_{a1} - \cos(2wt + \phi_{a1})] + \\ &= \frac{V_{mb} I_{b1}}{2} [\cos\phi_{b1} - \cos(2wt + \phi_{b1})] + \frac{V_{mc} I_{c1}}{2} [\cos\phi_{c1} - \cos(2wt + \phi_{c1})] \end{aligned} \quad (12)$$

Step 3:

The average fundamental power in 3-phase is given by

$$\begin{aligned} P_{av} &= \int_0^T P_{fund} dt \\ &= \frac{V_{ma} I_{a1}}{2} \cos\phi_{a1} + \frac{V_{mb} I_{b1}}{2} \cos\phi_{b1} + \frac{V_{mc} I_{c1}}{2} \cos\phi_{c1} \end{aligned} \quad (13)$$

For 3- phase balanced nonlinear load the followings can be written as

$$V_{ma} = V_{mb} = V_{mc} = V$$
$$I_{a1} = I_{b1} = I_{c1} = I$$
$$\phi_{a1} = \phi_{b1} = \phi_{c1} = \phi_1$$

$$P_{av} = \frac{3VI}{2}\cos\phi_1 \quad (14)$$

Step 4:
Using equation (14), avarage power per phase can be written as

$$(P_{av})_{ph} = \frac{VI}{2}\cos\phi_1 \quad (15)$$

Let $I\cos\phi_1 = I_m =$ Maximum amplitude of per phase fundamental current

$$I_m = \frac{2(P_{av})_{ph}}{V} \quad (16)$$

Step 5:
The fundamental current is given by

$$I_{Fa} = I_m \sin wt \quad (17)$$

$$I_{Fb} = I_m \sin(wt - 120^0) \quad (18)$$

$$I_{Fc} = I_m \sin(wt - 240^0) \quad (19)$$

The expression of reference current for shunt active power filter in each phase $(i^*_{ca,} i^*_{cb,} i^*_{cc})$

$$i^*_{ca} = I_{La} - I_{Fa}$$

$$i^*_{cb} = I_{Lb} - I_{Fb}$$

$$i^*_{cc} = I_{Lb} - I_{Fc}$$

After getting the reference current, it is compared with the actual current by using hysteresis current comparator to generate six switching pulses, which are used to control the IGBT either by turning ON or OFF.

5 Simulation Results and Discussions

The system considered for simulation, is a three-phase source supplying a diode bridge rectifier feeding an inductive load. This load draws a highly non-linear current rich in harmonics with a substantial reactive power requirement. A three-phase, VSI based shunt active power filter is connected to the system for reactive power compensation and harmonics elimination. Figure.3 shows simulation model of 3-phase SAPF done in the SIMULINK/MATLAB environment to examine the superiority in performance of SAPF by employing proposed HCC.

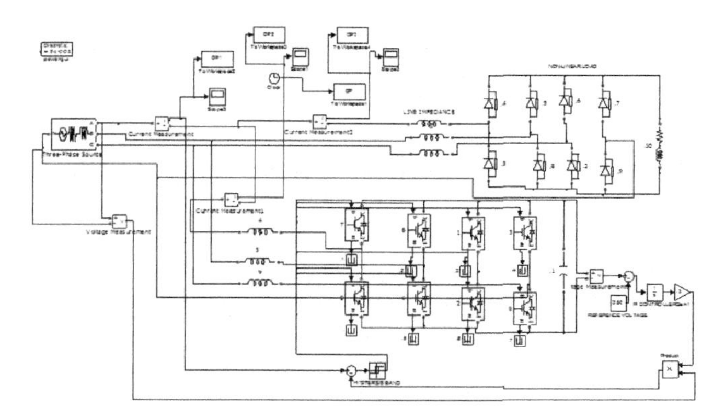

Fig. 3. Simulink model of 3-phase Shunt APF

The system parameters for the model developed are given below:

Table 1. System Parameters

Parameter	Value
Rms voltage	230v
frequency	50Hz
DClink capacitor	1000uF
Capacitor Voltage	350v
Filter inductance	1mH
Line impedance	0.001H

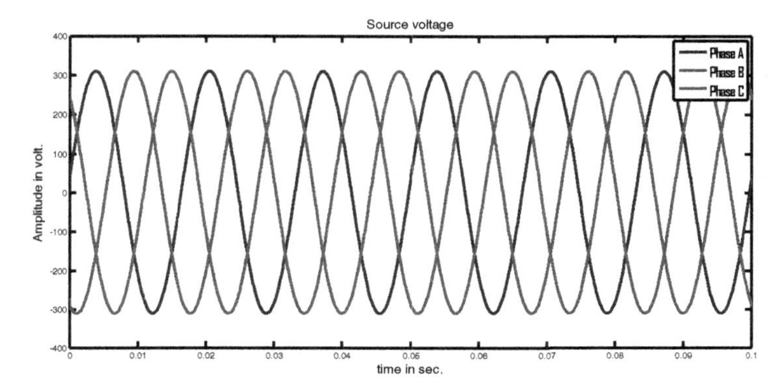

Fig. 4(a). Source voltages wave form

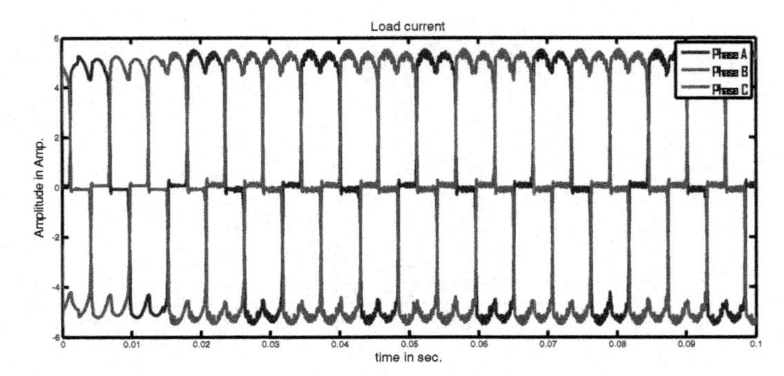

Fig. 4(b). Waveforms of current drawn by non linear load

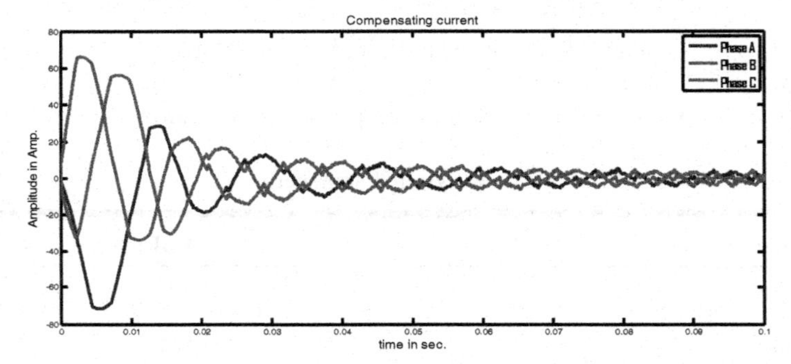

Fig. 4(c). Compensating signals generated by HCC based SAPF

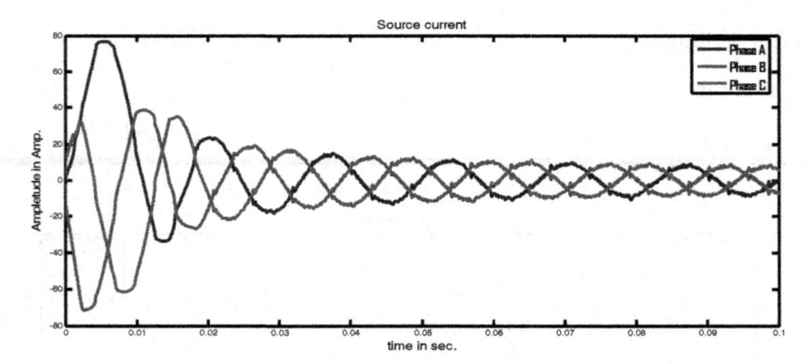

Fig. 4.(d). source currents after compensation

Fig. 4. Simulated waveforms with a shunt active power filter

Fig. 4 shows the supply voltage, load current iL, shunt active filter current ic, and supply current is of a nonlinear load connected to AC mains. The analysis of the steady-state waveforms shows close to full compensation of the source current. The harmonic spectrum of the source current using active power filter is shown in Fig. 5. The Total Harmonics Distortion (THD) of the source current is reduced from 29.59 % before compensation to 9.45% after compensation. Table 2. shows the % reduction of amplitude of individual harmonic in the system.

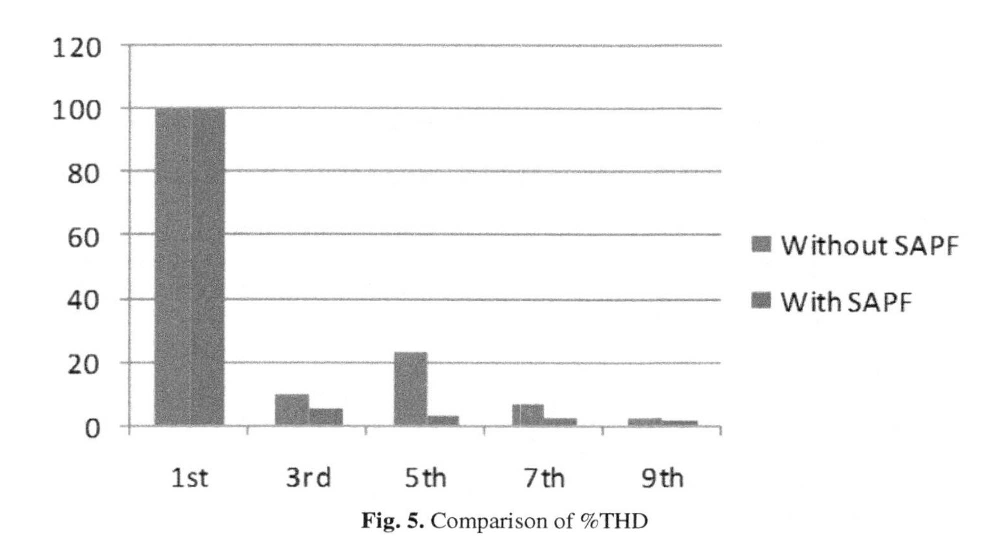

Fig. 5. Comparison of %THD

Table 2. Total Harmonic Distortion without SAPF and with SAPF

Harmonic order	Amplitude		
	Without SAPF (%THD= 29.59)	With SAPF (%THD= 9.45)	% of Reduction
5th	22.18	3.88	19
7th	11.48	2.81	8.67
9th	8.67	3.92	4.75
11th	6.47	2.30	4.44

6 Conclusion

By using different nonlinear loads in the power system harmonics are generated which is the main factor in reduction of performance of various system. Reduction of harmonics from the system directly improves the performance. From the simulation results given in this paper it is seen that total harmonic distortion is reduced up to great extent by the SAPF, leads to increase in system performance.

References

1. Hingorani, N.G.: Introducing Custom Power. IEEE Spectrum, 41–48 (June 1995)
2. Subjak Jr., J.S., Mcquilkin, J.S.: Harmonics- Causes, Effects, Measurements, and Analysis: An Update. IEEE Trans. on Industry Appl. 26(6), 1034–1042 (1990)

3. Ohnishi, T., Ueki, K., Hojo, M.: Source current detection control off activates filter for preventing current oscillation. In: 35th Annual IEEE Power Electronics Specialists Conference, Aachen, Germany, pp. 965–965 (2004)
4. Brod, D.M., Novotny, D.M.: Current control of VSI-PWM Inverter. IEEE Trans. on Industry Appl. 21, 562–570 (1985)
5. Kaunierkowski, M.P., Dzieniakowslu, M.A.: Review of Current Regulation Techniques For Three-phase PWM Inverters. In: Conference on IECON 1994, vol. 1, pp. 567–575 (1994)
6. Duffey, C.K., Stratford, R.P.: Update of Harmonic Standard IEEE-519: IEEE Recommended Practices and Requirements for Harmonic Control in Electric Power Systems. IEEE Trans. on Industry Appl. 25(6), 1025–1034 (1989)
7. Wu, J.C., Jou, H.L.: Simplified control for the individual phase activates power filter. IEE Proc. Electr. Power Appl. 143(3), 219–224 (1996)
8. Akagi, H.: Trends in Active Power Line Conditioners. IEEE Transactions on Power Electronics 9(3), 263–268 (1994)
9. Akagi, H.: New Trends in Active Filters for Power Conditioning. IEEE Transactions on Industry Applications 32(6), 1312–1322 (1996)
10. Terbobri, G.G.: Real time activates power filtering using sliding control mode. PhD Thesis, University off Wales, Swansea, the U.K. (1999)
11. Barrero, F., Martinez, S., Yeves, F., Martinez, P.M.: Active Power Filters for Line Conditioning: A Critical Evaluation. IEEE Trans. on Power Delivery 15(1), 319–325 (2000)
12. Singh, B., Al-Haddad, K., Chandrs, A.: A Review of Active Filters for Power Quality Improvements. IEEE Trans. on Industrial Electronics 46(5), 960–970 (1999)
13. Grady, W.M., Samotyj, M.J., Noyola, A.H.: Survey of Active Power Line Conditioning Methodologies. IEEE Trans. on Power Delivery 5(3), 1536–1542 (1990)
14. Montero, M.I.M., Cadaval, E.R., Gonzalez, F.B.: Comparison of Control Strategies for Shunt Active Power Filters in Three-Phase Four-Wire Systems. IEEE Trans. on Power Elects. 22(1) (2007)
15. Vodyakho, O., Kim, T.: Shunt active filter based on three-level inverter for 3-phase four-wire systems. IET Power Electron 2(3), 216–226 (2006)
16. Anand, V., Srivastava, S.K.: Performance Investigation of Shunt Active Power Filter Using Hysteresis Current Control Method. IJERT 1(4) (June 2012)
17. Moran, L.A., Dixon, J.W.: Using Active Power Filters to Improve Power Quality

Achieving Robust Network Centric Warfare with Multi Sensor Multi Target Data Association Techniques

Kanan Bala Sahoo and Aarti Dixit

Defence Institute of Advanced Technology (DIAT)
Girinagar, Pune – 411025
{mcse2011_kanan,aratidixit}@diat.ac.in

Abstract. A robust and secured Network Centric Warfare (NCW) environment can be sustained by real-time tracking of multiple objects in air, water and land, to identify them as friend or enemy. The accurate identification can help the commander to take timely appropriate decisions. The identification of objects has to be supported by Data Association (DA) process. In DA process multiple tracks received for multiple targets from a set of sensors are processed to correlate tracks/measurements to targets. In cluttered environment DA process becomes more challenging and needs appropriate algorithm to produce exact picture to the commander to perform combat control operations in time. The existing DA algorithms are compared and an advanced *Multiple Hypothesis Tracking* algorithm is implemented and tested. This algorithm contributes towards achievement of a secure and robust network centric Warfare environment.

Keywords: Network Centric Warfare, Data Association, Command and Control, Multiple Hypothesis Tracking.

1 Introduction

In conventional platform centric warfare, the linkages between services and Command and Control (C2) system are ineffective. In this, control flows unidirectional and delays the reaction time. The current advancement in technology has shifted the traditional centralized platform centric warfare to a highly distributed, agile, and self adapting Network Centric Warfare. The NCW is defined [1] as:

Network Centric Warfare is the conduct of military operations through the utilization of networked information systems, which supply the war fighter with the right information at the right time in the right form to the right person being put to the right use, in order to achieve desired effects across the physical, information, and cognitive domains of warfare.

Self synchronization in the view of NCW is defined as the ability of low level shooter groups to coordinate and execute the command without delaying for C2 commander's decision. This is achieved due to individual situational awareness at the shooter level and the shared situational awareness [2] of each weapon system node as shown in Fig.1. It helps in decision making. Finally mission objectives are met with the collaboration of decision making and self synchronization.

S. Unnikrishnan, S. Surve, and D. Bhoir (Eds.): ICAC3 2013, CCIS 361, pp. 693–704, 2013.

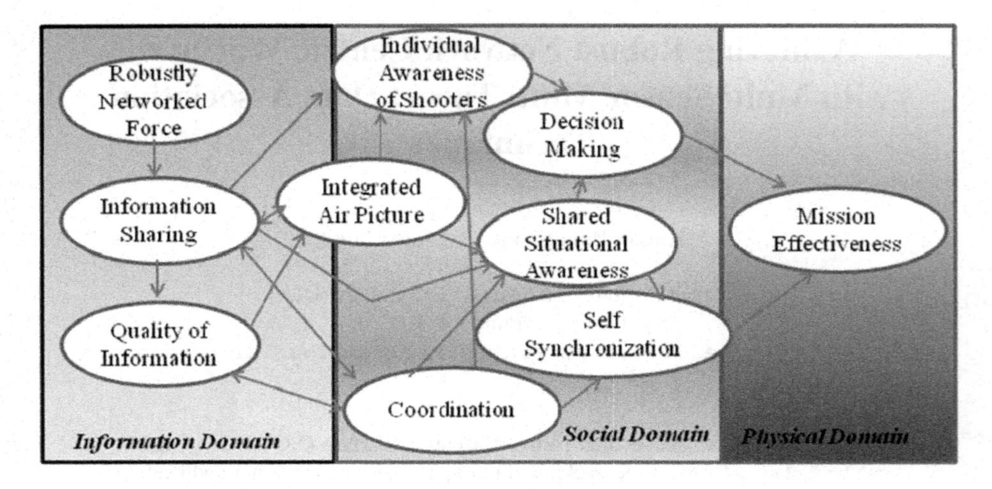

Fig. 1. Network Centric Operations

Network Centric Operation (NCO) is [2] defined as operations that are enabled by the networked infrastructure and with the collaboration of people. NCW refers to such operations when they are engaged in warfare or military operations. So, NCO is a superset of NCW. In NCO multiple Surface to Air Missile (SAM) systems comprising of multiple sensors are integrated to accomplish the mission objectives. In this scenario, data association of multiple targets tracked by multiple sensors plays a significant role in producing an Integrated Air Picture for the commander.

The automated C2 sequence of operations for an Air Defence SAM system as seen in Fig.2 involves the following major tasks:

a) Target Tracking by multiple sensors and sending data to C2 centre asynchronously.
b) An integrated Air Situation Picture Generation (ASPG) which involves bias correction, coordinate conversion of track data to a common frame, Filtering, DA, and Data Fusion.
c) Combat control module performs radar control, target identification, target classification, assigning the target to a suitable launcher, engaging the interceptor missile against the assigned target and performing kill assessment.
d) Provide real time display to the commander.

The details of NCW and the significance of DA in NCW are discussed in section-2. The section-3 describes different association techniques with their pros and cons. Multiple Hypothesis Tracking based DA technique with its subcategories are addressed in section-4. The proposed algorithm, experiment scenario and results are discussed in section-5. Finally section-6 concludes the paper.

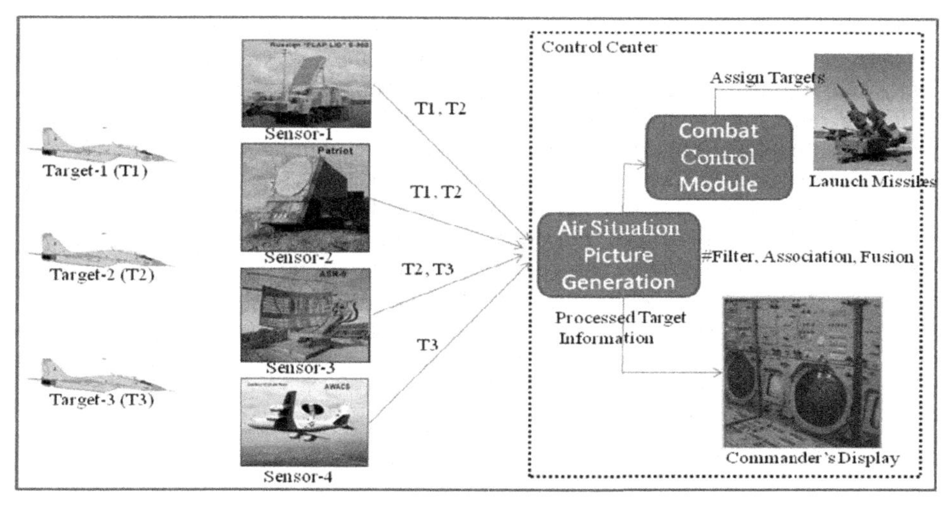

Fig. 2. C2 Sequence of Operations

2 Network Centric Warfare

NCW [2] comprises of three highly networked and coordinated layers of sensor grid in the forefront, C2 (information) grid in the middle for the control and the shooter grid at the bottom level to execute the commands which is depicted in Fig.3. In the diagram blue colour arrow indicates the link within the elements of a grid and the red colour arrows shows the information flow across the layers of NCW. In this data and control flows bidirectional and meets the real time requirements of sensor-to-shooter in handling time critical events and sensitive targets. It reduces ambiguities and provides a robust, secure and timely response in achieving mission objectives.

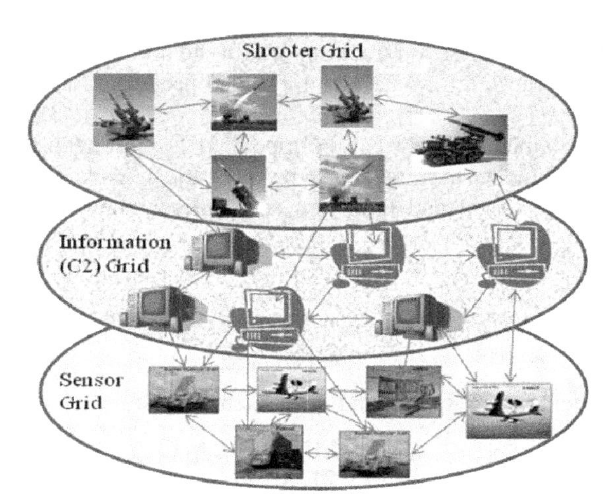

Fig. 3. Three layers of NCW

Today the war scenario has moved from platform centric to a highly complex network centric. Now the aim is to integrate all the three services (Army, Air Force and Navy) to cover land, air and water bodies associated with the region of interest. This integrated force will provide redundant, accurate and timely information to the control centre to take appropriate decision against the incoming threat.

Multiple sensors shall be deployed at various locations to provide early warning and subsequent target information to C2 centre. To achieve spatial and temporal coverage, these sensors are placed in an overlapped coverage fashion. The challenging task is to provide an integrated air picture to the commander as depicted in the air. In this context Multi Sensor Multi Target (MSMT) DA plays a significant role.

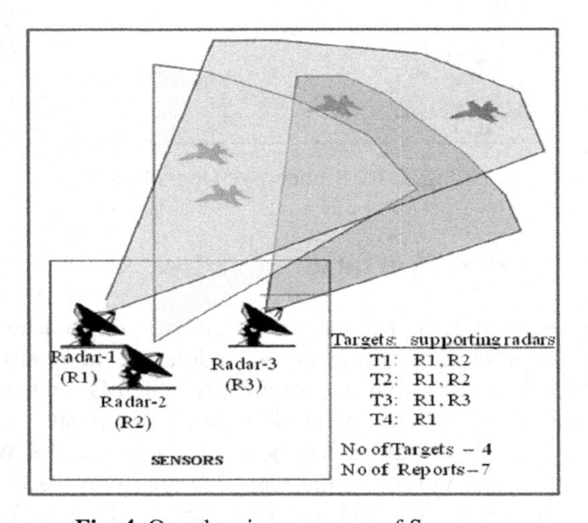

Fig. 4. Over lapping coverage of Sensors

A MSMT scenario is shown in Fig.4 where four targets are tracked by three sensors resulting altogether seven track reports in the commander's real time display. DA techniques take these information as input and produces four targets as output as seen in the air.

The system explained in Fig.4 can be modelled for DA with the help of targets, sensors and measurements. A target may be any object tracked continuously by the sensor. In this paper target and track are used interchangeably. Let the targets be denoted as:

$$T = \{T1, T2, T3, T4\} \tag{1}$$

Radars are denoted as:

$$R = \{R1, R2, R3\} \tag{2}$$

Let the measurements from R1, R2 and R3 be denoted as:

$$\left.\begin{array}{c}\{M11, M12, M13, M14\}\\ \{M21, M22\}\\ \{M33\}\end{array}\right\} \tag{3}$$

DA performs pairing of tracks with radar measurements and the output of the above discussed model is shown in table -1. It shows the associated measurements for each track.

Table 1. Data Association output

Targets	Measurements
T1	M11, M21
T2	M12, M22
T3	M13, M31
T4	M14

3 Data Association Techniques

DA is defined as a process of establishing a relation between the sensor reports or measurements with the tracks or targets. It is a technique for taking decision that which measurement can be paired with which track. It is also called correlation.

DA is required because received measurements are associated with the environmental or sensor noise, which produces clutter or false alarm. It helps in taking correct decisions in a scenario of dense or closely moving targets. This process is made simpler by using gating techniques. Gating is defined as a screening mechanism to determine which measurements are valid candidates to update the existing tracks or to create new tracks.

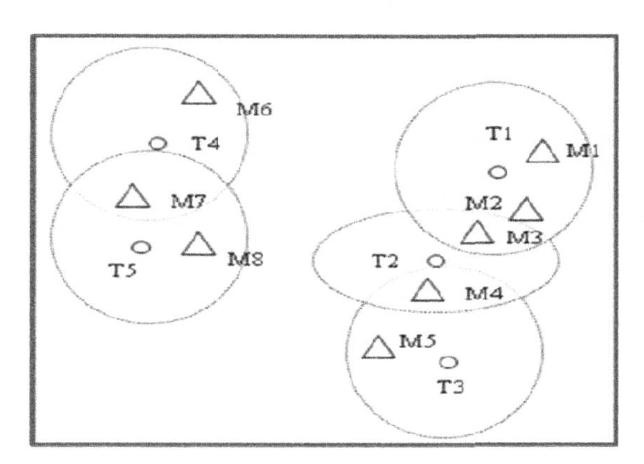

Fig. 5. Gating Mechanism

A gate is formed around the predicted position of the target as shown in Fig.5. In this targets are indicated as small circle and measurements are in triangle. All the measurements which are falling within the gate of the target are considered for target update. It reduces the computation and the whole data set can be divided into multiple groups or clusters for parallel processing. All those measurements which satisfy gating threshold are supplied as input to association algorithm. DA algorithms are broadly categories into two groups and as depicted in Fig.6:

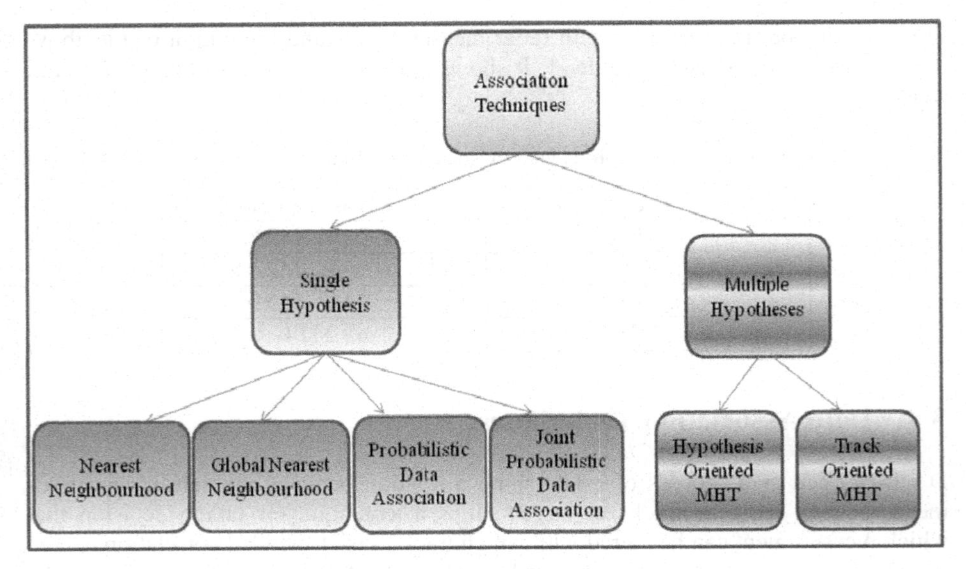

Fig. 6. Association Techniques

A hypothesis is formed by taking decision about tracks over n-measurements at mth -scan. Single hypothesis based DA algorithms are as follows:

3.1 Nearest Neighborhood
3.2 Global Nearest Neighborhood
3.3 Probabilistic Data Association
3.4 Joint Probabilistic Data Association

3.1 Nearest Neighborhood (NN)

NN approach is the most likely hood rule in which the best measurement from sensors is selected based on those measurements falling within the gate of the predicted target

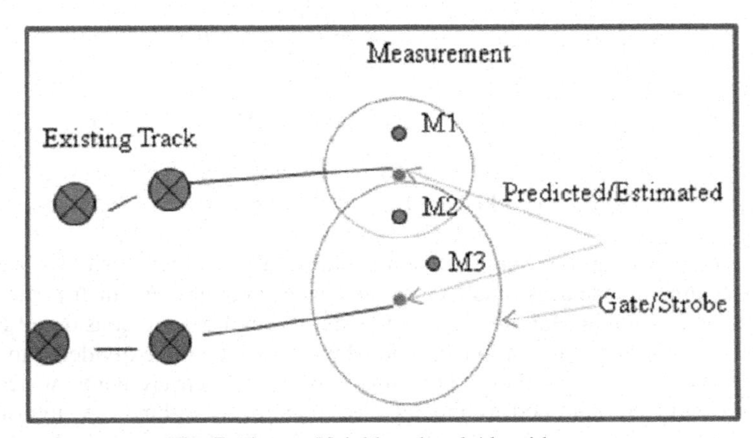

Fig. 7. Nearest Neighbourhood Algorithm

position. It [3] allows each track to be updated with its closet measurement and the distance between the target and measurement can be computed by using "Euclidian distance" or "Mahalanobis distance". The root mean square errors are used in dynamic gate computation.

NN is very simple, easy to implement and less computational intensive. But it cannot handle the complex conflict scenario is presented in Fig.7 where 'm2' is falling within both the target gates. So, one measurement may be used to update two tracks. The wrong association may end up with the real target drop or in creation of new track.

3.2 Global Nearest Neighborhood (GNN)

GNN considers all measurements-to-track pairings for taking decision. It determines a unique pair so that one measurement can be used to update at most a single track. This handles the conflict situation shown in Fig.6 but gives erroneous result in case of closely moving targets. This algorithm [3] works well only in the case of widely spaced targets, accurate measurements and few false tracks in the track gates.

3.3 Probabilistic Data Association (PDA)

In [4] PDA a single track is updated based on a probabilistically determined weighted sum of all measurements within the track gate. In this method all the neighbors are processed for a single target in clutter.

3.4 Joint Probabilistic Data Association (JPDA)

JPDA is an extension to PDA where the association probabilities are computed by considering all measurements and all tracks in a dense cluttered environment. It operates [4] on a scan by scan basis without maintaining all the alternate association hypotheses. In closely moving target scenario, JPDA may update two or more targets with one measurement. As a result the real target may get dropped.

4 Multiple Hypothesis Tracking

In Multiple Hypothesis Tracking (MHT) alternative DA hypotheses are formed whenever there is a conflict in measurement-to-track pairing as shown in Fig.7. These [5] hypotheses are propagated in expectation that the subsequent measurements will resolve the uncertainty. This approach provides a much higher correct decision probability than the single hypothesis based tracking. It can work in highly dense and closely spaced targets. MHT is broadly categorized in to two types:

4.1 Hypothesis Oriented MHT
4.2 Track Oriented MHT

4.1 Hypothesis Oriented MHT (HOMHT)

The original MHT algorithm was proposed by Donald B. Reid. In [6] this algorithm, hypotheses are generated for each measurement in each scan accounting to all possible combinations. A measurement may get associated with an existing target or may be a false track or end up with the creation of a new track. Reid's clustering mechanism was used to divide the entire set of targets and measurements into independent groups such that those can be processed parallel and reduce computation load.

Hypotheses [6] are represented as a tree and stored in a matrix form in the computer for recursive process. A cluster of two tracks and two measurements as shown in Fig. 8 are supplied for hypothesis generation. Tracks are numbered as 1 and 2. Measurements are numbered as 21 and 22. The first measurement 21 may be a false track (denoted as 0) or may get associated with track 1 or with track 2 or create a new track 3. The second measurement 22 will be processed similarly and tree branches are expanded. Hypotheses tree and hypotheses matrix for the cluster are shown in Fig.8.

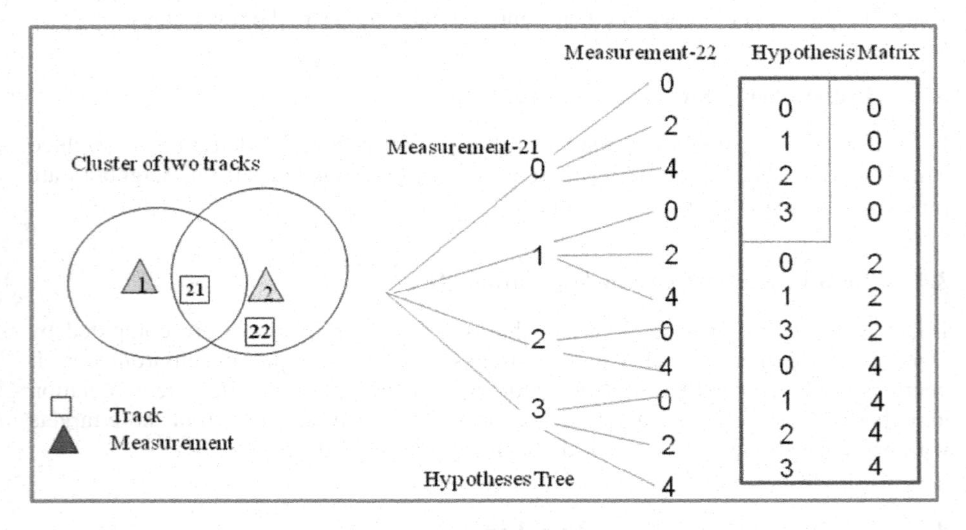

Fig. 8. Representation of Hypotheses

The probability of each hypothesis is calculated and target estimates are updated. In HOMHT, the number of hypotheses increases exponentially as the number of targets increases. Hypotheses are managed by applying pruning and merging mechanisms periodically.

HOMHT is the preferred method for DA. It has advantages over the conventional single hypothesis algorithms mentioned in section-3. But HOMHT is computationally intensive and needs huge memory to store all the hypotheses which makes it inappropriate for real time applications.

4.2 Track Oriented MHT (TOMHT)

The computational feasibility of HOMHT has been greatly enhanced by the use of Murty's K-best algorithm [7] to generate hypotheses more efficiently. TOMHT generates hypotheses only in [8] conflict situations where a track can be assigned to more than one measurement or one measurement may be paired with two or more tracks. It does not maintain hypotheses from scan to scan. Hypotheses are updated with the survived tracks after pruning and are updated in each scan. Probability of each hypothesis and tracks are computed. The tracks of probability lower than a defined threshold are deleted. Two or more tracks [8] can be merged if they share consecutive measurements in the past and are closer than a defined threshold.

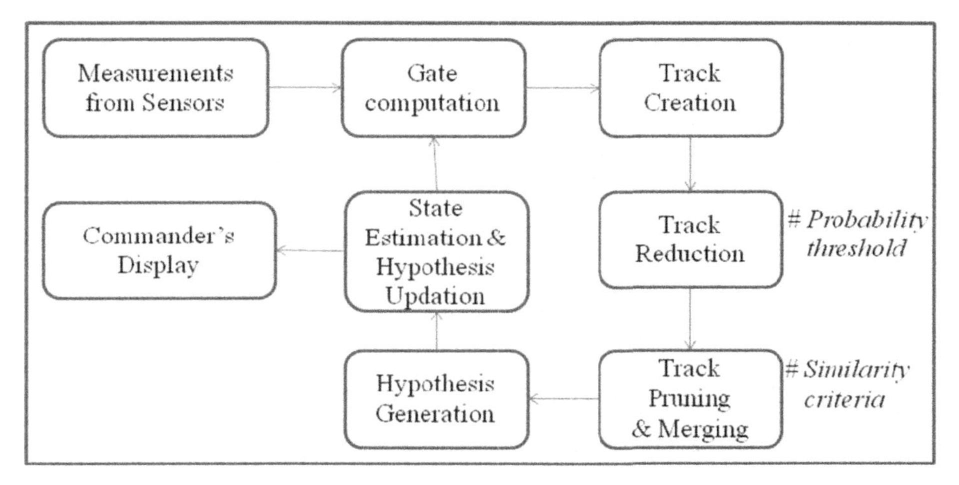

Fig. 9. Steps of TOMHT algorithm

The steps carried out in DA process by using TOMHT algorithm are explained in Fig.9, which involves measurement reception from sensors to integrated air picture generation for the commander's display in real time. TOMHT approach is implemented using multidimensional/multi frame assignment method [9, 10] which uses Lagrangian relaxation method. This makes the feasibility of real-time application of TOMHT.

5 Proposed Algorithm, Experiment and Results

The proposed algorithm of TOMHT consists of following steps:

a. Load track data comprising time, position, velocity and acceleration along three axes.
b. Load covariance data (root mean square error) associated to each track.
c. Extrapolate all the tracks to a common time.

d. Compute cost matrix by supplying extrapolated track and covariance information for all possible combinations of track and measurement pairings.
e. Supply the cost matrix to [7] Murty's assignment Algorithm to find the k-best hypotheses. (Here k = 2 for two targets)
f. Update the tracks by performing data fusion of associated information obtained as output from the previous step.
g. Continue this process from steps [c - f] until all the records over.

The possible pairings of tracks from two different sources are shown in Table 2. and Table 3. which are as follows:

Table 2. Possible combination-1

Targets	T1(R1)	T2(R1)
T3(R2)	1	0
T4(R2)	0	1

Table 3. Possible combination-2

Targets	T1(R1)	T2(R1)
T3(R2)	0	1
T4(R2)	1	0

Here one measurement of R1 can get associated with at most one measurement of R2. The cost matrix is defined as follows:

$$\text{Cost-mat} = \begin{bmatrix} C_{13} & C_{14} \\ C_{23} & C_{24} \end{bmatrix} \quad \text{Where,}$$

$$\left.\begin{aligned}
C_{13} &= \text{sqrt}\,[(X_1-X_3)^2/\,(VarX_1+VarX_3) + (Y_1-Y_3)^2/\,(VarY_1+VarY_3) + (Z_1-Z_3)^2/\,(VarZ_1+VarZ_3)] \\
C_{14} &= \text{sqrt}\,[(X_1-X_4)^2/\,(VarX_1+VarX_4) + (Y_1-Y_4)^2/\,(VarY_1+VarY_4) + (Z_1-Z_4)^2/\,(VarZ_1+VarZ_4)] \\
C_{23} &= \text{sqrt}\,[(X_2-X_3)^2/\,(VarX_2+VarX_3) + (Y_2-Y_3)^2/\,(VarY_2+VarY_3) + (Z_2-Z_3)^2/\,(VarZ_2+VarZ_3)] \\
C_{24} &= \text{sqrt}\,[(X_2-X_4)^2/\,(VarX_2+VarX_4) + (Y_2-Y_4)^2/\,(VarY_2+VarY_4) + (Z_2-Z_4)^2/\,(VarZ_2+VarZ_4)]
\end{aligned}\right\} \quad (4)$$

Here T1, T2 are two measurements from R1 and T3, T4 are two measurements from radar R2 respectively. The measurement positions are expressed as $<X_1, Y_1, Z_1>$, $<X_2, Y_2, Z_2>$, $<X_3, Y_3, Z_3>$, $<X_4, Y_4, Z_4>$ for T1, T2, T3 and T4 respectively. Similarly, the measurement variances (RMSE) values are expressed as $<VarX_1, VarY_1, VarZ_1>$, $<VarX_2, VarY_2, VarZ_2>$, $<VarX_3, VarY_3, VarZ_3>$, $<VarX_4, VarY_4, VarZ_4>$ for T1, T2, T3 and T4 respectively. The cost associated with each pair is denoted as C_{13}, C_{14}, C_{23} and C_{24} respectively.

TOMHT algorithm flowchart is explained in Fig. 10.

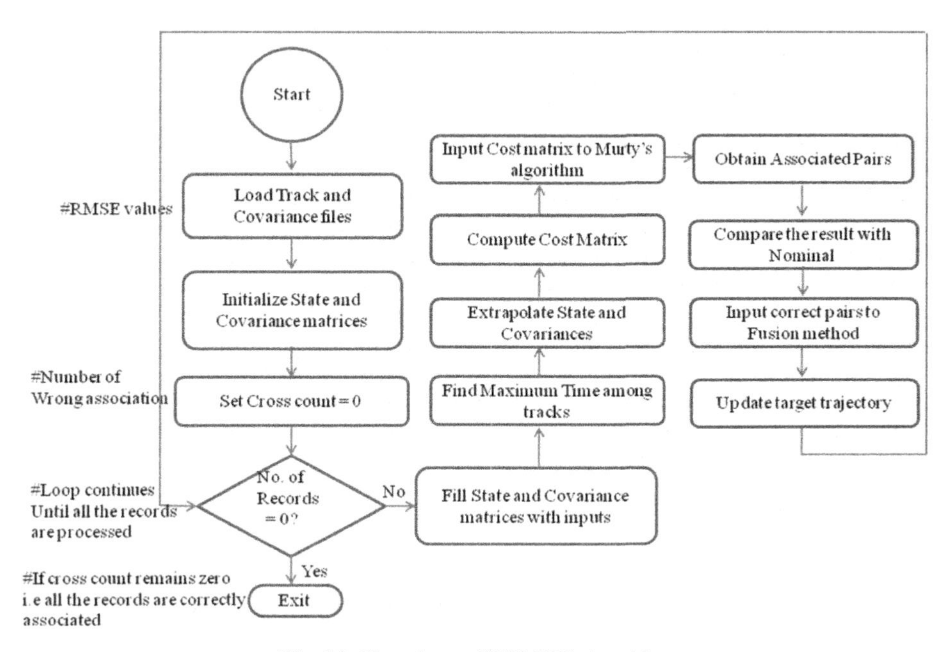

Fig. 10. Flowchart of TOMHT algorithm

The above algorithm is implemented using MATLAB for a limited number of sensors and targets. The Fig. 11 displays scenario of two radars tracking two tracks and supplying four measurements to the C2 system to perform DA.

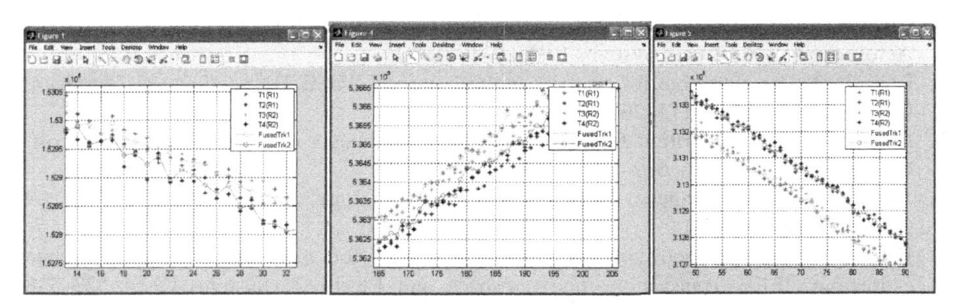

Fig. 11. TOMHT output for Two Targets and Four Measurements (along X, Y, Z directions)

These figures show the associated information as: T1 of R1 with T3 of R2 and T2 of R1 with T4 of R2. The data fusion of T1 and T3 updates Fusedtrack1 and fusion of T2 and T4 updates Fusedtrack2 respectively.

6 Conclusion

The NCW promises an agile, self healing, self-forming distributed environment focussing heterogeneous components to gain information superiority. The information is gathered with help of multiple sensors spread out geographically. These sensors generate humongous amount of data which has to be converted into information to generate a common picture for the commander to perform combat operations. This paper highlights different DA techniques with their pros and cons. MHT algorithm is explained in detail with its advantages over single hypothesis based algorithms which are suitable for simpler scenarios. Implementation and testing of TOMHT for limited number of targets is done. TOMHT algorithm supported with the use of High Performance computing will be the best method for MSMT data association contributing towards a secure and robust NCW environment. DA process concludes the supporting tracks from multiple sensors pertaining to a single target. Those supporting tracks go to data fusion process and the fused output updates the target. The resultant output of ASPG is sent to both combat control module and to the commander's display. Thus DA process will remain a crucial part of the automated C2 system and makes this area open for further research.

References

1. Honabarger, J.B.: Modeling Network Centric Warfare (NCW) With the System Effectiveness Analysis Simulation (SEAS). MS Thesis, Air Force Institute of Technology (2006)
2. Alberts, D.S., Garstka, J.J., Stein, F.P.: Network Centric Warfare: developing and leveraging information superiority. CCRP Publication, ISBN 1-57906-019-6
3. Blackman, S., Popoli, R.: Design and Analysis of Modern Tracking Systems. Artech House, Norwood (1999)
4. Bar-Shalom, Y., Li, X.-R.: Multitarget-Multisensor Tracking: Principles and Techniques. YBS Publishing, Storrs (1995)
5. Blackman, S.: Multiple Hypothesis Tracking for multiple target tracking. IEEE Trans. Aerospace and Electronic Systems 19(1), 5–18 (2003)
6. Reid, D.B.: An algorithm for tracking multiple targets. IEEE Transactions on Automatic Control 21(1), 101–104 (1976)
7. Murty, K.G.: An algorithm for ranking all the assignments in order of increasing cost. Operations Research 16, 682–687 (1968)
8. Apolinar Munoz Rodriguez, J.: Laser Scanner Technology, InTeO, ISBN: 9535102809 9789535102809 (2012)
9. Deb, S.: A generalized s-d assignment algorithm for multisensor-multitarget state estimation. IEEE Transactions on Aerospace and Electronic Systems 33(2), 523–538 (1997)
10. Poore, A.B., Robertson, A.J.: A new Lagrangian relaxation based algorithm for a class of multidimensional assignment problems. Computational Optimization and Applications 8(2), 129–150 (1997)

Eco-Friendly Horn Using Radio Frequencies

Raj Motwani[1] and Yash Arya[2]

[1] Department of IT, Thadomal Shahani Engineering College, Mumbai, Maharashtra, India
rajmotwani35@gmail.com
[2] Department of EXTC, Thadomal Shahani Engineering College, Mumbai, Maharashtra, India
yasharya1991@gmail.com

Abstract. In this paper we propose a new horn system, which if implemented in automobiles, along with the existing horn system, would reduce noise pollution caused due to the blowing of horns to a great extent. This new system does not compromise on the safety of people involved. The proposed system does not suggest the elimination of the existing horn system, rather operates in coexistence. In this paper, we will discuss the various ways of implementing the proposed system, its effectiveness and practicality.

Keywords: Noise pollution, green solutions, green technologies, pollution reduction horn systems.

1 Introduction

Today, the world is advancing rapidly, with new technologies and products developing every moment. But with this tremendous growth comes a cost that we have to bear; costs in terms of deteriorating environmental conditions. The present generation and the coming generation have to solve three grave problems- population, poverty and pollution, if they have to survive; Pollution being the most dangerous problem. Environmental pollution is assuming dangerous proportions all through the globe, and no country is free from its poisonous diseases.

Noise pollution is one of the deadly variants of pollution, and one of the major sources of noise pollution in cities is road traffic noise i.e. noise produced due to the continuous honking of automobiles [1]. To reduce this type of pollution, we propose a new mechanism of horns which can be heard only inside the transmitting automobile and the automobiles adjacent to it.

2 Proposed Technique

The proposed system would require all automobiles to have two separate horn buttons- one of them would act as the regular horn, which can be heard by all people in the surrounding, and the other will activate the proposed eco friendly horn that reduces noise pollution, which shall be referred to as –'Green Horn' from here on. This new system uses radio frequencies to communicate the horn signal between

S. Unnikrishnan, S. Surve, and D. Bhoir (Eds.): ICAC3 2013, CCIS 361, pp. 705–711, 2013.

automobiles. It involves fitting all automobiles with RF trans-receivers. These trans-receivers are to be mounted on the exterior body of the car.

Now let us have a look at the various possible scenarios, where a driver would require blowing a horn; to determine how many sensors are needed and where are they to be mounted.

The figures below depict the various scenarios where the driver of CAR-1 would require blowing a horn.

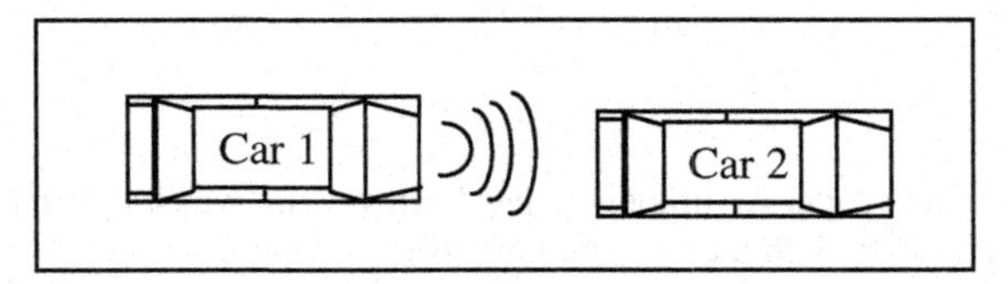

Fig. 1. Bumper to Bumper traffic

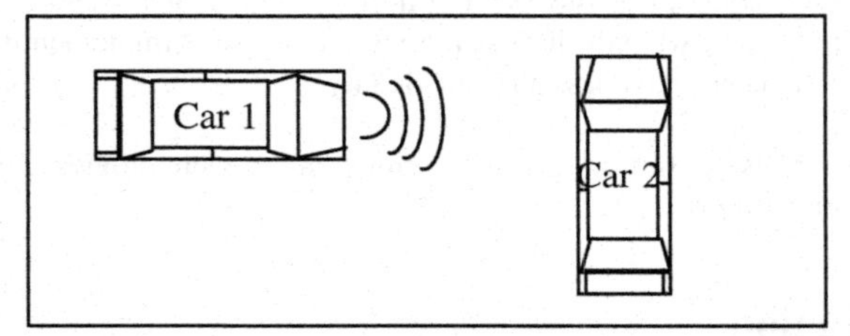

Fig. 2. Cross-junction (scene 1)

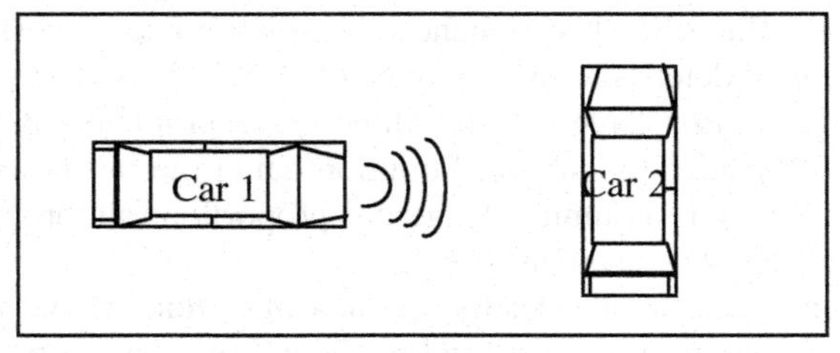

Fig. 3. Cross-junction (scene 2)

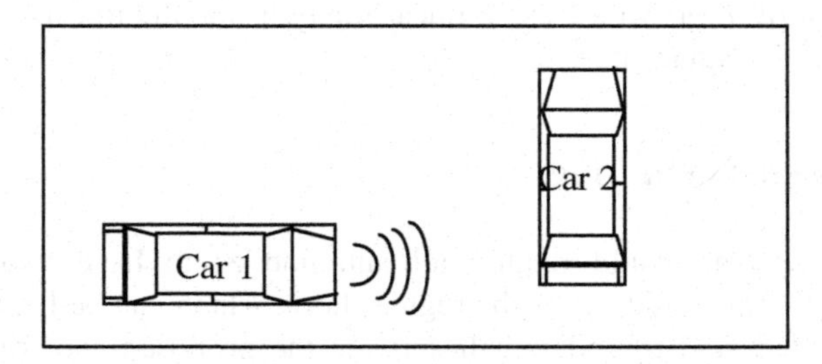

Fig. 4. Cross-junction (scene 3)

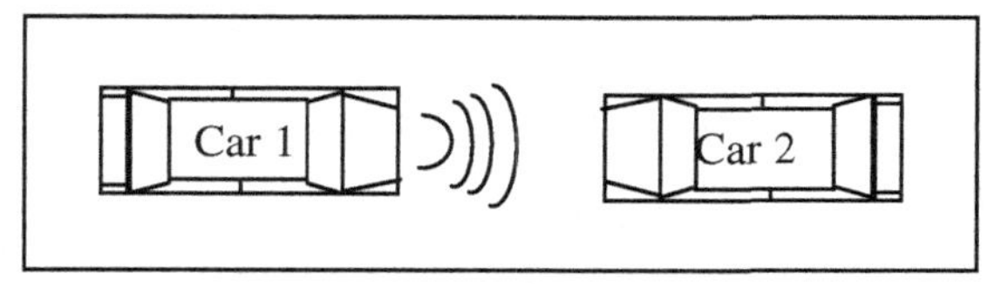

Fig. 5. Narrow Lane

The first four scenarios are regular overtaking or cross junction scenarios, while the last one occurs when a car enters the wrong side of the road or in a narrow lane.

3 Design Approach

Depending upon all these possibilities we have come up with three designs, depending upon the range and reliability of sensors. The first design consists of 1 transmitter and 6 receivers mounted as shown in the following figure.

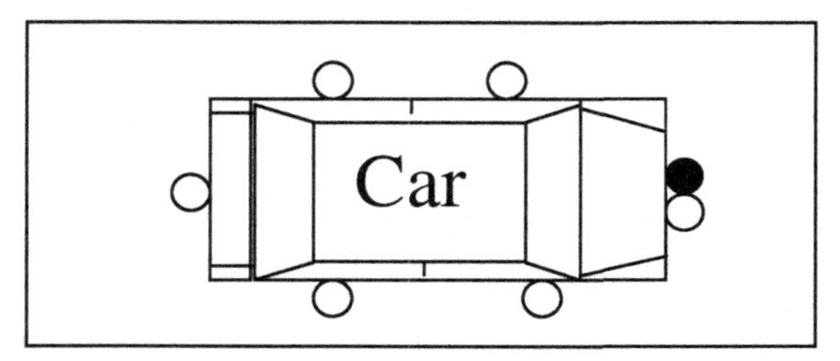

Fig. 6. First Design

The filled in circle represents the transmitter and the hollow circles represent the receivers.

Although this design satisfies all our requirements, it is not at all cost effective due to the high number of RF modules, which then led to an improvised second design.

The second design consists of 1 transmitter and 4 receivers mounted as shown in the following figure.

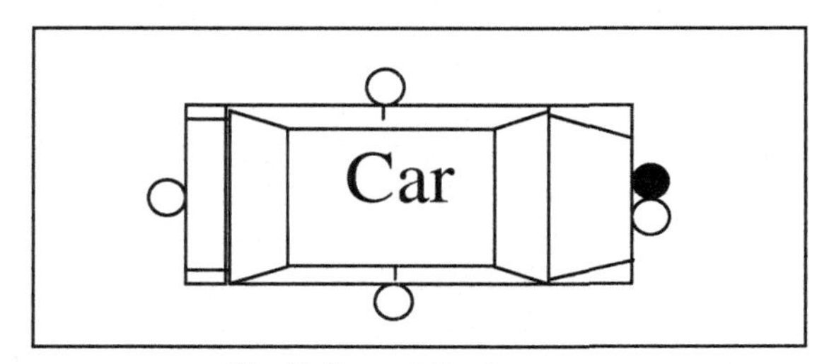

Fig. 7. Second Design

This design is more cost effective and satisfies all requirements of the system. It relies on the knowledge that radio waves have a spherical wave front [2]. But still it has a drawback, that the RF modules could be easily damaged in case of a minor crash or accident. Thus it was further improved to the third design.

The third design consists of only two trans-receivers and two aluminum plates at least 0.28 microns thick, which are mounted on the roof top of the vehicle. The thick lines represent the aluminum plates. This design reduces the possibility of physical damage to the RF modules in case of accidents. Also it requires less number of RF modules, so is less costly.

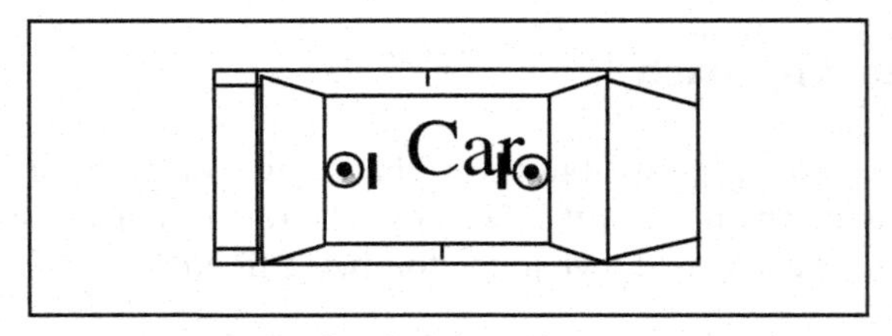

Fig. 8. Third Design

In this design, only one trans-receivers acts as a transmitter at a time, while the other operates as only a receiver. Normally, the trans-receiver mounted at the front will act as a transmitter unless, the vehicle is moving in the reverse direction, in which case the trans-receiver mounted at the rear end will act as a transmitter.

The aluminum plates are used to block radio signals from travelling in opposite direction of the desired direction.

4 Working Principle

Whenever the 'green horn' is pressed, an RF signal is transmitted by the transmitter. All receivers falling within its range will receive this signal. All receivers are connected to a special buzzer that is fitted inside the automobile to which the receiver is connected. On receiving this signal, the buzzer will blow, and thus will be heard only inside the vehicles, thus reducing noise pollution. Also, since a receiver is mounted right adjacent to the transmitter, the horn will also be heard inside the car blowing the horn. This will act as an acknowledgement to the driver, that the horn has been blown.

Using this system, all automobiles in the range of the transmitter will hear the horn. However, if we wish that the horn be transmitted only in a certain direction, then we can replace standard RF modules with directional antennas, but this would increase the cost drastically and does not give a major advantage.

Let us now have a look at the permissible noise levels given by MPCB (Maharashtra Pollution Control Board), and the prevailing noise levels in some of the major areas of Mumbai [3].

Table 1. Permissible Noise Levels by MPCB

Permissible Noise Levels (dB)			
Location	Day	Night	Contribution by Vehicular Traffic
Industrial	75	70	55
Commercial	65	55	45
Residential	55	45	35
Silence Zones	50	40	25

Table 2. Observed Noise Levels in the year 2011

Location	Highest Value (dB)	Lowest Value (dB)
Churchgate	80.6	64.5
Marine Lines	80.3	70
Charni Road	82.5	70.2
Grant Road	91.4	59.9
Mahim and Bandra	85.9	63.3
Bandra (Turner Rd.)	94.9	61
Lower Parel	104	68.8

As, we can clearly see, the prevailing noise levels are well above the permissible levels. Also it has been found that vehicular traffic alone contributes about 65dB in most of the above mentioned areas.

A detailed report is available in [3]. These levels are expected to reduce substantially once the green horn is put in use. However, the level to which it would be reduced depends upon the traffic composition in a particular area, at a particular time.

5 Steering Wheel Design

A very important concern is the usability and convenience of the system. Since, we are not replacing the traditional horn, there are now two different horns that the driver can blow. Thus it is very likely that the driver may face some difficulty or confusion when he is about to use the horn system. Also we need to make the horn system such that the driver mostly uses the green horn, as only then will we be able to use the system optimistically. Thus considering these various issues we propose the following design for the steering wheel, with the two horn buttons mounted as shown.

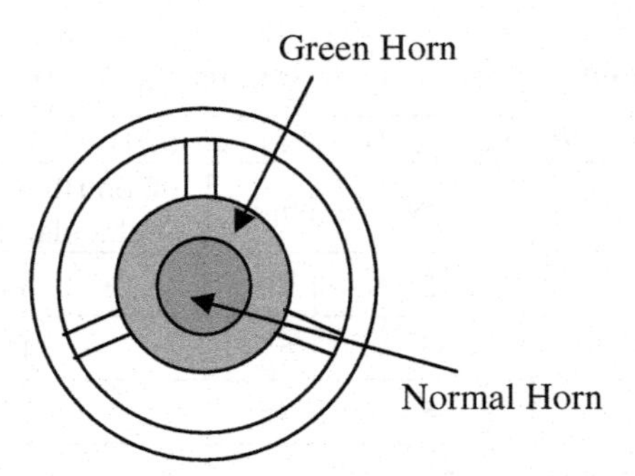

Fig. 9. Steering Wheel Design

As shown in the above figure, we propose the inner dial of the steering wheel be divided into an inner and outer region. The outer region when pressed activates the green horn, while the inner region activates the normal traditional horn. This is because it is observed, when driving, a driver tends to press the outer portion of the dial under normal conditions, and applies force at the central portion in emergency situations. Also we recommend the two regions be color coded to make the distinction clear. Also the inner/outer portion can be a little depressed with respect to the other portion. Thus by following the proposed design, we can make the decision process of the driver faster and almost involuntary, when using the horn.

6 Advantages

Using this system, the horn would only be heard by people for whom it might be intended. Thus it is not noise for them, instead it is critical information. Those for whom the horn is not intended, will not hear it, thus it is not noise to them. But since the regular horn has also been provided, which shall be used in case of emergencies and for two wheelers and pedestrians, during that time only, will there be some noise. Hence we can say that noise pollution caused due to the honking of vehicles will be reduced to a great extent.

7 Limitations

The green horn cannot be used to signal pedestrians and two wheelers. At such times, the driver will have to use the regular horn, which will cause some noise pollution.

Also, it cannot be used at blind turns or when one needs to signal a vehicle that is far off. Here again, we will have to use the regular horn.

Another problem is that of jammers. Since this new system uses radio frequencies, it is affected by jammers. To overcome this problem we can use the Direct Sequence Spread Spectrum technique with all transmitters having the same spreading code [4].

8 Practicality

The proposed system can very easily be implemented due to its following features:

8.1 Power and Size Requirement

We need a RF transmitter such that the transmitted signal covers at least a distance of 100 meters in the forward direction. Also the size of the module should be as small as possible. One can easily procure RF trans-receivers which are as small as 3cm squares and have a height of 1mm and satisfy the range and power requirements. Thus one would hardly notice these sensors.

8.2 Resistance to External Damage

It is required that the RF modules be protected in some form from external damage and exposure to liquids. This can be achieved by enclosing it in some waterproof casing that allows the RF signals to pass without introducing attenuation.

8.3 Cost Effective

The RF modules that meet the above requirements are pretty easy to procure and are economic as well [5].

9 Conclusion

Thus we conclude that the proposed system is effective in reducing noise pollution caused due to the honking of vehicles, which is a major source of noise pollution in cities, to a great extent without compromising on the safety of people. Also the system is practically implementable and its cost will be reduced when in mass production.

References

1. http://www.legalserviceindia.com/articles/noip.htm
2. Reinhold, L., Pavel, B.: RF Circuit Design – Theory and Applications. Prentice-Hall, New Jersey (2000)
3. Hindustan Times Daily Newspaper (October 24, 2011)
4. Behrouz, F.: Data Communications and Networking, 4th edn. Tata McGraw-Hill, New Delhi (2006)
5. http://www.robosoftsystems.co.in/roboshop/media/catalog/product/pdf/RFpro%20Module%20-%20how%20to%20use%20v4%20.pdf

Optimization of Sub-threshold Slope in Submicron MOSFET's for Switching Applications

Akhil Ulhas Masurkar and Swapnali Mahadik

Electronics Engineering, Fr.Conceicao Rodrigues College of Engineering,
Mumbai – 400050, India
swapnali@frcrce.ac.in, akhilmasurkar@gmail.com

Abstract. Scaling transistors into the nano-meter regime has resulted in a dramatic increase in MOS leakage current. Threshold voltages of transistors have scaled to maintain performance at reduced power supply voltages. Increased transistor leakages not only impact the overall power consumed by a CMOS system, but also reduce the margins available for design due to the strong relationship between process variation and leakage power. Therefore it is essential for circuit and system designers to understand the components of leakage, sensitivity of leakage to different design parameters, and leakage mitigation techniques by designing accurate models of short channel devices and simulating the same using the available standard CAD tools. This paper provides ways to optimize the sub-threshold slope in submicron devices which is an important parameter that defines transistor efficiency in switching applications using standard CAD tools like SILVACO and SPICE.

Keywords: Sub-threshold, slope, Model, SILVACO.

1 Introduction

Historically, VLSI designers have used circuit speed as the "performance" metric. Large gains, in terms of performance and silicon area, have been made for digital processors, microprocessors, DSPs (Digital Signal Processors), ASIC's (Application Specific IC's), etc. In general, "small area" and "high performance" are two conflicting constraints. The IC designer's activities have been involved in trading off these constraints. Power dissipation issue was not a design criterion but an afterthought [1][2]. In fact, power considerations have been the ultimate design criteria in special portable applications such as wristwatches and pacemakers for a long time. The objective in these applications was minimum power for maximum battery life time. Recently power dissipation is becoming an important constraint in a design [3]. Several reasons underlie the emerging of this issue.

Battery-powered systems such *as la*ptop, notebook computers, electronic organizers, etc, need an extended battery life. Many portable electronics use the rechargeable Nickel Cadmium (NiCd) batteries. Although the battery industry has been making efforts to develop batteries with higher energy capacity than that of NiCd, a strident increase does not seem imminent. The expected improvement of the energy density was 40% by the turn of the century [3][5]. With recent NiCd batteries, the energy

S. Unnikrishnan, S. Surve, and D. Bhoir (Eds.): ICAC3 2013, CCIS 361, pp. 712–720, 2013.

density is around 20 Watt-hour/ pound and the voltage is around 1.2 V. This means, for a notebook consuming a typical power of 10 Watts and using 1.5 pound of batteries, the time of operation between recharges is 3 hours. Even with the advanced battery technologies such as Nickel-Metal Hydride (Ni-MH) which provide large energy density characteristics (30 Watt-hour/pound), the life time of the battery is still low[4][6]. Since battery technology has offered a limited improvement. Low-power design techniques are essential for portable devices.

The key challenge in reducing the leakage power is the Sub-threshold current. This can be reduced by doping the substrate and channel implants with acceptor impurity (P type) or donor impurity (N type) in a particular proportion. Optimizing these proportions to a value independent of the technology under consideration will lead to reduction in the sub-threshold current, with a steep roll- off which is desirable in any transistor technology.

2 Model of the MOSFET

The figure below represents the equivalent circuit structure of a N-channel MOSFET. The MOSFET is voltage controlled current device.

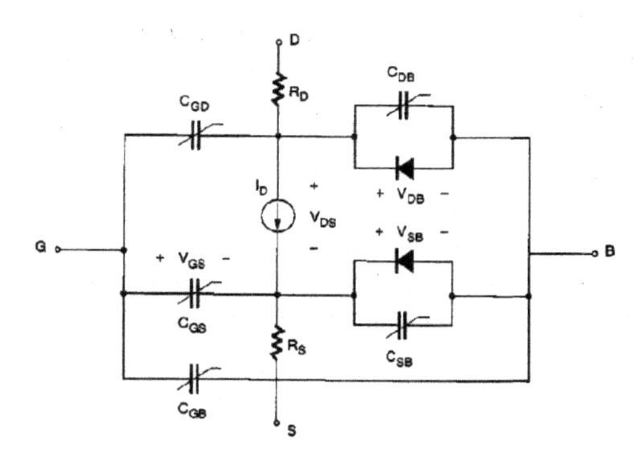

Fig. 1. Equivalent MOS Model

Note that the voltage-controlled current source I_D determines the steady-state current-voltage behaviour of the device, while the voltage controlled (nonlinear) capacitors connected between the terminals represent the parasitic oxide-related and junction capacitances. The source-substrate and the drain-substrate junctions, which are reverse-biased under normal operating conditions, are represented by ideal diodes in this equivalent circuit. Finally, the parasitic source and drain resistances are represented by the resistors R_D and R_S, respectively, connected between the drain current source and the respective terminals.

The basic geometry of an MOS transistor can be described by specifying the nominal channel (gate) length L and the channel width W both of which are indicated on the element description line. The channel width W is, by definition, the *width* of the area covered by the thin gate oxide. Note that the effective channel length L_{eff} *is* defined as the distance on the surface between the two (source and drain) diffusion regions. Thus, in order to find the effective channel length, the gate-source overlap distance and the gate-drain overlap distance must be subtracted from the nominal (mask) gate length specified on the device description line. The amount of gate overlap over the source and the drain can be specified by using the *lateral diffusion* coefficient L_D.

For modelling p-channel MOS transistors, the direction of the dependent current source, the polarities of the terminal voltages, and the directions of the two diodes representing the source-substrate and the drain-substrate junctions must be reversed. Otherwise, the equations to be presented in the following sections apply to p-channel MOSFETs as well.

The basic model implemented in SPICE calculates the drift current in the channel when the surface potential is equal to or larger than $2\Phi_F$ that is in strong surface inversion. In reality, a significant concentration of electrons exists near the surface for $V_{GS} < V_T$, therefore, there is a channel current even when the surface is not in strong inversion. This is known as *the sub-threshold current* and is mainly due to diffusion between the source and the channel.

3 Sub-threshold Dynamics in Submicron Devices

A closer inspection of the I_D-V_{GS} curves of Fig (2) reveals that the current does not drop abruptly to 0 at $V_{GS} = V_T$. It becomes apparent that the MOS transistor is already partially conducting for voltages below the threshold voltage. This effect is called *sub-threshold* or *weak-inversion* conduction [15][16][17]. The onset of strong inversion means that ample carriers are available for conduction, but by no means implies that no current will flow for gate-source voltages below V_T, although the current levels are small under those conditions. The transition from the ON to the OFF condition is thus not abrupt, but gradual.

When the $I_D$versus V_{GS} curve is drawn on a logarithmic scale as shown in Fig 3, confirms that the current does not drop to zero immediately for $V_{GS}< V_T$ but actually decays in an exponential fashion, similar to the operation of a bipolar transistor.

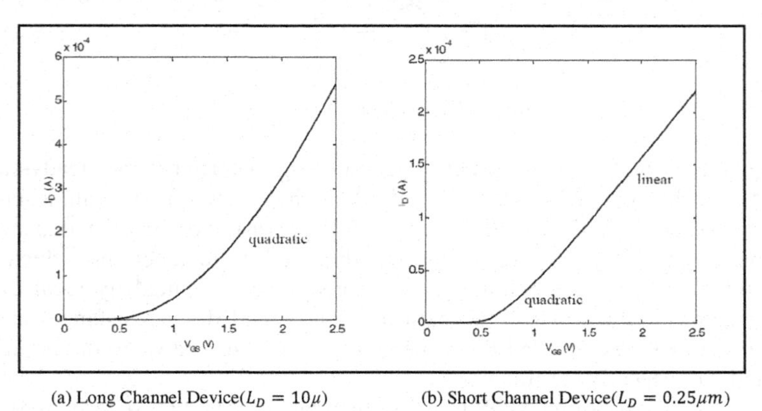

(a) Long Channel Device($L_D = 10\mu$) (b) Short Channel Device($L_D = 0.25\mu m$)

Fig. 2. NMOS transistor I_D-V_{GS} characteristic for long and short-channel devices $W/L = 1.5$ for both transistors and $V_{DS} = 2.5$ V

In the absence of a conducting channel, the *n*+ (source) - *p* (bulk) - *n*+ (drain) terminals actually form a parasitic bipolar transistor[11][12]. The current in this region can be approximated by the expression:

$$I_D = I_S e^{\frac{V_{GS}}{nKT/q}} \left(1 - e^{-\frac{V_{DS}}{KT/q}} \right) \qquad (1)$$

Where I_S and n are empirical parameters, with $n \geq 1$ and typically ranging around 1.5.

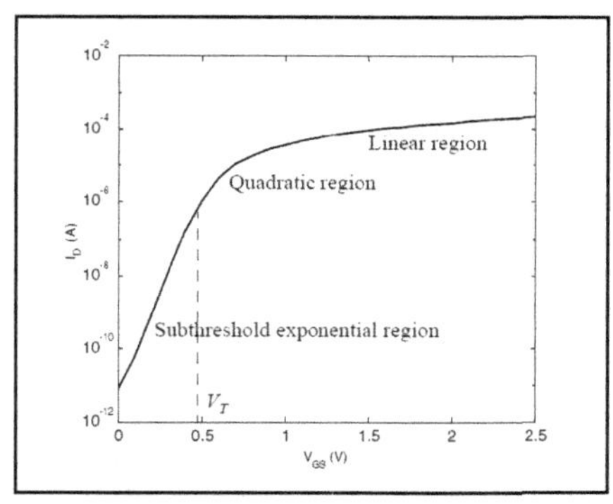

Fig. 3. I_D versus V_{GS} (on logarithmic scale), showing the exponential characteristic of the sub-threshold region

In most digital applications, the presence of sub threshold current is undesirable as it detracts from the ideal switch-like behaviour that we like to assume for the MOS transistor. We would rather have the current drop as fast as possible once the gate-source voltage falls belowV_T. The (inverse) rate of decline of the current with respect to V_{GS} below V_T hence is a quality measure of a device. It is often quantified by the *slope factor S,* which measures by how much V_{GS} has to be reduced for the drain current to drop by a factor of 10. From (1) we find

$$S = n\left(\frac{kT}{q}\right)\ln(10) \qquad (2)$$

$$\text{Where, } n = 1 + \frac{\gamma}{2\sqrt{(Vp+2\Phi f)}} \qquad (3)$$

Where '*S*' is expressed in mV/decade. V_p is the pinch off voltage. For an ideal transistor with the sharpest possible roll-off, n=1 and $(kT/q)\ln(10)$ evaluates to 60 mV/decade at room temperature, which means that the sub-threshold current drops by a factor of 10 for a reduction in V_{GS} of 60 mV.

Unfortunately, n is larger than 1 for actual devices and the current falls at a reduced rate (90 mV/decade for n = 1.5). The current roll-off is further affected in a negative sense by an increase in the operating temperature (most integrated circuits operate at temperatures considerably beyond room temperature). The value of n is determined by the intrinsic device topology and structure [10][13]. Reducing its value hence requires a different process technology, such as silicon-on-insulator [6].

Sub-threshold current has some important repercussions. In general, we want the current through the transistor to be as close as possible to zero at V_{GS} = 0. This is especially important in *dynamic circuits*, which rely on the storage of charge on a capacitor and whose operation can be severely degraded by sub-threshold leakage. Achieving this in the presence of sub-threshold current requires a firm lower bound on the value of the threshold voltage of the device.

4 Proposed Optimization Technique

Recent studies in the VLSI design area have proposed that the sub-threshold current can be reduced by

a. Changing the size of the drain and source regions.
b. Changing the doping concentrations of the source and drain implants depending on whether the MOS under fabrication is a PMOS or an NMOS.
c. Changing the doping concentrations of the Substrate used for fabrication.
d. Varying the Channel Length.

The following observations were made for a change in drain current in the sub-threshold region with variation in the doping concentrations of the P-substrate for 200nm length NMOS transistors in SILVACO.

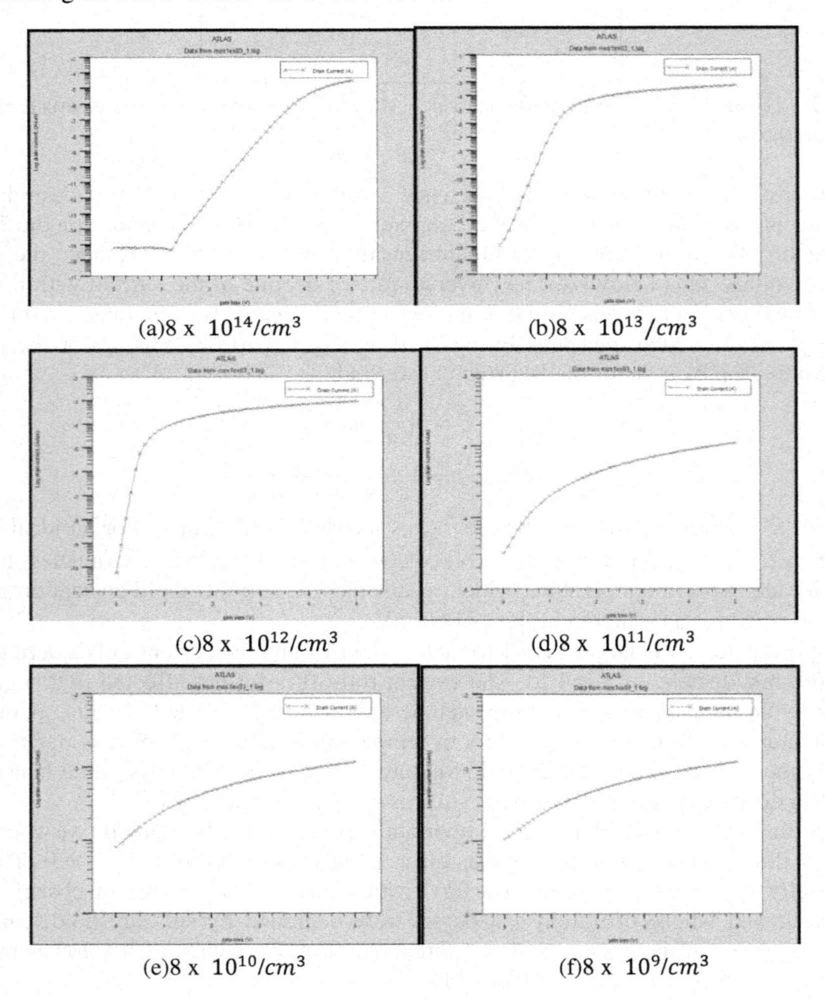

(a)8 x $10^{14}/cm^3$ (b)8 x $10^{13}/cm^3$

(c)8 x $10^{12}/cm^3$ (d)8 x $10^{11}/cm^3$

(e)8 x $10^{10}/cm^3$ (f)8 x $10^{9}/cm^3$

Fig. 4.

5 Observations

Channel Implant Doping Concentration	P-substrate Doping Concentrations	Sub-threshold Slope mv/decade	Threshold Voltage
3 x $10^{12}/cm^3$	8 x $10^{14}/cm^3$	30	4.483
3 x $10^{12}/cm^3$	8 x $10^{13}/cm^3$	60	1.41201
3 x $10^{12}/cm^3$	8 x $10^{12}/cm^3$	125	0.2457
3 x $10^{12}/cm^3$	8 x $10^{11}/cm^3$	NA	0
3 x $10^{12}/cm^3$	8 x $10^{10}/cm^3$	NA	0
3 x $10^{12}/cm^3$	8 x $10^{9}/cm^3$	NA	0

6 Results

From the above results in Fig 4 it can be noted that, when the doping concentration was set to a value of 8 x $10^{13}/cm^3$ at V_{GS}=0V, the off state current was equal to 1 x $10^{-15}\,A$. But the roll off slope is about 60mV/decade. The threshold voltage observed was 1V. To mitigate this effect the P-substrate doping concentration was reduced by 10 times and controlled to 8 x $10^{12}/cm^3$. The n+ implant doping concentration was controlled to 3 x $10^{12}/cm^3$. It was observed that the drain current at V_{GS}=0 increased to 1 x $10^{-12}A$, but the slope of the roll off increased to 125 mV/decade. The threshold voltage observed was 0.2457V. This value of the threshold voltage is 9.45 times the thermal equivalent value of temperature which is 26mV at 300 Kelvin and 8 times the thermal equivalent of temperature which is 31mV at 373 Kelvin. This value of the threshold voltage for the steepest sub-threshold roll-off is well with the prescribed limits which is 7.5 times the thermal equivalent of the ambient temperature.

From the results below it can also be observed that when the P-substrate and the n+ implant concentrations was set to the above values it was also observed that the gate to drain and gate to source overlaps are minimum (4nm), thus reducing the values of gate overlap parasitic capacitances. Also the shape of the source and drain implants have a shape close to a rectangular contour.

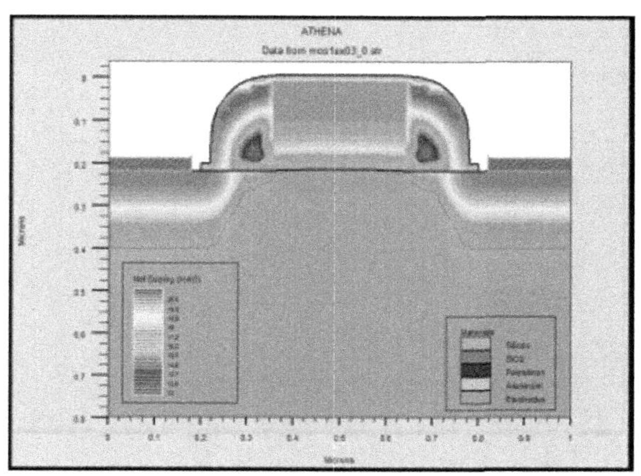

(5.a)Doping Profile

Fig. 5. Optimized Results for minimum sub-threshold leakage

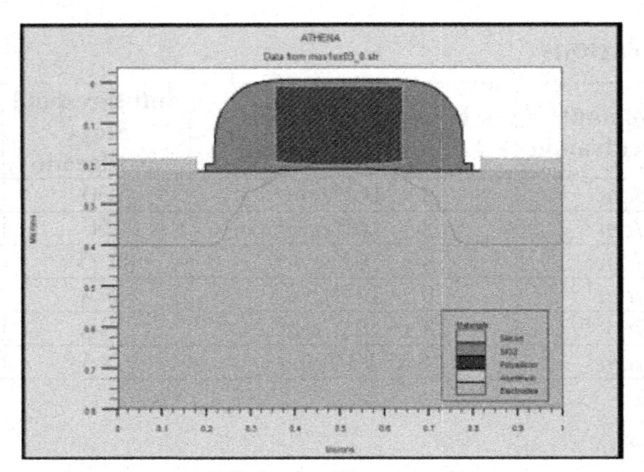

(5.b) Device Geometry

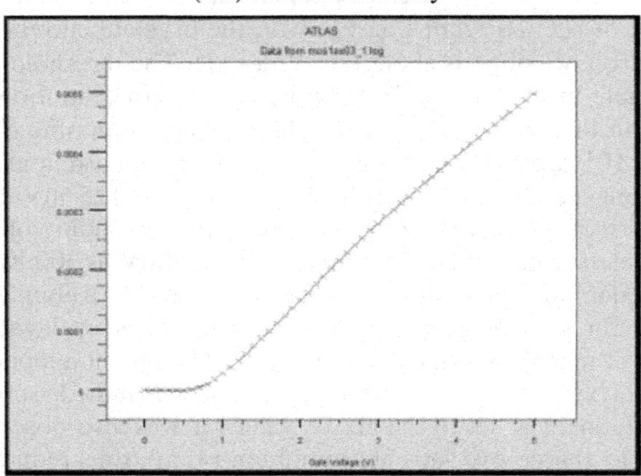

(5.c) I_D V/s V_{GS} Curve

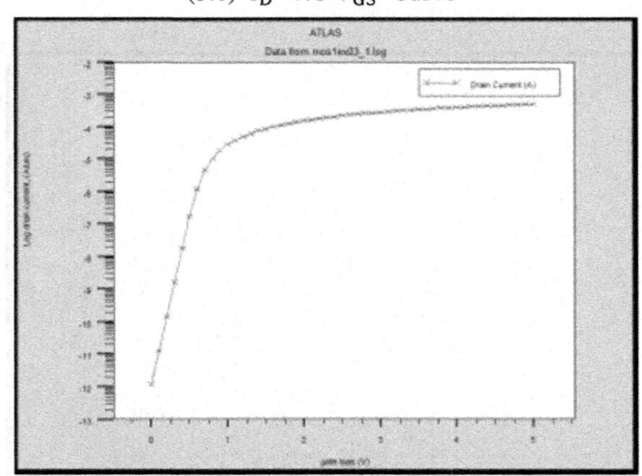

(5.d) I_D V/s V_{GS} Curve (Logarithmic Scale)

Fig. 5. (*Continued*)

7 Conclusion

Since majority of the transistors in a digital circuit are designed at the minimum channel length, the variation of the threshold voltage as a function of the length is almost uniform over the complete design, and is therefore not much of an issue except for the increased sub-threshold leakage currents. More troublesome is the sub-threshold current, as it varies with the operating voltage. This is, for instance, a problem in dynamic memories, where the leakage current of a cell (being the sub-threshold current of the access transistor) becomes a function of the voltage on the data-line, which depends upon the applied data patterns.

From the above results it can be observed that sub-threshold currents can be reduced to a minimum by keeping the P-substrate to N+ implants doping ratio to (8:3). By maintaining this doping ratio the threshold voltage is well within the design limits while still achieving low sub-threshold leakage currents and sharp roll off in the drain current below the value of the threshold voltage. It can also be observed that the gate overlaps are minimal which may have a reducing effect on the gate parasitic. Due to this there is a reduction in the rise time, the fall times and the dynamic power dissipation when the transistor is used as a switch. Hence the transistor may be used for low voltage as well as high speed applications.

Acknowledgment. I thank Prof. Swapnali Mahadik, and Prof. Sapna Prabhu(Associate Professor) with deep gratitude for their valuable guidance and encouragement which helped me throughout this venture. They suggested the necessary modifications which has greatly improved the quality and appearance of the text. I am also thankful to Dr. Deepak V. Bhoir, Head of Electronics Engineering Department and Dr. Srija Unnikrishnan, Principal who provided the necessary tools and help from time to time as required.

References

1. Moore, G.E.: Cramming more components onto Integrated circuits. Electronics 38(8) (April 19, 1965)
2. Piguet, C.: Are early computer architectures a source of ideas for low-power?. Invited Paper, Volta 1999, Como, Italy, March 4-5 (1999)
3. Piguet, C.: Histoire des ordinateurs. Invited Paper at FTFC 1999, Paris, May 26-28, pp. 7–16 (1999)
4. Baccarani, G., Wordeman, M., Dennard, R.: Generalized Scaling Theory and Its Application to 1/4 Micrometer MOSFET Design. IEEE Trans. Electron Devices ED-31(4), 452 (1984)
5. Meindl, P.J., Swanson, R.: Design of Ion-Implanted MOSFETS with Very Small Physical Dimensions. IEEE Journal of Solid-State Circuits SC-9, 256–258 (1972)
6. Ko, P.: Approaches to Scaling. In: VLSI Electronics: Microstructure Science, ch. 1, vol. 18, pp. 1–37. Academic Press (1989)
7. Nagel, L.: SPICE2: a Computer Program to Simulate Semiconductor Circuits, MemoERL-M520, Dept. Elect. and Computer Science, University of California at Berkeley (1975)
8. Thorpe, T.: Computerized Circuit Analysis with SPICE. John Wiley and Sons (1992)
9. Massobrio, G., Vladimirescu, P.: The SPICE Book. John Wiley and Sons (1993)

10. Antognetti, Semiconductor Device Modelling with SPICE, 2nd edn. McGraw-Hill, New York (1993)
11. Meyer, J.E.: MOS models and circuit simulation. RCA Review 32, 42–63 (1971)
12. Jeng, M.C., Lee, P.M., Kuo, M.M., Ko, P.K., Hu, C.: Theory, Algorithms, and User's Guide for BSIM and SCALP, Electronic Research Laboratory Memorandum, UCB/ERL M87/35. University of California, Berkeley (2008)
13. Sheu, B.J., Scharfetter, D.L., Ko, P.K., Jeng, M.C.: BSIM, Berkeley short-channel IGFET model. IEEE Journal of Solid-State Circuits SC-22, 558–566 (2005)
14. Yang, P., Chatterjee, P.K.: SPICE modelling for small geometry MOSFET circuits. IEEE Transactions on Computer-Aided Design CAD-1(4), 169–182 (1999)
15. Lee, S.-W., Rennick, R.C.: A compact IGFET model-ASIM. IEEE Transactions on Computer-Aided Design 7(9), 952–975 (2001)
16. Lee, K., Shur, M., Fjeldly, T.A., Ytterdal, Y.: Semiconductor Device Modelling for VLSI. Prentice-Hall, Inc., Englewood Cliffs (2009)
17. Cheng, Y., Chan, M., Hui, K., Jeng, M., Liu, Z., Huang, J., Chen, K., Tu, R., Ko, P.K., Hu, C.: BS1M3v3 Manual. Department of Electrical Engineering and Computer Science. University of California, Berkeley (2010)

Co-design Approach for Implementation of Decryption Block of Rijndael's Algorithm Using Soft Core Processor

Pradnya G. Ramteke[1], Meghana Hasamnis[1], and S.S. Limaye[2]

[1] Department of Electronics Engg.
Shri Ramdeobaba College of Engg. and Management
Nagpur, India
{pradnyag_ramteke,meghanahasamnis}@rediffmail.com
[2] Jhulelal Institute of Technology
Nagpur, India
shyam_limaye@hotmail.com

Abstract. The design process of embedded systems has changed in recent years. Because of increasing size of integrated circuits, increasing software complexity and decreasing time-to-market requirements and product costs, designing embedded systems becomes more and more complex. Such approaches combine modelling of software and ardware, and are called Co-Design approaches. Now-a–days systems are designed with some dedicated hardware unit and software unit on the same chip called SoC design and is the motivating factor for hardware/software co-design. Hardware/Software co-design improves the performance of the system.

Rijndael algorithm is a combination of encryption and decryption structure. In this paper decryption structure is designed and interfaced with NIOS II system. Using SOPC builder tool the NIOS II system is generated. Programming is done in C and NIOS II IDE is used to integrate the system. Decryption is also implemented separately as an accelerator and with different hardware/software partitions to improve the performance in terms of speed and area. By incorporating the co-design approach an optimized design for decryption is obtained.

Keywords: Decryption structure of Rijndael algorithm, NIOS II Processor, QUARTUS II, SOPC Builder, NIOS II IDE.

1 Introduction

In embedded systems, hardware and software component works together to perform a specific function. Approaches combine modelling of software and hardware, and are called Co-Design approaches. System-level objectives can be achieved by exploiting the trade-offs between hardware and software in a system through their concurrent design which is called Co-Design where hardware and software developed at the same time on parallel paths [1]. Co-design concept is advantageous in design flexibility to create systems that can meet performance requirements with a shorter design cycle. It Improves design quality, design cycle time, and cost.

S. Unnikrishnan, S. Surve, and D. Bhoir (Eds.): ICAC3 2013, CCIS 361, pp. 721–729, 2013.

Basic features of a co-design process are [2],

- It enables mutual influence of both HW and SW early in the design cycle
- Provides continual verification throughout the design cycle
- Separate HW/SW development paths can lead to costly modifications
- Enables evaluation of larger design space through tool interoperability and automation of co-design at abstract design levels
- Advances in key enabling technologies (e.g., logic synthesis and formal methods) make it easier to explore design trade offs

Important part of the co-design process is partitioning the system into dedicated hardware components and software components. Partitioning requires the use of performance to assist the petitioner in choosing from among several alternative hardware and software solutions. Partitioning is the process of deciding, for each subsystem, whether the required functionality is more advantageously implemented in hardware or software. There are two basic approaches that most designers use when performing partitioning. They either start with all operations in software or move some into hardware (when speed is critical) or they start with all operations in hardware and move some into software [10]. Different design environments support one or the other. Obviously designing hardware and software for a system in an integral manner is an extremely complex task with many aspects. There is a wide range of architectures and design methods, so the hardware/software Co-Design problem is treated in many different ways. Following Fig.1.shows general flow of co-design.

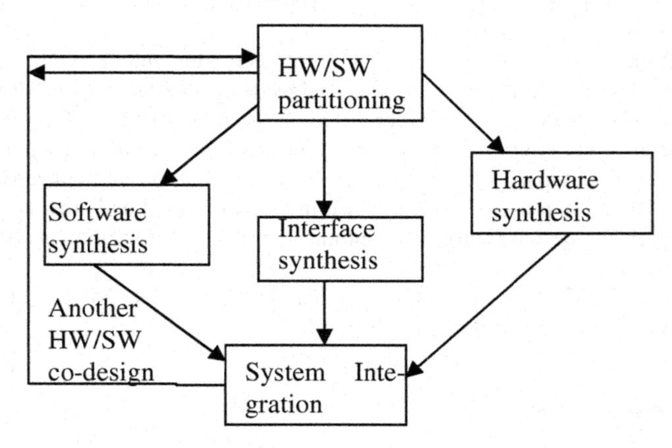

Fig. 1. General Flow of Co-design

Rijndeal's Algorithm

Rijndeal's algorithm is a cryptographic algorithm. Cryptography plays an important role in the security of data. It enables us to store sensitive information or transmit it across insecure networks so that unauthorized persons cannot read it. The algorithm originates from the initiative of the National Institute of Standards and Technology

(NIST) in 1997 to select a new symmetric key encryption algorithm. Rijndael algorithm was selected as the Advanced Encryption Standard (AES) [2, 3] due to the combination of security, performance, efficiency, ease of implementation and flexibility. The algorithm is composed of three main parts: Cipher, Inverse Cipher and Key Expansion. Cipher converts data to an unintelligible form called cipher text while Inverse Cipher converts data back into its original form called plaintext. Key Expansion generates a Key Schedule that is used in Cipher and Inverse Cipher procedure. Cipher and Inverse Cipher are composed of specific number of rounds. For the Rijndael algorithm, the number of rounds to be performed during the execution of the algorithm is dependent on the key length. Fig.2. gives structure of Rijndeal Algorithm [13].

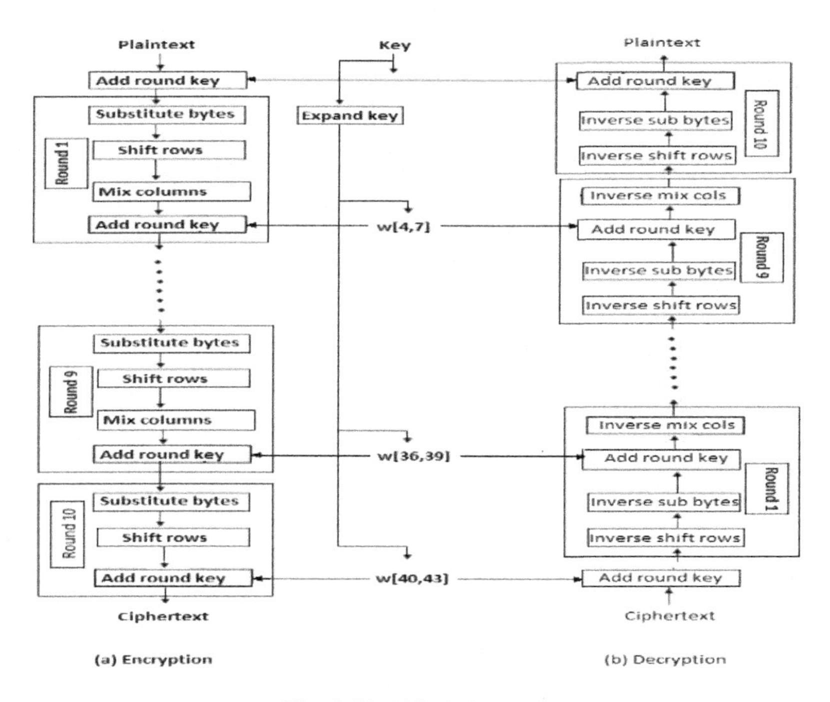

Fig. 2. The Rijndeal Algorithm

In the decryption of Rijndael algorithm, each round consists of four transformations which are listed below whereas final round doesn't include InvMixcolumns transformation.

AddRoundKey: The AddRoundKey transformation performs a simple bitwise XOR of the current state with a portion of expanded key.

InvSubBytes: The InvSubbytes transformation is a non-linear byte substitution that operates independently on each byte of the state. Each byte in the state is substituted by the corresponding byte in the substitution table (Inverse S-box).

InvShiftRows: The first row of State array is not altered. For the second row, a 1-byte circular right shift is performed. For the third row, a 2-byte circular right shift is performed. For the third row, a 3-byte circular right shift is performed.

4. InvMixcolumns: The InvMixcolumns transformation takes all of the columns of the state and mixes their data (independently of one another) to produce new columns [4].

2 Design Environment

The design environment consists of NIOS II processor and it is developed around NIOS II IDE, Cyclone II FPGA.

2.1 NIOS II Processor

The NIOS II processor is a general-purpose RISC processor core. A NIOS II processor system is equivalent to a microcontroller or "computer on a chip" that includes a processor and a combination of peripherals and memory on a single chip. The NIOS II processor is a configurable soft IP core, as opposed to a fixed, off-the-shelf microcontroller. NIOS II is offered in 3 different configurations: NIOS II/f (fast), NIOS II/s (standard), and NIOS II/e (economy) [5].

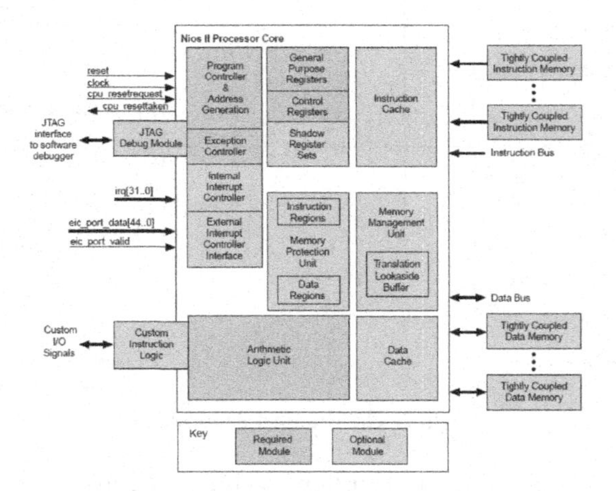

Fig. 3. NIOS II Core block diagram

2.2 NIOS II System

The NIOS II processor can be used with a variety of other components to form a complete system. These components include a number of standard peripherals, but it is also possible to define custom peripherals. Altera's DE2 Development and Education board contains several components that can be integrated into a NIOS II system.

Interfacing the partitioned hardware blocks of the algorithm as a custom hardware with NIOS II Processor is implemented through GPIO. The system is designed using the SOPC builder tool. Along with the processor the peripheral components used are SRAM, SRAM Controller, JTAG UART, Performance Counter, and GPIO's.

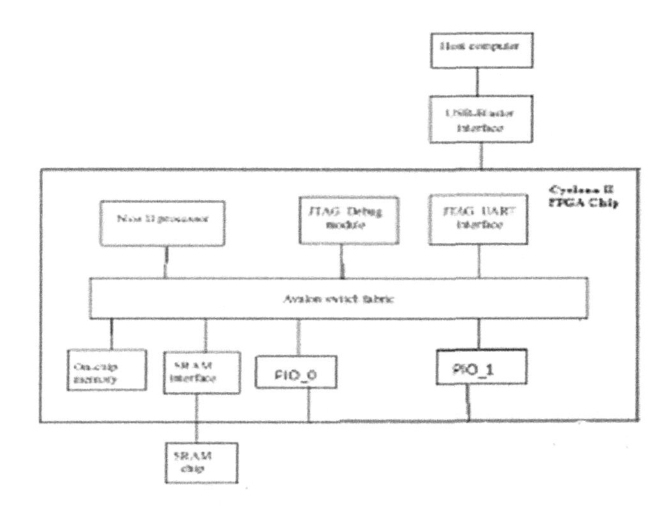

Fig. 4. NIOS II system implemented on the DE2 board

3 System Implementation

To have proper analysis, system implementation is done in three different ways, Hardware, Software and Co-Design Implementation. For the FPGA implementation we have to follow steps as,

1. Analyze system requirements.
2. Generate system in SOPC builder. SOPC builder is a tool provided in Quartus II environment [8]. Through the GPIO connect the custom hardware block or blocks which are partitioned using the co-design methodology in hardware with the system generated using SOPC.
3. Integrate SOPC builder system in Quartus II environment.
4. Compile hardware design for target board and download in FPGA.
5. Develop software with NIOS II IDE [10].
6. Build and run the software in NIOS II IDE.
7. Download software executable to NIOS II system on target board.

3.1 Software Implementation

For software implementation of AES algorithm on NIOS II processor, the algorithm is written in C language. The program written for decryption gets compiled by using

NIOS II IDE. After compilation of program, an executable file will be created, which is executed on NIOS II processor. Fig.5. shows the complete NIOS II system which is implemented on FPGA platform.

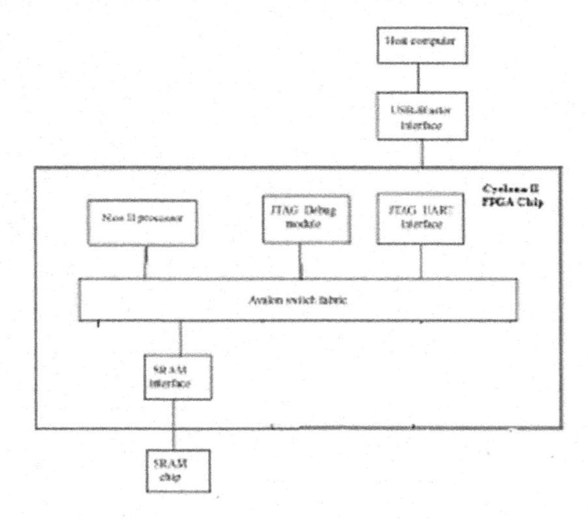

Fig. 5. NIOS II system for software implementation

3.2 Hardware Implementation

The system is designed around NIOS II Processor. Decryption is written in VHDL and connected as a custom hardware with the system generated using NIOS II processor. Control of the algorithm is kept in software and is written in C language.

NIOS II is a 32 bit processor and inputs, outputs of decryption are of 128 bits. General purpose registers are being used to send and receive the bits. So instead of giving 128 bits of input, at a time 16 bits of input is sending to the input of Decryption VHDL block. After collecting 128 bits of input and 128 bits of key, hardware will start to operate. Thus 16 bits of inputs are applied through DEMUX. Similarly, output is read from hardware block through MUX, as shown below.

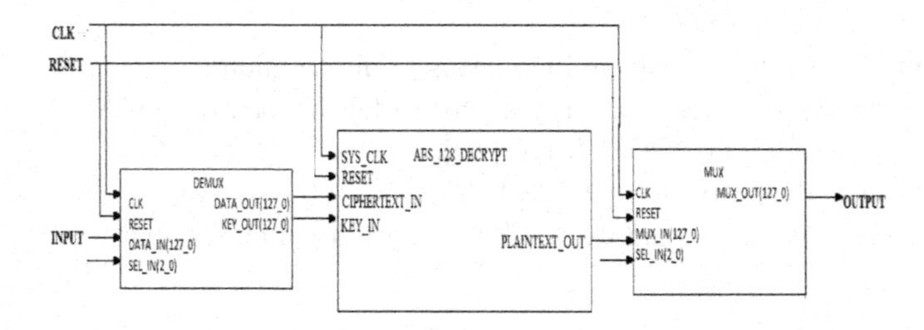

Fig. 6. Block Diagram of Decryption using MUX-DEMUX

3.3 HW/SW Co-design

Every block of the decryption (AddRoundkey , InvSubbytes, InvShiftRows, InvMix-column) is tested in terms for its number of clock cycles required and area in terms of number of Logic elements (LE's) on FPGA

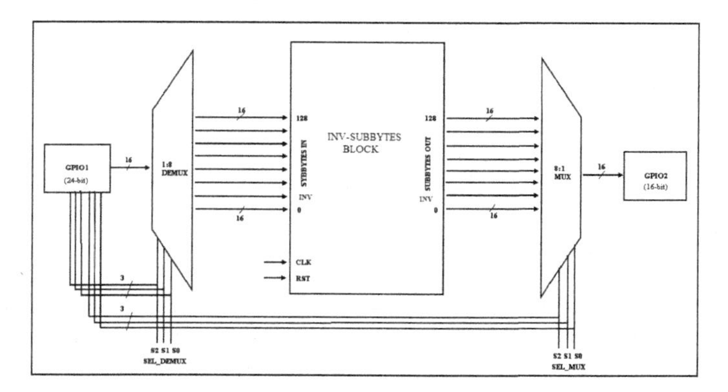

Fig. 7. InvSubbytes interfacing

4 Results

After integrating decryption with NIOS II Processor, results are compared with software. The plain text of 128 bits is given as the input along with the cipher key of 128 bits. Once the decryption algorithm is implemented the output plain text of 128 bits is obtained. This decrypted data is send to the PC through JTAG URAT to analyze the results. The system is designed around 50MHz clock. Cyclone II FPGA KIT EP2C35F672C6 is used.

Table 1. Input and output of Decryption

Input	Output
Cipher text	3925841D02DC09FBDC118597196A0B32
Cipher Key	2B7E151628AED2A6ABF7158809CF4F3C
Decrypted Data	3243F6A8885A308D313198A2E0370734

Table 2 shows the number of CPU clock cycles required for four functions of Decryption block, when these functions are separately implemented in software on NIOS II processor.

Table 2. CPU clock cycles required

AES algorithm	CPU clock cycles
	Software
AddRoundKey	2554
InvSubbytes	63912
InvShiftrows	383
InvMixcolumns	17421

From Table 2 it can be observed that, function InvSubbytes required more number of CPU clock cycles in software as compared to AddRoundKey, InvShiftrows and InvMixcolumn functions. Number of CPU clock cycles gets reduced when function InvSubbytes is put in hardware.

Table 3. CPU clock cycles required and Area for Decryption

AES algorithm		Total CPU clock cycles required for decryption	Area (Logic elements)
Hardware	Software		
_	Decryption	826235	2,515 / 33,216 (8 %)
Decryption as custom hardware	_	7520	18,196 / 33,216 (55 %)
AddRoundKey	InvSubbytes,InvShiftrows, InvMixcolumns	883617	3,092 / 33,216 (9 %)
InvSubbytes	AddRoundKey, InvShif-trows, Inv Mixcolumns	220287	2,480 / 33,216 (9 %)
InvShiftrows	AddRoundKey, InvSubbytes, InvMixcolumns	858675	2,893 / 33,216 (9 %)
InvMixcolumns	AddRoundKey, InvSubbytes, InvShiftrows	707308	3,045 / 33,216 (9 %)

Table 3 shows the total number of CPU clock cycles required for decryption, for three different cases i.e. when Decryption block is implemented completely in software, when interfaced as custom hardware with NIOS II processor and when individual block is kept in hardware and remaining part of the algorithm is kept in software i.e. using co-design approach.

5 Conclusion

Algorithm is implemented on CYCLONE II FPGA using NIOS II processor. Firstly decryption is implemented in software only. When one block of decryption (InvSubbytes) kept in hardware and remaining blocks in software, speed of the algorithm is increased by (73.33%) by the loss of area. Also, when Algorithm is implemented with NIOS II system as a custom hardware, speed of algorithm is increased by (99.08%) but at the same time area required for the implementation is also increased largely. Hence best optimized results for decryption block in terms of speed and area is obtained by co-design approach.

References

1. Ernst, R.: Co-design of embedded systems: status and trends. In: Proceedings of IEEE Design and Test, pp. 45–54 (April-June 1998)
2. Ghewari, P.B.: Efficient Hardware Design and Implementation of AES Cryptosystem. International Journal of Engineering Science and Technology 2(3), 213–219 (2010)
3. Bhargav, S., Chen, L., Majumdar, A., Ramudit, S.: 128-bit AES decryption. Project report, CSEE 4840 – Embedded System Design Spring. Columbia University (2008)
4. Daemen, J., Rijmen, V.: AES Proposal: Rijndael, The Rijndael Block Cipher, AES Proposal, pp. 1–45 (1999), `http://csrc.nist.gov/CryptoToolkit/aes/`
5. Zambreno, J., Nguyen, D., Choudhary, A.: Exploring Area/Delay Tradeoffs in an AES FPGA Implementation. In: Becker, J., Platzner, M., Vernalde, S. (eds.) FPL 2004. LNCS, vol. 3203, pp. 575–585. Springer, Heidelberg (2004)
6. NIOS II Processor Handbook, Altera Corporation (October 2008)
7. SOPC Builder User Guide, Altera Corporation (December 2010)
8. Altera Corporation, Introduction to Quartus II, (Online Document) (January 2004), `http://www.altera.com/literature/manual/intro_to_quartus2.pdf` (cited February 6, 2004)
9. Altera Corporation, Nios II IDE Help System (May 2007)
10. Wolf, W.H.: Hardware-Software Co-Design of Embedded System. In: Proceeding of the IEEE, Vol. 82(7) (July 1994)

Physical Design of Two Stage Ultra Low Power, High Gain Cmos OP-AMP for Portable Device Applications

Karandeep Singh, Vishal Mehta, and Mandeep Singh

Department of Electronics & Communication Engineering,
University Institute of Engineering & Technology (UIET),
Panjab University, Chandigarh
{Karandeep963,Vishal.2jan,Cme2016}@gmail.com

Abstract. A Two-Stage CMOS Op-Amp using P-Channel input differential stage suitable for portable device applications with ultra-low power, high swing, high gain is proposed. DC gain is increased by using Cascode Technique. A gain-stage implemented in Miller capacitor feedback path enhances the Unity-Gain Bandwidth. Topology selection, practical issues in designing micro-power Op-Amps and theoretical analysis of the design are discussed. The circuit is designed and simulated using TSMC 180nm technology. The circuit is simulated with 1.5V DC supply voltage. The proposed Op-amp provides 228MHz unity-gain bandwidth, 61.3 degree phase margin and a peak to peak output swing 1.15v. The circuit has 95.6dB gain. The maximum power dissipation of the designed Op-Amp is only 72μW. Suitable response in different temperature range is demonstrated by the designed system. Layouts of the proposed Op-Amp have been done in Cadence® Virtuoso Layout XL Design Environment.

Keywords: Two stage Op-Amp, Low voltage, Ultra low power, Cascode technique, Unity-Gain Bandwidth.

1 Introduction

Research in analog-circuit design is focused on low-voltage low-power battery operated equipment to be used as an example in portable equipment, wireless communication products and consumer electronics. A reduced supply voltage is necessary to decrease power consumption to ensure a reasonable battery lifetime in portable electronics. For the same reason, low-power circuits are also expected to reduce thermal dissipation, of increasing importance with general trend in miniaturization [1].

Advancements required by International Technology Roadmap for Semiconductors (ITRS), means that with current CMOS standard fabrication processes, circuits must work at supply voltages as low as 1.5V. Industry and academia are researching new circuits techniques that will make them operate at this voltage. For this fact, the

S. Unnikrishnan, S. Surve, and D. Bhoir (Eds.): ICAC3 2013, CCIS 361, pp. 730–739, 2013.

industry and the academia are doing research in new circuit's techniques that will make them able to operate at this voltage [1].

Also other point, as for various recently developed high-performance integrated electronic systems or subsystems, e.g. A/D converter, switched-capacitor filter, RF modulator and audio system, CMOS operational amplifier with high unity-gain bandwidth and large dynamic range are required [2].

Another important consideration from bio-medical applications typically requires high gain, high swing and ultra low power amplifiers that typically occupy the minimal chip real estate [10].

So, it can be deducted that designing of Op-Amp puts new challenges in low power applications with reduced channel length devices.

2 Circuit Topology Selection

The comparison of performance of various Op-Amp topologies is presented in Table 1. Each topology exposes its uniqueness of performance. Considering the design requirements the two-stage amplifier seems most suitable one. In a two-stage Op-Amp, the first stage provides high gain and the second provide high output swing.

Table 1. Topology Selection for CMOS Op-Amp[4]

Topology	*Gain*	*Output Swing*	*Speed*	*Power Dissipation*	*Noise*
Telescopic	Medium	Medium	Highest	Low	Low
Folded Cascode	Medium	Medium	High	Medium	Medium
Two Stage	High	Highest	Low	Medium	Low
Gain Boosted	High	Medium	Medium	High	Medium

3 Conventional RHP Zero Controlling and Improved Compensation Techniques

The RHP zero creates the un-stability problem as it boosts the magnitude but lags the phase which can be solved by eliminating one of the paths to the output. The effect of RHP zero on the stability or phase margin of Op-Amp is shown in Fig. 1 [4].

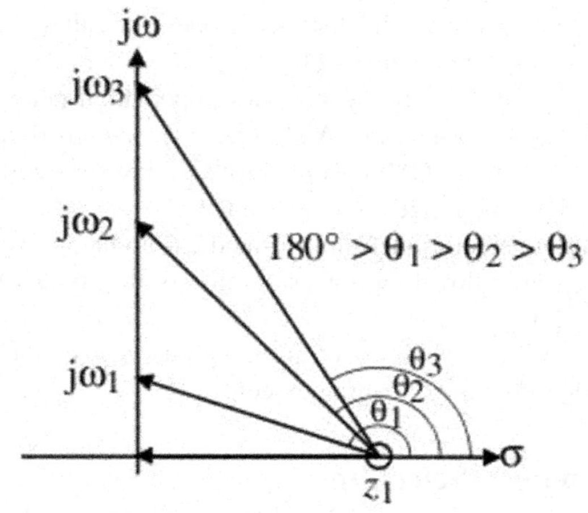

Fig. 1. RHP-Zero effect on stability

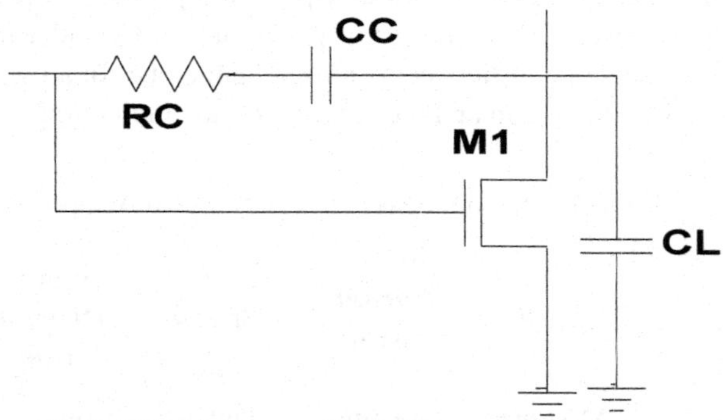

Fig. 2. RHP Zero control by nulling resistor approach

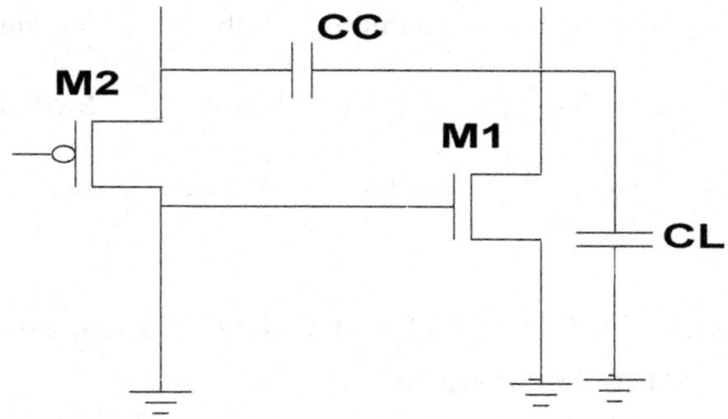

Fig. 3. RHP Zero control by gain stage approach

RHP zero control using nulling resistor and gain stage approach is illustrated in Fig.2. and Fig.3. respectively. RHP zero displacement form the basis of nulling resistor approach. On the other hand, as shown in Fig.4 ., the employed gain-stage M2 prevent the input current from going directly through the Miller capacitor, thus, the RHP zero will eliminate [7]. As observed in Fig. 2. , the compensation capacitor Cc can be ignored at frequencies near unity gain frequency. Therefore, the output resistance seen by C_L can be derived as given by

$$R_{out} = 1 / g_{m1} \tag{1}$$

while from Fig. 2. , the output resistance seen by C_L can be expressed as [5]

$$R_{out} = 1 / (g_{m1}\, g_{m2}\, r_{ds2}) \tag{2}$$

Also the output pole can be written as,

$$P_o = 1 / R_{out}\, C_{out} \tag{3}$$

Obviously, it can be deduced from (1) and (2) that noticeable reduction of R_{out} by a factor of gm2 r ds2 is the basis of improvement in "pole-splitting" effect which is valid only for Fig. 2. and not for Fig. 3. Unity-Gain Bandwidth is enhanced by increasing the magnitude of the non-dominant output pole by employed gain-stage M2.

As observed in Fig. 3. , the compensation result is to keep the dominant pole roughly the same as normal Miller compensation and to increase the output pole by approximately the gain of a single stage M6 the magnitude of the output pole can be given by

$$P_o = (g_{m12}\, g_{m6}\, r_{ds6}) / C_{out} \tag{4}$$

Consequently, the output pole Po has split from the dominant pole by a factor of gm6rds6 which leads to an enhancement in unity gain bandwidth, when compared to the nulling resistor approach.

4 Circuit Design and Implementation

The proposed schematic of a Two-Stage CMOS Op-Amp with p-channel input pair, with un-buffered output for Unity-Gain Bandwidth enhancement is shown in Fig. 4.

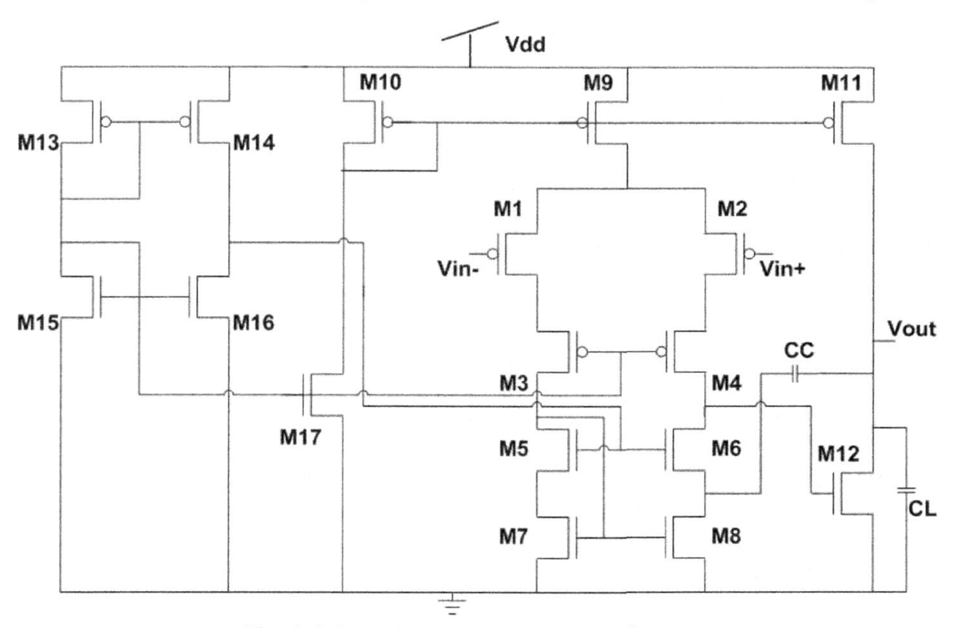

Fig. 4. Schematic proposed Two-Stage Op-Amp

In this work the various transconductance parameters are related as:

$$g_{m1} = g_{mI}\ ,\ g_{m2} = g_{mII}\ ,\ g_{ds2} + g_{ds4} = G_I \tag{5}$$

Also

$$g_{ds6} + g_{ds7} = G_{II} \tag{6}$$

$$g_m = \sqrt{2\mu_{n,p}\, C_{ox} \frac{W}{L}\, I_D} \tag{7}$$

$$g_m = 2\,\frac{I_D}{V_{eff}} \tag{8}$$

While the I_d current is given as

$$I_D = \frac{\mu_{n,p}\, C_{ox}\frac{W}{L} V_{eff}}{2} \tag{9}$$

from the expressions in relation with the compensation capacitor Cc, the slew rate is given by

$$SR= dv_o/dt \tag{10}$$

Therefore,

$$SR = I_{9(max)} / Cc \tag{11}$$

Requirements for the transconductance input transistors can be determined from Knowledge of Cc and GBW the transconductance g_{ml} can be calculated by the following equation

$$GBW = g_{m1} /Cc \tag{12}$$

The aspect ratio $(W/L)_1$ is directly obtainable from g_{ml} as shown below.

$$(W/L)_1 = g^2_{m1}/2\beta I_1 \tag{13}$$

5 Simulation Results

The circuit is simulated with BSIM3v3.1 model based on a standard 0.18 µm CMOS process. The OP AMP operates with the 1.5V power supply and consumes maximum 72µW power at maximum amplitude. Simulations resulted on considerable increase of unity-gain bandwidth to the value of 228 MHz, the improved DC gain of 95.6 dB, phase margin of 61.3 degree and output swing of 1.15V. Layout of the proposed Op-Amp has been done in Cadence® Virtuoso Layout XL Design Environment. LVS and DRC has been checked and compared with the corresponding circuits using Mentor Caliber, Synopsys Hercules and STAR RCXT respectively. The simulated results have been developed for the various performance characteristics and the superior features of the proposed Op-Amp are established.

Fig. 5. presents the simulated DC gain and Fig. 6. shows the phase margin of the OPAMP. Swing voltage is 1.15 v presented by Fig. 7. Additionally, with the load capacitor CL of 1 pF and with a compensation capacitor Cc of 208 fF. Fig. 8. demonstrates the performance metrics at different temperature range. The layout is depicted in Fig. 9. Simulation Results represent that the designed OP AMP has high voltage gain, low supply voltage, high swing voltage, high unity-gain bandwidth and ultra-low power dissipation.

Table 2. Aspect Ratios of proposed Op-Amp design

Transistor	*W*	*Transistor*	*W*
M1	15	M10	2.8
M2	15	M11	11
M3	10	M12	8
M4	10	M13	0.38
M5	2	M14	0.31
M6	2	M15	0.21
M7	3	M16	0.21
M8	3	M17	0.21
M9	6.8	*L*	***0.36***

* All the above measurements are in micro-meters.
**L is length of all the transistors

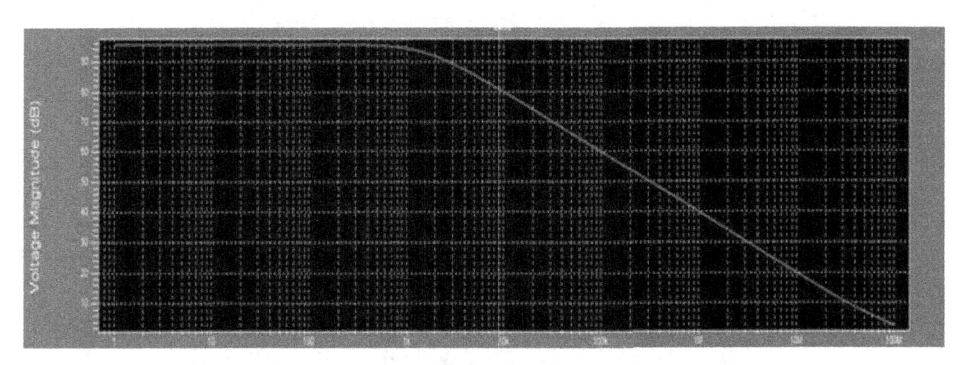

Fig. 5. Simulated DC Gain

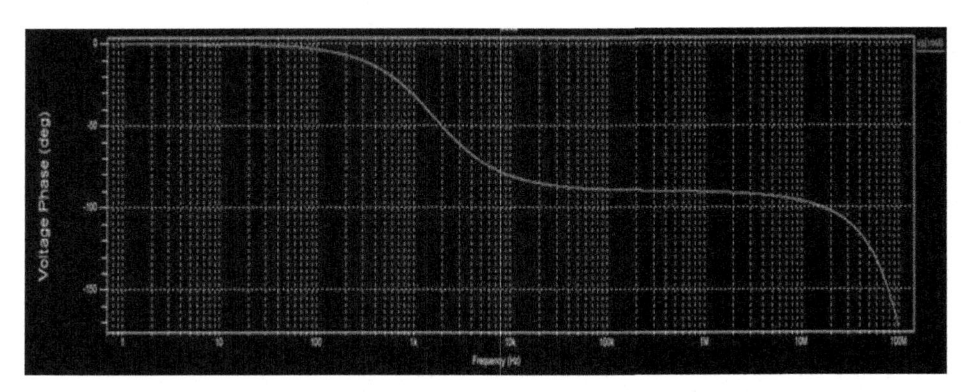

Fig. 6. Simulated Phase Margin

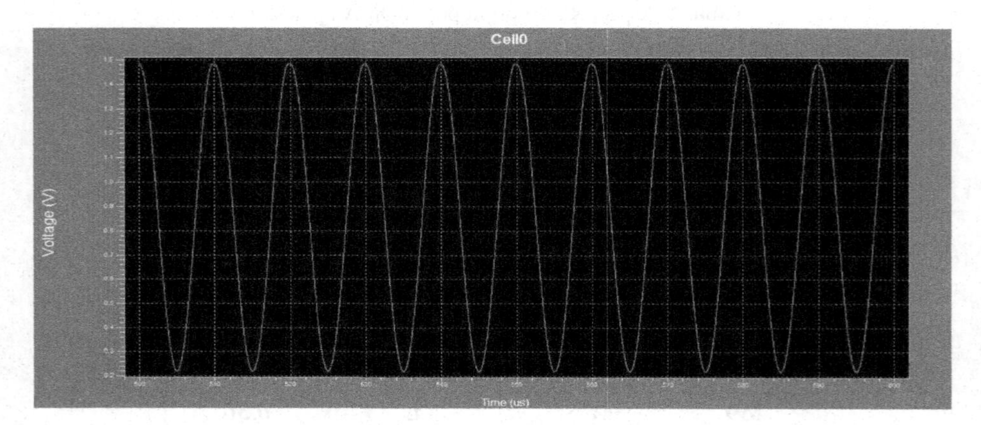

Fig. 7. Simulated Output-Swing

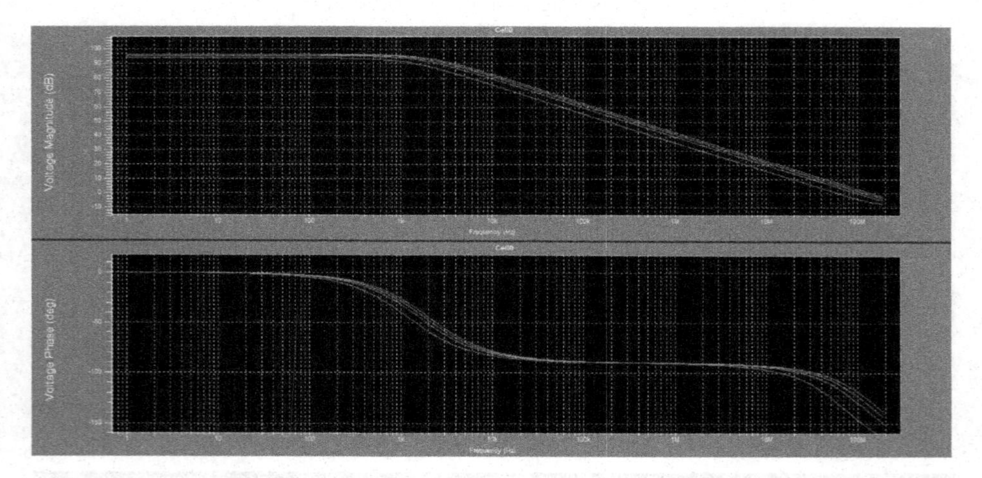

Fig. 8. Performance analysis at different temperature range of proposed Op-Amp

Table 3. The designed system demonstration in different temperature

Temp. (deg)	-20	0	27	100
Gain (db)	94	95.6	96.2	96.6
Phase Margin (deg)	12	61.3	63	66
UGB (MHz)	220	228	238	102

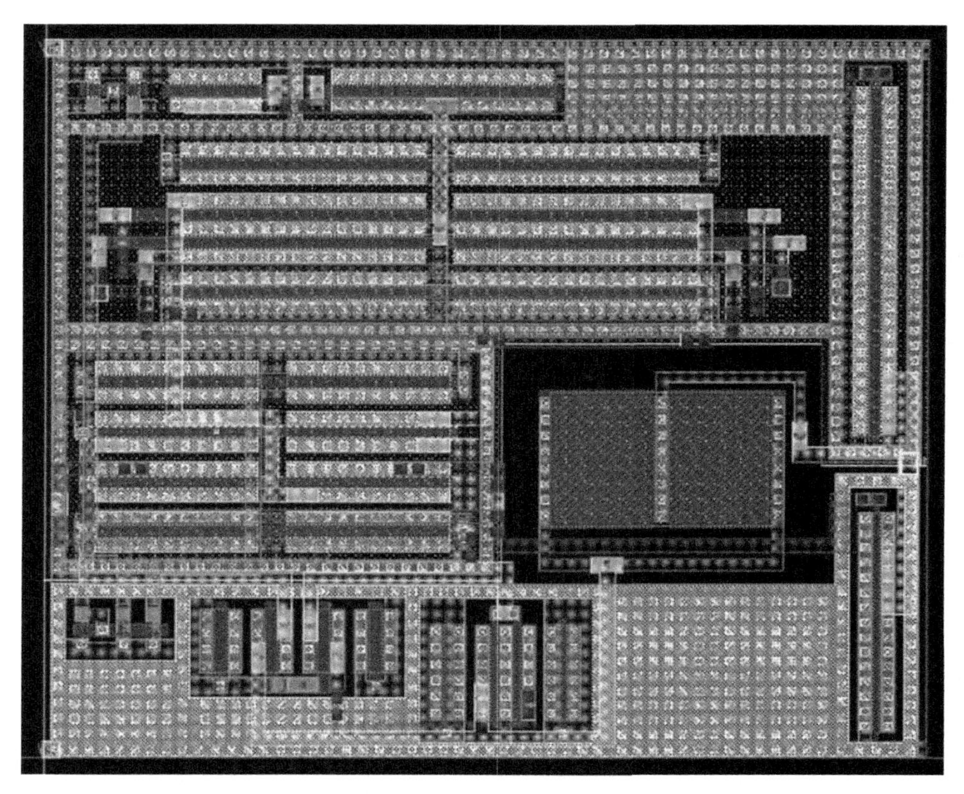

Fig. 9. Layout view of the proposed Op-Amp

Table 4. Performance summary of the proposed Op-Amp

Reference	Performance Metrics					
	DC Gain (db)	UGB (MHz)	PM (degrees)	Power Supply (V)	Power Diss. (μW)	Output Swing (V)
This Work	95.6	228	61.3	1.5	72	1.15

The OP AMP performance at different temperature is illustrated in Table 3. The performance summary of this proposed CMOS OPAMP is listed in Table 4.

6 Applications

Recent advances in IC technology as well as innovation in circuit design techniques, have led to system with processing capabilities that can supplement or even entirely replace complex circuits in the mixed signal system.

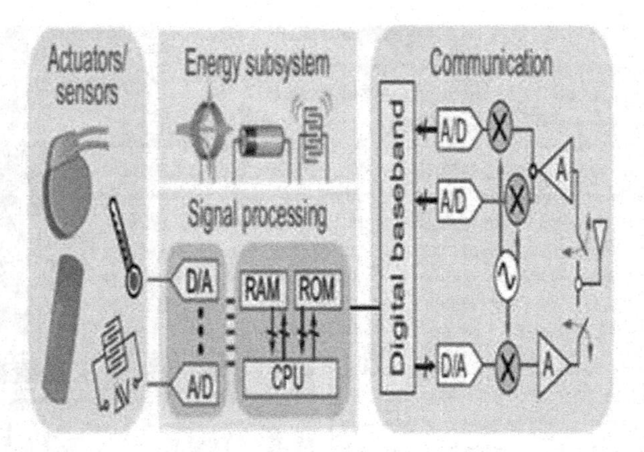

Fig. 10. Electronic Bio-Medical system

The electronic bio-medical system can be observed from Fig. 10. Due to the small voltage drops, the ultra-low power consumptions the system is very amenable too implantable, bio-medical and portable device applications, where battery size and capacity are limited.

7 Conclusion

The design procedure of low voltage, high gain, ultra-low power two stage CMOS Op-Amp is discussed in this paper. The unity-gain bandwidth enhancement using a gain-stage in the Miller capacitor feedback path is presented and verified by simulation achieving the UGB of 228MHz with peak to peak output swing of 1.15V. The concept of stability is presented and implemented with the phase margin of 61.3̊. The technique implemented in the circuit with 0.18µm CMOS technology and supply voltage of 1.5v. The circuit is power efficient with consumption of 72µW and DC gain of 95.6dB. A relatively suitable response in different temperature range is demonstrated by this system. Layout of the proposed Op-Amp have been done in Cadence® Virtuoso Layout XL Design Environment. LVS and DRC has been checked and compared with the corresponding circuits using Mentor Caliber, Synopsys Hercules and STAR RCXT respectively. Continued device scaling, reduction in power-supply voltage and integration of technology makes this design amenable to implantable, bio-medical and portable device applications, where battery size and capacity are limited.

References

1. Ramson, J., Steyaert, M.S.J.: Positive Feedback Frequency Compensation for Low Voltage Low Power Three Stage Amplifier. IEEE Transaction on Circuits and Systems 51(10), 1967–1974 (2004)

2. Ramirez-Angulo, J., Carvajal, R.G., Tombs, J., Torralba, A.: Low voltage CMOS Opamp with rail to rail input and output signal swing for continuous time signal processing using multiple input floating gate transistor. IEEE Transactions on Circuits and SystemsII 48, 110–116 (2001)
3. Razavi, B.: Design of Analog CMOS Integrated Circuits. CMOS operational amplifier, pp. 243–323. Mc Graw-Hill International Edition
4. Allen, P.E., Holberg, D.R.: CMOS Analog Circuit Design, 2nd edn. Operational Amplifier, pp. 324–396. Oxford University Press, London (2003)
5. Ahuja, B.K.: An improved frequency compensation technique for CMOS operational Amplifier. IEEE Journals of Solid-State SC-18(6), 629–633 (1983)
6. Yavari, M., Zare-Hoseini, H., Faeazian, M., Shoaei, O.: A new compensation technique for two-stage CMOS operational Transconductance Amplifier. In: Proc. IEEE Intl. Conf Electronics, Circuits, Systems, ICECS, pp. 539–542 (December 2003)
7. Liod, J., Lee, H.: A CMOS OP-AMP with Fully-Differential Gain-Enhancement. IEEE Transactions on Circuits and Systems-II: Analog and Digital Signal Processing 41(3) (March 1994)
8. Liu, Z., Bian, C., Wang, Z., Zhang, C.: Full custom design of two stage fully differential CMOS amplifier with high unity-gain bandwidth and large dynamic range at output. In: 48th IEEE International Midwest Symposium on Circuits and Systems (MWSCAS), pp. 25–28 (2007)
9. Angulo, J.R., Lopez-Martin, A., Garimella, A., Garimella, L.M.K., Carvajal, R.G.: Low voltage Low power Rail to Rail two stage op-amp with dynamic baising and no miller caompenastaion. In: 50th Midwest Symposium on Circuits and Systems (MWSCAS), pp. 35–41 (2009)
10. Fonderie, J., Huijsing, J.H.: Operational amplifier with 1-V rail-to-rail multipath driven output stage. IEEE J. Solid-State Circuits 26, 1817–1824 (1991)

Security Issues of Online Social Networks

M.A. Devmane[1] and N.K. Rana[2]

[1] P.V.P.P. College of Engineering, Mumbai
dmahavir@gmail.com
[2] Theem College of Engineering, Mumbai
ranank@rediffmail.com

Abstract. The number of users of the online social networks like facebook, twitter, google, linkedIn are going on increasing tremendously. Similarly the time spend by each user on such online social networks (OSN) is also increasing. These things clearly indicate that the popularity of these OSN is increasing like a wildfire. These OSN provide a very efficient platform for the user to establish contacts with others. These users are having frequent communication with each other. So the adversaries find the OSN as a soft target to attack easily and spread it to the large number of users in less time. In this paper we consider some of such threats to OSN as well countermeasures to some of these threats so as to make the OSN as well as the user secure in the digital world.

Keywords: Online Social network, cyber security, profile cloning, botnet.

1 Introduction

The OSNs are so popular that the number of users of facebook alone is more than 1000 million around the globe. To become member of the OSN the user has to create his profile by entering information like name ,photo, date of birth , Email ID, place of work , home town and so on [1][2]. Some of these fields are mandatory and remaining are optional and varies from one OSN to the other. To attract more and more friends or to let others to find easily the users try to provide the maximum information in the profile [3]. OSNs are mainly used for keeping in touch with friends, forming new contacts, as well as search for someone else on the OSN and establish contact with him by sending a friend request. Such contacts are used to share some information with each other as well as broadcast the information through a group.

Due to the availability of more number of OSNs some of the users are having their account on more than one OSN so that they can remain in touch with their friends in that OSN also. While creating such profiles or accounts most of the users doesn't go through the privacy policy of each such OSN and provide the information. In such cases there is high risk of leakage of information. As the OSN follow the client-server architecture all the data given by the user in the profile will remain with the server so no one can predict that how secure is the data.

This paper takes into consideration the various threats to the Online Social Networks and provides remedies to some of them in the following sections.

S. Unnikrishnan, S. Surve, and D. Bhoir (Eds.): ICAC3 2013, CCIS 361, pp. 740–746, 2013.

2 Information Leakage

Information leakage means the information stored by the user in his profile is accessed by someone else and use the same for malicious activities [4].Most of the OSN allow the friend of the user to have access to most of the fields from his profile. This must be taken into consideration while either sending friend request or accepting friend request from anybody. Whenever a user accepts friend request it is assumed that he is having trust in that user so such access is provided.

To avoid such malicious access it is necessary to be careful while selecting security settings for various fields in the profile. A field likes contact details can be set as "visible to me only" so that the adversary can't get access to it. Similarly before sending or accepting any friend request just check that the person can be trusted or not otherwise adversary may get direct access to your information and so many other things can happen. Most of the OSN users are careless in this matter which can be supported by the example of automated script which send friend requests to 250000 facebook users and out of that 75000 users accepted the request and him in the friend list without taking due care.

3 Spam in OSN

Spam is abuse of electronic messaging systems to send unsolicited bulk messages indiscriminately. Although the most widely recognized form of spam is E-mail spam the term is similarly applied to similar abuses in other media like online social networks [5]. There are two types of spamming.

3.1 Broadcast Spamming

In broadcast spamming the malicious user does not know the exact Email ID of the user so he normally does some combinations of words to generate some probable Email IDs and send Emails. Such spams are not that much effective as the user does not believe in it easily.

3.2 Context Aware Spamming

In the context aware spamming the Email may contain some information of the user which is taken from the profile of the user in one of the OSN. Such information is gathered just by searching the OSN for a name and if it exists find its Email ID and other credentials in the profile and use the same in the spam. As it contains some information about the user he will easily believe in the Email and may reply to it. It is observed that the context aware spams are more successful than the broadcast spam.

4 Profile Cloning

To have account with an OSN the user has to create his profile by providing his details. By using the credentials from this profile a malicious user can create a profile using the credentials and pretend that he is the legitimate owner of that profile and misguide others. This is called as profile cloning [6]. There are two types of profile cloning as given below.

4.1 Same-Site Cloning

In the same-site cloning the cloning will be done in the same OSN and impersonate like the legitimate user. He start using contacts of the legitimate user and start sending comments, uploading photos etc as the legitimate user.

4.2 Cross-Site Cloning

In the cross-site cloning the information from one profile of the user on one OSN will be used to create a clone profile on some other OSN and tried to find friends on that OSN. As the profile contains correct information of the user the target users doesn't know that they are dealing with a malicious user [6].

Such cloned profiles may be used to defame the legitimate user and sometimes he may send some messages containing links to some malware etc. which may be harmful to the recipient. As the recipient assume that as it comes from his friend he will just trust it and click on the link. Sometimes the adversary use the information of last log of the user's friend, which is provided by some of the OSN and try to contact the one who is not in touch with the user for longer duration. So that he will not be suspicious about the cloned profile of his friend.

Such profile cloning is difficult to avoid but it can be avoided up to certain extent by using proper privacy setting for various fields in the profile so that the adversary should not get information which he can use for profile cloning. Also the OSN on which the user is having profile as well as the other OSNs are to be checked frequently for existence of cloned profile.

5 Phishing Attacks

Phishing is a kind of identity theft that uses various social engineering techniques and tries to get secret and confidential information from the user. From the profile of user phisher get the Email ID of the profile owner. Such Email ID can be used for phishing [7]. A common factor among the phishers is that they maliciously mislead the recipients to believe that they received that Email from a legitimate website [8].

Some of the phishing attacks on OSN are as explained below.

5.1 Keylogger Attack

In the Keylogger attack [9] the spyware will be installed on the web browser or sometimes it may be installed as a device driver. This program will capture the keyboard inputs as well as mouse inputs and send it to the adversary without knowledge of the user.

5.2 Content Injection Phishing

In content injection phishing a malicious content is added into a legitimate website which will redirect the user to some other website where it will be tried to get the sensitive or confidential information from the user.

5.3 Session Hijacking

In session hijacking a malicious browser component monitors activities of the user and whenever the user logs into E-commerce related websites for some online transaction the session will be hijacked to perform some malicious actions [9].

5.4 Search Engine Phishing

In search engine phishing the phisher creates some fake WebPages get them indexed in the search engine. These web pages are designed by using various keywords in such a way that it will be having higher ranking in the search. These fake WebPages are mostly related with E commerce and will be used to get secret and confidential information of the user like his bank account number, credit/debit card details etc. such information will be further used for malicious purpose.

Countermeasures to the phishing attacks may be mainly to educate the users so that they will get protected at the first step only. Similarly blacklisting of the web pages which are used for phishing can be a better option where a dedicated server will keep track of all such blacklisted pages.

Apart from these a phishGaurd [10] browser plug-in is proposed to avoid phishing. Here the browser plug-in will exploit the method of sending random credentials to test the presence of phisher.

Another approach is a BogusBiter [11] which is a client side anti-phishing tool which transparently feeds a large number of bogus credentials into a suspicious phishing site. The BogusBiter hides the user's correct credentials into the bogus credentials and it enables the legitimate target website to identify the correct credentials of the user.

6 Sybil Attack

Sybil attack is an attack against identity in which an individual entity masquerades as multiple simultaneous identities. It poses a serious threat to the security of OSN. The malicious user creates new accounts in the OSN and link them in the network so when the anonymized network is released he can recover information using the particular topological feature introduced by the Sybil accounts to launch such Sybil attacks the requirement for the malicious user is that he should have some accounts in that OSN [5] .

Such sybil attacks can be avoided if the creation of fake accounts made impossible. Similarly a protocol like SybilGuard [12] can be used which limits the corruptive influence of sybil attacks as well as the sybil attacks which exploit the IP harvesting.

7 Friend-In-The-Middle Attack

Friend-in-the-middle (FITM) attacks are active eavesdropping attacks on the OSN. These attacks are using the missing protection of the communication link between the client and the OSN. By hijacking the session cookies an attacker can impersonate the victim and interact with the OSN as legitimate user though it doesn't have proper authorization. An attacker can access the communication link between the user and OSN either passively i.e. monitoring unencrypted wireless network or by actively i.e. using Address Resolution Protocol spoofing on the LAN [13]. The attacker may simply clone the HTTP header containing the authentication cookies and can interact with the OSN without the knowledge of OSN or the user. Sometimes the attacker may try to inject i.e. add himself as friend in the victims profile so now he can have a open access to the user as well as his contacts.

The countermeasures to effectively mitigate the FITM attacks are like the OSN provider should design the OSN in such a way that the communication between the user and OSN is done using the HTTPS instead of HTTP. Similarly sophisticated protection mechanism must be in place to limit the access to profile information to avoid the friend injection.

8 Botnet

In Botnet the adversary tries to install malicious application on the target computers which are there in network (internet) and get a complete control on that computer [14, 15]. Once a computer is infected it is called as zombie and now it will work according to the instructions given by the adversary. Though the authenticated user is working on the computer, a zombie in the computer may be working as per the instructions given by the adversary in the background. To create a botnet the adversary creates a large pool of such zombies to work for him.

The OSN is having some of the properties which make them ideal for attackers to attack and create a botnet [16]. Some of these properties are:-

A very large and distributed user base.

The group of users (friends) having trust with each other and seek access to so many common entities.

OSN platform openness for deploying application which be installed by the users. sometimes it may be a malicious or harmful application

In the OSN the adversary can have a very big pool of computers which can be converted into zombie. It also gets a platform to get spread just like a wildfire. The general procedure to use OSN to create a botnet is a malware is published as an application on someone's profile and lure the user to click on it, whenever the user clicks on the application the malware gets executed which converts the computer into zombie. Now this automated program will send this application to the contacts in the user's profile. As the application come through a trusted contact the destination user may click on it get infected and start working as zombie.

One of such example is koobface malware which was detected in late 2008 on the facebook [17]. The infection chain starts with one of the accounts on facebook which may be a compromised one may be generated by koobface to automatically befriend the victims [18].A zombie will regularly create fraudulent accounts by using a random profiles generated with photo, date of birth etc. further this account further this account joins some groups based on some randomly generated popular keywords. By using such groups the koobface sends spam to the members of these groups. The spams will be having some obfuscated URLs which are probably not blacklisted by the facebook. Such links will take the target user to some redirections and lure him to click on some popular downloads. So in this way the chain go on up to a large number of zombies.

9 Conclusion

The threats to the OSN lead to defaming, commercial loss and so on. To overcome most of the threats it is necessary to have the user education about importance of privacy of their data. To avoid fraudulent accounts the OSN should use some mandatory fields for PAN, Driving License Number etc. and it should be cross checked with the database and then only the account should be enabled. This will not only reduce the bogus and automatically created accounts but will also reduce various attacks which are taking place with help of such accounts. More efforts are required in this field to find new techniques to avoid and prevent such attacks.

References

[1] Gladbeck, J.: Trust and nuanced profile similarity in online social networks. ACM Transactions on the Web 3(4), Article 12 (September 2009)

[2] Nagy, J., Pecho, P.: Social network security. In: 2009 IEEE International Conference on Emergency Security Information Systems and Technologies, pp. 321–325 (2009)

[3] Irani, D., Web, S., et al.: Large online social footprints-an emerging threat. In: 2009 IEEE International Conference on Computational Science and Engineering, pp. 271–277 (2009)

[4] Irani, D., Web, S., et al.: Modeling unintended personal-information leakage from multiple online social networks, pp. 13–19. IEEE Computer Society

[5] Gao, H., Hu, J., et al.: The status quo of online social network security: A survey. IEEE Internet Computing, 1–6 (2011)

[6] Kontaxis, G., Polakis, I., et al.: Detecting social network profile cloning. In: 2011 IEEE Third International Conference on Security and Social Networking, pp. 295–299 (2011)

[7] Huang, H., Tan, J., et al.: Countermeasure techniques for deceptive phishing attack. In: 2009 IEEE International Conference on New Trends in Information and Service Science, pp. 636–641 (2009)

[8] Soni, P., Phirake, S., Meshram, B.B.: A phishing analysis of web based systems. In: ACM ICCCS, pp. 527–530 (February 2011)

[9] Badra, M., Ei-Sawda, S., Hajjeh, I.: Phishing attacks and solutions. In: Mobimedia (2007)

[10] Joshi, Y., Saklikar, S.: PhishGuard: A browser plug-in for protection from phishing. In: IMSAA (2008)
[11] Yue, C., Wang, H.: BogusBiter: A transparent protection against attacks. ACM Transactions on Internet Technology 10(2), Article 6 (May 2010)
[12] Yu, H., Michael, K., et al.: SybilGuard: Defending Against Sybil At-tacks Via Social Networks. In: ACM Conference on Applications, Technologies, Architectures, and Protocols for Computer Communications (2006)
[13] Huber, M., Mulazzani, M., et al.: Friend-in-the-middle attacks. IEEE Internet Computing (2011)
[14] Devmane, M.A., Meshram, B.B.: Attacks on wired and wireless networks and mobile devices. International Journal of Computer Applications in Engineering, Technology and Sciences 3(2), 208–215 (2011)
[15] Athanasopoulos, E., Makridakis, A.: Antisocial Networks: turning a social network into botnet. In: Proceedings of the 11th International Conference on Information Security (2008)
[16] Makridakis, A., Athanasopoulos, E., et al.: Understanding the behavior of malicious applications in social networks. IEEE Network (October 2010)
[17] Abu Rajah, M., Zarfoss, J., et al.: A multifaceted approach to understanding the botnet phenomenon. In: SIGCOMM Conference on Internet Measurement (2006)
[18] Thomas, K., Nicol, D.M.: The Koobface botnet and the rise of social malware. In: IEEE 5th International Conference on Malicious and Unwanted Software, pp. 63–70 (2010)

Author Index